W9-BTJ-965

4.2 Simplex Method

1. Determine the objective function.

2. Write all necessary constraints.

3. Convert each constraint into an equation by adding slack variables.

4. Set up the initial simplex tableau.

5. Locate the most negative indicator. If there are two such indicators, choose the one furthest to the left.

6. Form the necessary quotients to find the pivot. Disregard any quotients with 0 or a negative number in the denominator. The smallest nonnegative quotient gives the location of the pivot. If all quotients must be disregarded, no maximum solution exists. If two quotients are both equal and smallest, choose the pivot in the row nearest the top of the matrix.

7. Use row operations to change all other numbers in the pivot column to zero by adding a suitable multiple of the pivot row to a multiple of each row.

8. If the indicators are all positive or 0, this is the final tableau. If not, go back to Step 5 and repeat the process until a tableau with no negative indicators is obtained.

9. Read the solution from this final tableau.

5.1 Compound Amount

$$A = P(1 + i)^n,$$

where $i = \dfrac{r}{m}$ and $n = mt$,

A is the future (maturity) value;
P is the principal;
r is the annual interest rate;
m is the number of compounding periods per year;
t is the number of years;
n is the number of compounding periods;
i is the interest rate per period.

5.2 Future Value of an Ordinary Annuity

$$S = R\left[\frac{(1 + i)^n - 1}{i}\right] \qquad \text{or} \qquad S = Rs_{\overline{n}|i},$$

where
S is the future value;
R is the payment;
i is the interest rate per period;
n is the number of periods.

5.3 Present Value of an Ordinary Annuity

The present value P of an annuity of n payments of R dollars each at the end of each consecutive interest period, with interest compounded at a rate of interest i per period, is

$$P = R\left[\frac{1 - (1 + i)^{-n}}{i}\right] \qquad \text{or} \qquad P = R \cdot a_{\overline{n}|i}.$$

FINITE MATHEMATICS

SEVENTH EDITION

FINITE MATHEMATICS

SEVENTH EDITION

- **Margaret L. Lial**
  American River College

- **Raymond N. Greenwell**
  Hofstra University

- **Nathan P. Ritchey**
  Youngstown State University

Addison
Wesley

Boston San Francisco New York
London Toronto Sydney Tokyo Singapore Madrid
Mexico City Munich Paris Cape Town Hong Kong Montreal

Sponsoring Editor	Laurie Rosatone
Executive Project Manager	Christine O'Brien
Senior Production Supervisor	Karen Wernholm
Production Coordination and Text Design	Elm Street Publishing Services, Inc.
Marketing Manager	Michael Boezi
Senior Prepress Supervisor	Caroline Fell
Manufacturing Buyer	Evelyn Beaton
Compositor	The Beacon Group, Inc.
Illustrations	Precision Graphics
Cover Design	Barbara T. Atkinson
Cover Photos	© FPG, © 2001 Corbis, and © PhotoDisc

For permission to use copyrighted material, grateful acknowledgment is made to the copyright holders on page A-34, which is hereby made part of this copyright page.

Library of Congress Cataloging-in-Publication Data
Lial, Margaret L.
    Finite mathematics / Margaret L. Lial, Raymond N. Greenwell, Nathan P. Ritchey.—7th ed.
        p. cm.
    Includes index.
    ISBN 0-321-06714-2
    1. Mathematics.   I. Greenwell, Raymond N.   II. Ritchey, Nathan P.   III. Title.
    QA37.3.L53 2001
    510—dc21   2001022728

Copyright © 2002 Pearson Education, Inc. All rights reserved. No part of this publication may be reproduced, stored in a retrieval system, or transmitted, in any form or by any means, electronic, mechanical, photocopying, recording, or otherwise, without the prior written permission of the publisher. Printed in the United States of America.

6 7 8 9 10—QWT—0403

CONTENTS

■ CHAPTER 7   SETS AND PROBABILITY

■ CHAPTER 8   COUNTING PRINCIPLES; FURTHER PROBABILITY TOPICS

■ CHAPTER 9   STATISTICS

■ CHAPTER 10   MARKOV CHAINS

CHAPTER 11   GAME THEORY

SPECIAL TOPICS TO ACCOMPANY FINITE MATHEMATICS*

---

*These supplements will be provided free to adopters who request them.

PREFACE

Finite Mathematics is a solid, application-oriented text for students majoring in business, management, economics, or the life or social sciences. A prerequisite of two to three semesters of high school algebra is assumed. Many new features, including new exercises, new applications, increased use of spreadsheets, and multiple methods of solution, make this latest edition a richer, stronger learning resource for students.

■ NEW AND ENHANCED FEATURES

Multiple Methods of Solution As in the previous edition, we continue to emphasize multiple representations of a topic, whenever possible, by examining each topic symbolically, numerically, graphically, and verbally. In this edition, we have added multiple methods of solution to various examples. Some of these alternative methods involve technology, while others represent a different way of performing a computation. For example, see pages 31–33, 62–65, and 215–216. We believe that students will better understand a topic when they learn more than one approach. Furthermore, the multiple methods will give instructors greater flexibility to emphasize the methods they prefer.

Use of Spreadsheets and Graphing Calculators We continue the emphasis on graphing calculators introduced in the previous edition and have added the use of spreadsheets with this edition. Although students reading this book do not need a graphing calculator or a spreadsheet program, using either or both will enhance their learning. Exercises requiring technology are labeled with the ▨ icon; many of these can be done with either a spreadsheet or a graphing calculator. For example, see pages 38–43 and 74. We have used the TI-83 for our graphing calculator examples and Microsoft® Excel for our spreadsheet examples, but students can succeed in the course with any appropriate graphing calculator or spreadsheet program. Inclusion of both graphing calculators and spreadsheets in the text gives instructors the ability to incorporate technology into their course as they wish.

Increased Use of Referenced Real-Data Applications This edition includes even more application exercises and examples using real data than the previous edition, with references to articles appearing in newspapers, books, and journals. For example, see pages 18–20 and 120–122. These examples should help students learn how extensively mathematics is applied, as well as motivate those who wonder how the mathematics they study relates to the outside world and where they might use it. We believe the quantity and quality of real-data applications set this book apart from the others available for this course.

Increased Extended Applications   As in previous editions, we have included in-depth applied exercises, called Extended Applications, at the end of most chapters to stimulate student interest and for use as group projects. This edition includes many new Extended Applications, and a larger collection of them is available at the Web site for this book: www.aw.com/LGR. Topics for these Extended Applications include Statistics in the Law and Teacher Retention.

Full-Color Graphics and Photography   To make this book easier, livelier, and more enjoyable to read, we have used full-color graphics in this edition. Occasionally, color in the text enhances the exposition, such as on page 4. More often, we have used color to clarify the different parts of a figure, such as on pages 142 or 492. Most importantly, the use of different colors in the graphs and figures emphasizes their most significant aspects, ensuring that the students grasp and retain each illustration's key mathematical concept, as on pages 315 and 478. We have also added photographs illustrating many of the applications.

CONTINUING FEATURES

This edition continues to offer the many popular features of the previous edition:

Pedagogical Features

- careful explanation of the mathematics
- fully developed examples with explanatory comments in color on the side
- an algebra reference (Chapter R), designed to be used either in class or by individual students
- thought-provoking questions that open most sections, which are answered in an application within the section or in the section exercises
- "just in time" margin reviews giving short explanations or comments reminding students of skills or techniques learned earlier that are needed at this point
- common student difficulties and errors highlighted under the heading "Caution"
- important treatments and asides highlighted with the heading "Notes"
- summaries of rules or formulas for chapters where students may have trouble deciding which of several techniques to use
- an index of applications showing the abundant variety of real-data applications used in the text and allowing direct reference to particular topics

Exercises

- exercises carefully arranged according to the material in the section, with the more challenging exercises placed near the end
- applied exercises (labeled "Applications") that are grouped by subject, with subheadings indicating the specific topic

- writing exercises, labeled with the 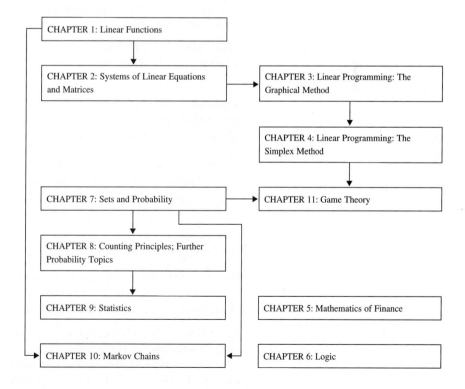 icon, to provide students with an opportunity to explain important mathematical ideas
- connections exercises, labeled with the ⟳ icon, that integrate topics presented in different sections or chapters
- exercises from actuarial or CPA exams to give students interested in these fields valuable exposure to such exams
- problems from entrance examinations for Japanese universities to give students a glimpse of international standards in math education

■KEY CONTENT CHANGES

The table of contents for this edition closely follows that of the previous edition. The main change is the addition of Chapter 6 on Logic, an important area of mathematics that instructors asked us to include. Also, Chapter 1 has been condensed to three sections, giving a more concise, cohesive presentation of the material on linear functions, and allowing the material on finite mathematics to be reached more quickly. We have made hundreds of minor changes to improve the exposition without changing the overall style of the book.

The flexibility of the text is indicated in the following chart of chapter prerequisites. As shown, the course could begin with either Chapter 1 or Chapter 7. Chapter 5 on the mathematics of finance and Chapter 6 on Logic could be covered at any time, although Chapter 6 makes a nice introduction to ideas covered in Chapter 7.

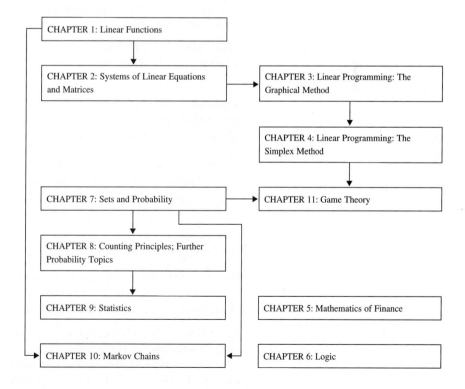

SUPPLEMENTS

For the Instructor The *Instructor's Resource Guide and Solutions Manual* (ISBN 0-321-09377-1) contains solutions to even-numbered exercises; a multiple-choice version and a short-answer version of a pretest and final exam, with answers; and a set of teaching tips.

CourseCompass (www.coursecompass.com) is a Web-based on-line course management tool powered by Blackboard. Content from the text's Web site and supplemental materials is provided, and the instructor has the option to post custom materials, forming an integrated package of course materials for the student. CourseCompass is free when bundled with the text.

TestGen-EQ with QuizMaster-EQ (dual platform for Windows and Macintosh, ISBN 0-321-09253-8) is a computerized test generator with algorithmically defined problems organized specifically for this textbook. Its user-friendly graphical interface enables instructors to select, view, edit, and add test items, then print tests in a variety of fonts and forms. A built-in question editor gives the user the power to create graphs, import graphics, insert mathematical symbols and templates, and insert variable numbers or text. An "Export to HTML" feature lets instructors create practice tests that can be posted to a Web site. Tests created with TestGen-EQ can also be displayed on Web pages when the free TestGen-EQ plug-in has been installed with Internet Explorer or Netscape Navigator. Tests created with TestGen-EQ can be used with QuizMaster-EQ, which enables students to take exams on a computer network. QuizMaster-EQ automatically grades the exams, stores results on disk, and allows the instructor to view or print a variety of reports for individual students, classes, or courses. Contact your Addison-Wesley sales consultant.

PowerPoint CD-ROM (dual platform for Windows and Macintosh, ISBN 0-321-09560-X) contains PowerPoint slides of selected key figures from each chapter of the text.

A supplementary section entitled *Digraphs and Networks* is available to you and your students at no charge should you desire to cover these topics. Contact your Addison-Wesley sales consultant.

For the Student The *Student's Solutions Manual* (ISBN 0-321-06717-7) provides solutions to odd-numbered exercises and sample chapter tests with answers.

The *Graphing Calculator Manual for Finite Mathematics and Calculus with Applications* (ISBN 0-321-09379-8) provides detailed information on using some of the more popular graphing calculators to work through examples in the text. Listings of programs for graphing calculators are also included.

The *Excel Spreadsheet Manual for Finite Mathematics and Calculus with Applications* (ISBN 0-321-09380-1) provides detailed information on using the Excel spreadsheet program to work through examples in the text.

The *AW Math Tutor Center* (www.aw.com/tutorcenter) provides assistance to students who take finite mathematics and purchase a mathematics textbook published by Addison-Wesley. Help is provided via phone, fax, or e-mail. Students who use the service will be assisted by experienced, qualified math instructors.

InterAct Math® Tutorial Software (Windows CD-ROM) (ISBN 0-321-09375-5) has been developed and designed by professional software engineers working closely with a team of experienced math educators. The software includes exercises that are linked with every objective in the textbook and require the same computational and problem-solving skills as their companion exercises in the text. Each exercise has an example and an interactive guided solution that are designed to involve students in the solution process and to help them identify precisely where they are having trouble. The software recognizes common student errors and provides students with appropriate customized feedback. With its sophisticated answer recognition capabilities, InterAct Math Tutorial Software recognizes appropriate forms of the same answer for any kind of input. It also tracks student activity and scores for each section, which can then be printed out. The software is free to qualifying adopters or can be bundled with the books for sale to students.

InterAct MathXL® 12-month registration (www.mathxl.com) (ISBN 0-201-71630-5) is a Web-based diagnostic tutorial and testing system that can be used to improve students' mathematics skills and understanding. Students take a test online, receive a customized study plan based on their test results, are directed to problems and sections covering the topics they need to study, and receive help for these topics. After completing the study plan, students can then take the test again to gauge their progress.

Companion Web site (www.aw.com/LGR) is a free site available to everyone with World Wide Web access. It provides additional text-specific resources for both students and instructors including additional Extended Applications, tutorials, quizzes, and more.

Acknowledgments

We wish to thank the following professors for their contributions in reviewing portions of this text.

Vic Akatsa, *Howard College*
Faiz Al-Rubaee, *University of North Florida*
Joe Apaloo, *Saint Francis Xavier University*
Kathryn Bollinger, *Texas A&M University*
Larry Bouldin, *Roane State Community College*
Debra Bryant, *Tennessee Technological University*
Linda Buchanan, *Howard College*
Charlie Cook, *University of South Carolina at Sumter*
Rattan Dupta, *The University of Texas at Brownsville* and *Texas Southmost College*
Irvin R. Hentzel, *Iowa State University*
J. Taylor Hollist, *State University of New York, College at Oneonta*
Jack Hoppin, *University of Arizona*
Karla Karstens, *University of Vermont*
Sandy Lofstock, *California Lutheran University*
Mary Treanor, *Valparaiso University*
Jackie Vogel, *Pellissippi State Technical College*
Daniel Wilshire, *Pennsylvania State University at Altoona*
Cathy Zucco-Teveloff, *Trinity College*

We are grateful to LaurelTech Integrated Publishing Services for doing an excellent job coordinating the Student's Solutions Manual and the Instructor's Resource Guide and Solutions Manual, an enormous and time-consuming task. We also thank Sheri Minkner and Judy Martinez for typesetting these manuals. Paul Van Erden, retired from American River College, has created an accurate and complete index for us, and Becky Troutman has compiled the extensive index of applications. Paula Grafton Young of Salem College, Jim Eckerman of American River College, and Thomas W. Polaski of Winthrop University have done a superb job of writing the graphing calculator manual, as have Paula Grafton Young and Todd Lee of Elon College in preparing the new spreadsheet manual. Frank Purcell of Twin Prime Editorial has been exceptionally helpful in writing new Extended Applications. We thank Vern E. Heeren and John Hornsby for permission to adapt and modify the chapter on logic from their book *Mathematical Ideas* (9th ed.), published by Addison-Wesley. For their invaluable help in maintaining high standards of accuracy in the answer section, we thank Tim Mogill, Julie Daberkow, S. K. Wyckoff, Thomas Wegleitner, and Pete Ciccarelli. Thomas Wakefield of Youngstown State University did a great job in finding new applications for exercises and examples. We are indebted to J. Laurie Snell of Dartmouth College whose electronic newsletter *Chance* alerted us to many applications in probability and statistics. We also want to thank Karla Harby and Mary Ann Ritchey for their editorial assistance. We especially appreciate the staff at Addison-Wesley, whose contributions have been very important in bringing this project to a successful conclusion: Greg Tobin, Christine O'Brien, and Laurie Rosatone. Finally, we wish to thank Susan Gallier of Elm Street Publishing Services for a great job as project editor.

Margaret L. Lial
Raymond N. Greenwell
Nathan P. Ritchey

INDEX OF APPLICATIONS

CHAPTER

R

Algebra Reference

In this chapter, we will review the most important topics in algebra. Knowing algebra is a fundamental prerequisite to success in higher mathematics. This algebra reference is designed for self-study; study it all at once or refer to it when needed throughout the course. Since this is a review, answers to all exercises are given in the answer section at the back of the book.

R.1 POLYNOMIALS

An expression such as $9p^4$ is a **term;** the number 9 is the **coefficient,** p is the **variable,** and 4 is the **exponent.** The expression p^4 means $p \cdot p \cdot p \cdot p$, while p^2 means $p \cdot p$, and so on. Terms having the same variable and the same exponent, such as $9x^4$ and $-3x^4$, are **like terms.** Terms that do not have both the same variable and the same exponent, such as m^2 and m^4, are **unlike terms.**

A **polynomial** is a term or a finite sum of terms in which all variables have whole number exponents, and no variables appear in denominators. Examples of polynomials include

$$5x^4 + 2x^3 + 6x, \qquad 8m^3 + 9m^2n - 6mn^2 + 3n^3, \qquad 10p, \qquad \text{and} \qquad -9.$$

Adding and Subtracting Polynomials The following properties of real numbers are useful for performing operations on polynomials.

PROPERTIES OF REAL NUMBERS

For all real numbers a, b, and c,

1. $a + b = b + a$;   **Commutative properties**
$ab = ba$;

2. $(a + b) + c = a + (b + c)$;   **Associative properties**
$(ab)c = a(bc)$;

3. $a(b + c) = ab + ac$.   **Distributive property**

EXAMPLE 1 Properties of Real Numbers

(a) $2 + x = x + 2$ Commutative property of addition

(b) $x \cdot 3 = 3x$ Commutative property of multiplication

(c) $(7x)x = 7(x \cdot x) = 7x^2$ Associative property of multiplication

(d) $3(x + 4) = 3x + 12$ Distributive property

The distributive property is used to add or subtract polynomials. Only like terms may be added or subtracted. For example,

$$12y^4 + 6y^4 = (12 + 6)y^4 = 18y^4,$$

and

$$-2m^2 + 8m^2 = (-2 + 8)m^2 = 6m^2,$$

but the polynomial $8y^4 + 2y^5$ cannot be further simplified. To subtract polynomials, use the facts that $-(a + b) = -a - b$ and $-(a - b) = -a + b$. In the next example, we show how to add and subtract polynomials.

EXAMPLE 2 Adding and Subtracting Polynomials
Add or subtract as indicated.

(a) $(8x^3 - 4x^2 + 6x) + (3x^3 + 5x^2 - 9x + 8)$

Solution Combine like terms.
$(8x^3 - 4x^2 + 6x) + (3x^3 + 5x^2 - 9x + 8)$
$$= (8x^3 + 3x^3) + (-4x^2 + 5x^2) + (6x - 9x) + 8$$
$$= 11x^3 + x^2 - 3x + 8$$

(b) $(-4x^4 + 6x^3 - 9x^2 - 12) + (-3x^3 + 8x^2 - 11x + 7)$

Solution Combining like terms as before yields
$-4x^4 + 3x^3 - x^2 - 11x - 5$.

(c) $(2x^2 - 11x + 8) - (7x^2 - 6x + 2)$

Solution Distributing the minus sign yields
$(2x^2 - 11x + 8) + (-7x^2 + 6x - 2)$
$$= -5x^2 - 5x + 6.$$

Multiplying Polynomials

The distributive property is also used to multiply polynomials, along with the fact that $a^m \cdot a^n = a^{m+n}$. For example,

$$x \cdot x = x^1 \cdot x^1 = x^2 \qquad \text{and} \qquad x^2 \cdot x^5 = x^7.$$

EXAMPLE 3 Multiplying Polynomials
Multiply.

(a) $8x(6x - 4)$

Solution
$8x(6x - 4) = 8x(6x) - 8x(4)$
$$= 48x^2 - 32x$$

(b) $(3p - 2)(p^2 + 5p - 1)$

Solution
$(3p - 2)(p^2 + 5p - 1)$
$$= 3p(p^2 + 5p - 1) - 2(p^2 + 5p - 1)$$
$$= 3p(p^2) + 3p(5p) + 3p(-1) - 2(p^2) - 2(5p) - 2(-1)$$
$$= 3p^3 + 15p^2 - 3p - 2p^2 - 10p + 2$$
$$= 3p^3 + 13p^2 - 13p + 2$$

(c) $(x + 2)(x + 3)(x - 4)$

Solution
$(x + 2)(x + 3)(x - 4)$
$$= [(x + 2)(x + 3)](x - 4)$$
$$= (x^2 + 2x + 3x + 6)(x - 4)$$
$$= (x^2 + 5x + 6)(x - 4)$$
$$= x^3 + 5x^2 + 6x - 4x^2 - 20x - 24$$
$$= x^3 + x^2 - 14x - 24$$

A **binomial** is a polynomial with exactly two terms, such as $2x + 1$ or $m + n$. When two binomials are multiplied, the FOIL method (First, Outer, Inner, Last) is used as a memory aid.

EXAMPLE 4 Multiplying Polynomials
Find $(2m - 5)(m + 4)$ using the FOIL method.

Solution

$$(2m - 5)(m + 4) = \overset{F}{(2m)(m)} + \overset{O}{(2m)(4)} + \overset{I}{(-5)(m)} + \overset{L}{(-5)(4)}$$
$$= 2m^2 + 8m - 5m - 20$$
$$= 2m^2 + 3m - 20$$

EXAMPLE 5 Multiplying Polynomials
Find $(2k - 5)^2$.

Solution Use FOIL.
$$(2k - 5)^2 = (2k - 5)(2k - 5)$$
$$= 4k^2 - 10k - 10k + 25$$
$$= 4k^2 - 20k + 25$$

Notice that the product of the square of a binomial is the square of the first term, $(2k)^2$, plus twice the product of the two terms, $(2)(2k)(-5)$, plus the square of the last term, $(-5)^2$.

CAUTION Avoid the common error of writing $(x + y)^2 = x^2 + y^2$. As Example 5 shows, the square of a binomial has three terms, so

$$(x + y)^2 = x^2 + 2xy + y^2.$$

Furthermore, higher powers of a binomial also result in more than two terms. For example, verify by multiplication that

$$(x + y)^3 = x^3 + 3x^2y + 3xy^2 + y^3.$$

Remember, for any value of $n \neq 1$,

$$(x + y)^n \neq x^n + y^n.$$

R.1 EXERCISES

Perform the indicated operations.

1. $(2x^2 - 6x + 11) + (-3x^2 + 7x - 2)$

2. $(-4y^2 - 3y + 8) - (2y^2 - 6y - 2)$

3. $-3(4q^2 - 3q + 2) + 2(-q^2 + q - 4)$

4. $2(3r^2 + 4r + 2) - 3(-r^2 + 4r - 5)$

5. $(.613x^2 - 4.215x + .892) - .47(2x^2 - 3x + 5)$

6. $.83(5r^2 - 2r + 7) - (7.12r^2 + 6.423r - 2)$

7. $-9m(2m^2 + 3m - 1)$

8. $(6k - 1)(2k - 3)$

9. $(5r - 3s)(5r + 4s)$

10. $(9k + q)(2k - q)$

11. $\left(\frac{2}{5}y + \frac{1}{8}z\right)\left(\frac{3}{5}y + \frac{1}{2}z\right)$

12. $\left(\frac{3}{4}r - \frac{2}{3}s\right)\left(\frac{5}{4}r + \frac{1}{3}s\right)$

13. $(12x - 1)(12x + 1)$

14. $(6m + 5)(6m - 5)$

15. $(3p - 1)(9p^2 + 3p + 1)$

17. $(2m + 1)(4m^2 - 2m + 1)$

19. $(m - n + k)(m + 2n - 3k)$

21. $(x + 1)(x + 2)(x + 3)$

23. $(3a + b)^2$

16. $(2p - 1)(3p^2 - 4p + 5)$

18. $(k + 2)(12k^3 - 3k^2 + k + 1)$

20. $(r - 3s + t)(2r - s + t)$

22. $(x - 1)(x + 2)(x - 3)$

24. $(x - 2y)^3$

■ R.2 FACTORING

Multiplication of polynomials relies on the distributive property. The reverse process, where a polynomial is written as a product of other polynomials, is called **factoring.** For example, one way to factor the number 18 is to write it as the product $9 \cdot 2$; both 9 and 2 are **factors** of 18. Usually, only integers are used as factors of integers. The number 18 can also be written with three integer factors as $2 \cdot 3 \cdot 3$.

The Greatest Common Factor   To factor the algebraic expression $15m + 45$, first note that both $15m$ and 45 are divisible by 15; $15m = 15 \cdot m$ and $45 = 15 \cdot 3$. By the distributive property,

$$15m + 45 = 15 \cdot m + 15 \cdot 3 = 15(m + 3).$$

Both 15 and $m + 3$ are factors of $15m + 45$. Since 15 divides into both terms of $15m + 45$ (and is the largest number that will do so), 15 is the **greatest common factor** for the polynomial $15m + 45$. The process of writing $15m + 45$ as $15(m + 3)$ is often called **factoring out** the greatest common factor.

EXAMPLE 1   Factoring

Factor out the greatest common factor.

(a) $12p - 18q$

Solution   Both $12p$ and $18q$ are divisible by 6. Therefore,

$$12p - 18q = 6 \cdot 2p - 6 \cdot 3q = 6(2p - 3q).$$

(b) $8x^3 - 9x^2 + 15x$

Solution   Each of these terms is divisible by x.

$$8x^3 - 9x^2 + 15x = (8x^2) \cdot x - (9x) \cdot x + 15 \cdot x$$
$$= x(8x^2 - 9x + 15) \quad \text{or} \quad (8x^2 - 9x + 15)x \quad ■$$

One can always check factorization by finding the product of the factors and comparing it to the original expression.

CAUTION   When factoring out the greatest common factor in an expression like $2x^2 + x$, be careful to remember the 1 in the second term.

$$2x^2 + x = 2x^2 + 1x = x(2x + 1), \text{ not } x(2x).$$

Factoring Trinomials A polynomial that has no greatest common factor (other than 1) may still be factorable. For example, the polynomial $x^2 + 5x + 6$ can be factored as $(x + 2)(x + 3)$. To see that this is correct, find the product $(x + 2)(x + 3)$; you should get $x^2 + 5x + 6$. A polynomial such as this with three terms is called a **trinomial.** To factor the trinomial $x^2 + 5x + 6$, where the coefficient of x^2 is 1, we use FOIL backwards.

EXAMPLE 2 Factoring a Trinomial
Factor $y^2 + 8y + 15$.

Solution Since the coefficient of y^2 is 1, factor by finding two numbers whose *product* is 15 and whose *sum* is 8. Since the constant and the middle term are positive, the numbers must both be positive. Begin by listing all pairs of positive integers having a product of 15. As you do this, also form the sum of each pair of numbers.

Products	Sums
$15 \cdot 1 = 15$	$15 + 1 = 16$
$5 \cdot 3 = 15$	$5 + 3 = 8$

The numbers 5 and 3 have a product of 15 and a sum of 8. Thus, $y^2 + 8y + 15$ factors as

$$y^2 + 8y + 15 = (y + 5)(y + 3).$$

The answer also can be written as $(y + 3)(y + 5)$.

If the coefficient of the squared term is *not* 1, work as shown below.

EXAMPLE 3 Factoring a Trinomial
Factor $2x^2 + 9xy - 5y^2$.

Solution The factors of $2x^2$ are $2x$ and x; the possible factors of $-5y^2$ are $-5y$ and y, or $5y$ and $-y$. Try various combinations of these factors until one works (if, indeed, any work). For example, try the product $(2x + 5y)(x - y)$.

$$(2x + 5y)(x - y) = 2x^2 - 2xy + 5xy - 5y^2$$
$$= 2x^2 + 3xy - 5y^2$$

This product is not correct, so try another combination.

$$(2x - y)(x + 5y) = 2x^2 + 10xy - xy - 5y^2$$
$$= 2x^2 + 9xy - 5y^2$$

Since this combination gives the correct polynomial,

$$2x^2 + 9xy - 5y^2 = (2x - y)(x + 5y).$$

Special Factorizations Four special factorizations occur so often that they are listed here for future reference.

SPECIAL FACTORIZATIONS

$x^2 - y^2 = (x + y)(x - y)$	**Difference of two squares**
$x^2 + 2xy + y^2 = (x + y)^2$	**Perfect square**
$x^3 - y^3 = (x - y)(x^2 + xy + y^2)$	**Difference of two cubes**
$x^3 + y^3 = (x + y)(x^2 - xy + y^2)$	**Sum of two cubes**

A polynomial that cannot be factored is called a **prime polynomial.**

EXAMPLE 4 Factoring

Factor each of the following.

(a) $64p^2 - 49q^2 = (8p)^2 - (7q)^2 = (8p + 7q)(8p - 7q)$

(b) $x^2 + 36$ is a prime polynomial.

(c) $x^2 + 12x + 36 = (x + 6)^2$

(d) $9y^2 - 24yz + 16z^2 = (3y - 4z)^2$

(e) $y^3 - 8 = y^3 - 2^3 = (y - 2)(y^2 + 2y + 4)$

(f) $m^3 + 125 = m^3 + 5^3 = (m + 5)(m^2 - 5m + 25)$

(g) $8k^3 - 27z^3 = (2k)^3 - (3z)^3 = (2k - 3z)(4k^2 + 6kz + 9z^2)$

> **CAUTION** In factoring, always look for a common factor first. Since $36x^2 - 4y^2$ has a common factor of 4,
>
> $$36x^2 - 4y^2 = 4(9x^2 - y^2) = 4(3x + y)(3x - y).$$
>
> It would be incomplete to factor it as
>
> $$36x^2 - 4y^2 = (6x + 2y)(6x - 2y)$$
>
> since each factor can be factored still further. To *factor* means to factor completely, so that each polynomial factor is prime.

R.2 EXERCISES

Factor each of the following. If a polynomial cannot be factored, write prime. *Factor out the greatest common factor as necessary.*

1. $8a^3 - 16a^2 + 24a$

2. $3y^3 + 24y^2 + 9y$

3. $25p^4 - 20p^3q + 100p^2q^2$

4. $60m^4 - 120m^3n + 50m^2n^2$

5. $m^2 + 9m + 14$

6. $x^2 + 4x - 5$

7. $z^2 + 9z + 20$

8. $b^2 - 8b + 7$

9. $a^2 - 6ab + 5b^2$

10. $s^2 + 2st - 35t^2$

11. $y^2 - 4yz - 21z^2$

12. $6a^2 - 48a - 120$

13. $3m^3 + 12m^2 + 9m$

14. $2x^2 - 5x - 3$

15. $3a^2 + 10a + 7$

16. $2a^2 - 17a + 30$

17. $15y^2 + y - 2$

18. $21m^2 + 13mn + 2n^2$

19. $24a^4 + 10a^3b - 4a^2b^2$

20. $32z^5 - 20z^4a - 12z^3a^2$

21. $x^2 - 64$

22. $9m^2 - 25$

23. $121a^2 - 100$

24. $9x^2 + 64$

25. $z^2 + 14zy + 49y^2$

26. $m^2 - 6mn + 9n^2$

27. $9p^2 - 24p + 16$

28. $a^3 - 216$

29. $8r^3 - 27s^3$

30. $64m^3 + 125$

31. $x^4 - y^4$

32. $16a^4 - 81b^4$

■ R.3 RATIONAL EXPRESSIONS

Many algebraic fractions are **rational expressions,** which are quotients of polynomials with nonzero denominators. Examples include

$$\frac{8}{x-1}, \qquad \frac{3x^2 + 4x}{5x - 6}, \qquad \text{and} \qquad \frac{2y + 1}{y^2}.$$

Properties for working with rational expressions are summarized next.

PROPERTIES OF RATIONAL EXPRESSIONS

For all mathematical expressions P, Q, R, and S, with Q and $S \neq 0$:

$$\frac{P}{Q} = \frac{PS}{QS} \qquad \qquad \textbf{Fundamental property}$$

$$\frac{P}{Q} + \frac{R}{Q} = \frac{P+R}{Q} \qquad \qquad \textbf{Addition}$$

$$\frac{P}{Q} - \frac{R}{Q} = \frac{P-R}{Q} \qquad \qquad \textbf{Subtraction}$$

$$\frac{P}{Q} \cdot \frac{R}{S} = \frac{PR}{QS} \qquad \qquad \textbf{Multiplication}$$

$$\frac{P}{Q} \div \frac{R}{S} = \frac{P}{Q} \cdot \frac{S}{R} \quad (R \neq 0) \qquad \textbf{Division}$$

When using the fundamental property to write a rational expression in lowest terms, we may need to use the fact that $\dfrac{a^m}{a^n} = a^{m-n}$. For example,

$$\frac{x^2}{3x} = \frac{1x^2}{3x} = \frac{1}{3} \cdot \frac{x^2}{x} = \frac{1}{3}x.$$

EXAMPLE 1 Reducing Rational Expressions

Write each rational expression in lowest terms, that is, reduce the expression as much as possible.

(a) $\dfrac{8x + 16}{4} = \dfrac{8(x + 2)}{4} = \dfrac{4 \cdot 2(x + 2)}{4} = 2(x + 2)$

Factor both the numerator and denominator in order to identify any common factors, which have a quotient of 1. The answer could also be written as $2x + 4$.

(b) $\dfrac{k^2 + 7k + 12}{k^2 + 2k - 3} = \dfrac{(k + 4)(k + 3)}{(k - 1)(k + 3)} = \dfrac{k + 4}{k - 1}$

The answer cannot be further reduced.

CAUTION One of the most common errors in algebra involves incorrect use of the fundamental property of rational expressions. Only common *factors* may be divided or "canceled." It is essential to factor rational expressions before writing them in lowest terms. In Example 1(b), for instance, it is not correct to "cancel" k^2 (or cancel k, or divide 12 by -3) because the additions and subtraction must be performed first. Here they cannot be performed, so it is not possible to divide. After factoring, however, the fundamental property can be used to write the expression in lowest terms.

EXAMPLE 2 Combining Rational Expressions

Perform each operation.

(a) $\dfrac{3y + 9}{6} \cdot \dfrac{18}{5y + 15}$

Solution Factor where possible, then multiply numerators and denominators and reduce to lowest terms.

$$\frac{3y + 9}{6} \cdot \frac{18}{5y + 15} = \frac{3(y + 3)}{6} \cdot \frac{18}{5(y + 3)}$$

$$= \frac{3 \cdot 18(y + 3)}{6 \cdot 5(y + 3)}$$

$$= \frac{3 \cdot 6 \cdot 3(y + 3)}{6 \cdot 5(y + 3)} = \frac{3 \cdot 3}{5} = \frac{9}{5}$$

(b) $\dfrac{m^2 + 5m + 6}{m + 3} \cdot \dfrac{m}{m^2 + 3m + 2}$

Solution Factor where possible.

$$\frac{(m + 2)(m + 3)}{m + 3} \cdot \frac{m}{(m + 2)(m + 1)}$$

$$= \frac{m(m + 2)(m + 3)}{(m + 3)(m + 2)(m + 1)} = \frac{m}{m + 1}$$

(c) $\dfrac{9p - 36}{12} \div \dfrac{5(p - 4)}{18}$

Solution Use the division property of rational expressions.

$$\frac{9p - 36}{12} \cdot \frac{18}{5(p - 4)} \quad \text{Invert and multiply.}$$

$$= \frac{9(p - 4)}{6 \cdot 2} \cdot \frac{6 \cdot 3}{5(p - 4)} = \frac{27}{10}$$

(d) $\dfrac{4}{5k} - \dfrac{11}{5k}$

Solution As shown in the list of properties, to subtract two rational expressions that have the same denominators, we subtract the numerators while keeping the same denominator.

$$\frac{4}{5k} - \frac{11}{5k} = \frac{4 - 11}{5k} = -\frac{7}{5k}$$

(e) $\dfrac{7}{p} + \dfrac{9}{2p} + \dfrac{1}{3p}$

Solution These three fractions cannot be added until their denominators are the same. A **common denominator** into which p, $2p$, and $3p$ all divide is $6p$. Note that $12p$ is also a common denominator, but $6p$ is the **least common denominator.** Use the fundamental property to rewrite each rational expression with a denominator of $6p$.

$$\dfrac{7}{p} + \dfrac{9}{2p} + \dfrac{1}{3p} = \dfrac{6 \cdot 7}{6 \cdot p} + \dfrac{3 \cdot 9}{3 \cdot 2p} + \dfrac{2 \cdot 1}{2 \cdot 3p}$$

$$= \dfrac{42}{6p} + \dfrac{27}{6p} + \dfrac{2}{6p}$$

$$= \dfrac{42 + 27 + 2}{6p}$$

$$= \dfrac{71}{6p}$$

(f) $\dfrac{x + 1}{x^2 + 5x + 6} - \dfrac{5x - 1}{x^2 - x - 12}$

Solution To find the least common denominator, first factor each denominator. Then change each fraction so they all have the same denominator, being careful to multiply only by quotients that equal 1.

$$\dfrac{x + 1}{x^2 + 5x + 6} - \dfrac{5x - 1}{x^2 - x - 12}$$

$$= \dfrac{x + 1}{(x + 2)(x + 3)} - \dfrac{5x - 1}{(x + 3)(x - 4)}$$

$$= \dfrac{x + 1}{(x + 2)(x + 3)} \cdot \dfrac{(x - 4)}{(x - 4)} - \dfrac{5x - 1}{(x + 3)(x - 4)} \cdot \dfrac{(x + 2)}{(x + 2)}$$

$$= \dfrac{(x^2 - 3x - 4) - (5x^2 + 9x - 2)}{(x + 2)(x + 3)(x - 4)}$$

$$= \dfrac{-4x^2 - 12x - 2}{(x + 2)(x + 3)(x - 4)}$$

$$= \dfrac{-2(2x^2 + 6x + 1)}{(x + 2)(x + 3)(x - 4)}$$

Because the numerator cannot be factored further, we leave our answer in this form. We could also multiply out the denominator, but factored form is usually more useful.

R.3 EXERCISES

Write each rational expression in lowest terms.

1. $\dfrac{7z^2}{14z}$

2. $\dfrac{25p^3}{10p^2}$

3. $\dfrac{8k + 16}{9k + 18}$

4. $\dfrac{3(t + 5)}{(t + 5)(t - 3)}$

5. $\dfrac{8x^2 + 16x}{4x^2}$

6. $\dfrac{36y^2 + 72y}{9y}$

7. $\dfrac{m^2 - 4m + 4}{m^2 + m - 6}$

8. $\dfrac{r^2 - r - 6}{r^2 + r - 12}$

9. $\dfrac{x^2 + 3x - 4}{x^2 - 1}$ **10.** $\dfrac{z^2 - 5z + 6}{z^2 - 4}$ **11.** $\dfrac{8m^2 + 6m - 9}{16m^2 - 9}$ **12.** $\dfrac{6y^2 + 11y + 4}{3y^2 + 7y + 4}$

Perform the indicated operations.

13. $\dfrac{9k^2}{25} \cdot \dfrac{5}{3k}$

14. $\dfrac{15p^3}{9p^2} \div \dfrac{6p}{10p^2}$

15. $\dfrac{a + b}{2p} \cdot \dfrac{12}{5(a + b)}$

16. $\dfrac{a - 3}{16} \div \dfrac{a - 3}{32}$

17. $\dfrac{2k + 8}{6} \div \dfrac{3k + 12}{2}$

18. $\dfrac{9y - 18}{6y + 12} \cdot \dfrac{3y + 6}{15y - 30}$

19. $\dfrac{4a + 12}{2a - 10} \div \dfrac{a^2 - 9}{a^2 - a - 20}$

20. $\dfrac{6r - 18}{9r^2 + 6r - 24} \cdot \dfrac{12r - 16}{4r - 12}$

21. $\dfrac{k^2 - k - 6}{k^2 + k - 12} \cdot \dfrac{k^2 + 3k - 4}{k^2 + 2k - 3}$

22. $\dfrac{m^2 + 3m + 2}{m^2 + 5m + 4} \div \dfrac{m^2 + 5m + 6}{m^2 + 10m + 24}$

23. $\dfrac{2m^2 - 5m - 12}{m^2 - 10m + 24} \div \dfrac{4m^2 - 9}{m^2 - 9m + 18}$

24. $\dfrac{6n^2 - 5n - 6}{6n^2 + 5n - 6} \cdot \dfrac{12n^2 - 17n + 6}{12n^2 - n - 6}$

25. $\dfrac{a + 1}{2} - \dfrac{a - 1}{2}$

26. $\dfrac{3}{p} + \dfrac{1}{2}$

27. $\dfrac{2}{y} - \dfrac{1}{4}$

28. $\dfrac{1}{6m} + \dfrac{2}{5m} + \dfrac{4}{m}$

29. $\dfrac{1}{m - 1} + \dfrac{2}{m}$

30. $\dfrac{6}{r} - \dfrac{5}{r - 2}$

31. $\dfrac{8}{3(a - 1)} + \dfrac{2}{a - 1}$

32. $\dfrac{2}{5(k - 2)} + \dfrac{3}{4(k - 2)}$

33. $\dfrac{2}{x^2 - 2x - 3} + \dfrac{5}{x^2 - x - 6}$

34. $\dfrac{2y}{y^2 + 7y + 12} - \dfrac{y}{y^2 + 5y + 6}$

35. $\dfrac{3k}{2k^2 + 3k - 2} - \dfrac{2k}{2k^2 - 7k + 3}$

36. $\dfrac{4m}{3m^2 + 7m - 6} - \dfrac{m}{3m^2 - 14m + 8}$

37. $\dfrac{2}{a + 2} + \dfrac{1}{a} + \dfrac{a - 1}{a^2 + 2a}$

38. $\dfrac{5x + 2}{x^2 - 1} + \dfrac{3}{x^2 + x} - \dfrac{1}{x^2 - x}$

■ R.4   EQUATIONS

Linear Equations Equations that can be written in the form $ax + b = 0$, where a and b are real numbers, with $a \neq 0$, are **linear equations.** Examples of linear equations include $5y + 9 = 16$, $8x = 4$, and $-3p + 5 = -8$. Equations that are *not* linear include absolute value equations such as $|x| = 4$. The following properties are used to solve linear equations.

PROPERTIES OF EQUALITY

For all real numbers a, b, and c:

1. If $a = b$, then $a + c = b + c$.   **Addition property of equality**
(The same number may be added
to both sides of an equation.)

2. If $a = b$, then $ac = bc$.   **Multiplication property of**
(Both sides of an equation may be   **equality**
multiplied by the same number.)

EXAMPLE 1 Solving Linear Equations

(a) If $x - 2 = 3$, then $x = 2 + 3 = 5$ Addition property of equality

(b) If $x/2 = 3$, then $x = 2 \cdot 3 = 6$ Multiplication property of equality

The following example shows how these properties are used to solve linear equations. Of course, the solutions should always be checked by substitution in the original equation.

EXAMPLE 2 Solving Linear Equations

Solve $2x - 5 + 8 = 3x + 2(2 - 3x)$.

Solution

$$2x - 5 + 8 = 3x + 4 - 6x \qquad \text{Distributive property}$$

$$2x + 3 = -3x + 4 \qquad \text{Combine like terms.}$$

$$5x + 3 = 4 \qquad \text{Add } 3x \text{ to both sides.}$$

$$5x = 1 \qquad \text{Add } -3 \text{ to both sides.}$$

$$x = \frac{1}{5} \qquad \text{Multiply both sides by } \tfrac{1}{5}.$$

Check by substituting in the original equation. The left side becomes $2(1/5) - 5 + 8$ and the right side becomes $3(1/5) + 2(2 - 3(1/5))$. Verify that both of these expressions simplify to $17/5$.

Quadratic Equations An equation with 2 as the highest exponent of the variable is a *quadratic equation*. A **quadratic equation** has the form $ax^2 + bx + c = 0$, where a, b, and c are real numbers and $a \neq 0$. A quadratic equation written in the form $ax^2 + bx + c = 0$ is said to be in **standard form.**

The simplest way to solve a quadratic equation, but one that is not always applicable, is by factoring. This method depends on the **zero-factor property.**

ZERO-FACTOR PROPERTY

If a and b are real numbers, with $ab = 0$, then

$$a = 0, b = 0, \text{ or both.}$$

EXAMPLE 3 Solving Quadratic Equations

Solve $6r^2 + 7r = 3$.

Solution First write the equation in standard form.

$$6r^2 + 7r - 3 = 0$$

Now factor $6r^2 + 7r - 3$ to get

$$(3r - 1)(2r + 3) = 0.$$

By the zero-factor property, the product $(3r - 1)(2r + 3)$ can equal 0 if and only if

$$3r - 1 = 0 \qquad \text{or} \qquad 2r + 3 = 0.$$

Solve each of these equations separately to find that the solutions are $1/3$ and $-3/2$. Check these solutions by substituting them in the original equation.

▌**CAUTION** Remember, the zero-factor property requires that the product of two (or more) factors be equal to *zero,* not some other quantity. It would be incorrect to use the zero-factor property with an equation in the form $(x + 3)(x - 1) = 4$, for example.

If a quadratic equation cannot be solved easily by factoring, use the *quadratic formula.* (The derivation of the quadratic formula is given in most algebra books.)

QUADRATIC FORMULA

The solutions of the quadratic equation $ax^2 + bx + c = 0$, where $a \neq 0$, are given by

$$x = \frac{-b \pm \sqrt{b^2 - 4ac}}{2a}.$$

EXAMPLE 4   The Quadratic Formula

Solve $x^2 - 4x - 5 = 0$ by the quadratic formula.

Solution The equation is already in standard form (it has 0 alone on one side of the equals sign), so the values of a, b, and c from the quadratic formula are easily identified. The coefficient of the squared term gives the value of a; here, $a = 1$. Also, $b = -4$ and $c = -5$. (Be careful to use the correct signs.) Substitute these values into the quadratic formula.

$$x = \frac{-(-4) \pm \sqrt{(-4)^2 - 4(1)(-5)}}{2(1)} \qquad \text{Let } a = 1,\, b = -4,\, c = -5.$$

$$x = \frac{4 \pm \sqrt{16 + 20}}{2} \qquad (-4)^2 = (-4)(-4) = 16$$

$$x = \frac{4 \pm 6}{2} \qquad \sqrt{16 + 20} = \sqrt{36} = 6$$

The $\pm$ sign represents the two solutions of the equation. To find both of the solutions, first use $+$ and then use $-$.

$$x = \frac{4 + 6}{2} = \frac{10}{2} = 5 \qquad \text{or} \qquad x = \frac{4 - 6}{2} = \frac{-2}{2} = -1$$

The two solutions are 5 and -1.

▌**CAUTION** Notice in the quadratic formula that the square root is added to or subtracted from the value of $-b$ *before* dividing by $2a$.

EXAMPLE 5   Quadratic Formula

Solve $x^2 + 1 = 4x$.

Solution First, add $-4x$ on both sides of the equals sign in order to get the equation in standard form.

$$x^2 - 4x + 1 = 0$$

Now identify the letters a, b, and c. Here $a = 1$, $b = -4$, and $c = 1$. Substitute these numbers into the quadratic formula.

$$x = \frac{-(-4) \pm \sqrt{(-4)^2 - 4(1)(1)}}{2(1)}$$

$$= \frac{4 \pm \sqrt{16 - 4}}{2}$$

$$= \frac{4 \pm \sqrt{12}}{2}$$

Simplify the solutions by writing $\sqrt{12}$ as $\sqrt{4 \cdot 3} = \sqrt{4} \cdot \sqrt{3} = 2\sqrt{3}$. Substituting $2\sqrt{3}$ for $\sqrt{12}$ gives

$$x = \frac{4 \pm 2\sqrt{3}}{2}$$

$$= \frac{2(2 \pm \sqrt{3})}{2} \qquad \text{Factor } 4 \pm 2\sqrt{3}.$$

$$= 2 \pm \sqrt{3}. \qquad \text{Reduce to lowest terms.}$$

The two solutions are $2 + \sqrt{3}$ and $2 - \sqrt{3}$.

The exact values of the solutions are $2 + \sqrt{3}$ and $2 - \sqrt{3}$. The $\sqrt{}$ key on a calculator gives decimal approximations of these solutions (to the nearest thousandth):

$$2 + \sqrt{3} \approx 2 + 1.732 = 3.732*$$

$$2 - \sqrt{3} \approx 2 - 1.732 = .268$$

NOTE Sometimes the quadratic formula will give a result with a negative number under the radical sign, such as $3 \pm \sqrt{-5}$. A solution of this type is not a real number. Since this text deals only with real numbers, such solutions cannot be used.

Equations with Fractions

When an equation includes fractions, first eliminate all denominators by multiplying both sides of the equation by a common denominator, a number that can be divided (with no remainder) by each denominator in the equation. When an equation involves fractions with variable denominators, it is *necessary* to check all solutions in the original equation to be sure that no solution will lead to a zero denominator.

EXAMPLE 6 Solving Rational Equations

Solve each equation.

(a) $\dfrac{r}{10} - \dfrac{2}{15} = \dfrac{3r}{20} - \dfrac{1}{5}$

Solution The denominators are 10, 15, 20, and 5. Each of these numbers can be divided into 60, so 60 is a common denominator. Multiply both sides of the equation by 60 and use the distributive property. (If a common denomi-

---

*The symbol $\approx$ means "is approximately equal to."

nator cannot be found easily, all the denominators in the problem can be multiplied together to produce one.)

$$\frac{r}{10} - \frac{2}{15} = \frac{3r}{20} - \frac{1}{5}$$

$$60\left(\frac{r}{10} - \frac{2}{15}\right) = 60\left(\frac{3r}{20} - \frac{1}{5}\right) \qquad \text{Multiply by the common denominator.}$$

$$60\left(\frac{r}{10}\right) - 60\left(\frac{2}{15}\right) = 60\left(\frac{3r}{20}\right) - 60\left(\frac{1}{5}\right) \qquad \text{Distributive property}$$

$$6r - 8 = 9r - 12$$

Add $-9r$ and 8 to both sides.

$$6r - 8 + (-9r) + 8 = 9r - 12 + (-9r) + 8$$

$$-3r = -4$$

$$r = \frac{4}{3} \qquad \text{Multiply each side by } -\frac{1}{3}.$$

Check by substituting into the original equation.

(b) $\dfrac{3}{x^2} - 12 = 0$

Solution Begin by multiplying both sides of the equation by x^2 to get $3 - 12x^2 = 0$. This equation could be solved by using the quadratic formula with $a = -12$, $b = 0$, and $c = 3$. Another method, which works well for the type of quadratic equation in which $b = 0$, is shown below.

$$3 - 12x^2 = 0$$

$$3 = 12x^2 \qquad \text{Add } 12x^2.$$

$$\frac{1}{4} = x^2 \qquad \text{Multiply by } \frac{1}{12}.$$

$$\pm\frac{1}{2} = x \qquad \text{Take square roots.}$$

Verify that there are two solutions, $-1/2$ and $1/2$.

(c) $\dfrac{2}{k} - \dfrac{3k}{k + 2} = \dfrac{k}{k^2 + 2k}$

Solution Factor $k^2 + 2k$ as $k(k + 2)$. The least common denominator for all the fractions is $k(k + 2)$. Multiplying both sides by $k(k + 2)$ gives the following.

$$2(k + 2) - 3k(k) = k$$

$$2k + 4 - 3k^2 = k \qquad \text{Distributive property}$$

$$-3k^2 + k + 4 = 0 \qquad \text{Add } -k\text{; rearrange terms.}$$

$$3k^2 - k - 4 = 0 \qquad \text{Multiply by } -1.$$

$$(3k - 4)(k + 1) = 0 \qquad \text{Factor.}$$

$$3k - 4 = 0 \qquad \text{or} \qquad k + 1 = 0$$

$$k = \frac{4}{3} \qquad\qquad\qquad k = -1$$

Verify that the solutions are $4/3$ and -1.

CAUTION It is possible to get, as a solution of a rational equation, a number that makes one or more of the denominators in the original equation equal to zero. That number is not a solution, so it is *necessary* to check all potential solutions of rational equations. These introduced solutions are called **extraneous solutions.**

EXAMPLE 7   Solving Rational Equations

Solve $\dfrac{2}{x - 3} + \dfrac{1}{x} = \dfrac{6}{x(x - 3)}$.

Solution The common denominator is $x(x - 3)$. Multiply both sides by $x(x - 3)$ and solve the resulting equation.

$$\frac{2}{x - 3} + \frac{1}{x} = \frac{6}{x(x - 3)}$$
$$2x + x - 3 = 6$$
$$3x = 9$$
$$x = 3$$

Checking this potential solution by substitution in the original equation shows that 3 makes two denominators 0. Thus 3 cannot be a solution, so there is no solution for this equation.

R.4 EXERCISES

Solve each equation.

1. $.2m - .5 = .1m + .7$

2. $\dfrac{5}{6}k - 2k + \dfrac{1}{3} = \dfrac{2}{3}$

3. $2x + 8 = x - 4$

4. $5x + 2 = 8 - 3x$

5. $3r + 2 - 5(r + 1) = 6r + 4$

6. $5(a + 3) + 4a - 5 = -(2a - 4)$

7. $2[m - (4 + 2m) + 3] = 2m + 2$

8. $4[2p - (3 - p) + 5] = -7p - 2$

Solve each of the following equations by factoring or by using the quadratic formula. If the solutions involve square roots, give both the exact solutions and the approximate solutions to three decimal places.

9. $x^2 + 5x + 6 = 0$

10. $x^2 = 3 + 2x$

11. $m^2 + 16 = 8m$

12. $2k^2 - k = 10$

13. $6x^2 - 5x = 4$

14. $m(m - 7) = -10$

15. $9x^2 - 16 = 0$

16. $z(2z + 7) = 4$

17. $12y^2 - 48y = 0$

18. $3x^2 - 5x + 1 = 0$

19. $2m^2 = m + 4$

20. $p^2 + p - 1 = 0$

21. $k^2 - 10k = -20$

22. $2x^2 + 12x + 5 = 0$

23. $2r^2 - 7r + 5 = 0$

24. $2x^2 - 7x + 30 = 0$

25. $3k^2 + k = 6$

26. $5m^2 + 5m = 0$

Solve each of the following equations.

27. $\dfrac{3x - 2}{7} = \dfrac{x + 2}{5}$

28. $\dfrac{x}{3} - 7 = 6 - \dfrac{3x}{4}$

29. $\dfrac{4}{x - 3} - \dfrac{8}{2x + 5} + \dfrac{3}{x - 3} = 0$

30. $\dfrac{5}{2p + 3} - \dfrac{3}{p - 2} = \dfrac{4}{2p + 3}$

31. $\dfrac{2}{m} + \dfrac{m}{m + 3} = \dfrac{3m}{m^2 + 3m}$

32. $\dfrac{2y}{y - 1} = \dfrac{5}{y} + \dfrac{10 - 8y}{y^2 - y}$

33. $\dfrac{1}{x - 2} - \dfrac{3x}{x - 1} = \dfrac{2x + 1}{x^2 - 3x + 2}$ **34.** $\dfrac{5}{a} + \dfrac{-7}{a + 1} = \dfrac{a^2 - 2a + 4}{a^2 + a}$ **35.** $\dfrac{2b^2 + 5b - 8}{b^2 + 2b} + \dfrac{5}{b + 2} = -\dfrac{3}{b}$

36. $\dfrac{2}{x^2 - 2x - 3} + \dfrac{5}{x^2 - x - 6} = \dfrac{1}{x^2 + 3x + 2}$ **37.** $\dfrac{2}{y^2 + 7y + 12} - \dfrac{1}{y^2 + 5y + 6} = \dfrac{5}{y^2 + 6y + 8}$

■ R.5 INEQUALITIES

To write that one number is greater than or less than another number, we use the following symbols.

> **INEQUALITY SYMBOLS**
>
> $<$ means *is less than* $\leq$ means *is less than or equal to*
> $>$ means *is greater than* $\geq$ means *is greater than or equal to*

Linear Inequalities An equation states that two expressions are equal; an **inequality** states that they are unequal. A **linear inequality** is an inequality that can be simplified to the form $ax < b$. (Properties introduced in this section are given only for $<$, but they are equally valid for $>$, $\leq$, or $\geq$.) Linear inequalities are solved with the following properties.

> **PROPERTIES OF INEQUALITY**
>
> For all real numbers a, b, and c:
>
> **1.** If $a < b$, then $a + c < b + c$.
> **2.** If $a < b$ and if $c > 0$, then $ac < bc$.
> **3.** If $a < b$ and if $c < 0$, then $ac > bc$.

Pay careful attention to property 3; it says that if both sides of an inequality are multiplied by a negative number, the direction of the inequality symbol must be reversed.

EXAMPLE 1 Solving Linear Inequalities
Solve $4 - 3y \leq 7 + 2y$.

Solution Use the properties of inequality.

$$4 - 3y + (-4) \leq 7 + 2y + (-4) \quad \text{Add } -4 \text{ to both sides.}$$
$$-3y \leq 3 + 2y$$

Remember that *adding* the same number to both sides never changes the direction of the inequality symbol.

$$-3y + (-2y) \leq 3 + 2y + (-2y) \quad \text{Add } -2y \text{ to both sides.}$$
$$-5y \leq 3$$

Multiply both sides by $-1/5$. Since $-1/5$ is negative, change the direction of the inequality symbol.

$$-\frac{1}{5}(-5y) \geq -\frac{1}{5}(3)$$

$$y \geq -\frac{3}{5}$$

CAUTION It is a common error to forget to reverse the direction of the inequality sign when multiplying or dividing by a negative number. For example, to solve $-4x \leq 12$, we must multiply by $-1/4$ on both sides *and* reverse the inequality symbol to get $x \geq -3$.

The solution $y \geq -3/5$ in Example 1 represents an interval on the number line. **Interval notation** often is used for writing intervals. With interval notation, $y \geq -3/5$ is written as $[-3/5, \infty)$. This is an example of a **half-open interval,** since one endpoint, $-3/5$, is included. The **open interval** $(2, 5)$ corresponds to $2 < x < 5$, with neither endpoint included. The **closed interval** $[2, 5]$ includes both endpoints and corresponds to $2 \leq x \leq 5$.

The **graph** of an interval shows all points on a number line that correspond to the numbers in the interval. To graph the interval $[-3/5, \infty)$, for example, use a solid circle at $-3/5$, since $-3/5$ is part of the solution. To show that the solution includes all real numbers greater than or equal to $-3/5$, draw a heavy arrow pointing to the right (the positive direction). See Figure 1.

FIGURE 1

EXAMPLE 2 Graphing Linear Inequalities

Solve $-2 < 5 + 3m < 20$. Graph the solution.

Solution The inequality $-2 < 5 + 3m < 20$ says that $5 + 3m$ is *between* -2 and 20. Solve this inequality with an extension of the properties given above. Work as follows, first adding -5 to each part.

$$-2 + (-5) < 5 + 3m + (-5) < 20 + (-5)$$
$$-7 < 3m < 15$$

Now multiply each part by $1/3$.

$$-\frac{7}{3} < m < 5$$

FIGURE 2

A graph of the solution is given in Figure 2; here open circles are used to show that $-7/3$ and 5 are *not* part of the graph.*

Quadratic Inequalities A **quadratic inequality** has the form $ax^2 + bx + c > 0$ (or $<$, or $\leq$, or $\geq$). The highest exponent is 2. The next few examples show how to solve quadratic inequalities.

---

*Some textbooks use brackets in place of solid circles for the graph of a closed interval, and parentheses in place of open circles for the graph of an open interval.

EXAMPLE 3 Solving Quadratic Inequalities

Solve the quadratic inequality $x^2 - x < 12$.

Solution Write the inequality with 0 on one side, as $x^2 - x - 12 < 0$. This inequality is solved with values of x that make $x^2 - x - 12$ negative (<0). The quantity $x^2 - x - 12$ changes from positive to negative or from negative to positive at the points where it equals 0. For this reason, first solve the *equation* $x^2 - x - 12 = 0$.

$$x^2 - x - 12 = 0$$
$$(x - 4)(x + 3) = 0$$
$$x = 4 \quad \text{or} \quad x = -3$$

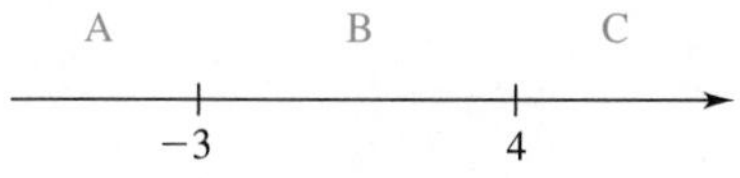

Locating -3 and 4 on a number line, as shown in Figure 3, determines three intervals A, B, and C. Decide which intervals include numbers that make $x^2 - x - 12$ negative by substituting any number from each interval in the polynomial. For example,

FIGURE 3

choose -4 from interval A: $(-4)^2 - (-4) - 12 = 8 > 0$;

choose 0 from interval B: $0^2 - 0 - 12 = -12 < 0$;

choose 5 from interval C: $5^2 - 5 - 12 = 8 > 0$.

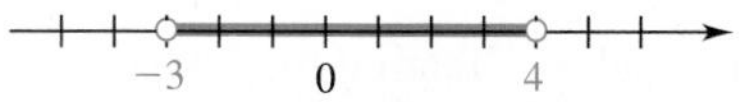

Only numbers in interval B satisfy the given inequality, so the solution is $(-3, 4)$. A graph of this solution is shown in Figure 4.

FIGURE 4

EXAMPLE 4 Solving Polynomial Inequalities

Solve the inequality $x(x - 1)(x + 3) \geq 0$.

Solution This is not a quadratic inequality. If the three factors are multiplied, the highest-degree term is x^3. However, it can be solved in the same way as a quadratic inequality because it is in factored form. First solve the corresponding equation.

$$x(x - 1)(x + 3) = 0$$
$$x = 0 \quad \text{or} \quad x - 1 = 0 \quad \text{or} \quad x + 3 = 0$$
$$x = 1 \qquad\qquad x = -3$$

These three solutions determine four intervals on the number line: $(-\infty, -3)$, $(-3, 0)$, $(0, 1)$, and $(1, \infty)$. Substitute a number from each interval into the original inequality to determine that the solution includes the numbers less than or equal to -3 and the numbers that are equal to or between 0 and 1. See Figure 5. In interval notation, the solution is

FIGURE 5

$$(-\infty, -3] \cup [0, 1].*$$

Inequalities with Fractions
Inequalities with fractions are solved in a similar manner as quadratic inequalities.

EXAMPLE 5 Solving Rational Inequalities

Solve $\dfrac{2x - 3}{x} \geq 1$.

---

*The symbol $\cup$ indicates the *union* of two sets, which includes all elements in either set.

Solution First solve the corresponding equation.

$$\frac{2x - 3}{x} = 1$$

$$2x - 3 = x$$

$$x = 3$$

The solution, $x = 3$, determines the intervals on the number line where the fraction may change from greater than 1 to less than 1. This change also may occur on either side of a number that makes the denominator equal 0. Here, the x-value that makes the denominator 0 is $x = 0$. Test each of the three intervals determined by the numbers 0 and 3.

$$\text{For } (-\infty, 0), \text{ choose } -1: \frac{2(-1) - 3}{-1} = 5 \geq 1.$$

$$\text{For } (0, 3), \quad \text{choose } \quad 1: \frac{2(1) - 3}{1} = -1 \ngeq 1.$$

$$\text{For } (3, \infty), \quad \text{choose } \quad 4: \frac{2(4) - 3}{4} = \frac{5}{4} \geq 1.$$

The symbol $\ngeq$ means "is *not* greater than or equal to." Testing the endpoints 0 and 3 shows that the solution is $(-\infty, 0) \cup [3, \infty)$. ■

> **CAUTION** A common error is to try to solve the inequality in Example 5 by multiplying both sides by x. The reason this is wrong is that we don't know in the beginning whether x is positive or negative. If x is negative, the $\geq$ would change to $\leq$ according to the third property of inequality listed at the beginning of this section.

EXAMPLE 6   Solving Rational Inequalities

Solve $\dfrac{(x - 1)(x + 1)}{x} \leq 0$.

Solution Solve the corresponding equation.

$$\frac{(x - 1)(x + 1)}{x} = 0$$

$$(x - 1)(x + 1) = 0 \qquad \text{Multiply both sides by } x.$$

$$x = 1 \qquad \text{or} \qquad x = -1 \qquad \text{Use the zero-factor property.}$$

Setting the denominator equal to 0 gives $x = 0$, so the intervals of interest are $(-\infty, -1)$, $(-1, 0)$, and $(0, \infty)$. Testing a number from each region in the original inequality and checking the endpoints, we find the solution is

$$(-\infty, -1] \cup (0, 1]. \qquad ■$$

> **CAUTION** Remember to solve the equation formed by setting the *denominator* equal to zero. Any number that makes the denominator zero always creates two intervals on the number line. For instance, in Example 6, 0 makes the denominator of the rational inequality equal to 0, so we know that there may be a sign change from one side of 0 to the other (as was indeed the case).

EXAMPLE 7   Solving Rational Inequalities

Solve $\dfrac{x^2 - 3x}{x^2 - 9} < 4$.

Solution   Solve the corresponding equation.

$$\frac{x^2 - 3x}{x^2 - 9} = 4$$

$$x^2 - 3x = 4x^2 - 36 \qquad \text{Multiply by } x^2 - 9.$$

$$0 = 3x^2 + 3x - 36 \qquad \text{Get 0 on one side.}$$

$$0 = x^2 + x - 12 \qquad \text{Multiply by } \tfrac{1}{3}.$$

$$0 = (x + 4)(x - 3) \qquad \text{Factor.}$$

$$x = -4 \qquad \text{or} \qquad x = 3$$

Now set the denominator equal to 0 and solve that equation.

$$x^2 - 9 = 0$$

$$(x - 3)(x + 3) = 0$$

$$x = 3 \qquad \text{or} \qquad x = -3$$

The intervals determined by the three (different) solutions are $(-\infty, -4)$, $(-4, -3)$, $(-3, 3)$, and $(3, \infty)$. Testing a number from each interval in the given inequality shows that the solution is

$$(-\infty, -4) \cup (-3, 3) \cup (3, \infty).$$

For this example, none of the endpoints are part of the solution because $x = 3$ and $x = -3$ force the denominator to be zero and $x = -4$ produces an equality.

R.5   EXERCISES

Write each expression in interval notation. Graph each interval.

1. $x < 0$ **2.** $x \geq -3$ **3.** $1 \leq x < 2$

4. $-5 < x \leq -4$ **5.** $-9 > x$ **6.** $6 \leq x$

Using the variable x, write each interval in Exercises 7–14 as an inequality.

7. $(-4, 3)$ **8.** $[2, 7)$ **9.** $(-\infty, -1]$ **10.** $(3, \infty)$

11. 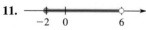   **12.**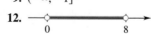

13.    **14.**

Solve each inequality and graph the solution.

15. $-3p - 2 \geq 1$ **16.** $6k - 4 < 3k - 1$

17. $m - (4 + 2m) + 3 < 2m + 2$ **18.** $-2(3y - 8) \geq 5(4y - 2)$

19. $3p - 1 < 6p + 2(p - 1)$ **20.** $x + 5(x + 1) > 4(2 - x) + x$

21. $-7 < y - 2 < 4$ **22.** $8 \leq 3r + 1 \leq 13$

23. $-4 \le \dfrac{2k-1}{3} \le 2$

24. $-1 \le \dfrac{5y+2}{3} \le 4$

25. $\dfrac{3}{5}(2p+3) \ge \dfrac{1}{10}(5p+1)$

26. $\dfrac{8}{3}(z-4) \le \dfrac{2}{9}(3z+2)$

Solve each of the following quadratic inequalities. Graph each solution.

27. $(m+2)(m-4) < 0$

28. $(t+6)(t-1) \ge 0$

29. $y^2 - 3y + 2 < 0$

30. $2k^2 + 7k - 4 > 0$

31. $q^2 - 7q + 6 \le 0$

32. $2k^2 - 7k - 15 \le 0$

33. $6m^2 + m > 1$

34. $10r^2 + r \le 2$

35. $2y^2 + 5y \le 3$

36. $3a^2 + a > 10$

37. $x^2 \le 25$

38. $p^2 - 16p > 0$

Solve the following inequalities.

39. $\dfrac{m-3}{m+5} \le 0$

40. $\dfrac{r+1}{r-1} > 0$

41. $\dfrac{k-1}{k+2} > 1$

42. $\dfrac{a-5}{a+2} < -1$

43. $\dfrac{2y+3}{y-5} \le 1$

44. $\dfrac{a+2}{3+2a} \le 5$

45. $\dfrac{7}{k+2} \ge \dfrac{1}{k+2}$

46. $\dfrac{5}{p+1} > \dfrac{12}{p+1}$

47. $\dfrac{3x}{x^2-1} < 2$

48. $\dfrac{8}{p^2+2p} > 1$

49. $\dfrac{z^2+z}{z^2-1} \ge 3$

50. $\dfrac{a^2+2a}{a^2-4} \le 2$

R.6 EXPONENTS

Integer Exponents   Recall that $a^2 = a \cdot a$, while $a^3 = a \cdot a \cdot a$, and so on. In this section a more general meaning is given to the symbol a^n.

DEFINITION OF EXPONENT

If n is a natural number, then

$$a^n = a \cdot a \cdot a \cdot \,\cdots\, \cdot a,$$

where a appears as a factor n times.

In the expression a^n, n is the **exponent** and a is the **base.** This definition can be extended by defining a^n for zero and negative integer values of n.

ZERO AND NEGATIVE EXPONENTS

If a is any nonzero real number, and if n is a positive integer, then

$$a^0 = 1 \quad \text{and} \quad a^{-n} = \dfrac{1}{a^n}.$$

(The symbol 0^0 is meaningless.)

EXAMPLE 1 Exponents

(a) $6^0 = 1$

(b) $(-9)^0 = 1$

(c) $3^{-2} = \dfrac{1}{3^2} = \dfrac{1}{9}$

(d) $9^{-1} = \dfrac{1}{9^1} = \dfrac{1}{9}$

(e) $\left(\dfrac{3}{4}\right)^{-1} = \dfrac{1}{(3/4)^1} = \dfrac{1}{3/4} = \dfrac{4}{3}$

The following properties follow from the definitions of exponents given above.

PROPERTIES OF EXPONENTS

For any integers m and n, and any real numbers a and b for which the following exist:

1. $a^m \cdot a^n = a^{m+n}$ **4.** $(ab)^m = a^m \cdot b^m$

2. $\dfrac{a^m}{a^n} = a^{m-n}$ **5.** $\left(\dfrac{a}{b}\right)^m = \dfrac{a^m}{b^m}.$

3. $(a^m)^n = a^{mn}$

EXAMPLE 2 Simplifying Exponential Expressions

Use the properties of exponents to simplify each of the following. Leave answers with positive exponents. Assume that all variables represent positive real numbers.

(a) $7^4 \cdot 7^6 = 7^{4+6} = 7^{10}$ (or 282,475,249)   Property 1

(b) $\dfrac{9^{14}}{9^6} = 9^{14-6} = 9^8$ (or 43,046,721)   Property 2

(c) $\dfrac{r^9}{r^{17}} = r^{9-17} = r^{-8} = \dfrac{1}{r^8}$ Property 2

(d) $(2m^3)^4 = 2^4 \cdot (m^3)^4 = 16m^{12}$ Properties 3 and 4

(e) $(3x)^4 = 3^4 \cdot x^4 = 81x^4$ Property 4

(f) $\left(\dfrac{x^2}{y^3}\right)^6 = \dfrac{(x^2)^6}{(y^3)^6} = \dfrac{x^{2\cdot6}}{y^{3\cdot6}} = \dfrac{x^{12}}{y^{18}}$ Properties 4 and 5

(g) $\dfrac{a^{-3}b^5}{a^4 b^{-7}} = \dfrac{b^{5-(-7)}}{a^{4-(-3)}} = \dfrac{b^{5+7}}{a^{4+3}} = \dfrac{b^{12}}{a^7}$ Property 2

(h) $p^{-1} + q^{-1} = \dfrac{1}{p} + \dfrac{1}{q} = \dfrac{1}{p} \cdot \dfrac{q}{q} + \dfrac{1}{q} \cdot \dfrac{p}{p} = \dfrac{q}{pq} + \dfrac{p}{pq} = \dfrac{p+q}{pq}$

(i) $\dfrac{x^{-2} - y^{-2}}{x^{-1} - y^{-1}} = \dfrac{\dfrac{1}{x^2} - \dfrac{1}{y^2}}{\dfrac{1}{x} - \dfrac{1}{y}}$ Definition of a^{-n}

$\qquad\qquad = \dfrac{\dfrac{y^2 - x^2}{x^2y^2}}{\dfrac{y - x}{xy}}$ Get common denominators and combine terms.

$\qquad\qquad = \dfrac{y^2 - x^2}{x^2y^2} \cdot \dfrac{xy}{y - x}$ Invert and multiply.

$\qquad\qquad = \dfrac{(y - x)(y + x)}{x^2y^2} \cdot \dfrac{xy}{y - x}$ Factor.

$\qquad\qquad = \dfrac{x + y}{xy}$ Simplify.

> **CAUTION** If Example 2(e) were written $3x^4$, the properties of exponents would not apply. When no parentheses are used, the exponent refers only to the factor closest to it. Also notice in Examples 2(c), 2(g), 2(h), and 2(i) that a negative exponent does *not* indicate a negative number.

Roots   For *even* values of n, the expression $a^{1/n}$ is defined to be the **positive nth root** of a or the **principal nth root** of a. For example, $a^{1/2}$ denotes the positive second root, or **square root,** of a, while $a^{1/4}$ is the positive fourth root of a. When n is *odd,* there is only one nth root, which has the same sign as a. For example, $a^{1/3}$, the **cube root** of a, has the same sign as a. By definition, if $b = a^{1/n}$, then $b^n = a$. On a calculator, a number is raised to a power using a key labeled x^y, y^x, or $\wedge$. For example, to take the fourth root of 6 on a TI-83 calculator, enter $6 \wedge (1/4)$, to get the result 1.56508458.

EXAMPLE 3   Calculations with Exponents

(a) $121^{1/2} = 11$, since 11 is positive and $11^2 = 121$.

(b) $625^{1/4} = 5$, since $5^4 = 625$.

(c) $256^{1/4} = 4$

(d) $64^{1/6} = 2$

(e) $27^{1/3} = 3$

(f) $(-32)^{1/5} = -2$

(g) $128^{1/7} = 2$

(h) $(-49)^{1/2}$ is not a real number.

Rational Exponents   In the following definition, the domain of an exponent is extended to include all rational numbers.

DEFINITION OF $a^{m/n}$

For all real numbers a for which the indicated roots exist, and for any rational number m/n,

$$a^{m/n} = (a^{1/n})^m.$$

EXAMPLE 4 Calculations with Exponents

(a) $27^{2/3} = (27^{1/3})^2 = 3^2 = 9$

(b) $32^{2/5} = (32^{1/5})^2 = 2^2 = 4$

(c) $64^{4/3} = (64^{1/3})^4 = 4^4 = 256$

(d) $25^{3/2} = (25^{1/2})^3 = 5^3 = 125$

NOTE $27^{2/3}$ could also be evaluated as $(27^2)^{1/3}$, but this is more difficult to perform without a calculator because it involves squaring 27, and then taking the cube root of this large number. On the other hand, when we evaluate it as $(27^{1/3})^2$, we know that the cube root of 27 is 3 without using a calculator, and squaring 3 is easy.

All the properties for integer exponents given in this section also apply to any rational exponent on a nonnegative real-number base.

EXAMPLE 5 Simplifying Exponential Expressions

(a) $\dfrac{y^{1/3}y^{5/3}}{y^3} = \dfrac{y^{1/3+5/3}}{y^3} = \dfrac{y^2}{y^3} = y^{2-3} = y^{-1} = \dfrac{1}{y}$

(b) $m^{2/3}(m^{7/3} + 2m^{1/3}) = m^{2/3+7/3} + 2m^{2/3+1/3} = m^3 + 2m$

(c) $\left(\dfrac{m^7 n^{-2}}{m^{-5} n^2}\right)^{1/4} = \left(\dfrac{m^{7-(-5)}}{n^{2-(-2)}}\right)^{1/4} = \left(\dfrac{m^{12}}{n^4}\right)^{1/4} = \dfrac{(m^{12})^{1/4}}{(n^4)^{1/4}} = \dfrac{m^{12/4}}{n^{4/4}} = \dfrac{m^3}{n}$

In calculus, it is often necessary to factor expressions involving fractional exponents.

EXAMPLE 6 Simplifying Exponential Expressions

Factor out the smallest power of the variable, assuming all variables represent positive real numbers.

(a) $4m^{1/2} + 3m^{3/2} = m^{1/2}(4 + 3m)$

Solution To check this result, multiply $m^{1/2}$ by $4 + 3m$.

(b) $9x^{-2} - 6x^{-3}$

Solution The smallest exponent here is -3. Since 3 is a common numerical factor, factor out $3x^{-3}$.

$$9x^{-2} - 6x^{-3} = 3x^{-3}(3x^{-2-(-3)} - 2x^{-3-(-3)}) = 3x^{-3}(3x - 2)$$

Check by multiplying. The factored form can be written without negative exponents as

$$\dfrac{3(3x - 2)}{x^3}.$$

(c) $(x^2 + 5)(3x - 1)^{-1/2}(2) + (3x - 1)^{1/2}(2x).$

Solution There is a common factor of 2. Also, $(3x - 1)^{-1/2}$ and $(3x - 1)^{1/2}$ have a common factor. Always factor out the quantity to the *smallest* exponent. Here $-1/2 < 1/2$, so the common factor is $2(3x - 1)^{-1/2}$ and the factored form is

$$2(3x - 1)^{-1/2}[(x^2 + 5) + (3x - 1)x] = 2(3x - 1)^{-1/2}(4x^2 - x + 5).$$

R.6 EXERCISES

Evaluate each expression. Write all answers without exponents.

1. 8^{-2}

2. 3^{-4}

3. 5^0

4. $(-12)^0$

5. $-(-3)^{-2}$

6. $-(-3^{-2})$

7. $\left(\dfrac{2}{7}\right)^{-2}$

8. $\left(\dfrac{4}{3}\right)^{-3}$

Simplify each expression. Assume that all variables represent positive real numbers. Write answers with only positive exponents.

9. $\dfrac{3^{-4}}{3^2}$

10. $\dfrac{8^9 \cdot 8^{-7}}{8^{-3}}$

11. $\dfrac{10^8 \cdot 10^{-10}}{10^4 \cdot 10^2}$

12. $\left(\dfrac{5^{-6} \cdot 5^3}{5^{-2}}\right)^{-1}$

13. $\dfrac{x^4 \cdot x^3}{x^5}$

14. $\dfrac{y^9 \cdot y^7}{y^{13}}$

15. $\dfrac{(4k^{-1})^2}{2k^{-5}}$

16. $\dfrac{(3z^2)^{-1}}{z^5}$

17. $\dfrac{2^{-1}x^3y^{-3}}{xy^{-2}}$

18. $\dfrac{5^{-2}m^2y^{-2}}{5^2m^{-1}y^{-2}}$

19. $\left(\dfrac{a^{-1}}{b^2}\right)^{-3}$

20. $\left(\dfrac{2c^2}{d^3}\right)^{-2}$

21. $\left(\dfrac{x^6y^{-3}}{x^{-2}y^5}\right)^{1/2}$

22. $\left(\dfrac{a^{-7}b^{-1}}{b^{-4}a^2}\right)^{1/3}$

Simplify each expression, writing the answer as a single term without negative exponents.

23. $a^{-1} + b^{-1}$

24. $b^{-2} - a$

25. $\dfrac{2n^{-1} - 2m^{-1}}{m + n^2}$

26. $\left(\dfrac{m}{3}\right)^{-1} + \left(\dfrac{n}{2}\right)^{-2}$

27. $(x^{-1} - y^{-1})^{-1}$

28. $(x^{-2} + y^{-2})^{-2}$

Write each number without exponents.

29. $81^{1/2}$

30. $27^{1/3}$

31. $32^{2/5}$

32. $-125^{2/3}$

33. $\left(\dfrac{4}{9}\right)^{1/2}$

34. $\left(\dfrac{64}{27}\right)^{1/3}$

35. $16^{-5/4}$

36. $625^{-1/4}$

37. $\left(\dfrac{27}{64}\right)^{-1/3}$

38. $\left(\dfrac{121}{100}\right)^{-3/2}$

Simplify each expression. Write all answers with only positive exponents. Assume that all variables represent positive real numbers.

39. $2^{1/2} \cdot 2^{3/2}$

40. $27^{2/3} \cdot 27^{-1/3}$

41. $\dfrac{4^{2/3} \cdot 4^{5/3}}{4^{1/3}}$

42. $\dfrac{3^{-5/2} \cdot 3^{3/2}}{3^{7/2} \cdot 3^{-9/2}}$

43. $\dfrac{7^{-1/3} \cdot 7r^{-3}}{7^{2/3} \cdot (r^{-2})^2}$

44. $\dfrac{12^{3/4} \cdot 12^{5/4} \cdot y^{-2}}{12^{-1} \cdot (y^{-3})^{-2}}$

45. $\dfrac{6k^{-4} \cdot (3k^{-1})^{-2}}{2^3 \cdot k^{1/2}}$

46. $\dfrac{8p^{-3} \cdot (4p^2)^{-2}}{p^{-5}}$

47. $\dfrac{a^{4/3} \cdot b^{1/2}}{a^{2/3} \cdot b^{-3/2}}$

48. $\dfrac{x^{1/3} \cdot y^{2/3} \cdot z^{1/4}}{x^{5/3} \cdot y^{-1/3} \cdot z^{3/4}}$

49. $\dfrac{k^{-3/5} \cdot h^{-1/3} \cdot t^{2/5}}{k^{-1/5} \cdot h^{-2/3} \cdot t^{1/5}}$

50. $\dfrac{m^{7/3} \cdot n^{-2/5} \cdot p^{3/8}}{m^{-2/3} \cdot n^{3/5} \cdot p^{-5/8}}$

Factor each expression.

51. $12x^2(x^2 + 2)^2 - 4x(4x^3 + 1)(x^2 + 2)$

52. $6x(x^3 + 7)^2 - 6x^2(3x^2 + 5)(x^3 + 7)$

53. $(x^2 + 2)(x^2 - 1)^{-1/2}(x) + (x^2 - 1)^{1/2}(2x)$

54. $9(6x + 2)^{1/2} + 3(9x - 1)(6x + 2)^{-1/2}$

55. $x(2x + 5)^2(x^2 - 4)^{-1/2} + 2(x^2 - 4)^{1/2}(2x + 5)$

56. $(4x^2 + 1)^2(2x - 1)^{-1/2} + 16x(4x^2 + 1)(2x - 1)^{1/2}$

R.7 RADICALS

We have defined $a^{1/n}$ as the positive or principal nth root of a for appropriate values of a and n. An alternative notation for $a^{1/n}$ uses radicals.

RADICALS

If n is an even natural number and $a > 0$, or n is an odd natural number, then

$$a^{1/n} = \sqrt[n]{a}.$$

The symbol $\sqrt{}$ is a **radical sign,** the number a is the **radicand,** and n is the **index** of the radical. The familiar symbol $\sqrt{a}$ is used instead of $\sqrt[2]{a}$.

EXAMPLE 1 Radical Calculations

(a) $\sqrt[4]{16} = 16^{1/4} = 2$ (b) $\sqrt[5]{-32} = -2$

(c) $\sqrt[3]{1000} = 10$ (d) $\sqrt[6]{\dfrac{64}{729}} = \dfrac{2}{3}$

With $a^{1/n}$ written as $\sqrt[n]{a}$, $a^{m/n}$ also can be written using radicals.

$$a^{m/n} = \left(\sqrt[n]{a}\right)^m \qquad \text{or} \qquad a^{m/n} = \sqrt[n]{a^m}$$

The following properties of radicals depend on the definitions and properties of exponents.

PROPERTIES OF RADICALS

For all real numbers a and b and natural numbers m and n such that $\sqrt[n]{a}$ and $\sqrt[n]{b}$ are real numbers:

1. $\left(\sqrt[n]{a}\right)^n = a$

2. $\sqrt[n]{a^n} = \begin{cases} |a| & \text{if } n \text{ is even} \\ a & \text{if } n \text{ is odd} \end{cases}$

3. $\sqrt[n]{a} \cdot \sqrt[n]{b} = \sqrt[n]{ab}$

4. $\dfrac{\sqrt[n]{a}}{\sqrt[n]{b}} = \sqrt[n]{\dfrac{a}{b}}$ $(b \neq 0)$

5. $\sqrt[m]{\sqrt[n]{a}} = \sqrt[mn]{a}$

Property 3 can be used to simplify certain radicals. For example, since $48 = 16 \cdot 3$,

$$\sqrt{48} = \sqrt{16 \cdot 3} = \sqrt{16} \cdot \sqrt{3} = 4\sqrt{3}.$$

To some extent, simplification is in the eye of the beholder, and $\sqrt{48}$ might be considered as simple as $4\sqrt{3}$. In this textbook, we will consider an expression to be simpler when we have removed as many factors as possible from under the radical.

EXAMPLE 2 Radical Calculations

(a) $\sqrt{1000} = \sqrt{100 \cdot 10} = \sqrt{100} \cdot \sqrt{10} = 10\sqrt{10}$

(b) $\sqrt{128} = \sqrt{64 \cdot 2} = 8\sqrt{2}$

(c) $\sqrt{108} = \sqrt{36 \cdot 3} = 6\sqrt{3}$

(d) $\sqrt[3]{54} = \sqrt[3]{27 \cdot 2} = \sqrt[3]{27} \cdot \sqrt[3]{2} = 3\sqrt[3]{2}$

(e) $\sqrt{288m^5} = \sqrt{144 \cdot m^4 \cdot 2m} = 12m^2\sqrt{2m}$

(f) $2\sqrt{18} - 5\sqrt{32} = 2\sqrt{9 \cdot 2} - 5\sqrt{16 \cdot 2}$
$$= 2\sqrt{9} \cdot \sqrt{2} - 5\sqrt{16} \cdot \sqrt{2}$$
$$= 2(3)\sqrt{2} - 5(4)\sqrt{2} = -14\sqrt{2}$$

Rationalizing Denominators The next example shows how to *rationalize* (remove all radicals from) the denominator in an expression containing radicals.

EXAMPLE 3 Rationalizing the Denominator

Simplify each of the following expressions by rationalizing the denominator.

(a) $\dfrac{4}{\sqrt{3}}$

Solution To rationalize the denominator, multiply by $\sqrt{3}/\sqrt{3}$ (or 1) so that the denominator of the product is a rational number.

$$\frac{4}{\sqrt{3}} \cdot \frac{\sqrt{3}}{\sqrt{3}} = \frac{4\sqrt{3}}{3}$$

(b) $\dfrac{2}{\sqrt[3]{x}}$

Solution Here, we need a perfect cube under the radical sign to rationalize the denominator. Multiplying by $\sqrt[3]{x^2}/\sqrt[3]{x^2}$ gives

$$\frac{2}{\sqrt[3]{x}} \cdot \frac{\sqrt[3]{x^2}}{\sqrt[3]{x^2}} = \frac{2\sqrt[3]{x^2}}{\sqrt[3]{x^3}} = \frac{2\sqrt[3]{x^2}}{x}.$$

(c) $\dfrac{1}{1 - \sqrt{2}}$

Solution The best approach here is to multiply both numerator and denominator by the number $1 + \sqrt{2}$. The expressions $1 + \sqrt{2}$ and $1 - \sqrt{2}$ are conjugates.* Thus,

$$\frac{1}{1 - \sqrt{2}} = \frac{1(1 + \sqrt{2})}{(1 - \sqrt{2})(1 + \sqrt{2})} = \frac{1 + \sqrt{2}}{1 - 2} = -1 - \sqrt{2}.$$ ■

Sometimes it is advantageous to rationalize the *numerator* of a rational expression. The following example arises in calculus when evaluating a *limit*.

EXAMPLE 4   Rationalizing the Numerator
Rationalize the numerator.

(a) $\dfrac{\sqrt{x} - 3}{x - 9}$.

Solution Multiply numerator and denominator by the conjugate of the numerator, $\sqrt{x} + 3$.

$$\frac{\sqrt{x} - 3}{x - 9} \cdot \frac{\sqrt{x} + 3}{\sqrt{x} + 3} = \frac{\sqrt{x} \cdot \sqrt{x} - 3\sqrt{x} + 3\sqrt{x} - 9}{(x - 9)(\sqrt{x} + 3)}$$

$$= \frac{x - 9}{(x - 9)(\sqrt{x} + 3)}$$

$$= \frac{1}{\sqrt{x} + 3}$$

(b) $\dfrac{\sqrt{3} + \sqrt{x + 3}}{\sqrt{3} - \sqrt{x + 3}}$

Solution Multiply the numerator and denominator by the conjugate of the numerator, $\sqrt{3} - \sqrt{x + 3}$.

$$\frac{\sqrt{3} + \sqrt{x + 3}}{\sqrt{3} - \sqrt{x + 3}} \cdot \frac{\sqrt{3} - \sqrt{x + 3}}{\sqrt{3} - \sqrt{x + 3}} = \frac{3 - (x + 3)}{3 - 2\sqrt{3}\sqrt{x + 3} + (x + 3)}$$

$$= \frac{-x}{6 + x - 2\sqrt{3(x + 3)}}$$ ■

When simplifying a square root, keep in mind that $\sqrt{x}$ is positive by definition. Also, $\sqrt{x^2}$ is not x, but $|x|$, the **absolute value of x,** defined as

$$|x| = \begin{cases} x \text{ if } x \geq 0 \\ -x \text{ if } x < 0. \end{cases}$$

For example, $\sqrt{(-5)^2} = |-5| = 5$.

EXAMPLE 5   Simplifying by Factoring
Simplify $\sqrt{m^2 - 4m + 4}$.

Solution Factor the polynomial as $m^2 - 4m + 4 = (m - 2)^2$. Then by property 2 of radicals, and the definition of absolute value,

$$\sqrt{(m - 2)^2} = |m - 2| = \begin{cases} m - 2 & \text{if } m - 2 \geq 0 \\ -(m - 2) = 2 - m & \text{if } m - 2 < 0. \end{cases}$$ ■

*If a and b are real numbers, the *conjugate* of $a + b$ is $a - b$.

> **CAUTION** Avoid the common error of writing $\sqrt{a^2 + b^2}$ as $\sqrt{a^2} + \sqrt{b^2}$. We must add a^2 and b^2 *before* taking the square root. For example, $\sqrt{16 + 9} = \sqrt{25} = 5$, *not* $\sqrt{16} + \sqrt{9} = 4 + 3 = 7$. This idea applies as well to higher roots. For example, in general,
>
> $$\sqrt[3]{a^3 + b^3} \neq \sqrt[3]{a^3} + \sqrt[3]{b^3},$$
> $$\sqrt[4]{a^4 + b^4} \neq \sqrt[4]{a^4} + \sqrt[4]{b^4}.$$
>
> Also, $$\sqrt{a + b} \neq \sqrt{a} + \sqrt{b}.$$

R.7   EXERCISES

Simplify each expression by removing as many factors as possible from under the radical. Assume that all variables represent positive real numbers.

1. $\sqrt[3]{125}$

2. $\sqrt[4]{1296}$

3. $\sqrt[5]{-3125}$

4. $\sqrt{50}$

5. $\sqrt{2000}$

6. $\sqrt{32y^5}$

7. $7\sqrt{2} - 8\sqrt{18} + 4\sqrt{72}$

8. $4\sqrt{3} - 5\sqrt{12} + 3\sqrt{75}$

9. $2\sqrt{5} - 3\sqrt{20} + 2\sqrt{45}$

10. $3\sqrt{28} - 4\sqrt{63} + \sqrt{112}$

11. $\sqrt[3]{2} - \sqrt[3]{16} + 2\sqrt[3]{54}$

12. $2\sqrt[3]{3} + 4\sqrt[3]{24} - \sqrt[3]{81}$

13. $\sqrt[3]{32} - 5\sqrt[3]{4} + 2\sqrt[3]{108}$

14. $\sqrt{2x^3y^2z^4}$

15. $\sqrt{98r^3s^4t^{10}}$

16. $\sqrt[3]{16x^8y^4z^5}$

17. $\sqrt[4]{x^8y^7z^{11}}$

18. $\sqrt{a^3b^5} - 2\sqrt{a^7b^3} + \sqrt{a^3b^9}$

19. $\sqrt{p^7q^3} - \sqrt{p^5q^9} + \sqrt{p^9q}$

Rationalize each denominator. Assume that all radicands represent positive real numbers.

20. $\dfrac{5}{\sqrt{7}}$

21. $\dfrac{-2}{\sqrt{3}}$

22. $\dfrac{-3}{\sqrt{12}}$

23. $\dfrac{4}{\sqrt{8}}$

24. $\dfrac{3}{1 - \sqrt{5}}$

25. $\dfrac{5}{2 - \sqrt{6}}$

26. $\dfrac{-2}{\sqrt{3} - \sqrt{2}}$

27. $\dfrac{1}{\sqrt{10} + \sqrt{3}}$

28. $\dfrac{1}{\sqrt{r} - \sqrt{3}}$

29. $\dfrac{5}{\sqrt{m} - \sqrt{5}}$

30. $\dfrac{y - 5}{\sqrt{y} - \sqrt{5}}$

31. $\dfrac{z - 11}{\sqrt{z} - \sqrt{11}}$

32. $\dfrac{\sqrt{x} + \sqrt{x + 1}}{\sqrt{x} - \sqrt{x + 1}}$

33. $\dfrac{\sqrt{p} + \sqrt{p^2 - 1}}{\sqrt{p} - \sqrt{p^2 - 1}}$

Rationalize each numerator. Assume that all radicands represent positive real numbers.

34. $\dfrac{1 + \sqrt{2}}{2}$

35. $\dfrac{1 - \sqrt{3}}{3}$

36. $\dfrac{\sqrt{x} + \sqrt{x + 1}}{\sqrt{x} - \sqrt{x + 1}}$

37. $\dfrac{\sqrt{p} + \sqrt{p^2 - 1}}{\sqrt{p} - \sqrt{p^2 - 1}}$

Simplify each root, if possible.

38. $\sqrt{16 - 8x + x^2}$

39. $\sqrt{4y^2 + 4y + 1}$

40. $\sqrt{4 - 25z^2}$

41. $\sqrt{9k^2 + h^2}$

Linear Functions

Over short time intervals, many changes in the economy are well modeled by linear functions. In an exercise in the first section of this chapter we will examine a linear model that predicts airline passenger traffic in the year 2005 at some of the fastest-growing airports in the United States. Such predictions are important tools for airline executives and airport planners.

Before using mathematics to solve a real-world problem, we must usually set up a **mathematical model,** a mathematical description of the situation. Constructing such a model requires a solid understanding of the situation to be modeled, as well as familiarity with relevant mathematical ideas and techniques.

Much mathematical theory is available for building models, but the very richness and diversity of contemporary mathematics often prevents people in other fields from finding the mathematical tools they need. There are so many useful parts of mathematics that it can be hard to know which to choose.

To avoid this problem, it is helpful to have a thorough understanding of the most basic and useful mathematical tools that are available for constructing mathematical models. In this chapter we look at some mathematics of *linear* models, which are used for data whose graphs can be approximated by straight lines.

■ 1.1 SLOPES AND EQUATIONS OF LINES

 THINK ABOUT IT How fast has tuition at public colleges been increasing in recent years, and how well can we predict tuition in the future?

In Example 15 of this section, we will answer these questions using the equation of a line.

There are many everyday situations in which two quantities are related. For example, if a bank account pays 6% simple interest per year, then the interest I that a deposit of P dollars would earn in one year is given by

$$I = .06 \cdot P, \qquad \text{or} \qquad I = .06P.$$

The formula $I = .06P$ describes the relationship between interest and the amount of money deposited.

Using this formula, we see, for example, that if $P = \$100$, then $I = \$6$, and if $I = \$12$, then $P = \$200$. These corresponding pairs of numbers can be written as **ordered pairs,** $(100, 6)$ and $(200, 12)$, pairs of numbers whose order is important. The first number denotes the value of P and the second number the value of I.

Ordered pairs are **graphed** with the perpendicular number lines of a **Cartesian coordinate system,** shown in Figure 1. The horizontal number line, or **x-axis,** represents the first components of the ordered pairs, while the vertical or **y-axis** represents the second components. The point where the number lines cross is the zero point on both lines; this point is called the **origin.**

The name "Cartesian" honors René Descartes (1596–1650), one of the greatest mathematicians of the seventeenth century. According to legend, Descartes was lying in bed when he noticed an insect crawling on the ceiling and realized that if he could determine the distance from the bug to each of two perpendicular walls, he could describe its position at any given moment. The same idea can be used to locate a point in a plane.

Each point on the xy-plane corresponds to an ordered pair of numbers, where the x-value is written first. From now on, we will refer to the point corresponding to the ordered pair (a, b) as "the point (a, b)."

Locate the point $(-2, 4)$ on the coordinate system by starting at the origin and counting 2 units to the left on the horizontal axis and 4 units upward, parallel to the vertical axis. This point is shown in Figure 1, along with several other sample points. The number -2 is the **x-coordinate** and the number 4 is the **y-coordinate** of the point $(-2, 4)$.

The x-axis and y-axis divide the plane into four parts, or **quadrants.** For example, quadrant I includes all those points whose x- and y-coordinates are both positive. The quadrants are numbered as shown in Figure 1. The points on the axes themselves belong to no quadrant. The set of points corresponding to the ordered pairs of an equation is the **graph** of the equation.

The x- and y-values of the points where the graph of an equation crosses the axes are called the **x-intercept** and **y-intercept,** respectively.* See Figure 2.

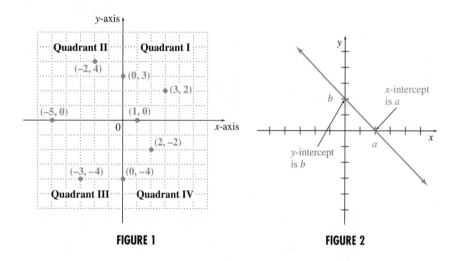

FIGURE 1　　　　　**FIGURE 2**

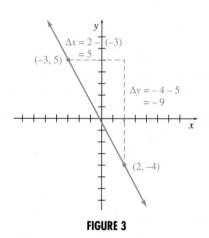

FIGURE 3

Slope of a Line An important characteristic of a straight line is its *slope,* a number that represents the "steepness" of the line. To see how slope is defined, look at the line in Figure 3. The line goes through the points $(x_1, y_1) = (-3, 5)$ and $(x_2, y_2) = (2, -4)$. The difference in the two x-values,

$$x_2 - x_1 = 2 - (-3) = 5$$

in this example, is called the **change in x.** The symbol Δx (read "delta x") is used to represent the change in x. In the same way, Δy represents the **change in y.** In our example,

$$\Delta y = y_2 - y_1$$
$$= -4 - 5$$
$$= -9.$$

These symbols, Δx and Δy, are used in the following definition of slope.

---

*Some people prefer to define the intercepts as ordered pairs, rather than as numbers.

SLOPE OF A LINE

The **slope** of a line is defined as the vertical change (the "rise") over the horizontal change (the "run") as one travels along the line. In symbols, taking two different points (x_1, y_1) and (x_2, y_2) on the line, the slope is

$$m = \frac{\text{Change in } y}{\text{Change in } x} = \frac{\Delta y}{\Delta x} = \frac{y_2 - y_1}{x_2 - x_1},$$

where $x_1 \neq x_2$.

By this definition, the slope of the line in Figure 3 is

$$m = \frac{\Delta y}{\Delta x} = \frac{-4 - 5}{2 - (-3)} = -\frac{9}{5}.$$

The slope of a line tells how fast y changes for each unit of change in x.

> **NOTE** Using similar triangles, it can be shown that the slope of a line is independent of the choice of points on the line. That is, the same slope will be obtained for *any* choice of two different points on the line.

EXAMPLE 1 Slope

Find the slope through each of the following pairs of points.

(a) $(-7, 6)$ and $(4, 5)$

   Solution Let $(x_1, y_1) = (-7, 6)$ and $(x_2, y_2) = (4, 5)$. Use the definition of slope.

$$m = \frac{\Delta y}{\Delta x} = \frac{5 - 6}{4 - (-7)} = -\frac{1}{11}$$

(b) $(5, -3)$ and $(-2, -3)$

   Solution Let $(x_1, y_1) = (5, -3)$ and $(x_2, y_2) = (-2, -3)$. Then

$$m = \frac{-3 - (-3)}{-2 - 5} = \frac{0}{-7} = 0.$$

   Lines with zero slope are horizontal (parallel to the x-axis).

(c) $(2, -4)$ and $(2, 3)$

   Solution Let $(x_1, y_1) = (2, -4)$ and $(x_2, y_2) = (2, 3)$. Then

$$m = \frac{3 - (-4)}{2 - 2} = \frac{7}{0},$$

   which is undefined. This happens when the line is vertical (parallel to the y-axis).

> **CAUTION** The phrase "no slope" should be avoided; specify instead whether the slope is zero or undefined.

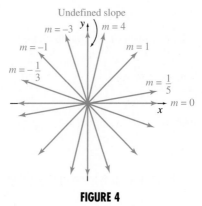

FIGURE 4

In finding the slope of the line in Example 1(a) we could have let $(x_1, y_1) = (4, 5)$ and $(x_2, y_2) = (-7, 6)$. In that case,

$$m = \frac{6 - 5}{-7 - 4} = \frac{1}{-11} = -\frac{1}{11},$$

the same answer as before. The order in which coordinates are subtracted does not matter, as long as it is done consistently.

Figure 4 shows examples of lines with different slopes. Lines with positive slopes go up from left to right, while lines with negative slopes go down from left to right.

It might help you to compare slope with the percent grade of a hill. If a sign says a hill has a 10% grade uphill, this means the slope is .10, or 1/10, so the hill rises 1 foot for every 10 feet horizontally. A 15% grade downhill means the slope is $-.15$.

FOR REVIEW ■——————

For review on solving a linear equation, see Section R.4.

Equations of a Line   An equation in two first-degree variables, such as $4x + 7y = 20$, has a line as its graph, so it is called a **linear equation.** In the rest of this section, we consider various forms of the equation of a line.

EXAMPLE 2   Equation of a Line
Find the equation of the line through $(0, -3)$ with slope 3/4.

Solution   We can use the definition of slope, letting $(x_1, y_1) = (0, -3)$ and (x, y) represent another point on the line.

$$m = \frac{y_2 - y_1}{x_2 - x_1}$$

$$\frac{3}{4} = \frac{y - (-3)}{x - 0} = \frac{y + 3}{x} \qquad \text{Substitute.}$$

$$3x = 4(y + 3) \qquad \text{Cross multiply.}$$

$$3x = 4y + 12$$

A generalization of the method of Example 2 can be used to find the equation of any line, given its y-intercept and slope. Assume that a line has y-intercept b, so that it goes through the point $(0, b)$. Let the slope of the line be represented by m. If (x, y) is any point on the line *other* than $(0, b)$, then the definition of slope can be used with the points $(0, b)$ and (x, y) to get

$$m = \frac{y - b}{x - 0}$$

$$m = \frac{y - b}{x}$$

$$mx = y - b$$

$$y = mx + b.$$

This result is called the *slope-intercept form* of the equation of a line, because b is the y-intercept of the graph of the line.

SLOPE-INTERCEPT FORM

If a line has slope m and y-intercept b, then the equation of the line in **slope-intercept form** is

$$y = mx + b.$$

When $b = 0$, we say that y is **proportional** to x.

EXAMPLE 3   Slope-Intercept Form

Find the equation of the line in slope-intercept form having y-intercept $7/2$ and slope $-5/2$.

Solution   Use the slope-intercept form with $b = 7/2$ and $m = -5/2$.

$$y = mx + b$$

$$y = -\frac{5}{2}x + \frac{7}{2}$$

The slope-intercept form shows that we can find the slope of a line by solving its equation for y. In that form the coefficient of x is the slope and the constant term is the y-intercept. For instance, in Example 2 the slope of the line $3x = 4y + 12$ was given as $3/4$. This slope also could be found by solving the equation for y.

$$4y + 12 = 3x$$

$$4y = 3x - 12$$

$$y = \frac{3}{4}x - 3$$

The coefficient of x, $3/4$, is the slope of the line. The y-intercept is -3.

The slope-intercept form of the equation of a line involves the slope and the y-intercept. Sometimes, however, the slope of a line is known, together with one point (perhaps *not* the y-intercept) that the line goes through. The *point-slope form* of the equation of a line is used to find the equation in this case. Let (x_1, y_1) be any fixed point on the line and let (x, y) represent any other point on the line. If m is the slope of the line, then by the definition of slope,

$$\frac{y - y_1}{x - x_1} = m,$$

or

$$y - y_1 = m(x - x_1).$$

POINT-SLOPE FORM

If a line has slope m and passes through the point (x_1, y_1), then an equation of the line is given by

$$y - y_1 = m(x - x_1),$$

the **point-slope form** of the equation of a line.

EXAMPLE 4   Point-Slope Form

Find an equation of the line that passes through the point $(3, -7)$ and has slope $m = 5/4$.

Solution Use the point-slope form.

$$y - y_1 = m(x - x_1)$$

$$y - (-7) = \frac{5}{4}(x - 3) \quad \text{Let } y_1 = -7, m = \tfrac{5}{4}, x_1 = 3.$$

$$y + 7 = \frac{5}{4}(x - 3)$$

$$4y + 28 = 5(x - 3) \qquad \text{Multiply both sides by 4.}$$

$$4y + 28 = 5x - 15$$

$$4y = 5x - 43 \qquad \text{Combine constants.}$$

FOR REVIEW ■

See Section R.4 for details on eliminating denominators in an equation.

The equation of the same line can be given in many forms. To avoid confusion, the linear equations used in the rest of this section will be written in slope-intercept form, $y = mx + b$, which is often the most useful form.

The point-slope form also can be useful to find an equation of a line if we know two different points that the line goes through. The procedure for doing this is shown in the next example.

EXAMPLE 5   Using Point-Slope Form to Find Equation

Find an equation of the line through $(5, 4)$ and $(-10, -2)$.

Solution Begin by using the definition of slope to find the slope of the line that passes through the given points.

$$\text{Slope} = m = \frac{-2 - 4}{-10 - 5} = \frac{-6}{-15} = \frac{2}{5}$$

Either $(5, 4)$ or $(-10, -2)$ can be used in the point-slope form with $m = 2/5$. If $(x_1, y_1) = (5, 4)$, then

$$y - y_1 = m(x - x_1)$$

$$y - 4 = \frac{2}{5}(x - 5) \quad \text{Let } y_1 = 4, m = \tfrac{2}{5}, x_1 = 5.$$

$$5y - 20 = 2(x - 5) \qquad \text{Multiply both sides by 5.}$$

$$5y - 20 = 2x - 10 \qquad \text{Distributive property}$$

$$5y = 2x + 10 \qquad \text{Add 20 to both sides.}$$

$$y = \frac{2}{5}x + 2 \qquad \text{Divide by 5 to put in slope-intercept form.}$$

Check that the same result is found if $(x_1, y_1) = (-10, -2)$.

EXAMPLE 6   Horizontal Line

Find an equation of the line through $(8, -4)$ and $(-2, -4)$.

Solution Find the slope.

$$m = \frac{-4 - (-4)}{-2 - 8} = \frac{0}{-10} = 0$$

Choose, say, $(8, -4)$ as (x_1, y_1).

$$y - y_1 = m(x - x_1)$$
$$y - (-4) = 0(x - 8) \qquad \text{Let } y_1 = -4, m = 0, x_1 = 8.$$
$$y + 4 = 0 \qquad\qquad 0(x - 8) = 0$$
$$y = -4$$

Plotting the given ordered pairs and drawing a line through the points, show that the equation $y = -4$ represents a horizontal line. See Figure 5(a). Every horizontal line has a slope of zero and an equation of the form $y = k$, where k is the y-value of all ordered pairs on the line.

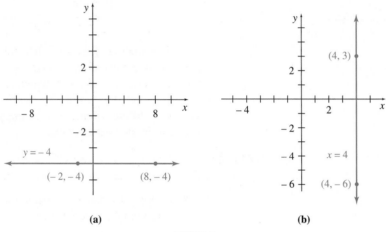

(a)   **(b)**

FIGURE 5

EXAMPLE 7   Vertical Line

Find an equation of the line through $(4, 3)$ and $(4, -6)$.

Solution The slope of the line is

$$m = \frac{-6 - 3}{4 - 4} = \frac{-9}{0},$$

which is undefined. Since both ordered pairs have x-coordinate 4, the equation is $x = 4$. Because the slope is undefined, the equation of this line cannot be written in the slope-intercept form.

Again, plotting the given ordered pairs and drawing a line through them show that the graph of $x = 4$ is a vertical line. See Figure 5(b).

> The slope of a horizontal line is 0.
> The slope of a vertical line is undefined.

The different forms of linear equations discussed in this section are summarized on the next page. The slope-intercept and point-slope forms are equivalent ways to express the equation of a nonvertical line. The slope-intercept form is simpler for a final answer, but you may find the point-slope form easier to use when you know the slope of a line and a point through which the line passes.

EQUATIONS OF LINES

Equation	Description
$y = mx + b$	**Slope-intercept form:** slope m, y-intercept b
$y - y_1 = m(x - x_1)$	**Point-slope form:** slope m, line passes through (x_1, y_1)
$x = k$	**Vertical line:** x-intercept k, no y-intercept (except when $k = 0$), undefined slope
$y = k$	**Horizontal line:** y-intercept k, no x-intercept (except when $k = 0$), slope 0

Parallel and Perpendicular Lines   One application of slope involves deciding whether two lines are parallel. Since two parallel lines are equally "steep," they should have the same slope. Also, two lines with the same "steepness" are parallel.

PARALLEL LINES

Two lines are parallel if and only if they have the same slope, or if they are both vertical.

EXAMPLE 8   Parallel Line

Find the equation of the line that passes through the point $(3, 5)$ and is parallel to the line $2x + 5y = 4$.

Solution   The slope of $2x + 5y = 4$ can be found by writing the equation in slope-intercept form.

$$2x + 5y = 4$$

$$y = -\frac{2}{5}x + \frac{4}{5}$$

This result shows that the slope is $-2/5$. Since the lines are parallel, $-2/5$ is also the slope of the line whose equation we want. This line passes through $(3, 5)$. Substituting $m = -2/5$, $x_1 = 3$, and $y_1 = 5$ into the point-slope form gives

$$y - y_1 = m(x - x_1)$$

$$y - 5 = -\frac{2}{5}x + \frac{6}{5}$$

$$y = -\frac{2}{5}x + \frac{6}{5} + 5$$

$$y = -\frac{2}{5}x + \frac{31}{5}$$

As already mentioned, two nonvertical lines are parallel if and only if they have the same slope. Two lines having slopes with a product of -1 are perpendicular. A proof of this fact, which depends on similar triangles from geometry, is given as Exercise 43 in this section.

PERPENDICULAR LINES

Two lines are perpendicular if and only if the product of their slopes is -1, or if one is vertical and the other horizontal.

EXAMPLE 9 Perpendicular Line

Find the slope of the line L perpendicular to the line having the equation $5x - y = 4$.

Solution To find the slope, write $5x - y = 4$ in slope-intercept form:

$$y = 5x - 4.$$

The slope is 5. Since the lines are perpendicular, if line L has slope m, then

$$5m = -1$$

$$m = -\frac{1}{5}.$$

Many real-world situations can be approximately described by a straight-line graph. One way to find the equation of such a straight line is to use two typical data points from the graph and the point-slope form of the equation of a line.

EXAMPLE 10 Work Force

In recent decades, the percentage of the U.S. civilian population age 16 and over that is in the work force has risen at a roughly constant rate, from 59.4% in 1960 to 67.1% in 1998.* Find the equation describing this linear relationship.

Solution For this example, let x represent time in years, with $x = 0$ for 1960. Such rescaling of a variable is often used to simplify the arithmetic, although computers and calculators have made rescaling less important than in the past. Here it allows us to work with smaller numbers, and, as you will see, find the y-intercept of the line more easily. We will use such rescaling on many examples throughout this book. When we do, it is important to be consistent. In this example, if we want to refer to the year 1975, we must let $x = 15$, and not $x = 1975$. Let y represent the percent of the population in the work force.

With 1960 corresponding to $x = 0$, 1998 corresponds to $x = 1998 - 1960 = 38$. The two ordered pairs representing the given information are $(0, 59.4)$ and $(38, 67.1)$. The slope of the line through these points is

$$m = \frac{67.1 - 59.4}{38 - 0} = \frac{7.7}{38} = .2026316 \approx .203.^\dagger$$

This means that, on average, the percent of the population in the work force has gone up by about .2% per year.

Using $m = .203$ and $(x_1, y_1) = (0, 59.4)$ in the point-slope form gives the required equation,

$$y - 59.4 = .203(x - 0)$$

$$y = .203x + 59.4.$$

This result could also have been obtained by observing that $(0, 59.4)$ is the y-intercept.

---

*U.S. Bureau of Labor Statistics, *Employment and Earnings.*
$\dagger$The symbol $\approx$ means "is approximately equal to."

Notice that if this formula is valid for all nonnegative x, then eventually y becomes 100:

$$.203x + 59.4 = 100$$

$$.203x = 40.6 \qquad \text{Subtract 59.4.}$$

$$x = 40.6/.203 = 200, \qquad \text{Divide by .203.}$$

which indicates that 200 years from 1960 (in the year 2160), 100% of the population will be in the work force.

In Example 10, of course, it is still possible that in 2160 there will be people who are not in the work force; the trend of recent decades may not continue. Most equations are valid for some specific set of numbers. It is highly speculative to extrapolate beyond those values. On the other hand, people in business and government often need to make some prediction about what will happen in the future, so a tentative conclusion based on past trends may be better than no conclusion at all. There are also circumstances, particularly in the physical sciences, in which theoretical reasons imply that the trend will continue.

EXAMPLE 11   Antibiotic Resistance

The linear equation $y = 1.5457x - 3067.7$, where x represents the year, can be used to estimate the percent of gonorrhea cases with antibiotic resistance diagnosed from 1985 to 1990.*

(a) Determine this percent in 1988.

Solution Substitute 1988 for x in the equation.

$$y = 1.5457x - 3067.7$$

$$= 1.5457(1988) - 3067.7$$

$$\approx 5.2$$

This means that 5.2% of gonorrhea cases in 1988 were antibiotic resistant.

(b) Find and interpret the slope of the line.

Solution The equation is given in slope-intercept form, so the slope is the coefficient of x, 1.5457. Since

$$m = 1.5457 = \frac{\text{change in } y}{\text{change in } x} = \frac{1.5457}{1},$$

the slope indicates the change in the percent of antibiotic-resistant cases per year from 1985 to 1990. Because the slope is positive, the percent of antibiotic-resistant cases increased by 1.5457% per year.

Graph of a Line   We can graph the linear equation defined by $y = x + 1$ by finding several ordered pairs. For example, if $x = 2$, then $y = 2 + 1 = 3$, giving the ordered pair $(2, 3)$. Also, $(0, 1)$, $(4, 5)$, $(-2, -1)$, $(-5, -4)$, $(-3, -2)$, among many others, are ordered pairs that satisfy the equation.

---

*Teutsch, S. and R. Churchhill, *Principles and Practice of Public Health Surveillance,* New York: Oxford University Press, 1994.

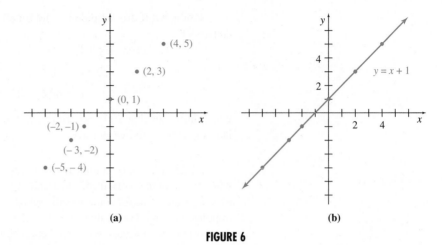

FIGURE 6

To graph $y = x + 1$, we begin by locating the ordered pairs obtained above, as shown in Figure 6(a). All the points of this graph appear to lie on a straight line, as in Figure 6(b). This straight line is the graph of $y = x + 1$.

It can be shown that every equation of the form $ax + by = c$ has a straight line as its graph. Although just two points are needed to determine a line, it is a good idea to plot a third point as a check. It is often convenient to use the x- and y-intercepts as the two points, as in the following example.

EXAMPLE 12 Graph of a Line

Graph $3x + 2y = 12$.

Solution To find the y-intercept, let $x = 0$.

$$3(0) + 2y = 12$$
$$2y = 12 \quad \text{Divide both sides by 2.}$$
$$y = 6$$

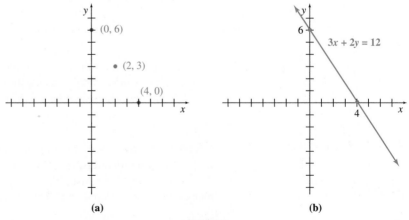

FIGURE 7

Similarly, find the *x*-intercept by letting $y = 0$, which gives $x = 4$. Verify that when $x = 2$, the result is $y = 3$. These three points are plotted in Figure 7(a). A line is drawn through them in Figure 7(b).

Not every line has two distinct intercepts; the graph in the next example does not cross the *x*-axis, and so it has no *x*-intercept.

EXAMPLE 13 Graph of a Horizontal Line
Graph $y = -3$.

Solution The equation $y = -3$, or equivalently, $y = 0x - 3$, always gives the same *y*-value, -3, for any value of *x*. Therefore, no value of *x* will make $y = 0$, so the graph has no *x*-intercept. As we saw in Example 6, the graph of such an equation is a horizontal line parallel to the *x*-axis. In this case the *y*-intercept is -3, as shown in Figure 8.

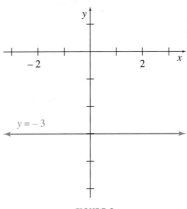

FIGURE 8

In general, the graph of $y = k$, where *k* is a real number, is the horizontal line having *y*-intercept *k*.

The graph in Example 13 had only one intercept. Another type of linear equation with coinciding intercepts is graphed in Example 14.

EXAMPLE 14 Graph of a Line Through the Origin
Graph $y = -3x$.

Solution Begin by looking for the *x*-intercept. If $y = 0$, then

$$y = -3x$$
$$0 = -3x \quad \text{Let } y = 0.$$
$$0 = x. \quad \text{Divide both sides by } -3.$$

We have the ordered pair $(0, 0)$. Starting with $x = 0$ gives exactly the same ordered pair, $(0, 0)$. Two points are needed to determine a straight line, and the intercepts have led to only one point. To get a second point, choose some other value of *x* (or *y*). For example, if $x = 2$, then

$$y = -3x = -3(2) = -6, \quad \text{Let } x = 2.$$

giving the ordered pair $(2, -6)$. These two ordered pairs, $(0, 0)$ and $(2, -6)$, were used to get the graph shown in Figure 9.

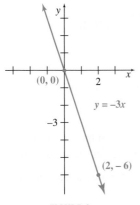

FIGURE 9

36. Find k so that the line through $(4, -1)$ and $(k, 2)$ is

    a. parallel to $2x + 3y = 6$

    b. perpendicular to $5x - 2y = -1$

37. Use slopes to show that the quadrilateral with vertices at $(1, 3)$, $(-5/2, 2)$, $(-7/2, 4)$, and $(2, 1)$ is a parallelogram.

38. Use slopes to show that the square with vertices at $(-2, 5)$, $(4, 5)$, $(4, -1)$, and $(-2, -1)$ has diagonals that are perpendicular.

For the lines in Exercises 39 and 40, which of the following is closest to the slope of the line? **(a)** 1   **(b)** 2   **(c)** 3   **(d)** -1 **(e)** -2 **(f)** -3

39.

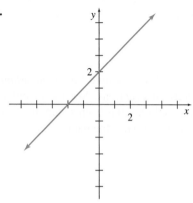

40.

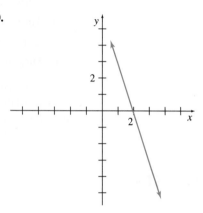

Estimate the slope of the lines in Exercises 41 and 42.

41.

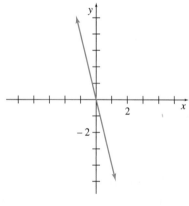

42.

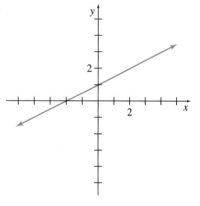

43. To show that two perpendicular lines, neither of which is vertical, have slopes with a product of -1, go through the following steps. Let line L_1 have equation $y = m_1x + b_1$, and let L_2 have equation $y = m_2x + b_2$. Assume that L_1 and L_2 are perpendicular, and use right triangle MPN shown in the figure. Prove each of the following statements.

    a. MQ has length m_1.

    b. QN has length $-m_2$.

    c. Triangles MPQ and PNQ are similar.

    d. $m_1/1 = 1/(-m_2)$ and $m_1m_2 = -1$

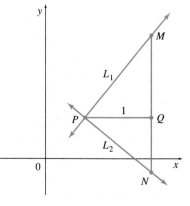

Graph the equations for Exercises 44–59.

44. $y = x - 1$

45. $y = 2x + 3$

46. $y = -4x + 9$

47. $y = -6x + 12$

48. $2x - 3y = 12$

49. $3x - y = -9$

50. $3y + 4x = 12$

51. $4y + 5x = 10$

52. $y = -2$

53. $x = 4$

54. $x + 5 = 0$

55. $y - 4 = 0$

56. $y = 2x$

57. $y = -5x$

58. $x + 4y = 0$

59. $x - 3y = 0$

Applications

BUSINESS AND ECONOMICS

60. *Sales* The sales of a small company were $27,000 in its second year of operation and $63,000 in its fifth year. Let y represent sales in the xth year of operation. Assume that the data can be approximated by a straight line.

  a. Find the slope of the sales line, and give an equation for the line in the form $y = mx + b$.

  b. Use your answer from part a to find out how many years must pass before the sales surpass $100,000.

61. *Federal Debt* The table lists the total federal debt (in trillions of dollars) from 1990 to 1996.*

Year	Federal Debt
1990	3.207
1991	3.599
1992	4.002
1993	4.351
1994	4.644
1995	4.921
1996	5.182

  a. Plot the data by letting $x = 0$ correspond to 1990. Discuss any trends of the federal debt over this time period.

  b. Find a linear equation that approximates the data, using the points $(0, 3.207)$ and $(6, 5.182)$. What does the slope of the graph represent? Graph the line and the data on the same coordinate axes.

  c. Use the equation from part b to predict the federal debt in the years 1997 and 1998. Compare your results to the actual values of 5.370 and 5.479 trillion dollars.

  d. Add the information for 1997 and 1998 to your graph from part b. Does the linear trend of the earlier years con-

tinue? Discuss the accuracy of using a linear approximation to model the federal debt over different domains.

62. *Airline Passenger Growth* The following table estimates the growth in the number of airline passengers (in millions) at some of the fastest-growing airports in the United States between 1992 and 2005.[†]

Airport	1992	2005
Harrisburg Intl.	.7	1.4
Dayton Intl.	1.1	2.4
Austin Robert Mueller	2.2	4.7
Milwaukee Gen. Mitchell Intl.	2.2	4.4
Sacramento Metropolitan	2.6	5.0
Fort Lauderdale–Hollywood	4.1	8.1
Washington Dulles Intl.	5.3	10.9
Greater Cincinnati Airport	5.8	12.3

  a. Determine a linear equation that approximates the data using the points $(.7, 1.4)$ and $(5.3, 10.9)$.

  b. In 1992, 4.9 million passengers used Raleigh–Durham International Airport. Using the equation from part a, approximate the number of passengers using this airport in 2005 and compare it with the Federal Aviation Administration's estimate of 10.3 million passengers.

LIFE SCIENCES

63. *HIV Infection* The time interval between a person's initial infection with HIV and that person's eventual development of AIDS symptoms is an important issue. The method of infection with HIV affects the time interval before AIDS develops. One study of HIV patients who were infected by intravenous drug use found that 17% of the patients had AIDS after 4 years, and 33% had developed the disease after 7 years. The relationship between the time interval

---

*Budget of the U.S. government, Fiscal Year 2000.
†Federal Aviation Administration

and the percentage of patients with AIDS can be modeled accurately with a linear equation.*

a. Write a linear equation $y = mx + b$ that models this data, using the ordered pairs (4, .17) and (7, .33).

b. Use your equation from part a to predict the number of years before half of these patients will have AIDS.

64. *Exercise Heart Rate* To achieve the maximum benefit for the heart when exercising, your heart rate (in beats per minute) should be in the target heart rate zone. The lower limit of this zone is found by taking 70% of the difference between 220 and your age. The upper limit is found by using 85%.†

a. Find formulas for the upper and lower limits (u and l) as linear equations involving the age x.

b. What is the target heart rate zone for a 20-year-old?

c. What is the target heart rate zone for a 40-year-old?

d. Two women in an aerobics class stop to take their pulse, and are surprised to find that they have the same pulse. One woman is 36 years older than the other and is working at the upper limit of her target heart rate zone. The younger woman is working at the lower limit of her target heart rate zone. What are the ages of the two women, and what is their pulse?

e. Run for 10 minutes, take your pulse, and see if it is in your target heart rate zone. (After all, this is listed as an exercise!)

65. *Ponies Trotting* A 1991 study found that the peak vertical force on a trotting Shetland pony increased linearly with the pony's speed, and that when the force reached a critical level, the pony switched from a trot to a gallop.‡ For one pony, the critical force was 1.16 times its body weight. It experienced a force of .75 times its body weight at a speed of 2 meters per second, and a force of .93 times its body weight at 3 meters per second. At what speed did the pony switch from a trot to a gallop?

66. *Life Span* Some scientists believe there is a limit to how long humans can live.§ One supporting argument is that during the last century, life expectancy from age 65 has increased more slowly than life expectancy from birth, so eventually these two will be equal, at which point, according to these scientists, life expectancy should increase no further. In 1900, life expectancy at birth was 46 years, and life expectancy at age 65 was 76. In 1975, these figures had risen to 75 and 80, respectively. In both cases, the increase in life expectancy has been linear. Using these assumptions and the data given, find the maximum life expectancy for humans.

67. *Deer Ticks* Deer ticks cause concern because they can carry Lyme disease. One study found a relationship between the density of acorns produced in the fall and the density of deer tick larvae the following spring.‖ The relationship can be approximated by the linear equation

$$y = 34x + 230,$$

where x is the number of acorns per square meter in the fall, and y is the number of deer tick larvae per 400 square meters the following spring. According to this formula, approximately how many acorns per square meter would result in 1000 deer tick larvae per 400 square meters?

SOCIAL SCIENCES

68. *Immigration* In 1974, 86,821 people from other countries immigrated to the state of California. In 1996, the number of immigrants was 199,483.#

a. If the change in foreign immigration to California is considered to be linear, write an equation expressing the number of immigrants, y, in terms of the number of years after 1974, x.

*Alcabes, P., A. Munoz, D. Vlahov, and G. Friedland, "Incubation Period of Human Immunodeficiency Virus," *Epidemiologic Review,* Vol. 15, No. 2, The Johns Hopkins University School of Hygiene and Public Health, 1993.

†Hockey, Robert V., *Physical Fitness: The Pathway to Healthful Living,* Times Mirror/Mosby College Publishing, 1989, pp. 85–87.

‡*Science,* July 19, 1991, pp. 306–8.

§*Science,* Nov. 15, 1991, pp. 936–38.

‖*Science,* Vol. 281, No. 5375, July 17, 1998, pp. 350–351.

#*Legal Immigration to California in Federal Fiscal Year 1996,* State of California Demographic Research Unit, June 1999.

b. Use your result in part a to predict the foreign immigration to California in the year 2010.

69. *Cohabitation* The number of unmarried couples in the United States who are living together has been rising at a roughly linear rate in recent years. The number of cohabiting adults was 1.1 million in 1977 and 4.9 million in 1997.*

   a. Write an equation expressing the number of cohabiting adults (in millions), y, in terms of the number of years after 1977, x.

   b. Use your result in part a to predict the number of cohabiting adults in the year 2010.

70. *Older College Students* The percentage of college students who are age 35 and older has been increasing at roughly a linear rate. In 1972 the percentage was 9%, and in 1998 it was 17%.[†]

   a. Find an equation giving the percentage of college students age 35 and older in terms of time t, where t represents the number of years since 1970.

   b. If this linear trend continues, what percentage of college students will be 35 and over in 2010?

   c. If this linear trend continues, in what year will the percentage of college students 35 and over reach 31%?

71. *Family Payments* The figure shows an idealized linear relationship for the average monthly family payment to families with dependent children in 1994 dollars.[‡]

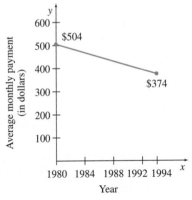

   a. Use this information to determine a linear function for this data, letting x be the years since 1980.

   b. Based on your function, what was the average payment in 1987?

   c. How is the average payment changing?

PHYSICAL SCIENCES

72. *Global Warming* In 1990, the Intergovernmental Panel on Climate Change predicted that the average temperature on Earth would rise .3°C per decade in the absence of international controls on greenhouse emissions.[§] Let t measure the time in years since 1970, when the average global temperature was 15°C.

   a. Find a linear equation giving the average global temperature in degrees Celsius in terms of t, the number of years since 1970.

   b. Scientists have estimated that the sea level will rise by 65 cm if the average global temperature rises to 19°C. According to your answer to part a, when would this occur?

73. *Galactic Distance* The table lists the distances (in megaparsecs where 1 megaparsec $\approx 3.1 \cdot 10^{19}$ km) and velocities (in kilometers per second) of four galaxies moving rapidly away from Earth.[‖]

Galaxy	Distance	Velocity
Virga	15	1600
Ursa Minor	200	15,000
Corona Borealis	290	24,000
Bootes	520	40,000

   a. Plot the data points letting x represent distance and y represent velocity. Do the points lie in an approximately linear pattern?

   b. Write a linear equation $y = mx$ to model this data, using the ordered pair (520, 40,000).

   c. The galaxy Hydra has a velocity of 60,000 km per sec. Use your equation to determine how far away it is from Earth.

   d. The value of m in the equation is called the *Hubble constant*. The Hubble constant can be used to estimate the age of the universe A (in years) using the formula

$$A = \frac{9.5 \times 10^{11}}{m}.$$

   Approximate A using your value of m.

---

*The New York Times, Feb. 15, 2000, p. F8.

[†]The New York Times, Aug. 17, 1994, p. B9, and Jan. 9, 2000, p. 4A.

[‡]Office of Financial Management, Administration for Children and Families.

[§]Science News, June 23, 1990, p. 391.

[‖]Acker, A. and C. Jaschek, *Astronomical Methods and Calculations,* John Wiley & Sons, 1986.

Karttunen, H. (editor), *Fundamental Astronomy,* Springer-Verlag, 1994.

74. *Radio Stations* The graph shows the number of U.S. radio stations on the air along with the graph of a linear equation that models the data.*

a. Use the two ordered pairs (1950, 2773) and (1998, 12,641) to find the approximate slope of the line shown. Interpret your answer.

b. Use the same two ordered pairs to write an equation of the line that models the data.

c. Estimate the year when it is expected that the number of stations will first exceed 15,000.

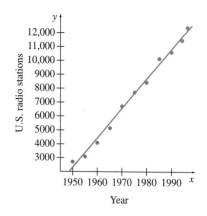

75. *Tuition* The table lists the average annual cost (in dollars) of tuition and fees at private four-year colleges for selected years.[†] See Example 15.

Year	Tuition and Fees
1986	6658
1988	8004
1990	9340
1992	10,449
1994	11,719
1996	12,994
1998	14,508

a. Sketch a graph of the data. Do the data appear to lie roughly along a straight line?

b. Let $x = 86$ correspond to 1986. Use the points (86, 6658) and (96, 12,994) to determine a linear equation that models the data. What does the slope of the graph of the equation indicate?

c. Discuss the accuracy of using this equation to estimate the cost of private college in 2020.

■ 1.2   LINEAR FUNCTIONS AND APPLICATIONS

THINK ABOUT IT   How many units must be sold for a firm to break even?

Later in this section, this question will be answered using a linear function.

As we saw in the previous section, many situations involve two variables related by a linear equation. For such a relationship, when we express the variable y in terms of x, we say that y is a **linear function** of x. This means that for any allowed value of x (the **independent variable**), we can use the equation to find the corresponding value of y (the **dependent variable**). Examples of linear functions include $y = 2x + 3$, $y = -5$, and $2x - 3y = 7$, which can be written as $y = (2/3)x - (7/3)$. Equations in the form $x = k$, where k is a constant, are not linear functions. All other linear equations define linear functions.

$f(x)$ **Notation**   `Letters such as f, g, or h are often used to name functions. For example, f might be used to name the function

$$y = 5 - 3x.$$

---

*National Association of Broadcasters
[†]The College Board

To show that this function is named f, it is common to replace y with $f(x)$ (read "f of x") to get

$$f(x) = 5 - 3x.$$

By choosing 2 as a value of x, $f(x)$ becomes $5 - 3 \cdot 2 = 5 - 6 = -1$, written

$$f(2) = -1.$$

The corresponding ordered pair is $(2, -1)$. In a similar manner,

$$f(-4) = 5 - 3(-4) = 17, \qquad f(0) = 5, \qquad f(-6) = 23,$$

and so on.

EXAMPLE 1   Function Notation

Let $g(x) = -4x + 5$. Find $g(3)$, $g(0)$, and $g(-2)$.

Solution To find $g(3)$, substitute 3 for x.

$$g(3) = -4(3) + 5 = -12 + 5 = -7$$

Similarly,

$$g(0) = -4(0) + 5 = 0 + 5 = 5,$$

and $$g(-2) = -4(-2) + 5 = 8 + 5 = 13.$$

We summarize the discussion below.

LINEAR FUNCTION

A relationship f defined by

$$y = f(x) = mx + b,$$

for real numbers m and b, is a **linear function.**

Supply and Demand   Linear functions are often good choices for **supply and demand curves.** Typically, as the price of an item increases, the demand for the item decreases, while the supply increases. On the other hand, when demand for an item increases, so does its price, causing the supply of the item to decrease.

For example, during the 1970s the price of gasoline increased rapidly. As the price continued to escalate, most buyers became more and more prudent in their use of gasoline in order to restrict their demand to an affordable amount. Consequently, the overall demand for gasoline decreased and the supply increased, to a point where there was an oversupply of gasoline. This caused prices to fall until supply and demand were approximately balanced. Many other factors were involved in the situation, but the relationship between price, supply, and demand was nonetheless typical. Some commodities, however, such as medical care, college education, and certain luxury items, may be exceptions to these typical relationships.

Although economists consider price to be the independent variable, they have the unfortunate habit of plotting price, usually denoted by p, on the vertical axis, while everyone else graphs the independent variable on the horizontal axis. This custom was started by the English economist Alfred Marshall (1842–1924).

In order to abide by this custom, we will write p, the price, as a function of q, the quantity produced, and plot p on the vertical axis. But remember, it is really *price* that determines how much consumers demand and producers supply, not the other way around.

Supply and demand functions are not necessarily linear, the simplest kind of function. Yet most functions are approximately linear if a small enough piece of the graph is taken, allowing applied mathematicians to often use linear functions for simplicity. That approach will be taken in this chapter.

EXAMPLE 2 Supply and Demand

Suppose that Greg Tobin, an economist, has studied the supply and demand for vinyl siding and has determined that the price (in dollars) per square yard, p, and the quantity demanded monthly (in thousands of square yards), q, are related by the linear function

$$p = D(q) = 60 - \frac{3}{4}q, \quad \text{Demand}$$

while the price p and the supply q are related by

$$p = S(q) = \frac{3}{4}q. \quad \text{Supply}$$

(a) Find the demand at a price of \$45 and at a price of \$18.

Solution Start with the demand function

$$p = 60 - \frac{3}{4}q,$$

and replace p with 45.

$$45 = 60 - \frac{3}{4}q$$

$$-15 = -\frac{3}{4}q \quad \text{Subtract 60 from both sides.}$$

$$20 = q \quad \text{Multiply both sides by } -\frac{4}{3}.$$

Thus, at a price of \$45, 20,000 square yards are demanded per month.

Similarly, replace p with 18 to find the demand when the price is \$18. Verify that this leads to $q = 56$. When the price is lowered from \$45 to \$18, the demand increases from 20,000 square yards to 56,000 square yards.

(b) Find the supply at a price of \$60 and at a price of \$12.

Solution Substitute 60 for p in the supply equation,

$$p = \frac{3}{4}q,$$

to find that $q = 80$, so the supply is 80,000 square yards. Similarly, replacing p with 12 in the supply equation gives a supply of 16,000 square yards. If the price decreases from \$60 to \$12, the supply also decreases, from 80,000 square yards to 16,000 square yards.

FOR REVIEW ■

In the second-to-last step of the solution in Example 2(a), q was multiplied by $-3/4$, so both sides of the equation had to be divided by $-3/4$. This was done by multiplying both sides by $-4/3$. The way to divide by a fraction is to multiply by its reciprocal. In other words, to divide by the fraction a/b, multiply by b/a.

(c) Graph both functions on the same axes.

> **Solution** The results of part (a) are written as the ordered pairs $(20, 45)$ and $(56, 18)$. The line through those points is the graph of $p = 60 - (3/4)q$, shown in red in Figure 11(a). We used the ordered pairs $(80, 60)$ and $(16, 12)$ from the work in part (b) to get the supply graph shown in blue in Figure 11(a).

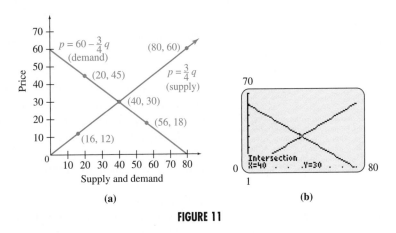

FIGURE 11

A calculator-generated graph of the lines representing the supply and demand in Example 2 is shown in Figure 11(b). The equation of each line, using x and y instead of q and p, was entered along with an appropriate window to get this graph. A special menu choice gives the coordinates of the intersection point, shown at the bottom of the graph.

> **NOTE** Not all supply/demand problems will have the same scale on both axes. It helps to consider the intercepts of both the supply and demand graphs to decide what scale to use. For example, in Figure 11, the y-intercept of the demand function is 60, so the scale should allow values from 0 to at least 60 on the vertical axis. The x-intercept of the supply function is 80, so values on the x-axis must go from 0 to 80. Letting each tick mark represent 10 gives a reasonable number of marks on each axis.

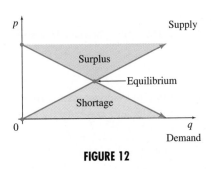

FIGURE 12

As shown in the graphs of Figure 11, both the supply and the demand graphs pass through the point $(40, 30)$. If the price of a square yard of siding is more than $30, the supply will exceed the demand. At a price less than $30, the demand will exceed the supply. Only at a price of $30 will demand and supply be equal. For this reason, $30 is called the *equilibrium price*. When the price is $30, demand and supply both equal 40,000 square yards, the *equilibrium quantity*. In general, the **equilibrium price** of a commodity is the price found at the point where the supply and demand graphs for that commodity intersect. The **equilibrium quantity** is the demand and the supply at that same point. Figure 12 illustrates a general supply and demand situation.

EXAMPLE 3 Equilibrium Quantity

Use algebra to find the equilibrium quantity for the vinyl siding in Example 2.

Solution The equilibrium quantity is found when the prices from both supply and demand are equal. Set the two expressions for p equal to each other and solve.

$$60 - \frac{3}{4}q = \frac{3}{4}q$$

$$240 - 3q = 3q \qquad \text{Multiply both sides by 4.}$$

$$240 = 6q \qquad \text{Add } 3q \text{ to both sides.}$$

$$40 = q$$

The equilibrium quantity is 40,000 square yards, the same answer found earlier.

You may prefer to find the equilibrium quantity by solving the equation with your calculator. Or, if your calculator has a TABLE feature, you can use it to find the value of q that makes the two expressions equal.

Another important issue is how, in practice, the equations of the supply and demand functions can be found. This issue is important for many problems involving linear functions in this section and the next. Data need to be collected, and if they lie perfectly along a line, then the equation can easily be found with any two points. What usually happens, however, is that the data are scattered, and there is no line that goes through all the points. In this case we must find a line that approximates the linear trend of the data as closely as possible (assuming the points lie approximately along a line) as in Example 15 in the previous section. This is usually done by the *method of least squares,* also referred to as *linear regression.* We will discuss this method in Section 1.3.

Cost Analysis   The cost of manufacturing an item commonly consists of two parts. The first is a **fixed cost** for designing the product, setting up a factory, training workers, and so on. Within broad limits, the fixed cost is constant for a particular product and does not change as more items are made. The second part is a *cost per item* for labor, materials, packing, shipping, and so on. The total value of this second cost *does* depend on the number of items made.

EXAMPLE 4   Cost Analysis

Suppose that the cost of producing videocassette tapes can be approximated by

$$C(x) = 12x + 100,$$

where $C(x)$ is the cost in dollars to produce x tapes. The cost to produce 0 tapes is

$$C(0) = 12(0) + 100 = 100,$$

or $100. This sum, $100, is the fixed cost.

Once the company has invested the fixed cost into the videocassette project, what will the additional cost per tape be? As an example, let's compare the costs of making 5 tapes and 6 tapes.

$$C(5) = 12(5) + 100 = 160 \qquad \text{and} \qquad C(6) = 12(6) + 100 = 172,$$

or $160 and $172, respectively.

So the 6th tape itself costs $172 - $160 = $12 to produce. In the same way, the 81st tape costs $C(81) - C(80) = $1072 - $1060 = 12 to produce. In fact,

the $(n + 1)$st tape costs

$$C(n + 1) - C(n) = [12(n + 1) + 100] - (12n + 100)$$
$$= 12,$$

or $12, to produce. The number 12 is also the slope of the graph of the cost function $C(x) = 12x + 100$; the slope gives us the cost to produce an additional item.

In economics, **marginal cost** is the rate of change of cost $C(x)$ at a level of production x and is equal to the slope of the cost function at x. It approximates the cost of producing one additional item. In fact, some books define the marginal cost to be the cost of producing one additional item. With *linear functions,* these two definitions are equivalent, and the marginal cost, which is equal to the slope of the cost function, is *constant.* For instance, in the videocassette example, the marginal cost of each tape is $12. For other types of functions, these two definitions are only approximately equal. Marginal cost is important to management in making decisions in areas such as cost control, pricing, and production planning.

The work in Example 4 can be generalized. Suppose the total cost to make x items is given by the linear cost function $C(x) = mx + b$. The fixed cost is found by letting $x = 0$:

$$C(0) = m \cdot 0 + b = b;$$

thus, the fixed cost is b dollars. The additional cost of the $(n + 1)$st item, the marginal cost, is m, the slope of the line $C(x) = mx + b$.

COST FUNCTION

In a cost function of the form $C(x) = mx + b$, m represents the marginal cost per item and b the fixed cost. Conversely, if the fixed cost of producing an item is b and the marginal cost is m, then the **cost function** $C(x)$ for producing x items is $C(x) = mx + b$.

EXAMPLE 5   Cost Function

The marginal cost to make x tablets of a prescription medication is $10 per batch, while the cost to produce 100 batches is $1500. Find the cost function $C(x)$, given that it is linear.

Solution Since the cost function is linear, it can be expressed in the form $C(x) = mx + b$. The marginal cost is $10 per batch, which gives the value for m, leading to $C(x) = 10x + b$. To find b, use the fact that the cost of producing 100 batches of tablets is $1500, or $C(100) = 1500$. Substituting $C(x) = 1500$ and $x = 100$ into $C(x) = 10x + b$ gives

$$1500 = 10 \cdot 100 + b$$
$$1500 = 1000 + b$$
$$500 = b.$$

The cost function is given by $C(x) = 10x + 500$, where the fixed cost is $500.

Break-Even Analysis   The **revenue** $R(x)$ from selling x units of a product is the product of the price per unit p and the number of units sold (demand) x, so that

$$R(x) = px.$$

The corresponding **profit** $P(x)$ is the difference between revenue $R(x)$ and cost $C(x)$. That is,

$$P(x) = R(x) - C(x).$$

A company can make a profit only if the revenue received from its customers exceeds the cost of producing and selling its goods and services. The number of units at which revenue just equals cost is the **break-even quantity;** the corresponding ordered pair gives the **break-even point.**

EXAMPLE 6   Break-Even Analysis

A firm producing poultry feed finds that the total cost $C(x)$ in dollars of producing and selling x units is given by

$$C(x) = 20x + 100.$$

Management plans to charge $24 per unit for the feed.

(a) How many units must be sold for the firm to break even?

Solution The firm will break even (no profit and no loss) as long as revenue just equals cost, or $R(x) = C(x)$. From the given information, since $R(x) = px$ and $p = \$24$,

$$R(x) = 24x.$$

Substituting for $R(x)$ and $C(x)$ in the equation $R(x) = C(x)$ gives

$$24x = 20x + 100,$$

from which $x = 25$. The firm breaks even by selling 25 units, which is the break-even quantity. The graphs of $C(x) = 20x + 100$ and $R(x) = 24x$ are shown in Figure 13. The break-even point (where $x = 25$) is shown on the graph. If the company sells more than 25 units (if $x > 25$), it makes a profit. If it sells less than 25 units, it loses money.

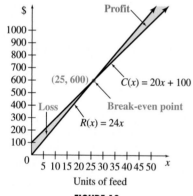

Units of feed

FIGURE 13

(b) What is the profit if 100 units of feed are sold?

Solution Use the formula for profit $P(x)$.

$$
\begin{aligned}
P(x) &= R(x) - C(x) \\
&= 24x - (20x + 100) \\
&= 4x - 100
\end{aligned}
$$

Then $P(100) = 4(100) - 100 = 300$. The firm will make a profit of $300 from the sale of 100 units of feed.

(c) How many units must be sold to produce a profit of $900?

Solution Let $P(x) = 900$ in the equation $P(x) = 4x - 100$ and solve for x.

$$
\begin{aligned}
900 &= 4x - 100 \\
1000 &= 4x \\
x &= 250
\end{aligned}
$$

Sales of 250 units will produce $900 profit.

Temperature   One of the most common linear relationships found in everyday situations deals with temperature. Recall that water freezes at 32° Fahrenheit and 0° Celsius, while it boils at 212° Fahrenheit and 100° Celsius.* The ordered pairs $(0, 32)$ and $(100, 212)$ are graphed in Figure 14 on axes showing Fahrenheit (F) as a function of Celsius (C). The line joining them is the graph of the function.

EXAMPLE 7   Temperature

Derive an equation relating F and C.

Solution   To derive the required linear equation, first find the slope using the given ordered pairs, $(0, 32)$ and $(100, 212)$.

$$m = \frac{212 - 32}{100 - 0} = \frac{9}{5}$$

The F-intercept of the graph is 32, so by the slope-intercept form, the equation of the line is

$$F = \frac{9}{5}C + 32.$$

With simple algebra this equation can be rewritten to give C in terms of F:

$$C = \frac{5}{9}(F - 32).$$

FIGURE 14

1.2   EXERCISES

In Exercises 1–4, decide whether the statement is true or false.

1. To find the x-intercept of the graph of a linear function, we solve $y = f(x) = 0$, and to find the y-intercept, we evaluate $f(0)$.

2. The graph of $f(x) = -3$ is a vertical line.

3. The slope of the graph of a linear function cannot be undefined.

4. The graph of $f(x) = ax$ is a straight line that passes through the origin.

5. In a few sentences, explain why the price of a commodity not already at its equilibrium price should move in that direction.

6. Explain why a linear function may not be adequate for describing the supply and demand functions.

Write a linear cost function for each situation. Identify all variables used.

7. A chain saw rental firm charges $12 plus $1 per hour.

8. A trailer-hauling service charges $45 plus $2 per mile.

9. A parking garage charges 50 cents plus 35 cents per half-hour.

10. For a one-day rental, a car rental firm charges $44 plus 28 cents per mile.

---

*Gabriel Fahrenheit (1686–1736), a German physicist, invented his scale with 0° representing the temperature of an equal mixture of ice and ammonium chloride (a type of salt), and 96° as the temperature of the human body. (It is often said, erroneously, that Fahrenheit set 100° as the temperature of the human body. Fahrenheit's own words are quoted in *A History of the Thermometer and Its Use in Meteorology* by W. E. Knowles, Middleton: The Johns Hopkins Press, 1966, p. 75.) The Swedish astronomer Anders Celsius (1701–1744) set 0° and 100° as the freezing and boiling points of water.

Assume that each of the following can be expressed as a linear cost function. Find the cost function in each case.

11. Fixed cost: $100; 50 items cost $1600 to produce.

12. Fixed cost: $400; 10 items cost $650 to produce.

13. Marginal cost: $90; 150 items cost $16,000 to produce.

14. Marginal cost: $120; 700 items cost $96,500 to produce.

 15. How is the average rate of change related to the graph of a function?

 16. In your own words describe the break-even quantity, how to find it, and what it indicates.

Applications

BUSINESS AND ECONOMICS

17. *Supply and Demand* Suppose that the demand and price for a certain model of electric can opener are related by

$$p = D(q) = 16 - \frac{5}{4}q,$$

where p is the price (in dollars) and q is the demand (in hundreds). Find the price at each of the following levels of demand.

a. 0 can openers   **b.** 400 can openers

c. 800 can openers

Find the demand for the electric can opener at each of the following prices.

d. $6   **e.** $11   **f.** $16

g. Graph $p = 16 - \frac{5}{4}q$.

Suppose the price and supply of the electric can opener are related by

$$p = S(q) = \frac{3}{4}q,$$

where p is the price (in dollars) and q is the supply (in hundreds) of can openers. Find the supply at each of the following prices.

h. $0   **i.** $10   **j.** $20

k. Graph $p = \frac{3}{4}q$ on the same axes used for part g.

l. Find the equilibrium quantity and the equilibrium price.

18. *Supply and Demand* Let the supply and demand functions for strawberry-flavored licorice be given by

$$p = S(q) = \frac{3}{2}q \quad \text{and} \quad p = D(q) = 81 - \frac{3}{4}q,$$

where p is the price in dollars and q is the number of batches.

a. Graph these on the same axes.

b. Find the equilibrium quantity and the equilibrium price.

19. *Supply and Demand* Let the supply and demand functions for butter pecan ice cream be given by

$$p = S(q) = \frac{2}{5}q \quad \text{and} \quad p = D(q) = 100 - \frac{2}{5}q,$$

where p is the price in dollars and q is the number of 10 gallon tubs.

a. Graph these on the same axes.

b. Find the equilibrium quantity and the equilibrium price.

20. *Supply and Demand* Let the supply and demand functions for sugar be given by

$$p = S(q) = 1.4q - .6 \quad \text{and}$$
$$p = D(q) = -2q + 3.2,$$

where p is the price per pound and q is the quantity in thousands of pounds.

a. Graph these on the same axes.

b. Find the equilibrium quantity and the equilibrium price.

21. *T-Shirt Cost* Yoshi Yamamura sells silk-screened T-shirts at community festivals and crafts fairs. Her marginal cost to produce one T-shirt is $3.50. Her total cost to produce 60 T-shirts is $300, and she sells them for $9 each.

a. Find the linear cost function for Yoshi's T-shirt production.

b. How many T-shirts must she produce and sell in order to break even?

c. How many T-shirts must she produce and sell to make a profit of $500?

22. *Publishing Costs* Enrique Gonzales owns a small publishing house specializing in Latin American poetry. His fixed cost to produce a typical poetry volume is $525, and his total cost to produce 1000 copies of the book is $2675. His books sell for $4.95 each.

a. Find the linear cost function for Enrique's book production.

b. How many poetry books must he produce and sell in order to break even?

c. How many books must he produce and sell to make a profit of $1000?

23. *Marginal Cost of Coffee* The manager of a restaurant found that the cost to produce 100 cups of coffee is $11.02, while the cost to produce 400 cups is $40.12. Assume the cost $C(x)$ is a linear function of x, the number of cups produced.

a. Find a formula for $C(x)$.

b. Find the total cost of producing 1000 cups.

c. Find the total cost of producing 1001 cups.

d. Find the marginal cost of the 1001st cup.

e. What is the marginal cost of *any* cup?

24. *Marginal Cost of a New Plant* In deciding whether to set up a new manufacturing plant, company analysts have decided that a linear function is a reasonable estimation for the total cost $C(x)$ in dollars to produce x items. They estimate the cost to produce 10,000 items as $547,500, and the cost to produce 50,000 items as $737,500.

a. Find a formula for $C(x)$.

b. Find the total cost to produce 100,000 items.

c. Find the marginal cost of the items to be produced in this plant.

25. *Bread Sales* Bread Boutiques, which sell freshly baked bread with no preservatives, are located in many malls around the United States and are growing rapidly. The Saint Louis Bread Company (now called Panera Bread) claims a sales growth of 5000% in its first five years.*

a. Suppose sales were $100,000 in 1991. What would they be in 1996 at that growth rate?

b. Let 1991 correspond to $x = 1$. Write two ordered pairs representing sales in 1991 and 1996.

c. Assuming sales increased linearly, write a linear sales function for this company.

d. If sales continue to increase at the same rate, when will they reach $20,000,000?

26. *Bread Sales* Au Bon Pain, another boutique bakery (see Exercise 25), had 1994 revenues of $45 million. Assume these revenues increase linearly with an average rate of change of $.5 million.*

a. Let 1994 correspond to $x = 4$ and write a linear function for revenue.

b. According to the revenue function from part a, when will revenue reach $50 million?

27. *Break-Even Analysis* Producing x units of tacos costs $C(x) = 5x + 20$; revenue is $R(x) = 15x$, where $C(x)$ and $R(x)$ are in dollars.

a. What is the break-even quantity?

b. What is the profit from 100 units?

c. How many units will produce a profit of $500?

28. *Break-Even Analysis* To produce x units of a religious medal costs $C(x) = 12x + 39$. The revenue is $R(x) = 25x$. Both $C(x)$ and $R(x)$ are in dollars.

a. Find the break-even quantity.

b. Find the profit from 250 units.

c. Find the number of units that must be produced for a profit of $130.

Break-Even Analysis You are the manager of a firm. You are considering the manufacture of a new product, so you ask the accounting department for cost estimates and the sales department for sales estimates. After you receive the data, you must decide whether to go ahead with production of the new product. Analyze the data in Exercises 29–32 (find a break-even quantity) and then decide what you would do in each case. Also write the profit function.

29. $C(x) = 85x + 900$; $R(x) = 105x$; no more than 38 units can be sold.

30. $C(x) = 105x + 6000$; $R(x) = 250x$; no more than 400 units can be sold.

31. $C(x) = 70x + 500$; $R(x) = 60x$ (*Hint*: What does a negative break-even quantity mean?)

32. $C(x) = 1000x + 5000$; $R(x) = 900x$

PHYSICAL SCIENCES

33. *Temperature* Use the formula for conversion between Fahrenheit and Celsius derived in Example 7 to convert each of the following temperatures.

a. 58°F to Celsius

b. −20°F to Celsius

c. 50°C to Fahrenheit

34. *Body Temperature* You may have heard that the average temperature of the human body is 98.6°. Recent experiments show that the actual figure is closer to 98.2°.[†] The figure of 98.6 comes from experiments done by Carl Wunderlich in 1868. But Wunderlich measured the temperatures in degrees Celsius and rounded the average to the nearest degree, giving 37°C as the average temperature.[‡]

---

*The New York Times, Nov. 18, 1995, pp. 19 and 21.
[†]Science News, Sept. 26, 1992, p. 195.
[‡]Science News, Nov. 7, 1992, p. 399.

a. What is the Fahrenheit equivalent of 37°C?

b. Given that Wunderlich rounded to the nearest degree Celsius, his experiments tell us that the actual average human body temperature is somewhere between 36.5°C and 37.5°C. Find what this range corresponds to in degrees Fahrenheit.

35. *Temperature* Find the temperature at which the Celsius and Fahrenheit temperatures are numerically equal.

■ 1.3 THE LEAST SQUARES LINE

 THINK ABOUT IT   How has the accidental death rate in the United States changed over time?

In this section, we show how to answer such questions using the method of least squares. We use past data to find trends and to make tentative predictions about the future. The only assumption we make is that the data are related linearly— that is, if we plot pairs of data, the resulting points will lie close to some line. This method cannot give exact answers. The best we can expect is that, if we are careful, we will get a reasonable approximation.

The table lists the number of accidental deaths per 100,000 population in the United States through the past century.* If you were a manager at an insurance company, these data could be very important. You might need to make some predictions about how much you will pay out next year in accidental death benefits, and even a very tentative prediction based on past trends is better than no prediction at all.

The first step is to draw a scatterplot, as we have done in Figure 15. Notice that the points lie approximately along a line, which means that a linear function may give a good approximation of the data. If we select two points and find the line that passes through them, as we did in Section 1.1, we will get a different line for each pair of points, and in some cases the lines will be very different. We want to draw one line that is simultaneously close to all the points on the graph, but many such lines are possible, depending upon how we define the phrase "simultaneously close to all the points." How do we decide on the best possible line? Before going on, you might want to try drawing the line you think is best on Figure 15.

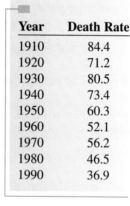

Year	Death Rate
1910	84.4
1920	71.2
1930	80.5
1940	73.4
1950	60.3
1960	52.1
1970	56.2
1980	46.5
1990	36.9

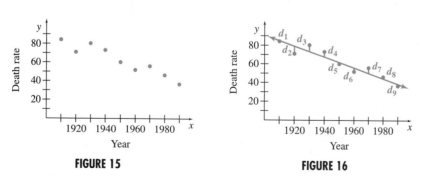

FIGURE 15

FIGURE 16

*U.S. Dept. of Health and Human Services, National Center for Health Statistics.

The line used most often in applications is that in which the sum of the squares of the vertical distances from the data points to the line is as small as possible. Such a line is called the **least squares line.** The least squares line for the data in Figure 15 is drawn in Figure 16. How does the line compare with the one you drew on Figure 15? It may not be exactly the same, but should appear similar.

In Figure 16, the vertical distances from the points to the line are indicated by d_1, d_2, and so on, up through d_9 (read "d-sub-one, d-sub-two, d-sub-three," and so on). For n points, corresponding to the n pairs of data, the least squares line is found by minimizing the sum $(d_1)^2 + (d_2)^2 + (d_3)^2 + \cdots + (d_n)^2$.

For the points $(x_1, y_1), (x_2, y_2), \ldots, (x_n, y_n)$, if the equation of the desired line is $Y = mx + b$, then

$$d_1 = |Y_1 - y_1| = |mx_1 + b - y_1|,$$
$$d_2 = |Y_2 - y_2| = |mx_2 + b - y_2|,$$

and so on. We use Y in the equation of the line instead of y to distinguish the predicted values (Y) from the y-values of the given data points. The sum to be minimized becomes

$$(mx_1 + b - y_1)^2 + (mx_2 + b - y_2)^2 + \cdots + (mx_n + b - y_n)^2,$$

where $(x_1, y_1), (x_2, y_2), \ldots, (x_n, y_n)$ are known and m and b are to be found.

The method of minimizing this sum requires advanced techniques and is not given here. The result gives equations that can be solved for the slope m and the y-intercept b of the least squares line.* In these equations, the symbol Σ, the Greek letter sigma, indicates "the sum of"; this notation is known as **summation notation.** For example, we write the sum $x_1 + x_2 + \cdots + x_n$, where n is the number of data points, as

$$x_1 + x_2 + \cdots + x_n = \Sigma x.$$

Similarly, Σxy means $x_1 y_1 + x_2 y_2 + \cdots + x_n y_n$, and so on.

CAUTION Note that Σx^2 means $x_1^2 + x_2^2 + \cdots + x_n^2$, which is *not* the same as squaring Σx.

LEAST SQUARES LINE

The least squares line $Y = mx + b$ that gives the best fit to the data points $(x_1, y_1), (x_2, y_2), \ldots, (x_n, y_n)$ has slope m and y-intercept b that satisfy the equations

$$nb + (\Sigma x)m = \Sigma y$$
$$(\Sigma x)b + (\Sigma x^2)m = \Sigma xy.$$

Method 1: Calculating by Hand To find the least squares line for the given data, we first find the required sums. To reduce the size of the numbers, let x represent the years since 1900, so that, for example, $x = 10$ for the year 1910. Let y represent the death rate.

---

*Equations for m and b are derived in Exercise 4.

x	y	xy	x^2	y^2
10	84.4	844	100	7123.36
20	71.2	1424	400	5069.44
30	80.5	2415	900	6480.25
40	73.4	2936	1600	5387.56
50	60.3	3015	2500	3636.09
60	52.1	3126	3600	2714.41
70	56.2	3934	4900	3158.44
80	46.5	3720	6400	2162.25
90	36.9	3321	8100	1361.61
$\Sigma x = 450$	$\Sigma y = 561.5$	$\Sigma xy = 24{,}735$	$\Sigma x^2 = 28{,}500$	$\Sigma y^2 = 37{,}093.41$

(The column headed y^2 will be used later.) Now we can calculate m and b by solving a system of equations. Our method is to solve the first equation for b in terms of m, and substitute this into the second equation. We then solve the second equation for m. Once we have m, we put this back into the equation for b. Here, $n = 9$ (the number of data points).

$$nb + (\Sigma x)m = \Sigma y$$

$$9b + 450m = 561.5 \qquad \text{Substitute from the table.}$$

$$9b = 561.5 - 450m$$

$$b = (561.5 - 450m)/9 \qquad \text{Divide by 9.}$$

$$(\Sigma x)b + (\Sigma x^2)m = \Sigma xy$$

$$450(561.5 - 450m)/9 + 28{,}500m = 24{,}735 \qquad \text{Substitute.}$$

$$28{,}075 - 22{,}500m + 28{,}500m = 24{,}735 \qquad \text{Multiply.}$$

$$6000m = -3340 \qquad \text{Combine terms.}$$

$$m \approx -.5566667 \approx -.557$$

The significance of m is that the death rate per 100,000 population is tending to drop (because of the negative) at a rate of .557 per year.

Now substitute the value of m into the equation for b.

$$b = \frac{561.5 - 450(-.5566667)}{9} \approx 90.2222 \approx 90.2$$

Substitute m and b into the least squares line equation, $Y = mx + b$; the least squares line that best fits the nine data points has equation $Y = -.557x + 90.2$. This gives a mathematical description of the relationship between the year and the number of accidental deaths per 100,000 population. The equation can be used to predict y from a given value of x, as we will show in Example 1. As we mentioned before, however, caution must be exercised when using the least squares equation to predict data points that are far from the range of points on which the equation was modeled.

CAUTION In computing m and b, we rounded the final answer to three digits because the original data were known only to three digits. It is important, however, *not* to round any of the intermediate results (such as Σx^2) because round-off error may have a detrimental effect on the accuracy of the answer. Similarly, it is important not to use a rounded-off value of m when computing b.

Method 2: Graphing Calculator

The calculations for finding the least squares line are often tedious, even with the aid of a calculator. Fortunately, many calculators can calculate the least squares line with just a few keystrokes. For purposes of illustration, we will show how the least squares line in the previous example is found with a TI-83 graphing calculator.

We begin by entering the data into the calculator. We will be using the first two lists, called L_1 and L_2. Choosing the STAT menu, then choosing the fourth entry ClrList, we enter L_1, L_2, to indicate the lists to be cleared. Now we press STAT again and choose the first entry EDIT, which brings up the blank lists. As before, we will only use the last two digits of the year, putting the numbers in L_1. We put the death rate in L_2, giving the two screens shown in Figure 17.

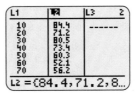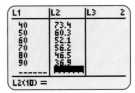

FIGURE 17

Press STAT again and choose CALC instead of EDIT. Then choose item 4 LinReg $(ax + b)$ to get the values of a (the slope) and b (the y-intercept) for the least squares line, as shown in Figure 18. With a and b rounded to three decimal places, the least squares line is $Y = -.557x + 90.2$. A graph of the data points and the line is shown in Figure 19.

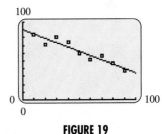

FIGURE 18   **FIGURE 19**

For more details on finding the least squares line with a graphing calculator, see *The Graphing Calculator Manual* that is available with this book.

Method 3: Spreadsheet

Many computer spreadsheet programs can also find the least squares line. Figure 20 shows the scatterplot and least squares line for the accidental death rate data using an Excel spreadsheet. The scatterplot was found using the XY(Scatter) command under Chart Wizard, and the line was found using the Add Trendline command under the Chart menu. For details, see *The Spreadsheet Manual* that is available with this book.

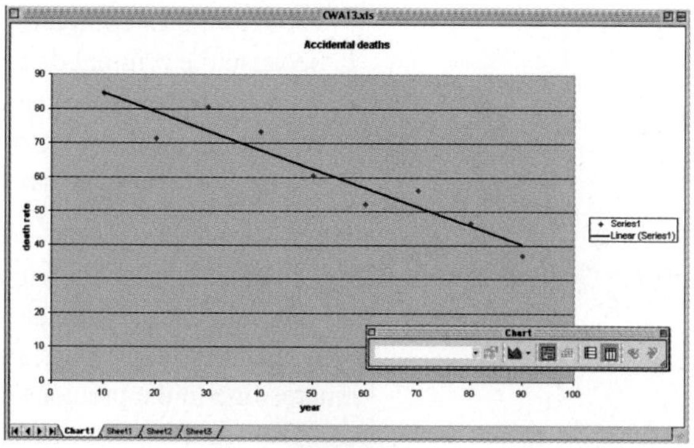

FIGURE 20

EXAMPLE 1 Least Squares Line

What do we predict the accidental death rate in 1997 to be?

Solution Use the least squares line equation given above with $x = 97$.

$$Y = -.557x + 90.2$$
$$= -.557(97) + 90.2$$
$$= 36.171$$

The death rate in 1997 is predicted to be about 36.2 per 100,000 population. In this case, we have the actual value for 1997. It happens to be 35.0, which is close to the predicted value.

EXAMPLE 2 Least Squares Line

In what year is the death rate predicted to drop below 26 per 100,000 population?

Solution Let $Y = 26$ in the equation above and solve for x.

$$26 = -.557x + 90.2$$
$$-64.2 = -.557x$$
$$x = 115.26$$

This means that after 115 years, the rate will not have quite reached 26 per 100,000, so we must wait 116 years for this to happen. This corresponds to the year 2016 (116 years after 1900), when our equation predicts the death rate to be $-.557(116) + 90.2 = 25.6$ per 100,000 population.

Correlation Once an equation is found for the least squares line, we need to have some way of judging just how good the equation is for predictive purposes. If the points from the data fit the line quite closely, then we have more reason to expect future data pairs to do so. But if the points are widely scattered about even the best-fitting line, then predictions are not likely to be accurate.

In order to have a quantitative basis for confidence in our predictions, we need a measure of the "goodness of fit" of the original data to the prediction line. One such measure is called the **coefficient of correlation,** denoted r.

COEFFICIENT OF CORRELATION

$$r = \frac{n(\Sigma xy) - (\Sigma x)(\Sigma y)}{\sqrt{n(\Sigma x^2) - (\Sigma x)^2} \cdot \sqrt{n(\Sigma y^2) - (\Sigma y)^2}}$$

Although the expression for r looks daunting, remember that each of the summations, Σx, Σy, Σxy, and so on, are just the totals from a table like the one we prepared for the data on accidental deaths. Also, with a calculator, the arithmetic is no problem!

The coefficient of correlation r is always equal to or between 1 and -1. Values of exactly 1 or -1 indicate that the data points lie *exactly* on the least squares line. If $r = 1$, the least squares line has a positive slope; $r = -1$ gives a negative slope. If $r = 0$, there is no linear correlation between the data points (but some *nonlinear* function might provide an excellent fit for the data). A correlation of zero may also indicate that the data fit a horizontal line. To investigate what is happening, it is always helpful to sketch a scatterplot of the data. Some scatterplots that correspond to these values of r are shown in Figure 21.

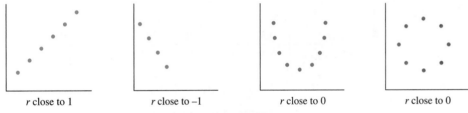

| r close to 1 | r close to -1 | r close to 0 | r close to 0 |

FIGURE 21

A value of r close to 1 or -1 indicates the presence of a linear relationship. The exact value of r necessary to conclude that there is a linear relationship depends upon n, the number of data points, as well as how confident we want to be of our conclusion. For details, consult a text on statistics.*

EXAMPLE 3 Coefficient of Correlation

Find r for the data on accidental death rates.

Solution

Method 1: Calculating by Hand

From the table on page 32, $\Sigma x = 450$, $\Sigma y = 561.5$, $\Sigma xy = 24{,}735$, $\Sigma x^2 = 28{,}500$, and $\Sigma y^2 = 37{,}093.41$. Also, $n = 9$. Substituting these values into the formula for r gives

$$r = \frac{9(24{,}735) - (450)(561.5)}{\sqrt{9(28{,}500) - (450)^2} \cdot \sqrt{9(37{,}093.41) - (561.5)^2}}$$

$$= \frac{-30{,}060}{\sqrt{54{,}000} \cdot \sqrt{18{,}558.44}}$$

$$= -.9495577064 \approx -.950.$$

This is a high correlation, which agrees with our observation that the data fit a line quite well.

---

*For example, see *Introductory Statistics,* 4th edition, by Neil A. Weiss, Reading, Mass.: Addison-Wesley, 1997.

Method 2: Graphing Calculator

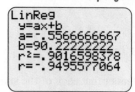

FIGURE 22

Most calculators that give the least squares line will also give the coefficient of correlation. To do this on the TI-83, we press the second function CATALOG and go down the list to the entry DiagnosticOn. Press ENTER at that point, then press STAT, CALC, and choose item 4 to get the display in Figure 22. The result is the same as we got by hand. The command DiagnosticOn need only be entered once, and the coefficient of correlation will always appear in the future.

Method 3: Spreadsheet

Many computer spreadsheet programs have a built-in command to find the coefficient of correlation. For example, in Excel, use the command "= CORREL(A1:A9, B1:B9)" to find the correlation of the 9 data points stored in columns A and B. For more details, see *The Spreadsheet Manual* that is available with this text.

EXAMPLE 4   Neediest Cases

The table shows donations to the *New York Times* Neediest Cases Fund (in millions) from 1985 through 1995.*

Year	'85	'86	'87	'88	'89	'90	'91	'92	'93	'94	'95
Donation	2.7	3.0	3.8	4.0	4.2	4.3	4.9	4.3	4.4	4.4	1.9

Find the correlation coefficient, as well as the line that best fits the data.

Solution A scatterplot of the data, along with the graph of the least squares line, is shown in Figure 23. Notice that the data appear to be linear for 1985 through 1994, but the last point ruins the linearity. Figure 24 shows the result of the LinReg command on the TI-83. The coefficient of correlation is only about .158, indicating that the data do not fit a line. In this case, the least squares line is of little value. When we remove the 1995 value, however, we get $r = .828$ and $Y = .184x - 12.5$. (Verify this on your calculator.) Therefore, the trend from 1985 through 1994 is roughly linear, with a correlation coefficient of .828 and a least squares line $Y = .184x - 12.5$. The figure for 1995 is an anomaly that does not fit the trend, and the managers of the Neediest Case Fund undoubtedly looked for the cause.

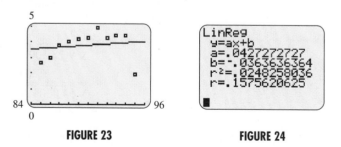

FIGURE 23   **FIGURE 24**

The final example illustrates why a plot of the data points is so important.

EXAMPLE 5   Silicone Implants

Silicones have long been used for fabricating medical devices on the presumption that they are biocompatible materials. This presumption is not entirely correct. Silicone prostheses, when implanted within the soft tissues of the breast, may evoke an inflammatory reaction. In response to silicone exposure, inflammatory mediator production was observed in experimental studies. After in vitro culture for 24 hours, the levels of four inflammatory mediators from ten patients with silicone breast implants were as shown in the following table.*

Patient	IL-2	TNF-α	IL-6	PGE$_2$
1	48	ND	231	68
2	219	78	308,287	2710
3	109	65	33,291	1804
4	2179	149	124,550	8053
5	219	451	17,075	7371
6	54	64	22,955	3418
7	6	79	95,102	9768
8	10	115	5649	441
9	42	618	840,585	9585
10	196	69	58,924	4536

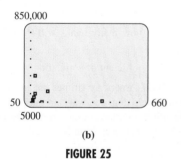

(a)

(b)

FIGURE 25

A graphing calculator was used to plot the last nine points in the table for TNF-$\alpha(x)$ and IL-6(y). (There is no data point for patient 1, since there is no TNF-α level.) The graph is shown in Figure 25(a). The correlation for these two mediators is .6936. Notice that one data point is way off by itself. As shown in Example 4, sometimes, by removing such a point from the graph, we can achieve a higher correlation.[†] Figure 25(b) shows the scatterplot with the remaining eight points. However, the correlation is now −.2493![‡] With the ninth point, the points were closer to a line with a positive slope. Without the point, there is little linear correlation and it has become negative.

1.3 EXERCISES

1. Suppose a positive linear correlation is found between two quantities. Does this mean that one of the quantities increasing causes the other to increase? If not, what does it mean?

2. Given a set of points, the least squares line formed by letting x be the independent variable will not necessarily be the same as the least squares line formed by letting y be the independent variable. Give an example to show why this is true.

---

*Mena, et al., "Inflammatory Intermediates Produced by Tissues Encasing Silicone Breast Prostheses," *Journal of Investigative Surgery,* Vol. 8, 1995, p. 33. Copyright © 1995. Reproduced by permission of Taylor & Francis, Inc., http://www.routledge-ny.com.
[†]Before discarding a point, we should investigate the reason it is an outlier.
[‡]The observation that removing one point changes the correlation from positive to negative was made by Patrick Fleury, *Chance News* (an electronic newsletter), Vol. 4, No. 16, Dec. 1995.

The following problem is reprinted from the November 1989 Actuarial Examination on Applied Statistical Methods.

3. You are given

X	6.8	7.0	7.1	7.2	7.4
Y	.8	1.2	.9	.9	1.5

Determine r^2, the coefficient of determination for the regression of Y on X. (*Note:* The coefficient of determination is defined as the square of the coefficient of correlation.)

a. .3   **b.** .4   **c.** .5   **d.** .6   **e.** .7

4. Follow the steps outlined in this section to solve the least squares line equations

$$nb + (\Sigma x)m = \Sigma y$$
$$(\Sigma x)b + (\Sigma x^2)m = \Sigma xy$$

for m and b to get

$$m = \frac{n\Sigma(xy) - (\Sigma x)(\Sigma y)}{n(\Sigma x^2) - (\Sigma x)^2}$$
$$b = \frac{\Sigma y - m(\Sigma x)}{n}.$$

◢ Applications

BUSINESS AND ECONOMICS

5. *New Jobs* Since 1993, California has generated new jobs to replace those lost—mainly in the defense and aerospace industries. Many of these new jobs are in the computer field. The results of a survey in which x represented the last two digits of the year and y represented the number of new jobs (in thousands) generated in computers, gave the following summations.[†]

$$n = 6 \qquad \Sigma x^2 = 51,355$$
$$\Sigma x = 555 \qquad \Sigma xy = 69,869$$
$$\Sigma y = 754 \qquad \Sigma y^2 = 95,740$$

a. Find an equation for the least squares line.

b. Predict the number of new jobs in 1997.

c. If this growth in jobs continues linearly, when will the number of new jobs reach 170,000?

d. Find and interpret the coefficient of correlation.

6. *Decrease in Banks* The number of banks in the United States has dropped more than 27% since 1985. The following data are from a survey in which x corresponded to the last two digits of the year and y corresponded to the number of banks, in thousands, in the U.S.[‡]

$$n = 10 \qquad \Sigma x^2 = 80,185$$
$$\Sigma x = 895 \qquad \Sigma xy = 10,540.6$$
$$\Sigma y = 118.2 \qquad \Sigma y^2 = 1415.04$$

a. Find an equation of the least squares line.

b. If the trend continues, how many banks will there be in 1999?

c. Find and interpret the coefficient of correlation.

7. *Air Fares* In January 2000, American Airlines ran an ad in *The New York Times* advertising one-way air fares from New York to various cities.[§] Fourteen of the cities are listed on the next page, with the distances from New York to the cities added.

a. Plot the data. Do the data points lie in a linear pattern?

b. Find the correlation coefficient. Combining this with your answer to part a, does the cost of a ticket tend to go up with the distance flown?

c. Find the equation of the least squares line, and use it to find the approximate marginal cost per mile to fly.

d. For similar data in an October 1993 ad, the equation of the least squares line was $y = 91.9 + .0313x$. Use this information and your answer to part b to compare the cost of flying American Airlines for these two time periods.

*"November 1989 Course 120 Examination Applied Statistical Methods" of the *Education and Examination Committee of The Society of Actuaries.* Reprinted by permission of The Society of Actuaries.
†UCLA, Center for Continuing Study of the California Economy. From *Chicago Tribune,* Dec. 24, 1995, Sec. 1, p. 13.
‡Federal Deposit Insurance Corp., *Chicago Tribune,* Jan. 2, 1996.
§*The New York Times,* Jan. 7, 2000, p. A9.

City	Distance (x) in miles	Price (y) in dollars
Boston	206	109
Chicago	802	124
Denver	1771	154
Kansas City	1198	144
Little Rock	1238	144
Los Angeles	2786	179
Minneapolis	1207	144
Nashville	892	144
Phoenix	2411	179
Portland	2885	179
Reno	2705	179
St. Louis	948	144
San Diego	2762	179
Seattle	2815	179

8. *Consumer Debt* Bank credit card debt has risen steadily over the years. The table gives debt per household in 1994 dollars.* In the table the years are coded so that 1975 corresponds to 75, 1980 corresponds to 80, and so on.

Year (x)	75	80	85	90	95
Debt (y)	270	650	1100	1800	3100

a. Plot the data. Does the graph show a linear pattern?

b. Find the equation of the least squares line and graph it on the same axes. Does the line appear to be a good fit?

c. Find and interpret the coefficient of correlation.

d. If this linear trend continues, when will household debt reach $3300?

9. *Used Car Sales* As cars are becoming more expensive, used car sales have increased at a faster rate since 1984 than new car sales.[†] Sales in millions from 1984 to 1996 are given in the table in the next column.

a. Find the equation of the least squares line and the coefficient of correlation.

b. Find the equation of the least squares line using only the data for every other year starting with 1985, 1987, and so on. Find the coefficient of correlation.

c. Compare your answers for parts a and b. What do you find? Why do you think this happens?

Year	Sales	Year	Sales
84	12.3	91	12.3
85	13.2	92	12.8
86	13.6	93	13.9
87	13.2	94	15.0
88	14.5	95	14.7
89	14.5	96	14.6
90	13.8		

LIFE SCIENCES

10. *Medical School Admissions* The number of applications to medical schools in the United States increased rapidly from 1989 to 1994 as indicated by the data in the table.[‡] Years are represented by their last two digits and applications are given in thousands.

Year (x)	88	89	90	91	92	93	94
Applications (y)	27	27	29	33	37	43	45

a. Plot the data. Do the data points lie in a linear pattern?

b. Determine the least squares line for this data and graph it on the same coordinate axes. Does the line fit the data reasonably well?

c. Find the coefficient of correlation. Does it agree with your estimate of the fit in part b?

d. Explain why the coefficient of correlation is close to 1, even though the data points do not appear to be linear.

*The New York Times, Dec. 28, 1995, p. C1.
[†]The New York Times, March 3, 1996.
[‡]The New York Times, Friday, Nov. 17, 1995, p. C2.

11. *Crickets Chirping* Biologists have observed a linear relationship between the temperature and the frequency with which a cricket chirps. The following data were measured for the striped ground cricket.*

Temperature °F (x)	Chirps per second (y)
88.6	20.0
71.6	16.0
93.3	19.8
84.3	18.4
80.6	17.1
75.2	15.5
69.7	14.7
82.0	17.1
69.4	15.4
83.3	16.2
79.6	15.0
82.6	17.2
80.6	16.0
83.5	17.0
76.3	14.4

a. Find the equation for the least squares line for the data.

b. Use the results of part a to determine how many chirps per second you would expect to hear from the striped ground cricket if the temperature were 73°F.

c. Use the results of part a to determine what the temperature is when the striped ground crickets are chirping at a rate of 18 times per second.

d. Find the coefficient of correlation.

SOCIAL SCIENCES

12. *Educational Expenditures* A 1991 report issued by the U.S. Department of Education listed the expenditure per pupil and the average mathematics proficiency in grade 8 for 37 states and the District of Columbia. Letting x equal the expenditure per pupil (ranging from $2838 in Idaho to $7850 in Washington, D.C.) and y equal the average mathematics proficiency score (ranging from 231 in Washington, D.C., to 281 in North Dakota), the data can be summarized as follows:[†]

$$\Sigma x = 175{,}878 \qquad \Sigma x^2 = 872{,}066{,}218$$
$$\Sigma y = 9989 \qquad \Sigma y^2 = 2{,}629{,}701$$
$$\Sigma xy = 46{,}209{,}266 \qquad n = 38$$

Compute the coefficient of correlation for the given data. Does there appear to be a trend in the amount of money spent per pupil and the proficiency of eighth graders in mathematics?

13. *Poverty Levels* The following table lists how poverty level income cutoffs (in dollars) for a family of four have changed over time.[‡]

Year	Income
1970	3968
1975	5500
1980	8414
1985	10,989
1990	13,359
1995	15,569

Let x be the year, with $x = 0$ corresponding to 1970, and y be the income in thousands of dollars. (*Note:* $\Sigma x = 75$, $\Sigma x^2 = 1375$, $\Sigma y = 57.799$, $\Sigma y^2 = 658.405183$, $\Sigma xy = 932.88$.)

a. Sketch a graph of the data. Do the data appear to lie along a straight line?

b. Calculate the coefficient of correlation. Does your result agree with your answer to part a?

c. Find the equation of the least squares line.

d. Use your answer from part c to predict the poverty level in the year 2015.

14. *Library Budget Decline* The budget for books in New York Public Library branches has declined recently as shown in the table. Budgets are given in millions of dollars.[§]

Year (x)	91	92	93	94	95	96
Budget (y)	7.1	6.3	7.8	7.6	7.5	6.6

a. Plot the data with a graphing calculator in the window $[90, 97]$ by $[6, 8]$. Do the data show a linear pattern?

b. Plot the data with a graphing calculator in the window $[90, 97]$ by $[4, 10]$. Do the data show a more linear pattern? Do you think a line would be a good fit?

c. Plot the data with a graphing calculator in the window $[90, 97]$ by $[0, 12]$. Is a line a good fit for this set of points?

---

*Pierce, George W., *The Songs of Insects,* Cambridge, Mass., Harvard University Press, Copyright © 1948 by the President and Fellows of Harvard College. Reprinted by permission of the publishers.
[†]*Newsday,* June 7, 1991.
[‡]U.S. Bureau of the Census, *Current Population Reports.*
[§]*New York Observer,* Feb. 12, 1996. Vol. 10, No. 6, p. 1.

d. What can you conclude from a comparison of the graphs in parts a–c?

e. What is the coefficient of correlation?

✎ **f.** Which scatterplot in parts a–c describes the data most accurately? How does the coefficient of correlation support your answer?

15. *SAT Scores* At Hofstra University, all students take the math SAT before entrance, and most students take a mathematics placement test before registration. Recently, one professor collected the following data for 19 students in his Finite Mathematics class:

Math SAT	Placement Test	Math SAT	Placement Test	Math SAT	Placement Test
540	20	580	8	440	10
510	16	680	15	520	11
490	10	560	8	620	11
560	8	560	13	680	8
470	12	500	14	550	8
600	11	470	10	620	7
540	10				

a. Find an equation for the least squares line. Let x be the math SAT and y be the placement test score.

b. Use your answer from part a to predict the mathematics placement test score for a student with a math SAT score of 420.

c. Use your answer from part a to predict the mathematics placement test score for a student with a math SAT score of 620.

d. Calculate the coefficient of correlation.

e. Based on your answer to part d, what can you conclude about the relationship between a student's math SAT and mathematics placement test score?

PHYSICAL SCIENCES

16. *Air Conditioning* While shopping for an air conditioner, Adam Bryer consulted the following table giving a machine's BTUs and the square footage (ft^2) that it would cool.

a. Find the equation for the least squares line for the data.

b. To check the fit of the data to the line, use the results from part a to find the BTUs required to cool a room of

ft^2 (x)	BTUs (y)
150	5000
175	5500
215	6000
250	6500
280	7000
310	7500
350	8000
370	8500
420	9000
450	9500

150 ft^2, 280 ft^2, and 420 ft^2. How well does the actual data agree with the predicted values?

c. Suppose Adam's room measures 230 ft^2. Use the results from part a to decide how many BTUs it requires. If air conditioners are available only with the BTU choices in the table, which would Adam choose?

✎ **d.** Why do you think the table gives ft^2 instead of ft^3, which would give the volume of the room?

17. *Length of a Pendulum* Grandfather clocks use pendulums to keep accurate time. The relationship between the length of a pendulum L and the time T for one complete oscillation can be determined from the data in the table.*

L (ft)	T (sec)
1.0	1.11
1.5	1.36
2.0	1.57
2.5	1.76
3.0	1.92
3.5	2.08
4.0	2.22

a. Plot the data from the table with L as the horizontal axis and T as the vertical axis.

b. Find the least squares line equation and graph it simultaneously, if possible, with the data points. Does it seem to fit the data?

c. Find the coefficient of correlation and interpret it. Does it confirm your answer to part b?†

---

*Data provided by Gary Rockswold, Mankato State University, Minnesota.
†The actual relationship is $L = .81T^2$, which is not a linear equation. This illustrates that even if the relationship is not linear, a line can give a good approximation.

GENERAL INTEREST

18. *Athletic Records* The following table shows the men's and women's world records (in seconds) in the 800-meter run.*

Year	Men's Record	Women's Record
1905	113.4	—
1915	111.9	—
1925	111.9	144.0
1935	109.7	135.6
1945	106.6	132.0
1955	105.7	125.0
1965	104.3	118.0
1975	103.7	117.48
1985	101.73	113.28
1995	101.11	113.28

a. Find the equation for the least squares line for the men's record (y) in terms of the year (x). Use 5 for 1905, 15 for 1915, and so on.

b. Find the equation for the least squares line for the women's record.

c. Suppose the men's and women's records continue to improve as predicted by the equations found in parts a and b. In what year will the women's record catch up with the men's record? Do you believe that will happen? Why or why not?

d. Calculate the coefficient of correlation for both the men's and the women's record. What do these numbers tell you?

19. *Football* The following data give the expected points for a football team with first down and 10 yards to go from various points on the field.[†] (*Note:* $\Sigma x = 500$, $\Sigma x^2 = 33{,}250$, $\Sigma y = 20.668$, $\Sigma y^2 = 91.927042$, $\Sigma xy = 399.16$.)

Yards from Goal (x)	Expected Points (y)
5	6.041
15	4.572
25	3.681
35	3.167
45	2.392
55	1.538
65	.923
75	.236
85	−.637
95	−1.245

a. Calculate the coefficient of correlation. Does there appear to be a linear correlation?

b. Find the equation of the least squares line.

c. Use your answer from part a to predict the expected points when a team is at the 50 yard line.

20. *Baseball* Some baseball fans are concerned about the recent increase in time to complete the game. The following table shows the average time (in hours and minutes) to complete baseball games in recent years.[‡]

Year	Average Completion Time
1981	2:33
1982	2:34
1983	2:36
1984	2:35
1985	2:40
1986	2:44
1987	2:48
1988	2:45
1989	2:46
1990	2:48
1991	2:49
1992	2:49
1993	2:48
1994	2:54
1995	2:57

Let x be the number of years since 1980, and let y be the number of minutes beyond 2 hours. (*Note:* $\Sigma x = 120$, $\Sigma x^2 = 1240$, $\Sigma y = 666$, $\Sigma xy = 5765$.)

a. Find the equation of the least squares line.

b. If the trend in the data continues, in what year will the average completion time be 3 hours and 15 minutes?

---

*Whipp, Brian J., and Susan A. Ward, "Will Women Soon Outrun Men?" *Nature,* Vol. 355 (Jan. 2, 1992), p. 25. The data are from Peter Matthews, *Track and Field Athletics: The Records,* Guinness, 1986, pp. 11, 44, and from Robert W. Schutz and Yuanlong Liu in *Statistics in Sport,* edited by Jay Bennett, Arnold, 1998, p. 189.

[†]Carter, Virgil, and Robert E. Machol, *Operations Research,* Vol. 19, 1971, pp. 541–545.

[‡]*The New York Times,* May 30, 1995, p. B9.

21. *Running* If you think a marathon is a long race, consider the Hardrock 100, a 101.1 mile running race held in southwestern Colorado. The chart lists the times that the 1994 winner, Scott Hirst, arrived at various mileage points along the way.*

  a. What was Hirst's average speed?

  b. Graph the data, plotting time on the *x*-axis and distance on the *y*-axis. You will need to convert the time from hours and minutes into hours. Do the data appear to lie approximately on a straight line?

  c. Find the equation for the least squares line, fitting distance as a linear function of time.

  d. Calculate the coefficient of correlation. Does it indicate a good fit of the least squares line to the data?

  e. Based on your answer to part d, what is a good value for Hirst's average speed? Compare this with your answer to part a. Which answer do you think is better? Explain your reasoning.

Miles	Time (hours:minutes)
0	0
12.2	2:57
19.0	5:03
28.0	7:22
33.4	9:26
36.6	10:10
43.8	11:16
59.4	16:01
69.6	21:40
78.4	24:00
85.0	26:48
91.5	28:59
101.1	32:00

CHAPTER SUMMARY

In this chapter, we have seen how to find the equation of a line, given a point and the slope or given two points. We have also seen how to express the result as a linear function. Equations of lines have a broad range of applications, as demonstrated in this chapter. They are used through the rest of this book, so fluency in their use is important. The method of least squares shows how mathematical models, such as many of those used throughout this book, are derived.

KEY TERMS

To understand the concepts presented in this chapter, you should know the meaning and use of the following terms. For easy reference, the section in the chapter where a term was first used is provided.

mathematical model	graph	**1.2** linear function	marginal cost
1.1 ordered pair	intercepts	independent variable	cost function
Cartesian coordinate	slope	dependent variable	break-even quantity
system	linear equation	supply curve	break-even point
axes	slope-intercept form	demand curve	**1.3** least squares line
origin	proportional	equilibrium price	summation notation
coordinates	point-slope form	equilibrium quantity	coefficient of correlation
quadrants	scatterplot	fixed cost	

---

*Frantz, Marny, and Sylvia Lazarnick, "Data Analysis and the Hardrock 100," *The Mathematics Teacher,* Vol. 90, No. 4, April 1997, pp. 274–276. Reprinted with permission from *The Mathematics Teacher,* copyright 1970 by the National Council of Teachers of Mathematics.

CHAPTER 1 REVIEW EXERCISES

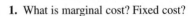 1. What is marginal cost? Fixed cost?

2. What six quantities are needed to compute a coefficient of correlation?

Find the slope for each line in Exercises 3–12 that has a slope.

3. Through $(-2, 5)$ and $(4, 7)$

4. Through $(4, -1)$ and $(3, -3)$

5. Through the origin and $(11, -2)$

6. Through the origin and $(0, 7)$

7. $2x + 3y = 15$

8. $4x - y = 7$

9. $y + 4 = 9$

10. $3y - 1 = 14$

11. $y = -3x$

12. $x = 5y$

Find an equation in the form $y = mx + b$ (where possible) for each line.

13. Through $(5, -1)$; slope $= 2/3$

14. Through $(8, 0)$; slope $= -1/4$

15. Through $(5, -2)$ and $(1, 3)$

16. Through $(2, -3)$ and $(-3, 4)$

17. Through $(-1, 4)$; undefined slope

18. Through $(-2, 5)$; slope $= 0$

19. Through $(2, -1)$, parallel to $3x - y = 1$

20. Through $(0, 5)$, perpendicular to $8x + 5y = 3$

21. Through $(2, -10)$, perpendicular to a line with undefined slope

22. Through $(3, -5)$, parallel to $y = 4$

23. Through $(-7, 4)$, perpendicular to $y = 8$

Graph each linear equation defined as follows.

24. $y = 4x + 3$

25. $y = 6 - 2x$

26. $3x - 5y = 15$

27. $2x + 7y = 14$

28. $x + 2 = 0$

29. $y = 1$

30. $y = 2x$

31. $x + 3y = 0$

Applications

BUSINESS AND ECONOMICS

32. *Profit* To manufacture x thousand computer chips requires fixed expenditures of $352 plus $42 per thousand chips. Receipts from the sale of x thousand chips amount to $130 per thousand.

a. Write an expression for expenditures.

b. Write an expression for receipts.

c. For profit to be made, receipts must be greater than expenditures. How many chips must be sold to produce a profit?

33. *Supply and Demand* The supply and demand for crabmeat in a local fish store are related by the equations

$$\text{Supply: } p = S(q) = 6q + 3$$

and $\text{Demand: } p = D(q) = 19 - 2q,$

where p represents the price in dollars per pound and q represents the quantity of crabmeat in pounds per day. Find the supply and demand at each of the following prices.

a. $10 b. $15 c. $18

d. Graph both the supply and the demand functions on the same axes.

e. Find the equilibrium price.

f. Find the equilibrium quantity.

34. *Supply* For a new diet pill, 60 will be supplied at a price of $40, while 100 pills will be supplied at a price of $60. Write a linear supply function for this product.

35. *Demand* The demand for the diet pills in Exercise 34 is 50 at a price of $47.50 and 80 at a price of $32.50. Determine a linear demand function for these pills.

36. *Supply and Demand* Find the equilibrium price and quantity for the diet pills in Exercises 34 and 35.

Cost Find a linear cost function in Exercises 37–40.

37. Eight units cost $300; fixed cost is $60.

38. Fixed cost is $2000; 36 units cost $8480.

39. Twelve units cost $445; 50 units cost $1585.

40. Thirty units cost $1500; 120 units cost $5640.

41. *Break-Even Analysis* The cost of producing x hundred cartons of CDs is $C(x)$ dollars, where $C(x) = 200x + 1000$. The CDs sell for $400 per carton.

a. Find the break-even quantity.

b. What revenue will the company receive if it sells just that number of cartons?

42. *Break-Even Analysis* The cost function for flavored coffee at an upscale coffeehouse is given in dollars by $C(x) = 3x + 160$, where x is in pounds. The coffee sells for $7 per pound.

a. Find the break-even quantity.

b. What will the revenue be at that point?

43. *U.S. Imports from China* The U.S. is China's largest export market. Imports from China have grown from about 8 billion dollars in 1988 to 39 billion dollars in 1995.* This growth has been approximately linear. Use the given data pairs to write a linear equation that describes this growth in imports over the years. Let $x = 88$ represent 1988 and $x = 95$ represent 1995.

44. *U.S. Exports to China* U.S. exports to China have grown (although at a slower rate than imports) since 1988. In 1988, about 8 billion dollars of goods were exported to China. By 1995, this amount had grown to 15.9 billion dollars.* Write a linear equation describing the number of exports each year, with $x = 88$ representing 1988 and $x = 95$ representing 1995.

45. *Pilot salaries* An ad for Northwest Airlines published in *The New York Times* showed the pay of the typical Northwest pilot to be a linear function of time.[†] In 1992, the pay was $120,000. In 1997, the pay was $165,000. Find a formula giving the typical pay, P, as a function of the year x, where $x = 0$ corresponds to 1990.

46. *New Car Cost* The average new car cost for the years from 1975 to 1995 is given in the table.[‡]

Year (x)	75	80	85	90	95
Cost (y)	6000	7500	12,000	16,000	20,400

a. Find an equation for the least squares line.

b. Use your equation from part a to predict the average cost of a new car in the year 2000 ($x = 100$).

c. Find and interpret the coefficient of correlation. Does it indicate that the line is a good fit for the data?

d. Plot the data. Does the scatterplot suggest the trend might not be linear?

LIFE SCIENCES

47. *World Health* In general, people tend to live longer in countries that have a greater supply of food. Listed below is the 1992 daily calorie supply and 1994 life expectancy at birth for 10 randomly selected countries.[§]

Country	Calories (x)	Life Expectancy (y)
Afghanistan	1523	44
Belize	2670	74
Cuba	2833	76
France	3465	79
India	2395	61
Mexico	3181	72
North Korea	2834	71
Peru	1883	67
Sweden	2960	78
United States	3671	76

a. Find the coefficient of correlation. Do the data seem to fit a straight line?

b. Draw a scatterplot of the data. Combining this with your results from part a, do the data seem to fit a straight line?

c. Find the equation for the least squares line.

d. Use your answer from part c to predict the life expectancy in the United Kingdom, which has a daily

---

Economist, U.S.-China Business Council, *China Business Review,* U.S. Commerce Department.
[†]*The New York Times,* Aug. 27, 1998, p. A16.
[‡]*Chicago Tribune,* Feb. 4, 1996, Sec. 5, p. 4.
[§]*The New York Times 2000 Almanac,* pp. 490–492.

calorie supply of 3149. Compare your answer with the actual value of 77 years.

e. Briefly explain why countries with a higher daily calorie supply might tend to have a longer life expectancy.

f. (For the ambitious!) Find the coefficient of correlation and least squares line using the data for a larger sample of countries, as found in an almanac or other reference. Is the result in general agreement with the previous results?

48. *Blood Sugar and Cholesterol Levels* The following data show the connection between blood sugar levels and cholesterol levels for 8 different patients.

Patient	Blood Sugar Level (x)	Cholesterol Level (y)
1	130	170
2	138	160
3	142	173
4	159	181
5	165	201
6	200	192
7	210	240
8	250	290

For the data given in the preceding table, $\Sigma x = 1394$, $\Sigma y = 1607$, $\Sigma xy = 291,990$, $\Sigma x^2 = 255,214$, and $\Sigma y^2 = 336,155$.

a. Find the equation of the least squares line, $Y = mx + b$.

b. Predict the cholesterol level for a person whose blood sugar level is 190.

c. Find r.

SOCIAL SCIENCES

49. *Red Meat Consumption* The per capita consumption of red meat in the United States decreased from 131.7 pounds in 1970 to 114.1 pounds in 1992.* Assume a linear function describes the decrease. Write a linear equation defining the function. Let x represent the number of years since 1900 and y represent the number of pounds of red meat consumed.

50. *Marital Status* More people are staying single longer in the United States. In 1970, the number of never-married adults, age 18 and over, was 21.4 million. By 1993, it was 42.3 million.† Assume the data increase linearly, and write an equation that defines a linear function for this data. Let x represent the number of years since 1900.

51. *Governors' salaries* In general, the larger a state's population, the more its governor earns. Listed below are the estimated 1998 populations (in millions) and the salary of the governor (in thousands of dollars) for 8 randomly selected states.‡

a. Find the coefficient of correlation. Do the data seem to fit a straight line?

b. Draw a scatterplot of the data. Compare this with your answer from part a.

c. Find the equation for the least squares line.

d. Based on your answer to part c, how much does a governor's salary increase, on average, for each additional million in population?

e. Use your answer from part c to predict the governor's salary in your state. Based on your answers from parts a and b, would this prediction be very accurate? Compare with the actual salary, as listed in an almanac or other reference.

f. (For the ambitious!) Find the coefficient of correlation and least squares line using the data for all 50 states, as found in an almanac or other reference. Is the result in general agreement with the previous results?

State	AZ	DE	MD	MA	NY	PA	TN	WY
Population (x)	4.67	.74	5.14	6.15	18.18	12.00	5.43	.48
Governor's Salary (y)	75	107	120	75	130	105	85	95

*U.S. Department of Agriculture, Economic Research Service, *Food Consumption, Price, and Expenditures,* annual.
†U.S. Bureau of the Census, *1970 Census of Population,* Vol. 1, Part 1, and *Current Population Reports,* pp. 20, 450.
‡*The New York Times 2000 Almanac,* pp. 183, 220.

▣ EXTENDED APPLICATION: Using Extrapolation to Predict Life Expectancy

One reason for developing a mathematical model is to make predictions. If your model is a least squares line, you can predict the y value corresponding to some new x by substituting this x into an equation of the form $Y = mx + b$. (We use a capital Y to remind us that we're getting a predicted value rather than an actual data value.) Data analysts distinguish two very different kinds of prediction, *interpolation* and *extrapolation*. An interpolation uses a new x inside the x range of your original data. For example, if you have inflation data at five-year intervals from 1950 to 2000, estimating the rate of inflation in 1957 is an interpolation problem. But if you use the same data to estimate what the inflation rate was in 1920, or what it will be in 2020, you are extrapolating.

In general, interpolation is much safer than extrapolation, because data that are approximately linear over a short interval may be nonlinear over a larger interval. One way to detect nonlinearity is to look at *residuals,* which are the differences between the actual data values and the values predicted by the line of best fit. Here is a simple example:

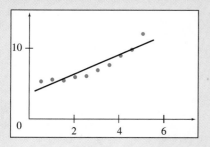

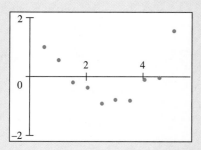

The regression equation for the linear fit on the top is $Y = 3.431 + 1.334x$. Since the r value for this regression line is 0.93, our linear model fits the data very well. But we might notice that the predictions are a bit low at the ends and high in the middle. We can get a better look at this pattern by plotting the residuals. To find them, we put each value of the independent variable into the regression equation, calculate the predicted value Y, and subtract it from the actual y value. The residual plot is below the linear fit graph, with the vertical axis rescaled to

exaggerate the pattern. The residuals indicate that our data has a nonlinear, U-shaped component that isn't captured by the linear fit. Extrapolating from this data set is probably not a good idea; our linear prediction for the value of y when x is 10 may be much too low.

Exercises

The following table gives the life expectancy at birth of females born in the United States in various years from 1950 to 1995.*

Year of birth	Life expectancy in years
1950	71.3
1960	73.1
1970	74.7
1980	77.4
1985	78.2
1990	78.8
1995	78.9

1. Find an equation for the least squares line for this data, using year of birth as the independent variable.

2. Use your regression equation to guess a value for the life expectancy of females born in 1900.

3. Compare your answer with the actual life expectancy for females born in 1900, which was 48.3 years. Are you surprised?

4. Find the life expectancy predicted by your regression equation for each year in the table, and subtract it from the actual value in the second column. This gives you a table of residuals. Plot your residuals as points on a graph.

5. Now look at the residuals as a fresh data set and see if you can sketch the graph of a smooth function that fits the residuals well. How easy do you think it will be to predict the life expectancy at birth of females born in 2010?

6. What will happen if you try linear regression on the *residuals?* If you're not sure, use your calculator or software to find the regression equation for the residuals. Why does this result make sense?

7. Since most of the females born in 1985 are still alive, how did the Public Health Service come up with a life expectancy of 78.2 years for these women?

**Health, United States, 1998,* Centers for Disease Control.

CHAPTER

2

Systems of Linear Equations and Matrices

The synchronized movements of band members marching on a field can be modeled using matrix arithmetic. An exercise in Section 5 in this chapter shows how multiplication by a matrix inverse transforms the original positions of the marchers into their new coordinates as they change direction.

Many mathematical models require finding the solutions of two or more equations. The solutions must satisfy *all* of the equations in the model. A set of equations related in this way is called a **system of equations.** In this chapter we will discuss systems of equations, introduce the idea of a *matrix,* and then show how matrices are used to solve systems of equations.

■ 2.1 SOLUTION OF LINEAR SYSTEMS BY THE ECHELON METHOD

 THINK ABOUT IT   How much of each ingredient should be used in an animal feed to meet dietary requirements?

Suppose that an animal feed is made from three ingredients: corn, soybeans, and cottonseed. One unit of each ingredient provides the number of units of protein, fat, and fiber shown in the table. For example, the entries in the first column, .25, .4, and .3, indicate that one unit of corn provides twenty-five hundredths (one-fourth) of a unit of protein, four-tenths of a unit of fat, and three-tenths of a unit of fiber.

	Corn	Soybeans	Cottonseed
Protein	.25	.4	.2
Fat	.4	.2	.3
Fiber	.3	.2	.1

    Now suppose we need to know the number of units of each ingredient that should be used to make a feed that contains 22 units of protein, 28 units of fat, and 18 units of fiber. To find out, we let x represent the required number of units of corn, y the number of units of soybeans, and z the number of units of cottonseed. Since the total amount of protein is to be 22 units,

$$.25x + .4y + .2z = 22.$$

The feed must supply 28 units of fat, so

$$.4x + .2y + .3z = 28,$$

and 18 units of fiber, so

$$.3x + .2y + .1z = 18.$$

To solve this problem, we must find values of x, y, and z that satisfy this system of equations. Verify that $x = 40$, $y = 15$, and $z = 30$ is a solution of the system, since these numbers satisfy all three equations. In fact, this is the only solution of this system. Many practical problems lead to such systems of *first-degree equations.*

A **first-degree equation** in n unknowns is any equation of the form

$$a_1x_1 + a_2x_2 + \cdots + a_nx_n = k,$$

where $a_1, a_2, \ldots, a_n$ and k are all real numbers.* Each of the three equations from the animal feed problem is a first-degree equation in three unknowns. A *solution* of the first-degree equation

$$a_1x_1 + a_2x_2 + \cdots + a_nx_n = k$$

is a sequence of numbers $s_1, s_2, \ldots, s_n$ such that

$$a_1s_1 + a_2s_2 + \cdots + a_ns_n = k.$$

A solution of an equation is usually written between parentheses as $(s_1, s_2, \ldots, s_n)$. For example, $(1, 6, 2)$ is a solution of the equation $3x_1 + 2x_2 - 4x_3 = 7$, since $3(1) + 2(6) - 4(2) = 7$. This is an extension of the idea of an ordered pair, which was introduced in Chapter 1. A solution of a first-degree equation in two unknowns is an ordered pair, and the graph of the equation is a straight line. For this reason all first-degree equations are also called linear equations. In this section we develop a method for solving a system of first-degree equations. Although the discussion will be confined to equations with only a few variables, the method of solution can be extended to systems with many variables.

Because the graph of a linear equation in two unknowns is a straight line, there are three possibilities for the solutions of a system of two linear equations in two unknowns.

1. The two graphs are lines intersecting at a single point. The system has a **unique solution,** and it is given by the coordinates of this point. See Figure 1(a).

2. The graphs are distinct parallel lines. When this is the case, the system is **inconsistent;** that is, there is no solution common to both equations. See Figure 1(b).

3. The graphs are the same line. In this case, the equations are said to be **dependent,** since any solution of one equation is also a solution of the other. There are infinitely many solutions. See Figure 1(c).

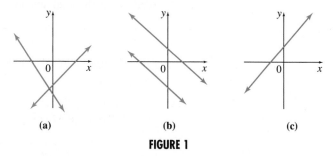

(a)   (b)   (c)

FIGURE 1

In larger systems, with more equations and more variables, there also may be exactly one solution, no solutions, or infinitely many solutions. If no solution satisfies every equation in the system, the system is inconsistent, and if there are infinitely many solutions that satisfy all the equations in the system, the equations are dependent.

---

*a_1 is read "a-sub-one." The notation $a_1, a_2, \ldots, a_n$ represents n real-number coefficients (some of which may be equal), and the notation $x_1, x_2, \ldots, x_n$ represents n different variables, or unknowns.

Transformations   To solve a linear system of equations, we use properties of algebra to change, or transform, the system into a simpler *equivalent* system. An **equivalent system** is one that has the same solutions as the given system. Algebraic properties are the basis of the following transformations.

> **TRANSFORMATIONS OF A SYSTEM**
>
> The following transformations can be applied to a system of equations to get an equivalent system:
>
> **1.** exchanging any two equations;
> **2.** multiplying both sides of an equation by any nonzero real number;
> **3.** replacing any equation by a nonzero multiple of that equation plus a nonzero multiple of any other equation.

Use of these transformations leads to an equivalent system because each transformation can be reversed or "undone," allowing a return to the original system.

The Echelon Method   A systematic approach for solving systems of equations using the three transformations is called the **echelon method.** The goal of the echelon method is to use the transformations to rewrite the equations of the system until the system has a triangular form. For a system of three equations in three variables, for example, the system should have the form

$$x + ay + bz = c$$
$$y + dz = e$$
$$z = f,$$

where a, b, c, d, e, and f are constants. Then the value of z from the third equation can be substituted into the second equation to find y, and the values of y and z can be substituted into the first equation to find x. This is called **back-substitution.**

EXAMPLE 1   Solving Systems of Equations
Solve the system

$$2x + 3y = 12 \tag{1}$$
$$3x - 4y = 1. \tag{2}$$

Solution   We will first use transformation 3 to eliminate the x-term from equation (2). Multiply equation (1) by 3 and add the results to -2 times equation (2).

$$6x + 9y + (-6x + 8y) = 36 + (-2)$$
$$17y = 34$$

We will indicate this process by the notation $3R_1 + (-2)R_2 \rightarrow R_2$. (R stands for row.) The new system is

$$2x + 3y = 12 \tag{1}$$
$$3R_1 + (-2)R_2 \rightarrow R_2 \qquad 17y = 34. \tag{3}$$

Now use transformation 2 to make the coefficient of the first term in each row equal to 1. Here, we must multiply equation (1) by $1/2$ and equation (3) by $1/17$ to accomplish this.

We get the system

$$\frac{1}{2}R_1 \rightarrow R_1 \quad x + \frac{3}{2}y = 6$$

$$\frac{1}{17}R_2 \rightarrow R_2 \qquad\qquad y = 2.$$

Back-substitution gives

$$x + \frac{3}{2}(2) = 6$$

$$x + 3 = 6$$

$$x = 3.$$

The solution of the system is $(3, 2)$. The graphs of the two equations in Figure 2 suggest that $(3, 2)$ satisfies both equations in the system. Verify that $(3, 2)$ does indeed satisfy both original equations.

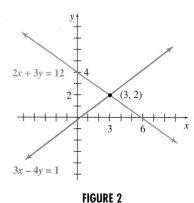

FIGURE 2   **FIGURE 3**

EXAMPLE 2   Systems of Equations with No Solution
Solve the system

$$3x - 2y = 4 \tag{1}$$

$$-6x + 4y = 7. \tag{2}$$

Solution Eliminate x in equation (2) to get the system

$$3x - 2y = \;\; 4 \tag{1}$$

$$2R_1 + R_2 \rightarrow R_2 \qquad 0 = 15. \tag{3}$$

In equation (3), both variables have been eliminated, leaving a false statement. This is a signal that these two equations have no common solution. This system is inconsistent and has no solution. As Figure 3 shows, the graph of the system is made up of two distinct parallel lines.

EXAMPLE 3   Systems of Equations with an Infinite Number of Solutions
Solve the system

$$-4x + \;\; y = \;\;\; 2 \tag{1}$$

$$8x - 2y = -4. \tag{2}$$

Use transformation 3 to eliminate x in equation (2), getting the system

$$-4x + y = 2 \tag{1}$$

$$2R_1 + R_2 \rightarrow R_2 \qquad 0 = 0. \tag{3}$$

The system becomes

$$x - \frac{1}{4}y = -\frac{1}{2} \quad \text{Multiply by } -\tfrac{1}{4}. \tag{4}$$

$$0 = 0. \tag{5}$$

The true statement in equation (5) indicates that the two equations have the same graph, which means that there is an infinite number of solutions for the system. See Figure 4.

We will express the solutions in terms of y, where y can be any real number. The variable y in this case is called a **parameter.** (We could also let x be the parameter. In this text, we will follow the common practice of letting the rightmost variable be the parameter.) Solving equation (4) for x gives $x = (1/4)y - (1/2) = (y - 2)/4$, and all the ordered pairs of the form

$$\left(\frac{y - 2}{4}, y \right)$$

are solutions. For example, letting $y = 6$, we have $x = (6 - 2)/4 = 1$, so one solution is $(1, 6)$. Similarly, letting $y = 3$ and $y = -10$ gives the solutions $(1/4, 3)$ and $(-3, -10)$.

In some applications, x and y must be nonnegative integers. For instance, in Example 3, if x and y represent the number of men and women workers in a factory, it makes no sense to have $x = 1/4$ or $x = -3$. To make both x and y nonnegative, we solve the inequalities

$$\frac{y - 2}{4} \geq 0 \qquad \text{and} \qquad y \geq 0,$$

yielding

$$y \geq 2 \qquad \text{and} \qquad y \geq 0.$$

To make these last two inequalities true, we require $y \geq 2$, from which $y \geq 0$ automatically follows. Furthermore, to ensure that $(y - 2)/4$ is an integer, it is necessary that y be 2 more than a whole number multiple of 4. Therefore, the possible values of y are 2, 6, 10, 14, etc., and the corresponding values of x are 0, 1, 2, 3, etc.

In Section 1.3, we solved systems of two equations in two unknowns when finding the least squares line. Our method was to solve one equation for one unknown and substitute the result into the other equation. The echelon method is not only slightly easier to perform, but also generalizes to systems with more equations and unknowns. Because systems with three or more unknowns are complicated, however, we will only do a few in this section. In the next section, we will show a procedure based upon the echelon method that is useful for solving large systems of equations. Meanwhile, the following example illustrates the additional steps needed to solve a system with three equations in three unknowns by the echelon method.

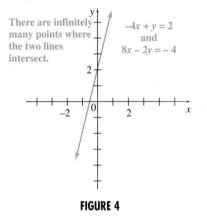

There are infinitely many points where the two lines intersect.

$-4x + y = 2$
and
$8x - 2y = -4$

FIGURE 4

EXAMPLE 4 Solving Systems of Equations

Solve the system

$$2x + y - z = 2 \qquad (1)$$
$$x + 3y + 2z = 1 \qquad (2)$$
$$x + y + z = 2. \qquad (3)$$

Solution As in the previous examples, we begin by eliminating the term with x, this time from equations (2) and (3), as follows.

$$2x + y - z = 2 \qquad (1)$$
$$R_1 + (-2)R_2 \rightarrow R_2 \qquad -5y - 5z = 0 \qquad (4)$$
$$R_1 + (-2)R_3 \rightarrow R_3 \qquad -y - 3z = -2 \qquad (5)$$

In the same way, use equation (4) to eliminate y in equation (5). The new system is

$$2x + y - z = 2$$
$$-5y - 5z = 0$$
$$R_2 + (-5)R_3 \rightarrow R_3 \qquad 10z = 10.$$

Make the coefficient of the first term in each equation equal 1.

$$\tfrac{1}{2}R_1 \rightarrow R_1 \qquad x + \frac{1}{2}y - \frac{1}{2}z = 1 \qquad (6)$$

$$\left(-\tfrac{1}{5}\right)R_2 \rightarrow R_2 \qquad y + z = 0 \qquad (7)$$

$$\tfrac{1}{10}R_3 \rightarrow R_3 \qquad z = 1$$

Substitute 1 for z in equation (7) to get $y = -1$. Finally, substitute 1 for z and -1 for y in equation (6) to get $x = 2$. The solution of the system is $(2, -1, 1)$. Note the triangular form of the last system. This is the typical echelon form.　■

In summary, to solve a linear system in n variables by the echelon method, perform the following steps using the three transformations given earlier.

THE ECHELON METHOD OF SOLVING A LINEAR SYSTEM

1. If possible, arrange the equations so that there is an x_1 term in the first equation, an x_2 term in the second equation, and so on.
2. Eliminate the x_1 term in all equations after the first equation.
3. Eliminate the x_2 term in all equations after the second equation.
4. Eliminate the x_3 term in all equations after the third equation.
5. Continue in this way until the last equation has the form $ax_n = k$, for constants a and k.
6. Multiply each equation by the reciprocal of the coefficient of its first term.
7. Use back-substitution to find the value of each variable.

Applications The mathematical techniques in this text will be useful to you only if you are able to apply them to practical problems. To do this, always begin by reading the problem carefully. Next, identify what must be found. Let each unknown quantity be represented by a variable. (It is a good idea to *write down* exactly what each variable represents.) Now reread the problem, looking for all necessary data. Write those down, too. Finally, look for one or more sentences that lead to equations or inequalities. The next example illustrates these steps.

EXAMPLE 5   Flight Time

A flight leaves New York at 8 P.M. and arrives in Paris at 9 A.M. (Paris time). This 13-hour difference includes the flight time plus the change in time zones. The return flight leaves Paris at 1 P.M. and arrives in New York at 3 P.M. (New York time). This 2-hour difference includes the flight time *minus* time zones, plus an extra hour due to the fact that flying westward is against the wind. Find the actual flight time eastward and the difference in time zones.

Solution Let x be the flight time and y be the difference in time zones. For the trip east, we have

$$x + y = 13.$$

The flight time for the trip westward is $x + 1$ hours due to the wind, so

$$(x + 1) - y = 2.$$

Subtract 1 from both sides of this equation, and then solve the system

$$x + y = 13$$
$$x - y = 1$$

using the echelon method.

$$x + y = 13$$
$$\text{R}_1 + (-1)\text{R}_2 \rightarrow \text{R}_2 \qquad 2y = 12$$

Dividing the last equation by 2 gives $y = 6$. Substituting this into the first equation gives $x + 6 = 13$, so $x = 7$. Therefore, the flight time eastward is 7 hours, and the difference in time zones is 6 hours.

EXAMPLE 6   Integral Solutions

A restaurant owner orders a replacement set of knives, forks, and spoons. The box arrives containing 40 utensils and weighing 141.3 ounces (ignoring the weight of the box). A knife, fork, and spoon weigh 3.9 ounces, 3.6 ounces, and 3.0 ounces, respectively.

(a) How many solutions are there for the number of knives, forks, and spoons in the box?

Solution Let

$$x = \text{the number of knives;}$$
$$y = \text{the number of forks;}$$
$$z = \text{the number of spoons.}$$

A chart is useful for organizing the information in a problem of this type.

	Knives	Forks	Spoons	Total
Number	x	y	z	40
Weight	3.9	3.6	3.0	141.3

Because the box contains 40 utensils,

$$x + y + z = 40.$$

The x knives weigh $3.9x$ ounces, the y forks weigh $3.6y$ ounces, and the z spoons weigh $3.0z$ ounces. Since the total weight is 141.3 ounces, we have the system

$$\begin{aligned} x + \quad y + \quad z &= \quad 40 \\ 3.9x + 3.6y + 3.0z &= 141.3. \end{aligned}$$

Solve using the echelon method.

$$\begin{aligned} x + \quad y + \quad z &= 40 \\ 3.9R_1 + (-1)R_2 \to R_2 \qquad .3y + .9z &= 14.7 \end{aligned}$$

We do not have a third equation to solve for z, as we did in Example 4. This system, then, has an infinite number of solutions. Letting z be the parameter, solve the second equation for y to get

$$y = \frac{14.7 - .9z}{.3} = 49 - 3z.$$

Substitute this into the first equation to get

$$x + (49 - 3z) + z = 40.$$

Solving this for x gives

$$x = 2z - 9.$$

Thus, the solutions are $(2z - 9, 49 - 3z, z)$ where z is any real number.

Now that we have solved for x and y in terms of z, let us investigate what values z can take on. This application demands that the solutions be nonnegative integers. The number of forks cannot be negative, so set

$$49 - 3z \geq 0.$$

Solving for z gives

$$z \leq \frac{49}{3} \approx 16.33.$$

Also, the number of knives cannot be negative, so set

$$2z - 9 \geq 0.$$

Solving for z gives

$$z \geq \frac{9}{2} = 4.5.$$

Therefore, the permissible values of z are 5, 6, 7, …, 16, for a total of 12 solutions.

(b) Find the solution with the smallest number of spoons.

Solution The smallest value of z is $z = 5$, from which we find $x = 2(5) - 9 = 1$ and $y = 49 - 3(5) = 34$. This solution has 1 knife, 34 forks, and 5 spoons.

2.1 EXERCISES

Use the echelon method to solve each system of two equations in two unknowns. Check your answers.

1. $x + y = 9$
$2x - y = 0$

2. $4x + y = 9$
$3x - y = 5$

3. $5x + 3y = 7$
$7x - 3y = -19$

4. $2x + 7y = -8$
$-2x + 3y = -12$

5. $3x + 2y = -6$
$5x - 2y = -10$

6. $-6x + 2y = 8$
$5x - 2y = -8$

7. $2x - 3y = -7$
$5x + 4y = 17$

8. $4m + 3n = -1$
$2m + 5n = 3$

9. $5p + 7q = 6$
$10p - 3q = 46$

10. $12s - 5t = 9$
$3s - 8t = -18$

11. $6x + 7y = -2$
$7x - 6y = 26$

12. $2a + 9b = 3$
$5a + 7b = -8$

13. $3x + 2y = 5$
$6x + 4y = 8$

14. $9x - 5y = 1$
$-18x + 10y = 1$

15. $4x - y = 9$
$-8x + 2y = -18$

16. $3x + 5y + 2 = 0$
$9x + 15y + 6 = 0$

17. An inconsistent system has _____ solutions.

18. The solution of a system with two dependent equations in two variables is

_____ .

Use the echelon method to solve each system. Check your answers.

19. $\dfrac{x}{2} + \dfrac{y}{3} = 8$

$\dfrac{2x}{3} + \dfrac{3y}{2} = 17$

20. $\dfrac{x}{5} + 3y = 31$

$2x - \dfrac{y}{5} = 8$

21. $\dfrac{x}{2} + y = \dfrac{3}{2}$

$\dfrac{x}{3} + y = \dfrac{1}{3}$

22. $x + \dfrac{y}{3} = -6$

$\dfrac{x}{5} + \dfrac{y}{4} = -\dfrac{7}{4}$

Use the echelon method to solve each system of three equations in three unknowns. Check your answers.

23. $x + y + z = 2$
$2x + y - z = 5$
$x - y + z = -2$

24. $2x + y + z = 9$
$-x - y + z = 1$
$3x - y + z = 9$

25. $x + 3y + 4z = 14$
$2x - 3y + 2z = 10$
$3x - y + z = 9$

26. $4x - y + 3z = -2$
$3x + 5y - z = 15$
$-2x + y + 4z = 14$

27. In your own words, describe the echelon method as used to solve a system of three equations in three variables.

Solve each of the following systems of equations. Let z be the parameter.

28. $5x + 3y + 4z = 19$
$3x - y + z = -4$

29. $3x + y - z = 0$
$2x - y + 3z = -7$

30. $x + 2y + 3z = 11$
$2x - y + z = 2$

31. $-x + y - z = -7$
$2x + 3y + z = 7$

32. In an exercise in Section 1.3, you were asked to solve the system of least squares line equations

$$nb + (\Sigma x)m = \Sigma y$$
$$(\Sigma x)b + (\Sigma x^2)m = \Sigma xy$$

by the method of substitution. Now solve the system by the echelon method to get

$$m = \frac{n\Sigma(xy) - (\Sigma x)(\Sigma y)}{n(\Sigma x^2) - (\Sigma x)^2}$$
$$b = \frac{\Sigma y - m(\Sigma x)}{n}.$$

33. The examples in this section did not use the first transformation. How might this transformation be used in the echelon method?

Applications

BUSINESS AND ECONOMICS

34. *Production* The Gonzalez Company makes ROM chips and RAM chips for computers. Both require time on two assembly lines. Each unit of ROM chips requires 1 hr on line A and 2 hr on line B. Each unit of RAM chips requires 3 hr on line A and 1 hr on line B. Both assembly lines operate 15 hr per day. How many units of each product can be produced in a day under these conditions?

35. *Groceries* If 20 lb of rice and 10 lb of potatoes cost $16.20, and 30 lb of rice and 12 lb of potatoes cost $23.04, how much will 10 lb of rice and 50 lb of potatoes cost?

36. *Sales* An apparel shop sells skirts for $45 and blouses for $35. Its entire stock is worth $51,750. But sales are slow and only half the skirts and two-thirds of the blouses are sold, for a total of $30,600. How many skirts and blouses are left in the store?

37. *Sales* A theater charges $8 for main floor seats and $5 for balcony seats. If all seats are sold, the ticket income is $4200. At one show, 25% of the main floor seats and 40% of the balcony seats were sold and ticket income was $1200. How many seats are on the main floor and how many are in the balcony?

38. *Stock* Shirley Cicero has $16,000 invested in Disney and Intel stock. The Disney stock currently sells for $30 a share and the Intel stock for $70 a share. Her stockbroker points out that if Disney stock goes up 50% and Intel stock goes up by $35 a share, her stock will be worth $25,500. Is this possible? If so, tell how many shares of each stock she owns. If not, explain why not.

39. *Production* A company produces two models of bicycles, model 201 and model 301. Model 201 requires 2 hours of assembly time and model 301 requires 3 hours of assembly time. The parts for model 201 cost $18 per bike and the parts for model 301 cost $27 per bike. If the company has a total of 34 hours of assembly time and $335 available per

day for these two models, how many of each should be made in a day to use up all available time and money? If it is not possible, explain why not.

40. *Banking* A bank teller has a total of 70 bills in five-, ten-, and twenty-dollar denominations. The number of fives is three times the number of tens, while the total value of the money is $960. Find the number of each type of bill.

41. *Investments* Nui invests $10,000, received from her grandmother, in three ways. With one part, she buys mutual funds that offer a return of 6.5% per year. She uses the second part, which amounts to twice the first, to buy government bonds at 6% per year. She puts the rest in the bank at 5% annual interest. The first year her investments bring a return of $605. How much did she invest in each way?

42. *Production* Turley Tailor Inc. makes long-sleeve, short-sleeve, and sleeveless blouses. A long-sleeve blouse requires 1.5 hours of cutting and 1.2 hours of sewing. A short-sleeve blouse requires 1 hour of cutting and .9 hours of sewing. A sleeveless blouse requires .5 hours of cutting and .6 hours of sewing. There are 380 hours of labor available in the cutting department each day and 330 hours in the sewing department. If the plant is to run at full capacity, how many of each type of blouse should be made each day?

43. *Production* Felsted Furniture makes dining room furniture. A buffet requires 30 hours for construction and 10 hours for finishing. A chair requires 10 hours for construction and 10 hours for finishing. A table requires 10 hours for construction and 30 hours for finishing. The construction department has 350 hours of labor and the finishing department has 150 hours of labor available each week. How many pieces of each type of furniture should be produced each week if the factory is to run at full capacity?

44. *Rug Cleaning Machines* Kelly Karpet Kleaners sells rug cleaning machines. The EZ model weighs 10 lb and comes in a 10 cubic ft box. The compact model weighs 20 lb and comes in an 8 cubic ft box. The commercial model weighs

60 lb and comes in a 28 cubic ft box. Each of their delivery vans has 248 cubic ft of space and can hold a maximum of 440 lb. In order for a van to be fully loaded, how many of each model should it carry?

PHYSICAL SCIENCES

45. *Stopping Distance* The stopping distance of a car traveling 25 mph is 61.7 ft, and for a car traveling 35 mph it is 106 ft.* The stopping distance in feet can be described by the equation $y = ax^2 + bx + c$, where x is the speed in mph.

a. Find the values of a and b.

b. Use your answers from part a to find the stopping distance for a car traveling 55 mph.

GENERAL INTEREST

46. *The 24 Game* The object of the 24 Game™, created by Robert Sun, is to combine four numbers, using addition, subtraction, multiplication, and/or division, to get the number 24.[†] For example, the numbers 2, 5, 5, 4 can be combined as $2(5 + 5) + 4 = 24$. For the algebra edition of the game and the game card below, the object is to find single-digit positive integer values x and y so the four numbers $x + y$, $3x + 2y$, 8, and 9 can be combined to make 24.

a. Using the game card, write a system of equations that, when solved, can be used to make 24 from the game card. What is the solution to this system, and how can these be used to make 24 on the game card?

b. Repeat part a and develop a second system of equations.

■ 2.2   SOLUTION OF LINEAR SYSTEMS BY THE GAUSS-JORDAN METHOD

 THINK **A**BOUT **I**T   How can an auto manufacturer with more than one factory and several dealers decide how many cars to send to each dealer from each factory?

Questions like this are called *transportation problems;* they frequently lead to a system of equations that must be satisfied. In this section we use a further refinement of the echelon method to answer this question. When we use the echelon method, since the variables are in the same order in each equation, we really need to keep track of just the coefficients and the constants. For example, look at the system solved in Example 4 of the previous section.

$$2x + y - z = 2$$
$$x + 3y + 2z = 1$$
$$x + y + z = 2$$

This system can be written in an abbreviated form as

$$\text{Rows} \begin{bmatrix} 2 & 1 & -1 & 2 \\ 1 & 3 & 2 & 1 \\ 1 & 1 & 1 & 2 \end{bmatrix}.$$

Columns

---

*National Traffic Safety Institute Student Workbook, 1993, p. 7.
[†]Copied with permission by Suntex Inc., Easton, PA, www.Math24.com.

Such a rectangular array of numbers enclosed by brackets is called a **matrix** (plural: **matrices**).* Each number in the array is an **element** or **entry.** To separate the constants in the last column of the matrix from the coefficients of the variables, we use a vertical line, producing the following **augmented matrix.**

$$\left[\begin{array}{ccc|c} 2 & 1 & -1 & 2 \\ 1 & 3 & 2 & 1 \\ 1 & 1 & 1 & 2 \end{array}\right]$$

The rows of the augmented matrix can be transformed in the same way as the equations of the system, since the matrix is just a shortened form of the system. The following **row operations** on the augmented matrix correspond to the transformations of systems of equations given earlier.

ROW OPERATIONS

For any augmented matrix of a system of equations, the following operations produce the augmented matrix of an equivalent system:

1. interchanging any two rows;
2. multiplying the elements of a row by any nonzero real number;
3. adding a nonzero multiple of the elements of one row to the corresponding elements of a nonzero multiple of some other row.

In steps 2 and 3, we are replacing a row with a new, modified row which the old row helped to form, just as we replaced an equation with a new, modified equation in the previous section.

Row operations, like the transformations of systems of equations, are reversible. If they are used to change matrix A to matrix B, then it is possible to use row operations to transform B back into A. In addition to their use in solving equations, row operations are very important in the simplex method to be described in Chapter 4.

In the examples in this section, we will use the same notation as in Section 1 to show the row operation used. The notation R_1 indicates row 1 of the previous matrix, for example, and $-3R_1 + R_2$ means that row 1 is multiplied by -3 and added to row 2.

By the first row operation, interchanging two rows, the matrix

$$\left[\begin{array}{ccc|c} 1 & 3 & 5 & 6 \\ 0 & 1 & 2 & 3 \\ 2 & 1 & -2 & -5 \end{array}\right] \text{ becomes } \left[\begin{array}{ccc|c} 0 & 1 & 2 & 3 \\ 1 & 3 & 5 & 6 \\ 2 & 1 & -2 & -5 \end{array}\right] \quad \text{Interchange } R_1 \text{ and } R_2.$$

by interchanging the first two rows. Row 3 is left unchanged.

The second row operation, multiplying a row by a number, allows us to change

$$\left[\begin{array}{ccc|c} 1 & 3 & 5 & 6 \\ 0 & 1 & 2 & 3 \\ 2 & 1 & -2 & -5 \end{array}\right] \text{ to } \left[\begin{array}{ccc|c} -2 & -6 & -10 & -12 \\ 0 & 1 & 2 & 3 \\ 2 & 1 & -2 & -5 \end{array}\right] \quad -2R_1 \rightarrow R_1$$

*The word matrix, Latin for "womb," was coined by James Joseph Sylvester (1814–1897) and made popular by his friend Arthur Cayley (1821–1895). Both mathematicians were English, although Sylvester spent much of his life in the United States.

by multiplying the elements of row 1 of the original matrix by -2. Note that rows 2 and 3 are left unchanged.

Using the third row operation, adding a multiple of one row to another, we change

$$\begin{bmatrix} 1 & 3 & 5 & | & 6 \\ 0 & 1 & 2 & | & 3 \\ 2 & 1 & -2 & | & -5 \end{bmatrix} \text{ to } \begin{bmatrix} -1 & 2 & 7 & | & 11 \\ 0 & 1 & 2 & | & 3 \\ 2 & 1 & -2 & | & -5 \end{bmatrix} \quad -1R_3 + R_1 \to R_1$$

by first multiplying each element in row 3 of the original matrix by -1 and then adding the results to the corresponding elements in the first row of that matrix. Work as follows.

$$\begin{bmatrix} 1 + 2(-1) & 3 + 1(-1) & 5 + (-2)(-1) & | & 6 + (-5)(-1) \\ 0 & 1 & 2 & | & 3 \\ 2 & 1 & -2 & | & -5 \end{bmatrix} = \begin{bmatrix} -1 & 2 & 7 & | & 11 \\ 0 & 1 & 2 & | & 3 \\ 2 & 1 & -2 & | & -5 \end{bmatrix}$$

Again rows 2 and 3 are left unchanged, *even though the elements of row 3 were used to transform row 1.*

The Gauss-Jordan Method The **Gauss-Jordan method** is an extension of the echelon method of solving systems.* Before the Gauss-Jordan method can be used, the system must be in proper form: the terms with variables should be on the left and the constants on the right in each equation, with the variables in the same order in each equation. The following example illustrates the use of the Gauss-Jordan method to solve a system of equations.

EXAMPLE 1 Gauss-Jordan

Solve the system

$$3x - 4y = 1 \tag{1}$$
$$5x + 2y = 19. \tag{2}$$

Solution The system is already in the proper form for use of the Gauss-Jordan method. The steps used are essentially the same as in the echelon method, but we go further by eliminating the term with y in equation (1) before using transformation 2. By doing this, the final solution can be read directly from the last matrix. To begin, change the 5 in row 2 to 0. (Notice that the same notation is used to indicate each transformation, as in the previous section.)

$$\begin{bmatrix} 3 & -4 & | & 1 \\ 5 & 2 & | & 19 \end{bmatrix} \quad \text{Given matrix}$$

$$5R_1 + (-3)R_2 \to R_2 \quad \begin{bmatrix} 3 & -4 & | & 1 \\ 0 & -26 & | & -52 \end{bmatrix}$$

---

*The great German mathematician Carl Friedrich Gauss (1777–1855), sometimes referred to as the "Prince of Mathematicians," originally developed his elimination method for use in finding least squares coefficients. (See Section 1.3.) The German geodesist Wilhelm Jordan (1842–1899) improved his method and used it in surveying problems. Gauss's method had been known to the Chinese at least 1800 years earlier and was described in the *Jiuahang Suanshu (Nine Chapters on the Mathematical Art)*.

Change the -4 in row 1 to 0.

$$-4R_2 + 26R_1 \rightarrow R_1 \quad \begin{bmatrix} 78 & 0 & | & 234 \\ 0 & -26 & | & -52 \end{bmatrix}$$

Change the first nonzero number in each row to 1.

$$\begin{array}{c} \frac{1}{78}R_1 \rightarrow R_1 \\ -\frac{1}{26}R_2 \rightarrow R_2 \end{array} \quad \begin{bmatrix} 1 & 0 & | & 3 \\ 0 & 1 & | & 2 \end{bmatrix}$$

The last matrix corresponds to the system

$$x = 3$$
$$y = 2,$$

so the solution can be read directly from the last column of the final matrix. Check the solution by substitution in the equations of the original system.

> **NOTE** If your solution does not check, the most efficient way to find the error is to substitute back through the equations that correspond to each matrix, starting with the last matrix. When you find a system that is not satisfied by your (incorrect) answers, you have probably reached the matrix just before the error occurred. Look for the error in the transformation to the next matrix.

When the Gauss-Jordan method is used to solve a system, the final matrix always will have zeros above and below the diagonal of 1's on the left of the vertical bar. To transform the matrix, it is best to work column by column from left to right. Such an orderly method avoids confusion and going around in circles. For each column, first perform the steps that give the zeros. When all columns have zeros in place, multiply each row by the reciprocal of the coefficient of the remaining nonzero number in that row to get the required 1's. With dependent equations or inconsistent systems, it will not be possible to get the complete diagonal of 1's.

EXAMPLE 2 Gauss-Jordan

Use the Gauss-Jordan method to solve the system

$$x + 5z = -6 + y$$
$$3x + 3y = 10 + z$$
$$x + 3y + 2z = 5.$$

Solution

Method 1: Calculating by Hand   First, rewrite the system in proper form as follows.

$$x - y + 5z = -6$$
$$3x + 3y - z = 10$$
$$x + 3y + 2z = 5$$

Begin to find the solution by writing the augmented matrix of the linear system.

$$\begin{bmatrix} 1 & -1 & 5 & | & -6 \\ 3 & 3 & -1 & | & 10 \\ 1 & 3 & 2 & | & 5 \end{bmatrix}$$

Row transformations will be used to rewrite this matrix in the form

$$\begin{bmatrix} 1 & 0 & 0 & | & m \\ 0 & 1 & 0 & | & n \\ 0 & 0 & 1 & | & p \end{bmatrix},$$

where m, n, and p are real numbers. From this final form of the matrix, the solution can be read: $x = m$, $y = n$, $z = p$, or (m, n, p).

In the first column, we need zeros in the second and third rows. Multiply the first row by -3 and add to the second row to get a zero there. Then multiply the first row by -1 and add to the third row to get that zero.

$$\begin{matrix} \\ -3R_1 + R_2 \to R_2 \\ -1R_1 + R_3 \to R_3 \end{matrix} \begin{bmatrix} 1 & -1 & 5 & | & -6 \\ 0 & 6 & -16 & | & 28 \\ 0 & 4 & -3 & | & 11 \end{bmatrix}$$

Now get zeros in the second column in a similar way. We want zeros in the first and third rows. Row two will not change.

$$\begin{matrix} R_2 + 6R_1 \to R_1 \\ \\ 2R_2 + (-3)R_3 \to R_3 \end{matrix} \begin{bmatrix} 6 & 0 & 14 & | & -8 \\ 0 & 6 & -16 & | & 28 \\ 0 & 0 & -23 & | & 23 \end{bmatrix}$$

In transforming the third row, you may have used the operation $4R_2 + (-6)R_3 \to R_3$ instead of $2R_2 + (-3)R_3 \to R_3$. This is perfectly fine; the last row would then have -46 and 46 in place of -23 and 23. To avoid errors, it helps to keep the numbers as small as possible. We observe at this point that all of the numbers can be reduced in size by multiplying each row by an appropriate constant. This next step is not essential, but it simplifies the arithmetic.

$$\begin{matrix} \frac{1}{2}R_1 \to R_1 \\ \frac{1}{2}R_2 \to R_2 \\ -\frac{1}{23}R_3 \to R_3 \end{matrix} \begin{bmatrix} 3 & 0 & 7 & | & -4 \\ 0 & 3 & -8 & | & 14 \\ 0 & 0 & 1 & | & -1 \end{bmatrix}$$

Next, we want zeros in the first and second rows of the third column. Row three will not change.

$$\begin{matrix} -7R_3 + R_1 \to R_1 \\ 8R_3 + R_2 \to R_2 \\ \\ \end{matrix} \begin{bmatrix} 3 & 0 & 0 & | & 3 \\ 0 & 3 & 0 & | & 6 \\ 0 & 0 & 1 & | & -1 \end{bmatrix}$$

Finally, get 1's in each row by multiplying the row by the reciprocal of (or dividing the row by) the number in the diagonal position.

$$\begin{matrix} \frac{1}{3}R_1 \to R_1 \\ \frac{1}{3}R_2 \to R_2 \\ \\ \end{matrix} \begin{bmatrix} 1 & 0 & 0 & | & 1 \\ 0 & 1 & 0 & | & 2 \\ 0 & 0 & 1 & | & -1 \end{bmatrix}$$

The linear system associated with the final augmented matrix is

$$\begin{aligned} x &= 1 \\ y &= 2 \\ z &= -1, \end{aligned}$$

and the solution is $(1, 2, -1)$. Verify that this is the solution to the original system of equations.

CAUTION Notice that we have performed two or three operations on the same matrix in one step. This is permissible as long as we do not use a row that we are changing as part of another row operation. For example, when we changed row 2 in the first step, we could not use row 2 to transform row 3 in the same step. To avoid difficulty, use *only* row 1 to get zeros in column 1, row 2 to get zeros in column 2, and so on.

Method 2: Graphing Calculators

The row operations of the Gauss-Jordan method can also be done on a graphing calculator. For example, Figure 5 shows the result when the augmented matrix is entered into a TI-83. Figures 6 and 7 show how row operations can be used to get zeros in rows 2 and 3 of the first column.

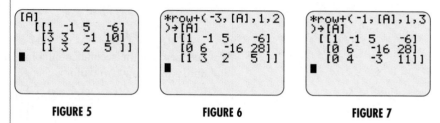

FIGURE 5   FIGURE 6   FIGURE 7

Calculators typically do not allow any multiple of a row to be added to any multiple of another row, such as in the operation $2R_2 + 6R_1 \rightarrow R_1$. They normally allow a multiple of a row to be added only to another unmodified row. To get around this restriction, we can convert the diagonal element to a 1 before changing the other elements in the column to 0. In this example, we change the 6 in row 2, column 2, to a 1 by dividing by 6. The result is shown in Figure 8. (The right side of the matrix is not visible, but can be seen by pressing the right arrow key.) Notice that this operation introduces decimals. Converting to fractions is preferable on calculators that have that option; 1/3 is certainly more concise than .3333333333. Figure 9 shows such a conversion on the TI-83.

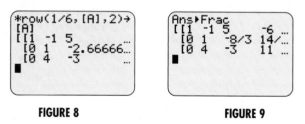

FIGURE 8   FIGURE 9

When performing row operations without a graphing calculator, it is best to avoid fractions and decimals, because these make the operations more difficult and more prone to error. A calculator, on the other hand, encounters no such difficulties.

Continuing in the same manner, the solution $(1, 2, -1)$ is found as shown in Figure 10.

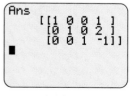

FIGURE 10

Some calculators can do the entire Gauss-Jordan process with a single command; on the TI-83, for example, this is done with the `rref` command. This is very useful in practice, although it does not show any of the intermediate steps.

Method 3: Spreadsheets The Gauss-Jordan method can be done using a spreadsheet either by using a macro or by developing the pivot steps using formulas with the copy and paste commands. However, spreadsheets also have built-in methods to solve systems of equations. Although these solvers do not usually employ the Gauss-Jordan method for solving systems of equations, they are, nonetheless, efficient and practical to use.

The Solver that is included with Excel can solve systems of equations that are both linear and nonlinear. The Solver is located in the Tools menu and requires that cells be identified ahead of time for each variable in the problem. It also requires that the left-hand side of each equation be placed in the spreadsheet as a formula. For example, to solve the above problem, we could identify cells A1, B1, and C1 for the variables x, y, and z, respectively. The Solver requires that we place a guess for the answer in these cells. Thus, it is convenient to place a zero in each of these cells. The left-hand side of each equation must be placed in a cell. Thus, we could choose A3, A4, and A5 to hold each of these formulas. Thus, in cell A3, we would type "=A1 − B1 + 5*C1" and put the other two equations in cells A4 and A5.

We now click on the Tools menu and choose Solver. Since this solver attempts to find a solution that is best in some way, we are required to identify a cell with a formula in it that we want to optimize. In this case, it is convenient to use the cell with the left-hand side of the first constraint in it, A3. Figure 11 illustrates the Solver box and the items placed in it.

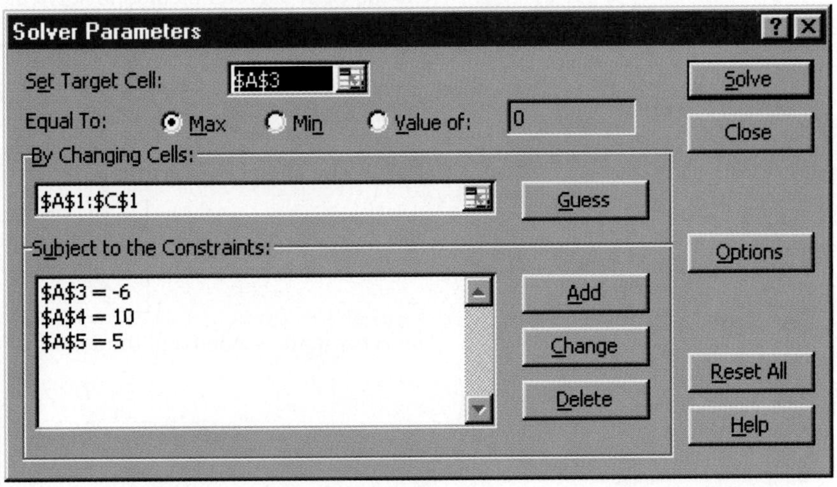

FIGURE 11

To obtain a solution, click on Solve. The solution is located in cells A1, B1, and C1, and these correspond to x, y, and z, respectively.

Systems with Nonunique Solutions If some of the given steps cannot be performed, the solution cannot be completed with the Gauss-Jordan method. The next examples illustrate how such cases may be handled.

EXAMPLE 3 Systems of Equations with No Solution
Use the Gauss-Jordan method to solve the system

$$x + y = 2$$
$$2x + 2y = 5.$$

Solution Begin by writing the augmented matrix.

$$\begin{bmatrix} 1 & 1 & | & 2 \\ 2 & 2 & | & 5 \end{bmatrix}$$

To get a zero for the second element in column 1, multiply the numbers in row 1 by -2 and add the results to the corresponding elements in row 2.

$$-2R_1 + R_2 \rightarrow R_2 \quad \begin{bmatrix} 1 & 1 & | & 2 \\ 0 & 0 & | & 1 \end{bmatrix}$$

This matrix corresponds to the system

$$x + y = 2$$
$$0x + 0y = 1.$$

Since the second equation is $0 = 1$, the system is inconsistent and therefore has no solution. The row $[0 \quad 0 \,|\, 1]$ is a signal that the given system is inconsistent.

EXAMPLE 4 Systems of Equations with an Infinite Number of Solutions
Use the Gauss-Jordan method to solve the system

$$x + 2y - z = 0$$
$$3x - y + z = 6$$
$$-2x - 4y + 2z = 0.$$

Solution The augmented matrix is

$$\begin{bmatrix} 1 & 2 & -1 & | & 0 \\ 3 & -1 & 1 & | & 6 \\ -2 & -4 & 2 & | & 0 \end{bmatrix}.$$

Get zeros in the second and third rows of column 1.

$$\begin{array}{c} -3R_1 + R_2 \rightarrow R_2 \\ 2R_1 + R_3 \rightarrow R_3 \end{array} \quad \begin{bmatrix} 1 & 2 & -1 & | & 0 \\ 0 & -7 & 4 & | & 6 \\ 0 & 0 & 0 & | & 0 \end{bmatrix}$$

To continue, get a zero in the first row of column 2 using the second row, as usual.

$$2R_2 + 7R_1 \rightarrow R_1 \quad \begin{bmatrix} 7 & 0 & 1 & | & 12 \\ 0 & -7 & 4 & | & 6 \\ 0 & 0 & 0 & | & 0 \end{bmatrix}$$

We cannot get a zero for the first-row, third-column element without changing the form of the first two columns. We must multiply each of the first two rows by the reciprocal of the first nonzero number.

$$\begin{array}{c} \frac{1}{7}R_1 \rightarrow R_1 \\ -\frac{1}{7}R_2 \rightarrow R_2 \end{array} \quad \left[\begin{array}{ccc|c} 1 & 0 & \frac{1}{7} & \frac{12}{7} \\ 0 & 1 & -\frac{4}{7} & -\frac{6}{7} \\ 0 & 0 & 0 & 0 \end{array} \right]$$

To complete the solution, write the equations that correspond to the first two rows of the matrix.

$$x + \frac{1}{7}z = \frac{12}{7}$$

$$y - \frac{4}{7}z = -\frac{6}{7}$$

Because both equations involve z, let z be the parameter. There are an infinite number of solutions, corresponding to the infinite number of values of z. Solve the first equation for x and the second for y to get

$$x = \frac{12 - z}{7} \quad \text{and} \quad y = \frac{4z - 6}{7}.$$

As shown in the previous section, the general solution is written

$$\left(\frac{12 - z}{7}, \frac{4z - 6}{7}, z \right),$$

where z is any real number. For example, $z = 2$ and $z = 12$ lead to the solutions $(10/7, 2/7, 2)$ and $(0, 6, 12)$.

EXAMPLE 5   Systems of Equations with an Infinite Number of Solutions
Consider the following system of equations.

$$x + 2y + 3z - w = 4$$
$$2x + 3y + w = -3$$
$$3x + 5y + 3z = 1$$

(a) Set this up as an augmented matrix, and verify that the result after the Gauss-Jordan method is

$$\left[\begin{array}{cccc|c} 1 & 0 & -9 & 5 & -18 \\ 0 & 1 & 6 & -3 & 11 \\ 0 & 0 & 0 & 0 & 0 \end{array} \right]$$

(b) Find the solution to this system of equations.

Solution To complete the solution, write the equations that correspond to the first two rows of the matrix.

$$x \quad - 9z + 5w = -18$$
$$y + 6z - 3w = \quad 11$$

Because both equations involve both z and w, let z and w be parameters. There are an infinite number of solutions, corresponding to the infinite number of values of z and w. Solve the first equation for x and the second for y to get

$$x = -18 + 9z - 5w \quad \text{and} \quad y = 11 - 6z + 3w.$$

In an analogous manner to problems with a single parameter, the general solution is written

$$(-18 + 9z - 5w, 11 - 6z + 3w, z, w),$$

where z and w are any real numbers. For example, $z = 1$ and $w = -2$ leads to the solution $(1, -1, 1, -2)$.

Although the examples have used only systems with two equations in two unknowns, three equations in three unknowns, or three equations in four unknowns, the Gauss-Jordan method can be used for any system with n equations and m unknowns. The method becomes tedious with more than three equations in three unknowns; on the other hand, it is very suitable for use by graphing calculators and computers, which can solve fairly large systems quickly. Sophisticated computer programs modify the method to reduce round-off error. Other methods used for special types of large matrices are studied in a course on numerical analysis.

EXAMPLE 6 Soda Sales

A convenience store sells 23 sodas one summer afternoon in 12-, 16-, and 20-ounce cups (small, medium, and large). The total volume of soda sold was 376 ounces.

(a) Suppose that the prices for a small, medium, and large soda are $1, $1.25, and $1.40, respectively, and that the total sales were $28.45. How many of each size did the store sell?

Solution As in Example 6 of the previous section, we will organize the information in a table.

	Small	**Medium**	**Large**	**Total**
Number	x	y	z	23
Volume	12	16	20	376
Price	1.00	1.25	1.40	28.45

The three rows of the table lead to three equations: one for the total number of sodas, one for the volume, and one for the price.

$$\begin{aligned} x + y + z &= 23 \\ 12x + 16y + 20z &= 376 \\ 1.00x + 1.25y + 1.40z &= 28.45 \end{aligned}$$

Set this up as an augmented matrix, and verify that the result after the Gauss-Jordan method is

$$\begin{bmatrix} 1 & 0 & 0 & | & 6 \\ 0 & 1 & 0 & | & 9 \\ 0 & 0 & 1 & | & 8 \end{bmatrix}.$$

The store sold 6 small, 9 medium, and 8 large sodas.

(b) Suppose the prices for small, medium, and large sodas are changed to $1, $2, and $3, respectively, but all other information is kept the same. How many of each size did the store sell?

Solution Change the third equation to

$$x + 2y + 3z = 28.45$$

and go through the Gauss-Jordan method again. The result is

$$\begin{bmatrix} 1 & 0 & -1 & | & 2 \\ 0 & 1 & 2 & | & 25 \\ 0 & 0 & 0 & | & -19.55 \end{bmatrix}.$$

(If you do the row operations in a different order in this example, you will have different numbers in the last column.) The last row of this matrix says that $0 = -19.55$, so the system is inconsistent and has no solution. (In retrospect, this is clear, because each soda sells for a whole number of dollars, and the total amount of money is not a whole number of dollars. In general, however, it is not easy to tell whether a system of equations has a solution or not by just looking at it.)

(c) Suppose the prices are the same as in part (b), but the total revenue is $48. Now how many of each size did the store sell?

Solution The third equation becomes

$$x + 2y + 3z = 48,$$

and the Gauss-Jordan method leads to

$$\begin{bmatrix} 1 & 0 & -1 & | & -2 \\ 0 & 1 & 2 & | & 25 \\ 0 & 0 & 0 & | & 0 \end{bmatrix}.$$

The system is dependent, similar to Example 4. Let z be the parameter, and solve the first two equations for x and y, yielding

$$x = z - 2 \quad \text{and} \quad y = 25 - 2z.$$

Remember that in this problem, x, y, and z must be nonnegative integers. From the equation for x, we must have

$$z \geq 2,$$

and from the equation for y, we must have

$$25 - 2z \geq 0,$$

from which we find

$$z \leq 12.5.$$

We therefore have 11 solutions corresponding to $z = 2, 3, \ldots, 12$.

(d) Give the solutions from part (c) that have the smallest and largest numbers of large sodas.

Solution For the smallest number of large sodas, let $z = 2$, giving $x = 2 - 2 = 0$ and $y = 25 - 2(2) = 21$. There are 0 small, 21 medium, and 2 large sodas.
For the largest number of large sodas, let $z = 12$, giving $x = 12 - 2 = 10$ and $y = 25 - 2(12) = 1$. There are 10 small, 1 medium, and 12 large sodas.

In summary, the Gauss-Jordan method of solving a linear system requires the following steps.

THE GAUSS-JORDAN METHOD OF SOLVING A LINEAR SYSTEM

1. Write each equation so that variable terms are in the same order on the left side of the equals sign and constants are on the right.

2. Write the augmented matrix that corresponds to the system.

3. Use row operations to transform the first column so that all elements except the element in the first row are zero.

4. Use row operations to transform the second column so that all elements except the element in the second row are zero.

5. Use row operations to transform the third column so that all elements except the element in the third row are zero.

6. Continue in this way until the last row is written in the form

$$[0 \quad 0 \quad 0 \quad \cdots \quad 0 \quad j \,|\, k],$$

where j and k are constants.

7. Multiply each row by the reciprocal of the nonzero element in that row.

2.2 EXERCISES

Write the augmented matrix for each of the following systems. **Do not solve.**

1. $2x + 3y = 11$
   $x + 2y = 8$

2. $3x + 5y = -13$
   $2x + 3y = -9$

3. $2x + y + z = 3$
   $3x - 4y + 2z = -7$
   $x + y + z = 2$

4. $4x - 2y + 3z = 4$
   $3x + 5y + z = 7$
   $5x - y + 4z = 7$

Write the system of equations associated with each of the following augmented matrices.

5. $\begin{bmatrix} 1 & 0 & | & 2 \\ 0 & 1 & | & 3 \end{bmatrix}$

6. $\begin{bmatrix} 1 & 0 & | & 5 \\ 0 & 1 & | & -3 \end{bmatrix}$

7. $\begin{bmatrix} 1 & 0 & 0 & | & 2 \\ 0 & 1 & 0 & | & 3 \\ 0 & 0 & 1 & | & -2 \end{bmatrix}$

8. $\begin{bmatrix} 1 & 0 & 0 & | & 4 \\ 0 & 1 & 0 & | & 2 \\ 0 & 0 & 1 & | & 3 \end{bmatrix}$

9. _____ on a matrix correspond to transformations of a system of equations.

10. Describe in your own words what $2R_1 + R_3 \rightarrow R_3$ means.

Use the indicated row operations to change each matrix.

11. Replace R_2 by $R_1 + (-2)R_2$.
$$\begin{bmatrix} 2 & 3 & 8 & | & 20 \\ 1 & 4 & 6 & | & 12 \\ 0 & 3 & 5 & | & 10 \end{bmatrix}$$

12. Replace R_3 by $-1R_1 + 3R_3$.
$$\begin{bmatrix} 3 & 2 & 6 & | & 18 \\ 2 & -2 & 5 & | & 7 \\ 1 & 0 & 5 & | & 20 \end{bmatrix}$$

13. Replace R_1 by $-4R_2 + R_1$.
$$\begin{bmatrix} 1 & 4 & 2 & | & 9 \\ 0 & 1 & 5 & | & 14 \\ 0 & 3 & 8 & | & 16 \end{bmatrix}$$

14. Replace R_1 by $R_3 + (-3)R_1$.
$$\begin{bmatrix} 1 & 0 & 4 & | & 21 \\ 0 & 6 & 5 & | & 30 \\ 0 & 0 & 12 & | & 15 \end{bmatrix}$$

15. Replace R_1 by $\frac{1}{3}R_1$.

$$\begin{bmatrix} 3 & 0 & 0 & | & 18 \\ 0 & 5 & 0 & | & 9 \\ 0 & 0 & 4 & | & 8 \end{bmatrix}$$

16. Replace R_3 by $\frac{1}{4}R_3$.

$$\begin{bmatrix} 1 & 0 & 0 & | & 6 \\ 0 & 1 & 0 & | & 5 \\ 0 & 0 & 4 & | & 12 \end{bmatrix}$$

Use the Gauss-Jordan method to solve the following systems of equations.

17. $x + y = 5$
$x - y = -1$

18. $x + 2y = 5$
$2x + y = -2$

19. $2x - 5y = 10$
$4x - 5y = 15$

20. $4x - 2y = 3$
$-2x + 3y = 1$

21. $2x - 3y = 2$
$4x - 6y = 1$

22. $x + 2y = 1$
$2x + 4y = 3$

23. $6x - 3y = 1$
$-12x + 6y = -2$

24. $x - y = 1$
$-x + y = -1$

25. $y = x - 1$
$y = 6 + z$
$z = -1 - x$

26. $x = 1 - y$
$2x = z$
$2z = -2 - y$

27. $2x - 2y = -2$
$y + z = 4$
$x + z = 1$

28. $x - z = -3$
$y + z = 9$
$-x + z = 3$

29. $4x + 4y - 4z = 24$
$2x - y + z = -9$
$x - 2y + 3z = 1$

30. $x + 3y - 6z = 7$
$2x - y + 2z = 0$
$x + y + 2z = -1$

31. $3x + 5y - z = 0$
$4x - y + 2z = 1$
$-6x - 10y + 2z = 0$

32. $3x - 6y + 3z = 15$
$2x + y - z = 2$
$-2x + 4y - 2z = 2$

33. $5x - 4y + 2z = 4$
$5x + 3y - z = 17$
$15x - 5y + 3z = 25$

34. $3x + 2y - z = -16$
$6x - 4y + 3z = 12$
$3x + 3y + z = -11$

35. $2x + 3y + z = 9$
$4x + 6y + 2z = 18$
$-\frac{1}{2}x - \frac{3}{4}y - \frac{1}{4}z = -\frac{9}{4}$

36. $4x - 2y - 3z = -23$
$-4x + 3y + z = 11$
$8x - 5y + 4z = 6$

37. $x + 2y - w = 3$
$2x + 4z + 2w = -6$
$x + 2y - z = 6$
$2x - y + z + w = -3$

38. $x + 3y - 2z - w = 9$
$2x + 4y + 2w = 10$
$-3x - 5y + 2z - w = -15$
$x - y - 3z + 2w = 6$

39. $x + y - z + 2w = -20$
$2x - y + z + w = 11$
$3x - 2y + z - 2w = 27$

40. $4x - 3y + z + w = 21$
$-2x - y + 2z + 7w = 2$
$10x - 5z - 20w = 15$

41. $10.47x + 3.52y + 2.58z - 6.42w = 218.65$
$8.62x - 4.93y - 1.75z + 2.83w = 157.03$
$4.92x + 6.83y - 2.97z + 2.65w = 462.3$
$2.86x + 19.10y - 6.24z - 8.73w = 398.4$

42. $28.6x + 94.5y + 16.0z - 2.94w = 198.3$
$16.7x + 44.3y - 27.3z + 8.9w = 254.7$
$12.5x - 38.7y + 92.5z + 22.4w = 562.7$
$40.1x - 28.3y + 17.5z - 10.2w = 375.4$

Applications

BUSINESS AND ECONOMICS

43. *Surveys* The president of Sam's Supermarkets plans to hire two public relations firms to survey 500 customers by phone, 750 by mail, and 250 by in-person interviews. The Garcia firm has personnel to do 10 phone surveys, 30 mail surveys, and 5 interviews per hour. The Wong firm can handle 20 phone surveys, 10 mail surveys, and 10 interviews per hour. For how many hours should each firm be hired to produce the exact number of surveys needed?

44. *Transportation* A knitting shop orders yarn from three suppliers in Toronto, Montreal, and Ottawa. One month the shop ordered a total of 100 units of yarn from these suppliers. The delivery costs were $80, $50, and $65 per unit for the orders from Toronto, Montreal, and Ottawa, respectively, with total delivery costs of $5990. The shop ordered the same amount from Toronto and Ottawa. How many units were ordered from each supplier?

45. *Manufacturing* Fred's Furniture Factory has 1950 machine hours available each week in the cutting department, 1490 hours in the assembly department, and 2160 in the finishing department. Manufacturing a chair requires .2 hours of cutting, .3 hours of assembly, and .1 hours of finishing. A cabinet requires .5 hours of cutting, .4 hours of assembly, and .6 hours of finishing. A buffet requires .3 hours of cutting, .1 hours of assembly, and .4 hours of finishing. How many chairs, cabinets, and buffets should be produced in order to use all the available production capacity?

46. *Manufacturing* Nadir Inc. produces three models of television sets: deluxe, super-deluxe, and ultra. Each deluxe set requires 2 hours of electronics work, 2 hours of assembly time, and 1 hour of finishing time. Each super-deluxe requires 1, 3, and 1 hours of electronics, assembly, and finishing time, respectively. Each ultra requires 3, 2, and 2 hours of the same work, respectively. There are 100 hours available for electronics, 100 hours available for assembly, and 65 hours available for finishing per week. How many of each model should be produced each week if all available time is to be used?

47. *Transportation* An electronics company produces three models of stereo speakers, models A, B, and C, and can deliver them by truck, van, or station wagon. A truck holds 2 boxes of model A, 1 of model B, and 3 of model C. A van holds 1 box of model A, 3 boxes of model B, and 2 boxes of model C. A station wagon holds 1 box of model A, 3 boxes of model B, and 1 box of model C.

  a. If 15 boxes of model A, 20 boxes of model B, and 22 boxes of model C are to be delivered, how many vehicles of each type should be used so that all operate at full capacity?

  b. Model C has been discontinued. Each kind of delivery vehicle can now carry one more box of model B than previously and the same number of boxes of model A. If 16 boxes of model A and 22 boxes of model B are to be delivered, how many vehicles of each type should be used so that all operate at full capacity?

48. *Truck Rental* The U-Drive Rent-A-Truck Company plans to spend $6 million on 200 new vehicles. Each van will cost $20,000, each small truck $30,000, and each large truck $50,000. Past experience shows that they need twice as many vans as small trucks. How many of each kind of vehicle can they buy?

49. *Loans* To get the necessary funds for a planned expansion, a small company took out three loans totaling $25,000. Company owners were able to borrow some of the money at 13%. They borrowed $2000 more than one-half the amount of the 13% loan at 14%, and the rest at 12%. The total annual interest on the loans was $3240.

  a. How much did they borrow at each rate?

  b. Suppose we drop the condition that the amount borrowed at 14% is $2000 more than one-half the amount borrowed at 13%. What can you say about the amount borrowed at 12%? What is the solution if the amount borrowed at 12% is $5000?

  c. Suppose the company can borrow only $6000 at 12%. Is a solution possible that still meets the conditions given in part a?

50. *Transportation* An auto manufacturer sends cars from two plants, I and II, to dealerships A and B located in a midwestern city. Plant I has a total of 28 cars to send, and plant II has 8. Dealer A needs 20 cars, and dealer B needs 16. Transportation costs per car, based on the distance of each dealership from each plant, are $220 from I to A, $300 from I to B, $400 from II to A, and $180 from II to B. The manufacturer wants to limit transportation costs to $10,640. How many cars should be sent from each plant to each of the two dealerships?

51. *Transportation* A manufacturer purchases a part for use at both of its plants—one at Roseville, California, the other at Akron, Ohio. The part is available in limited quantities from two suppliers. Each supplier has 75 units available. The Roseville plant needs 40 units, and the Akron plant requires 75 units. The first supplier charges $70 per unit delivered to Roseville and $90 per unit delivered to Akron. Corresponding costs from the second supplier are $80 and $120. The manufacturer wants to order a total of 75 units from the first, less expensive supplier, with the remaining 40 units to come from the second supplier. If the company spends $10,750 to purchase the required number of units for the two plants, find the number of units that should be purchased from each supplier.

52. *Packaging* A company produces three combinations of mixed vegetables that sell in 1-kg packages. Italian style combines .3 kg of zucchini, .3 of broccoli, and .4 of carrots. French style combines .6 kg of broccoli and .4 of carrots. Oriental style combines .2 kg of zucchini, .5 of broccoli, and .3 of carrots. The company has a stock of 16,200 kg of zucchini, 41,400 kg of broccoli, and 29,400 kg of carrots. How many packages of each style should it prepare to use up existing supplies?

53. *Broadway Economics* When Neil Simon opens a new play, he has to decide whether to open the show on Broadway or Off Broadway. For example, in his play *London Suite,* he decided to open it Off Broadway. From information provided by Emanuel Azenberg, his producer, the following equations were developed:

$$43,500x - y = 1,295,000$$
$$27,000x - y = 440,000,$$

where x represents the number of weeks that the show has run and y represents the profit or loss from the show (first equation is Broadway and second equation is Off Broadway).*

a. Solve this system of equations to determine when the profit/loss from the show will be equal for each venue. What is the profit at that point?

b. Discuss which venue is favorable for the show.

SOCIAL SCIENCES

54. *Traffic Control* At rush hours, substantial traffic congestion is encountered at the traffic intersections shown in the figure. (The streets are one-way, as shown by the arrows.)

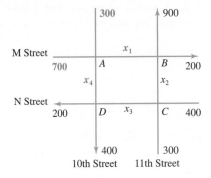

The city wishes to improve the signals at these corners so as to speed the flow of traffic. The traffic engineers first gather data. As the figure shows, 700 cars per hour come down M Street to intersection A, and 300 cars per hour come down 10th Street to intersection A. A total of x_1 of these cars leave A on M Street, and x_4 cars leave A on 10th Street. The number of cars entering A must equal the number leaving, so that

$$x_1 + x_4 = 700 + 300$$

or

$$x_1 + x_4 = 1000.$$

For intersection B, x_1 cars enter on M Street and x_2 on 11th Street. The figure shows that 900 cars leave B on 11th and 200 on M. We have

$$x_1 + x_2 = 900 + 200$$
$$x_1 + x_2 = 1100.$$

a. Write two equations representing the traffic entering and leaving intersections C and D.

b. Use the four equations to set up an augmented matrix, and solve the system by the Gauss-Jordan method, using x_4 as the parameter.

c. Based on your solution to part b, what are the largest and smallest possible values for the number of cars leaving intersection A on 10th Street?

d. Answer the question in part c for the other three variables.

e. Verify that you could have discarded any one of the four original equations without changing the solution. What does this tell you about the original problem?

55. *Modeling War* One of the factors that contribute to the success or failure of a particular army during war is its ability to get new troops ready for service. It is possible to analyze the rate of change in the number of troops of two hypothetical armies with the following simplified model,

Rate of increase (RED ARMY) $= 200{,}000 - .5r - .3b$
Rate of increase (BLUE ARMY) $= 350{,}000 - .5r - .7b$,

where r is the number of soldiers in the Red Army at a given time and b is the number of soldiers in the Blue Army at a given time. The factors .5 and .7 represent each army's efficiency of bringing new soldiers to the fight.[†]

a. Solve this system of equations to determine the number of soldiers in each army when the rate of increase for each is zero.

b. Describe what might be going on in a war when the rate of increase is zero.

LIFE SCIENCES

56. *Birds* The date of the first sighting of robins has been occurring earlier each spring over the past 25 years at the Rocky Mountain Biological Laboratory. Scientists from this laboratory have developed two linear equations that estimate the date of the first sighting of robins:

$$y = 759 - .338x$$
$$y = 1637 - .779x,$$

*Goetz, Albert, "Basic Economics: Calculating Against Theatrical Disaster," *The Mathematics Teacher,* Vol. 89, No. 1, Jan. 1996, pp. 30–32.
†Bellany, Ian, "Modeling War," *Journal of Peace Research,* Vol. 36, No. 6, 1999, pp. 729–739.

where x is the year and y is the estimated number of days into the year when a robin can be expected.*

a. Compare the date of first sighting in 1980 for each of these equations.

b. Solve this system of equations to find the year in which the two estimates agree.

57. *Dietetics* A hospital dietician is planning a special diet for a certain patient. The total amount per meal of food groups A, B, and C must equal 400 grams. The diet should include one-third as much of group A as of group B, and the sum of the amounts of group A and group C should equal twice the amount of group B.

a. How many grams of each food group should be included?

b. Suppose we drop the requirement that the diet include one-third as much of group A as of group B. Describe the set of all possible solutions.

c. Suppose that, in addition to the conditions given in the original problem, foods A and B cost 2 cents per gram and food C costs 3 cents per gram, and that a meal must cost $8. Is a solution possible?

58. *Animal Breeding* An animal breeder can buy four types of food for Vietnamese pot-bellied pigs. Each case of Brand A contains 25 units of fiber, 30 units of protein, and 30 units of fat. Each case of Brand B contains 50 units of fiber, 30 units of protein, and 20 units of fat. Each case of Brand C contains 75 units of fiber, 30 units of protein, and 20 units of fat. Each case of Brand D contains 100 units of fiber, 60 units of protein, and 30 units of fat. How many cases of each should the breeder mix together to obtain a food that provides 1200 units of fiber, 600 units of protein, and 400 units of fat?

59. *Mixing Plant Foods* Natural Brand plant food is made from three chemicals. The mix must include 10.8% of the first chemical, and the other two chemicals must be in a ratio of 4 to 3 as measured by weight. How much of each chemical is required to make 750 kg of the plant food?

60. *Bacterial Food Requirements* Three species of bacteria are fed three foods, I, II, and III. A bacterium of the first species consumes 1.3 units each of foods I and II and 2.3 units of food III each day. A bacterium of the second species consumes 1.1 units of food I, 2.4 units of food II, and 3.7 units of food III each day. A bacterium of the third species consumes 8.1 units of I, 2.9 units of II, and 5.1 units of III each day. If 16,000 units of I, 28,000 units of II, and 44,000 units of III are supplied each day, how many of each species can be maintained in this environment?

61. *Fish Food Requirements* A lake is stocked each spring with three species of fish, A, B, and C. Three foods, I, II, and III, are available in the lake. Each fish of species A requires an average of 1.32 units of food I, 2.9 units of food II, and 1.75 units of food III each day. Species B fish each require 2.1 units of food I, .95 unit of food II, and .6 unit of food III daily. Species C fish require .86, 1.52, and 2.01 units of I, II, and III per day, respectively. If 490 units of food I, 897 units of food II, and 653 units of food III are available daily, how many of each species should be stocked?

62. *Agriculture* According to data from a 1984 Texas agricultural report, the amount of nitrogen (in lb/acre), phosphate (in lb/acre), and labor (in hr/acre) needed to grow honeydews, yellow onions, and lettuce is given by the following table.[†]

	Honeydews	Yellow Onions	Lettuce
Nitrogen	120	150	180
Phosphate	180	80	80
Labor	4.97	4.45	4.65

a. If the farmer has 220 acres, 29,100 lb of nitrogen, 32,600 lb of phosphate, and 480 hours of labor, is it possible to use all resources completely? If so, how many acres should he allot for each crop?

b. Suppose everything is the same as in part a, except that 1061 hours of labor are available. Is it possible to use all resources completely? If so, how many acres should he allot for each crop?

63. *Archimedes' Problem Bovinum* Archimedes is credited with the authorship of a famous problem involving the number of cattle of the sun god. A simplified version of the problem has been stated as follows:[‡]

> *The sun god had a herd of cattle consisting of bulls and cows, one part of which was white, a second black, a third spotted, and a fourth brown.*
>
> *Among the bulls, the number of white ones was one half plus one third the number of the black greater than the brown; the number of the black, one quarter plus one fifth the number of the spotted greater than the brown; the number of the spotted, one sixth and one seventh the number of the white greater than the brown.*

*Inouye, David, Billy Barr, Kenneth Armitage, and Brian Inouye, "Climate Change Is Affecting Altitudinal Migrants and Hibernating Species," *Proceedings of the National Academy of Science,* Vol. 97, No. 4, Feb. 15, 2000, pp. 1630–1633.

†Paredes, Miguel, Mohammad Fatehi, and Richard Hinthorn, "The Transformation of an Inconsistent Linear System into a Consistent System," *The AMATYC Review,* Vol. 13, No. 2, Spring 1992.

‡Dorrie, Heinrich, *100 Great Problems of Elementary Mathematics, Their History and Solution,* Dover Publications, New York, 1965, pp. 3–7.

Among the cows, the number of white ones was one third plus one quarter of the total black cattle; the number of the black, one quarter plus one fifth the total of the spotted cattle; the number of the spotted, one fifth plus one sixth the total of the brown cattle; the number of the brown, one sixth plus one seventh the total of the white cattle.

What was the composition of the herd?

The problem can be solved by converting the statements into two systems of equations, using X, Y, Z, and T for the number of white, black, spotted, and brown bulls, respectively, and x, y, z, and t for the number of white, black, spotted, and brown cows, respectively. For example, the first statement can be written as $X = (1/2 + 1/3)Y + T$ and then reduced. The result is the following two systems of equations:

$$
\begin{aligned}
6X - 5Y &= 6T \\
20Y - 9Z &= 20T \\
42Z - 13X &= 42T
\end{aligned}
\qquad \text{and} \qquad
\begin{aligned}
12x - 7y &= 7Y \\
20y - 9z &= 9Z \\
30z - 11t &= 11T \\
-13x + 42t &= 13X
\end{aligned}
$$

a. Show that these two systems of equations represent Archimedes' Problem Bovinum.

b. If it is known that the number of brown bulls, T, is 4,149,387, use Gauss-Jordan elimination to first find a solution to the 3×3 system and then use these values and Gauss-Jordan elimination to find a solution to the 4×4 system of equations.

PHYSICAL SCIENCES

64. *Lead Emissions* Estimates of the total amount of lead emissions in the United States are made by the Environmental Protection Agency's Office of Air Quality Planning and Standards each year. The lead estimates decreased in the early 1990s but did not continue to fall. The table lists lead emissions estimates for four years.*

Year	Lead Emissions (in short tons)
1988	7053
1992	3808
1994	4043
1997	3915

a. If the relationship between the lead emissions L and the year t is expressed as $L = at^2 + bt + c$, where $t = 0$ corresponds to 1988, use data from 1988, 1992, and 1997 and a linear system of equations to determine the constants a, b, and c.

b. Use the equation from part a to predict the emissions in 1994, and compare the result with the actual data.

c. If the relationship between the lead emissions L and the year t is expressed as $L = at^3 + bt^2 + ct + d$, where $t = 0$ corresponds to 1988, use all four data points and a linear system of equations to determine the constants a, b, c, and d.

d. Discuss the appropriateness of the functions used in parts a and b to model this data.

GENERAL INTEREST

65. *Toys* 100 toys are to be given out to a group of children. A ball costs \$2, a doll costs \$3, and a car costs \$4. \$295 was spent.

a. A ball weighs 12 oz, a doll 16 oz, and a car 18 oz. The total weight of all the toys was 1542 oz. Find how many of each toy there are.

b. Now suppose the weight of a ball, doll, and car are 11, 15, and 19 oz, respectively. If the total weight is still 1542 oz, how many solutions are there now?

c. Keep the weights as in part b, but change the total weight to 1480 oz. How many solutions are there?

d. Give the solution to part c that has the smallest number of cars.

e. Give the solution to part c that has the largest number of cars.

66. *Ice Cream* Researchers have determined that the amount of sugar contained in ice cream helps to determine the overall "degree of like" that a consumer has toward that particular flavor. They have also determined that too much or too little sugar will have the same negative affect on the "degree of like" and that this relationship follows a quadratic function. In an experiment conducted at Pennsylvania State University, the following condensed table was obtained.[†]

Percentage of Sugar	Degree of Like
8	5.4
13	6.3
18	5.6

a. Use this information and Gauss-Jordan elimination to determine the coefficients a, b, c, of the quadratic equation

$$y = ax^2 + bx + c,$$

where y is the degree of like and x is the percentage of sugar in the ice cream mix.

The World Almanac and Book of Facts 2000, World Almanac Books, p. 169.
[†]Guinard, J., C. Zoumas-Morse, L. Mori, B. Uatoni, D. Panyam, and A. Kilar, "Sugar and Fat Effects on Sensory Properties of Ice Cream," *Journal of Food Science,* Vol. 62, No. 4, Sept./Oct. 1997, pp. 1087–1094.

b. Repeat part a by using the quadratic regression feature on a graphing calculator. Compare your answers.

67. *Lights Out* The Tiger Electronics' game, Lights Out, consists of five rows of five lighted buttons. When a button is pushed, it changes the on/off status of it and the status of all of its vertical and horizontal neighbors. For any given situation where some of the lights are on and some are off, the goal of the game is to push buttons until all of the lights are turned off. Thus, for any given array of lights, solving a system of equations can be used to develop a strategy for turning the lights out.* The following system of equations can be used to solve the problem for a simplified, 2×2, version of the game where all of the lights are initially turned on.

$$x_{11} + x_{12} + x_{21} = 1$$
$$x_{11} + x_{12} + x_{22} = 1$$
$$x_{11} + x_{21} + x_{22} = 1$$
$$x_{12} + x_{21} + x_{22} = 1,$$

where $x_{ij} = 1$ if the light in row i, column j is on and $x_{ij} = 0$ when it is off. The order in which the buttons are pushed does not matter, so we are only seeking which buttons should be pushed.

a. Solve this system of equations and determine a strategy to turn the lights out. (*Hint:* While doing pivot operations, if an odd number is found, immediately replace this value with a 1; if an even number is found, then immediately replace that number with a zero. This is called modulo 2 arithmetic, and it is necessary in problems dealing with on/off switches.)

b. Resolve the equation with the right side changed to $(0, 1, 1, 0)$.

2.3   ADDITION AND SUBTRACTION OF MATRICES

? THINK ABOUT IT   A company sends monthly shipments to its warehouses in several cities. How might the company keep track of the shipments to each warehouse most efficiently?

In the previous section, matrices were used to store information about systems of linear equations. In this section, we begin a study of matrices and show additional uses of matrix notation that will answer the question posed above. The use of matrices has gained increasing importance in the fields of management, natural science, and social science because matrices provide a convenient way to organize data, as Example 1 demonstrates.

EXAMPLE 1   Furniture Shipments

The EZ Life Company manufactures sofas and armchairs in three models, A, B, and C. The company has regional warehouses in New York, Chicago, and San Francisco. In its August shipment, the company sends 10 model-A sofas, 12 model-B sofas, 5 model-C sofas, 15 model-A chairs, 20 model-B chairs, and 8 model-C chairs to each warehouse.

  Solution To organize this data, we might first list it as follows.

| Sofas | 10 model-A | 12 model-B | 5 model-C |
| Chairs | 15 model-A | 20 model-B | 8 model-C |

---

*Anderson, Marlow, and Todd Feil, "Turning Lights Out with Linear Algebra," *Mathematics Magazine,* Vol. 71, No. 4, 1998, pp. 300–303.

Alternatively, we might tabulate the data in a chart.

		Model		
		A	**B**	**C**
Furniture Type	*Sofas*	10	12	5
	Chairs	15	20	8

With the understanding that the numbers in each row refer to the furniture type (sofa, chair) and the numbers in each column refer to the model (A, B, C), the same information can be given by a matrix, as follows.

$$M = \begin{bmatrix} 10 & 12 & 5 \\ 15 & 20 & 8 \end{bmatrix}$$

Matrices often are named with capital letters, as in Example 1. Matrices are classified by **size;** that is, by the number of rows and columns they contain. For example, matrix M above has two rows and three columns. This matrix is a 2×3 (read "2 by 3") matrix. By definition, a matrix with m rows and n columns is an $m \times n$ matrix. The number of rows is always given first.

EXAMPLE 2   Matrix Size

(a) The matrix $\begin{bmatrix} 6 & 5 \\ 3 & 4 \\ 5 & -1 \end{bmatrix}$ is a 3×2 matrix.

(b) $\begin{bmatrix} 5 & 8 & 9 \\ 0 & 5 & -3 \\ -4 & 0 & 5 \end{bmatrix}$ is a 3×3 matrix.

(c) $\begin{bmatrix} 1 & 6 & 5 & -2 & 5 \end{bmatrix}$ is a 1×5 matrix.

(d) $\begin{bmatrix} 3 \\ -5 \\ 0 \\ 2 \end{bmatrix}$ is a 4×1 matrix.

A matrix with the same number of rows as columns is called a **square matrix.** The matrix in Example 2(b) is a square matrix.

A matrix containing only one row is called a **row matrix** or a **row vector.** The matrix in Example 2(c) is a row matrix, as are

$$[5 \ \ 8], \quad [6 \ -9 \ \ 2], \quad \text{and} \quad [-4 \ \ 0 \ \ 0 \ \ 0].$$

A matrix of only one column, as in Example 2(d), is a **column matrix** or a **column vector.**

Equality for matrices is defined as follows.

> **MATRIX EQUALITY**
>
> Two matrices are equal if they are the same size and if each pair of corresponding elements is equal.

By this definition,

$$\begin{bmatrix} 2 & 1 \\ 3 & -5 \end{bmatrix} \quad \text{and} \quad \begin{bmatrix} 1 & 2 \\ -5 & 3 \end{bmatrix}$$

are not equal (even though they contain the same elements and are the same size) since the corresponding elements differ.

EXAMPLE 3   Matrix Equality

(a) From the definition of matrix equality given above, the only way that the statement

$$\begin{bmatrix} 2 & 1 \\ p & q \end{bmatrix} = \begin{bmatrix} x & y \\ -1 & 0 \end{bmatrix}$$

can be true is if $2 = x$, $1 = y$, $p = -1$, and $q = 0$.

(b) The statement

$$\begin{bmatrix} x \\ y \end{bmatrix} = \begin{bmatrix} 1 \\ 4 \\ 0 \end{bmatrix}$$

can never be true, since the two matrices are different sizes. (One is 2×1 and the other is 3×1.) ▪

Addition   The matrix given in Example 1,

$$M = \begin{bmatrix} 10 & 12 & 5 \\ 15 & 20 & 8 \end{bmatrix},$$

shows the August shipment from the EZ Life plant to each of its warehouses. If matrix N below gives the September shipment to the New York warehouse, what is the total shipment of each item of furniture to the New York warehouse for these two months?

$$N = \begin{bmatrix} 45 & 35 & 20 \\ 65 & 40 & 35 \end{bmatrix}$$

If 10 model-A sofas were shipped in August and 45 in September, then altogether $10 + 45 = 55$ model-A sofas were shipped in the two months. The other corresponding entries can be added in a similar way to get a new matrix Q, that represents the total shipment for the two months.

$$Q = \begin{bmatrix} 55 & 47 & 25 \\ 80 & 60 & 43 \end{bmatrix}$$

It is convenient to refer to Q as the sum of M and N.

The way these two matrices were added illustrates the following definition of addition of matrices.

ADDITION OF MATRICES

The sum of two $m \times n$ matrices X and Y is the $m \times n$ matrix $X + Y$ in which each element is the sum of the corresponding elements of X and Y.

█ **CAUTION** It is important to remember that only matrices that are the same size can be added.

EXAMPLE 4   Addition of Matrices
Find each sum if possible.

Solution

(a) $\begin{bmatrix} 5 & -6 \\ 8 & 9 \end{bmatrix} + \begin{bmatrix} -4 & 6 \\ 8 & -3 \end{bmatrix} = \begin{bmatrix} 5 + (-4) & -6 + 6 \\ 8 + 8 & 9 + (-3) \end{bmatrix} = \begin{bmatrix} 1 & 0 \\ 16 & 6 \end{bmatrix}$

(b) The matrices

$$A = \begin{bmatrix} 5 & 8 \\ 6 & 2 \end{bmatrix} \quad \text{and} \quad B = \begin{bmatrix} 3 & 9 & 1 \\ 4 & 2 & 5 \end{bmatrix}$$

are different sizes. Therefore, the sum $A + B$ does not exist. ▪

EXAMPLE 5   Furniture Shipments
The September shipments from the EZ Life Company to the New York, San Francisco, and Chicago warehouses are given in matrices N, S, and C below.

$$N = \begin{bmatrix} 45 & 35 & 20 \\ 65 & 40 & 35 \end{bmatrix} \quad S = \begin{bmatrix} 30 & 32 & 28 \\ 43 & 47 & 30 \end{bmatrix} \quad C = \begin{bmatrix} 22 & 25 & 38 \\ 31 & 34 & 35 \end{bmatrix}$$

What was the total amount shipped to the three warehouses in September?

?

Solution The total of the September shipments is represented by the sum of the three matrices N, S, and C.

$$N + S + C = \begin{bmatrix} 45 & 35 & 20 \\ 65 & 40 & 35 \end{bmatrix} + \begin{bmatrix} 30 & 32 & 28 \\ 43 & 47 & 30 \end{bmatrix} + \begin{bmatrix} 22 & 25 & 38 \\ 31 & 34 & 35 \end{bmatrix}$$

$$= \begin{bmatrix} 97 & 92 & 86 \\ 139 & 121 & 100 \end{bmatrix}$$

For example, this sum shows that the total number of model-C sofas shipped to the three warehouses in September was 86. ▪

The additive inverse of the real number a is $-a$; a similar definition applies to matrices.

ADDITIVE INVERSE

The **additive inverse** (or **negative**) of a matrix X is the matrix $-X$ in which each element is the additive inverse of the corresponding element of X.

FOR REVIEW ▰▰▬

Compare this with the identity property for real numbers: for any real number a, $a + 0 = 0 + a = a$. Exercises 34–37 give other properties of matrices that are parallel to the properties of real numbers.

If

$$A = \begin{bmatrix} 1 & 2 & 3 \\ 0 & -1 & 5 \end{bmatrix} \quad \text{and} \quad B = \begin{bmatrix} -2 & 3 & 0 \\ 1 & -7 & 2 \end{bmatrix},$$

then by the definition of the additive inverse of a matrix,

$$-A = \begin{bmatrix} -1 & -2 & -3 \\ 0 & 1 & -5 \end{bmatrix} \quad \text{and} \quad -B = \begin{bmatrix} 2 & -3 & 0 \\ -1 & 7 & -2 \end{bmatrix}.$$

By the definition of matrix addition, for each matrix X the sum $X + (-X)$ is a **zero matrix,** O, whose elements are all zeros. For the matrix A above,

$$A - A = \begin{bmatrix} 0 & 0 & 0 \\ 0 & 0 & 0 \end{bmatrix}.$$

There is an $m \times n$ zero matrix for each pair of values of m and n. Such a matrix serves as an $m \times n$ **additive identity,** similar to the additive identity 0 for any real number. Zero matrices have the following identity property.

ZERO MATRIX

If O is an $m \times n$ zero matrix, and A is any $m \times n$ matrix, then

$$A + O = O + A = A.$$

Subtraction The subtraction of matrices is defined in a manner comparable to subtraction of real numbers.

SUBTRACTION OF MATRICES

For two $m \times n$ matrices X and Y, the difference $X - Y$ is the $m \times n$ matrix defined by

$$X - Y = X + (-Y).$$

This definition means that matrix subtraction can be performed by subtracting corresponding elements. For example, with A, B, and $-B$ as defined above,

$$A - B = A + (-B) = \begin{bmatrix} 1 & 2 & 3 \\ 0 & -1 & 5 \end{bmatrix} + \begin{bmatrix} 2 & -3 & 0 \\ -1 & 7 & -2 \end{bmatrix}$$

$$= \begin{bmatrix} 3 & -1 & 3 \\ -1 & 6 & 3 \end{bmatrix}.$$

Matrix operations are easily performed on a graphing calculator. Figure 12 shows the previous operation; the matrices A and B were already entered into the calculator.

Spreadsheet programs are designed to effectively organize data that can be represented in rows and columns. Accordingly, matrix operations are also easily performed on spreadsheets. See *The Spreadsheet Manual* that is available with this book for details.

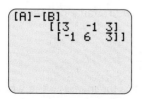

FIGURE 12

EXAMPLE 6 Subtraction of Matrices

(a) $[8 \quad 6 \quad -4] - [3 \quad 5 \quad -8] = [5 \quad 1 \quad 4]$

(b) The matrices

$$\begin{bmatrix} -2 & 5 \\ 0 & 1 \end{bmatrix} \quad \text{and} \quad \begin{bmatrix} 3 \\ 5 \end{bmatrix}$$

are different sizes and cannot be subtracted.

EXAMPLE 7 Furniture Shipments

During September the Chicago warehouse of the EZ Life Company shipped out the following numbers of each model.

$$K = \begin{bmatrix} 5 & 10 & 8 \\ 11 & 14 & 15 \end{bmatrix}$$

What was the Chicago warehouse inventory on October 1, taking into account only the number of items received and sent out during the month?

Solution The number of each kind of item received during September is given by matrix C from Example 5; the number of each model sent out during September is given by matrix K. The October 1 inventory will be represented by the matrix $C - K$:

$$\begin{bmatrix} 22 & 25 & 38 \\ 31 & 34 & 35 \end{bmatrix} - \begin{bmatrix} 5 & 10 & 8 \\ 11 & 14 & 15 \end{bmatrix} = \begin{bmatrix} 17 & 15 & 30 \\ 20 & 20 & 20 \end{bmatrix}.$$

2.3 EXERCISES

Mark each statement as true or false. If false, tell why.

1. $\begin{bmatrix} 1 & 3 \\ 5 & 7 \end{bmatrix} = \begin{bmatrix} 1 & 5 \\ 3 & 7 \end{bmatrix}$

2. $\begin{bmatrix} 1 \\ 2 \\ 3 \end{bmatrix} = [1 \quad 2 \quad 3]$

3. $\begin{bmatrix} x \\ y \end{bmatrix} = \begin{bmatrix} 3 \\ 5 \end{bmatrix}$ if $x = 3$ and $y = 5$.

4. $\begin{bmatrix} 3 & 5 & 2 & 8 \\ 1 & -1 & 4 & 0 \end{bmatrix}$ is a 4×2 matrix.

5. $\begin{bmatrix} 1 & 9 & -4 \\ 3 & 7 & 2 \\ -1 & 1 & 0 \end{bmatrix}$ is a square matrix.

6. $\begin{bmatrix} 2 & 4 & -1 \\ 3 & 7 & 5 \\ 0 & 0 & 0 \end{bmatrix} = \begin{bmatrix} 2 & 4 & -1 \\ 3 & 7 & 5 \end{bmatrix}$

Find the size of each matrix. Identify any square, column, or row matrices. Give the additive inverse of each matrix.

7. $\begin{bmatrix} -4 & 8 \\ 2 & 3 \end{bmatrix}$

8. $\begin{bmatrix} -9 & 6 & 2 \\ 4 & 1 & 8 \end{bmatrix}$

9. $\begin{bmatrix} -6 & 8 & 0 & 0 \\ 4 & 1 & 9 & 2 \\ 3 & -5 & 7 & 1 \end{bmatrix}$

10. $[8 \quad -2 \quad 4 \quad 6 \quad 3]$

11. $\begin{bmatrix} 2 \\ 4 \end{bmatrix}$

12. $[-9]$

13. The sum of an $n \times m$ matrix and its additive inverse is _____ .

14. If A is a 5×2 matrix and $A + K = A$, what do you know about K?

Find the values of the variables in each equation.

15. $\begin{bmatrix} 2 & 1 \\ 4 & 8 \end{bmatrix} = \begin{bmatrix} x & 1 \\ y & z \end{bmatrix}$

16. $\begin{bmatrix} -5 \\ y \end{bmatrix} = \begin{bmatrix} -5 \\ 8 \end{bmatrix}$

17. $\begin{bmatrix} x+6 & y+2 \\ 8 & 3 \end{bmatrix} = \begin{bmatrix} -9 & 7 \\ 8 & k \end{bmatrix}$

18. $\begin{bmatrix} 9 & 7 \\ r & 0 \end{bmatrix} = \begin{bmatrix} m-3 & n+5 \\ 8 & 0 \end{bmatrix}$

19. $\begin{bmatrix} -7+z & 4r & 8s \\ 6p & 2 & 5 \end{bmatrix} + \begin{bmatrix} -9 & 8r & 3 \\ 2 & 5 & 4 \end{bmatrix} = \begin{bmatrix} 2 & 36 & 27 \\ 20 & 7 & 12a \end{bmatrix}$

20. $\begin{bmatrix} a+2 & 3z+1 & 5m \\ 4k & 0 & 3 \end{bmatrix} + \begin{bmatrix} 3a & 2z & 5m \\ 2k & 5 & 6 \end{bmatrix} = \begin{bmatrix} 10 & -14 & 80 \\ 10 & 5 & 9 \end{bmatrix}$

Perform the indicated operations where possible.

21. $\begin{bmatrix} 1 & 2 & 5 & -1 \\ 3 & 0 & 2 & -4 \end{bmatrix} + \begin{bmatrix} 8 & 10 & -5 & 3 \\ -2 & -1 & 0 & 0 \end{bmatrix}$

22. $\begin{bmatrix} 1 & 5 \\ 2 & -3 \\ 3 & 7 \end{bmatrix} + \begin{bmatrix} 2 & 3 \\ 8 & 5 \\ -1 & 9 \end{bmatrix}$

23. $\begin{bmatrix} 1 & 3 & -2 \\ 4 & 7 & 1 \end{bmatrix} + \begin{bmatrix} 3 & 0 \\ 6 & 4 \\ -5 & 2 \end{bmatrix}$

24. $\begin{bmatrix} 1 & 3 & -2 \\ 4 & 7 & 1 \end{bmatrix} - \begin{bmatrix} 3 & 6 & -5 \\ 0 & 4 & 2 \end{bmatrix}$

25. $\begin{bmatrix} 2 & 8 & 12 & 0 \\ 7 & 4 & -1 & 5 \\ 1 & 2 & 0 & 10 \end{bmatrix} - \begin{bmatrix} 1 & 3 & 6 & 9 \\ 2 & -3 & -3 & 4 \\ 8 & 0 & -2 & 17 \end{bmatrix}$

26. $\begin{bmatrix} 2 & 1 \\ 5 & -3 \\ -7 & 2 \\ 9 & 0 \end{bmatrix} + \begin{bmatrix} 1 & -8 & 0 \\ 5 & 3 & 2 \\ -6 & 7 & -5 \\ 2 & -1 & 0 \end{bmatrix}$

27. $\begin{bmatrix} 2 & 3 \\ -2 & 4 \end{bmatrix} + \begin{bmatrix} 4 & 3 \\ 7 & 8 \end{bmatrix} - \begin{bmatrix} 3 & 2 \\ 1 & 4 \end{bmatrix}$

28. $\begin{bmatrix} 4 & 3 \\ 1 & 2 \end{bmatrix} - \begin{bmatrix} 1 & 1 \\ 1 & 0 \end{bmatrix} + \begin{bmatrix} 1 & 1 \\ 1 & 4 \end{bmatrix}$

29. $\begin{bmatrix} 1 & 5 \\ -3 & 7 \end{bmatrix} - \begin{bmatrix} 6 & 3 \\ 2 & 4 \end{bmatrix} + \begin{bmatrix} 8 & 10 \\ -1 & 0 \end{bmatrix}$

30. $\begin{bmatrix} -1 & -1 \\ -1 & 0 \end{bmatrix} + \begin{bmatrix} 4 & 3 \\ 1 & 2 \end{bmatrix} + \begin{bmatrix} 1 & 1 \\ 1 & 4 \end{bmatrix}$

31. $\begin{bmatrix} -4x+2y & -3x+y \\ 6x-3y & 2x-5y \end{bmatrix} + \begin{bmatrix} -8x+6y & 2x \\ 3y-5x & 6x+4y \end{bmatrix}$

32. $\begin{bmatrix} 4k-8y \\ 6z-3x \\ 2k+5a \\ -4m+2n \end{bmatrix} - \begin{bmatrix} 5k+6y \\ 2z+5x \\ 4k+6a \\ 4m-2n \end{bmatrix}$

33. For matrix X given below, find the matrix $-X$.

Using matrices $O = \begin{bmatrix} 0 & 0 \\ 0 & 0 \end{bmatrix}$, $P = \begin{bmatrix} m & n \\ p & q \end{bmatrix}$, $T = \begin{bmatrix} r & s \\ t & u \end{bmatrix}$, *and* $X = \begin{bmatrix} x & y \\ z & w \end{bmatrix}$, *verify the statements in Exercises 34–37.*

34. $X + T = T + X$ (commutative property of addition of matrices)

35. $X + (T + P) = (X + T) + P$ (associative property of addition of matrices)

36. $X + (-X) = O$ (inverse property of addition of matrices)

37. $P + O = P$ (identity property of addition of matrices)

38. Which of the above properties are valid for matrices that are not square?

Applications

BUSINESS AND ECONOMICS

39. *Management* A toy company has plants in Boston, Chicago, and Seattle that manufacture toy phones and calculators. The following matrix gives the production costs (in dollars) for each item at the Boston plant:

$$\begin{array}{c} \\ \text{Material} \\ \text{Labor} \end{array} \begin{array}{c} \text{Phones} \quad \text{Calculators} \\ \begin{bmatrix} 4.27 & 6.94 \\ 3.45 & 3.65 \end{bmatrix} \end{array}$$

a. In Chicago, a phone costs $4.05 for material and $3.27 for labor; a calculator costs $7.01 for material and $3.51 for labor. In Seattle, material costs are $4.40 for a phone and $6.90 for a calculator; labor costs are $3.54 for a phone and $3.76 for a calculator. Write the production cost matrices for Chicago and Seattle.

b. Suppose labor costs increase by $.11 per item in Chicago and material costs there increase by $.37 for a phone and $.42 for a calculator. What is the new production cost matrix for Chicago?

40. *Management* There are three convenience stores in Folsom. This week, Store I sold 88 loaves of bread, 48 quarts of milk, 16 jars of peanut butter, and 112 pounds of cold cuts. Store II sold 105 loaves of bread, 72 quarts of milk, 21 jars of peanut butter, and 147 pounds of cold cuts. Store III sold 60 loaves of bread, 40 quarts of milk, no peanut butter, and 50 pounds of cold cuts.

a. Use a 4×3 matrix to express the sales information for the three stores.

b. During the following week, sales on these products at Store I increased by 25%; sales at Store II increased by 1/3; and sales at Store III increased by 10%. Write the sales matrix for that week.

c. Write a matrix that represents total sales over the two-week period.

LIFE SCIENCES

41. *Dietetics* A dietician prepares a diet specifying the amounts a patient should eat of four basic food groups: group I, meats; group II, fruits and vegetables; group III, breads and starches; group IV, milk products. Amounts are given in "exchanges" that represent 1 oz (meat), 1/2 cup (fruits and vegetables), 1 slice (bread), 8 oz (milk), or other suitable measurements.

a. The number of "exchanges" for breakfast for each of the four food groups, respectively, are 2, 1, 2, and 1; for lunch, 3, 2, 2, and 1; and for dinner, 4, 3, 2, and 1. Write a 3×4 matrix using this information.

b. The amounts of fat, carbohydrates, and protein (in appropriate units) in each food group, respectively, are

as follows.

Fat: 5, 0, 0, 10
Carbohydrates: 0, 10, 15, 12
Protein: 7, 1, 2, 8

Use this information to write a 4×3 matrix.

c. There are 8 calories per exchange of fat, 4 calories per exchange of carbohydrates, and 5 calories per exchange of protein; summarize this data in a 3×1 matrix.

42. *Animal Growth* At the beginning of a laboratory experiment, five baby rats measured 5.6, 6.4, 6.9, 7.6, and 6.1 centimeters in length, and weighed 144, 138, 149, 152, and 146 grams, respectively.

a. Write a 2×5 matrix using this information.

b. At the end of two weeks, their lengths in centimeters were 10.2, 11.4, 11.4, 12.7, and 10.8 and their weights (in grams) were 196, 196, 225, 250, and 230. Write a 2×5 matrix with this information.

c. Use matrix subtraction and the matrices found in parts a and b to write a matrix that gives the amount of change in length and weight for each rat.

d. During the third week, the rats grew by the amounts shown in the matrix below.

$$\begin{array}{c} \text{Length} \\ \text{Weight} \end{array} \begin{bmatrix} 1.8 & 1.5 & 2.3 & 1.8 & 2.0 \\ 25 & 22 & 29 & 33 & 20 \end{bmatrix}$$

What were their lengths and weights at the end of this week?

43. *Testing Medication* A drug company is testing 200 patients to see if Painfree (a new headache medicine) is effective. Half the patients receive Painfree and half receive a placebo. The data on the first 50 patients is summarized in this matrix:

$$\begin{array}{c} \\ \text{Painfree} \\ \text{Placebo} \end{array} \begin{array}{c} \text{Pain Relief Obtained} \\ \begin{array}{cc} \text{Yes} & \text{No} \end{array} \\ \begin{bmatrix} 22 & 3 \\ 8 & 17 \end{bmatrix} \end{array}.$$

a. Of those who took the placebo, how many got relief?

b. Of those who took the new medication, how many got no relief?

c. The test was repeated on three more groups of 50 patients each, with the results summarized by these matrices.

$$\begin{bmatrix} 21 & 4 \\ 6 & 19 \end{bmatrix} \quad \begin{bmatrix} 19 & 6 \\ 10 & 15 \end{bmatrix} \quad \begin{bmatrix} 23 & 2 \\ 3 & 22 \end{bmatrix}$$

Find the total results for all 200 patients.

d. On the basis of these results, does it appear that Painfree is effective?

44. *Car Accidents* The following tables give the death rates, per million person trips, for male and female drivers for various ages and number of passengers.*

Male Drivers	Number of Passengers			
Age	0	1	2	≥ 3
16	2.61	4.39	6.29	9.08
17	1.63	2.77	4.61	6.92
30–59	.92	.75	.62	.54

Female Drivers	Number of Passengers			
Age	0	1	2	≥ 3
16	1.38	1.72	1.94	3.31
17	1.26	1.48	2.82	2.28
30–59	.41	.33	.27	.40

a. Write a matrix for the death rate of male drivers.

b. Write a matrix for the death rate of female drivers.

c. Use the matrices from parts a and b to write a matrix showing the difference between the death rates of males and females.

d. Analyze the results of part c and make some suggestions on how to solve this problem.

45. *Life Expectancy* The following table gives the life expectancy for African American males and females and white American males and females for various times in the last 30 years.[†]

	Black		White	
Year	M	F	M	F
1970	60.0	68.3	68.0	75.6
1980	63.8	72.5	70.7	78.1
1990	64.5	73.6	72.7	79.4
1997	67.2	74.7	74.3	79.9

a. Write a matrix for the life expectancy of African Americans.

b. Write a matrix for the life expectancy of white Americans.

c. Use the matrices from parts a and b to write a matrix showing the difference between the life expectancy between the two groups.

d. Analyze the results of part c and make some suggestions on how to solve this problem.

SOCIAL SCIENCES

46. *Educational Attainment* The following table gives the educational attainment of the U.S. population 25 years and older.[‡]

	Male		Female	
	Percent with Four Years of High School or More	Percent with Four Years of College or More	Percent with Four Years of High School or More	Percent with Four Years of College or More
1960	39.5	9.7	42.5	5.8
1970	51.9	13.5	52.8	8.1
1980	67.3	20.1	65.8	12.8
1990	77.7	24.4	77.5	18.4
1995	81.7	26.0	81.6	20.2
1998	82.8	26.5	82.9	22.4

a. Write a matrix for the educational attainment of males.

b. Write a matrix for the educational attainment of females.

c. Use the matrices from parts a and b to write a matrix showing how much more (or less) education males have attained than females.

---

*Chen, Li-Hui, Susan Baker, Elisa Braver, and Guohua Li, "Carrying Passengers as a Risk Factor for Crashes Fatal to 16- and 17-Year-Old Drivers," *JAMA,* Vol. 283, No. 12, Mar. 22/29, 2000, pp. 1578–1582.
[†]*National Vital Statistics Reports,* Vol. 47, No. 19, June 30, 1999, Table 6.
[‡]"Educational Attainment, by Race, Hispanic Origin and Sex: 1960–1998," *Statistical Abstracts of the United States: 1999,* Table 264.

47. *Educational Attainment* The following table gives the educational attainment of African Americans 25 years and older.*

	Male		Female	
	Percent with Four Years of High School or More	Percent with Four Years of College or More	Percent with Four Years of High School or More	Percent with Four Years of College or More
1960	18.2	2.8	21.8	3.3
1970	30.1	4.2	32.5	4.6
1980	50.8	8.4	51.5	8.3
1990	65.8	11.9	66.5	10.8
1995	73.4	13.6	74.1	12.9
1998	75.2	13.9	76.7	15.4

a. Write a matrix for the educational attainment of African American males.

b. Write a matrix for the educational attainment of African American females.

c. Use the matrices from parts a and b to write a matrix showing how more (or less) education African American males have attained than African American females.

GENERAL INTEREST

48. *Animal Interactions* When two kittens named Cauchy and Cliché were introduced into a household with Jamie (an older cat) and Musk (a dog), the interactions between animals were complicated. The two kittens liked each other and Jamie, but didn't like Musk. Musk liked everybody, but Jamie didn't like any of the other animals.

a. Write a 4 × 4 matrix in which rows (and columns) 1, 2, 3, and 4 refer to Musk, Jamie, Cauchy, and Cliché. Make an element a 1 if the animal for that row likes the animal for that column, and otherwise make the element a 0. Assume every animal likes herself.

b. Within a few days, Cauchy and Cliché decided that they liked Musk after all. Write a 4 × 4 matrix, as you did in part a, representing the new situation.

■ 2.4 MULTIPLICATION OF MATRICES

THINK ABOUT IT   What is a contractor's total cost for materials required for various types of model homes?

Matrix multiplication will be used to answer this question in Example 5. We begin by defining the product of a real number and a matrix. In work with matrices, a real number is called a **scalar.**

> **PRODUCT OF A MATRIX AND A SCALAR**
> The product of a scalar k and a matrix X is the matrix kX, each of whose elements is k times the corresponding element of X.

---

*"Educational Attainment, by Race, Hispanic Origin and Sex: 1960–1998," *Statistical Abstracts of the United States: 1999,* Table 264.

For example,

$$(-3)\begin{bmatrix} 2 & -5 \\ 1 & 7 \end{bmatrix} = \begin{bmatrix} -6 & 15 \\ -3 & -21 \end{bmatrix}.$$

Finding the product of two matrices is more involved, but such multiplication is important in solving practical problems. To understand the reasoning behind matrix multiplication, it may be helpful to consider another example concerning EZ Life Company discussed in the previous section. Suppose sofas and chairs of the same model are often sold as sets. Matrix W shows the number of sets of each model in each warehouse.

$$
\begin{array}{c}
 \\
\text{New York} \\
\text{Chicago} \\
\text{San Francisco}
\end{array}
\begin{array}{ccc}
\text{A} & \text{B} & \text{C} \\
\begin{bmatrix} 10 & 7 & 3 \\ 5 & 9 & 6 \\ 4 & 8 & 2 \end{bmatrix} = W
\end{array}
$$

If the selling price of a model-A set is $800, of a model-B set $1000, and of a model-C set $1200, the total value of the sets in the New York warehouse is found as follows.

Type	Number of Sets		Price of Set		Total
A	10	×	$800	=	$8000
B	7	×	$1000	=	$7000
C	3	×	$1200	=	$3600
		(Total for New York)			$18,600

The total value of the three kinds of sets in New York is $18,600.

The work done in the table above is summarized as follows:

$$10(\$800) + 7(\$1000) + 3(\$1200) = \$18,600.$$

In the same way, we find that the Chicago sets have a total value of

$$5(\$800) + 9(\$1000) + 6(\$1200) = \$20,200,$$

and in San Francisco, the total value of the sets is

$$4(\$800) + 8(\$1000) + 2(\$1200) = \$13,600.$$

The selling prices can be written as a column matrix P, and the total value in each location as another column matrix, V.

$$\begin{bmatrix} 800 \\ 1000 \\ 1200 \end{bmatrix} = P \qquad \begin{bmatrix} 18,600 \\ 20,200 \\ 13,600 \end{bmatrix} = V$$

Look at the elements of W and P below; multiplying the first, second, and third elements of the first row of W by the first, second, and third elements, respectively, of the column matrix P and then adding these products gives the first element in V. Doing the same thing with the second row of W gives the second element of V; the third row of W leads to the third element of V, suggesting that it is reasonable to write the product of matrices

$$W = \begin{bmatrix} 10 & 7 & 3 \\ 5 & 9 & 6 \\ 4 & 8 & 2 \end{bmatrix} \qquad \text{and} \qquad P = \begin{bmatrix} 800 \\ 1000 \\ 1200 \end{bmatrix}$$

as

$$WP = \begin{bmatrix} 10 & 7 & 3 \\ 5 & 9 & 6 \\ 4 & 8 & 2 \end{bmatrix} \begin{bmatrix} 800 \\ 1000 \\ 1200 \end{bmatrix} = \begin{bmatrix} 18,600 \\ 20,200 \\ 13,600 \end{bmatrix} = V.$$

The product was found by multiplying the elements of *rows* of the matrix on the left and the corresponding elements of the *column* of the matrix on the right, and then finding the sum of these separate products. Notice that the product of a 3×3 matrix and a 3×1 matrix is a 3×1 matrix.

The product AB of an $m \times n$ matrix A and an $n \times k$ matrix B is found as follows. Multiply each element of the *first row* of A by the corresponding element of the *first column* of B. The sum of these n products is the *first-row, first-column* element of AB. Similarly, the sum of the products found by multiplying the elements of the *first row* of A by the corresponding elements of the *second column* of B gives the *first-row, second-column* element of AB, and so on.

PRODUCT OF TWO MATRICES

Let A be an $m \times n$ matrix and let B be an $n \times k$ matrix. To find the element in the ith row and jth column of the **product matrix** AB, multiply each element in the ith row of A by the corresponding element in the jth column of B, and then add these products. The product matrix AB is an $m \times k$ matrix.

EXAMPLE 1   Matrix Product

Find the product AB of matrices

$$A = \begin{bmatrix} 2 & 3 & -1 \\ 4 & 2 & 2 \end{bmatrix} \quad \text{and} \quad B = \begin{bmatrix} 1 \\ 8 \\ 6 \end{bmatrix}.$$

Solution

Step 1 Multiply the elements of the first row of A and the corresponding elements of the column of B.

$$\begin{bmatrix} \mathbf{2} & \mathbf{3} & \mathbf{-1} \\ 4 & 2 & 2 \end{bmatrix} \begin{bmatrix} \mathbf{1} \\ \mathbf{8} \\ \mathbf{6} \end{bmatrix} \quad 2 \cdot \mathbf{1} + 3 \cdot \mathbf{8} + (\mathbf{-1}) \cdot \mathbf{6} = 20$$

Thus, 20 is the first-row entry of the product matrix AB.

Step 2 Multiply the elements of the second row of A and the corresponding elements of B.

$$\begin{bmatrix} 2 & 3 & -1 \\ \mathbf{4} & \mathbf{2} & \mathbf{2} \end{bmatrix} \begin{bmatrix} \mathbf{1} \\ \mathbf{8} \\ \mathbf{6} \end{bmatrix} \quad 4 \cdot \mathbf{1} + 2 \cdot \mathbf{8} + 2 \cdot \mathbf{6} = 32$$

The second-row entry of the product matrix AB is 32.

Step 3 Write the product as a column matrix using the two entries found above.

$$AB = \begin{bmatrix} 2 & 3 & -1 \\ 4 & 2 & 2 \end{bmatrix} \begin{bmatrix} 1 \\ 8 \\ 6 \end{bmatrix} = \begin{bmatrix} 20 \\ 32 \end{bmatrix}$$

EXAMPLE 2 Matrix Product

Find the product CD of matrices

$$C = \begin{bmatrix} -3 & 4 & 2 \\ 5 & 0 & 4 \end{bmatrix} \quad \text{and} \quad D = \begin{bmatrix} -6 & 4 \\ 2 & 3 \\ 3 & -2 \end{bmatrix}.$$

Solution

Step 1

$$\begin{bmatrix} -3 & 4 & 2 \\ 5 & 0 & 4 \end{bmatrix} \begin{bmatrix} -6 & 4 \\ 2 & 3 \\ 3 & -2 \end{bmatrix} \quad (-3) \cdot (-6) + 4 \cdot 2 + 2 \cdot 3 = 32$$

Step 2

$$\begin{bmatrix} -3 & 4 & 2 \\ 5 & 0 & 4 \end{bmatrix} \begin{bmatrix} -6 & 4 \\ 2 & 3 \\ 3 & -2 \end{bmatrix} \quad (-3) \cdot 4 + 4 \cdot 3 + 2 \cdot (-2) = -4$$

Step 3

$$\begin{bmatrix} -3 & 4 & 2 \\ 5 & 0 & 4 \end{bmatrix} \begin{bmatrix} -6 & 4 \\ 2 & 3 \\ 3 & -2 \end{bmatrix} \quad 5 \cdot (-6) + 0 \cdot 2 + 4 \cdot 3 = -18$$

Step 4

$$\begin{bmatrix} -3 & 4 & 2 \\ 5 & 0 & 4 \end{bmatrix} \begin{bmatrix} -6 & 4 \\ 2 & 3 \\ 3 & -2 \end{bmatrix} \quad 5 \cdot 4 + 0 \cdot 3 + 4 \cdot (-2) = 12$$

Step 5 The product is

$$CD = \begin{bmatrix} -3 & 4 & 2 \\ 5 & 0 & 4 \end{bmatrix} \begin{bmatrix} -6 & 4 \\ 2 & 3 \\ 3 & -2 \end{bmatrix} = \begin{bmatrix} 32 & -4 \\ -18 & 12 \end{bmatrix}.$$

Here the product of a 2×3 matrix and a 3×2 matrix is a 2×2 matrix. ▪

NOTE One way to avoid errors in matrix multiplication is to lower the first matrix so it is below and to the left of the second matrix, and then write the product in the space between the two matrices. For example, to multiply the matrices in Example 2, we could rewrite the product as shown below.

$$\begin{array}{c} \downarrow \\ \begin{bmatrix} -6 & 4 \\ 2 & 3 \\ 3 & -2 \end{bmatrix} \\ \rightarrow \begin{bmatrix} -3 & 4 & 2 \\ 5 & 0 & 4 \end{bmatrix} \begin{bmatrix} \\ * \end{bmatrix} \end{array}$$

To find the entry where the $*$ is, for example, multiply the row and the column indicated by the arrows: $5 \cdot (-6) + 0 \cdot 2 + 4 \cdot 3 = -18$.

As the definition of matrix multiplication shows,

the product AB of two matrices A and B can be found only if the number of columns of A is the same as the number of rows of B.

The final product will have as many rows as A and as many columns as B.

EXAMPLE 3 Matrix Product

Suppose matrix A is 2×2 and matrix B is 2×4. Can the products AB and BA be calculated? If so, what is the size of each product?

Solution The following diagram helps decide the answers to these questions.

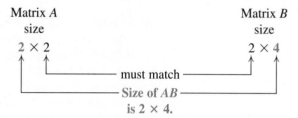

The product of A and B can be found because A has two columns and B has two rows. The size of the product is 2×4.

Matrix B size   Matrix A size

2×4 2×2

do not match

The product BA cannot be found because B has 4 columns and A has 2 rows.

EXAMPLE 4 Comparing Matrix Products AB and BA

Find AB and BA, given

$$A = \begin{bmatrix} 1 & -3 \\ 7 & 2 \\ -2 & 5 \end{bmatrix} \quad \text{and} \quad B = \begin{bmatrix} 1 & 0 & -1 \\ 3 & 1 & 4 \end{bmatrix}.$$

Method 1: Calculating by Hand

Solution

$$AB = \begin{bmatrix} 1 & -3 \\ 7 & 2 \\ -2 & 5 \end{bmatrix} \begin{bmatrix} 1 & 0 & -1 \\ 3 & 1 & 4 \end{bmatrix}$$

$$= \begin{bmatrix} -8 & -3 & -13 \\ 13 & 2 & 1 \\ 13 & 5 & 22 \end{bmatrix}$$

$$BA = \begin{bmatrix} 1 & 0 & -1 \\ 3 & 1 & 4 \end{bmatrix} \begin{bmatrix} 1 & -3 \\ 7 & 2 \\ -2 & 5 \end{bmatrix}$$

$$= \begin{bmatrix} 3 & -8 \\ 2 & 13 \end{bmatrix}$$

Method 2: Graphing Calculators

Matrix multiplication is easily performed on a graphing calculator. Figure 13 shows the results. The matrices A and B were already entered into the calculator.

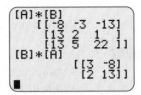

```
[A]*[B]
   [[-8 -3 -13]
    [13  2   1 ]
    [13  5  22 ]]
[B]*[A]
          [[3 -8]
           [2 13]]
■
```

FIGURE 13

Matrix multiplication can also be easily done with a spreadsheet. See *The Spreadsheet Manual* that is available with this textbook for details.

Notice in Example 4 that $AB \neq BA$; AB and BA aren't even the same size. In Example 3, we showed that they may not both exist. This means that matrix multiplication is *not* commutative. Even if both A and B are square matrices, in general, matrices AB and BA are not equal. (See Exercise 31.) Of course, there may be special cases in which they are equal, but this is not true in general.

CAUTION Since matrix multiplication is not commutative, always be careful to multiply matrices in the correct order.

Matrix multiplication *is* associative, however. For example, if

$$C = \begin{bmatrix} 2 & 1 \\ 3 & 4 \\ 1 & 5 \end{bmatrix},$$

then $(AB)C = A(BC)$, where A and B are the matrices given in Example 4. (Verify this.) Also, there is a distributive property of matrices such that, for appropriate matrices A, B, and C,

$$A(B + C) = AB + AC.$$

(See Exercises 32 and 33.) Other properties of matrix multiplication involving scalars are included in the exercises. Multiplicative inverses and multiplicative identities are defined in the next section.

EXAMPLE 5 Home Construction

A contractor builds three kinds of houses, models A, B, and C, with a choice of two styles, Spanish and contemporary. Matrix P shows the number of each kind of house planned for a new 100-home subdivision. The amounts for each of the exterior materials depend primarily on the style of the house. These amounts are shown in matrix Q. (Concrete is in cubic yards, lumber in units of 1000 board feet, brick in 1000s, and shingles in units of 100 square feet.) Matrix R gives the cost in dollars for each kind of material.

$$\begin{array}{c} \\ \text{Model A} \\ \text{Model B} \\ \text{Model C} \end{array} \begin{array}{cc} \text{Spanish} & \text{Contemporary} \end{array} \\ \begin{bmatrix} 0 & 30 \\ 10 & 20 \\ 20 & 20 \end{bmatrix} = P$$

$$\begin{array}{c} \text{Concrete} \quad \text{Lumber} \quad \text{Brick} \quad \text{Shingles} \\ \begin{array}{c} \text{Spanish} \\ \text{Contemporary} \end{array} \left[\begin{array}{cccc} 10 & 2 & 0 & 2 \\ 50 & 1 & 20 & 2 \end{array} \right] = Q \end{array}$$

$$\begin{array}{c} \text{Cost per Unit} \\ \begin{array}{c} \text{Concrete} \\ \text{Lumber} \\ \text{Brick} \\ \text{Shingles} \end{array} \left[\begin{array}{c} 20 \\ 180 \\ 60 \\ 25 \end{array} \right] = R \end{array}$$

(a) What is the total cost of these materials for each model?

Solution To find the cost for each model, first find PQ, which shows the amount of each material needed for each model.

$$PQ = \begin{bmatrix} 0 & 30 \\ 10 & 20 \\ 20 & 20 \end{bmatrix} \begin{bmatrix} 10 & 2 & 0 & 2 \\ 50 & 1 & 20 & 2 \end{bmatrix}$$

$$= \begin{array}{c} \text{Concrete} \quad \text{Lumber} \quad \text{Brick} \quad \text{Shingles} \\ \begin{bmatrix} 1500 & 30 & 600 & 60 \\ 1100 & 40 & 400 & 60 \\ 1200 & 60 & 400 & 80 \end{bmatrix} \begin{array}{c} \text{Model A} \\ \text{Model B} \\ \text{Model C} \end{array} \end{array}$$

Now multiply PQ and R, the cost matrix, to get the total cost of the exterior materials for each model.

$$\begin{array}{c} \quad\quad\quad\quad\quad\quad\quad\quad\quad \text{Cost} \\ \begin{bmatrix} 1500 & 30 & 600 & 60 \\ 1100 & 40 & 400 & 60 \\ 1200 & 60 & 400 & 80 \end{bmatrix} \begin{bmatrix} 20 \\ 180 \\ 60 \\ 25 \end{bmatrix} = \begin{bmatrix} 72{,}900 \\ 54{,}700 \\ 60{,}800 \end{bmatrix} \begin{array}{c} \text{Model A} \\ \text{Model B} \\ \text{Model C} \end{array} \end{array}$$

The total cost of materials is $72,900 for Model A, $54,700 for Model B, and $60,800 for Model C.

(b) How much of each of the four kinds of material must be ordered?

Solution The totals of the columns of matrix PQ will give a matrix whose elements represent the total amounts of each material needed for the subdivision. Call this matrix T, and write it as a row matrix.

$$T = [3800 \quad 130 \quad 1400 \quad 200]$$

Thus, 3800 cubic yards of concrete, 130,000 board feet of lumber, 1,400,000 bricks, and 20,000 square feet of shingles are needed.

(c) What is the total cost for exterior materials?

Solution For the total cost of all the exterior materials, find the product of matrix T, the matrix showing the total amount of each material, and matrix R, the cost matrix. (To multiply these and get a 1×1 matrix representing total cost, we need a 1×4 matrix multiplied by a 4×1 matrix. This is why T was written as a row matrix in (b) above.)

$$TR = [3800 \quad 130 \quad 1400 \quad 200] \begin{bmatrix} 20 \\ 180 \\ 60 \\ 25 \end{bmatrix} = [188{,}400]$$

The total cost for exterior materials is $188,400.

(d) Suppose the contractor builds the same number of homes in five subdivisions. Calculate the total amount of each exterior material for each model for all five subdivisions.

Solution Multiply PQ by the scalar 5, as follows.

$$5(PQ) = 5 \begin{bmatrix} 1500 & 30 & 600 & 60 \\ 1100 & 40 & 400 & 60 \\ 1200 & 60 & 400 & 80 \end{bmatrix} = \begin{bmatrix} 7500 & 150 & 3000 & 300 \\ 5500 & 200 & 2000 & 300 \\ 6000 & 300 & 2000 & 400 \end{bmatrix}$$

The total amount of concrete needed for Model A homes, for example, is 7500 cubic yards.

Choosing Matrix Notation It is helpful to use a notation that keeps track of the quantities a matrix represents. We will use the notation

meaning of the rows/meaning of the columns,

that is, writing the meaning of the rows first, followed by the meaning of the columns. In Example 5, we would use the notation models/styles for matrix P, styles/materials for matrix Q, and materials/cost for matrix R. In multiplying PQ, we are multiplying models/styles by styles/materials. The result is models/materials. Notice that styles, the common quantity in both P and Q, was eliminated in the product PQ. By this method, the product $(PQ)R$ represents models/cost.

In practical problems this notation helps us decide in which order to multiply matrices so that the results are meaningful. In Example 5(c) either RT or TR can be calculated. Since T represents subdivisions/materials and R represents materials/cost, the product TR gives subdivisions/cost, while the product RT is meaningless.

2.4 EXERCISES

Let $A = \begin{bmatrix} -2 & 4 \\ 0 & 3 \end{bmatrix}$ and $B = \begin{bmatrix} -6 & 2 \\ 4 & 0 \end{bmatrix}$. Find each of the following.

1. $2A$

2. $-3B$

3. $-4B$

4. $5A$

5. $-4A + 5B$

6. $3A - 10B$

In Exercises 7–12, the sizes of two matrices A and B are given. Find the sizes of the product AB and the product BA, whenever these products exist.

7. A is 2×2, and B is 2×2.

8. A is 3×3, and B is 3×3.

9. A is 3×5, and B is 5×2.

10. A is 4×3, and B is 3×6.

11. A is 4×2, and B is 3×4.

12. A is 7×3, and B is 2×7.

13. To find the product matrix AB, the number of _____ of A must be the same as the number of _____ of B.

14. The product matrix AB has the same number of _____ as A and the same number of _____ as B.

Find each of the matrix products in Exercises 15–30 if possible.

15. $\begin{bmatrix} 1 & 2 \\ 3 & 4 \end{bmatrix} \begin{bmatrix} -1 \\ 7 \end{bmatrix}$

16. $\begin{bmatrix} -1 & 5 \\ 7 & 0 \end{bmatrix} \begin{bmatrix} 6 \\ 2 \end{bmatrix}$

17. $\begin{bmatrix} 1 & 5 & 3 \\ -1 & 2 & 7 \end{bmatrix} \begin{bmatrix} 4 \\ 2 \\ -3 \end{bmatrix}$

18. $\begin{bmatrix} 5 & 2 \\ 7 & 6 \\ 1 & 0 \end{bmatrix} \begin{bmatrix} 1 & 4 & 0 \\ 2 & -1 & 2 \end{bmatrix}$

19. $\begin{bmatrix} 5 & 1 \\ 2 & 3 \end{bmatrix} \begin{bmatrix} 3 & -1 & 0 \\ 1 & 0 & 2 \end{bmatrix}$

20. $\begin{bmatrix} 6 & 0 & -4 \\ 1 & 2 & 5 \\ 10 & -1 & 3 \end{bmatrix} \begin{bmatrix} 1 \\ 2 \\ 0 \end{bmatrix}$

21. $\begin{bmatrix} 2 & 2 & -1 \\ 3 & 0 & 1 \end{bmatrix} \begin{bmatrix} 0 & 2 \\ -1 & 4 \\ 0 & 2 \end{bmatrix}$

22. $\begin{bmatrix} -9 & 2 & 1 \\ 3 & 0 & 0 \end{bmatrix} \begin{bmatrix} 2 \\ -1 \\ 4 \end{bmatrix}$

23. $\begin{bmatrix} 1 & 2 \\ 3 & 4 \end{bmatrix} \begin{bmatrix} -1 & 5 \\ 7 & 0 \end{bmatrix}$

24. $\begin{bmatrix} -1 & 5 \\ 7 & 0 \end{bmatrix} \begin{bmatrix} 1 & 2 \\ 3 & 4 \end{bmatrix}$

25. $\begin{bmatrix} -2 & -3 & 7 \\ 1 & 5 & 6 \end{bmatrix} \begin{bmatrix} 1 \\ 2 \\ 3 \end{bmatrix}$

26. $\begin{bmatrix} 6 \\ 5 \\ 4 \end{bmatrix} \begin{bmatrix} -1 & 1 & 1 \end{bmatrix}$

27. $\left(\begin{bmatrix} 4 & 3 \\ 1 & 2 \\ 0 & -5 \end{bmatrix} \begin{bmatrix} 2 & -2 \\ 1 & -1 \end{bmatrix} \right) \begin{bmatrix} 10 \\ 0 \end{bmatrix}$

28. $\begin{bmatrix} 4 & 3 \\ 1 & 2 \\ 0 & -5 \end{bmatrix} \left(\begin{bmatrix} 2 & -2 \\ 1 & -1 \end{bmatrix} \begin{bmatrix} 10 \\ 0 \end{bmatrix} \right)$

29. $\begin{bmatrix} 2 & -2 \\ 1 & -1 \end{bmatrix} \left(\begin{bmatrix} 4 & 3 \\ 1 & 2 \end{bmatrix} + \begin{bmatrix} 7 & 0 \\ -1 & 5 \end{bmatrix} \right)$

30. $\begin{bmatrix} 2 & -2 \\ 1 & -1 \end{bmatrix} \begin{bmatrix} 4 & 3 \\ 1 & 2 \end{bmatrix} + \begin{bmatrix} 2 & -2 \\ 1 & -1 \end{bmatrix} \begin{bmatrix} 7 & 0 \\ -1 & 5 \end{bmatrix}$

31. Let $A = \begin{bmatrix} -2 & 4 \\ 1 & 3 \end{bmatrix}$ and $B = \begin{bmatrix} -2 & 1 \\ 3 & 6 \end{bmatrix}$.

 a. Find AB.

 b. Find BA.

 c. Did you get the same answer in parts a and b?

 d. In general, for matrices A and B such that AB and BA both exist, does AB always equal BA?

Given matrices $P = \begin{bmatrix} m & n \\ p & q \end{bmatrix}$, $X = \begin{bmatrix} x & y \\ z & w \end{bmatrix}$, and $T = \begin{bmatrix} r & s \\ t & u \end{bmatrix}$, verify that the statements in Exercises 32–35 are true. The statements are valid for any matrices whenever matrix multiplication and addition can be carried out. This, of course, depends on the size of the matrices.

32. $(PX)T = P(XT)$ (associative property: see Exercises 27 and 28)

33. $P(X + T) = PX + PT$ (distributive property: see Exercises 29 and 30)

34. $k(X + T) = kX + kT$ for any real number k.

35. $(k + h)P = kP + hP$ for any real numbers k and h.

36. Let I be the matrix $I = \begin{bmatrix} 1 & 0 \\ 0 & 1 \end{bmatrix}$, and let matrices P, X, and T be defined as for Exercises 32–35.

 a. Find IP, PI, and IX.

 b. Without calculating, guess what the matrix IT might be.

 c. Suggest a reason for naming a matrix such as I an *identity matrix*.

37. Show that the system of linear equations

$$2x_1 + 3x_2 + x_3 = 5$$
$$x_1 - 4x_2 + 5x_3 = 8$$

can be written as the matrix equation

$$\begin{bmatrix} 2 & 3 & 1 \\ 1 & -4 & 5 \end{bmatrix} \begin{bmatrix} x_1 \\ x_2 \\ x_3 \end{bmatrix} = \begin{bmatrix} 5 \\ 8 \end{bmatrix}.$$

38. Let $A = \begin{bmatrix} 1 & 2 \\ -3 & 5 \end{bmatrix}$, $X = \begin{bmatrix} x_1 \\ x_2 \end{bmatrix}$, and $B = \begin{bmatrix} -4 \\ 12 \end{bmatrix}$. Show that the equation $AX = B$ represents a linear system of two equations in two unknowns. Solve the system and substitute into the matrix equation to check your results.

Use a computer or graphing calculator and the following matrices to find the matrix products and sums in Exercises 39–41.

$$A = \begin{bmatrix} 2 & 3 & -1 & 5 & 10 \\ 2 & 8 & 7 & 4 & 3 \\ -1 & -4 & -12 & 6 & 8 \\ 2 & 5 & 7 & 1 & 4 \end{bmatrix} \quad B = \begin{bmatrix} 9 & 3 & 7 & -6 \\ -1 & 0 & 4 & 2 \\ -10 & -7 & 6 & 9 \\ 8 & 4 & 2 & -1 \\ 2 & -5 & 3 & 7 \end{bmatrix}$$

$$C = \begin{bmatrix} -6 & 8 & 2 & 4 & -3 \\ 1 & 9 & 7 & -12 & 5 \\ 15 & 2 & -8 & 10 & 11 \\ 4 & 7 & 9 & 6 & -2 \\ 1 & 3 & 8 & 23 & 4 \end{bmatrix} \quad D = \begin{bmatrix} 5 & -3 & 7 & 9 & 2 \\ 6 & 8 & -5 & 2 & 1 \\ 3 & 7 & -4 & 2 & 11 \\ 5 & -3 & 9 & 4 & -1 \\ 0 & 3 & 2 & 5 & 1 \end{bmatrix}$$

39. a. Find *AC*.   **b.** Find *CA*.   **c.** Does *AC = CA*?

40. a. Find *CD*.   **b.** Find *DC*.   **c.** Does *CD = DC*?

41. a. Find *C + D*.   **b.** Find *(C + D)B*.   **c.** Find *CB*.

   d. Find *DB*.   **e.** Find *CB + DB*.   **f.** Does *(C + D)B = CB + DB*?

42. Which property of matrices does Exercise 41 illustrate?

Applications

BUSINESS AND ECONOMICS

43. *Cost Analysis* The four departments of Spangler Enterprises need to order the following amounts of the same products.

	Paper	Tape	Printer Ribbon	Memo Pads	Pens
Department 1	10	4	3	5	6
Department 2	7	2	2	3	8
Department 3	4	5	1	0	10
Department 4	0	3	4	5	5

The unit price (in dollars) of each product is given in the next column for two suppliers.

	Supplier A	Supplier B
Paper	2	3
Tape	1	1
Printer Ribbon	4	3
Memo Pads	3	3
Pens	1	2

a. Use matrix multiplication to get a matrix showing the comparative costs for each department for the products from the two suppliers.

b. Find the total cost over all departments to buy products from each supplier. From which supplier should the company make the purchase?

44. *Cost Analysis* The Mundo Candy Company makes three types of chocolate candy: Cheery Cherry, Mucho Mocha, and Almond Delight. The company produces its products in San Diego, Mexico City, and Managua using two main ingredients: chocolate and sugar.

  a. Each kilogram of Cheery Cherry requires .5 kg of sugar and .2 kg of chocolate; each kilogram of Mucho Mocha requires .4 kg of sugar and .3 kg of chocolate; and each kilogram of Almond Delight requires .3 kg of sugar and .3 kg of chocolate. Put this information into a 2 × 3 matrix, labeling the rows and columns.

  b. The cost of 1 kg of sugar is $4 in San Diego, $2 in Mexico City, and $1 in Managua. The cost of 1 kg of chocolate is $3 in San Diego, $5 in Mexico City, and $7 in Managua. Put this information into a matrix in such a way that when you multiply it with your matrix from part a, you get a matrix representing the ingredient cost of producing each type of candy in each city.

  c. Multiply the matrices in parts a and b, labeling the product matrix.

  d. From part c, what is the combined sugar-and-chocolate cost to produce 1 kg of Mucho Mocha in Managua?

  e. Mundo Candy needs to quickly produce a special shipment of 100 kg of Cheery Cherry, 200 kg of Mucho Mocha, and 500 kg of Almond Delight, and it decides to select one factory to fill the entire order. Use matrix multiplication to determine in which city the total sugar-and-chocolate cost to produce the order is the smallest.

45. *Management* In Exercise 39 from Section 2.3, consider the matrices $\begin{bmatrix} 4.27 & 6.94 \\ 3.45 & 3.65 \end{bmatrix}$, $\begin{bmatrix} 4.05 & 7.01 \\ 3.27 & 3.51 \end{bmatrix}$, and $\begin{bmatrix} 4.40 & 6.90 \\ 3.54 & 3.76 \end{bmatrix}$ for the production costs at the Boston, Chicago, and Seattle plants, respectively.

  a. Assume each plant makes the same number of each item. Write a matrix that expresses the average production costs for all three plants.

  b. In part b of Exercise 39 in Section 2.3, cost increases for the Chicago plant resulted in a new production cost matrix $\begin{bmatrix} 4.42 & 7.43 \\ 3.38 & 3.62 \end{bmatrix}$. Following those cost increases the Boston plant was closed and production divided evenly between the Chicago and Seattle plants. What is the matrix that now expresses the average production cost for the entire country?

46. *House Construction* Consider the matrices P, Q, and R given in Example 5.

  a. Find and interpret the matrix product QR.

  b. Verify that P(QR) is equal to (PQ)R calculated in Example 5.

47. *Shoe Sales* Sal's Shoes and Fred's Footwear both have outlets in California and Arizona. Sal's sells shoes for $80, sandals for $40, and boots for $120. Fred's prices are $60, $30, and $150 for shoes, sandals, and boots, respectively. Half of all sales in California stores are shoes, 1/4 are sandals, and 1/4 are boots. In Arizona the fractions are 1/5 shoes, 1/5 sandals, and 3/5 boots.

  a. Write a 2 × 3 matrix called P representing prices for the two stores and three types of footwear.

  b. Write a 3 × 2 matrix called F representing the fraction of each type of footwear sold in each state.

  c. Only one of the two products PF and FP is meaningful. Determine which one it is, calculate the product, and describe what the entries represent.

48. *Management* In Exercise 40 from Section 2.3, consider the matrix

$$\begin{bmatrix} 88 & 105 & 60 \\ 48 & 72 & 40 \\ 16 & 21 & 0 \\ 112 & 147 & 50 \end{bmatrix}$$

expressing the sales information for the three stores.

  a. Write a 3 × 1 matrix expressing the factors by which sales in each store should be multiplied to reflect the fact that sales increased during the following week by 25%, 1/3, and 10% in Stores I, II, and III, respectively, as described in part b of Exercise 40 from Section 2.3.

  b. Multiply the matrix expressing sales information by the matrix found in part a of this exercise to find the sales for all three stores in the second week.

SOCIAL SCIENCES

49. *World Population* The 1998 birth and death rates per million for several regions and the world population (in millions) by region are given below.*

	Births	Deaths
Africa	.038	.014
Asia	.022	.008
Latin America	.023	.007
North America	.014	.009
Europe	.010	.011

*"Vital Events and Rates by Region and Development Category, 1998" and "World Population by Region and Development Category, 1950–2050" from U.S. Bureau of the Census, *World Population Profile: 1998.*

	Africa	Asia	Latin America	North America	Europe
1960	283	1627	218	199	425
1970	360	2038	286	227	460
1980	468	2498	362	252	484
1990	621	2987	443	278	498
2000	798	3451	523	306	508

a. Write the information in each table as a matrix.

b. Use the matrices from part a to find the total number (in millions) of births and deaths in each year.

c. Using the results of part b, compare the number of births in 1960 and in 2000. Also compare the birth rates from part a. Which gives better information?

d. Using the results of part b, compare the number of deaths in 1980 and in 2000. Discuss how this comparison differs from comparing death rates from part a.

LIFE SCIENCES

50. *Dietetics* In Exercise 41 from Section 2.3, label the matrices

$$\begin{bmatrix} 2 & 1 & 2 & 1 \\ 3 & 2 & 2 & 1 \\ 4 & 3 & 2 & 1 \end{bmatrix}, \quad \begin{bmatrix} 5 & 0 & 7 \\ 0 & 10 & 1 \\ 0 & 15 & 2 \\ 10 & 12 & 8 \end{bmatrix}, \quad \text{and} \quad \begin{bmatrix} 8 \\ 4 \\ 5 \end{bmatrix}$$

found in parts a, b, and c, respectively, X, Y, and Z.

a. Find the product matrix XY. What do the entries of this matrix represent?

b. Find the product matrix YZ. What do the entries represent?

c. Find the products $(XY)Z$ and $X(YZ)$ and verify that they are equal. What do the entries represent?

51. *Car Accidents* In Exercise 44 of the previous section, you constructed matrices that represent the death rates, per million person trips, for both male and female drivers for various ages and number of passengers. Use matrix operations to combine these two matrices to form one matrix that represents the combined death rates for males and females, per million person trips, of drivers of various ages and number of passengers. (*Hint:* Add the two matrices together and then multiply the resulting matrix by the scalar $1/2$.)

52. *Life Expectancy* In Exercise 45 of the previous section, you constructed matrices that represent the life expectancy of African American and white American males and females. Use matrix operations to combine these two matrices to form one matrix that represents the combined life expectancy for African American and white Americans from 1970 through 1997. See the hint for Exercise 51.

53. *Northern Spotted Owl Population** In an attempt to save the endangered northern spotted owl, the U.S. Fish and Wildlife Service imposed strict guidelines for the use of 12 million acres of Pacific Northwest forest. This decision led to a national debate between the logging industry and environmentalists. Mathematical ecologists have created a mathematical model to analyze population dynamics of the northern spotted owl by dividing the female owl population into three categories: juvenile (up to 1 year old), subadult (1 to 2 years), and adult (over 2 years old). By analyzing these three subgroups, it is possible to use the number of females in each subgroup at time n to estimate the number of females in each group at any time $n + 1$ with the following matrix equation:

$$\begin{bmatrix} j_{n+1} \\ s_{n+1} \\ a_{n+1} \end{bmatrix} = \begin{bmatrix} 0 & 0 & .33 \\ .18 & 0 & 0 \\ 0 & .71 & .94 \end{bmatrix} \begin{bmatrix} j_n \\ s_n \\ a_n \end{bmatrix},$$

where j_n is the number of juveniles, s_n is the number of subadults, and a_n is the number of adults at time n.[†]

a. If there are currently 4000 female northern spotted owls made up of 900 juveniles, 500 subadults, and 2600 adults, use a graphing calculator or spreadsheet and matrix operations to determine the total number of female owls for each of the next 5 years. (*Hint:* Round each answer to the nearest whole number after each matrix multiplication.)

*This problem was created by David I. Schneider, University of Maryland.
[†]Lamberson, R., R. McKelvey, B. Noon, and C. Voss, "A Dynamic Analysis of Northern Spotted Owl Viability in a Fragmented Forest Landscape," *Conservation Biology,* Vol. 6, No. 4, Dec. 1992, pp. 505–512.

b. With advanced techniques from linear algebra, it is possible to show that in the long run, the following holds.

$$\begin{bmatrix} j_{n+1} \\ s_{n+1} \\ a_{n+1} \end{bmatrix} \approx .98359 \begin{bmatrix} j_n \\ s_n \\ a_n \end{bmatrix}$$

What can we conclude about the long-term survival of the northern spotted owl?

c. Notice that only 18 percent of the juveniles become subadults. Assuming that, through better habitat management, this number could be increased to 40 percent, rework part a. Discuss possible reasons why only 18 percent of the juveniles become subadults. Under the new assumption, what can you conclude about the long-term survival of the northern spotted owl?

■ 2.5 MATRIX INVERSES

 THINK ABOUT IT

An investor wants to invest a specific amount of money in three types of bonds with restrictions on each investment. How much money should be invested in each type?

This question is answered in Exercise 61 for this section by solving a matrix equation. In this section, we introduce the idea of a matrix inverse, which is comparable to the reciprocal of a real number. This will allow us to solve a matrix equation.

Earlier, we defined a zero matrix as an additive identity matrix with properties similar to those of the real number 0, the additive identity for real numbers. The real number 1 is the *multiplicative* identity for real numbers: for any real number a, $a \cdot 1 = 1 \cdot a = a$. In this section, we define a *multiplicative identity matrix I* that has properties similar to those of the number 1. We then use the definition of matrix I to find the *multiplicative inverse* of any square matrix that has an inverse.

If I is to be the identity matrix, both of the products AI and IA must equal A. This means that an identity matrix exists only for square matrices. The 2×2 **identity matrix** that satisfies these conditions is

$$I = \begin{bmatrix} 1 & 0 \\ 0 & 1 \end{bmatrix}.$$

To check that I, as defined above, is really the 2×2 identity, let

$$A = \begin{bmatrix} a & b \\ c & d \end{bmatrix}.$$

Then AI and IA should both equal A.

$$AI = \begin{bmatrix} a & b \\ c & d \end{bmatrix} \begin{bmatrix} 1 & 0 \\ 0 & 1 \end{bmatrix} = \begin{bmatrix} a(1) + b(0) & a(0) + b(1) \\ c(1) + d(0) & c(0) + d(1) \end{bmatrix} = \begin{bmatrix} a & b \\ c & d \end{bmatrix} = A$$

$$IA = \begin{bmatrix} 1 & 0 \\ 0 & 1 \end{bmatrix} \begin{bmatrix} a & b \\ c & d \end{bmatrix} = \begin{bmatrix} 1(a) + 0(c) & 1(b) + 0(d) \\ 0(a) + 1(c) & 0(b) + 1(d) \end{bmatrix} = \begin{bmatrix} a & b \\ c & d \end{bmatrix} = A$$

This verifies that I has been defined correctly.

It is easy to verify that the identity matrix I is unique. Suppose there is another identity; call it J. Then IJ must equal I, because J is an identity, and IJ must also equal J, because I is an identity. Thus $I = J$.

The identity matrices for 3×3 matrices and 4×4 matrices, respectively, are

$$I = \begin{bmatrix} 1 & 0 & 0 \\ 0 & 1 & 0 \\ 0 & 0 & 1 \end{bmatrix} \quad \text{and} \quad I = \begin{bmatrix} 1 & 0 & 0 & 0 \\ 0 & 1 & 0 & 0 \\ 0 & 0 & 1 & 0 \\ 0 & 0 & 0 & 1 \end{bmatrix}.$$

By generalizing, we can find an $n \times n$ identity matrix for any value of n.

Recall that the multiplicative inverse of the nonzero real number a is $1/a$. The product of a and its multiplicative inverse $1/a$ is 1. Given a matrix A, can a **multiplicative inverse matrix A^{-1}** (read "A-inverse") be found satisfying both

$$AA^{-1} = I \quad \text{and} \quad A^{-1}A = I?$$

For a given matrix, we often can find an inverse matrix by using the row operations of Section 2.2.

> **NOTE** A^{-1} does not mean $1/A$; here, A^{-1} is just the notation for the multiplicative inverse of matrix A. Also, only square matrices can have inverses because both $A^{-1}A$ and AA^{-1} must exist and be equal to I.

If an inverse exists, it is unique. That is, any given square matrix has no more than one inverse. The proof of this is left to Exercise 50 in this section.

As an example, let us find the inverse of

$$A = \begin{bmatrix} 2 & 4 \\ 1 & -1 \end{bmatrix}.$$

Let the unknown inverse matrix be

$$A^{-1} = \begin{bmatrix} x & y \\ z & w \end{bmatrix}.$$

By the definition of matrix inverse, $AA^{-1} = I$, or

$$AA^{-1} = \begin{bmatrix} 2 & 4 \\ 1 & -1 \end{bmatrix} \begin{bmatrix} x & y \\ z & w \end{bmatrix} = \begin{bmatrix} 1 & 0 \\ 0 & 1 \end{bmatrix}.$$

By matrix multiplication,

$$\begin{bmatrix} 2x + 4z & 2y + 4w \\ x - z & y - w \end{bmatrix} = \begin{bmatrix} 1 & 0 \\ 0 & 1 \end{bmatrix}.$$

Setting corresponding elements equal gives the system of equations

$$2x + 4z = 1 \tag{1}$$
$$2y + 4w = 0 \tag{2}$$
$$x - z = 0 \tag{3}$$
$$y - w = 1. \tag{4}$$

Since equations (1) and (3) involve only x and z, while equations (2) and (4) involve only y and w, these four equations lead to two systems of equations,

$$\begin{array}{cc} 2x + 4z = 1 \\ x - z = 0 \end{array} \quad \text{and} \quad \begin{array}{cc} 2y + 4w = 0 \\ y - w = 1. \end{array}$$

Writing the two systems as augmented matrices gives

$$\begin{bmatrix} 2 & 4 & \bigg| & 1 \\ 1 & -1 & \bigg| & 0 \end{bmatrix} \quad \text{and} \quad \begin{bmatrix} 2 & 4 & \bigg| & 0 \\ 1 & -1 & \bigg| & 1 \end{bmatrix}.$$

Each of these systems can be solved by the Gauss-Jordan method. Notice, however, that the elements to the left of the vertical bar are identical. The two systems can be combined into the single matrix

$$\begin{bmatrix} 2 & 4 & \bigg| & 1 & 0 \\ 1 & -1 & \bigg| & 0 & 1 \end{bmatrix}.$$

This is of the form $[A \,|\, I]$. It is solved simultaneously as follows.

$$R_1 + (-2)R_2 \rightarrow R_2 \quad \begin{bmatrix} 2 & 4 & \bigg| & 1 & 0 \\ 0 & 6 & \bigg| & 1 & -2 \end{bmatrix}$$ Get 0 in the second-row, first-column position.

$$-2R_2 + 3R_1 \rightarrow R_1 \quad \begin{bmatrix} 6 & 0 & \bigg| & 1 & 4 \\ 0 & 6 & \bigg| & 1 & -2 \end{bmatrix}$$ Get 0 in the first-row, second-column position.

$$\tfrac{1}{6}R_1 \rightarrow R_1 \quad \begin{bmatrix} 1 & 0 & \bigg| & \tfrac{1}{6} & \tfrac{2}{3} \\ 0 & 1 & \bigg| & \tfrac{1}{6} & -\tfrac{1}{3} \end{bmatrix}$$ Get ones down the diagonal.
$$\tfrac{1}{6}R_2 \rightarrow R_2$$

The numbers in the first column to the right of the vertical bar give the values of x and z. The second column gives the values of y and w. That is,

$$\begin{bmatrix} 1 & 0 & \bigg| & x & y \\ 0 & 1 & \bigg| & z & w \end{bmatrix} = \begin{bmatrix} 1 & 0 & \bigg| & \tfrac{1}{6} & \tfrac{2}{3} \\ 0 & 1 & \bigg| & \tfrac{1}{6} & -\tfrac{1}{3} \end{bmatrix}$$

so that

$$A^{-1} = \begin{bmatrix} x & y \\ z & w \end{bmatrix} = \begin{bmatrix} \tfrac{1}{6} & \tfrac{2}{3} \\ \tfrac{1}{6} & -\tfrac{1}{3} \end{bmatrix}.$$

To check, multiply A by A^{-1}. The result should be I.

$$AA^{-1} = \begin{bmatrix} 2 & 4 \\ 1 & -1 \end{bmatrix} \begin{bmatrix} \tfrac{1}{6} & \tfrac{2}{3} \\ \tfrac{1}{6} & -\tfrac{1}{3} \end{bmatrix} = \begin{bmatrix} \tfrac{1}{3} + \tfrac{2}{3} & \tfrac{4}{3} - \tfrac{4}{3} \\ \tfrac{1}{6} - \tfrac{1}{6} & \tfrac{2}{3} + \tfrac{1}{3} \end{bmatrix} = \begin{bmatrix} 1 & 0 \\ 0 & 1 \end{bmatrix} = I$$

Verify that $A^{-1}A = I$, also. Finally,

$$A^{-1} = \begin{bmatrix} \tfrac{1}{6} & \tfrac{2}{3} \\ \tfrac{1}{6} & -\tfrac{1}{3} \end{bmatrix}.$$

FINDING A MULTIPLICATIVE INVERSE MATRIX

To obtain A^{-1} for any $n \times n$ matrix A for which A^{-1} exists, follow these steps.

1. Form the augmented matrix $[A \,|\, I]$, where I is the $n \times n$ identity matrix.

2. Perform row operations on $[A \,|\, I]$ to get a matrix of the form $[I \,|\, B]$.

3. Matrix B is A^{-1}.

EXAMPLE 1   Inverse Matrix

Find A^{-1} if $A = \begin{bmatrix} 1 & 0 & 1 \\ 2 & -2 & -1 \\ 3 & 0 & 0 \end{bmatrix}.$

Solution

Method 1: Calculating by Hand   Write the augmented matrix $[A \mid I]$.

$$[A \mid I] = \left[\begin{array}{ccc|ccc} 1 & 0 & 1 & 1 & 0 & 0 \\ 2 & -2 & -1 & 0 & 1 & 0 \\ 3 & 0 & 0 & 0 & 0 & 1 \end{array}\right]$$

Begin by selecting the row operation that produces a zero for the first element in row 2.

$$\begin{array}{c} -2R_1 + R_2 \rightarrow R_2 \\ -3R_1 + R_3 \rightarrow R_3 \end{array} \left[\begin{array}{ccc|ccc} 1 & 0 & 1 & 1 & 0 & 0 \\ 0 & -2 & -3 & -2 & 1 & 0 \\ 0 & 0 & -3 & -3 & 0 & 1 \end{array}\right]$$ Get zeros in the first column.

Column 2 already has zeros in the required positions, so we work on column 3.

$$\begin{array}{c} R_3 + 3R_1 \rightarrow R_1 \\ R_3 + (-1)R_2 \rightarrow R_2 \end{array} \left[\begin{array}{ccc|ccc} 3 & 0 & 0 & 0 & 0 & 1 \\ 0 & 2 & 0 & -1 & -1 & 1 \\ 0 & 0 & -3 & -3 & 0 & 1 \end{array}\right]$$ Get zeros in the third column.

Now get 1's down the main diagonal.

$$\begin{array}{c} \frac{1}{3}R_1 \rightarrow R_1 \\ \frac{1}{2}R_2 \rightarrow R_2 \\ -\frac{1}{3}R_3 \rightarrow R_3 \end{array} \left[\begin{array}{ccc|ccc} 1 & 0 & 0 & 0 & 0 & \frac{1}{3} \\ 0 & 1 & 0 & -\frac{1}{2} & -\frac{1}{2} & \frac{1}{2} \\ 0 & 0 & 1 & 1 & 0 & -\frac{1}{3} \end{array}\right]$$ Get ones down the diagonal.

From the last transformation, the desired inverse is

$$A^{-1} = \left[\begin{array}{ccc} 0 & 0 & \frac{1}{3} \\ -\frac{1}{2} & -\frac{1}{2} & \frac{1}{2} \\ 1 & 0 & -\frac{1}{3} \end{array}\right].$$

Confirm this by forming the products $A^{-1}A$ and AA^{-1}, both of which should equal I.

Method 2: Graphing Calculators   The inverse of A can also be found with a graphing calculator, as shown in Figure 14. (The matrix A had previously been entered into the calculator.) The entire answer can be viewed by pressing the right and left arrow keys on the calculator.

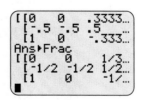

FIGURE 14

Spreadsheets also have the capability of calculating the inverse of a matrix with a simple command. See *The Spreadsheet Manual* that is available with this book for details.

EXAMPLE 2 Inverse Matrix

Find A^{-1} if $A = \begin{bmatrix} 2 & -4 \\ 1 & -2 \end{bmatrix}$.

Solution Using row operations to transform the first column of the augmented matrix

$$\begin{bmatrix} 2 & -4 & | & 1 & 0 \\ 1 & -2 & | & 0 & 1 \end{bmatrix}$$

gives the following results.

$$\mathbf{R_1 + (-2)R_2 \rightarrow R_2} \quad \begin{bmatrix} 2 & -4 & | & 1 & 0 \\ 0 & 0 & | & 1 & -2 \end{bmatrix}$$

Because the last row has all zeros to the left of the vertical bar, there is no way to complete the process of finding the inverse matrix. What is wrong? Just as the real number 0 has no multiplicative inverse, some matrices do not have inverses. Matrix A is an example of a matrix that has no inverse: there is no matrix A^{-1} such that $AA^{-1} = A^{-1}A = I$. ◼

Solving Systems of Equations with Inverses

We used matrices to solve systems of linear equations by the Gauss-Jordan method in Section 2.2. Another way to use matrices to solve linear systems is to write the system as a matrix equation $AX = B$, where A is the matrix of the coefficients of the variables of the system, X is the matrix of the variables, and B is the matrix of the constants. Matrix A is called the **coefficient matrix.**

To solve the matrix equation $AX = B$, first see if A^{-1} exists. Assuming A^{-1} exists and using the facts that $A^{-1}A = I$ and $IX = X$ gives

$$AX = B$$
$$A^{-1}(AX) = A^{-1}B \qquad \text{Multiply both sides by } A^{-1}.$$
$$(A^{-1}A)X = A^{-1}B \qquad \text{Associative property}$$
$$IX = A^{-1}B \qquad \text{Multiplicative inverse property}$$
$$X = A^{-1}B. \qquad \text{Identity property}$$

CAUTION When multiplying by matrices on both sides of a matrix equation, be careful to multiply in the same order on both sides of the equation, since multiplication of matrices is not commutative (unlike multiplication of real numbers).

The work above leads to the following method of solving a system of equations written as a matrix equation.

SOLVING A SYSTEM $AX = B$ USING MATRIX INVERSES

To solve a system of equations $AX = B$, where A is the matrix of coefficients, X is the matrix of variables, and B is the matrix of constants, first find A^{-1}. Then $X = A^{-1}B$.

This method is most practical in solving several systems that have the same coefficient matrix but different constants, as in Example 4 in this section. Then just one inverse matrix must be found.

EXAMPLE 3 Inverse Matrices and Systems of Equations

Use the inverse of the coefficient matrix to solve the linear system

$$2x - 3y = 4.$$
$$x + 5y = 2.$$

Solution To represent the system as a matrix equation, use the coefficient matrix of the system together with the matrix of variables and the matrix of constants:

$$A = \begin{bmatrix} 2 & -3 \\ 1 & 5 \end{bmatrix}, \quad X = \begin{bmatrix} x \\ y \end{bmatrix}, \quad \text{and} \quad B = \begin{bmatrix} 4 \\ 2 \end{bmatrix}.$$

The system can now be written in matrix form as the equation $AX = B$ since

$$AX = \begin{bmatrix} 2 & -3 \\ 1 & 5 \end{bmatrix} \begin{bmatrix} x \\ y \end{bmatrix} = \begin{bmatrix} 2x - 3y \\ x + 5y \end{bmatrix} = \begin{bmatrix} 4 \\ 2 \end{bmatrix} = B.$$

To solve the system, first find A^{-1}. Do this by using row operations on matrix $[A\,|\,I]$ to get

$$\begin{bmatrix} 1 & 0 & \bigg| & \frac{5}{13} & \frac{3}{13} \\ 0 & 1 & \bigg| & -\frac{1}{13} & \frac{2}{13} \end{bmatrix}.$$

From this result,

$$A^{-1} = \begin{bmatrix} \frac{5}{13} & \frac{3}{13} \\ -\frac{1}{13} & \frac{2}{13} \end{bmatrix}.$$

Next, find the product $A^{-1}B$.

$$A^{-1}B = \begin{bmatrix} \frac{5}{13} & \frac{3}{13} \\ -\frac{1}{13} & \frac{2}{13} \end{bmatrix} \begin{bmatrix} 4 \\ 2 \end{bmatrix} = \begin{bmatrix} 2 \\ 0 \end{bmatrix}$$

Since $X = A^{-1}B$,

$$X = \begin{bmatrix} x \\ y \end{bmatrix} = \begin{bmatrix} 2 \\ 0 \end{bmatrix}.$$

The solution of the system is $(2, 0)$.

EXAMPLE 4 Fertilizer

Three brands of fertilizer are available that provide nitrogen, phosphoric acid, and soluble potash to the soil. One bag of each brand provides the following units of each nutrient.

Brand Nutrient	Fertifun	Big Grow	Soakem
Nitrogen	1	2	3
Phosphoric Acid	3	1	2
Potash	2	0	1

For ideal growth, the soil on a Michigan farm needs 18 units of nitrogen, 23 units of phosphoric acid, and 13 units of potash per acre. The corresponding numbers for a California farm are 31, 24, and 11, and for a Kansas farm are 20, 19, and 15. How many bags of each brand of fertilizer should be used per acre for ideal growth on each farm?

Solution Rather than solve three separate systems, consider the single system

$$
\begin{aligned}
x + 2y + 3z &= a \\
3x + y + 2z &= b \\
2x + z &= c,
\end{aligned}
$$

where a, b, and c represent the units of nitrogen, phosphoric acid, and potash needed for the different farms. The system of equations is then of the form $AX = B$, where

$$
A = \begin{bmatrix} 1 & 2 & 3 \\ 3 & 1 & 2 \\ 2 & 0 & 1 \end{bmatrix} \quad \text{and} \quad X = \begin{bmatrix} x \\ y \\ z \end{bmatrix}.
$$

B has different values for the different farms. Find A^{-1} first, then use it to solve all three systems.

To find A^{-1}, we start with the matrix

$$
[A|I] = \begin{bmatrix} 1 & 2 & 3 & | & 1 & 0 & 0 \\ 3 & 1 & 2 & | & 0 & 1 & 0 \\ 2 & 0 & 1 & | & 0 & 0 & 1 \end{bmatrix}
$$

and use row operations to get $[I|A^{-1}]$. The result is

$$
A^{-1} = \begin{bmatrix} -\frac{1}{3} & \frac{2}{3} & -\frac{1}{3} \\ -\frac{1}{3} & \frac{5}{3} & -\frac{7}{3} \\ \frac{2}{3} & -\frac{4}{3} & \frac{5}{3} \end{bmatrix}.
$$

Now we can solve each of the three systems by using $X = A^{-1}B$.

For the Michigan farm, $B = \begin{bmatrix} 18 \\ 23 \\ 13 \end{bmatrix}$, and

$$
X = \begin{bmatrix} -\frac{1}{3} & \frac{2}{3} & -\frac{1}{3} \\ -\frac{1}{3} & \frac{5}{3} & -\frac{7}{3} \\ \frac{2}{3} & -\frac{4}{3} & \frac{5}{3} \end{bmatrix} \begin{bmatrix} 18 \\ 23 \\ 13 \end{bmatrix} = \begin{bmatrix} 5 \\ 2 \\ 3 \end{bmatrix}.
$$

Therefore, $x = 5$, $y = 2$, and $z = 3$. Buy 5 bags of Fertifun, 2 bags of Big Grow, and 3 bags of Soakem.

For the California farm, $B = \begin{bmatrix} 31 \\ 24 \\ 11 \end{bmatrix}$, and

$$
X = \begin{bmatrix} -\frac{1}{3} & \frac{2}{3} & -\frac{1}{3} \\ -\frac{1}{3} & \frac{5}{3} & -\frac{7}{3} \\ \frac{2}{3} & -\frac{4}{3} & \frac{5}{3} \end{bmatrix} \begin{bmatrix} 31 \\ 24 \\ 11 \end{bmatrix} = \begin{bmatrix} 2 \\ 4 \\ 7 \end{bmatrix}.
$$

Buy 2 bags of Fertifun, 4 bags of Big Grow, and 7 bags of Soakem.

For the Kansas farm, $B = \begin{bmatrix} 20 \\ 19 \\ 15 \end{bmatrix}$. Verify that this leads to $x = 1$, $y = -10$, and $z = 13$. We cannot have a negative number of bags, so this solution is impossible. In buying enough bags to meet all of the nutrient requirements the farmer must purchase an excess of some nutrients. In the next two chapters, we will study a method of solving such problems at a minimum cost. ▪

In Example 4, using the matrix inverse method of solving the systems involved considerably less work than using row operations for each of the three systems.

EXAMPLE 5 Inconsistent Systems of Equations
Use the inverse of the coefficient matrix to solve the system

$$2x - 4y = 13$$
$$x - 2y = 1.$$

Solution We saw in Example 2 that the coefficient matrix $\begin{bmatrix} 2 & -4 \\ 1 & -2 \end{bmatrix}$ does not have an inverse. This means that the given system either has no solution or has an infinite number of solutions. Verify that this system is inconsistent and has no solution. ▪

EXAMPLE 6 Cryptography
Throughout the Cold War and as the Internet has grown and developed, the need for sophisticated methods of coding and decoding messages has increased. Although there are many methods of encrypting messages, one fairly sophisticated method uses matrix operations. This method first assigns a number to each letter of the alphabet. The simplest way to do this is to assign the number 1 to A, 2 to B, and so on, with the number 27 used to represent a space between words.

For example, the message *math is cool* can be divided into groups of three letters each and then converted into numbers as follows

$$\begin{bmatrix} m \\ a \\ t \end{bmatrix} = \begin{bmatrix} 13 \\ 1 \\ 20 \end{bmatrix}.$$

The entire message would then consist of four 3×1 columns of numbers:

$$\begin{bmatrix} 13 \\ 1 \\ 20 \end{bmatrix}, \quad \begin{bmatrix} 8 \\ 27 \\ 9 \end{bmatrix}, \quad \begin{bmatrix} 19 \\ 27 \\ 3 \end{bmatrix}, \quad \begin{bmatrix} 15 \\ 15 \\ 12 \end{bmatrix}.$$

This code is easy to break, so we further complicate the code by choosing a matrix that has an inverse (in this case a 3×3 matrix) and calculate the products of the matrix and each of the column vectors above.

If we choose the coding matrix

$$A = \begin{bmatrix} 1 & 3 & 4 \\ 2 & 1 & 3 \\ 4 & 2 & 1 \end{bmatrix},$$

then the products of A with each of the column vectors above produce a new set of vectors

$$\begin{bmatrix} 96 \\ 87 \\ 74 \end{bmatrix}, \quad \begin{bmatrix} 125 \\ 70 \\ 95 \end{bmatrix}, \quad \begin{bmatrix} 112 \\ 74 \\ 133 \end{bmatrix}, \quad \begin{bmatrix} 108 \\ 81 \\ 102 \end{bmatrix}.$$

This set of vectors represents our coded message and it will be transmitted as 96, 87, 74, 125 and so on.

When the intended person receives the message, it is divided into groups of three numbers, and each group is formed into a column matrix. The message is easily decoded if the receiver knows the inverse of the original matrix. The inverse of matrix A is

$$A^{-1} = \begin{bmatrix} -.2 & .2 & .2 \\ .4 & -.6 & .2 \\ 0 & .4 & -.2 \end{bmatrix}.$$

Thus, the message is decoded by taking the product of the inverse matrix with each column vector of the received message. For example,

$$A^{-1} \begin{bmatrix} 96 \\ 87 \\ 74 \end{bmatrix} = \begin{bmatrix} 13 \\ 1 \\ 20 \end{bmatrix}.$$

Unless the original matrix or its inverse is known, this type of code can be difficult to break. In fact, very large matrices can be used to encrypt data. It is interesting to note that many mathematicians are employed by the National Security Agency to develop encryption methods that are virtually unbreakable.

2.5 EXERCISES

In Exercises 1–8, decide whether the given matrices are inverses of each other. (Check to see if their product is the identity matrix I.)

1. $\begin{bmatrix} 2 & 3 \\ 1 & 1 \end{bmatrix}$ and $\begin{bmatrix} -1 & 3 \\ 1 & -2 \end{bmatrix}$

2. $\begin{bmatrix} 5 & 7 \\ 2 & 3 \end{bmatrix}$ and $\begin{bmatrix} 3 & -7 \\ -2 & 5 \end{bmatrix}$

3. $\begin{bmatrix} 2 & 1 \\ 3 & 2 \end{bmatrix}$ and $\begin{bmatrix} 2 & 1 \\ -3 & 2 \end{bmatrix}$

4. $\begin{bmatrix} -1 & 2 \\ 3 & -5 \end{bmatrix}$ and $\begin{bmatrix} -5 & -2 \\ -3 & -1 \end{bmatrix}$

5. $\begin{bmatrix} 1 & 2 & 0 \\ 0 & 1 & 0 \\ 0 & 1 & 0 \end{bmatrix}$ and $\begin{bmatrix} 1 & -2 & 0 \\ 0 & 1 & 0 \\ 0 & -1 & 1 \end{bmatrix}$

6. $\begin{bmatrix} 0 & 1 & 0 \\ 0 & 0 & -2 \\ 1 & -1 & 0 \end{bmatrix}$ and $\begin{bmatrix} 1 & 0 & 1 \\ 1 & 0 & 0 \\ 0 & -1 & 0 \end{bmatrix}$

7. $\begin{bmatrix} 1 & 3 & 3 \\ 1 & 4 & 3 \\ 1 & 3 & 4 \end{bmatrix}$ and $\begin{bmatrix} 7 & -3 & -3 \\ -1 & 1 & 0 \\ -1 & 0 & 1 \end{bmatrix}$

8. $\begin{bmatrix} -1 & 0 & 2 \\ 3 & 1 & 0 \\ 0 & 2 & -3 \end{bmatrix}$ and $\begin{bmatrix} -\frac{1}{5} & \frac{4}{15} & -\frac{2}{15} \\ \frac{3}{5} & \frac{1}{5} & \frac{2}{5} \\ \frac{2}{5} & \frac{2}{15} & -\frac{1}{15} \end{bmatrix}$

9. Does a matrix with a row of all zeros have an inverse? Why?

10. Matrix A has A^{-1} as its inverse. What does $(A^{-1})^{-1}$ equal? (*Hint:* Experiment with a few matrices to see what you get.)

Find the inverse, if it exists, for each matrix.

11. $\begin{bmatrix} 1 & -1 \\ 2 & 0 \end{bmatrix}$
12. $\begin{bmatrix} -1 & 2 \\ -2 & -1 \end{bmatrix}$
13. $\begin{bmatrix} 3 & -1 \\ -5 & 2 \end{bmatrix}$
14. $\begin{bmatrix} -1 & -2 \\ 3 & 4 \end{bmatrix}$

15. $\begin{bmatrix} -6 & 4 \\ -3 & 2 \end{bmatrix}$
16. $\begin{bmatrix} 5 & 10 \\ -3 & -6 \end{bmatrix}$
17. $\begin{bmatrix} 1 & 0 & 0 \\ 0 & -1 & 0 \\ 1 & 0 & 1 \end{bmatrix}$
18. $\begin{bmatrix} 1 & 0 & 1 \\ 0 & -1 & 0 \\ 2 & 1 & 1 \end{bmatrix}$

19. $\begin{bmatrix} -1 & -1 & -1 \\ 4 & 5 & 0 \\ 0 & 1 & -3 \end{bmatrix}$
20. $\begin{bmatrix} 2 & 0 & 4 \\ 3 & 1 & 5 \\ -1 & 1 & -2 \end{bmatrix}$
21. $\begin{bmatrix} 1 & 2 & 3 \\ -3 & -2 & -1 \\ -1 & 0 & 1 \end{bmatrix}$
22. $\begin{bmatrix} 2 & 0 & 4 \\ 1 & 0 & -1 \\ 3 & 0 & -2 \end{bmatrix}$

23. $\begin{bmatrix} 2 & 4 & 6 \\ -1 & -4 & -3 \\ 0 & 1 & -1 \end{bmatrix}$
24. $\begin{bmatrix} 2 & 2 & -4 \\ 2 & 6 & 0 \\ -3 & -3 & 5 \end{bmatrix}$
25. $\begin{bmatrix} 1 & -2 & 3 & 0 \\ 0 & 1 & -1 & 1 \\ -2 & 2 & -2 & 4 \\ 0 & 2 & -3 & 1 \end{bmatrix}$
26. $\begin{bmatrix} 1 & 1 & 0 & 2 \\ 2 & -1 & 1 & -1 \\ 3 & 3 & 2 & -2 \\ 1 & 2 & 1 & 0 \end{bmatrix}$

Solve each system of equations by using the inverse of the coefficient matrix.

27. $\begin{aligned} 2x + 3y &= 10 \\ x - y &= -5 \end{aligned}$

28. $\begin{aligned} -x + 2y &= 15 \\ -2x - y &= 20 \end{aligned}$

29. $\begin{aligned} 2x + y &= 5 \\ 5x + 3y &= 13 \end{aligned}$

30. $\begin{aligned} -x - 2y &= 8 \\ 3x + 4y &= 24 \end{aligned}$

31. $\begin{aligned} -x + y &= 1 \\ 2x - y &= 1 \end{aligned}$

32. $\begin{aligned} 3x - 6y &= 1 \\ -5x + 9y &= -1 \end{aligned}$

33. $\begin{aligned} -x - 8y &= 12 \\ 3x + 24y &= -36 \end{aligned}$

34. $\begin{aligned} x + 3y &= -14 \\ 2x - y &= 7 \end{aligned}$

Solve each system of equations by using the inverse of the coefficient matrix. (The inverses for the first four problems were found in Exercises 19, 20, 23, and 24 above.)

35. $\begin{aligned} -x - y - z &= 1 \\ 4x + 5y &= -2 \\ y - 3z &= 3 \end{aligned}$

36. $\begin{aligned} 2x + 4z &= -8 \\ 3x + y + 5z &= 2 \\ -x + y - 2z &= 4 \end{aligned}$

37. $\begin{aligned} 2x + 4y + 6z &= 4 \\ -x - 4y - 3z &= 8 \\ y - z &= -4 \end{aligned}$

38. $\begin{aligned} 2x + 2y - 4z &= 12 \\ 2x + 6y &= 16 \\ -3x - 3y + 5z &= -20 \end{aligned}$

39. $\begin{aligned} 2x - 2y &= 5 \\ 4y + 8z &= 7 \\ x + 2z &= 1 \end{aligned}$

40. $\begin{aligned} x + z &= 3 \\ y + 2z &= 8 \\ -x + y &= 4 \end{aligned}$

Solve the systems of equations in Exercises 41 and 42 by using the inverse of the coefficient matrix. (The inverses were found in Exercises 25 and 26.)

41. $\begin{aligned} x - 2y + 3z &= 4 \\ y - z + w &= -8 \\ -2x + 2y - 2z + 4w &= 12 \\ 2y - 3z + w &= -4 \end{aligned}$

42. $\begin{aligned} x + y + 2w &= 3 \\ 2x - y + z - w &= 3 \\ 3x + 3y + 2z - 2w &= 5 \\ x + 2y + z &= 3 \end{aligned}$

Let $A = \begin{bmatrix} a & b \\ c & d \end{bmatrix}$ in Exercises 43–48.

43. Show that $IA = A$.

44. Show that $AI = A$.

45. Show that $A \cdot 0 = 0$.

46. Find A^{-1}.
(Assume $ad - bc \neq 0$.)

47. Show that $A^{-1}A = I$.

48. Show that $AA^{-1} = I$.

49. Using the definition and properties listed in this section, show that for square matrices A and B of the same size, if $AB = O$ and if A^{-1} exists, then $B = O$.

50. Prove that, if it exists, the inverse of a matrix is unique. (*Hint:* Assume there are two inverses B and C for some matrix A, so that $AB = BA = I$ and $AC = CA = I$. Multiply the first equation by C and the second by B.)

Use matrices C and D in Exercises 51–55.

$$C = \begin{bmatrix} -6 & 8 & 2 & 4 & -3 \\ 1 & 9 & 7 & -12 & 5 \\ 15 & 2 & -8 & 10 & 11 \\ 4 & 7 & 9 & 6 & -2 \\ 1 & 3 & 8 & 23 & 4 \end{bmatrix}, \quad D = \begin{bmatrix} 5 & -3 & 7 & 9 & 2 \\ 6 & 8 & -5 & 2 & 1 \\ 3 & 7 & -4 & 2 & 11 \\ 5 & -3 & 9 & 4 & -1 \\ 0 & 3 & 2 & 5 & 1 \end{bmatrix}$$

51. Find C^{-1}.   **52.** Find $(CD)^{-1}$.   **53.** Find D^{-1}.   **54.** Is $C^{-1}D^{-1} = (CD)^{-1}$?   **55.** Is $D^{-1}C^{-1} = (CD)^{-1}$?

Solve the matrix equation $AX = B$ for X by finding A^{-1}, given A and B as follows.

56. $A = \begin{bmatrix} 2 & 3 & 5 \\ 1 & 7 & 9 \\ -3 & 2 & 10 \end{bmatrix}$, $B = \begin{bmatrix} 3 \\ 4 \\ 1 \end{bmatrix}$

57. $A = \begin{bmatrix} 2 & 5 & 7 & 9 \\ 1 & 3 & -4 & 6 \\ -1 & 0 & 5 & 8 \\ 2 & -2 & 4 & 10 \end{bmatrix}$, $B = \begin{bmatrix} 3 \\ 7 \\ -1 \\ 5 \end{bmatrix}$

58. $A = \begin{bmatrix} 3 & 2 & -1 & -2 & 6 \\ -5 & 17 & 4 & 3 & 15 \\ 7 & 9 & -3 & -7 & 12 \\ 9 & -2 & 1 & 4 & 8 \\ 1 & 21 & 9 & -7 & 25 \end{bmatrix}$, $B = \begin{bmatrix} -2 \\ 5 \\ 3 \\ -8 \\ 25 \end{bmatrix}$

Applications

BUSINESS AND ECONOMICS

Solve the following exercises by using the inverse of the coefficient matrix to solve a system of equations.

59. *Analysis of Orders* The Bread Box Bakery sells three types of cakes, each requiring the amounts of the basic ingredients shown in the following matrix.

		Type of Cake		
		I	II	III
	Flour (in cups)	2	4	2
Ingredient	Sugar (in cups)	2	1	2
	Eggs	2	1	3

To fill its daily orders for these three kinds of cake, the bakery uses 72 cups of flour, 48 cups of sugar, and 60 eggs.

a. Write a 3×1 matrix for the amounts used daily.

b. Let the number of daily orders for cakes be a 3×1 matrix X with entries x_1, x_2, and x_3. Write a matrix equation that can be solved for X, using the given matrix and the matrix from part a.

c. Solve the equation from part b to find the number of daily orders for each type of cake.

60. *Production Requirements* An electronics company produces transistors, resistors, and computer chips. Each transistor requires 3 units of copper, 1 unit of zinc, and 2 units of glass. Each resistor requires 3, 2, and 1 units of the three materials, and each computer chip requires 2, 1, and 2 units of these materials, respectively. How many of each product can be made with the following amounts of materials?

a. 810 units of copper, 410 units of zinc, and 490 units of glass

b. 765 units of copper, 385 units of zinc, and 470 units of glass

c. 1010 units of copper, 500 units of zinc, and 610 units of glass

61. *Investments* An investment firm recommends that a client invest in AAA, A, and B rated bonds. The average yield on AAA bonds is 6%, on A bonds 7%, and on B bonds 10%. The client wants to invest twice as much in AAA bonds as in B bonds. How much should be invested in each type of bond under the following conditions?

a. The total investment is $25,000, and the investor wants an annual return of $1810 on the three investments.

b. The values in part a are changed to $30,000 and $2150, respectively.

c. The values in part a are changed to $40,000 and $2900, respectively.

62. *Production* Pretzels cost $3 per pound, dried fruit $4 per pound, and nuts $8 per pound. The three ingredients are to be combined in a trail mix containing twice the weight of pretzels as dried fruit. How many pounds of each should be used to produce the following amounts at the given cost?

a. 140 lb at $6 per lb

b. 112 lb at $6.50 per lb

c. 126 lb at $5 per lb

LIFE SCIENCES

63. *Vitamins* Greg Tobin mixes together three types of vitamin tablets. Each Super Vim tablet contains, among other things, 15 mg of niacin and 12 I.U. of Vitamin E. The figures for a Multitab tablet are 20 mg and 15 I.U., and for a Mighty Mix are 25 mg and 35 I.U. How many of each tablet are there if the total number of tablets, total amount of niacin, and total amount of Vitamin E are as follows?

a. 225 tablets, 4750 mg of niacin, and 5225 I.U. of Vitamin E

b. 185 tablets, 3625 mg of niacin, and 3750 I.U. of Vitamin E

c. 230 tablets, 4450 mg of niacin, and 4210 I.U. of Vitamin E

GENERAL INTEREST

64. *Encryption* Use the matrices presented in Example 6 of this section to do the following:

a. Encode the message, "All is fair in love and war."

b. Decode the message 138, 81, 102, 101, 67, 109, 162, 124, 173, 210, 150, 165.

65. *Music* During a marching band's half-time show, the band members generally line up in such a way that a common shape is recognized by the fans. For example, as illustrated in the figure, a band might form a letter T, where an x represents a member of the band. As the music is played the band will either create a new shape or rotate the original shape. In doing this, each member of the band will need to move from one point on the field to another. For larger bands, keeping track of who goes where can be a daunting task. However, it is possible to use matrix inverses to make the process a bit easier.* The entire process is calculated by knowing how three band members, all of whom cannot be in a straight line, will move from the current position to a new position. For example, in the figure, we can see that there are band members at $(50, 0)$, $(50, 15)$, and $(45, 20)$. We will assume that these three band members move to $(40, 10)$, $(55, 10)$, and $(60, 15)$, respectively.

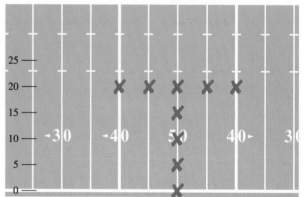

a. Find the inverse of $B = \begin{bmatrix} 50 & 50 & 45 \\ 0 & 15 & 20 \\ 1 & 1 & 1 \end{bmatrix}$.

b. Find $A = \begin{bmatrix} 40 & 55 & 60 \\ 10 & 10 & 15 \\ 1 & 1 & 1 \end{bmatrix} B^{-1}$.

c. Use the result of part b to find the new position of the other band members. What is the shape of the new position? (*Hint:* Multiply the matrix A by a 3×1 column vector with the first two components equal to the original position of each band member and the third component equal to 1. The new position of the band member is in the first two components of the product.)

■ 2.6 INPUT-OUTPUT MODELS

THINK ABOUT IT What production levels are needed to keep an economy going and to supply demands from outside the economy?

A method for solving such questions is developed in this section.

---

*Isaksen, Daniel, "Linear Algebra on the Gridiron," *The College Mathematics Journal,* Vol. 26, No. 5, Nov. 1995, pp. 358–360.

Wassily Leontief (1906–) developed an interesting and powerful application of matrix theory to economics and was recognized for this contribution with the Nobel prize in economics in 1973. His matrix models for studying the interdependencies in an economy are called *input-output* models. In practice these models are very complicated, with many variables. Only simple examples with a few variables are discussed here.

Input-output models are concerned with the production and flow of goods (and perhaps services). In an economy with n basic commodities, or sectors, the production of each commodity uses some (perhaps all) of the commodities in the economy as inputs. The amounts of each commodity used in the production of one unit of each commodity can be written as an $n \times n$ matrix A, called the **technological matrix** or **input-output matrix** of the economy.

EXAMPLE 1   Input-Output Matrix

Suppose a simplified economy involves just three commodity categories: agriculture, manufacturing, and transportation, all in appropriate units. Production of 1 unit of agriculture requires $1/2$ unit of manufacturing and $1/4$ unit of transportation; production of 1 unit of manufacturing requires $1/4$ unit of agriculture and $1/4$ unit of transportation; and production of 1 unit of transportation requires $1/3$ unit of agriculture and $1/4$ unit of manufacturing. Give the input-output matrix for this economy.

Solution

$$
\begin{array}{c}
\\
\text{Agriculture} \\
\text{Manufacturing} \\
\text{Transportation}
\end{array}
\begin{array}{ccc}
\text{Agriculture} & \text{Manufacturing} & \text{Transportation}
\end{array}
$$

$$
\begin{array}{c}
\text{Agriculture} \\
\text{Manufacturing} \\
\text{Transportation}
\end{array}
\left[
\begin{array}{ccc}
0 & \frac{1}{4} & \frac{1}{3} \\
\frac{1}{2} & 0 & \frac{1}{4} \\
\frac{1}{4} & \frac{1}{4} & 0
\end{array}
\right] = A
$$

The first column of the input-output matrix represents the amount of each of the three commodities consumed in the production of 1 unit of agriculture. The second column gives the amounts required to produce 1 unit of manufacturing, and the last column gives the amounts required to produce 1 unit of transportation. (Although it is perhaps unrealistic that production of a unit of each commodity requires none of that commodity, the simpler matrix involved is useful for our purposes.)

> **NOTE** Notice that for each commodity produced, the various units needed are put in a column. Each column corresponds to a commodity produced, and the rows correspond to what is needed to produce the commodity.

Another matrix used with the input-output matrix is the matrix giving the amount of each commodity produced, called the **production matrix,** or the matrix of gross output. In an economy producing n commodities, the production matrix can be represented by a column matrix X with entries $x_1, x_2, x_3, \ldots, x_n$.

EXAMPLE 2   Production Matrix

In Example 1, suppose the production matrix is

$$
X = \begin{bmatrix} 60 \\ 52 \\ 48 \end{bmatrix}.
$$

Then 60 units of agriculture, 52 units of manufacturing, and 48 units of transportation are produced. Because 1/4 unit of agriculture is used for each unit of manufacturing produced, $1/4 \times 52 = 13$ units of agriculture must be used in the "production" of manufacturing. Similarly, $1/3 \times 48 = 16$ units of agriculture will be used in the "production" of transportation. Thus, $13 + 6 = 29$ units of agriculture are used for production in the economy. Look again at the matrices A and X. Since X gives the number of units of each commodity produced and A gives the amount (in units) of each commodity used to produce 1 unit of each of the various commodities, the matrix product AX gives the amount of each commodity used in the production process.

$$AX = \begin{bmatrix} 0 & \frac{1}{4} & \frac{1}{3} \\ \frac{1}{2} & 0 & \frac{1}{4} \\ \frac{1}{4} & \frac{1}{4} & 0 \end{bmatrix} \begin{bmatrix} 60 \\ 52 \\ 48 \end{bmatrix} = \begin{bmatrix} 29 \\ 42 \\ 28 \end{bmatrix}$$

From this result, 29 units of agriculture, 42 units of manufacturing, and 28 units of transportation are used to produce 60 units of agriculture, 52 units of manufacturing, and 48 units of transportation.

The matrix product AX represents the amount of each commodity used in the production process. The remainder (if any) must be enough to satisfy the demand for the various commodities from outside the production system. In an n-commodity economy, this demand can be represented by a **demand matrix** D with entries $d_1, d_2, \ldots, d_n$. If no production is to remain unused, the difference between the production matrix X and the amount AX used in the production process must equal the demand D, or

$$D = X - AX.$$

In Example 2,

$$D = \begin{bmatrix} 60 \\ 52 \\ 48 \end{bmatrix} - \begin{bmatrix} 29 \\ 42 \\ 28 \end{bmatrix} = \begin{bmatrix} 31 \\ 10 \\ 20 \end{bmatrix},$$

so production of 60 units of agriculture, 52 units of manufacturing, and 48 units of transportation would satisfy a demand of 31, 10, and 20 units of each commodity, respectively.

In practice, A and D usually are known and X must be found. That is, we need to decide what amounts of production are needed to satisfy the required demands. Matrix algebra can be used to solve the equation $D = X - AX$ for X.

$$D = X - AX$$
$$D = IX - AX \qquad \text{Identity property}$$
$$D = (I - A)X \qquad \text{Distributive property}$$

If the matrix $I - A$ has an inverse, then

$$X = (I - A)^{-1}D.$$

If the production matrix is large or complicated, we could use a graphing calculator. On the TI-83, for example, we would enter the command `(identity(3) - [A])`$^{-1}$`*[D]` for a 3×3 matrix A. It is also practical to do these calculations on a spreadsheet.

EXAMPLE 3   Demand Matrix

Suppose, in the three-commodity economy from Examples 1 and 2, there is a demand for 516 units of agriculture, 258 units of manufacturing, and 129 units of transportation. What should production of each commodity be?

Solution The demand matrix is

$$D = \begin{bmatrix} 516 \\ 258 \\ 129 \end{bmatrix}.$$

To find the production matrix X, first calculate $I - A$.

$$I - A = \begin{bmatrix} 1 & 0 & 0 \\ 0 & 1 & 0 \\ 0 & 0 & 1 \end{bmatrix} - \begin{bmatrix} 0 & \frac{1}{4} & \frac{1}{3} \\ \frac{1}{2} & 0 & \frac{1}{4} \\ \frac{1}{4} & \frac{1}{4} & 0 \end{bmatrix} = \begin{bmatrix} 1 & -\frac{1}{4} & -\frac{1}{3} \\ -\frac{1}{2} & 1 & -\frac{1}{4} \\ -\frac{1}{4} & -\frac{1}{4} & 1 \end{bmatrix}$$

Use row operations to find the inverse of $I - A$ (the entries are rounded to two decimal places).

$$(I - A)^{-1} = \begin{bmatrix} 1.40 & .50 & .59 \\ .84 & 1.36 & .62 \\ .56 & .47 & 1.30 \end{bmatrix}$$

Since $X = (I - A)^{-1}D$,

$$X = \begin{bmatrix} 1.40 & .50 & .59 \\ .84 & 1.36 & .62 \\ .56 & .47 & 1.30 \end{bmatrix} \begin{bmatrix} 516 \\ 258 \\ 129 \end{bmatrix} = \begin{bmatrix} 928 \\ 864 \\ 578 \end{bmatrix}.$$

(Each entry in X has been rounded to the nearest whole number.)

The last result shows that production of 928 units of agriculture, 864 units of manufacturing, and 578 units of transportation are required to satisfy demands of 516, 258, and 129 units, respectively.

The entries in the matrix $(I - A)^{-1}$ are often called *multipliers*, and they have important economic interpretations. For example, every \$1 increase in total agricultural demand will result in an increase in agricultural production by \$1.40, an increase in manufacturing production by \$.84, and an increase in transportation production by \$.56. Similarly, every \$3 increase in total manufacturing demand will result in an increase of $3(.50) = 1.50$, $3(1.36) = 4.08$, and $3(.47) = 1.41$ dollars in agricultural production, manufacturing production, and transportation production, respectively.

EXAMPLE 4   Wheat and Oil Production

An economy depends on two basic products, wheat and oil. To produce 1 metric ton of wheat requires .25 metric tons of wheat and .33 metric tons of oil. Production of 1 metric ton of oil consumes .08 metric tons of wheat and .11 metric tons of oil. Find the production that will satisfy a demand for 500 metric tons of wheat and 1000 metric tons of oil.

Solution The input-output matrix is

$$A = \begin{bmatrix} .25 & .08 \\ .33 & .11 \end{bmatrix}.$$

Also,

$$I - A = \begin{bmatrix} .75 & -.08 \\ -.33 & .89 \end{bmatrix}.$$

Next, calculate $(I - A)^{-1}$.

$$(I - A)^{-1} = \begin{bmatrix} 1.39 & .13 \\ .51 & 1.17 \end{bmatrix} \quad \text{(rounded)}$$

To find the production matrix X, use the equation $X = (I - A)^{-1}D$, with

$$D = \begin{bmatrix} 500 \\ 1000 \end{bmatrix}.$$

The production matrix is

$$X = \begin{bmatrix} 1.39 & .13 \\ .51 & 1.17 \end{bmatrix} \begin{bmatrix} 500 \\ 1000 \end{bmatrix} = \begin{bmatrix} 825 \\ 1425 \end{bmatrix}.$$

Production of 825 metric tons of wheat and 1425 metric tons of oil is required to satisfy the indicated demand. ▮

The input-output model discussed above is referred to as an **open model**, since it allows for a surplus from the production equal to D. In the **closed model**, all the production is consumed internally in the production process, so that $X = AX$. There is nothing left over to satisfy any outside demands from other parts of the economy or from other economies. In this case, the sum of each column in the input-output matrix equals 1.

To solve the closed model, set $D = O$ in the equation derived earlier.

$$(I - A)X = D = O$$

The system of equations that corresponds to $(I - A)X = O$ does not have a single unique solution, but it can be solved in terms of a parameter. (It can be shown that if the columns of a matrix A sum to 1, then the equation $(I - A)X = O$ has an infinite number of solutions.)

FOR REVIEW ▬

Parameters were discussed in the first section of this chapter. As mentioned there, parameters are required when a system has infinitely many solutions.

EXAMPLE 5   Closed Input-Output Model

Use matrix A below to find the production of each commodity in a closed model.

$$A = \begin{bmatrix} \frac{1}{2} & \frac{1}{4} & \frac{1}{3} \\ 0 & \frac{1}{4} & \frac{1}{3} \\ \frac{1}{2} & \frac{1}{2} & \frac{1}{3} \end{bmatrix}$$

Solution Find the value of $I - A$, then set $(I - A)X = O$ to find X.

$$I - A = \begin{bmatrix} \frac{1}{2} & -\frac{1}{4} & -\frac{1}{3} \\ 0 & \frac{3}{4} & -\frac{1}{3} \\ -\frac{1}{2} & -\frac{1}{2} & \frac{2}{3} \end{bmatrix}$$

$$(I - A)X = \begin{bmatrix} \frac{1}{2} & -\frac{1}{4} & -\frac{1}{3} \\ 0 & \frac{3}{4} & -\frac{1}{3} \\ -\frac{1}{2} & -\frac{1}{2} & \frac{2}{3} \end{bmatrix} \begin{bmatrix} x_1 \\ x_2 \\ x_3 \end{bmatrix} = \begin{bmatrix} 0 \\ 0 \\ 0 \end{bmatrix}$$

Multiply to get

$$\begin{bmatrix} \frac{1}{2}x_1 - \frac{1}{4}x_2 - \frac{1}{3}x_3 \\ 0x_1 + \frac{3}{4}x_2 - \frac{1}{3}x_3 \\ -\frac{1}{2}x_1 - \frac{1}{2}x_2 + \frac{2}{3}x_3 \end{bmatrix} = \begin{bmatrix} 0 \\ 0 \\ 0 \end{bmatrix}.$$

The last matrix equation corresponds to the following system.

$$\frac{1}{2}x_1 - \frac{1}{4}x_2 - \frac{1}{3}x_3 = 0$$
$$\frac{3}{4}x_2 - \frac{1}{3}x_3 = 0$$
$$-\frac{1}{2}x_1 - \frac{1}{2}x_2 + \frac{2}{3}x_3 = 0$$

Solving the system with x_3 as the parameter gives the solution of the system

$$\left(\frac{8}{9}x_3, \frac{4}{9}x_3, x_3\right).$$

For example, if we let $x_3 = 9$ (a choice that eliminates fractions in the answer), then $x_1 = 8$ and $x_2 = 4$, so the production of the three commodities should be in the ratio $8:4:9$.

Production matrices for actual economies are much larger than those shown in this section. An analysis of the U.S. economy in 1967 has close to 200 commodity categories.* Such matrices require large human and computer resources for their analysis. Some of the exercises at the end of this section use actual data in which categories have been combined to simplify the work.

FINDING THE PRODUCTION MATRIX

To obtain the production matrix, X, for an open input-output model, follow these steps:

1. Form the $n \times n$ input-output matrix, A, by placing in each column the amount of the various commodities required to produce 1 unit of a particular commodity.
2. Calculate $I - A$, where I is the $n \times n$ identity matrix.
3. Find the inverse, $(I - A)^{-1}$.
4. Multiply the inverse on the right by the demand matrix, D, to obtain $X = (I - A)^{-1}D$.

To obtain a production matrix, X, for a closed input-output model, solve the system $(I - A)X = 0$.

2.6 EXERCISES

Find the production matrix for the following input-output and demand matrices using the open model.

1. $A = \begin{bmatrix} .5 & .4 \\ .25 & .2 \end{bmatrix}$, $D = \begin{bmatrix} 2 \\ 4 \end{bmatrix}$

2. $A = \begin{bmatrix} .2 & .04 \\ .6 & .05 \end{bmatrix}$, $D = \begin{bmatrix} 3 \\ 10 \end{bmatrix}$

---

*Input-Output Structure of U.S. Economy: 1967, U.S. Department of Commerce, 1974.

3. $A = \begin{bmatrix} .1 & .03 \\ .07 & .6 \end{bmatrix}$, $D = \begin{bmatrix} 5 \\ 10 \end{bmatrix}$

4. $A = \begin{bmatrix} .01 & .03 \\ .05 & .05 \end{bmatrix}$, $D = \begin{bmatrix} 100 \\ 200 \end{bmatrix}$

5. $A = \begin{bmatrix} .4 & 0 & .3 \\ 0 & .8 & .1 \\ 0 & .2 & .4 \end{bmatrix}$, $D = \begin{bmatrix} 1 \\ 3 \\ 2 \end{bmatrix}$

6. $A = \begin{bmatrix} .1 & .5 & 0 \\ 0 & .3 & .4 \\ .1 & .2 & .1 \end{bmatrix}$, $D = \begin{bmatrix} 10 \\ 4 \\ 2 \end{bmatrix}$

Find the ratios of products A, B, *and* C *using a closed model.*

7.
	A	B	C
A	.3	.1	.8
B	.5	.6	.1
C	.2	.3	.1

8.
	A	B	C
A	.2	.1	.5
B	.4	.3	.4
C	.4	.6	.1

Use a graphing calculator or computer to find the production matrix X, given the following input-output and demand matrices.

9. $A = \begin{bmatrix} .25 & .25 & .25 & .05 \\ .01 & .02 & .01 & .1 \\ .3 & .3 & .01 & .1 \\ .2 & .01 & .3 & .01 \end{bmatrix}$, $D = \begin{bmatrix} 2930 \\ 3570 \\ 2300 \\ 580 \end{bmatrix}$

10. $A = \begin{bmatrix} .2 & .1 & .2 & .2 \\ .3 & .05 & .07 & .02 \\ .1 & .03 & .02 & .01 \\ .05 & .3 & .05 & .03 \end{bmatrix}$, $D = \begin{bmatrix} 5000 \\ 1000 \\ 8000 \\ 500 \end{bmatrix}$

■ Applications

BUSINESS AND ECONOMICS

Input-Output Open Model In Exercises 11 and 12, refer to Example 4.

11. If the demand is changed to 690 metric tons of wheat and 920 metric tons of oil, how many units of each commodity should be produced?

12. Change the technological matrix so that production of 1 ton of wheat requires 1/5 metric ton of oil (and no wheat), and production of 1 metric ton of oil requires 1/3 metric ton of wheat (and no oil). To satisfy the same demand matrix, how many units of each commodity should be produced?

Input-Output Open Model In Exercises 13–16, refer to Example 3.

13. If the demand is changed to 516 units of each commodity, how many units of each commodity should be produced?

14. Suppose 1/3 unit of manufacturing (no agriculture or transportation) is required to produce 1 unit of agriculture, 1/4 unit of transportation is required to produce 1 unit of manufacturing, and 1/2 unit of agriculture is required to produce 1 unit of transportation. How many units of each commodity should be produced to satisfy a demand of 1000 units of each commodity?

15. Suppose 1/4 unit of manufacturing and 1/2 unit of transportation are required to produce 1 unit of agriculture, 1/2 unit of agriculture and 1/4 unit of transportation to produce 1 unit of manufacturing, and 1/4 unit of agricul-ture and 1/4 unit of manufacturing to produce 1 unit of transportation. How many units of each commodity should be produced to satisfy a demand of 1000 units for each commodity?

16. If the input-output matrix is changed so that 1/4 unit of manufacturing and 1/2 unit of transportation are required to produce 1 unit of agriculture, 1/2 unit of agriculture and 1/4 unit of transportation are required to produce 1 unit of manufacturing, and 1/4 unit each of agriculture and manufacturing are required to produce 1 unit of transporta-tion, find the number of units of each commodity that should be produced to satisfy a demand for 500 units of each commodity.

Input-Output Open Model

17. A primitive economy depends on two basic goods, yams and pork. Production of 1 bushel of yams requires 1/4 bushel of yams and 1/2 of a pig. To produce 1 pig requires 1/6 bushel of yams. Find the amount of each commodity that should be produced to get the following.

a. 1 bushel of yams and 1 pig

b. 100 bushels of yams and 70 pigs

18. A simple economy depends on three commodities: oil, corn, and coffee. Production of 1 unit of oil requires .1 unit of oil, .2 unit of corn, and no units of coffee. To produce 1 unit of corn requires .2 unit of oil, .1 unit of corn, and .05 unit of coffee. To produce 1 unit of coffee requires .1 unit of oil, .05 unit of corn, and .1 unit of coffee. Find the production required to meet a demand of 1000 units each of oil, corn, and coffee.

19. In his work *Input-Output Economics,* Leontief provides an example of a simplified economy with just three sectors: agriculture, manufacturing, and households (i.e., the sector of the economy that produces labor).* It has the following input-output matrix:

	Agriculture	Manufacturing	Households
Agriculture	.25	.40	.133
Manufacturing	.14	.12	.100
Households	.80	3.60	.133

He also gives the demand matrix

$$D = \begin{bmatrix} 35 \\ 38 \\ 40 \end{bmatrix}.$$

Find the amount of each commodity that should be produced.

20. A much simplified version of Leontief's 42-sector analysis of the 1947 American economy has the following input-output matrix.[†]

	Agriculture	Manufacturing	Households
Agriculture	.245	.102	.051
Manufacturing	.099	.291	.279
Households	.433	.372	.011

The demand matrix (in billions of dollars) is

$$D = \begin{bmatrix} 2.88 \\ 31.45 \\ 30.91 \end{bmatrix}.$$

Find the amount of each commodity that should be produced.

21. An analysis of the 1958 Israeli economy is simplified here by grouping the economy into three sectors, with the following input-output matrix:[‡]

	Agriculture	Manufacturing	Energy
Agriculture	.293	0	0
Manufacturing	.014	.207	.017
Energy	.044	.010	.216

The demand (in thousands of Israeli pounds) as measured by exports is

$$D = \begin{bmatrix} 138,213 \\ 17,597 \\ 1786 \end{bmatrix}.$$

Find the amount of each commodity that should be produced.

22. The 1981 Chinese economy can be simplified to three sectors: agriculture, industry and construction, and transportation and commerce.[§] The input-output matrix is given below.

	Agriculture	Industry/Constr.	Trans./Commerce
Agriculture	.158	.156	.009
Industry/Constr.	.136	.432	.071
Trans./Commerce	.013	.041	.011

The demand (in 100,000 RMB, the unit of money in China) is

$$D = \begin{bmatrix} 106,674 \\ 144,739 \\ 26,725 \end{bmatrix}.$$

a. Find the amount of each commodity that should be produced.

b. Interpret the economic value of an increase in demand of 1 RMB in agricultural exports.

23. *Washington* The 1987 economy of the state of Washington has been simplified to four sectors: natural resource, manufacturing, trade and services, and personal consumption. The input-output matrix is given below.[‖]

	Natural Resources	Manufacturing	Trade & Services	Personal Consumption
Natural Resources	.1045	.0428	.0029	.0031
Manufacturing	.0826	.1087	.0584	.0321
Trade & Services	.0867	.1019	.2032	.3555
Personal Consumption	.6253	.3448	.6106	.0798

Suppose the demand (in millions of dollars) is

$$D = \begin{bmatrix} 450 \\ 300 \\ 125 \\ 100 \end{bmatrix}.$$

Find the amount of each commodity that should be produced.

*Leontief, Wassily, *Input-Output Economics,* 2nd ed., Oxford University Press, 1966, pp. 20–27.
†Ibid, pp. 6–9.
‡Ibid, pp. 174–177.
§*Input-Output Tables of China, 1981,* China Statistical Information and Consultancy Service Centre, 1987, pp. 17–19.
‖Chase, Robert, Philip Bourque, and Richard Conway Jr., "The 1987 Washington State Input-Output Study," Report to the Graduate School Business Administration, University of Washington, Sept. 1993.

24. *Washington* In addition to solving the previous input-output model, most models of this nature also include an employment equation. For the previous model, the employment equation is added and a new system of equations is obtained as follows.*

$$\begin{bmatrix} x_1 \\ x_2 \\ x_3 \\ x_4 \\ N \end{bmatrix} = (I - B)^{-1}C,$$

where x_1, x_2, x_3, x_4 represent the amount, in millions of dollars, that must be produced to satisfy internal and external demands of the four sectors; N is the total work force required for a particular set of demands; and

$$B = \begin{bmatrix} .1045 & .0428 & .0029 & .0031 & 0 \\ .0826 & .1087 & .0584 & .0321 & 0 \\ .0867 & .1019 & .2032 & .3555 & 0 \\ .6253 & .3448 & .6106 & .0798 & 0 \\ 21.6 & 6.6 & 20.2 & 0 & 0 \end{bmatrix}.$$

a. Suppose that a $50 million change in manufacturing occurs. How will this increase in demand affect the economy? (*Hint:* Find $(I - B)^{-1}C$ where $C = \begin{bmatrix} 0 \\ 50 \\ 0 \\ 0 \\ 0 \end{bmatrix}$.)

b. Interpret the meaning of the bottom row in the matrix $(I - B)^{-1}$.

25. *Community Links* The use of input-output analysis can also be used to model how changes in one city can affect cities that are connected with it in some way.[†] For example, if a large manufacturing company shuts down in one city, it is very likely that the economic welfare of all of the cities around it will suffer. Consider three Pennsylvania communities: Sharon, Farrell, and Hermitage. Due to their proximity to each other, residents of these three communities regularly spend time and money in the other communities. Suppose that we have gathered information in the form

of an input-output matrix.

$$A = \begin{array}{c} \\ S \\ F \\ H \end{array} \begin{array}{c} \begin{array}{ccc} S & F & H \end{array} \\ \begin{bmatrix} .2 & .1 & .1 \\ .1 & .1 & 0 \\ .5 & .6 & .7 \end{bmatrix} \end{array}$$

This matrix can be thought of as the likelihood that a person from a particular community will spend money in each of the communities.

a. Treat this matrix like an input-output matrix and calculate $(I - A)^{-1}$.

b. Interpret the entries of this inverse matrix.

Input-Output Closed Model

26. Use the input-output matrix

$$\begin{array}{c} \\ \text{Yams} \\ \text{Pigs} \end{array} \begin{array}{c} \begin{array}{cc} \text{Yams} & \text{Pigs} \end{array} \\ \begin{bmatrix} \frac{1}{4} & \frac{1}{2} \\ \frac{3}{4} & \frac{1}{2} \end{bmatrix} \end{array}$$

and the closed model to find the ratio of yams to pigs produced.

27. Use the input-output matrix

$$\begin{array}{c} \\ \text{Steel} \\ \text{Coal} \end{array} \begin{array}{c} \begin{array}{cc} \text{Steel} & \text{Coal} \end{array} \\ \begin{bmatrix} \frac{1}{3} & \frac{3}{5} \\ \frac{2}{3} & \frac{2}{5} \end{bmatrix} \end{array}$$

to find the ratio of coal to steel produced.

28. Suppose that production of 1 unit of agriculture requires 1/3 unit of agriculture, 1/3 unit of manufacturing, and 1/3 unit of transportation. To produce 1 unit of manufacturing requires 1/2 unit of agriculture, 1/4 unit of manufacturing, and 1/4 unit of transportation. To produce 1 unit of transportation requires 0 units of agriculture, 1/4 unit of manufacturing, and 3/4 unit of transportation. Find the ratio of the three commodities in the closed model.

29. Suppose that production of 1 unit of mining requires 1/5 unit of mining, 2/5 unit of manufacturing, and 2/5 unit of communication. To produce 1 unit of manufacturing requires 3/5 unit of mining, 1/5 unit of manufacturing, and 1/5 unit of communication. To produce 1 unit of communication requires 0 units of mining, 4/5 unit of manufacturing, and 1/5 unit of communication. Find the ratio of the three commodities in the closed model.

---

*Chase, Robert, Philip Bourque, and Richard Conway Jr., "The 1987 Washington State Input-Output Study," Report to the Graduate School Business Administration, University of Washington, Sept. 1993.

[†]The idea for this problem came from an example created by Thayer Watkins, Department of Economics, San Jose State University, www.sjsu/faculty/watkins/inputoutput.htm.

■ CHAPTER SUMMARY

In this chapter we extended our study of linear functions to include finding solutions of systems of linear equations. Techniques such as the echelon method and the Gauss-Jordan method were developed and used to solve systems of linear equations. It is convenient to store mathematical information in rows and columns called a matrix. We saw that matrices can be combined and manipulated using addition, subtraction, scalar multiplication, and matrix multiplication. The concept of multiplicative inverse of real numbers was then extended to matrices, and matrix inverses were explored. The fact that matrix inverses can be used to solve systems of equations was introduced and then applied to solve a wide range of problems. One application of particular importance was the Leontief input-output models, which are used to study interdependencies in an economy.

KEY TERMS

To understand the concepts presented in this chapter, you should know the meaning and use of the following terms. For easy reference, the section in the chapter where a word (or expression) was first used is given with each item.

system of equations	**2.2 matrix (matrices)**	**additive inverse**	**2.6 input-output**
2.1 first-degree equation in	**element (entry)**	**(negative) of a matrix**	**(technological)**
***n* unknowns**	**augmented matrix**	**zero matrix**	**matrix**
unique solution	**row operations**	**additive identity**	**production matrix**
inconsistent system	**Gauss-Jordan method**	**2.4 scalar**	**demand matrix**
dependent equations	**2.3 size**	**product matrix**	**open model**
equivalent system	**square matrix**	**2.5 identity matrix**	**closed model**
echelon method	**row matrix (row vector)**	**multiplicative inverse**	
back-substitution	**column matrix (column**	**matrix**	
parameter	**vector)**	**coefficient matrix**	

CHAPTER 2 REVIEW EXERCISES

1. What is true about the number of solutions to a system of m linear equations in n unknowns if $m = n$? If $m < n$? If $m > n$?

2. Suppose someone says that a more reasonable way to multiply two matrices than the method presented in the text is to multiply corresponding elements. For example, the result of

$$\begin{bmatrix} 1 & 2 \\ 3 & 4 \end{bmatrix} \cdot \begin{bmatrix} 3 & 5 \\ 7 & 11 \end{bmatrix} \text{ should be } \begin{bmatrix} 3 & 10 \\ 21 & 44 \end{bmatrix},$$

according to this person. How would you respond?

Solve each system by the echelon method.

3.
$$\begin{aligned} 2x + 3y &= 10 \\ -3x + y &= 18 \end{aligned}$$

4.
$$\begin{aligned} \frac{x}{2} + \frac{y}{4} &= 3 \\ \frac{x}{4} - \frac{y}{2} &= 4 \end{aligned}$$

5.
$$\begin{aligned} 2x - 3y + z &= -5 \\ x + 4y + 2z &= 13 \\ 5x + 5y + 3z &= 14 \end{aligned}$$

6.
$$\begin{aligned} x - y &= 3 \\ 2x + 3y + z &= 13 \\ 3x - 2z &= 21 \end{aligned}$$

Solve each system by the Gauss-Jordan method.

7. $2x + 4y = -6$
$-3x - 5y = 12$

8. $x + 2y = -9$
$4x + 9y = 41$

9. $x - y + 3z = 13$
$4x + y + 2z = 17$
$3x + 2y + 2z = 1$

10. $x - 2z = 5$
$3x + 2y = 8$
$-x + 2z = 10$

11. $3x - 6y + 9z = 12$
$-x + 2y - 3z = -4$
$x + y + 2z = 7$

Find the size of each matrix, find the values of any variables, and identify any square, row, or column matrices.

12. $\begin{bmatrix} 2 & 3 \\ 5 & q \end{bmatrix} = \begin{bmatrix} a & b \\ c & 9 \end{bmatrix}$

13. $\begin{bmatrix} 2 & x \\ y & 6 \\ 5 & z \end{bmatrix} = \begin{bmatrix} a & -1 \\ 4 & 6 \\ p & 7 \end{bmatrix}$

14. $[m \quad 4 \quad z \quad -1] = [12 \quad k \quad -8 \quad r]$

15. $\begin{bmatrix} a+5 & 3b & 6 \\ 4c & 2+d & -3 \\ -1 & 4p & q-1 \end{bmatrix} = \begin{bmatrix} -7 & b+2 & 2k-3 \\ 3 & 2d-1 & 4l \\ m & 12 & 8 \end{bmatrix}$

Given the matrices

$$A = \begin{bmatrix} 4 & 10 \\ -2 & -3 \\ 6 & 9 \end{bmatrix}, \quad B = \begin{bmatrix} 2 & 3 & -2 \\ 2 & 4 & 0 \\ 0 & 1 & 2 \end{bmatrix}, \quad C = \begin{bmatrix} 5 & 0 \\ -1 & 3 \\ 4 & 7 \end{bmatrix},$$

$$D = \begin{bmatrix} 6 \\ 1 \\ 0 \end{bmatrix}, \quad E = [1 \quad 3 \quad -4], \quad F = \begin{bmatrix} -1 & 4 \\ 3 & 7 \end{bmatrix}, \quad G = \begin{bmatrix} 2 & 5 \\ 1 & 6 \end{bmatrix},$$

find each of the following, if it exists.

16. $A + C$

17. $2G - 4F$

18. $3C + 2A$

19. $B - A$

20. $2A - 5C$

21. AF

22. AC

23. DE

24. ED

25. BD

26. EA

27. F^{-1}

28. B^{-1}

29. $(A + C)^{-1}$

Find the inverse of each of the following matrices that has an inverse.

30. $\begin{bmatrix} 2 & 1 \\ 5 & 3 \end{bmatrix}$

31. $\begin{bmatrix} -4 & 2 \\ 0 & 3 \end{bmatrix}$

32. $\begin{bmatrix} 2 & 0 \\ -1 & 5 \end{bmatrix}$

33. $\begin{bmatrix} 6 & 4 \\ 3 & 2 \end{bmatrix}$

34. $\begin{bmatrix} 2 & -1 & 0 \\ 1 & 0 & 1 \\ 1 & -2 & 0 \end{bmatrix}$

35. $\begin{bmatrix} 2 & 0 & 4 \\ 1 & -1 & 0 \\ 0 & 1 & -2 \end{bmatrix}$

36. $\begin{bmatrix} 1 & 3 & 6 \\ 4 & 0 & 9 \\ 5 & 15 & 30 \end{bmatrix}$

37. $\begin{bmatrix} 2 & 3 & 5 \\ -2 & -3 & -5 \\ 1 & 4 & 2 \end{bmatrix}$

Solve the matrix equation $AX = B$ for X using the given matrices.

38. $A = \begin{bmatrix} 2 & 4 \\ -1 & -3 \end{bmatrix}, \quad B = \begin{bmatrix} 8 \\ 3 \end{bmatrix}$

39. $A = \begin{bmatrix} 1 & 2 \\ 2 & 4 \end{bmatrix}, \quad B = \begin{bmatrix} 5 \\ 10 \end{bmatrix}$

40. $A = \begin{bmatrix} 1 & 0 & 2 \\ -1 & 1 & 0 \\ 3 & 0 & 4 \end{bmatrix}, \quad B = \begin{bmatrix} 8 \\ 4 \\ -6 \end{bmatrix}$

41. $A = \begin{bmatrix} 2 & 4 & 0 \\ 1 & -2 & 0 \\ 0 & 0 & 3 \end{bmatrix}, \quad B = \begin{bmatrix} 72 \\ -24 \\ 48 \end{bmatrix}$

Solve each of the following systems of equations by inverses.

42. $2x + y = 5$
$3x - 2y = 4$

43. $5x + 10y = 80$
$3x - 2y = 120$

44. $x + y + z = 1$
$2x + y = -2$
$3y + z = 2$

45. $x + 4y - z = 6$
$2x - y + z = 3$
$3x + 2y + 3z = 16$

Find each production matrix, given the following input-output and demand matrices.

46. $A = \begin{bmatrix} .01 & .05 \\ .04 & .03 \end{bmatrix}$, $D = \begin{bmatrix} 200 \\ 300 \end{bmatrix}$

47. $A = \begin{bmatrix} .2 & .1 & .3 \\ .1 & 0 & .2 \\ 0 & 0 & .4 \end{bmatrix}$, $D = \begin{bmatrix} 500 \\ 200 \\ 100 \end{bmatrix}$

48. The following system of equations is given.

$$x + 2y + z = 7$$
$$2x - y - z = 2$$
$$3x - 3y + 2z = -5$$

a. Solve by the echelon method.

b. Solve by the Gauss-Jordan method. Compare with the echelon method.

c. Write the system as a matrix equation, $AX = B$.

d. Find the inverse of matrix A from part c.

e. Solve the system using A^{-1} from part d.

Applications

BUSINESS AND ECONOMICS

Write each of Exercises 49–52 as a system of equations and solve.

49. *Scheduling Production* An office supply manufacturer makes two kinds of paper clips, standard and extra large. To make 1000 standard paper clips requires 1/4 hr on a cutting machine and 1/2 hr on a machine that shapes the clips. One thousand extra large paper clips require 1/3 hr on each machine. The manager of paper clip production has 4 hr per day available on the cutting machine and 6 hr per day on the shaping machine. How many of each kind of clip can he make?

50. *Investment* Gretchen Schmidt plans to buy shares of two stocks. One costs $32 per share and pays dividends of $1.20 per share. The other costs $23 per share and pays dividends of $1.40 per share. She has $10,100 to spend and wants to earn dividends of $540. How many shares of each stock should she buy?

51. *Production Requirements* The Waputi Indians make woven blankets, rugs, and skirts. Each blanket requires 24 hr for spinning the yarn, 4 hr for dyeing the yarn, and 15 hr for weaving. Rugs require 30, 5, and 18 hr and skirts 12, 3, and 9 hr, respectively. If there are 306, 59, and 201 hr available for spinning, dyeing, and weaving, respectively, how many of each item can be made? (*Hint:* Simplify the equations you write, if possible, before solving the system.)

52. *Distribution* An oil refinery in Tulsa sells 50% of its production to a Chicago distributor, 20% to a Dallas distributor, and 30% to an Atlanta distributor. Another refinery in New Orleans sells 40% of its production to the Chicago distributor, 40% to the Dallas distributor, and 20% to the Atlanta distributor. A third refinery in Ardmore sells the same distributors 30%, 40%, and 30% of its production. The three distributors received 219,000, 192,000, and 144,000 gallons of oil, respectively. How many gallons of oil were produced at each of the three plants?

53. *Stock Reports* The New York Stock Exchange reports in daily newspapers give the dividend, price-to-earnings ratio, sales (in hundreds of shares), last price, and change in price for each company. Write the following stock reports as a 4×5 matrix: American Telephone & Telegraph: .14, 11, 333,675, 20.13, +1.88; General Electric: .64, 39, 390,591, 47.81, +4.06; Lucent: .08, 41, 436,351, 15.19, +1.88; Sara Lee: .54, 17, 27,077, 23.13, −1.50.

54. *Filling Orders* A printer has three orders for pamphlets that require three kinds of paper, as shown in the following matrix.

		Order		
		I	II	III
	High-grade	10	5	8
Paper	Medium-grade	12	0	4
	Coated	0	10	5

The printer has on hand 3170 sheets of high-grade paper, 2360 sheets of medium-grade paper, and 1800 sheets of coated paper. All the paper must be used in preparing the order.

a. Write a 3 × 1 matrix for the amounts of paper on hand.

b. Write a matrix of variables to represent the number of pamphlets that must be printed in each of the three orders.

c. Write a matrix equation using the given matrix and your matrices from parts a and b.

d. Solve the equation from part c.

55. *Input-Output* An economy depends on two commodities, goats and cheese. It takes 2/3 of a unit of goats to produce 1 unit of cheese and 1/2 unit of cheese to produce 1 unit of goats.

a. Write the input-output matrix for this economy.

b. Find the production required to satisfy a demand of 400 units of cheese and 800 units of goats.

56. *Nebraska* The 1970 economy of the state of Nebraska has been condensed to six sectors: livestock, crops, food products, mining and manufacturing, households, and other. The input-output matrix is given below.*

$$\begin{bmatrix} .178 & .018 & .411 & 0 & .005 & 0 \\ .143 & .018 & .088 & 0 & .001 & 0 \\ .089 & 0 & .035 & 0 & .060 & .003 \\ .001 & .010 & .012 & .063 & .007 & .014 \\ .141 & .252 & .088 & .089 & .402 & .124 \\ .188 & .156 & .103 & .255 & .008 & .474 \end{bmatrix}$$

a. Find the matrix $(I - A)^{-1}$ and interpret the value in row 2, column 1 of this matrix.

b. Suppose the demand (in millions of dollars) is

$$D = \begin{bmatrix} 1980 \\ 650 \\ 1750 \\ 1000 \\ 2500 \\ 3750 \end{bmatrix}.$$

Find the dollar amount of each commodity that should be produced.

LIFE SCIENCES

57. *Animal Activity* The activities of a grazing animal can be classified roughly into three categories: grazing, moving,

and resting. Suppose horses spend 8 hr grazing, 8 moving, and 8 resting; cattle spend 10 grazing, 5 moving, and 9 resting; sheep spend 7 grazing, 10 moving, and 7 resting; and goats spend 8 grazing, 9 moving, and 7 resting. Write this information as a 4 × 3 matrix.

58. *CAT Scans* Computer Aided Tomography (CAT) scanners take X-rays of a part of the body from different directions, and put the information together to create a picture of a cross section of the body.[†] The amount by which the energy of the X-ray decreases, measured in linear-attenuation units, tells whether the X-ray has passed through healthy tissue, tumorous tissue, or bone, based on the following table.

Type of Tissue	Linear-Attenuation Values
Healthy tissue	.1625–.2977
Tumorous tissue	.2679–.3930
Bone	.3857–.5108

The part of the body to be scanned is divided into cells. If an X-ray passes through more than one cell, the total linear-attenuation value is the sum of the values for the cells. For example, in the figure, let *a*, *b*, and *c* be the values for cells A, B, and C. The attenuation value for beam 1 is $a + b$ and for beam 2 is $a + c$.

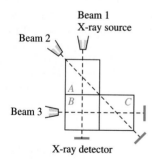

a. Find the attenuation value for beam 3.

b. Suppose that the attenuation values are .8, .55, and .65 for beams 1, 2, and 3, respectively. Set up and solve the system of three equations for *a*, *b*, and *c*. What can you conclude about cells A, B, and C?

c. Find the inverse of the coefficient matrix from part b to find *a*, *b*, and *c* for the following three cases, and make conclusions about cells A, B, and C for each.

---

*Lamphear, F. Charles, and Theodore Roesler, "1970 Nebraska Input-Output Tables," *Nebraska Economic and Business Report No. 10,* Bureau of Business Research, University of Nebraska-Lincoln, 1971.
[†]Exercises 58 and 59 are based on the article "Medical Applications of Linear Equations" by Jabon, David, Gail Nord, Bryce W. Wilson, and Penny Coffman, *The Mathematics Teacher,* Vol. 89, No. 5, May 1996, p. 398.

	Linear-Attenuation Values		
Patient	Beam 1	Beam 2	Beam 3
X	.54	.40	.52
Y	.65	.80	.75
Z	.51	.49	.44

59. *CAT Scans* (Refer to Exercise 58.)* Four X-ray beams are aimed at four cells. as shown in the following figure.

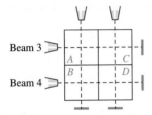

Beam 1 Beam 2

Beam 3

Beam 4

a. Suppose the attenuation values for beams 1, 2, 3, and 4 are .60, .75, .65, and .70, respectively. Do we have enough information to determine the values of a, b, c, and d? Explain.

b. Suppose we have the data from part a, as well as the following values for d. Find the values for a, b, and c, and make conclusions about cells A, B, C, and D in each case.

(i) .33   **(ii)** .43

c. Two X-ray beams are added, as shown in the figure. In addition to the data in part a, we now have attenuation values for beams 5 and 6 of .85 and .50. Find the values for a, b, c, and d, and make conclusions about cells A, B, C, and D.

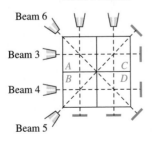

Beam 1 Beam 2

Beam 6

Beam 3

Beam 4

Beam 5

d. Six X-ray beams are not necessary because four appropriately chosen beams are sufficient. Give two examples of four beams (chosen from beams 1–6 in part c) that will give the solution. (*Note:* There are 12 possible solutions.)

e. Discuss what properties the four beams selected in part d must have in order to provide a unique solution.

60. *Hockey* In a recent study, the number of head and neck injuries among hockey players wearing full face shields and half face shields were compared. The following table provides the rates per 1000 athlete-exposures for specific injuries that caused a player wearing either shield to miss one or more events.[†]

	Half Shield	Full Shield
Head and Face Injuries (Excluding Concussions)	3.54	1.41
Concussions	1.53	1.57
Neck Injuries	.34	.29
Other	7.53	6.21

If an equal number of players in a large league wear each type of shield and the total number of athlete-exposures for the league in a season is 8000, use matrix operations to estimate the total number of injuries of each type.

PHYSICAL SCIENCES

61. *Roof Trusses* Linear systems occur in the design of roof trusses for new homes and buildings. The simplest type of roof truss is a triangle. The truss shown in the figure on the next page is used to frame roofs of small buildings. If

*Exercises 58 and 59 are based on the article "Medical Applications of Linear Equations" by Jabon, David, Gail Nord, Bryce W. Wilson, and Penny Coffman, *The Mathematics Teacher,* Vol. 89, No. 5, May 1996, p. 398.
†Benson, Brian, Nicholas Nohtadi, Sarah Rose, and Willem Meeuwisse, "Head and Neck Injuries Among Ice Hockey Players Wearing Full Face Shields vs. Half Face Shields," *JAMA,* Vol. 282, No. 24, Dec. 22/29 1999, pp. 2328–2332.

a 100-lb force is applied at the peak of the truss, then the forces or weights W_1 and W_2 exerted parallel to each rafter of the truss are determined by the following linear system of equations.

$$\frac{\sqrt{3}}{2}(W_1 + W_2) = 100$$

$$W_1 - W_2 = 0$$

Solve the system to find W_1 and W_2.*

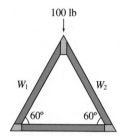

100 lb

62. *Roof Trusses* (Refer to Exercise 61.) Use the following system of equations to determine the force or weights W_1 and W_2 exerted on each rafter for the truss shown in the figure.

$$\frac{1}{2}W_1 + \frac{\sqrt{2}}{2}W_2 = 150$$

$$\frac{\sqrt{3}}{2}W_1 - \frac{\sqrt{2}}{2}W_2 = 0$$

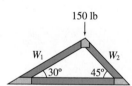

150 lb

63. *Carbon Dioxide* Determining the amount of carbon dioxide in the atmosphere is important because carbon dioxide is known to be a greenhouse gas. Carbon dioxide concentrations (in parts per million) have been measured at Mauna Loa, Hawaii, over the past 30 years. The concentrations have increased quadratically.[†] The table lists readings for three years.

Year	CO_2
1959	316
1979	337
1999	368

a. If the relationship between the carbon dioxide concentration C and the year t is expressed as $C = at^2 + bt + c$, where $t = 0$ corresponds to 1959, use a linear system of equations to determine the constants a, b, and c.

b. Predict the year when the amount of carbon dioxide in the atmosphere will double from its 1959 level. (*Hint:* This requires solving a quadratic equation. For review on how to do this, see Section R.4.)

64. *Chemistry* When carbon monoxide (CO) reacts with oxygen (O_2), carbon dioxide (CO_2) is formed. This can be written as $CO + (1/2)O_2 = CO_2$ and as a matrix equation.[‡] If we form a 2×1 column matrix by letting the first element be the number of carbon atoms and the second element be the number of oxygen atoms, then CO would have the column matrix

$$\begin{bmatrix} 1 \\ 1 \end{bmatrix}.$$

Similarly, O_2 and CO_2 would have the column matrices $\begin{bmatrix} 0 \\ 2 \end{bmatrix}$ and $\begin{bmatrix} 1 \\ 2 \end{bmatrix}$, respectively.

a. Use Gauss-Jordan to find numbers x and y (known as *stoichiometric numbers*) that solve the system of equations

$$\begin{bmatrix} 1 \\ 1 \end{bmatrix} x + \begin{bmatrix} 0 \\ 2 \end{bmatrix} y = \begin{bmatrix} 1 \\ 2 \end{bmatrix}.$$

Compare your answers to the equation written above.

b. Repeat the process for $xCO_2 + yH_2 + zCO = H_2O$, where H_2 is hydrogen, and H_2O is water. In words, what does this mean?

GENERAL INTEREST

65. *Students* Suppose 20% of the boys and 30% of the girls in a high school like tennis, and 60% of the boys and 90% of the girls like math. If 500 students like tennis and 1500 like math, how many boys and girls are in the school? Find all possible solutions.

---

*Hibbeler, R., *Structural Analysis,* Prentice Hall, 1995.

[†]Atmospheric Carbon Dioxide Record from Mauna Loa, University of California, La Jolla, http://cdiac.esd.ornl.gov/trends/co2/slo-mlo.htm.

[‡]Alberty, Robert, "Chemical Equations Are Actually Matrix Equations," *Journal of Chemical Education,* Vol. 68, No. 12, Dec. 1991, p. 984.

EXTENDED APPLICATION: Contagion

Suppose that three people have contracted a contagious disease.* A second group of five people may have been in contact with the three infected persons. A third group of six people may have been in contact with the second group. We can form a 3 × 5 matrix P with rows representing the first group of three and columns representing the second group of five. We enter a one in the corresponding position if a person in the first group has contact with a person in the second group. These direct contacts are called *first-order contacts*. Similarly, we form a 5 × 6 matrix Q representing the first-order contacts between the second and third group. For example, suppose

$$P = \begin{bmatrix} 1 & 0 & 0 & 1 & 0 \\ 0 & 0 & 1 & 1 & 0 \\ 1 & 1 & 0 & 0 & 0 \end{bmatrix} \text{ and }$$

$$Q = \begin{bmatrix} 1 & 1 & 0 & 1 & 1 & 1 \\ 0 & 0 & 0 & 0 & 1 & 0 \\ 0 & 0 & 0 & 0 & 0 & 0 \\ 0 & 1 & 0 & 1 & 0 & 0 \\ 1 & 0 & 0 & 0 & 1 & 0 \end{bmatrix}.$$

From matrix P we see that the first person in the first group had contact with the first and fourth persons in the second group. Also, none of the first group had contact with the last person in the second group.

A *second-order contact* is an indirect contact between persons in the first and third groups through some person in the second group. The product matrix PQ indicates these contacts. Verify that the second-row, fourth-column entry of PQ is 1. That is, there is one second-order contact between the second person in group one and the fourth person in group three. Let

a_{ij} denote the element in the i-th row and j-th column of the matrix PQ. By looking at the products that form a_{24} below, we see that the common contact was with the fourth individual in group two. (The p_{ij} are entries in P, and the q_{ij} are entries in Q.)

$$a_{24} = p_{21}q_{14} + p_{22}q_{24} + p_{23}q_{34} + p_{24}q_{44} + p_{25}q_{54}$$
$$= 0 \cdot 1 + 0 \cdot 0 + 1 \cdot 0 + 1 \cdot 1 + 0 \cdot 1$$
$$= 1$$

The second person in group one and the fourth person in group three both had contact with the fourth person in group two.

This idea could be extended to third-, fourth-, and larger-order contacts. It indicates a way to use matrices to trace the spread of a contagious disease. It could also pertain to the dispersal of ideas or anything that might pass from one individual to another.

Exercises

1. Find the second-order contact matrix PQ mentioned in the text.

2. How many second-order contacts were there between the second contagious person and the third person in the third group?

3. Is there anyone in the third group who has had no contacts at all with the first group?

4. The totals of the columns in PQ give the total number of second-order contacts per person, while the column totals in P and Q give the total number of first-order contacts per person. Which person(s) in the third group had the most contacts, counting first- and second-order contacts?

*Grossman, Stanley, "First and Second Order Contact to a Contagious Disease," *Finite Mathematics with Applications to Business, Life Sciences, and Social Sciences,* WCB/McGraw-Hill, 1993.

Linear Programming: The Graphical Method

An oil refinery turns crude oil into many different products, including gasoline and fuel oil. Efficient management requires matching the output of each product to the demand and the available shipping capacity. In an exercise in Section 3 we explore the use of linear programming to allocate refinery production for maximum profit.

Many realistic problems involve inequalities—a factory can manufacture *no more* than 12 items on a shift, or a medical researcher must interview *at least* a hundred patients to be sure that a new treatment for a disease is better than the old treatment. *Linear inequalities* of the form $ax + by \le c$ (or with $\ge$, $<$, or $>$ instead of $\le$) can be used in a process called *linear programming* to *optimize* (find the maximum or minimum value of a quantity) for a given situation.

In this chapter we introduce some *linear programming* problems that can be solved by graphical methods. Then, in Chapter 4, we discuss the simplex method, a general method for solving linear programming problems with many variables.

■ 3.1   GRAPHING LINEAR INEQUALITIES

  THINK ABOUT IT   How can a company determine the feasible number of units of each product to manufacture in order to meet all production requirements?

We can answer this question by graphing a set of inequalities.

As mentioned above, a linear inequality is defined as follows.

LINEAR INEQUALITY

A **linear inequality** in two variables has the form

$$ax + by \le c$$
$$ax + by < c,$$
$$ax + by \ge c,$$
$$\text{or} \quad ax + by > c,$$

for real numbers a, b, and c, with a and b not both 0.

EXAMPLE 1   Graphing an Inequality
Graph the linear inequality $3x - 2y \le 6$.

Solution Because of the "=" portion of $\le$, the points of the line $3x - 2y = 6$ satisfy the linear inequality $3x - 2y \le 6$ and are part of its graph. As in Chapter 1, find the intercepts by first letting $x = 0$ and then letting $y = 0$; use these points to get the graph of $3x - 2y = 6$ shown in Figure 1.

FOR REVIEW ■

Recall from Chapter 1 that one way to sketch a line is to first let $x = 0$ to find the y-intercept, then let $y = 0$ to find the x-intercept. For example, given $3x - 2y = 6$, letting $x = 0$ yields $-2y = 6$, so $y = -3$, and the corresponding point is $(0, -3)$. Letting $y = 0$ yields $3x = 6$, so $x = 2$ and the point is $(2, 0)$. Plot these two points, as in Figure 1, then use a straightedge to draw a line through them.

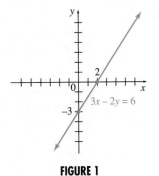

FIGURE 1

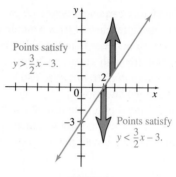

Points satisfy $y > \frac{3}{2}x - 3.$

Points satisfy $y < \frac{3}{2}x - 3.$

FIGURE 2

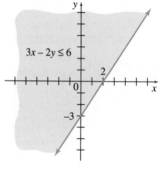

$3x - 2y \leq 6$

FIGURE 3

The points on the line satisfy "$3x - 2y$ *equals* 6." To locate the points satisfying "$3x - 2y$ *is less than* or equal to 6," first solve $3x - 2y \leq 6$ for y.

$$3x - 2y \leq 6$$
$$-2y \leq -3x + 6$$
$$y \geq \frac{3}{2}x - 3$$

(Recall that multiplying both sides of an inequality by a negative number reverses the direction of the inequality symbol.)

As shown in Figure 2, the points *above* the line $3x - 2y = 6$ satisfy

$$y > \frac{3}{2}x - 3,$$

while those below the line satisfy

$$y < \frac{3}{2}x - 3.$$

In summary, the inequality $3x - 2y \leq 6$ is satisfied by all points *on or above* the line $3x - 2y = 6$. Indicate the points above the line by shading, as in Figure 3. The line and shaded region in Figure 3 make up the graph of the linear inequality $3x - 2y \leq 6$.

CAUTION In this chapter, be sure to use a straightedge to draw lines, and to plot the points with care. A sloppily drawn line could give a deceptive picture of the region being considered.

In Example 1, the line $3x - 2y = 6$, which separates the points in the solution from the points that are not in the solution, is called the **boundary.**

EXAMPLE 2 Graphing an Inequality
Graph $x + 4y < 4$.

Solution The boundary here is the line $x + 4y = 4$. Since the points on this line do not satisfy $x + 4y < 4$, the line is drawn dashed, as in Figure 4. To decide whether to shade the region above the line or the region below the line, solve for y.

$$x + 4y < 4$$
$$4y < -x + 4$$
$$y < -\frac{1}{4}x + 1$$

Since y is less than $(-1/4)x + 1$, the solution is the region below the boundary, as shown by the shaded region in Figure 4.

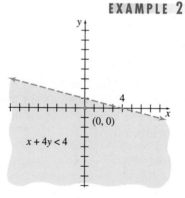

$x + 4y < 4$

$(0, 0)$

FIGURE 4

There is an alternative way to find the correct region to shade, or to check the method shown above. Choose as a test point any point not on the boundary line. For example, in Example 2 we could choose the point $(0, 0)$, which is not on the line $x + 4y = 4$. Substitute 0 for x and 0 for y in the given inequality.

$$x + 4y < 4$$
$$0 + 4(0) < 4 \quad \text{Let } x = 0, y = 0.$$
$$0 < 4 \quad \text{True}$$

Since the result $0 < 4$ is true, the test point $(0,0)$ belongs on the side of the boundary line where all the points satisfy $x + 4y < 4$. For this reason, we shade the side containing $(0,0)$, as in Figure 4. Choosing a point on the other side of the line, such as $(1,5)$, would produce a false result when the values $x = 1$ and $y = 5$ were substituted into the given inequality. In such a case, we would shade the side of the line *not including* the test point.

CAUTION Be careful. If the point $(0,0)$ is on the boundary line, it cannot be used as a test point.

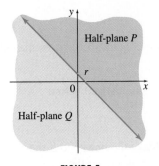

FIGURE 5

As the examples above suggest, the graph of a linear inequality is represented by a shaded region in the plane, perhaps including the line that is the boundary of the region. Each shaded region is an example of a **half-plane,** a region on one side of a line. For example, in Figure 5 line r divides the plane into half-planes P and Q. The points on r belong neither to P nor to Q. Line r is the boundary of each half-plane.

Graphing calculators can shade regions on the plane. Casio has an inequality mode which offers options for $y >$, $y <$, $y \geq$, or $y \leq$. Sharp has a fill screen which allows you to shade above and/or below selected functions. Refer to your instruction book for details.

TI calculators have a DRAW menu that includes an option to shade above or below a line. For instance, to graph the inequality in Example 2, first use your calculator to graph the line $y = -(1/4)x + 1$. Select the DRAW feature, then the Shade option, which requires an upper and a lower boundary for the region to be shaded. To match Figure 5, choose for the lower boundary a horizontal line that lies below the bottom of the graphing calculator screen. We will use a standard window with $-10 \leq y \leq 10$, and so we let $y = -20$ be the lower boundary. For the upper boundary, use $y = -(1/4)x + 1$. Then the command Shade(-20, $-(1/4)$X + 1) produces Figure 6(a).

The TI-83 calculator offers another way to graph the region above or below a line. Press the $y =$ key. Note the slanted line to the right of Y_1, Y_2, and so on. Use the left arrow key to move the cursor to that position for Y_1. Press ENTER until you see the symbol ◣. This indicates that the calculator will shade below the line whose equation is entered in Y_1. (The symbol ◥ operates similarly to shade above a line.) We used this method to get the graph in Figure 6(b).

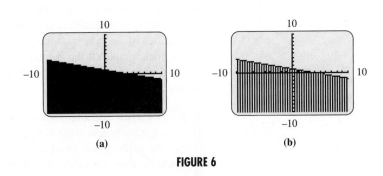

(a)   (b)

FIGURE 6

Notice that you cannot tell from the calculator graph whether the boundary line is solid or dashed. It is important to understand the concepts in order to interpret the graph correctly. In this case, the points on the line are not part of the solution, because of the strict inequality, $<$.

See *The Spreadsheet Manual* that is available with this book for information on graphing linear inequalities with a spreadsheet.

EXAMPLE 3 Graphing an Inequality

Graph $x \leq -1$.

Solution Recall that the graph of $x = -1$ is the vertical line through $(-1, 0)$. To decide which half-plane belongs to the solution, choose a test point. Choosing $(0, 0)$ and replacing x with 0 gives a false statement:

$$x < -1$$
$$0 < -1. \quad \text{False}$$

The correct half-plane is the one that does *not* contain $(0, 0)$; it is shaded in Figure 7.

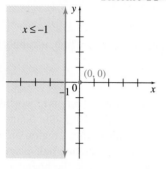

FIGURE 7

The steps in graphing a linear inequality are summarized below.

GRAPHING A LINEAR INEQUALITY

1. Draw the graph of the boundary line. Make the line solid if the inequality involves $\leq$ or $\geq$; make the line dashed if the inequality involves $<$ or $>$.

2. Decide which half-plane to shade. Use either of the following methods.

    a. Solve the inequality for y; shade the region above the line if the inequality is of the form $y > $ or $y \geq$; shade the region below the line if the inequality is of the form $y < $ or $y \leq$.

    b. Choose any point not on the line as a test point. Shade the half-plane that includes the test point if the test point satisfies the original inequality; otherwise, shade the half-plane on the other side of the boundary line.

Systems of Inequalities Realistic problems often involve many inequalities. For example, a manufacturing problem might produce inequalities resulting from production requirements, as well as inequalities about cost requirements. A collection of at least two inequalities is called a **system of inequalities.** The solution of a system of inequalities is made up of all those points that satisfy all the inequalities of the system at the same time. To graph the solution of a system of inequalities, graph all the inequalities on the same axes and identify, by heavy shading, the region common to all graphs. The next example shows how this is done.

> **NOTE** When shading regions by hand, it may be difficult to tell what is shaded heavily and what is shaded only lightly, particularly when more than two inequalities are involved. In such cases, an alternative technique is to shade the region *opposite* that of the inequality. In other words, the region that is *not* wanted can be shaded. Then, when the various regions are shaded, whatever is not shaded is the desired region. We will not use this technique in this text, but you may wish to try it on your own.

EXAMPLE 4 Graphing a System of Inequalities

Graph the system

$$y < -3x + 12$$
$$x < 2y.$$

Solution The graph of the first inequality has the line $y = -3x + 12$ as its boundary. Because of the $<$ symbol, we use a dotted line and shade *below* the line. The second inequality should first be solved for y to get $y > (1/2)x$ to see that the graph is the region *above* the dotted boundary line $y = (1/2)x$.

The heavily shaded region in Figure 8(a) shows all the points that satisfy both inequalities of the system. Since the points on the boundary lines are not in the solution, the boundary lines are dashed.

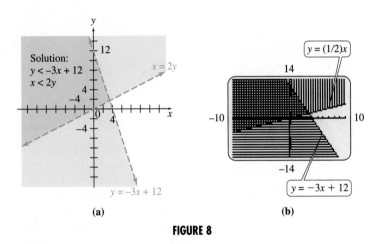

(a)   **(b)**

FIGURE 8

A calculator graph of the system in Example 4 is shown in Figure 8(b). You can also graph this system on your calculator using Shade(Y_2, Y_1).

A region consisting of the overlapping parts of two or more graphs of inequalities in a system, such as the heavily shaded region in Figure 8, is sometimes called the **region of feasible solutions** or the **feasible region,** since it is made up of all the points that satisfy (are feasible for) all inequalities of the system.

EXAMPLE 5 Feasible Region

Graph the feasible region for the system

$$2x - 5y \leq 10$$
$$x + 2y \leq 8$$
$$x \geq 0$$
$$y \geq 0.$$

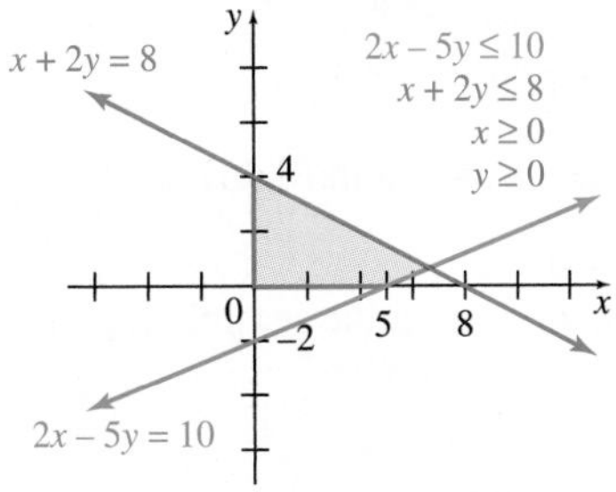

Solution On the same axes, graph each inequality by graphing the boundary and choosing the appropriate half-plane. Then find the feasible region by locating the overlap of all the half-planes. This feasible region is shaded in Figure 9.

FIGURE 9

▌ **NOTE** The inequalities $x \geq 0$ and $y \geq 0$ restrict the feasible region to the first quadrant.

Applications As shown in the rest of this chapter, many realistic problems lead to systems of linear inequalities. The next example is typical of such problems.

EXAMPLE 6 Manufacturing

Midtown Manufacturing Company makes plastic plates and cups, both of which require time on two machines. Manufacturing a run of plates requires one hour on machine A and two on machine B, and producing a run of cups requires three hours on machine A and one on machine B. Each machine is operated for at most 15 hours per day. Write a system of inequalities that expresses these restrictions, and sketch the feasible region.

Solution Start by making a chart that summarizes the given information.

	Plates	**Cups**	**Total**
Number of Units Made	x	y	
Hours on A	1	3	≤ 15
Hours on B	2	1	≤ 15

Here x represents the number of runs of plates to be made, and y represents the number of runs of cups. Since the machines are available at most 15 hours per day, we put the total number of hours as ≤ 15. Putting the inequality ($\leq$ or $\geq$) next to the number in the chart may help you remember which way to write the inequality.

On machine A, x runs of plates require a total of $1 \cdot x = x$ hours, and y runs of cups require $3 \cdot y = 3y$ hours. Since machine A is available no more than 15 hours per day,

$$x + 3y \leq 15. \quad \text{Machine A}$$

We translated "no more than" as "less than or equal to." Notice how this inequality corresponds to the row in the table for time on machine A. Similarly, the row corresponding to time on machine B gives

$$2x + y \leq 15. \quad \text{Machine B}$$

Since it is not possible to produce a negative number of cups or plates,

$$x \geq 0 \quad \text{and} \quad y \geq 0.$$

The feasible region for this system of inequalities is shown in Figure 10.

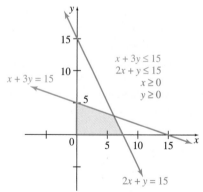

FIGURE 10

The following steps summarize the process of finding the feasible region.

1. Form a table that summarizes the information.

2. Convert the table into a set of linear inequalities.

3. Graph each linear inequality.

4. Graph the region that is common to all the regions graphed in step 3.

3.1 EXERCISES

Graph the following linear inequalities.

1. $x + y \leq 2$

2. $y \leq x + 1$

3. $x \geq 3 + y$

4. $y \geq x - 3$

5. $4x - y < 6$

6. $3y + x > 4$

7. $3x + y < 6$

8. $2x - y > 2$

9. $x + 3y \geq -2$

10. $2x + 3y \leq 6$

11. $x \leq 5y$

12. $2x \geq y$

13. $x + y \leq 0$

14. $3x + 2y \geq 0$

15. $y < x$

16. $y > -2x$

17. $x < 4$

18. $y > 5$

19. $y \leq -2$

20. $x \geq 3$

Graph the feasible region for each system of inequalities.

21. $x + y \leq 1$
$x - y \geq 2$

22. $2x - y < 1$
$3x + y < 6$

23. $x + 3y \leq 6$
$2x + 4y \geq 7$

24. $-x - y < 5$
$2x - y < 4$

25. $x + y \leq 4$
$x - y \leq 5$
$4x + y \leq -4$

26. $3x - 2y \geq 6$
$x + y \leq -5$
$y \leq 4$

27. $-2 < x < 3$
$-1 \leq y \leq 5$
$2x + y < 6$

28. $-2 < x < 2$
$y > 1$
$x - y > 0$

29. $2y + x \geq -5$
$y \leq 3 + x$
$x \geq 0$
$y \geq 0$

30. $2x + 3y \leq 12$
$2x + 3y > -6$
$3x + y < 4$
$x \geq 0$
$y \geq 0$

31. $3x + 4y > 12$
$2x - 3y < 6$
$0 \leq y \leq 2$
$x \geq 0$

32. $0 \leq x \leq 9$
$x - 2y \geq 4$
$3x + 5y \leq 30$
$y \geq 0$

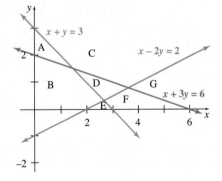 *Use a graphing calculator to graph the following.*

33. $2x - 4y > 3$

34. $4x - 3y < 12$

35. $3x - 4y < 6$
$2x + 5y > 15$

36. $6x - 4y > 8$
$3x + 2y > 4$

37. The regions A through G in the figure can be described by the inequalities

$$x + 3y \;?\; 6$$
$$x + y \;?\; 3$$
$$x - 2y \;?\; 2$$
$$x \geq 0$$
$$y \geq 0,$$

where ? can be either $\leq$ or $\geq$. For each region, tell what the ? should be in the three inequalities. For example, for region A, the ? should be $\geq$, $\leq$, and $\leq$, because region A is described by the inequalities

$$x + 3y \geq 6$$
$$x + y \leq 3$$
$$x - 2y \leq 2$$
$$x \geq 0$$
$$y \geq 0.$$

Applications

BUSINESS AND ECONOMICS

38. *Production Scheduling* A small pottery shop makes two kinds of planters, glazed and unglazed. The glazed type requires 1/2 hr to throw on the wheel and 1 hr in the kiln. The unglazed types takes 1 hr to throw on the wheel and 6 hr in the kiln. The wheel is available for at most 8 hr per day, and the kiln for at most 20 hr per day.

a. Complete this chart.

	Glazed	Unglazed	Total
Number Made	x	y	
Time on Wheel			
Time in Kiln			

b. Set up a system of inequalities and graph the feasible region.

c. Using your graph from part b, can 5 glazed and 2 unglazed planters be made? Can 10 glazed and 2 unglazed planters be made?

39. *Time Management* Carmella and Walt produce handmade shawls and afghans. They spin the yarn, dye it, and then weave it. A shawl requires 1 hr of spinning, 1 hr of dyeing, and 1 hr of weaving. An afghan needs 2 hr of spinning, 1 hr of dyeing, and 4 hr of weaving. Together, they spend at most 8 hr spinning, 6 hr dyeing, and 14 hr weaving.

a. Complete this chart.

	Shawls	Afghans	Total
Number Made	x	y	
Spinning Time			
Dyeing Time			
Weaving Time			

b. Set up a system of inequalities and graph the feasible region.

c. Using your graph from part b, can 3 shawls and 2 afghans be made? Can 4 shawls and 3 afghans be made?

For Exercises 40–45, perform each of the following steps.
(a) *Write a system of inequalities to express the conditions of the problem.*
(b) *Graph the feasible region of the system.*

40. *Transportation* Southwestern Oil supplies two distributors located in the Northwest. One distributor needs at least 3000 barrels of oil, and the other needs at least 5000 barrels. Southwestern can send out at most 10,000 barrels. Let $x =$ the number of barrels of oil sent to distributor 1 and $y =$ the number sent to distributor 2.

41. *Finance* The loan department in a bank will use at most $25 million for commercial and home loans. The bank's policy is to allocate at least four times as much money to home loans as to commercial loans. The bank's return is 12% on a home loan and 10% on a commercial loan. The manager of the loan department wants to earn a return of at least $2.8 million on these loans. Let $x =$ the amount for home loans and $y =$ the amount for commercial loans.

42. *Transportation* The California Almond Growers have at most 2400 boxes of almonds to be shipped from their plant in Sacramento to Des Moines and San Antonio. The Des Moines market needs at least 1000 boxes, while the San Antonio market must have at least 800 boxes. Let $x =$ the number of boxes to be shipped to Des Moines and $y =$ the number of boxes to be shipped to San Antonio.

43. *Management* An electric shaver manufacturer makes two models, the regular and the flex. Because of demand, the number of regular shavers made is never more than half the number of flex shavers. The factory's production cannot exceed 1000 shavers per week. Let $x =$ the number of regular shavers and $y =$ the number of flex shavers produced per week.

44. *Production Scheduling* A cement manufacturer produces at least 3.2 million barrels of cement annually. He is told by the Environmental Protection Agency (EPA) that his operation emits 2.5 lb of dust for each barrel produced. The EPA has ruled that annual emissions must be reduced to no more than 1.8 million pounds. To do this the manufacturer plans to replace the present dust collectors with two types of electronic precipitators. One type would reduce emissions to .5 lb per barrel and operating costs would be 16¢ per barrel.

The other would reduce the dust to .3 lb per barrel and operating costs would be 20¢ per barrel. The manufacturer does not want to spend more than .8 million dollars in operating costs on the precipitators. He needs to know how many barrels he could produce with each type. Let $x =$ the number of barrels (in millions) produced with the first type and $y =$ the number of barrels (in millions) produced with the second type.

LIFE SCIENCES

45. *Nutrition* A dietician is planning a snack package of fruit and nuts. Each ounce of fruit will supply 1 unit of protein, 2 units of carbohydrates, and 1 unit of fat. Each ounce of nuts will supply 1 unit of protein, 1 unit of carbohydrates, and 1 unit of fat. Every package must provide at least 7 units of protein, at least 10 units of carbohydrates, and no more than 9 units of fat. Let $x =$ the ounces of fruit and $y =$ the ounces of nuts to be used in each package.

■ 3.2 SOLVING LINEAR PROGRAMMING PROBLEMS GRAPHICALLY

Many mathematical models designed to solve problems in business, biology, and economics involve finding optimum value (maximum or minimum) of a function, subject to certain restrictions. In a **linear programming** problem, we must find the maximum or minimum value of a function, called the **objective function,** and also satisfy a set of restrictions, or **constraints,** given by linear inequalities. When only two variables are involved, the solution to a linear programming problem can be found by first graphing the set of constraints, then finding the feasible region as discussed in the previous section. This method is explained in the following example.

EXAMPLE 1 Maximization

Find the maximum value of the objective function $z = 2x + 5y$, subject to the following constraints.

$$3x + 2y \leq 6$$
$$-x + 2y \leq 4$$
$$x \geq 0$$
$$y \geq 0$$

Solution The feasible region is graphed in Figure 11. We can find the coordinates of point A, $(1/2, 9/4)$, by solving the system

$$3x + 2y = 6$$
$$-x + 2y = 4.$$

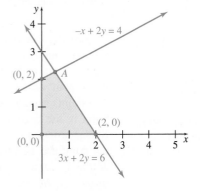

FIGURE 11

Every point in the feasible region satisfies all the constraints; however, we want to find those points that produce the maximum possible value of the objective function. To see how to find this maximum value, change the graph of Figure 11 by adding lines that represent the objective function $z = 2x + 5y$ for various sample values of z. By choosing the values 0, 5, 10, and 15 for z, the objective function becomes (in turn)

$$0 = 2x + 5y, \quad 5 = 2x + 5y, \quad 10 = 2x + 5y, \quad \text{and} \quad 15 = 2x + 5y.$$

These four lines are graphed in Figure 12. (Why are the lines parallel?) The figure shows that z cannot take on the value 15 because the graph for $z = 15$ is entirely outside the feasible region. The maximum possible value of z will be obtained from a line parallel to the others and between the lines representing the objective function when $z = 10$ and $z = 15$. The value of z will be as large as possible and all constraints will be satisfied if this line just touches the feasible region. This occurs at point A. We find that A has coordinates $(1/2, 9/4)$. (See the review in the margin.) The value of z at this point is

$$z = 2x + 5y = 2\left(\frac{1}{2}\right) + 5\left(\frac{9}{4}\right) = 12\frac{1}{4}.$$

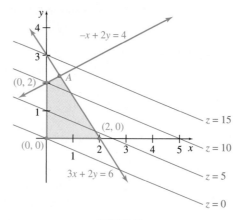

FIGURE 12

FOR REVIEW ◼

Recall from Chapter 2 that two equations in two unknowns can be solved by using row operations to eliminate one variable. For example, to solve the system

$$3x + 2y = 6$$
$$-x + 2y = 4,$$

we could take the first equation plus 3 times the second to eliminate x. (This is equivalent to $R_1 + 3R_2 \rightarrow R_2$ in the Gauss-Jordan method.) The result is $8y = 18$, so $y = 18/8 = 9/4$. We can then substitute this value of y into either equation and solve for x. For example, substitution into the first equation yields

$$3x + 2\left(\frac{9}{4}\right) = 6$$

$$3x = 6 - \frac{9}{2} = \frac{3}{2}$$

$$x = \frac{1}{2}.$$

We instead could have subtracted the two original equations to eliminate y, yielding $4x = 2$, or $x = 1/2$.

The maximum possible value of z is $12\frac{1}{4}$. Of all the points in the feasible region, A leads to the largest possible value of z. ◼

A graphing calculator is particularly useful for finding the coordinates of intersection points such as point A. We do this by solving each equation for y, graphing each line, and then using the capability of the calculator to find the coordinates of the point of intersection.

Points such as A in Example 1 are called corner points. A **corner point** is a point in the feasible region where the boundary lines of two constraints cross. Since corner points occur where two straight lines cross, the coordinates of a corner point are the solution of a system of two linear equations. As we saw in Example 1, corner points play a key role in the solution of linear programming problems. We will make this explicit after the following example.

EXAMPLE 2 Minimization

Solve the following linear programming problem.

$$\begin{aligned}
\text{Minimize} \quad & z = 2x + 4y \\
\text{subject to:} \quad & x + 2y \geq 10 \\
& 3x + y \geq 10 \\
& x \geq 0 \\
& y \geq 0.
\end{aligned}$$

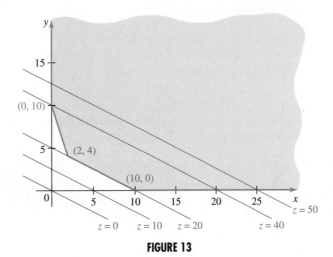

FIGURE 13

Solution Figure 13 shows the feasible region and the lines that result when z in the objective function is replaced by 0, 10, 20, 40, and 50. The line representing the objective function touches the region of feasible solutions when $z = 20$. Two corner points, $(2, 4)$ and $(10, 0)$, lie on this line; both $(2, 4)$ and $(10, 0)$, as well as all the points on the boundary line between them, give the same optimum value of z. There are infinitely many equally good values of x and y that will give the same minimum value of the objective function $z = 2x + 4y$. This minimum value is 20.

The feasible region in Example 1 is **bounded,** since the region is enclosed by boundary lines on all sides. Linear programming problems with bounded regions always have solutions. On the other hand, the feasible region in Example 2 is **unbounded,** and no solution will *maximize* the value of the objective function.

Some general conclusions can be drawn from the method of solution used in Examples 1 and 2. Figure 14 on the next page shows various feasible regions and the lines that result from various values of z. (We assume the lines are in order from left to right as z increases.) In Figure 14(a), the objective function takes on its minimum value at corner point Q and its maximum value at P. The minimum is again at Q in part (b), but the maximum occurs at P_1 or P_2, or any point on the line segment connecting them. Finally, in part (c), the minimum value occurs at Q, but the objective function has no maximum value because the feasible region is unbounded. As long as the objective function increases as x and y increase, the objective function will have no maximum over an unbounded region.

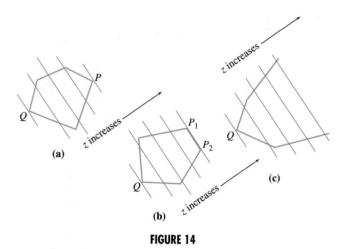

FIGURE 14

The preceding discussion suggests the truth of the **corner point theorem.**

CORNER-POINT THEOREM

If an optimum value (either a maximum or a minimum) of the objective function exists, it will occur at one or more of the corner points of the feasible region.

This theorem simplifies the job of finding an optimum value. First, we graph the feasible region and find all corner points. Then we test each corner point in the objective function. Finally, we identify the corner point producing the optimum solution. For unbounded regions, we must decide whether the required optimum can be found (see Example 2).

With the theorem, we can solve the problem in Example 1 by first identifying the four corner points in Figure 11: $(0,0)$, $(0,2)$, $(1/2, 9/4)$, and $(2,0)$. Then we substitute each of the four points into the objective function $z = 2x + 5y$ to identify the corner point that produces the maximum value of z.

Corner Point	Value of $z = 2x + 5y$	
$(0,0)$	$2(0) + 5(0) = 0$	
$(0,2)$	$2(0) + 5(2) = 10$	
$\left(\frac{1}{2}, \frac{9}{4}\right)$	$2\left(\frac{1}{2}\right) + 5\left(\frac{9}{4}\right) = 12\frac{1}{4}$	Maximum
$(2,0)$	$2(2) + 5(0) = 4$	

From these results, the corner point $(1/2, 9/4)$ yields the maximum value of $12\frac{1}{4}$. This is the same as the result found earlier.

The following summary gives the steps to use in solving a linear programming problem by the graphical method.

SOLVING A LINEAR PROGRAMMING PROBLEM

1. Write the objective function and all necessary constraints.

2. Graph the feasible region.

3. Identify all corner points.

4. Find the value of the objective function at each corner point.

5. For a bounded region, the solution is given by the corner point producing the optimum value of the objective function.

6. For an unbounded region, check that a solution actually exists. If it does, it will occur at a corner point.

EXAMPLE 3 Maximization and Minimization

Sketch the feasible region for the following set of constraints, and then find the maximum and minimum values of the objective function $z = 5x + 2y$.

$$3y - 2x \geq 0$$

$$y + 8x \leq 52$$

$$y - 2x \leq 2$$

$$x \geq 3$$

Solution The graph in Figure 15 shows that the feasible region is bounded. Use the corner points from the graph to find the maximum and minimum values of the objective function.

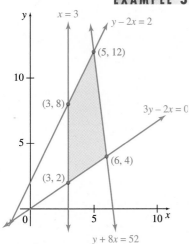

FIGURE 15

Corner Point	Value of $z = 5x + 2y$	
$(3, 2)$	$5(3) + 2(2) = 19$	Minimum
$(6, 4)$	$5(6) + 2(4) = 38$	
$(5, 12)$	$5(5) + 2(12) = 49$	Maximum
$(3, 8)$	$5(3) + 2(8) = 31$	

The minimum value of $z = 5x + 2y$ is 19 at the corner point $(3, 2)$. The maximum value is 49 at $(5, 12)$.

To verify that the minimum or maximum is correct in a linear programming problem, you might want to add the graph of the line $z = 0$ to the graph of the feasible region. For instance, in Example 3, the result of adding the line $5x + 2y = 0$ is shown in Figure 16. Now imagine moving a straightedge through the feasible region parallel to this line. It appears that the first place the line touches the feasible region is at $(3, 2)$, where we found the minimum. Similarly, the last place the line touches is at $(5, 12)$, where we found the maximum. In Figure 16, these parallel lines, labeled $z = 19$ and $z = 49$, are also shown.

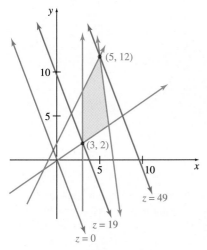

FIGURE 16

3.2 EXERCISES

The graphs in Exercises 1–6 show regions of feasible solutions. Use these regions to find maximum and minimum values of the given objective functions.

1. $z = 3x + 5y$

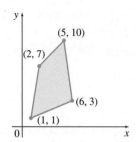

2. $z = 6x - y$

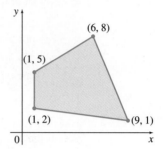

3. $z = .40x + .75y$

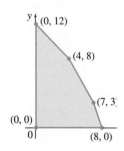

4. $z = .35x + 1.25y$

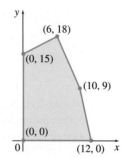

5. a. $z = 4x + 2y$
   b. $z = 2x + 3y$
   c. $z = 2x + 4y$
   d. $z = x + 4y$

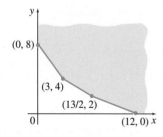

6. a. $z = 4x + y$
   b. $z = 5x + 6y$
   c. $z = x + 2y$
   d. $z = x + 6y$

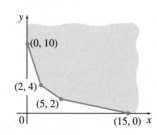

Use graphical methods to solve the linear programming problems in Exercises 7–14.

7. Minimize $z = 5x + 2y$
    subject to: $2x + 3y \geq 6$
                $4x + y \geq 6$
                $x \geq 0$
                $y \geq 0.$

8. Minimize $z = x + 3y$
    subject to: $x + y \leq 10$
                $5x + 2y \geq 20$
                $-x + 2y \geq 0$
                $x \geq 0$
                $y \geq 0.$

9. Maximize $z = 2x + 2y$
    subject to: $3x - y \geq 12$
                $x + y \leq 15$
                $x \geq 2$
                $y \geq 5.$

10. Maximize $\quad z = 10x + 8y$
  subject to: $\quad 2x + 3y \le 100$
  $\quad\quad\quad\quad 5x + 4y \le 200$
  $\quad\quad\quad\quad\quad\quad x \ge 10$
  $\quad\quad\quad\quad 0 \le y \le 20.$

11. Maximize $\quad z = 4x + 2y$
  subject to: $\quad x - y \le 10$
  $\quad\quad\quad\quad 5x + 3y \le 75$
  $\quad\quad\quad\quad x + y \le 20$
  $\quad\quad\quad\quad\quad x \ge 0$
  $\quad\quad\quad\quad\quad y \ge 0.$

12. Maximize $\quad z = 4x + 5y$
  subject to: $\quad 10x - 5y \le 100$
  $\quad\quad\quad\quad 20x + 10y \ge 150$
  $\quad\quad\quad\quad x + y \ge 12$
  $\quad\quad\quad\quad\quad x \ge 0$
  $\quad\quad\quad\quad\quad y \ge 0.$

13. Find values of $x \ge 0$ and $y \ge 0$ that maximize $z = 10x + 12y$ subject to each of the following sets of constraints.

a. $x + y \le 20$
   $x + 3y \le 24$

b. $3x + y \le 15$
   $x + 2y \le 18$

c. $2x + 5y \ge 22$
   $4x + 3y \le 28$
   $2x + 2y \le 17$

14. Find values of $x \ge 0$ and $y \ge 0$ that minimize $z = 3x + 2y$ subject to each of the following sets of constraints.

a. $10x + 7y \le 42$
   $4x + 10y \ge 35$

b. $6x + 5y \ge 25$
   $2x + 6y \ge 15$

c. $x + 2y \ge 10$
   $2x + y \ge 12$
   $x - y \le 8$

15. You are given the following linear programming problem:*

Maximize $\quad z = c_1 x_1 + c_2 x_2$
subject to: $\quad 2x_1 + x_2 \le 11$
$\quad\quad\quad\quad -x_1 + 2x_2 \le 2$
$\quad\quad\quad\quad x_1 \ge 0, x_2 \ge 0.$

If $c_2 > 0$, determine the range of c_1/c_2 for which $(x_1, x_2) = (4, 3)$ is an optimal solution.

a. $[-2, 1/2]$ **b.** $[-1/2, 2]$ **c.** $[-11, -1]$

d. $[1, 11]$ **e.** $[-11, 11]$

3.3 APPLICATIONS OF LINEAR PROGRAMMING

 THINK ABOUT IT

How many filing cabinets of different types should an office manager purchase, given a limited budget and limited office space?

We will use linear programming to answer this question in Example 1.

EXAMPLE 1 Filing Cabinets

An office manager needs to purchase new filing cabinets. He knows that Ace cabinets cost $40 each, require 6 square feet of floor space, and hold 24 cubic feet of files. On the other hand, each Excello cabinet costs $80, requires 8 square feet of floor space, and holds 36 cubic feet. His budget permits him to spend no more than $560 on cabinets, while the office has space for no more than 72 square feet of cabinets. The manager desires the greatest storage capacity within the limitations imposed by funds and space. How many of each type of cabinet should he buy?

---

*Problem 5 from "November 1989 Course 130 Examination Operations Research" of the *Education and Examination Committee of The Society of Actuaries*. Reprinted by permission of The Society of Actuaries.

Solution Let x represent the number of Ace cabinets to be bought and let y represent the number of Excello cabinets. Summarize the given information in a table.

	Ace	Excello	Total
Number of Cabinets	x	y	
Cost of Each	$40	$80	$\leq$ $560
Space Required	6 ft²	8 ft²	$\leq$ 72 ft²
Storage Capacity	24 ft³	36 ft³	

The constraints imposed by cost and space correspond to the rows in the table as follows.

$$40x + 80y \leq 560 \quad \text{Cost}$$
$$6x + 8y \leq 72 \quad \text{Floor space}$$

Since the number of cabinets cannot be negative, $x \geq 0$ and $y \geq 0$. The objective function to be maximized gives the amount of storage capacity provided by some combination of Ace and Excello cabinets. If the variable z represents the total storage space used, the objective function is

$$z = 24x + 36y.$$

In summary, the mathematical model for the given linear programming problem is as follows:

Maximize	$z = 24x + 36y$		(1)
subject to:	$40x + 80y \leq 560$		(2)
	$6x + 8y \leq 72$		(3)
	$x \geq 0$		(4)
	$y \geq 0.$		(5)

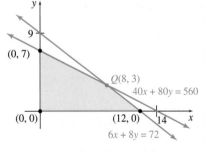

FIGURE 17

Using the methods described in the previous section, graph the feasible region for the system of inequalities (2)–(5), as in Figure 17. Three of the corner points can be identified from the graph as $(0,0)$, $(0,7)$, and $(12,0)$. The fourth corner point, labeled Q in the figure, can be found by solving the system of equations

$$40x + 80y = 560$$
$$6x + 8y = 72.$$

Solve this system to find that Q is the point $(8,3)$. Now test these four points in the objective function to determine the maximum value of z. The results are shown in the table.

Corner Point	Value of $z = 24x + 36y$
$(0,0)$	0
$(0,7)$	252
$(12,0)$	288
$(8,3)$	300 Maximum

The objective function, which represents storage space, is maximized when $x = 8$ and $y = 3$. The manager should buy 8 Ace cabinets and 3 Excello cabinets.

Fortunately, the answer to the linear programming problem in Example 1 is a point with integer coordinates, as the number of each type of cabinet must be an integer. Unfortunately, there is no guarantee that this will always happen. When the solution to a linear programming problem is restricted to integers, it is an *integer programming* problem, which is more difficult to solve than a linear programming problem. In this text, all problems in which fractional solutions are meaningless are contrived to have integer solutions.

EXAMPLE 2 Farm Animals

A 4-H member raises only geese and pigs. She wants to raise no more than 16 animals, including no more than 10 geese. She spends $5 to raise a goose and $15 to raise a pig, and she has $180 available for this project. The 4-H member wishes to maximize her profits. Each goose produces $6 in profit and each pig $20 in profit.

Solution First, set up a table that shows the information given in the problem.

	Geese	Pigs	Total
Number Raised	x	y	$\leq$ 16
Cost to Raise	$5	$15	$\leq$ $180
Profit (each)	$6	$20	

Use the table to write the necessary constraints. Since the total number of animals cannot exceed 16, the first constraint is

$$x + y \leq 16.$$

The cost to raise x geese at $5 per goose is $5x$ dollars, while the cost for y pigs at $15 each is $15y$ dollars. Since only $180 is available,

$$5x + 15y \leq 180.$$

Dividing both sides by 5 gives the equivalent inequality

$$x + 3y \leq 36.$$

"No more than 10 geese" means

$$x \leq 10.$$

The number of geese and pigs cannot be negative, so

$$x \geq 0 \quad \text{and} \quad y \geq 0.$$

The 4-H member wants to know how many geese and pigs to raise in order to produce maximum profit. Each goose yields $6 profit and each pig $20. If z represents total profit, then

$$z = 6x + 20y.$$

In summary, we have the following linear programming problem:

$$\text{Maximize} \quad z = 6x + 20y$$
$$\text{subject to:} \quad x + y \le 16$$
$$x + 3y \le 36$$
$$x \le 10$$
$$x \ge 0$$
$$y \ge 0.$$

A graph of the feasible region is shown in Figure 18. The corner points $(0, 12)$, $(0, 0)$, and $(10, 0)$ can be read directly from the graph. The coordinates of each of the other corner points can be found by solving a system of linear equations.

Test each corner point in the objective function to find the maximum profit.

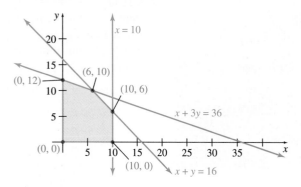

FIGURE 18

Corner Point	Value of $z = 6x + 20y$	
$(0, 12)$	$6(0) + 20(12) = 240$	Maximum
$(6, 10)$	$6(6) + 20(10) = 236$	
$(10, 6)$	$6(10) + 20(6) = 180$	
$(10, 0)$	$6(10) + 20(0) = 60$	
$(0, 0)$	$6(0) + 20(0) = 0$	

The maximum of 240 occurs at $(0, 12)$. Thus, 12 pigs and no geese will produce a maximum profit of $240.

EXAMPLE 3 Nutrition

Certain laboratory animals must have at least 30 grams of protein and at least 20 grams of fat per feeding period. These nutrients come from food A, which costs 18 cents per unit and supplies 2 grams of protein and 4 of fat; and food B, which costs 12 cents per unit and has 6 grams of protein and 2 of fat. Food B is bought under a long-term contract requiring that at least 2 units of B be used per serving. How much of each food must be bought to produce the minimum cost per serving?

Solution Let x represent the required amount of food A and y the amount of food B. Use the given information to prepare the following table.

	Food A	Food B	Total
Number of Units	x	y	
Grams of Protein	2	6	≥ 30
Grams of Fat	4	2	≥ 20
Cost	18¢	12¢	

Since the animals must have *at least* 30 grams of protein and 20 grams of fat, we use $\geq$ in the inequality. If the animals needed *at most* a certain amount of some nutrient, we would use $\leq$.

The linear programming problem can be stated as follows.

$$\text{Minimize} \qquad z = .18x + .12y$$
$$\text{subject to:} \qquad 2x + 6y \geq 30 \qquad \text{Protein}$$
$$4x + 2y \geq 20 \qquad \text{Fat}$$
$$y \geq 2$$
$$x \geq 0.$$

(The usual constraint $y \geq 0$ is redundant because of the constraint $y \geq 2$.) A graph of the feasible region is shown in Figure 19. The corner points are $(0, 10)$, $(3, 4)$, and $(9, 2)$.

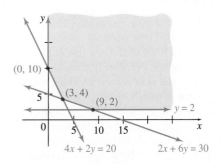

FIGURE 19

Test each corner point in the objective function to find the minimum cost.

Corner Point	Value of $z = .18x + .12y$	
$(0, 10)$	$.18(0) + .12(10) = 1.20$	
$(3, 4)$	$.18(3) + .12(4) = 1.02$	Minimum
$(9, 2)$	$.18(9) + .12(2) = 1.86$	

The minimum of 1.02 occurs at $(3, 4)$. Thus, 3 units of food A and 4 units of food B will produce a minimum cost of $1.02 per serving.

The feasible region in Figure 19 is an *unbounded* feasible region—the region extends indefinitely to the upper right. With this region it would not be possible to *maximize* the objective function, because the total cost of the food could always be increased by encouraging the animals to eat more.

3.3 EXERCISES

Write Exercises 1–6 as linear inequalities. Identify all variables used. (Note: Not all of the given information is used in Exercises 5 and 6.)

1. Product A requires 2 hr on machine I, while product B needs 3 hr on the same machine. The machine is available for at most 45 hr per week.

2. A cow requires a third of an acre of pasture and a sheep needs a quarter acre. A rancher wants to use at least 120 acres of pasture.

3. Jesus Garcia needs at least 25 units of vitamin A per day. Green pills provide 4 units and red pills provide 1.

4. Pauline Wong spends 3 hr selling a small computer and 5 hr selling a larger model. She works no more than 45 hr per week.

5. Coffee costing $6 per pound is to be mixed with coffee costing $5 per pound to get at least 50 lb of a blended coffee.

6. A tank in an oil refinery holds 120 gal. The tank contains a mixture of light oil worth $1.25 per gallon and heavy oil worth $.80 per gallon.

Applications

BUSINESS AND ECONOMICS

7. *Transportation* The Miers Company produces small engines for several manufacturers. The company receives orders from two assembly plants for their Top-flight engine. Plant I needs at least 50 engines, and plant II needs at least 27 engines. The company can send at most 85 engines to these two assembly plants. It costs $20 per engine to ship to plant I and $35 per engine to ship to plant II. Plant I gives Miers $15 in rebates towards its products for each engine they buy, while plant II gives similar $10 rebates. Miers estimates that they need at least $1110 in rebates to cover products they plan to buy from the two plants. How many engines should be shipped to each plant to minimize shipping costs? What is the minimum cost?

8. *Transportation* A manufacturer of refrigerators must ship at least 100 refrigerators to its two West Coast warehouses. Each warehouse holds a maximum of 100 refrigerators. Warehouse A holds 25 refrigerators already, and warehouse B has 20 on hand. It costs $12 to ship a refrigerator to warehouse A and $10 to ship one to warehouse B. Union rules require that at least 300 workers be hired. Shipping a refrigerator to warehouse A requires 4 workers, while shipping a refrigerator to warehouse B requires 2 workers. How many refrigerators should be shipped to each warehouse to minimize costs? What is the minimum cost?

9. *Insurance Premiums* A company is considering two insurance plans with the types of coverage and premiums shown in the following table.

	Policy A	Policy B
Fire/Theft	$10,000	$15,000
Liability	$180,000	$120,000
Premium	$50	$40

(For example, this means that $50 buys one unit of plan A, consisting of $10,000 fire and theft insurance and $180,000 of liability insurance.)

a. The company wants at least $300,000 fire/theft insurance and at least $3,000,000 liability insurance from these plans. How many units should be purchased from each plan to minimize the cost of the premiums? What is the minimum premium?

b. Suppose the premium for policy A is reduced to $25. Now how many units should be purchased from each plan to minimize the cost of the premiums? What is the minimum premium?

10. *Profit* Seall Manufacturing Company makes color television sets. It produces a bargain set that sells for $100 profit and a deluxe set that sells for $150 profit. On the assembly line the bargain set requires 3 hr, and the deluxe set takes 5 hr. The cabinet shop spends 1 hr on the cabinet for the bargain set and 3 hr on the cabinet for the deluxe set. Both sets require 2 hr of time for testing and packing. On a particular production run, the Seall Company has available 3900 work hours on the assembly line, 2100 work hours in

the cabinet shop, and 2200 work hours in the testing and packing department.

a. How many sets of each type should it produce to make a maximum profit? What is the maximum profit?

b. Suppose the profit on deluxe sets goes up to $170. Now how many sets of each type should be produced to make a maximum profit? What is the maximum profit?

11. *Revenue* A machine shop manufacturers two types of bolts. The bolts require time on each of three groups of machines, but the time required on each group differs, as shown in the table below.

		Machine Group		
		I	**II**	**III**
Bolts	Type 1	.1 min	.1 min	.1 min
	Type 2	.1 min	.4 min	.02 min

Production schedules are made up one day at a time. In a day there are 240, 720, and 160 min available, respectively, on these machines. Type 1 bolts sell for 10¢ and type 2 bolts for 12¢. How many of each type of bolt should be manufactured per day to maximize revenue? What is the maximum revenue?

12. *Revenue* The manufacturing process requires that oil refineries must manufacture at least 2 gal of gasoline for every gallon of fuel oil. To meet the winter demand for fuel oil, at least 3 million gallons a day must be produced. The demand for gasoline is no more than 6.4 million gallons per day. It takes .25 hour to ship each million gallons of gasoline and 1 hour to ship each million gallons of fuel oil out of the warehouse. No more than 4.65 hours are available for shipping. If the refinery sells gasoline for $1.25 per gallon and fuel oil for $1 per gallon, how much of each should be produced to maximize revenue? Find the maximum revenue.

13. *Revenue* A candy company has 100 kg of chocolate-covered nuts and 125 kg of chocolate-covered raisins to be sold as two different mixes. One mix will contain half nuts and half raisins and will sell for $6 per kilogram. The other mix will contain 1/3 nuts and 2/3 raisins and will sell for $4.80 per kilogram.

a. How many kilograms of each mix should the company prepare for maximum revenue? Find the maximum revenue.

b. The company raises the price of the half-and-half mix to $8 per kilogram. Now how many kilograms of each mix should the company prepare for maximum revenue? Find the maximum revenue.

14. *Profit* A small country can grow only two crops for export, coffee and cocoa. The country has 500,000 hectares of land available for the crops. Long-term contracts require that at least 100,000 hectares be devoted to coffee and at least 200,000 hectares to cocoa. Cocoa must be processed locally, and production bottlenecks limit cocoa to 270,000 hectares. Coffee requires two workers per hectare, with cocoa requiring five. No more than 1,750,000 people are available for working with these crops. Coffee produces a profit of $220 per hectare and cocoa a profit of $550 per hectare. How many hectares should the country devote to each crop in order to maximize profit? Find the maximum profit.

15. *Blending* The Mostpure Milk Company gets milk from two dairies and then blends the milk to get the desired amount of butterfat for the company's premier product. Milk from dairy I costs $.60 per gallon, and milk from dairy II costs $.20 per gallon. At most $36 is available for purchasing milk. Dairy I can supply at most 50 gallons of milk averaging 3.7% butterfat. Dairy II can supply at most 80 gallons of milk averaging 3.2% butterfat. How much milk from each supplier should Mostpure use to get at most 100 gallons of milk with the maximum total amount of butterfat? What is the maximum amount of butterfat?

16. *Transportation* A greeting card manufacturer has 370 boxes of a particular card in warehouse I and 290 boxes of the same card in warehouse II. A greeting card shop in San Jose orders 350 boxes of the card, and another shop in Memphis orders 300 boxes. The shipping costs per box to these shops from the two warehouses are shown in the following table.

		Destination	
		San Jose	**Memphis**
Warehouse	I	$.25	$.22
	II	$.23	$.21

How many boxes should be shipped to each city from each warehouse to minimize shipping costs? What is the minimum cost? (*Hint:* Use x, $350 - x$, y, and $300 - y$ as the variables.)

17. *Finance* A pension fund manager decides to invest a total of at most $40 million in U.S. Treasury bonds paying 12% annual interest and in mutual funds paying 6% annual interest. He plans to invest at least $20 million in bonds and at least $14 million in mutual funds. Bonds have an initial fee of $200 per million dollars, while the fee for mutual funds is $100 per million. The fund manager is allowed to spend no more than $6200 on fees. How much should be invested in each to maximize annual interest? What is the maximum annual interest?

Manufacturing (Note: Exercises 18–20 are from qualification examinations for Certified Public Accountants.*) The Random Company manufactures two products, Zeta and Beta. Each product must pass through two processing operations. All materials are introduced at the start of Process No. 1. There are no work-in-process inventories. Random may produce either one product exclusively or various combinations of both products subject to the following constraints:

	Process No. 1	Process No. 2	Contribution Margin (per unit)
Hours Required to Produce One Unit:			
Zeta	1 hr	1 hr	$4.00
Beta	2 hr	3 hr	$5.25
Total Capacity (in hours per day)	1000 hr	1275 hr	

A shortage of technical labor has limited Beta production to 400 units per day. There are no constraints on the production of Zeta other than the hour constraints in the above schedule. Assume that all relationships between capacity and production are linear.

18. Given the objective to maximize total contribution margin, what is the production constraint for Process No. 1?

   a. Zeta + Beta $\leq$ 1000   **b.** Zeta + 2 Beta $\leq$ 1000

   c. Zeta + Beta $\geq$ 1000   **d.** Zeta + 2 Beta $\geq$ 1000

19. Given the objective to maximize total contribution margin, what is the labor constraint for production of Beta?

   a. Beta $\leq$ 400   **b.** Beta $\geq$ 400

   c. Beta $\leq$ 425   **d.** Beta $\geq$ 425

20. What is the objective function of the data presented?

   a. Zeta + 2 Beta = $9.25

   b. $4.00 Zeta + 3($5.25) Beta = Total Contribution Margin

   c. $4.00 Zeta + $5.25 Beta = Total Contribution Margin

   d. 2($4.00) Zeta + 3($5.25) Beta = Total Contribution Margin

LIFE SCIENCES

21. *Health Care* Mark, who is ill, takes vitamin pills. Each day he must have at least 16 units of vitamin A, 5 units of vitamin B_1, and 20 units of vitamin C. He can choose between pill #1, which contains 8 units of A, 1 of B_1, and 2 of C; and pill #2, which contains 2 units of A, 1 of B_1, and 7 of C. Pill #1 costs 15¢, and pill #2 costs 30¢. How many of each pill should he buy in order to minimize his cost? What is the minimum cost?

22. *Predator Food Requirements* A certain predator requires at least 10 units of protein and 8 units of fat per day. One prey of species I provides 5 units of protein and 2 units of fat; one prey of species II provides 3 units of protein and 4 units of fat. Capturing and digesting each species II prey requires 3 units of energy, and capturing and digesting each species of prey I requires 2 units of energy. How many of each prey would meet the predator's daily food requirements with the least expenditure of energy? Are the answers reasonable? How could they be interpreted?

23. *Dietetics* Sing, who is dieting, requires two food supplements, I and II. He can get these supplements from two different products, A and B, as shown in the table below.

		Grams of Supplement per Serving	
		I	II
Product	A	3	2
	B	2	4

Sing's physician has recommended that he include at least 15 g of each supplement in his daily diet. If product A costs 25¢ per serving and product B costs 40¢ per serving, how can he satisfy his dietetic requirements most economically? Find the minimum cost.

24. *Health Care* Ms. Oliveras was given the following advice. She should supplement her daily diet with at least 6000 USP units of vitamin A, at least 195 milligrams of vitamin C, and at least 600 USP units of vitamin D. Ms. Oliveras finds that Mason's Pharmacy carries Brand X

*Material from *Uniform CPA Examinations and Unofficial Answers,* copyright © 1973, 1974, 1975 by the American Institute of Certified Public Accountants, Inc., is reprinted with permission.

vitamin pills at 5¢ each and Brand Y vitamins at 4¢ each. Each Brand X pill contains 3000 USP units of A, 45 milligrams of C, and 75 USP units of D, while Brand Y pills contain 1000 USP units of A, 50 milligrams of C, and 200 USP units of D. What combination of vitamin pills should she buy to obtain the least possible cost? What is the least possible cost per day?

SOCIAL SCIENCES

25. *Anthropology* An anthropology article presents a hypothetical situation that could be described by a linear programming model.* Suppose a population gathers plants and animals for survival. They need at least 360 units of energy, 300 units of protein, and 8 hides during some time period. One unit of plants provides 30 units of energy, 10 units of protein, and no hides. One animal provides 20 units of energy, 25 units of protein, and 1 hide. Only 25 units of plants and 25 animals are available. It costs the population 30 hours of labor to gather one unit of a plant and 15 hours for an animal. Find how many units of plants and how many animals should be gathered to meet the requirements with a minimum number of hours of labor.

26. *Construction* In a small town in South Carolina, zoning rules require that the window space (in square feet) in a house be at least one-sixth of the space used up by solid walls. The cost to build windows is $10 per square foot, while the cost to build solid walls is $20 per square foot. The total amount available for building walls and windows is no more than $12,000. The estimated monthly cost to heat the house is $.32 for each square foot of windows and $.20 for each square foot of solid walls. Find the maximum total area (windows plus walls) if no more than $160 per month is available to pay for heat.

27. *Farming* An agricultural advisor looks at the results of Example 2 and claims that it cannot possibly be correct. After all, the 4-H member is able to raise 16 animals, and she is only raising 12 animals. Surely she can earn more profit by raising all 16 animals. How would you respond?

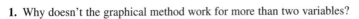

▪ CHAPTER SUMMARY

In this chapter, we have studied maximization and minimization problems with linear constraints that arise in realistic situations. The graphical method allows us to solve such problems by graphing the region described by the linear constraints. The corner point theorem assures us that the optimal solution, if it exists, must lie at one of the corner points. The graphical method is restricted to problems with two variables. In the next chapter, we will study a method that does not have this restriction.

▪ KEY TERMS

3.1	linear inequality	system of inequalities	3.2	linear programming	corner point
	boundary	region of feasible		objective function	bounded
	half-plane	solutions		constraints	unbounded

▪ CHAPTER 3 REVIEW EXERCISES

1. Why doesn't the graphical method work for more than two variables?

2. How many constraints are we limited to in the graphical method?

---

*Reidhead, Van A., "Linear Programming Models in Archaeology," *Annual Review of Anthropology,* Vol. 8, 1979, pp. 543–578.

Graph each linear inequality.

3. $y \geq 2x + 3$ **4.** $3x - y \leq 5$ **5.** $3x + 4y \leq 12$

6. $2x - 6y \geq 18$ **7.** $y \geq x$ **8.** $y \leq 3$

Graph the solution of each system of inequalities. Find all corner points.

9. $x + y \leq 6$
    $2x - y \geq 3$

10. $4x + y \geq 8$
    $2x - 3y \leq 6$

11. $-4 \leq x \leq 2$
    $-1 \leq y \leq 3$
    $x + y \leq 4$

12. $2 \leq x \leq 5$
    $1 \leq y \leq 7$
    $x - y \leq 3$

13. $x + 3y \geq 6$
    $4x - 3y \leq 12$
    $x \geq 0$
    $y \geq 0$

14. $x + 2y \leq 4$
    $2x - 3y \leq 6$
    $x \geq 0$
    $y \geq 0$

Use the given regions to find the maximum and minimum values of the objective function $z = 2x + 4y$.

15.

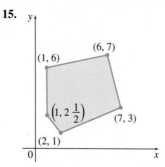

16.

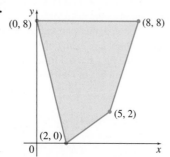

Use the graphical method to solve each linear programming problem.

17. Maximize $z = 2x + 4y$
    subject to: $3x + 2y \leq 12$
        $5x + y \geq 5$
        $x \geq 0$
        $y \geq 0.$

18. Minimize $z = 3x + 2y$
    subject to: $8x + 9y \geq 72$
        $6x + 8y \geq 72$
        $x \geq 0$
        $y \geq 0.$

19. Minimize $z = 4x + 2y$
    subject to: $x + y \leq 50$
        $2x + y \geq 20$
        $x + 2y \geq 30$
        $x \geq 0$
        $y \geq 0.$

20. Maximize $z = 4x + 3y$
    subject to: $2x + 7y \leq 14$
        $2x + 3y \leq 10$
        $x \geq 0$
        $y \geq 0.$

21. Why must the solution to a linear programming problem always occur at a corner point of the feasible region?

22. Is there necessarily a unique point in the feasible region where the maximum or minimum occurs? Why or why not?

23. It is not necessary to check all corner points in a linear programming problem. This exercise illustrates an alternative procedure, which is essentially an expansion of ideas illustrated in Example 1 of Section 3.2.

Maximize $z = 2x + 5y$
subject to: $3x + 2y \leq 6$
    $-x + 2y \leq 4$
        $x \geq 0$
        $y \geq 0.$

a. Sketch the feasible region, and add the line $z = 10$. (*Note:* 10 is chosen because the numbers work out simply, but the chosen value of z is arbitrary.)

b. Draw a line parallel to the line $z = 10$ that is as far from the origin as possible but still touches the feasible region.

c. The line you drew in part b should go through the point $(1/2, 9/4)$. Explain how you know the maximum must be located at this point.

24. Use the method described in the previous exercise to solve Exercise 20.

Applications

BUSINESS AND ECONOMICS

25. *Time Management* A bakery makes both cakes and cookies. Each batch of cakes requires 2 hr in the oven and 3 hr in the decorating room. Each batch of cookies needs $1\frac{1}{2}$ hr in the oven and 2/3 hr in the decorating room. The oven is available no more than 15 hr per day, and the decorating room can be used no more than 13 hr per day. Set up a system of inequalities, and then graph the solution of the system.

26. *Cost Analysis* A company makes two kinds of pizza, special and basic. The special has toppings of cheese, tomatoes, and vegetables. Basic has just cheese and tomatoes. The company sells at least 6 units a day of the special pizza and 4 units a day of the basic. The cost of the vegetables (including tomatoes) is $2 per unit for special and $1 per unit for basic. No more than $32 per day can be spent on vegetables (including tomatoes). The cheese used for the special is $5 per unit, and the cheese for the basic is $4 per unit. The company can spend no more than $100 per day on cheese. Set up a system of inequalities, and then graph the solution of the system.

27. *Profit* How many batches of cakes and cookies should the bakery in Exercise 25 make in order to maximize profits if cookies produce a profit of $20 per batch and cakes produce a profit of $30 per batch?

28. *Profit* How many units of each kind of pizza should the company in Exercise 26 make in order to maximize rev-

enue if the special sells for $20 per unit and the basic for $15 per unit?

29. *Planting* In Karla's garden shop, she makes two kinds of mixtures for planting. A package of gardening mixture requires 2 lbs of soil, 1 lb of peat moss, and 1 lb of fertilizer. A package of potting mixture requires 1 lb of soil, 2 lb of peat moss, and 3 lbs of fertilizer. She has 16 lbs of soil, 11 lbs of peat moss, and 15 lbs of fertilizer. If a package of gardening mixture sells for $3 and a package of potting mixture for $5, how many of each should she make in order to maximize her income? What is the maximum income?

30. *Construction* A contractor builds boathouses in two basic models, the Atlantic and the Pacific. Each Atlantic model requires 1000 feet of framing lumber, 3000 cubic feet of concrete, and $2000 for advertising. Each Pacific model requires 2000 feet of framing lumber, 3000 cubic feet of concrete, and $3000 for advertising. Contracts call for using at least 8000 feet of framing lumber, 18,000 cubic feet of concrete, and $15,000 worth of advertising. If the total spent on each Atlantic model is $3000 and the total spent on each Pacific model is $4000, how many of each model should be built to minimize costs?

31. *Steel* A steel company produces two types of alloys. A run of type I requires 3000 pounds of molybdenum and 2000 tons of iron ore pellets as well as $2000 in advertising. A run of type II requires 3000 pounds of molybdenum and 1000 tons of iron ore pellets as well as $3000 in advertising. Total costs are $15,000 on a run of type I and

$6000 on a run of type II. Because of various contracts, the company must use at least 18,000 pounds of molybdenum and 7000 tons of iron ore pellets and spend at least $14,000 on advertising. How much of each type should be produced to minimize costs?

LIFE SCIENCES

32. *Nutrition* A dietician in a hospital is to arrange a special diet containing two foods, Health Trough and Power Gunk. Each ounce of Health Trough contains 30 mg of calcium, 10 mg of iron, 10 IU of vitamin A, and 8 mg of cholesterol. Each ounce of Power Gunk contains 10 mg of calcium, 10 mg of iron, 30 IU of vitamin A, and 4 mg of cholesterol. If the minimum daily requirements are 360 mg of calcium, 160 mg of iron, and 240 IU of vitamin A, how many ounces of each food should be used to meet the minimum requirements and at the same time minimize the cholesterol intake? Also, what is the minimum cholesterol intake?

SOCIAL SCIENCES

33. *Anthropology* A simplified model of the Mountain Fur economy of central Africa has been proposed.* In this model, two crops can be grown, millet and wheat, which produce 400 lb and 800 lb per acre, respectively. Millet requires 36 days to harvest one acre, while wheat requires only 8 days. There are 2 acres of land and 48 days of harvest labor available. How many acres should be devoted to each crop to maximize the pounds of grain harvested?

GENERAL INTEREST

34. *Studying* Ron Hampton is trying to allocate his study time this weekend. He can spend time working with either his math tutor or his accounting tutor to prepare for exams in both classes the following Monday. His math tutor charges $20 per hour, and his accounting tutor charges $40 per hour. He has $220 to spend on tutoring. Each hour that he spends working with his math tutor requires 1 aspirin and 1 hr of sleep to recover, while each hour he spends with his accounting tutor requires 1/2 aspirin and 3 hr of sleep. The maximum dosage of aspirin that he can safely take during his study time is 8 tablets, and he can only afford 15 hr of sleep this weekend. He expects that each hour with his math tutor will increase his score on the math exam by 3 points, while each hour with his accounting tutor will increase his score on the accounting exam by 5 points. How many hours should he spend with each tutor in order to maximize the number of points he will get on the two tests combined?

---

*Joy, Leonard, "Barth's Presentation of Economic Spheres in Darfur," in *Themes in Economic Anthropology,* edited by Raymond Firth, Tavistock Publications, 1967, pp. 175–189.

Linear Programming: The Simplex Method

Each type of beer has its own recipe and an associated cost per unit, and brings in a specific revenue per unit. The brewery manager must meet a revenue target with minimum production costs. An exercise in Section 3 formulates the manager's goal as a linear programming problem and solves for the optimum production schedule when there are two beer varieties.

In the previous chapter, we discussed solving linear programming problems by the graphical method. This method illustrates the basic ideas of linear programming, but it is practical only for problems with two variables. For problems with more than two variables, or problems with two variables and many constraints, the *simplex method* is used. This method grew out of a practical problem faced by George B. Dantzig in 1947. Dantzig was concerned with finding the least expensive way to allocate supplies for the United States Air Force.

The **simplex method** starts with the selection of one corner point (often the origin) from the feasible region. Then, in a systematic way, another corner point is found that attempts to improve the value of the objective function. Finally, an optimum solution is reached, or it can be seen that no such solution exists.

The simplex method requires a number of steps. In this chapter we divide the presentation of these steps into two parts. First, a problem is set up in Section 4.1 and the method started, and then, in Section 4.2, the method is completed. Special situations are discussed in the remainder of the chapter.

4.1 SLACK VARIABLES AND THE PIVOT

Because the simplex method is used for problems with many variables, it usually is not convenient to use letters such as x, y, z, or w as variable names. Instead, the symbols x_1 (read "x-sub-one"), x_2, x_3, and so on, are used. These variable names lend themselves easily to use on the computer. In the simplex method, all constraints must be expressed in the linear form

$$a_1x_1 + a_2x_2 + a_3x_3 + \cdots \leq b,$$

where $x_1, x_2, x_3, \ldots$ are variables and $a_1, a_2, \ldots,$ and b are constants.

In this section we will use the simplex method only for problems such as the following:

$$\begin{aligned}
\text{Maximize} \quad & z = 2x_1 - 3x_2 \\
\text{subject to:} \quad & 2x_1 + x_2 \leq 10 \\
& x_1 - 3x_2 \leq 5 \\
\text{with} \quad & x_1 \geq 0, \quad x_2 \geq 0.
\end{aligned}$$

This type of problem is said to be in *standard maximum form.*

STANDARD MAXIMUM FORM

A linear programming problem is in **standard maximum form** if the following conditions are satisfied.

1. The objective function is to be maximized.
2. All variables are nonnegative ($x_i \geq 0$).
3. All remaining constraints are stated in the form

$$a_1x_1 + a_2x_2 + \cdots + a_nx_n \leq b \qquad \text{with } b \geq 0.$$

(Problems that do not meet all of these conditions are discussed in Sections 4.3 and 4.4.)

To use the simplex method, we start by converting the constraints, which are linear inequalities, into linear equations. We do this by adding a nonnegative variable, called a **slack variable,** to each constraint. For example, the inequality $x_1 + x_2 \leq 10$ is converted into an equation by adding the slack variable s_1 to get

$$x_1 + x_2 + s_1 = 10, \quad \text{where } s_1 \geq 0.$$

The inequality $x_1 + x_2 \leq 10$ says that the sum $x_1 + x_2$ is less than or perhaps equal to 10. The variable s_1 "takes up any slack" and represents the amount by which $x_1 + x_2$ fails to equal 10. For example, if $x_1 + x_2$ equals 8, then s_1 is 2. If $x_1 + x_2 = 10$, then s_1 is 0.

■ **CAUTION** A different slack variable must be used for each constraint.

EXAMPLE 1 Slack Variables

Restate the following linear programming problem by introducing slack variables.

$$\begin{array}{ll} \text{Maximize} & z = 3x_1 + 2x_2 + x_3 \\ \text{subject to:} & 2x_1 + x_2 + x_3 \leq 150 \\ & 2x_1 + 2x_2 + 8x_3 \leq 200 \\ & 2x_1 + 3x_2 + x_3 \leq 320 \\ \text{with} & x_1 \geq 0, \quad x_2 \geq 0, \quad x_3 \geq 0. \end{array}$$

Solution Rewrite the three constraints as equations by adding slack variables s_1, s_2, and s_3, one for each constraint. Then the problem can be restated as follows.

$$\begin{array}{ll} \text{Maximize} & z = 3x_1 + 2x_2 + x_3 \\ \text{subject to:} & 2x_1 + x_2 + x_3 + s_1 = 150 \\ & 2x_1 + 2x_2 + 8x_3 + s_2 = 200 \\ & 2x_1 + 3x_2 + x_3 + s_3 = 320 \\ \text{with} & x_1 \geq 0, \quad x_2 \geq 0, \quad x_3 \geq 0, \quad s_1 \geq 0, \quad s_2 \geq 0, \quad s_3 \geq 0. \end{array}$$

Adding slack variables to the constraints converts a linear programming problem into a system of linear equations. In each of these equations, all variables should be on the left side of the equals sign and all constants on the right. All the equations in Example 1 satisfy this condition except for the objective function, $z = 3x_1 + 2x_2 + x_3$, which may be written with all variables on the left as

$$-3x_1 - 2x_2 - x_3 + z = 0.$$

Now the equations in Example 1 can be written as the following augmented matrix.

x_1	x_2	x_3	s_1	s_2	s_3	z	
2	1	1	1	0	0	0	150
2	2	8	0	1	0	0	200
2	3	1	0	0	1	0	320
−3	−2	−1	0	0	0	1	0

Indicators

This matrix is called the initial **simplex tableau.** The numbers in the bottom row, which are from the objective function, are called **indicators** (except for the 1 and 0 at the far right).

EXAMPLE 2 Initial Simplex Tableau

Set up the initial simplex tableau for the following problem:

A farmer has 100 acres of available land that he wishes to plant with a mixture of potatoes, corn, and cabbage. It costs him $400 to produce an acre of potatoes, $160 to produce an acre of corn, and $280 to produce an acre of cabbage. He has a maximum of $20,000 to spend. He makes a profit of $120 per acre of potatoes, $40 per acre of corn, and $60 per acre of cabbage. How many acres of each crop should he plant to maximize his profit?

Solution Begin by summarizing the given information as follows.

	Potatoes	**Corn**	**Cabbage**		**Total**
Number of Acres	x_1	x_2	x_3	$\leq$	100
Cost (per acre)	$400	$160	$280	$\leq$	$20,000
Profit (per acre)	$120	$40	$60		

If the number of acres allotted to each of the three crops is represented by x_1, x_2, and x_3, respectively, then the constraints of the example can be expressed as

$$x_1 + \quad x_2 + \quad x_3 \leq \quad 100 \quad \text{Number of acres}$$
$$400x_1 + 160x_2 + 280x_3 \leq 20{,}000 \quad \text{Production costs}$$

where x_1, x_2, and x_3 are all nonnegative. The first of these constraints says that $x_1 + x_2 + x_3$ is less than or perhaps equal to 100. Use s_1 as the slack variable, giving the equation

$$x_1 + x_2 + x_3 + s_1 = 100.$$

Here s_1 represents the amount of the farmer's 100 acres that will not be used (s_1 may be 0 or any value up to 100).

The constraint $400x_1 + 160x_2 + 280x_3 \leq 20{,}000$ can be simplified by dividing by 40, to get

$$10x_1 + 4x_2 + 7x_3 \leq 500.$$

This inequality can also be converted into an equation by adding a slack variable, s_2.

$$10x_1 + 4x_2 + 7x_3 + s_2 = 500$$

If we had not divided by 40, the slack variable would have represented any unused portion of the farmer's $20,000 capital. Instead, the slack variable represents 1/40 of that unused portion. (Note that s_2 may be any value from 0 to 500.)

The objective function represents the profit. The farmer wants to maximize

$$z = 120x_1 + 40x_2 + 60x_3.$$

The linear programming problem can now be stated as follows:

Maximize $\qquad z = 120x_1 + 40x_2 + 60x_3$

subject to: $\qquad x_1 + x_2 + x_3 + s_1 \qquad = 100$

$\qquad 10x_1 + 4x_2 + 7x_3 \qquad + s_2 = 500$

with $\qquad x_1 \geq 0, \quad x_2 \geq 0, \quad x_3 \geq 0, \quad s_1 \geq 0, \quad s_2 \geq 0.$

Rewrite the objective function as $-120x_1 - 40x_2 - 60x_3 + z = 0$, and complete the initial simplex tableau as follows.

$$
\begin{array}{c}
\begin{array}{cccccc} x_1 & x_2 & x_3 & s_1 & s_2 & z \end{array} \\
\left[\begin{array}{cccccc|c}
1 & 1 & 1 & 1 & 0 & 0 & 100 \\
10 & 4 & 7 & 0 & 1 & 0 & 500 \\
\hline
-120 & -40 & -60 & 0 & 0 & 1 & 0
\end{array} \right]
\end{array}
$$

The maximization problem in Example 2 consists of a system of two equations (describing the constraints) in five variables, together with the objective function. As with the graphical method, it is necessary to solve this system to find corner points of the region of feasible solutions. To produce a single, distinct solution, the number of variables must equal the number of independent equations in the system.

For example, the system of equations in Example 2 has an infinite number of solutions since there are more variables than equations. To see this, solve the system for s_1 and s_2.

$$s_1 = 100 - x_1 - x_2 - x_3$$
$$s_2 = 500 - 10x_1 - 4x_2 - 7x_3$$

Each choice of values for x_1, x_2, and x_3 gives corresponding values for s_1 and s_2 that produce a solution of the system. But only some of these solutions are feasible. In a feasible solution all variables must be nonnegative. To get a unique feasible solution, we set three of the five variables equal to 0. In general, if there are m equations, then m variables can be nonzero. These m nonzero variables are called **basic variables,** and the corresponding solutions are called **basic feasible solutions.** Each basic feasible solution corresponds to a corner point. In particular, if we choose the solution with $x_1 = 0$, $x_2 = 0$, and $x_3 = 0$, then $s_1 = 100$ and $s_2 = 500$ are the basic variables. This solution, which corresponds to the corner point at the origin, is hardly optimal. It produces a profit of $0 for the farmer, since the equation that corresponds to the objective function becomes

$$-120(0) - 40(0) - 60(0) + 0s_1 + 0s_2 + z = 0.$$

In the next section we will use the simplex method to start with this solution and improve it to find the maximum possible profit.

Each step of the simplex method produces a solution that corresponds to a corner point of the region of feasible solutions. These solutions can be read directly from the matrix, as shown in the next example.

EXAMPLE 3 Basic Variables

Read a solution from the matrix below.

$$
\begin{array}{c}
\begin{array}{cccccc} x_1 & x_2 & x_3 & s_1 & s_2 & z \end{array} \\
\left[\begin{array}{cccccc|c}
2 & 0 & 8 & 5 & 2 & 0 & 17 \\
9 & 5 & 3 & 12 & 0 & 0 & 45 \\
\hline
-2 & 0 & -4 & 0 & 0 & 3 & 90
\end{array}\right]
\end{array}
$$

Solution In this solution, the variables x_2 and s_2 are basic variables. They can be identified quickly because the columns for these variables have all zeros except for one nonzero entry. All variables that are not basic variables have the value 0. This means that in the matrix just shown, x_2 and s_2 are the basic variables, while x_1, x_3, and s_1 have the value 0. The nonzero entry for x_2 is 5 in the second row. Since x_1, x_3, and s_1 are zero, the second row of the matrix represents the equation $5x_2 = 45$, so $x_2 = 9$. Similarly, from the top row, $2s_2 = 17$, so $s_2 = 17/2$. From the bottom row, $3z = 90$, so $z = 30$. The solution is thus $x_1 = 0$, $x_2 = 9$, $x_3 = 0$, $s_1 = 0$, and $s_2 = 17/2$, with $z = 30$.

FOR REVIEW ■■

We discussed three row operations in Chapter 2:

1. interchanging any two rows;

2. multiplying the elements of a row by any nonzero real number; and

3. adding a multiple of the elements of one row to the corresponding elements of a multiple of any other row.

In this chapter we will only use operations 2 and 3; we will never interchange two rows.

Pivots Solutions read directly from the initial simplex tableau are seldom optimal. It is necessary to proceed to other solutions (corresponding to other corner points of the feasible region) until an optimum solution is found. To get these other solutions, we use the row operations from Chapter 2 to change the tableau by using one of the nonzero entries of the tableau as a **pivot.** The row operations are performed to change to 0 all entries in the column containing the pivot (except for the pivot itself, which remains unchanged). Pivoting, explained in the next example, produces a new tableau leading to another solution of the system of equations obtained from the original problem.

CAUTION In this chapter, when adding a multiple of one row to a multiple of another, we will never take a negative multiple of the row being changed. For example, when changing row 2, we might use $-2R_1 + 3R_2 \rightarrow R_2$, but we will never use $2R_1 - 3R_2 \rightarrow R_2$. If you get a negative number in the rightmost column, you will know immediately that you have made an error. The reason for this restriction is that violating it turns negative numbers into positive, and vice versa. This is disastrous in the bottom row, where we will seek negative numbers when we choose our pivot column. It will also cause problems with choosing pivots, particularly in the algorithm for solving nonstandard problems in Section 4.4.

EXAMPLE 4 Pivot

Pivot about the indicated 2 of the initial simplex tableau given below.

$$
\begin{array}{c}
\begin{array}{ccccccc} x_1 & x_2 & x_3 & s_1 & s_2 & s_3 & z \end{array} \\
\left[\begin{array}{ccccccc|c}
\mathbf{2} & 1 & 1 & 1 & 0 & 0 & 0 & 150 \\
1 & 2 & 8 & 0 & 1 & 0 & 0 & 200 \\
2 & 3 & 1 & 0 & 0 & 1 & 0 & 320 \\
\hline
-3 & -2 & -1 & 0 & 0 & 0 & 1 & 0
\end{array}\right]
\end{array}
$$

Method 1: Calculating by Hand

Solution

Using the row operations indicated in color below to get zeros in the column with the pivot, we arrive at the following matrix.

$$
\begin{array}{c}
\\
\\
-R_1 + 2R_2 \rightarrow R_2 \\
-R_1 + R_3 \rightarrow R_3 \\
3R_1 + 2R_4 \rightarrow R_4
\end{array}
\begin{array}{c}
\begin{array}{ccccccc}
x_1 & x_2 & x_3 & s_1 & s_2 & s_3 & z
\end{array} \\
\left[
\begin{array}{ccccccc|c}
2 & 1 & 1 & 1 & 0 & 0 & 0 & 150 \\
0 & 3 & 15 & -1 & 2 & 0 & 0 & 250 \\
0 & 2 & 0 & -1 & 0 & 1 & 0 & 170 \\
\hline
0 & -1 & 1 & 3 & 0 & 0 & 2 & 450
\end{array}
\right]
\end{array}
$$

This simplex tableau gives the solution $x_1 = 75$, $x_2 = 0$, $x_3 = 0$, $s_1 = 0$, $s_2 = 125$, and $s_3 = 170$. Substituting these results into the objective function gives

$$0(75) - 1(0) + 1(0) + 3(0) + 0(125) + 0(170) + 2z = 450,$$

or $z = 225$. (This shows that the value of z can always be found using the number in the bottom row of the z column and the number in the lower right-hand corner.)

Finally, to be able to read the solution directly from the matrix, we multiply rows 1, 2, and 4 by $1/2$, getting the following matrix.

$$
\begin{array}{c}
\\
\\
\frac{1}{2}R_1 \rightarrow R_1 \\
\frac{1}{2}R_2 \rightarrow R_2 \\
\\
\frac{1}{2}R_4 \rightarrow R_4
\end{array}
\begin{array}{c}
\begin{array}{ccccccc}
x_1 & x_2 & x_3 & s_1 & s_2 & s_3 & z
\end{array} \\
\left[
\begin{array}{ccccccc|c}
1 & \frac{1}{2} & \frac{1}{2} & \frac{1}{2} & 0 & 0 & 0 & 75 \\
0 & \frac{3}{2} & \frac{15}{2} & -\frac{1}{2} & 1 & 0 & 0 & 125 \\
0 & 2 & 0 & -1 & 0 & 1 & 0 & 170 \\
\hline
0 & -\frac{1}{2} & \frac{1}{2} & \frac{3}{2} & 0 & 0 & 1 & 225
\end{array}
\right]
\end{array}
$$

Method 2: Graphing Calculator

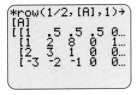

FIGURE 1

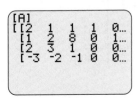

FIGURE 2

The row operations of the simplex method can also be done on a graphing calculator, as we saw in Chapter 2. Figure 1 shows the result when the matrix in this example is entered into a TI-83. The right side of the matrix is not visible, but can be seen by pressing the right arrow key.

Recall that we must change the pivot to 1 before performing row operations with a graphing calculator. Figure 2 shows the result of multiplying row 1 of matrix A by $1/2$. In Figure 3 we show the same result with the decimal numbers changed to fractions.

We can now modify column 1, using the commands described in Chapter 2, to agree with the tableau under Method 1. The result is shown in Figure 4.

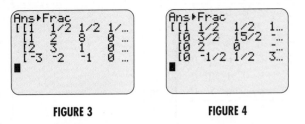

FIGURE 3   **FIGURE 4**

In the simplex method, the pivoting process (without the final step of getting a 1 in each basic variable column when using Method 1) is repeated until an optimum solution is found, if one exists. In the next section we will see how to decide where to pivot to improve the value of the objective function and how to tell when an optimum solution either has been reached or does not exist.

▪ 4.1 EXERCISES

Convert each of the following inequalities into an equation by adding a slack variable.

1. $x_1 + 2x_2 \leq 6$

2. $3x_1 + 5x_2 \leq 100$

3. $2x_1 + 4x_2 + 3x_3 \leq 100$

4. $8x_1 + 6x_2 + 5x_3 \leq 250$

*For Exercises 5–8, (**a**) determine the number of slack variables needed, (**b**) name them, and (**c**) use slack variables to convert each constraint into a linear equation.*

5. Maximize $z = 10x_1 + 12x_2$

subject to: $4x_1 + 2x_2 \leq 20$
$5x_1 + x_2 \leq 50$
$2x_1 + 3x_2 \leq 25$

with $x_1 \geq 0, \quad x_2 \geq 0.$

6. Maximize $z = 1.2x_1 + 3.5x_2$

subject to: $2.4x_1 + 1.5x_2 \leq 10$
$1.7x_1 + 1.9x_2 \leq 15$

with $x_1 \geq 0, \quad x_2 \geq 0.$

7. Maximize $z = 8x_1 + 3x_2 + x_3$

subject to: $7x_1 + 6x_2 + 8x_3 \leq 118$
$4x_1 + 5x_2 + 10x_3 \leq 220$

with $x_1 \geq 0, \quad x_2 \geq 0, \quad x_3 \geq 0.$

8. Maximize $z = 12x_1 + 15x_2 + 10x_3$

subject to: $2x_1 + 2x_2 + x_3 \leq 8$
$x_1 + 4x_2 + 3x_3 \leq 12$

with $x_1 \geq 0, \quad x_2 \geq 0, \quad x_3 \geq 0.$

Write the solutions that can be read from Exercises 9–12.

9.

x_1	x_2	x_3	s_1	s_2	z	
2	2	0	3	1	0	15
3	4	1	6	0	0	20
−2	−1	0	1	0	1	10

10.

x_1	x_2	x_3	s_1	s_2	z	
0	2	1	1	3	0	5
1	5	0	1	2	0	8
0	−2	0	1	1	1	10

11.

x_1	x_2	x_3	s_1	s_2	s_3	z	
6	2	2	3	0	0	0	16
2	2	0	1	0	5	0	35
2	1	0	3	1	0	0	6
−3	−2	0	2	0	0	3	36

12.

x_1	x_2	x_3	s_1	s_2	s_3	z	
0	2	0	5	2	2	0	15
0	3	1	0	1	2	0	2
7	4	0	0	3	5	0	35
0	−4	0	0	4	3	2	40

Pivot once as indicated in each simplex tableau. Read the solution from the result.

13.

x_1	x_2	x_3	s_1	s_2	z	
1	2	4	1	0	0	56
2	**2**	1	0	1	0	40
−1	−3	−2	0	0	1	0

14.

x_1	x_2	x_3	s_1	s_2	z	
5	4	1	1	0	0	50
3	3	**2**	0	1	0	40
−1	−2	−4	0	0	1	0

15.

x_1	x_2	x_3	s_1	s_2	s_3	z	
2	2	**1**	1	0	0	0	12
1	2	3	0	1	0	0	45
3	1	1	0	0	1	0	20
−2	−1	−3	0	0	0	1	0

16.

x_1	x_2	x_3	s_1	s_2	s_3	z	
4	2	3	1	0	0	0	22
2	2	**5**	0	1	0	0	28
1	3	2	0	0	1	0	45
−3	−2	−4	0	0	0	1	0

17.

x_1	x_2	x_3	s_1	s_2	s_3	z	
1	1	1	1	0	0	0	60
3	1	**2**	0	1	0	0	100
1	2	3	0	0	1	0	200
−1	−1	−2	0	0	0	1	0

18.

x_1	x_2	x_3	x_4	s_1	s_2	s_3	z	
1	2	3	1	1	0	0	0	115
2	1	8	5	0	1	0	0	200
1	0	1	0	0	0	1	0	50
−2	−1	−1	−1	0	0	0	1	0

Introduce slack variables as necessary, then write the initial simplex tableau for each of the following linear programming problems.

19. Find $x_1 \geq 0$ and $x_2 \geq 0$ such that

$$2x_1 + 3x_2 \leq 6$$
$$4x_1 + x_2 \leq 6$$

and $z = 5x_1 + x_2$ is maximized.

20. Find $x_1 \geq 0$ and $x_2 \geq 0$ such that

$$2x_1 + 3x_2 \leq 100$$
$$5x_1 + 4x_2 \leq 200$$

and $z = x_1 + 3x_2$ is maximized.

21. Find $x_1 \geq 0$ and $x_2 \geq 0$ such that

$$x_1 + x_2 \leq 10$$
$$5x_1 + 2x_2 \leq 20$$
$$x_1 + 2x_2 \leq 36$$

and $z = x_1 + 3x_2$ is maximized.

22. Find $x_1 \geq 0$ and $x_2 \geq 0$ such that

$$x_1 + x_2 \leq 10$$
$$5x_1 + 3x_2 \leq 75$$

and $z = 4x_1 + 2x_2$ is maximized.

23. Find $x_1 \geq 0$ and $x_2 \geq 0$ such that

$$3x_1 + x_2 \leq 12$$
$$x_1 + x_2 \leq 15$$

and $z = 2x_1 + x_2$ is maximized.

24. Find $x_1 \geq 0$ and $x_2 \geq 0$ such that

$$10x_1 + 4x_2 \leq 100$$
$$20x_1 + 10x_2 \leq 150$$

and $z = 4x_1 + 5x_2$ is maximized.

Applications

Set up Exercises 25–29 for solution by the simplex method. First express the linear constraints and objective function, then add slack variables, and then set up the initial simplex tableau. The solutions of some of these problems will be completed in the exercises for the next section.

BUSINESS AND ECONOMICS

25. *Royalties* The authors of a best-selling textbook in finite mathematics are told that, for the next edition of their book, each simple figure would cost the project $20, each figure with additions would cost $35, and each computer-drawn sketch would cost $60. They are limited to 400 figures, for which they are allowed to spend up to $2200. The number of computer-drawn sketches must be no more than the number of the other two types combined, and there must be at least twice as many simple figures as there are figures with additions. If each simple figure increases the royalties by $95, each figure with additions increases royalties by $200, and each computer-drawn figure increases royalties by $325, how many of each type of figure should be included to maximize royalties, assuming that all art costs are borne by the publisher?

26. *Manufacturing Bicycles* A manufacturer of bicycles builds racing, touring, and mountain models. The bicycles are made of both aluminum and steel. The company has available 91,800 units of steel and 42,000 units of aluminum. The racing, touring, and mountain models need 17, 27, and 34 units of steel, and 12, 21, and 15 units of aluminum, respectively. How many of each type of bicycle should be made in order to maximize profit if the company

makes $8 per racing bike, $12 per touring bike, and $22 per mountain bike? What is the maximum possible profit?

27. *Production—Picnic Tables* The manager of a large park has received many complaints about the insufficient number of picnic tables available. At the end of the park season, she has surplus cash and labor resources available and decides to make as many tables as possible. She considers three possible models: redwood, stained Douglas fir, and stained white spruce (all of which last equally well). She has carpenters available for assembly work for a maximum of 90 eight-hour days, while laborers for staining work are available for no more than 60 eight-hour days. Each redwood table requires 8 hours to assemble but no staining, and it costs $159 (including all labor and materials). Each Douglas fir table requires 7 hours to assemble and 2 hours to stain, and it costs $138.85. Each white spruce table requires 8 hours to assemble and 2 hours to stain, and it costs $129.35. If no more than $15,000 is available for this project, what is the maximum number of tables which can be made, and how many of each type should be made?*

28. *Production—Knives* The Cut-Right Company sells sets of kitchen knives. The Basic Set consists of 2 utility knives and 1 chef's knife. The Regular Set consists of 2 utility knives, 1 chef's knife, and 1 slicer. The Deluxe Set consists of 3 utility knives, 1 chef's knife, and 1 slicer. Their profit is $30 on a Basic Set, $40 on a Regular Set, and $60 on a Deluxe Set. The factory has on hand 800 utility knives, 400 chef's knives, and 200 slicers. Assuming that all sets will be sold, how many of each type should be produced in order to maximize profit? What is the maximum profit?

29. *Advertising* The Fancy Fashions Store has $8000 available each month for advertising. Newspaper ads cost $400 each and no more than 20 can be run per month. Radio ads cost $200 each and no more than 30 can run per month. TV ads cost $1200 each, with a maximum of 6 available each month. Approximately 2000 women will see each newspaper ad, 1200 will hear each radio commercial, and 10,000 will see each TV ad. How much of each type of advertising should be used if the store wants to maximize its ad exposure?

■ 4.2 MAXIMIZATION PROBLEMS

 THINK ABOUT IT   How many racing, touring, and mountain bicycles should a bicycle manufacturer make to maximize profit?

We will answer this question in Exercise 21 of this section using the simplex algorithm.

In the previous section we showed how to prepare a linear programming problem for solution. First, we converted the constraints to linear equations with slack variables; then we used the coefficients of the variables from the linear equation to write an augmented matrix. Finally, we used the pivot to go from one vertex to another vertex in the region of feasible solutions.

Now we are ready to put all this together and produce an optimum value for the objective function. To see how this is done, let us complete Example 2 from Section 4.1, the example about the farmer.

In the previous section we set up the following simplex tableau.

$$\begin{array}{c}
\begin{array}{cccccc} x_1 & x_2 & x_3 & s_1 & s_2 & z \end{array} \\
\left[\begin{array}{cccccc|c}
1 & 1 & 1 & 1 & 0 & 0 & 100 \\
10 & 4 & 7 & 0 & 1 & 0 & 500 \\
\hline
-120 & -40 & -60 & 0 & 0 & 1 & 0
\end{array}\right]
\end{array}$$

This tableau leads to the solution $x_1 = 0$, $x_2 = 0$, $x_3 = 0$, $s_1 = 100$, and $s_2 = 500$, with s_1 and s_2 as the basic variables. These values produce a value of 0 for z. Since a value of 0 for the farmer's profit is not an optimum, we will try to improve this value.

The coefficients of x_1, x_2, and x_3 in the objective function are negative, so the profit could be improved by making any one of these variables take on a nonzero value in a solution. To decide which variable to use, look at the indicators in the initial simplex tableau above. (Recall that the indicators are the numbers in the bottom row, except for the two numbers at the far right.) The coefficient of x_1, -120, is the most negative of the indicators. This means that x_1 has the largest coefficient in the objective function, so increasing x_1 by a certain amount will

---

*This exercise was provided by Dr. Karl K. Norton, Husson College.

increase profit more than increasing any other variable by the same amount. Since each variable can be increased by a different amount, the most negative indicator is not always the best choice. On average, though, this is the best choice, and so we will always choose the most negative indicator.

As we saw earlier, because there are two equations in the system (for the two constraints), only two of the five variables can be basic variables (and be nonzero). If x_1 is nonzero in the solution, then x_1 will be a basic variable. This means that either s_1 or s_2 no longer will be a basic variable. To decide which variable will no longer be basic, start with the equations of the system,

$$x_1 + x_2 + x_3 + s_1 \qquad = 100$$
$$10x_1 + 4x_2 + 7x_3 \qquad + s_2 = 500,$$

and solve for s_1 and s_2, respectively.

$$s_1 = 100 - x_1 - x_2 - x_3$$
$$s_2 = 500 - 10x_1 - 4x_2 - 7x_3$$

Only x_1 is being changed to a nonzero value; both x_2 and x_3 keep the value 0. Replacing x_2 and x_3 with 0 gives

$$s_1 = 100 - x_1$$
$$s_2 = 500 - 10x_1.$$

Since both s_1 and s_2 must remain nonnegative, there is a limit to how much the value of x_1 can be increased. The equation $s_1 = 100 - x_1$ (or $s_1 = 100 - 1x_1$) shows that x_1 cannot exceed $100/1$, or 100. The second equation, $s_2 = 500 - 10x_1$, shows that x_1 cannot exceed $500/10$, or 50. To satisfy both these conditions, x_1 cannot exceed 50, the smaller of 50 and 100. If we let x_1 take the value of 50, then $x_1 = 50$, $x_2 = 0$, $x_3 = 0$, and $s_2 = 0$. Since $s_1 = 100 - x_1$, then

$$s_1 = 100 - 50 = 50.$$

Therefore, s_1 is still a basic variable, while s_2 is no longer a basic variable, having been replaced in the set of basic variables by x_1. This solution gives a profit of

$$z = 120x_1 + 40x_2 + 60x_3 + 0s_1 + 0s_2$$
$$= 120(50) + 40(0) + 60(0) + 0(50) + 0(0) = 6000,$$

or \$6000.

The same result could have been found from the initial simplex tableau given above. To use the tableau, select the most negative indicator. (If no indicator is negative, then the value of the objective function cannot be improved.)

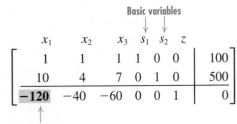

The most negative indicator identifies the variable whose value is to be made nonzero, if possible. To find the variable that is now basic and will become nonbasic, calculate the quotients that were found above. Do this by dividing each number from the right side of the tableau by the corresponding number from the column with the most negative indicator.

$$
\begin{array}{r}
\text{Quotients} \\[4pt]
100/1 = 100 \\[4pt]
\text{Smaller} \rightarrow 500/10 = 50
\end{array}
\qquad
\begin{array}{ccccccc}
 & & & \text{Basic variables} & & \\
 & & & \downarrow & \downarrow & \\
x_1 & x_2 & x_3 & s_1 & s_2 & z \\
\left[\begin{array}{cccccc|c}
1 & 1 & 1 & 1 & 0 & 0 & 100 \\
\mathbf{10} & 4 & 7 & 0 & 1 & 0 & 500 \\
-120 & -40 & -60 & 0 & 0 & 1 & 0
\end{array}\right]
\end{array}
$$

Notice that we do not form a quotient for the bottom row. Of the two quotients found, the smallest is 50 (from the second row), so 10 is the pivot. Using 10 as the pivot, perform the appropriate row operations to get zeros in the rest of the column. We will use Method 1 of Section 4.1 (calculating by hand) to perform the pivoting, but Method 2 (graphing calculator) could just as well be used. The new tableau is as follows.

$$
\begin{array}{r}
\\[4pt]
-R_2 + 10R_1 \rightarrow R_1 \\[10pt]
\\[4pt]
12R_2 + R_3 \rightarrow R_3
\end{array}
\qquad
\begin{array}{cccccc}
 & & \text{Basic variables} & & \\
\downarrow & & & \downarrow & & \\
x_1 & x_2 & x_3 & s_1 & s_2 & z \\
\left[\begin{array}{cccccc|c}
0 & 6 & 3 & 10 & -1 & 0 & 500 \\
10 & 4 & 7 & 0 & 1 & 0 & 500 \\
0 & 8 & 24 & 0 & 12 & 1 & 6000
\end{array}\right]
\end{array}
$$

The solution read from this tableau is

$$x_1 = 50, \quad x_2 = 0, \quad x_3 = 0, \quad s_1 = 50, \quad s_2 = 0,$$

with $z = 6000$, the same as the result found above.

None of the indicators in the final simplex tableau are negative, which means that the value of z cannot be improved beyond \$6000. To see why, recall that the last row gives the coefficients of the objective function so that

$$0x_1 + 8x_2 + 24x_3 + 0s_1 + 12s_2 + z = 6000,$$

or $\qquad z = 6000 - 0x_1 - 8x_2 - 24x_3 - 0s_1 - 12s_2.$

Since x_2, x_3, and s_2 are zero, $z = 6000$, but if any of these three variables were to increase, z would decrease.

This result suggests that the optimal solution has been found as soon as no indicators are negative. As long as an indicator is negative, the value of the objective function may be improved. If any indicators are negative, we just find a new pivot and use row operations, repeating the process until no negative indicators remain.

Once there are no longer any negative numbers in the final row, create a 1 in the columns corresponding to the basic variables and z. In the previous example, this is accomplished by dividing rows 1 and 2 by 10.

$$
\begin{array}{r}
\\[4pt]
R_1/10 \rightarrow R_1 \\[4pt]
R_2/10 \rightarrow R_2
\end{array}
\qquad
\begin{array}{cccccc}
x_1 & x_2 & x_3 & s_1 & s_2 & z \\
\left[\begin{array}{cccccc|c}
0 & \frac{6}{10} & \frac{3}{10} & 1 & -\frac{1}{10} & 0 & 50 \\
1 & \frac{4}{10} & \frac{7}{10} & 0 & \frac{1}{10} & 0 & 50 \\
0 & 8 & 24 & 0 & 12 & 1 & 6000
\end{array}\right]
\end{array}
$$

It is now easy to read the solution from this tableau:

$$x_1 = 50, \quad x_2 = 0, \quad x_3 = 0, \quad s_1 = 50, \quad s_2 = 0,$$

with $z = 6000$.

We can finally state the solution to the problem about the farmer. The farmer will make a maximum profit of $6000 by planting 50 acres of potatoes, no acres of corn, and no acres of cabbage. The other 50 acres should be left unplanted. It may seem strange that leaving assets unused can produce a maximum profit, but such results actually occur often.

In summary, the following steps are involved in solving a standard maximum linear programming problem by the simplex method.

SIMPLEX METHOD

1. Determine the objective function.

2. Write all the necessary constraints.

3. Convert each constraint into an equation by adding slack variables.

4. Set up the initial simplex tableau.

5. Locate the most negative indicator. If there are two such indicators, choose the one farther to the left.

6. Form the necessary quotients to find the pivot. Disregard any quotients with 0 or a negative number in the denominator. The smallest nonnegative quotient gives the location of the pivot. If all quotients must be disregarded, no maximum solution exists. If two quotients are both equal and smallest, choose the pivot in the row nearest the top of the matrix.

7. Use row operations to change all other numbers in the pivot column to zero by adding a suitable multiple of the pivot row to a multiple of each row.

8. If the indicators are all positive or 0, this is the final tableau. If not, go back to Step 5 and repeat the process until a tableau with no negative indicators is obtained.

9. Read the solution from the final tableau.

In Steps 5 and 6, the choice of the column farthest to the left or the row closest to the top is arbitrary. You may choose another row or column in case of a tie, and you will get the same final answer, but your intermediate results will be different.

CAUTION In performing the simplex method, a negative number in the right-hand column signals that a mistake has been made. One possible error is using a negative value for c_2 in the operation $c_1 R_i + c_2 R_j \rightarrow R_j$.

EXAMPLE 1 Simplex Method

To compare the simplex method with the graphical method, we use the simplex method to solve the problem in Example 1, Section 3.3. The graph is shown again in Figure 5. The objective function to be maximized was

$$z = 24x_1 + 36x_2. \quad \text{Storage space}$$

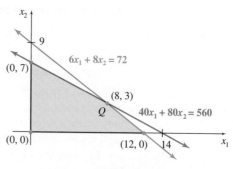

FIGURE 5

(Since we are using the simplex method, we use x_1 and x_2 as variables instead of x and y.) The constraints were as follows:

$$40x_1 + 80x_2 \leq 560 \quad \text{Cost}$$
$$6x_1 + 8x_2 \leq 72 \quad \text{Floor space}$$

with
$$x_1 \geq 0, \quad x_2 \geq 0.$$

To simplify the arithmetic, divide the first inequality on both sides by 40 and the second by 2, which gives

$$x_1 + 2x_2 \leq 14$$
$$3x_1 + 4x_2 \leq 36$$

with
$$x_1 \geq 0, \quad x_2 \geq 0.$$

Add a slack variable to each constraint:

$$x_1 + 2x_2 + s_1 \qquad = 14$$
$$3x_1 + 4x_2 \qquad + s_2 = 36$$

with
$$x_1 \geq 0, \quad x_2 \geq 0, \quad s_1 \geq 0, \quad s_2 \geq 0.$$

Write the initial tableau.

$$
\begin{array}{ccccc}
x_1 & x_2 & s_1 & s_2 & z \\
\end{array}
$$
$$
\left[
\begin{array}{ccccc|c}
1 & 2 & 1 & 0 & 0 & 14 \\
3 & 4 & 0 & 1 & 0 & 36 \\
\hline
-24 & -36 & 0 & 0 & 1 & 0
\end{array}
\right]
$$

This tableau leads to the solution $x_1 = 0$, $x_2 = 0$, $s_1 = 14$, and $s_2 = 36$, with $z = 0$, which corresponds to the origin in Figure 5. The most negative indicator is -36, which is in column 2 of row 3. The quotients of the numbers in the right-hand column and in column 2 are

$$\frac{14}{2} = 7 \quad \text{and} \quad \frac{36}{4} = 9.$$

The smaller quotient is 7, giving 2 as the pivot. Use row operations to get the new tableau. For clarity, we will continue to label the columns with x_1, x_2, etc., although this is not necessary in practice.

$$
\begin{array}{c}
\\
-2R_1 + R_2 \rightarrow R_2 \\
18R_1 + R_3 \rightarrow R_3
\end{array}
\begin{array}{ccccc}
x_1 & x_2 & s_1 & s_2 & z \\
\left[\begin{array}{ccccc|c}
1 & 2 & 1 & 0 & 0 & 14 \\
1 & 0 & -2 & 1 & 0 & 8 \\
-6 & 0 & 18 & 0 & 1 & 252
\end{array}\right]
\end{array}
$$

The solution from this tableau is $x_1 = 0$ and $x_2 = 7$, with $z = 252$. (From now on, we will only list the original variables when giving the solution.) This corresponds to the corner point $(0, 7)$ in Figure 5. Because of the indicator -6, the value of z might be improved. We compare quotients as before and choose the 1 in row 2, column 1 as pivot to get the final tableau.

$$
\begin{array}{c}
\\
-R_2 + R_1 \rightarrow R_1 \\
\\
6R_2 + R_3 \rightarrow R_3
\end{array}
\begin{array}{ccccc}
x_1 & x_2 & s_1 & s_2 & z \\
\left[\begin{array}{ccccc|c}
0 & 2 & 3 & -1 & 0 & 6 \\
1 & 0 & -2 & 1 & 0 & 8 \\
0 & 0 & 6 & 6 & 1 & 300
\end{array}\right]
\end{array}
$$

There are no more negative indicators, so the optimum solution has been achieved. Create a 1 in column 2 by dividing row 1 by 2.

$$
\begin{array}{c}
\\
R_1/2 \rightarrow R_1 \\
\\
\end{array}
\begin{array}{ccccc}
x_1 & x_2 & s_1 & s_2 & z \\
\left[\begin{array}{ccccc|c}
0 & 1 & \frac{3}{2} & -\frac{1}{2} & 0 & 3 \\
1 & 0 & -2 & 1 & 0 & 8 \\
0 & 0 & 6 & 6 & 1 & 300
\end{array}\right]
\end{array}
$$

Here the solution is $x_1 = 8$ and $x_2 = 3$, with $z = 300$. This solution, which corresponds to the corner point $(8, 3)$ in Figure 5, is the same as the solution found earlier.

Each simplex tableau above gave a solution corresponding to one of the corner points of the feasible region. As shown in Figure 6, the first solution corresponded to the origin, with $z = 0$. By choosing the appropriate pivot, we moved systematically to a new corner point, $(0, 7)$, which improved the value of z to 252. The next tableau took us to $(8, 3)$, producing the optimum value of $z = 300$. There was no reason to test the last corner point, $(12, 0)$, since the optimum value z was found before that point was reached.

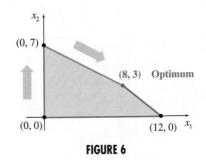

FIGURE 6

▌ **CAUTION** Never choose a zero or a negative number as the pivot. The reason for this is explained in the next example.

EXAMPLE 2 Pivot

Find the pivot for the following initial simplex tableau.

$$
\begin{array}{cccccc}
x_1 & x_2 & s_1 & s_2 & s_3 & z \\
\left[\begin{array}{cccccc|c}
1 & -2 & 1 & 0 & 0 & 0 & 100 \\
3 & 4 & 0 & 1 & 0 & 0 & 200 \\
5 & 0 & 0 & 0 & 1 & 0 & 150 \\
-10 & -25 & 0 & 0 & 0 & 1 & 0
\end{array}\right]
\end{array}
$$

Solution The most negative indicator is -25. To find the pivot, find the quotients formed by the entries in the rightmost column and in the x_2 column: $100/(-2)$, $200/4$, and $150/0$. The quotients predict the value of a variable in the solution. Thus, since we want all variables to be nonnegative, we must reject a negative quotient. Furthermore, we cannot choose 0 as the pivot, because no multiple of the row with 0, when added to the other rows, will cause the other entries in the x_2 column to become 0.

The only usable quotient is $200/4 = 50$, making 4 the pivot. If all the quotients either are negative or have zero denominators, no unique optimum solution will be found. Such a situation indicates an unbounded feasible region. The quotients, then, determine whether an optimum solution exists. ■

CAUTION If there is a 0 in the right-hand column, do not disregard that row, unless the corresponding number in the pivot column is negative or zero. In fact, such a row gives a quotient of 0, so it will automatically have the smallest ratio. It will not cause an increase in z, but it may lead to another tableau in which z can be further increased.

We saw earlier that graphing calculators can be used to perform row operations. A program to solve a linear programming problem with a graphing calculator is given in *The Graphing Calculator Manual* that is available with this book. Spreadsheets often have such a program built in. Figure 7 shows the Solver feature of Microsoft Excel (under the Tools menu) for Example 1.

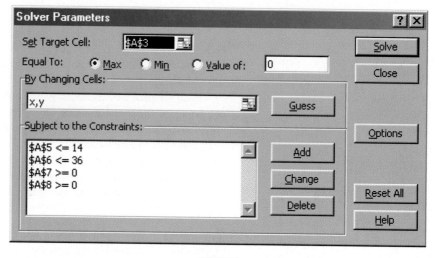

FIGURE 7

In addition, Solver provides a **sensitivity analysis,** which allows us to see how much the constraints could be varied without changing the solution. Figure 8 shows a sensitivity analysis for Example 1. Notice that the value of the first coefficient in the objective function is 24, with an allowable increase of 3 and an allowable decrease of 6. This means that, while keeping the second coefficient at 36, the first coefficient of 24 could be increased by 3 (to 27) or decreased by 6 (to 18), and $(8, 3)$ would still be a solution to the maximization problem. Similarly, for the second coefficient of 36, increasing it by 12 (to 48) or decreasing it by 4 (to 32) would still leave $(8, 3)$ as a solution to the maximization problem. This would be useful to a manufacturer who decides on the solution of $(8, 3)$ (8 Ace

Adjustable Cells						
Cell	Name	Final Value	Reduced Cost	Objective Coefficient	Allowable Increase	Allowable Decrease
A1	x	8	0	24	3	6
B1	y	3	0	36	12	4

Constraints						
Cell	Name	Final Value	Shadow Price	Constraint R. H. Side	Allowable Increase	Allowable Decrease
A5		14	6	14	4	2
A6		36	6	36	6	8
A7		8	0	0	8	1E+30
A8		3	0	0	3	1E+30

FIGURE 8

cabinets and 3 Excello cabinets in the original problem), and wonders how much the objective function would have to change before the solution would no longer be optimal. The original storage capacity for an Ace cabinet was 24 cubic feet, which is the source of the first coefficient in the objective function. Assuming that everything else stays the same, the manufacturer of Ace cabinets could change the storage capacity to anything from 18 to 27 cubic feet, and the office manager's original decision would still be optimal. Notice, however, that any change in the storage capacity for one of the cabinets will change the total storage in the optimal solution. For example, if the first coefficient of 24 is increased by 3 to 27, then the optimal objective value will increase by $3 \times 8 = 24$. One can perform similar changes to other parameters of the problem, but that is beyond the scope of this text.

In many real-life problems, the number of variables and constraints may be in the hundreds, if not the thousands, in which case a computer is used to implement the simplex algorithm. Computer programs for the simplex algorithm differ in some ways from the algorithm we have shown. For example, it is not necessary for a computer to divide common factors out of inequalities to simplify the arithmetic. In fact, computer versions of the algorithm do not necessarily keep all the numbers as integers. As we saw in the previous section, dividing a row by a number may introduce decimals, which makes the arithmetic more difficult to do by hand, but creates no problem for a computer other than round-off error.

If you use a graphing calculator to perform the simplex algorithm, we suggest that you review the pivoting procedure described in Method 2 of the previous section. It differs slightly from Method 1, because it converts each pivot element into a 1, but it works nicely with a calculator to keep track of the arithmetic details.

NOTE Sometimes the simplex method cycles and returns to a previously visited solution, rather than making progress. Methods are available for handling cycling. In this text, we will avoid examples with this behavior. For more details, see Alan Sultan's *Linear Programming: An Introduction with Applications,* Academic Press, 1993. In real applications, cycling is rare and tends not to come up because of computer rounding.

4.2 EXERCISES

In Exercises 1–6, the initial tableau of a linear programming problem is given. Use the simplex method to solve each problem.

1.
$$\begin{array}{cccccc} x_1 & x_2 & x_3 & s_1 & s_2 & z \end{array}$$
$$\left[\begin{array}{cccccc|c} 1 & 2 & 4 & 1 & 0 & 0 & 8 \\ 2 & 2 & 1 & 0 & 1 & 0 & 10 \\ \hline -2 & -5 & -1 & 0 & 0 & 1 & 0 \end{array}\right]$$

2.
$$\begin{array}{cccccc} x_1 & x_2 & x_3 & s_1 & s_2 & z \end{array}$$
$$\left[\begin{array}{cccccc|c} 2 & 2 & 1 & 1 & 0 & 0 & 10 \\ 1 & 2 & 3 & 0 & 1 & 0 & 15 \\ \hline -3 & -2 & -1 & 0 & 0 & 1 & 0 \end{array}\right]$$

3.
$$\begin{array}{cccccc} x_1 & x_2 & s_1 & s_2 & s_3 & z \end{array}$$
$$\left[\begin{array}{cccccc|c} 1 & 3 & 1 & 0 & 0 & 0 & 12 \\ 2 & 1 & 0 & 1 & 0 & 0 & 10 \\ 1 & 1 & 0 & 0 & 1 & 0 & 4 \\ \hline -2 & -1 & 0 & 0 & 0 & 1 & 0 \end{array}\right]$$

4.
$$\begin{array}{ccccccc} x_1 & x_2 & x_3 & s_1 & s_2 & s_3 & z \end{array}$$
$$\left[\begin{array}{ccccccc|c} 2 & 2 & 1 & 1 & 0 & 0 & 0 & 50 \\ 1 & 1 & 3 & 0 & 1 & 0 & 0 & 40 \\ 4 & 2 & 5 & 0 & 0 & 1 & 0 & 80 \\ \hline -2 & -3 & -5 & 0 & 0 & 0 & 1 & 0 \end{array}\right]$$

5.
$$\begin{array}{ccccccc} x_1 & x_2 & x_3 & s_1 & s_2 & s_3 & z \end{array}$$
$$\left[\begin{array}{ccccccc|c} 2 & 2 & 8 & 1 & 0 & 0 & 0 & 40 \\ 4 & -5 & 6 & 0 & 1 & 0 & 0 & 60 \\ 2 & -2 & 6 & 0 & 0 & 1 & 0 & 24 \\ \hline -14 & -10 & -12 & 0 & 0 & 0 & 1 & 0 \end{array}\right]$$

6.
$$\begin{array}{cccccc} x_1 & x_2 & x_3 & s_1 & s_2 & z \end{array}$$
$$\left[\begin{array}{cccccc|c} 3 & 2 & 4 & 1 & 0 & 0 & 18 \\ 2 & 1 & 5 & 0 & 1 & 0 & 8 \\ \hline -1 & -4 & -2 & 0 & 0 & 1 & 0 \end{array}\right]$$

Use the simplex method to solve each linear programming problem.

7. Maximize $z = 4x_1 + 3x_2$
subject to: $2x_1 + 3x_2 \le 11$
$x_1 + 2x_2 \le 6$
with $x_1 \ge 0, \quad x_2 \ge 0.$

8. Maximize $z = 2x_1 + 3x_2$
subject to: $3x_1 + 5x_2 \le 29$
$2x_1 + x_2 \le 10$
with $x_1 \ge 0, \quad x_2 \ge 0.$

9. Maximize $z = 10x_1 + 12x_2$
subject to: $4x_1 + 2x_2 \le 20$
$5x_1 + x_2 \le 50$
$2x_1 + 2x_2 \le 24$
with $x_1 \ge 0, \quad x_2 \ge 0.$

10. Maximize $z = 1.2x_1 + 3.5x_2$
subject to: $2.4x_1 + 1.5x_2 \le 10$
$1.7x_1 + 1.9x_2 \le 15$
with $x_1 \ge 0, \quad x_2 \ge 0.$

11. Maximize $z = 8x_1 + 3x_2 + x_3$
subject to: $x_1 + 6x_2 + 8x_3 \le 118$
$x_1 + 5x_2 + 10x_3 \le 220$
with $x_1 \ge 0, \quad x_2 \ge 0, \quad x_3 \ge 0.$

12. Maximize $z = 12x_1 + 15x_2 + 5x_3$
subject to: $2x_1 + 2x_2 + x_3 \le 8$
$x_1 + 4x_2 + 3x_3 \le 12$
with $x_1 \ge 0, \quad x_2 \ge 0, \quad x_3 \ge 0.$

13. Maximize $z = x_1 + 2x_2 + x_3 + 5x_4$
subject to: $x_1 + 2x_2 + x_3 + x_4 \le 50$
$3x_1 + x_2 + 2x_3 + x_4 \le 100$
with $x_1 \ge 0, \quad x_2 \ge 0, \quad x_3 \ge 0, \quad x_4 \ge 0.$

14. Maximize $z = x_1 + x_2 + 4x_3 + 5x_4$
subject to: $x_1 + 2x_2 + 3x_3 + x_4 \le 115$
$2x_1 + x_2 + 8x_3 + 5x_4 \le 200$
$x_1 + x_3 \le 50$
with $x_1 \ge 0, \quad x_2 \ge 0, \quad x_3 \ge 0, \quad x_4 \ge 0.$

15. The simplex algorithm still works if an indicator other than the most negative one is chosen. (Try it!) List the disadvantages that might occur if this is done.

16. What goes wrong if a quotient other than the smallest nonnegative quotient is chosen in the simplex algorithm? (Try it!)

Applications

Set up and solve Exercises 17–23 by the simplex method.

BUSINESS AND ECONOMICS

17. *Charitable Contributions* Jayanta is working to raise money for the homeless by sending information letters and making follow-up calls to local labor organizations and church groups. She discovers that each church group requires 2 hr of letter writing and 1 hr of follow-up, while for each labor union she needs 2 hr of letter writing and 3 hr of follow-up. Jayanta can raise $100 from each church group and $200 from each union local, and she has a maximum of 16 hr of letter-writing time and a maximum of 12 hr of follow-up time available per month. Determine the most profitable mixture of groups she should contact and the most money she can raise in a month.

18. *Profit* Seall Manufacturing Company makes color television sets. It produces a bargain set that sells for $100 profit and a deluxe set that sells for $150 profit. On the assembly line the bargain set requires 3 hr, and the deluxe set takes 5 hr. The cabinet shop spends 1 hr on the cabinet for the bargain set and 3 hr on the cabinet for the deluxe set. Both sets require 2 hr of time for testing and packing. On a particular production run the Seall Company has available 3900 work hours on the assembly line, 2100 work hours in the cabinet shop, and 2200 work hours in the testing and packing department. How many sets of each type should it produce to make maximum profit? What is the maximum profit? (See Exercise 10 in Section 3.3.)

19. *Profit* The Soul Sounds Recording Company produces three main types of musical CDs: jazz, blues, and reggae. Each jazz CD requires 4 hr of recording, 2 hr of mixing, and 6 hr of editing; each blues CD requires 4 hr of recording, 8 hr of mixing, and 2 of editing; and each reggae CD requires 10 hr of recording, 4 hr of mixing, and 6 hr of editing. The recording studio is available 80 hr per week, staff to operate the mixing board are available 52 hr a week,

and the editing crew has at most 54 hr per week for work. Soul Sounds makes 80¢ per jazz CD, 60¢ per blues CD, and $1.20 for each reggae CD. Use the simplex method to determine how many of each type of CD the company should produce to maximize weekly profit. Find the maximum profit.

20. *Income* A baker has 150 units of flour, 90 of sugar, and 150 of raisins. A loaf of raisin bread requires 1 unit of flour, 1 of sugar, and 2 of raisins, while a raisin cake needs 5, 2, and 1 units, respectively. If raisin bread sells for $1.75 a loaf and raisin cake for $4.00 each, how many of each should be baked so that gross income is maximized? What is the maximum gross income?

21. *Manufacturing Bicycles* A manufacturer of bicycles builds racing, touring, and mountain models. The bicycles are made of both aluminum and steel. The company has available 91,800 units of steel and 42,000 units of aluminum. The racing, touring, and mountain models need 17, 27, and 34 units of steel, and 12, 21, and 15 units of aluminum, respectively. (See Exercise 26 in Section 4.1.)

  a. How many of each type of bicycle should be made in order to maximize profit if the company makes $8 per racing bike, $12 per touring bike, and $22 per mountain bike?

  b. What is the maximum possible profit?

  c. There are many unstated assumptions in the problem given above. Even if the mathematical solution is to make only one or two types of the bicycles, there may be demand for the type(s) not being made, which would create problems for the company. Discuss this and other difficulties that would arise in a real situation.

22. *Production* The Cut-Right Company sells sets of kitchen knives. The Basic Set consists of 2 utility knives and 1 chef's knife. The Regular Set consists of 2 utility knives, 1 chef's knife, and 1 slicer. The Deluxe Set consists of 3 utility knives, 1 chef's knife, and 1 slicer. Their profit is $30 on a Basic Set, $40 on a Regular Set, and $60 on a Deluxe Set. The factory has on hand 800 utility knives, 400 chef's knives, and 200 slicers. (See Exercise 28 in Section 4.1.)

  a. Assuming that all sets will be sold, how many of each type should be made up in order to maximize profit? What is the maximum profit?

  b. A consultant for the Cut-Right Company notes that more profit is made on a Regular Set of knives than on a Basic Set, yet the result from part a recommends making up 100 Basic Sets but no Regular Sets. She is puzzled how this can be the best solution. How would you respond?

23. *Advertising* The Fancy Fashions Store had $8000 available each month for advertising. Newspaper ads cost $400 each and no more than 20 can be run per month. Radio ads cost $200 each and no more than 30 can run per month. TV ads cost $1200 each, with a maximum of 6 available each month. Approximately 2000 women will see each newspaper ad, 1200 will hear each radio commercial, and 10,000 will see each TV ad. (See Exercise 29 in Section 4.1.)

a. How much of each type of advertising should be used if the store wants to maximize its ad exposure?

b. A marketing analyst is puzzled by the results of part a. More women see each newspaper ad than hear each radio commercial, he reasons, so it makes no sense to use radio commercials and no newspaper ads. How would you respond?

24. *Profit* A manufacturer makes two products, toy trucks and toy fire engines. Both are processed in four different departments, each of which has a limited capacity. The sheet metal department can handle at least $1\frac{1}{2}$ times as many trucks as fire engines; the truck assembly department can handle at most 6700 trucks per week; and the fire engine assembly department assembles at most 5500 fire engines weekly. The painting department, which finishes both toys, has a maximum capacity of 12,000 per week.

a. If the profit is $8.50 for a toy truck and $12.10 for a toy fire engine, how many of each item should the company produce to maximize profit?

b. Find a value for the profit for a toy truck and a value for the profit for a toy fire engine that would result in no fire engines being manufactured to maximize profit, given the constraints in part a.

c. Find a value for the profit for a toy truck and a value for the profit for a toy fire engine that would result in no toy trucks being manufactured to maximize the profit, given the constraints in part a.

Exercises 25 and 26 come from past CPA examinations.*
Select the appropriate answer for each question.

25. *Profit* The Ball Company manufactures three types of lamps, labeled A, B, and C. Each lamp is processed in two departments, I and II. Total available work-hours per day for departments I and II are 400 and 600, respectively. No additional labor is available. Time requirements and profit per unit for each lamp type is as follows:

	A	B	C
Work-hours in I	2	3	1
Work-hours in II	4	2	3
Profit per Unit	$5	$4	$3

The company has assigned you as the accounting member of its profit planning committee to determine the numbers of types of A, B, and C lamps that it should produce in order to maximize its total profit from the sale of lamps. The following questions relate to a linear programming model that your group has developed.

a. The coefficients of the objective function would be

(1) 4, 2, 3.   **(2)** 2, 3, 1.

(3) 5, 4, 3.   **(4)** 400, 600.

b. The constrains in the model would be

(1) 2, 3, 1.   **(2)** 5, 4, 3.

(3) 4, 2, 3.   **(4)** 400, 600.

c. The constraint imposed by the available work-hours in department I could be expressed as

(1) $4X_1 + 2X_2 + 3X_3 \leq 400.$

(2) $4X_1 + 2X_2 + 3X_3 \geq 400.$

(3) $2X_1 + 3X_2 + 1X_3 \leq 400.$

(4) $2X_1 + 3X_2 + 1X_3 \geq 400.$

26. *Profit* The Golden Hawk Manufacturing Company wants to maximize the profits on products A, B, and C. The contribution margin for each product follows:

Product	Contribution Margin
A	$2
B	$5
C	$4

The production requirements and departmental capacities, by departments, are as follows:

Department	Production Requirements by Product (hours)			Departmental Capacity (total hours)
	A	B	C	
Assembling	2	3	2	30,000
Painting	1	2	2	38,000
Finishing	2	3	1	28,000

a. What is the profit-maximization formula for the Golden Hawk Company?

(1) $2A + $5B + $4C = X$ (where X = profit)

(2) $5A + 8B + 5C \leq 96,000$

---

*Material from *Uniform CPA Examination Questions and Unofficial Answers,* copyright © 1973, 1974, 1975 by the American Institute of Certified Public Accountants, Inc., is reprinted with permission.

    (3) $2A + $5B + $4C \leq X$

    (4) $2A + $5B + $4C = 96,000$

b. What is the constraint for the painting department of the Golden Hawk Company?

    (1) $1A + 2B + 2C \geq 38,000$

    (2) $2A + $5B + $4C \geq 38,000$

    (3) $1A + 2B + 2C \leq 38,000$

    (4) $2A + 3B + 2C \leq 30,000$

27. *Sensitivity Analysis* Using a computer spreadsheet, perform a sensitivity analysis for the objective function in Exercise 17. What are the highest and lowest possible values for the amount raised from each church group that would yield the same solution as the original problem? Answer the same question for the amount raised from each union local.

28. *Sensitivity Analysis* Using a computer spreadsheet, perform a sensitivity analysis for the objective function in Exercise 18. What are the highest and lowest possible values for profit on a bargain set that would yield the same solution as the original problem? Answer the same question for a deluxe set.

Set up and solve Exercises 29–32 by the simplex method.

LIFE SCIENCES

29. *Blending Nutrients* A biologist has 500 kg of nutrient A, 600 kg of nutrient B, and 300 kg of nutrient C. These nutrients will be used to make four types of food, whose contents (in percent of nutrient per kilogram of food) and whose "growth values" are as shown in the table.

	P	Q	R	S
A	0	0	37.5	62.5
B	0	75	50	37.5
C	100	25	12.5	0
Growth Value	90	70	60	50

How many kilograms of each food should be produced in order to maximize total growth value? Find the maximum growth value.

30. *Resource Management* The average weights of the three species stocked in the lake referred to in Section 2.2, Exercise 61, are 1.62, 2.14, and 3.01 kg for species A, B, and C, respectively.

a. If the largest amounts of food that can be supplied each day are as given in Exercise 61, how should the lake be stocked to maximize the weight of the fish supported by the lake?

b. Find a value for each of the average weights of the three species that would result in none of species B or C being

stocked to maximize the weight of the fish supported by the lake, given the constraints in part a.

c. Find a value for each of the average weights of the three species that would result in none of species A or B being stocked to maximize the weight of the fish supported by the lake, given the constraints in part a.

SOCIAL SCIENCES

31. *Politics* A political party is planning a half-hour television show. The show will have 3 min of direct requests for money from viewers. Three of the party's politicians will be on the show—a senator, a congresswoman, and a governor. The senator, a party "elder statesman," demands that he be on screen at least twice as long as the governor. The total time taken by the senator and the governor must be at least twice the time taken by the congresswoman. Based on a pre-show survey, it is believed that 40, 60, and 50 (in thousands) viewers will watch the program for each minute the senator, congresswoman, and governor, respectively, are on the air. Find the time that should be allotted to each politician in order to get the maximum number of viewers. Find the maximum number of viewers.

32. *Fund Raising* The political party in Exercise 31 is planning its fund-raising activities for a coming election. It plans to raise money through large fund-raising parties, letters requesting funds, and dinner parties where people can meet the candidate personally. Each large fund-raising party costs $3000, each mailing costs $1000, and each dinner party costs $12,000. The party can spend up to $102,000 for these activities. From experience, the planners know that each large party will raise $200,000, each letter campaign will raise $100,000, and each dinner party will raise $600,000. They are able to carry out a total of 25 of these activities.

a. How many of each should the party plan in order to raise the maximum amount of money? What is the maximum amount?

b. Dinner parties are more expensive than letter campaigns, yet the optimum solution found in part a includes dinner parties but no letter campaigns. Explain how this is possible.

■ 4.3 MINIMIZATION PROBLEMS; DUALITY

THINK ABOUT IT How many units of different types of feed should a dog breeder purchase to meet the nutrient requirements of her beagles at a minimum cost?

Using the method of duals, we will learn to answer this and other questions.

Minimization Problems The definition of a problem in standard maximum form was given earlier in this chapter. Now we can define a linear programming problem in *standard minimum form,* as follows.

> **STANDARD MINIMUM FORM**
>
> A linear programming problem is in **standard minimum form** if the following conditions are satisfied.
>
> 1. The objective function is to be minimized.
> 2. All variables are nonnegative.
> 3. All remaining constraints are stated in the form
>
> $$a_1y_1 + a_2y_2 + \cdots + a_ny_n \geq b, \qquad \text{with } b \geq 0.$$

The difference between maximization and minimization problems is in conditions 1 and 3: in problems stated in standard minimum form the objective function is to be *minimized,* rather than maximized, and all constraints must have $\geq$ instead of $\leq$.

We use y_1, y_2, etc., for the variables and w for the objective function as a reminder that these are minimizing problems. Thus, $w = c_1y_1 + c_2y_2 + \cdots + c_ny_n$.

NOTE In this section, we require that all coefficients in the objective function be positive, so $c_1 \geq 0, c_2 \geq 0, \ldots, c_n \geq 0$.

Duality An interesting connection exists between standard maximization and standard minimization problems: any solution of a standard maximization problem produces the solution of an associated standard minimization problem, and vice versa. Each of these associated problems is called the **dual** of the other. One advantage of duals is that standard minimization problems can be solved by the simplex method discussed in the first two sections of this chapter. Let us explain the idea of a dual with an example.

EXAMPLE 1 Duality

Minimize $w = 8y_1 + 16y_2$
subject to: $y_1 + 5y_2 \geq 9$
                   $2y_1 + 2y_2 \geq 10$
with $y_1 \geq 0, \quad y_2 \geq 0.$

Without considering slack variables just yet, write the augmented matrix of the system of inequalities, and include the coefficients of the objective function (not their negatives) as the last row in the matrix.

$$\begin{array}{c} \\ \\ \text{objective function} \longrightarrow \end{array} \overbrace{\begin{bmatrix} 1 & 5 & \vline & 9 \\ 2 & 2 & \vline & 10 \\ 8 & 16 & \vline & 0 \end{bmatrix}}^{\text{constants}}$$

Now look at the following matrix, which we obtain from the one above by interchanging rows and columns.

$$\begin{array}{c} \\ \\ \text{objective function} \longrightarrow \end{array} \overbrace{\begin{bmatrix} 1 & 2 & \vline & 8 \\ 5 & 2 & \vline & 16 \\ 9 & 10 & \vline & 0 \end{bmatrix}}^{\text{constants}}$$

The *rows* of the first matrix (for the minimization problem) are the *columns* of the second matrix.

The entries in this second matrix could be used to write the following maximization problem in standard form (again ignoring the fact that the numbers in the last row are not negative):

$$\text{Maximize} \qquad z = 9x_1 + 10x_2$$

$$\text{subject to:} \qquad x_1 + 2x_2 \le 8$$

$$5x_1 + 2x_2 \le 16$$

with all variables nonnegative.

Figure 9(a) shows the region of feasible solutions for the minimization problem just given, while Figure 9(b) shows the region of feasible solutions for the maximization problem produced by exchanging rows and columns. The solutions of the two problems are given on the next page.

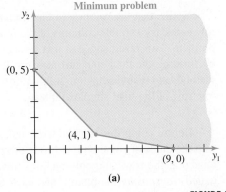

(a)

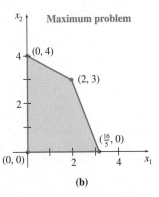

(b)

FIGURE 9

Corner Point	$w = 8y_1 + 16y_2$	
$(0, 5)$	80	
$(4, 1)$	48	Minimum
$(9, 0)$	72	

The minimum is 48 when $y_1 = 4$ and $y_2 = 1$.

Corner Point	$z = 9x_1 + 10x_2$	
$(0, 0)$	0	
$(0, 4)$	40	
$(2, 3)$	48	Maximum
$(16/5, 0)$	28.8	

The maximum is 48 when $x_1 = 2$ and $x_2 = 3$.

The two feasible regions in Figure 9 are different and the corner points are different, but the values of the objective functions are equal—both are 48. An even closer connection between the two problems is shown by using the simplex method to solve this maximization problem.

Maximization Problem

$$
\begin{array}{ccccc}
x_1 & x_2 & s_1 & s_2 & z \\
\end{array}
$$

$$
\left[\begin{array}{ccccc|c}
1 & \mathbf{2} & 1 & 0 & 0 & 8 \\
5 & 2 & 0 & 1 & 0 & 16 \\
\hline
-9 & -10 & 0 & 0 & 1 & 0
\end{array}\right]
$$

$$
\begin{array}{c}
-R_1 + R_2 \rightarrow R_2 \\
5R_1 + R_3 \rightarrow R_3
\end{array}
\left[\begin{array}{ccccc|c}
1 & 2 & 1 & 0 & 0 & 8 \\
\mathbf{4} & 0 & -1 & 1 & 0 & 8 \\
\hline
-4 & 0 & 5 & 0 & 1 & 40
\end{array}\right]
$$

$$
\begin{array}{c}
-R_2 + 4R_1 \rightarrow R_1 \\
\\
R_2 + R_3 \rightarrow R_3
\end{array}
\left[\begin{array}{ccccc|c}
0 & 8 & 5 & -1 & 0 & 24 \\
4 & 0 & -1 & 1 & 0 & 8 \\
\hline
0 & 0 & 4 & 1 & 1 & 48
\end{array}\right]
$$

$$
\begin{array}{c}
R_1/8 \rightarrow R_1 \\
R_2/4 \rightarrow R_2
\end{array}
\left[\begin{array}{ccccc|c}
0 & 1 & \frac{5}{8} & -\frac{1}{8} & 0 & 3 \\
1 & 0 & -\frac{1}{4} & \frac{1}{4} & 0 & 2 \\
\hline
0 & 0 & \mathbf{4} & \mathbf{1} & 1 & 48
\end{array}\right]
$$

The maximum is 48 when
$x_1 = 2$ and $x_2 = 3$.

Notice that the solution to the *minimization problem* is found in the bottom row and slack variable columns of the final simplex tableau for the maximization problem. This result suggests that standard minimization problems can be solved by forming the dual standard maximization problem, solving it by the simplex method, and then reading the solution for the minimization problem from the bottom row of the final simplex tableau.

Before using this method to actually solve a minimization problem, let us find the duals of some typical linear programming problems. The process of exchanging the rows and columns of a matrix, which is used to find the dual, is called *transposing* the matrix, and each of the two matrices is the **transpose** of the other.

EXAMPLE 2 Transpose

Find the transpose of each matrix.

(a) $A = \begin{bmatrix} 2 & -1 & 5 \\ 6 & 8 & 0 \\ -3 & 7 & -1 \end{bmatrix}$

Solution Write the rows of matrix A as the columns of the transpose.

$$\text{Transpose of } A = A^T = \begin{bmatrix} 2 & 6 & -3 \\ -1 & 8 & 7 \\ 5 & 0 & -1 \end{bmatrix}$$

(b) $B = \begin{bmatrix} 1 & 2 & 4 & 0 \\ 2 & 1 & 7 & 6 \end{bmatrix}$

Solution

$$\text{Transpose of } B = B^T = \begin{bmatrix} 1 & 2 \\ 2 & 1 \\ 4 & 7 \\ 0 & 6 \end{bmatrix}$$

EXAMPLE 3 Dual

Write the dual of each of the following standard linear programming problems.

(a) Maximize $z = 2x_1 + 5x_2$
   subject to: $x_1 + x_1 \le 10$
                   $2x_1 + x_2 \le 8$
   with $x_1 \ge 0, \quad x_2 \ge 0.$

Solution Begin by writing the augmented matrix for the given problem.

$$\begin{bmatrix} 1 & 1 & | & 10 \\ 2 & 1 & | & 8 \\ 2 & 5 & | & 0 \end{bmatrix}$$

Form the transpose of this matrix:

$$\begin{bmatrix} 1 & 2 & | & 2 \\ 1 & 1 & | & 5 \\ 10 & 8 & | & 0 \end{bmatrix}.$$

The dual problem is stated from this second matrix as follows (using y instead of x):

   Minimize $w = 10y_1 + 8y_2$
   subject to: $y_1 + 2y_2 \ge 2$
                  $y_1 + y_2 \ge 5$
   with $y_1 \ge 0, \quad y_2 \ge 0.$

(b) Minimize $w = 7y_1 + 5y_2 + 8y_3$
   subject to: $3y_1 + 2y_2 + y_3 \ge 10$
                  $y_1 + y_2 + y_3 \ge 8$
                  $4y_1 + 5y_2 \ge 25$
   with $y_1 \ge 0, \quad y_2 \ge 0, \quad y_3 \ge 0.$

Solution The dual problem is stated as follows.

$$\text{Maximize} \quad z = 10x_1 + 8x_2 + 25x_3$$
$$\text{subject to:} \quad 3x_1 + x_2 + 4x_3 \le 7$$
$$2x_1 + x_2 + 5x_3 \le 5$$
$$x_1 + x_2 \qquad \le 8$$
$$\text{with} \quad x_1 \ge 0, \quad x_2 \ge 0, \quad x_3 \ge 0.$$

In Example 3, all the constraints of the given standard maximization problems were $\le$ inequalities, while all those in the dual minimization problems were $\ge$ inequalities. This is generally the case; inequalities are reversed when the dual problem is stated.

The following table shows the close connection between a problem and its dual.

Given Problem	Dual Problem
m variables	n variables
n constraints	m constraints
Coefficients from objective function	Constraint constants
Constraint constants	Coefficients from objective function

NOTE To solve a minimization problem with duals, all of the coefficients in the objective function must be positive. Otherwise, negative numbers will appear on the right side of the constraints in the dual problem, which we do not allow. For a method that does not have this restriction, see the next section.

The next theorem, whose proof requires advanced methods, guarantees that a standard minimization problem can be solved by forming a dual standard maximization problem.

THEOREM OF DUALITY

The objective function w of a minimization linear programming problem takes on a minimum value if and only if the objective function z of the corresponding dual maximization problem takes on a maximum value. The maximum value of z equals the minimum value of w.

This method is illustrated in the following example.

EXAMPLE 4 Duality

$$\text{Minimize} \quad w = 3y_1 + 2y_2$$
$$\text{subject to:} \quad y_1 + 3y_2 \ge 6$$
$$2y_1 + y_2 \ge 3$$
$$\text{with} \quad y_1 \ge 0, \quad y_2 \ge 0.$$

Solution Use the given information to write the matrix.

$$\begin{bmatrix} 1 & 3 & | & 6 \\ 2 & 1 & | & 3 \\ \hline 3 & 2 & | & 0 \end{bmatrix}$$

Transpose to get the following matrix for the dual problem.

$$\left[\begin{array}{cc|c} 1 & 2 & 3 \\ 3 & 1 & 2 \\ \hline 6 & 3 & 0 \end{array}\right]$$

Write the dual problem from this matrix, as follows:

$$\begin{aligned} \text{Maximize} \quad & z = 6x_1 + 3x_2 \\ \text{subject to:} \quad & x_1 + 2x_2 \le 3 \\ & 3x_1 + x_2 \le 2 \\ \text{with} \quad & x_1 \ge 0, \quad x_2 \ge 0. \end{aligned}$$

Solve this standard maximization problem using the simplex method. Start by introducing slack variables to give the system

$$\begin{aligned} x_1 + 2x_2 + s_1 \qquad\qquad &= 3 \\ 3x_1 + x_2 \qquad + s_2 \qquad &= 2 \\ -6x_1 - 3x_2 - 0s_1 - 0s_2 + z &= 0 \end{aligned}$$

with $x_1 \ge 0, \quad x_2 \ge 0, \quad s_1 \ge 0, \quad s_2 \ge 0.$

The first tableau for this system is given below, with the pivot as indicated.

Quotients		x_1	x_2	s_1	s_2	z	
$3/1 = 3$		1	2	1	0	0	3
$2/3$		**3**	1	0	1	0	2
		-6	-3	0	0	1	0

The simplex method gives the following as the final tableau.

$$\left[\begin{array}{ccccc|c} x_1 & x_2 & s_1 & s_2 & z & \\ 0 & 1 & \frac{3}{5} & -\frac{1}{5} & 0 & \frac{7}{5} \\ 1 & 0 & -\frac{1}{5} & \frac{2}{5} & 0 & \frac{1}{5} \\ 0 & 0 & \frac{3}{5} & \frac{9}{5} & 1 & \frac{27}{5} \end{array}\right]$$

Since a 1 has been created in the z column, the last row of this final tableau gives the solution to the minimization problem. The minimum value of $w = 3y_1 + 2y_2$, subject to the given constraints, is $27/5$ and occurs when $y_1 = 3/5$ and $y_2 = 9/5$. The minimum value of w, $27/5$, is the same as the maximum value of z.

Let us summarize the steps in solving a standard minimization linear programming problem by the method of duals.

SOLVING STANDARD MINIMUM PROBLEMS WITH DUALS

1. Find the dual standard maximization problem.
2. Solve the maximization problem using the simplex method.
3. The minimum value of the objective function w is the maximum value of the objective function z.
4. The optimum solution is given by the entries in the bottom row of the columns corresponding to the slack variables, so long as the entry in the z column is equal to 1.

CAUTION (1) If the final entry in the z column is a value other than 1, divide the bottom row through by that value so that it will become 1. Only then can the solution of the minimization problem be found in the bottom row of the columns corresponding to the slack variables.

(2) Do not simplify an inequality in the dual by dividing out a common factor. For example, if an inequality in the dual is $3x_1 + 3x_2 \leq 6$, do not simplify to $x_1 + x_2 \leq 2$ by dividing out the 3. Doing so will give an incorrect solution to the original problem.

NOTE If the objective function is written below the constraints (except those stating that each variable must be greater than or equal to 0), and the variables are lined up vertically, it is easy to go from the original problem to the dual: the coefficients in any column of the original problem become the coefficients of the corresponding row in the dual. In Example 4, for instance, if the objective function is written below the constraints, the coefficients in the first column are 1, 2, 3, leading to the first row of the dual: $x_1 + 2x_2 \leq 3$. The last column is 6, 3, so the last row of the dual is this: Maximize $z = 6x_1 + 3x_2$.

 Further Uses of the Dual The dual is useful not only in solving minimization problems, but also in seeing how small changes in one variable will affect the value of the objective function. For example, suppose a dog breeder needs at least 6 units per day of nutrient A and at least 3 units of nutrient B for her beagles, and that the breeder can choose between two different feeds, feed 1 and feed 2. Find the minimum cost for the breeder if each bag of feed 1 costs $3 and provides 1 unit of nutrient A and 2 units of B, while each bag of feed 2 costs $2 and provides 3 units of nutrient A and 1 of B.

If y_1 represents the number of bags of feed 1 and y_2 represents the number of bags of feed 2, the given information leads to the following problem.

$$\begin{aligned} \text{Minimize} \quad & w = 3y_1 + 2y_2 \\ \text{subject to:} \quad & y_1 + 3y_2 \geq 6 \\ & 2y_1 + y_2 \geq 3 \\ \text{with} \quad & y_1 \geq 0, \quad y_2 \geq 0. \end{aligned}$$

This standard minimization linear programming problem is the one solved in Example 4 of this section. In that example, the dual was formed and the following tableau was found.

$$\begin{array}{ccccc} x_1 & x_2 & s_1 & s_2 & z \\ \left[\begin{array}{ccccc|c} 0 & 1 & \frac{3}{5} & -\frac{1}{5} & 0 & \frac{7}{5} \\ 1 & 0 & -\frac{1}{5} & \frac{2}{5} & 0 & \frac{1}{5} \\ 0 & 0 & \frac{3}{5} & \frac{9}{5} & 1 & \frac{27}{5} \end{array}\right] \end{array}$$

This final tableau shows that the breeder will obtain minimum feed costs by using $3/5$ bag of feed 1 and $9/5$ bags of feed 2 per day, for a daily cost of $27/5 = \$5.40$.

Now look at the data from the problem shown in the following table.

	Unit of Nutrient (per bag):		Cost (per bag)
	A	B	
Feed 1	1	2	$3
Feed 2	3	1	$2
Requirement	6	3	

If x_1 and x_2 are the costs per unit of nutrients A and B, the constraints of the dual problem can be stated as follows.

$$\text{Cost of feed 1:} \quad x_1 + 2x_2 \leq 3$$
$$\text{Cost of feed 2:} \quad 3x_1 + x_2 \leq 2$$

The solution of the dual problem, which maximizes nutrients, can be read from the final tableau:

$$x_1 = \frac{1}{5} = .20 \quad \text{and} \quad x_2 = \frac{7}{5} = 1.40,$$

which means that a unit of nutrient A costs $.20, while a unit of nutrient B costs $1.40. The minimum daily cost, $5.40, is found by the following procedure.

$$\underline{\begin{aligned} (\$.20 \text{ per unit of A}) \times (6 \text{ units of A}) &= \$1.20 \\ + (\$1.40 \text{ per unit of B}) \times (3 \text{ units of B}) &= \$4.20 \end{aligned}}$$
$$\text{Minimum daily cost} = \$5.40$$

The numbers .20 and 1.40 are called the **shadow costs** of the nutrients. These two numbers from the dual, $.20 and $1.40, also allow the breeder to calculate feed costs for small changes in nutrient requirements. For example, an increase of one unit in the requirement for each nutrient would produce a total cost of $7.00:

$5.40	6 units of A, 3 of B
.20	1 extra unit of A
+ 1.40	1 extra unit of B
$7.00	Total cost per day

Shadow costs only give the exact answer for a limited range. Unfortunately, finding that range is somewhat complicated. In the dog feed example, we can add up to 3 units or delete up to 4 units of A, and shadow costs will give the exact answer. If, however, we add 4 units of A, shadow costs give an answer of $6.20, while the true cost is $6.67.

CAUTION If you wish to use shadow costs, do not simplify an inequality in the original problem by dividing out a common factor. For example, if an inequality in the original problem is $3y_1 + 3y_2 \geq 6$, do not simplify to $y_1 + y_2 \geq 2$ by dividing out the 3. Doing so will give incorrect shadow costs.

NOTE Shadow costs become shadow values in maximization problems. For example, see Exercises 16 and 17.

The Solver in Microsoft Excel provides the values of the dual variables. See *The Spreadsheet Manual* that is available with this book for more details.

4.3 EXERCISES

Find the transpose of each matrix.

1. $\begin{bmatrix} 1 & 2 & 3 \\ 3 & 2 & 1 \\ 1 & 10 & 0 \end{bmatrix}$

2. $\begin{bmatrix} 2 & 5 & 8 & 6 & 0 \\ 1 & -1 & 0 & 12 & 14 \end{bmatrix}$

3. $\begin{bmatrix} -1 & 4 & 6 & 12 \\ 13 & 25 & 0 & 4 \\ -2 & -1 & 11 & 3 \end{bmatrix}$

4. $\begin{bmatrix} 1 & 11 & 15 \\ 0 & 10 & -6 \\ 4 & 12 & -2 \\ 1 & -1 & 13 \\ 2 & 25 & -1 \end{bmatrix}$

State the dual problem for each of the following.

5. Maximize $\quad z = 4x_1 + 3x_2 + 2x_3$
   subject to: $\quad x_1 + x_2 + x_3 \le 5$
   $\quad\quad\quad\quad x_1 + x_2 \quad\quad \le 4$
   $\quad\quad\quad\quad 2x_1 + x_2 + 3x_3 \le 15$
   with $\quad x_1 \ge 0, \quad x_2 \ge 0, \quad x_3 \ge 0.$

6. Maximize $\quad z = 8x_1 + 3x_2 + x_3$
   subject to: $\quad 7x_1 + 6x_2 + 8x_3 \le 18$
   $\quad\quad\quad\quad 4x_1 + 5x_2 + 10x_3 \le 20$
   with $\quad x_1 \ge 0, \quad x_2 \ge 0, \quad x_3 \ge 0.$

7. Minimize $\quad w = y_1 + 2y_2 + y_3 + 5y_4$
   subject to: $\quad y_1 + y_2 + y_3 + y_4 \ge 50$
   $\quad\quad\quad\quad 3y_1 + y_2 + 2y_3 + y_4 \ge 100$
   with $\quad y_1 \ge 0, \quad y_2 \ge 0, \quad y_3 \ge 0, \quad y_4 \ge 0.$

8. Minimize $\quad w = y_1 + y_2 + 4y_3$
   subject to: $\quad y_1 + 2y_2 + 3y_3 \ge 115$
   $\quad\quad\quad\quad 2y_1 + y_2 + 8y_3 \ge 200$
   $\quad\quad\quad\quad y_1 \quad\quad + y_3 \ge 50$
   with $\quad y_1 \ge 0, \quad y_2 \ge 0, \quad y_3 \ge 0.$

Use the simplex method to solve Exercises 9–14.

9. Find $y_1 \ge 0$ and $y_2 \ge 0$ such that
   $$2y_1 + 3y_2 \ge 6$$
   $$2y_1 + y_2 \ge 7$$
   and $w = 5y_1 + 2y_2$ is minimized.

10. Find $y_1 \ge 0$ and $y_2 \ge 0$ such that
    $$3y_1 + y_2 \ge 12$$
    $$y_1 + 4y_2 \ge 16$$
    and $w = 2y_1 + y_2$ is minimized.

11. Find $y_1 \ge 0$ and $y_2 \ge 0$ such that
    $$10y_1 + 5y_2 \ge 100$$
    $$20y_1 + 10y_2 \ge 150$$
    and $w = 4y_1 + 5y_2$ is minimized.

12. Minimize $\quad w = 3y_1 + 2y_2$
    subject to: $\quad 2y_1 + 3y_2 \ge 60$
    $\quad\quad\quad\quad y_1 + 4y_2 \ge 40$
    with $\quad y_1 \ge 0, \quad y_2 \ge 0.$

13. Minimize $\quad w = 2y_1 + y_2 + 3y_3$
    subject to: $\quad y_1 + y_2 + y_3 \ge 100$
    $\quad\quad\quad\quad 2y_1 + y_2 \quad\quad \ge 50$
    with $\quad y_1 \ge 0, \quad y_2 \ge 0, \quad y_3 \ge 0.$

14. Minimize $\quad w = 3y_1 + 2y_2$
    subject to: $\quad y_1 + 2y_2 \ge 10$
    $\quad\quad\quad\quad y_1 + y_2 \ge 8$
    $\quad\quad\quad\quad 2y_1 + y_2 \ge 12$
    with $\quad y_1 \ge 0, \quad y_2 \ge 0.$

15. You are given the following linear programming problem (P):*

Minimize $z = x_1 + 2x_2$
subject to: $-2x_1 + x_2 \geq 1$
$x_1 - 2x_2 \geq 1$
$x_1 \geq 0, \quad x_2 \geq 0.$

The dual of (P) is (D). Which of the statements in the next column is true?

a. (P) has no feasible solution and the objective function of (D) is unbounded.

b. (D) has no feasible solution and the objective function of (P) is unbounded.

c. The objective functions of both (P) and (D) are unbounded.

d. Both (P) and (D) have optimal solutions.

e. Neither (P) nor (D) has feasible solutions.

Applications

BUSINESS AND ECONOMICS

16. *Agriculture* Refer to the original information in Example 2, Section 4.1.

   a. Give the dual problem.

   b. Use the shadow values to estimate the farmer's profit if land is cut to 90 acres but capital increases to $21,000.

   c. Suppose the farmer has 110 acres but only $19,000. Find the optimum profit and the planting strategy that will produce this profit.

17. *Toy Manufacturing* A small toy manufacturing firm has 200 squares of felt, 600 oz of stuffing, and 90 ft of trim available to make two types of toys, a small bear and a monkey. The bear requires 1 square of felt and 4 oz of stuffing. The monkey requires 2 squares of felt, 3 oz of stuffing, and 1 ft of trim. The firm makes $1 profit on each bear and $1.50 profit on each monkey.

   a. Set up the linear programming problem to maximize profit.

   b. Solve the linear programming problem in part a.

   c. What is the corresponding dual problem?

   d. What is the optimal solution to the dual problem?

   e. Use the shadow values to calculate the profit the firm will make if its supply of felt increases to 210 squares.

   f. How much profit will the firm make if its supply of stuffing is cut to 590 oz and its supply of trim is cut to 80 ft?

   g. Explain why it makes sense that the shadow cost for trim is 0.

18. *Production Costs* A brewery produces regular beer and a lower-carbohydrate "light" beer. Steady customers of the brewery buy 12 units of regular beer and 10 units of light beer monthly. While setting up the brewery to produce the beers, the management decides to produce extra beer, beyond that needed to satisfy the customers. The cost per unit of regular beer is $36,000 and the cost per unit of light beer is $48,000. Every unit of regular beer brings in $100,000 in revenue, while every unit of light beer brings in $300,000 in revenue. The brewery wants at least $7,000,000 in revenue. At least 20 additional units of beer can be sold.

   a. How much of each type of beer should be made so as to minimize total production costs?

   b. Suppose the minimum revenue is increased to $7,500,000. Use shadow costs to calculate the total production cost in this case.

19. *Supply Costs* The chemistry department at a local college decides to stock at least 800 small test tubes and 500 large test tubes. It wants to buy at least 2100 test tubes to take advantage of a special price. Since the small tubes are broken twice as often as the large, the department will order at least twice as many small tubes as large.

   a. If the small test tubes cost 15¢ each and large ones, made of a cheaper glass, cost 12¢ each, how many of each size should be ordered to minimize cost?

   b. Suppose the minimum number of test tubes is increased to 2300. Use shadow costs to calculate the total cost in this case.

20. *Interview Time* Joan McKee has a part-time job conducting public opinion interviews. She has found that a political interview takes 45 min and a market interview takes 55 min. She needs to minimize the time she spends doing interviews to allow more time for her full-time job. Unfortunately, to keep her part-time job, she must complete at least 8 interviews each week. Also, she must earn at least $60 per

---

*Problem 2 from "November 1989 Course 130 Examination Operations Research" of the *Education and Examination Committee of The Society of Actuaries.* Reprinted by permission of The Society of Actuaries.

week at this job; she earns $8 for each political interview and $10 for each market interview. Finally, to stay in good standing with her supervisor, she must earn at least 40 bonus points per week; she receives 6 bonus points for each political interview and 5 points for each market interview. How many of each interview should she do each week to minimize the time spent?

21. *Animal Food* An animal food must provide at least 54 units of vitamins and 60 calories per serving. One gram of soybean meal provides at least 2.5 units of vitamins and 5 calories. One gram of meat byproducts provides at least 4.5 units of vitamins and 3 calories. One gram of grain provides at least 5 units of vitamins and 10 calories. If a gram of soybean meal costs 8¢, a gram of meat byproducts 9¢, and a gram of grain 10¢, what mixture of these three ingredients will provide the required vitamins and calories at minimum cost? What is the minimum cost?

22. *Feed Costs* Refer to the example at the end of this section on minimizing the daily cost of feeds.

 a. Find a combination of feeds that will cost $7 and give 7 units of A and 4 units of B.

 b. Use the dual variables to predict the daily cost of feed if the requirements change to 5 units of A and 4 units of B. Find a combination of feeds to meet these requirements at the predicted price.

23. *Pottery* Karla Harby makes three items in her pottery shop: large bowls, small bowls, and pots for plants. A large bowl requires 3 lb of clay and 6 fl oz of glaze. A small bowl requires 2 lb of clay, and 6 fl oz of glaze. A pot requires 4 lbs of clay and 2 fl oz of glaze. She must use up 72 lbs of old clay and 108 fl oz of old glaze; she can order more if necessary. If Karla can make a large bowl in 5 hours, a small bowl in 6 hours, and a pot in 4 hours, how many of each should she make to minimize her time? What is the minimum time?

LIFE SCIENCES

24. *Health Care* Mark, who is ill, takes vitamin pills. Each day he must have at least 16 units of vitamin A, 5 units of vitamin B_1, and 20 units of vitamin C. He can choose between pill #1, which costs 10¢ and contains 8 units of A, 1 of B_1, and 2 of C; and pill #2, which costs 20¢ and contains 2 units of A, 1 of B_1, and 7 of C. How many of each pill should he buy in order to minimize his cost?

25. *Dietetics* Sing, who is dieting, requires two food supplements, I and II. He can get these supplements from two different products, A and B, as shown below.

$$\begin{array}{c} \text{Grams of} \\ \text{Supplement} \\ \text{(per serving)} \end{array}$$

$$\text{Product} \quad \begin{array}{c} \\ A \\ B \end{array} \begin{array}{cc} I & II \\ \begin{bmatrix} 3 & 2 \\ 2 & 4 \end{bmatrix} \end{array}$$

Sing's physician has recommended that he include at least 15 g of each supplement in his daily diet. If product A costs 25¢ per serving and product B costs 40¢ per serving, how can he satisfy his requirements most economically? Find the minimum cost.

26. *Blending Nutrients* A biologist must make a nutrient of her algae. The nutrient must contain the three basic elements D, E, and F, and must contain at least 10 kg of D, 12 kg of E, and 20 kg of F. The nutrient is made from three ingredients, I, II, and III. The quantity of D, E, and F in one unit of each of the ingredients is as given in the following chart.

Ingredient		I	II	III
Kilograms of	D	4	1	10
Elements (per	E	3	2	1
unit of ingredient)	F	0	4	5
Cost per unit (in $)		4	7	5

How many units of each ingredient are required to meet the biologist's needs at minimum cost?

◼ 4.4 NONSTANDARD PROBLEMS

? THINK ABOUT IT How many cars should an auto manufacturer send from each of its two plants to each of two dealerships in order to minimize the cost while meeting each dealership's needs?

We will learn techniques in this section for answering questions like the one above.

So far we have used the simplex method to solve linear programming problems in standard maximum or minimum form only. Now, this work is extended to include linear programming problems with mixed $\leq$ and $\geq$ constraints.

For example, suppose a new constraint is added to the farmer problem in Example 2 of Section 4.1: to satisfy orders from regular buyers, the farmer must plant a total of at least 60 acres of the three crops. This constraint introduces the new inequality

$$x_1 + x_2 + x_3 \geq 60.$$

As before, this inequality must be rewritten as an equation in which the variables all represent nonnegative numbers. The inequality $x_1 + x_2 + x_3 \geq 60$ means that

$$x_1 + x_2 + x_3 - s_3 = 60$$

for some nonnegative variable s_3. (Remember that s_1 and s_2 are the slack variables in the problem.)

The new variable, s_3, is called a **surplus variable.** The value of this variable represents the excess number of acres (over 60) that may be planted. Since the total number of acres planted is to be no more than 100 but at least 60, the value of s_3 can vary from 0 to 40.

We must now solve the system of equations

$$
\begin{aligned}
x_1 + x_2 + x_3 + s_1 \phantom{{}+ s_2 - s_3 + z} &= 100 \\
10x_1 + 4x_2 + 7x_3 \phantom{{}+ s_1} + s_2 \phantom{{}- s_3 + z} &= 500 \\
x_1 + x_2 + x_3 \phantom{{}+ s_1 + s_2} - s_3 \phantom{{}+ z} &= 60 \\
-120x_1 - 40x_2 - 60x_3 \phantom{{}+ s_1 + s_2 - s_3} + z &= 0,
\end{aligned}
$$

with $x_1, x_2, x_3, s_1, s_2,$ and s_3 all nonnegative.

Set up the initial simplex tableau.

$$
\begin{array}{ccccccc}
x_1 & x_2 & x_3 & s_1 & s_2 & s_3 & z \\
\end{array}
$$

$$
\left[
\begin{array}{ccccccc|c}
1 & 1 & 1 & 1 & 0 & 0 & 0 & 100 \\
10 & 4 & 7 & 0 & 1 & 0 & 0 & 500 \\
1 & 1 & 1 & 0 & 0 & -1 & 0 & 60 \\
\hline
-120 & -40 & -60 & 0 & 0 & 0 & 1 & 0
\end{array}
\right]
$$

This tableau gives the solution

$$x_1 = 0, \quad x_2 = 0, \quad x_3 = 0, \quad s_1 = 100, \quad s_2 = 500, \quad s_3 = -60.$$

But this is not a feasible solution, since s_3 is negative. All the variables in any feasible solution must be nonnegative if the solution is to correspond to a corner point of the region of feasible solutions.

When a negative value of a variable appears in the solution, row operations are used to transform the matrix until a solution is found in which all variables are nonnegative. Here the problem is the -1 in a column corresponding to a basic variable. If the number in that row of the right-hand column were 0, we could simply multiply this row by -1 to remove the negative from the column. But we cannot do this with 60 in the right-hand column. Instead, we find the positive entry that is farthest to the left in the third row (the row containing the -1); namely, the 1 in row 3, column 1. We will pivot using this column. (Actually, any column with a positive entry in row 3 will do; we chose the column farthest to the left arbitrarily.) Use quotients as before to find the pivot, which is the 10 in row 2, column 1. Then use row operations to get the following tableau.

	x_1	x_2	x_3	s_1	s_2	s_3	z	
$-R_2 + 10R_1 \rightarrow R_1$	0	6	3	10	-1	0	0	500
	10	4	7	0	1	0	0	500
$-R_2 + 10R_3 \rightarrow R_3$	0	6	3	0	-1	-10	0	100
$12R_2 + R_4 \rightarrow R_4$	0	8	24	0	12	0	1	6000

Notice from the s_3 column that $-10s_3 = 100$, so s_3 is still negative. We therefore apply the procedure again. The 6 in row 3, column 2, is the positive entry farthest to the left in row 3, and by investigating quotients, we see that it is also the pivot. This leads to the following tableau.

	x_1	x_2	x_3	s_1	s_2	s_3	z	
$-R_3 + R_1 \rightarrow R_1$	0	0	0	10	0	10	0	400
$-2R_3 + 3R_2 \rightarrow R_2$	30	0	15	0	5	20	0	1300
	0	6	3	0	-1	-10	0	100
$-4R_3 + 3R_4 \rightarrow R_4$	0	0	60	0	40	40	3	17,600

The value of s_3 is now 0. If there were any negative numbers in the bottom row, we would continue as before. Since there are none, we have merely to create a 1 in each column corresponding to a basic variable or z.

	x_1	x_2	x_3	s_1	s_2	s_3	z	
$R_1/10 \rightarrow R_1$	0	0	0	1	0	1	0	40
$R_2/30 \rightarrow R_2$	1	0	$\frac{1}{2}$	0	$\frac{1}{6}$	$\frac{2}{3}$	0	$\frac{130}{3}$
$R_3/6 \rightarrow R_3$	0	1	$\frac{1}{2}$	0	$-\frac{1}{6}$	$-\frac{5}{3}$	0	$\frac{50}{3}$
$R_4/3 \rightarrow R_4$	0	0	20	0	$\frac{40}{3}$	$\frac{40}{3}$	1	$\frac{17,600}{3}$

The solution is

$$x_1 = \frac{130}{3} = 43\frac{1}{3}, \; x_2 = \frac{50}{3} = 16\frac{2}{3}, \; x_3 = 0, \; z = \frac{17,600}{3} = 5866.67.$$

For maximum profit with this new constraint, the farmer should plant $43\frac{1}{3}$ acres of potatoes, $16\frac{2}{3}$ acres of corn, and no cabbage. The profit will be $5866.67, less than the $6000 profit if the farmer were to plant only 50 acres of potatoes. Because of the additional constraint that at least 60 acres must be planted, the profit is reduced.

NOTE If we ever reach a point where a surplus variable still has a negative solution, but there are no positive elements left in the row, then the problem has no feasible solution.

The procedure we have followed is a simplified version of the **two-phase method,** which is widely used for solving problems with mixed constraints. To see the complete method, including how to handle some complications that may arise, see *Linear Programming: An Introduction with Applications* by Alan Sultan, Academic Press, 1993.

In the previous section we solved standard minimum problems using duals. If a minimizing problem has mixed $\leq$ and $\geq$ constraints, the dual method cannot be used. We solve such problems with the method presented in this section. To see how, consider the simple fact: when a number t gets smaller, then $-t$ gets larger, and vice versa. For instance, if t goes from 6 to 1 to 0 to -8, then $-t$ goes from -6 to -1 to 0 to 8. Thus, if w is the objective function of a minimizing linear programming problem, the feasible solution that produces the minimum value of w also produces the maximum value of $z = -w$, and vice versa. Therefore, to solve a minimization problem with objective function w, we need only solve the maximization problem with the same constraints and objective function $z = -w$.

In summary, the following steps are involved in solving the nonstandard problems in this section.

SOLVING NONSTANDARD PROBLEMS

1. If necessary, convert the problem to a maximization problem.
2. Add slack variables and subtract surplus variables as needed.
3. Write the initial simplex tableau.
4. If any basic variable has a negative value, locate the nonzero number in that variable's column, and note what row it is in.
5. In the row located in Step 4, find the positive entry that is farthest to the left, and note what column it is in.
6. In the column found in Step 5, choose a pivot by investigating quotients.
7. Use row operations to change the other numbers in the pivot column to 0.
8. Continue Steps 4 through 7 until all basic variables are nonnegative. If it ever becomes impossible to continue, then the problem has no feasible solution.
9. Once a feasible solution has been found, continue to use the simplex method until the optimal solution is found.

In the next example, we use this method to solve a minimization problem with mixed constraints.

EXAMPLE 1 Minimization

Minimize $\quad w = 3y_1 + 2y_2$
subject to: $\quad y_1 + 3y_2 \leq 6$
$\qquad\qquad 2y_1 + \ y_2 \geq 3$
with $\qquad\quad y_1 \geq 0, \quad y_2 \geq 0.$

Solution Change this to a maximization problem by letting z equal the *negative* of the objective function: $z = -w$. Then find the *maximum* value of

$$z = -w = -3y_1 - 2y_2.$$

The problem can now be stated as follows.

$$\text{Maximize} \qquad z = -3y_1 - 2y_2$$
$$\text{subject to:} \qquad y_1 + 3y_2 \leq 6$$
$$2y_1 + y_2 \geq 3$$
$$\text{with} \qquad y_1 \geq 0, \quad y_2 \geq 0.$$

To begin, we add slack and surplus variables and set up the first tableau.

$$\begin{array}{ccccc} y_1 & y_2 & s_1 & s_2 & z \\ \end{array}$$
$$\left[\begin{array}{ccccc|c} 1 & 3 & 1 & 0 & 0 & 6 \\ 2 & 1 & 0 & -1 & 0 & 3 \\ 3 & 2 & 0 & 0 & 1 & 0 \end{array} \right]$$

The solution $y_1 = 0$, $y_2 = 0$, $s_1 = 6$, and $s_2 = -3$, is not feasible. Row operations must be used to get a feasible solution. We start with s_2 which has a -1 in row 2. The positive entry farthest to the left in row 2 is the 2 in column 1. The element in column 1 which gives the smallest quotient is 2, so it becomes the pivot. Pivoting produces the following matrix.

$$\begin{array}{ccccc} & y_1 & y_2 & s_1 & s_2 & z \\ \end{array}$$
$$\begin{array}{c} -R_2 + 2R_1 \rightarrow R_1 \\ \\ -3R_2 + 2R_3 \rightarrow R_3 \end{array} \left[\begin{array}{ccccc|c} 0 & 5 & 2 & 1 & 0 & 9 \\ 2 & 1 & 0 & -1 & 0 & 3 \\ 0 & 1 & 0 & 3 & 2 & -9 \end{array} \right]$$

Now $s_2 = 0$, so the solution is feasible. Furthermore, there are no negative numbers in the bottom row to the left of the vertical bar. Divide row 1 by 2, row 2 by 2, and row 3 by 2 to find the final solution: $y_1 = 3/2$ and $y_2 = 0$. Since $z = -w = -9/2$, the minimum value is $w = 9/2$. ∎

An important application of linear programming is the problem of minimizing the cost of transporting goods. This type of problem is often referred to as a *transportation problem* or *warehouse problem.* Some problems of this type were included in the exercise sets in previous chapters. The next example is based on an exercise from Section 2.2, in which the transportation costs were set equal to $10,640. We will now use the simplex method to minimize the transportation costs.

EXAMPLE 2 Transportation Problem

An auto manufacturer sends cars from two plants, I and II, to dealerships A and B located in a midwestern city. Plant I has a total of 28 cars to send, and plant II has 8. Dealer A needs 20 cars, and dealer B needs 16. Transportation costs per car based on the distance of each dealership from each plant are $220 from I to A, $300 from I to B, $400 from II to A, and $180 from II to B. How many cars should be sent from each plant to each of the two dealerships to minimize transportation costs? Use the simplex method to find the solution.

Solution To begin, let

$$y_1 = \text{the number of cars shipped from I to A;}$$
$$y_2 = \text{the number of cars shipped from I to B;}$$
$$y_3 = \text{the number of cars shipped from II to A;}$$
and
$$y_4 = \text{the number of cars shipped from II to B.}$$

Plant I has only 28 cars to ship, so

$$y_1 + y_2 \leq 28.$$

Similarly,

$$y_3 + y_4 \leq 8.$$

Since dealership A needs 20 cars and dealership B needs 16 cars,

$$y_1 + y_3 \geq 20 \qquad \text{and} \qquad y_2 + y_4 \geq 16.$$

The manufacturer wants to minimize transportation costs, so the objective function is

$$w = 220y_1 + 300y_2 + 400y_3 + 180y_4.$$

Now write the problem as a system of linear equations, adding slack or surplus variables as needed, and let $z = -w$.

$$
\begin{array}{rcrcrcrcrcrcrcrcr}
y_1 & + & y_2 & & & & & + & s_1 & & & & & & & = & 28 \\
& & & & y_3 & + & y_4 & & & + & s_2 & & & & & & 8 \\
y_1 & & & + & y_3 & & & & & & & - & s_3 & & & = & 20 \\
& & y_2 & & & + & y_4 & & & & & & & - & s_4 & = & 16 \\
220y_1 & + & 300y_2 & + & 400y_3 & + & 180y_4 & & & & & & & & + z & = & 0
\end{array}
$$

Set up the first simplex tableau.

$$
\begin{array}{ccccccccc|c}
y_1 & y_2 & y_3 & y_4 & s_1 & s_2 & s_3 & s_4 & z & \\
\hline
1 & 1 & 0 & 0 & 1 & 0 & 0 & 0 & 0 & 28 \\
0 & 0 & 1 & 1 & 0 & 1 & 0 & 0 & 0 & 8 \\
1 & 0 & 1 & 0 & 0 & 0 & -1 & 0 & 0 & 20 \\
0 & 1 & 0 & 1 & 0 & 0 & 0 & -1 & 0 & 16 \\
\hline
220 & 300 & 400 & 180 & 0 & 0 & 0 & 0 & 1 & 0
\end{array}
$$

Because $s_3 = -20$, we choose the positive entry farthest to the left in row 3, which is the 1 in column 1. After forming the necessary quotients, we find that the 1 is also the pivot, leading to the following tableau.

$$
\begin{array}{ccccccccc|c}
y_1 & y_2 & y_3 & y_4 & s_1 & s_2 & s_3 & s_4 & z & \\
\hline
0 & 1 & -1 & 0 & 1 & 0 & 1 & 0 & 0 & 8 \\
0 & 0 & 1 & 1 & 0 & 1 & 0 & 0 & 0 & 8 \\
1 & 0 & 1 & 0 & 0 & 0 & -1 & 0 & 0 & 20 \\
0 & 1 & 0 & 1 & 0 & 0 & 0 & -1 & 0 & 16 \\
\hline
0 & 300 & 180 & 180 & 0 & 0 & 220 & 0 & 1 & -4400
\end{array}
$$

$-R_3 + R_1 \rightarrow R_1$ (row 1)
$-220R_3 + R_5 \rightarrow R_5$ (row 5)

We still have $s_4 = -16$. Verify that the 1 in row 1, column 2, is the next pivot, leading to the following tableau.

$$
\begin{array}{ccccccccc|c}
y_1 & y_2 & y_3 & y_4 & s_1 & s_2 & s_3 & s_4 & z & \\
\hline
0 & 1 & -1 & 0 & 1 & 0 & 1 & 0 & 0 & 8 \\
0 & 0 & 1 & 1 & 0 & 1 & 0 & 0 & 0 & 8 \\
1 & 0 & 1 & 0 & 0 & 0 & -1 & 0 & 0 & 20 \\
0 & 0 & 1 & 1 & -1 & 0 & -1 & -1 & 0 & 8 \\
\hline
0 & 0 & 480 & 180 & -300 & 0 & -80 & 0 & 1 & -6800
\end{array}
$$

$-R_1 + R_4 \rightarrow R_4$ (row 4)
$-300R_1 + R_5 \rightarrow R_5$ (row 5)

We still have $s_4 = -8$. Choosing column 3 to pivot, there is a tie between rows 2 and 4. Observe that if we choose row 4, we will remove s_4 from the set of basic variables. This would be a smart next move. But our algorithm says to choose the row nearest the top, so we will do this and see where it leads.

$$
\begin{array}{c}
\\
R_2 + R_1 \rightarrow R_1 \\
\\
-R_2 + R_3 \rightarrow R_3 \\
-R_2 + R_4 \rightarrow R_4 \\
-480R_2 + R_5 \rightarrow R_5
\end{array}
\begin{array}{c}
y_1 \;\; y_2 \;\; y_3 \qquad y_4 \quad\;\; s_1 \quad\;\; s_2 \quad\;\; s_3 \quad\; s_4 \;\; z \\
\left[
\begin{array}{ccccccccc|c}
0 & 1 & 0 & 1 & 1 & 1 & 1 & 0 & 0 & 16 \\
0 & 0 & 1 & 1 & 0 & 1 & 0 & 0 & 0 & 8 \\
1 & 0 & 0 & -1 & 0 & -1 & -1 & 0 & 0 & 12 \\
0 & 0 & 0 & 0 & -1 & -1 & -1 & -1 & 0 & 0 \\
0 & 0 & 0 & -300 & -300 & -480 & -80 & 0 & 1 & -10,640
\end{array}
\right]
\end{array}
$$

Now $s_4 = 0$. There is still a -1 in column 8, but this can be removed by multiplying row 4 by -1. We then have the feasible solution

$$y_1 = 12, \quad y_2 = 16, \quad y_3 = 8, \quad y_4 = 0, \quad s_1 = 0, \quad s_2 = 0, \quad s_3 = 0, \quad s_4 = 0,$$

with $w = 10,640$. But there are still negatives in the bottom row, so we can keep going. After three more tableaus, we find that

$$y_1 = 20, \quad y_2 = 8, \quad y_3 = 0, \quad y_4 = 8,$$

with $w = 8240$. Therefore, the manufacturer should send 20 cars from plant I to dealership A and 8 cars to dealership B. From plant II, 8 cars should be sent to dealership B and none to dealership A. The transportation cost will then be $8240, a savings of $2400 over the original stated cost of $10,640. ∎

When one or more of the constraints in a linear programming problem is an equation, rather than an inequality, there is no need for a slack or surplus variable. The simplex method requires an additional variable, however, for *each* constraint. To meet this condition, an **artificial variable** is added to each equation. These variables are called artificial variables because they have no meaning in the context of the original problem. The first goal of the simplex method is to eliminate any artificial variables as basic variables, since they must have a value of 0 in the solution.

EXAMPLE 3 Artificial Variables

In the transportation problem discussed in Example 2, it would be more realistic for the dealerships to order exactly 20 and 16 cars, respectively. Solve the problem with these two equality constraints.

Solution Using the same variables, we can state the problem as follows.

$$
\begin{array}{ll}
\text{Minimize} & w = 220y_1 + 300y_2 + 400y_3 + 180y_4 \\
\text{subject to:} & y_1 + y_2 \leq 28 \\
& y_3 + y_4 \leq 8 \\
& y_1 + y_3 = 20 \\
& y_2 + y_4 = 16
\end{array}
$$

with all variables nonnegative.

The corresponding system of equations requires slack variables s_1 and s_2 and two artificial variables that we shall call a_1 and a_2, to remind us that they require special handling. The system

$$
\begin{array}{rcl}
y_1 + y_2 & + s_1 & = 28 \\
y_3 + y_4 & + s_2 & = 8 \\
y_1 + y_3 & + a_1 & = 20 \\
y_2 + y_4 & + a_2 & = 16 \\
220y_1 + 300y_2 + 400y_3 + 180y_4 & + z & = 0
\end{array}
$$

produces a tableau exactly the same as in Example 2, except that the columns labeled s_3 and s_4 in that example are now labeled a_1 and a_2. We proceed as we did in Example 2, except for one difference: as soon as we get to the point where $a_1 = 0$, we drop the a_1 column. Similarly, as soon as $a_2 = 0$, we drop the a_2 column. The solution will proceed as in Example 2 and will give the same result. In other problems, however, equality constraints can result in a higher cost. ■

CAUTION If the artificial variables cannot be made equal to zero, the problem has no feasible solution.

NOTE Another way to handle this situation is by solving for y_3 and y_4 in terms of y_1 and y_2. Then proceed with the usual method for standard problems.

Several linear programming models in actual use are presented on the web site for this textbook. These models illustrate the usefulness of linear programming. In most real applications, the number of variables is so large that these problems could not be solved without using methods (like the simplex method) that can be adapted to computers.

4.4 EXERCISES

Rewrite each system of inequalities, adding slack variables or subtracting surplus variables as necessary.

1. $2x_1 + 3x_2 \le 8$
$x_1 + 4x_2 \ge 7$

2. $5x_1 + 8x_2 \le 10$
$6x_1 + 2x_2 \ge 7$

3. $x_1 + x_2 + x_3 \le 100$
$x_1 + x_2 + x_3 \ge 75$
$x_1 + x_2 \ge 27$

4. $2x_1 + x_3 \le 40$
$x_1 + x_2 \ge 18$
$x_1 + x_3 \ge 20$

Convert the following problems into maximization problems.

5. Minimize $w = 4y_1 + 3y_2 + 2y_3$
subject to: $y_1 + y_2 + y_3 \ge 5$
$y_1 + y_2 \ge 4$
$2y_1 + y_2 + 3y_3 \ge 15$
with $y_1 \ge 0, \quad y_2 \ge 0, \quad y_3 \ge 0.$

6. Minimize $w = 8y_1 + 3y_2 + y_3$
subject to: $7y_1 + 6y_2 + 8y_3 \ge 18$
$4y_1 + 5y_2 + 10y_3 \ge 20$
with $y_1 \ge 0, \quad y_2 \ge 0, \quad y_3 \ge 0.$

7. Minimize $w = y_1 + 2y_2 + y_3 + 5y_4$
subject to: $y_1 + y_2 + y_3 + y_4 \ge 50$
$3y_1 + y_2 + 2y_3 + y_4 \ge 100$
with $y_1 \ge 0, \quad y_2 \ge 0, \quad y_3 \ge 0, \quad y_4 \ge 0.$

8. Minimize $w = y_1 + y_2 + 4y_3$
subject to: $y_1 + 2y_2 + 3y_3 \ge 115$
$2y_1 + y_2 + y_3 \le 200$
$y_1 + y_3 \ge 50$
with $y_1 \ge 0, \quad y_2 \ge 0, \quad y_3 \ge 0.$

Use the simplex method to solve each of the following.

9. Find $x_1 \geq 0$ and $x_2 \geq 0$ such that

$$x_1 + 2x_2 \geq 24$$
$$x_1 + x_2 \leq 40$$

and $z = 12x_1 + 10x_2$ is maximized.

10. Find $x_1 \geq 0$ and $x_2 \geq 0$ such that

$$3x_1 + 4x_2 \geq 48$$
$$2x_1 + 4x_2 \leq 60$$

and $z = 6x_1 + 8x_2$ is maximized.

11. Find $x_1 \geq 0$, $x_2 \geq 0$, and $x_3 \geq 0$ such that

$$x_1 + x_2 + x_3 \leq 150$$
$$x_1 + x_2 + x_3 \geq 100$$

and $z = 2x_1 + 5x_2 + 3x_3$ is maximized.

12. Find $x_1 \geq 0$, $x_2 \geq 0$, and $x_3 \geq 0$ such that

$$x_1 + x_2 + 2x_3 \leq 38$$
$$2x_1 + x_2 + x_3 \geq 24$$

and $z = 3x_1 + 2x_2 + 2x_3$ is maximized.

13. Find $x_1 \geq 0$ and $x_2 \geq 0$ such that

$$x_1 + x_2 \leq 100$$
$$x_1 + x_2 \geq 50$$
$$2x_1 + x_2 \leq 110$$

and $z = -2x_1 + 3x_2$ is maximized.

14. Find $x_1 \geq 0$ and $x_2 \geq 0$ such that

$$x_1 + 2x_2 \leq 18$$
$$x_1 + 3x_2 \geq 12$$
$$2x_1 + 2x_2 \leq 24$$

and $z = 5x_1 - 10x_2$ is maximized.

15. Find $y_1 \geq 0$ and $y_2 \geq 0$ such that

$$10y_1 + 15y_2 \leq 150$$
$$20y_1 + 5y_2 \geq 100$$

and $w = 4y_1 + 5y_2$ is minimized.

16. Minimize $w = 3y_1 + 2y_2$

subject to: $2y_1 + 3y_2 \leq 60$

$$y_1 + 4y_2 \geq 40$$

with $y_1 \geq 0, \quad y_2 \geq 0.$

Solve each of the following using artificial variables.

17. Maximize $z = 3x_1 + 2x_2$

subject to: $x_1 + x_2 = 50$

$$4x_1 + 2x_2 \geq 120$$
$$5x_1 + 2x_2 \leq 200$$

with $x_1 \geq 0, \quad x_2 \geq 0.$

18. Maximize $z = 10x_1 + 9x_2$

subject to: $x_1 + x_2 = 30$

$$x_1 + x_2 \geq 25$$
$$2x_1 + x_2 \leq 40$$

with $x_1 \geq 0, \quad x_2 \geq 0.$

19. Minimize $w = 32y_1 + 40y_2$

subject to: $20y_1 + 10y_2 = 200$

$$25y_1 + 40y_2 \leq 500$$
$$18y_1 + 24y_2 \geq 300$$

with $y_1 \geq 0, \quad y_2 \geq 0.$

20. Minimize $w = 15y_1 + 12y_2$

subject to: $y_1 + 2y_2 \leq 12$

$$3y_1 + y_2 \geq 18$$
$$y_1 + y_2 = 10$$

with $y_1 \geq 0, \quad y_2 \geq 0.$

21. Explain how, in any linear programming problem, the value of the objective function can be found without using the number in the lower right-hand corner of the final tableau.

22. Explain why, for a maximization problem, you write the negative of the coefficients of the objective function on the bottom row, while, for a minimization problem, you write the coefficients themselves.

Applications

BUSINESS AND ECONOMICS

23. *Transportation* Southwestern Oil supplies two distributors in the Northwest from two outlets, S_1 and S_2. Distributor D_1 needs at least 3000 barrels of oil, and distributor D_2 needs at least 5000 barrels. The two outlets can each furnish up to 5000 barrels of oil. The costs per barrel to ship the oil are given in the table.

		Distributors	
		D_1	D_2
Outlets	S_1	$30	$20
	S_2	$25	$22

There is also a shipping tax per barrel as given in the table below. Southwestern Oil is determined to spend no more than $40,000 on shipping tax.

	D_1	D_2
S_1	$2	$6
S_2	$5	$4

How should the oil be supplied to minimize shipping costs?

24. *Transportation* Change Exercise 23 so that the two outlets each furnish exactly 5000 barrels of oil, with everything else the same. Use artificial variables to solve the problem, following the steps outlined in Example 3.

25. *Finance* A bank has set aside a maximum of $25 million for commercial and home loans. Every million dollars in commercial loans requires 2 lengthy application forms, while every million dollars in home loans requires 3 lengthy application forms. The bank cannot process more than 72 application forms at this time. The bank's policy is to loan at least four times as much for home loans as for commercial loans. Because of prior commitments, at least $10 million will be used for these two types of loans. The bank earns 12% on home loans and 10% on commercial loans. What amount of money should be allotted for each type of loan to maximize the interest income?

26. *Blending Seed* Topgrade Turf lawn seed mixture contains three types of seeds: bluegrass, rye, and Bermuda. The costs per pound of the three types of seed are 12¢, 15¢, and 5¢, respectively. In each batch there must be at least 20% bluegrass seed, and the amount of Bermuda must be no more than 2/3 the amount of rye. To fill current orders, the company must make at least 5000 pounds of the mixture. How much of each kind of seed should be used to minimize cost?

27. *Blending Seed* Change Exercise 26 so that the company must make exactly 5000 pounds of the mixture. Use artificial variables to solve the problem, following the steps outlined in Example 3.

28. *Investments* Karen Guardino has decided to invest a $100,000 inheritance in government securities that earn 7% per year, municipal bonds that earn 6% per year, and mutual funds that earn an average of 10% per year. She will spend at least $40,000 on government securities, and she wants at least half the inheritance to go to bonds and mutual funds. Government securities have an initial fee of 2%, municipal bonds have an initial fee of 1%, and mutual funds have an initial fee of 3%. Karen has $2400 available to pay initial fees. How much should be invested in each way to maximize the interest yet meet the constraints? What is the maximum interest she can earn?

29. *Transportation* The manufacturer of a popular personal computer has orders from two dealers. Dealer D_1 wants at least 32 computers, and dealer D_2 wants at least 20 computers. The manufacturer can fill the orders from either of two warehouses, W_1 or W_2. W_1 has 25 of the computers on hand, and W_2 has 30. The costs (in dollars) to ship one computer to each dealer from each warehouse are given below.

		Dealer	
		D_1	D_2
Warehouse	W_1	$14	$12
	W_2	$12	$10

How should the orders be filled to minimize shipping costs?

30. *Blending Chemicals* Natural Brand plant food is made from three chemicals, labeled I, II, and III. In each batch of the plant food, the amounts of chemicals II and III must be in the ratio of 4 to 3. The amount of nitrogen must be at least 30 kg. The percent of nitrogen in the three chemicals is 9%, 4%, and 3%, respectively. If the three chemicals cost $1.09, $.87, and $.65 per kilogram, respectively, how much of each should be used to minimize the cost of producing at least 750 kg of the plant food?

31. *Blending Gasoline* A company is developing a new additive for gasoline. The additive is a mixture of three liquid ingredients, I, II, and III. For proper performance, the total amount of additive must be at least 10 oz per gallon of gasoline. However, for safety reasons, the amount of additive should not exceed 15 oz per gallon of gasoline. At least 1/4 oz of ingredient I must be used for every ounce of ingredient II, and at least 1 oz of ingredient III must be used for every ounce of ingredient I. If the costs of I, II, and III are $.30, $.09, and $.27 per ounce, respectively, find the mixture of the three ingredients that produces the minimum cost of the additive. How much of the additive should be used per gallon of gasoline?

32. *Blending a Soft Drink* A popular soft drink called Sugarlo, which is advertised as having a sugar content of no more than 10%, is blended from five ingredients, each of which has some sugar content. Water may also be added to dilute the mixture. The sugar content of the ingredients and their costs per gallon are given below.

			Ingredient			
	1	2	3	4	5	Water
Sugar Content (%)	.28	.19	.43	.57	.22	0
Cost ($/gal)	.48	.32	.53	.28	.43	.04

At least .01 of the content of Sugarlo must come from ingredients 3 or 4, .01 must come from ingredients 2 or 5, and .01 from ingredients 1 or 4. How much of each ingredient should be used in preparing 15,000 gal of Sugarlo to minimize the cost?

CHAPTER SUMMARY

The simplex method is a procedure for solving any linear programming problem. In the first two sections of this chapter, we saw how to write a standard maximization problem in matrix form, how to choose a pivot, and how to perform the steps leading to the optimal solution. Standard minimization problems may be solved by duality, which links each standard minimization problem with a corresponding standard maximization problem, known as its *dual*. Finally, problems that are not standard because they have inequalities going in both directions (and perhaps equalities as well) may be solved using a variation on the basic simplex method described in the last section of this chapter.

KEY TERMS

simplex method	indicators	**4.3** standard minimum	**4.4** surplus variable
4.1 standard maximum	basic variable	form	two-phase method
form	basic feasible solution	dual	artificial variable
slack variable	pivot	transpose	
simplex tableau	**4.2** sensitivity analysis	shadow costs	

CHAPTER 4 REVIEW EXERCISES

1. When is it necessary to use the simplex method rather than the graphical method?

2. What can you conclude if a surplus variable cannot be made nonnegative?

For Exercises 3–6, (a) add slack variables or subtract surplus variables, and (b) set up the initial simplex tableau.

3. Maximize $z = 5x_1 + 3x_2$
   subject to: $2x_1 + 5x_2 \le 50$
   $x_1 + 3x_2 \le 25$
   $4x_1 + x_2 \le 18$
   $x_1 + x_2 \le 12$
   with $x_1 \ge 0, \quad x_2 \ge 0.$

4. Maximize $z = 25x_1 + 30x_2$
   subject to: $3x_1 + 5x_2 \le 47$
   $x_1 + x_2 \le 25$
   $5x_1 + 2x_2 \le 35$
   $2x_1 + x_2 \le 30$
   with $x_1 \ge 0, \quad x_2 \ge 0.$

5. Maximize $z = 5x_1 + 8x_2 + 6x_3$
   subject to: $x_1 + x_2 + x_3 \le 90$
   $2x_1 + 5x_2 + x_3 \le 120$
   $x_1 + 3x_2 \ge 80$
   with $x_1 \ge 0, \quad x_2 \ge 0, \quad x_3 \ge 0.$

6. Maximize $z = 2x_1 + 3x_2 + 4x_3$
   subject to: $x_1 + x_2 + x_3 \ge 100$
   $2x_1 + 3x_2 \le 500$
   $x_1 + 2x_3 \le 350$
   with $x_1 \ge 0, \quad x_2 \ge 0, \quad x_3 \ge 0.$

Use the simplex method to solve the maximization linear programming problems with initial tableaus as given in Exercises 7–10.

7.

$$\begin{array}{cccccc} x_1 & x_2 & x_3 & s_1 & s_2 & z \end{array}$$
$$\begin{bmatrix} 1 & 2 & 3 & 1 & 0 & 0 & | & 28 \\ 2 & 4 & 1 & 0 & 1 & 0 & | & 32 \\ \hline -5 & -2 & -3 & 0 & 0 & 1 & | & 0 \end{bmatrix}$$

8.

$$\begin{array}{ccccc} x_1 & x_2 & s_1 & s_2 & z \end{array}$$
$$\begin{bmatrix} 2 & 1 & 1 & 0 & 0 & | & 10 \\ 1 & 3 & 0 & 1 & 0 & | & 16 \\ \hline -2 & -3 & 0 & 0 & 1 & | & 0 \end{bmatrix}$$

9.

$$\begin{array}{ccccccc} x_1 & x_2 & x_3 & s_1 & s_2 & s_3 & z \end{array}$$
$$\begin{bmatrix} 1 & 2 & 2 & 1 & 0 & 0 & 0 & | & 50 \\ 3 & 1 & 0 & 0 & 1 & 0 & 0 & | & 20 \\ 1 & 0 & 2 & 0 & 0 & -1 & 0 & | & 15 \\ \hline -5 & -3 & -2 & 0 & 0 & 0 & 1 & | & 0 \end{bmatrix}$$

10.

$$\begin{array}{cccccc} x_1 & x_2 & s_1 & s_2 & s_3 & z \end{array}$$
$$\begin{bmatrix} 3 & 6 & -1 & 0 & 0 & 0 & | & 28 \\ 1 & 1 & 0 & 1 & 0 & 0 & | & 12 \\ 2 & 1 & 0 & 0 & 1 & 0 & | & 16 \\ \hline -1 & -2 & 0 & 0 & 0 & 1 & | & 0 \end{bmatrix}$$

Convert the problems of Exercises 11–13 into maximization problems, using both the dual method and the method of Section 4.4.

11. Minimize $w = 10y_1 + 15y_2$
   subject to: $y_1 + y_2 \geq 17$
                   $5y_1 + 8y_2 \geq 42$
   with $y_1 \geq 0, \quad y_2 \geq 0.$

12. Minimize $w = 20y_1 + 15y_2 + 18y_3$
   subject to: $2y_1 + y_2 + y_3 \geq 112$
                   $y_1 + y_2 + y_3 \geq 80$
                   $y_1 + y_2 \quad\ \geq 45$
   with $y_1 \geq 0, \quad y_2 \geq 0, \quad y_3 \geq 0.$

13. Minimize $w = 7y_1 + 2y_2 + 3y_3$
   subject to: $y_1 + y_2 + 2y_3 \geq 48$
                   $y_1 + y_2 \quad\ \geq 12$
                   $y_3 \geq 10$
                   $3y_1 \quad + y_3 \geq 30$
   with $y_1 \geq 0, \quad y_2 \geq 0, \quad y_3 \geq 0.$

The tableau in Exercises 14–16 are the final tableaus of minimization problems. State the solution and the minimum value of the objective function for each problem.

14.

$$\begin{array}{ccccccc} y_1 & y_2 & y_3 & s_1 & s_2 & s_3 & z \end{array}$$
$$\begin{bmatrix} 1 & 0 & 0 & 3 & 1 & 2 & 0 & | & 12 \\ 0 & 0 & 1 & 4 & 5 & 3 & 0 & | & 5 \\ 0 & 1 & 0 & -2 & 7 & -6 & 0 & | & 8 \\ \hline 0 & 0 & 0 & 5 & 7 & 3 & 1 & | & -172 \end{bmatrix}$$

15.

$$\begin{array}{cccccccc} y_1 & y_2 & s_1 & s_2 & s_3 & s_4 & z \end{array}$$
$$\begin{bmatrix} 0 & 0 & 3 & 0 & 1 & 1 & 0 & | & 2 \\ 1 & 0 & -2 & 0 & 2 & 0 & 0 & | & 8 \\ 0 & 1 & 7 & 0 & 0 & 0 & 0 & | & 12 \\ 0 & 0 & 1 & 1 & -4 & 0 & 0 & | & 1 \\ \hline 0 & 0 & 5 & 0 & 8 & 0 & 1 & | & -62 \end{bmatrix}$$

16.

$$\begin{array}{cccccc} y_1 & y_2 & y_3 & s_1 & s_2 & z \end{array}$$
$$\begin{bmatrix} 5 & 1 & 0 & 7 & -1 & 0 & | & 100 \\ -2 & 0 & 1 & 1 & 3 & 0 & | & 27 \\ \hline 12 & 0 & 0 & 7 & 2 & 1 & | & -640 \end{bmatrix}$$

17. What types of problems can be solved using slack, surplus, and artificial variables?

18. What kind of problems can be solved using the method of duals?

19. In solving a linear programming problem, you are given the following initial tableau.

$$\begin{bmatrix} 4 & 2 & 3 & 1 & 0 & 0 & | & 9 \\ 5 & 4 & 1 & 0 & 1 & 0 & | & 10 \\ \hline -6 & -7 & -5 & 0 & 0 & 1 & | & 0 \end{bmatrix}$$

a. What is the problem being solved?

b. If the 1 in row 1, column 4 were a -1 rather than a 1, how would it change your answer to part a?

c. After several steps of the simplex algorithm, the following tableau results.

$$\begin{bmatrix} 3 & 0 & 5 & 2 & -1 & 0 & | & 8 \\ 11 & 10 & 0 & -1 & 3 & 0 & | & 21 \\ \hline 47 & 0 & 0 & 13 & 11 & 10 & | & 227 \end{bmatrix}$$

What is the solution? (List only the values of the original variables and the objective function. Do not include slack or surplus variables.)

d. What is the dual of the problem you found in part a?

e. What is the solution of the dual you found in part d? (Do not perform any steps of the simplex algorithm; just examine the tableau given in part c.)

20. In Chapter 2, we wrote a system of linear equations using matrix notation. We can do the same thing for the system of linear inequalities in this chapter.

a. Find matrices A, B, C, and X such that the maximization problem in Example 1 of Section 4.1 can be written as

Maximize CX
subject to: $AX \le B$
with $X \ge 0$.

(*Hint:* Let B and X be column matrices, and C a row matrix.)

b. Show that the dual of the problem in part a can be written as

Minimize YB
subject to: $YA \ge C$
with $Y \ge 0$,

where Y is a row matrix.

c. Show that for any feasible solutions X and Y to the original and dual problems, respectively, $CX \le YB$. (*Hint:* Multiply both sides of $AX \le B$ by Y on the left. Then substitute for YA.)

d. For the solution X to the maximization problem and Y to the dual, it can be shown that

$$CX = YB$$

is always true. Verify this for Example 1 of Section 4.1. What is the significance of the value in CX (or YB)?

Applications

For Exercises 21–24, **(a)** *select appropriate variables;* **(b)** *write the objective functions;* **(c)** *write the constraints as inequalities.*

BUSINESS AND ECONOMICS

21. *Production* Roberta Hernandez sells three items, A, B, and C, in her gift shop. Each unit of A costs her $5 to buy, $1 to sell, and $2 to deliver. For each unit of B, the costs are $3, $2, and $1, respectively, and for each unit of C, the costs are $6, $2, and $5, respectively. The profit on A is $4; on B, $3; and on C, $3. How many of each should she order to maximize her profit if she can spend $1200 on buying costs, $800 on selling costs, and $500 on delivery costs?

22. *Investments* An investor is considering three types of investments: a high-risk venture into oil leases with a potential return of 15%, a medium-risk investment in stocks with a 9% return, and a relatively safe bond investment with a 5% return. He has $50,000 to invest. Because of the risk, he will limit his investment in oil leases and stocks to 30% and his investment in oil leases and bonds to 50%. How much should he invest in each to maximize his return, assuming investment returns are as expected?

23. *Profit* The Aged Wood Winery makes two white wines, Fruity and Crystal, from two kinds of grapes and sugar. A gallon of Fruity wine requires 2 bushels of Grape A, 2 bushels of Grape B, 2 pounds of sugar, and produces a profit of $12. A gallon of Crystal wine requires 1 bushel of Grape A, 3 bushels of Grape B, 1 pound of sugar, and produces a profit of $15. The winery has available 110 bushels of grape A, 125 bushels of grape B, and 90 lb of sugar. How much of each wine should be made to maximize profit?

24. *Production Costs* Cauchy Canners produces canned whole tomatoes and tomato sauce. This season, the company has available 3,000,000 kg of tomatoes for these two products. To meet the demands of regular customers, it must produce at least 80,000 kg of sauce and 800,000 kg of whole tomatoes. The cost per kilogram is $4 to produce canned whole tomatoes and $3.25 to produce tomato sauce. Labor agreements require that at least 110,000 person-hours be used. Each can of sauce requires 3 minutes for one worker, and each can of whole tomatoes requires 6 minutes for one worker. How many kilograms of tomatoes should Cauchy use for each product to minimize cost? (For simplicity, assume production of y_1 kg of canned whole tomatoes and y_2 kilograms of tomato sauce requires $y_1 + y_2$ kg of tomatoes.)

25. Solve Exercise 21.   **26.** Solve Exercise 22.

27. Solve Exercise 23.   **28.** Solve Exercise 24.

29. *Canning* Cauchy Canners produces canned corn, beans, and carrots. Demand for vegetables requires it to produce at least 1000 cases per month. Based on past sales, it should produce at least twice as many cases of corn as of beans, and at least 340 cases of carrots. It costs $10 to produce a case of corn, $15 to produce a case of beans, and $25 to produce a case of carrots.

a. Using the method of surplus variables, find how many cases of each vegetable should be produced to minimize costs. What is the minimum cost?

b. Using the method of duals, find how many cases of each vegetable should be produced to minimize costs. What is the minimum cost?

30. *Food Cost* A store sells two brands of snacks. A package of Sun Hill costs $3 and contains 10 oz of peanuts, 4 oz of raisins, and 2 oz of rolled oats. A package of Bear Valley costs $2 and contains 2 oz of peanuts, 4 oz of raisins, and 8 oz of rolled oats. Suppose you wish to make a mixture that contains at least 20 oz of peanuts, 24 oz of raisins, and 24 oz of rolled oats.

a. Using the method of surplus variables, find how many packages of each you should buy to minimize the cost. What is the minimum cost?

b. Using the method of duals, find how many packages of each you should buy to minimize the cost. What is the minimum cost?

c. Suppose the minimum amount of peanuts is increased to 28. Use shadow costs to calculate the total cost in this case.

d. Explain why it makes sense that the shadow cost for rolled oats is 0.

EXTENDED APPLICATION: Using Integer Programming in the Stock-Cutting Problem*

In Chapter 3 Section 3 we noted that some problems require solutions in integers because the resources to be allocated are items that can't be split into pieces, like cargo containers or airplanes. These *integer programming* problems are generally harder than the linear programming problems we have been solving by the simplex method, but often linear programming can be combined with other techniques to solve integer problems. Even if the number of variables and constraints is small, some help from software is usually required. We will introduce integer programming with the basic but important *stock-cutting problem*. (To get a feeling for the issues involved, you may want to try the simple stock-cutting problem given in Exercise 1.)

A paper mill produces rolls of paper that are much wider than most customers require, often as wide as 200 inches. The mill then cuts these wide rolls into smaller widths to fill orders for paper rolls to be used in printing and packaging and other applications. The stock-cutting problem is the following:

> *Given a list of roll widths and the number of rolls ordered for each width, decide how to cut the raw rolls that come from the paper-making machine into smaller rolls so as to fill all the orders with a minimum amount of waste.*

Another way to state the problem is: What is the minimum number of raw rolls required to fill the orders? This is an integer problem because the customers have ordered whole numbers of rolls, and each roll is cut in a single piece from one of the raw rolls.

As an example, suppose the paper machine produces rolls 100 inches wide. The manufacturer offers rolls in the following six widths: 14 inches, 17 inches, 31 inches, 33 inches, 36 inches, and 45 inches. (We'll call these the standard widths.) The current orders to be filled are as follows:

Width in inches	14	17	31	33	36	45
Number ordered	100	123	239	121	444	87

The cutting machine can make four simultaneous cuts, so a raw roll can be cut into as many as five pieces. With luck, all five pieces might be usable for filling orders, but there will usually be unusable waste on the end, and we also might end up with more rolls of some standard width than we need. We'll consider both the end pieces that are too narrow and any unused standard-width rolls as waste, and this is the waste we want to minimize.

The first question is, what are the possible cutting patterns? We're restricted to at most five standard rolls from any given raw roll, and we'll elect to use as much as possible in each raw roll so that the waste remaining at the end will always be less than 14 inches. So, for example, $14 \,|\, 36 \,|\, 45 \,|$ is a possible pattern, but $14 \,|\, 14 \,|\, 14 \,|\, 14 \,|\, 14 \,|$ is not, because it has too many cuts, and $45 \,|\, 36 \,|$ is not, because more than 14 inches is left at the end. (Each vertical bar represents a cut; if the piece on the end happens to be a standard width, then we don't need a cut after it, since we've reached the end of the roll.) This is already a tricky problem, and variations of it appear in many industrial applications involving packing objects of different sizes into a fixed space (for example, packing crates into a container for shipment overseas). In the Exercises we'll ask you to write down some more possible patterns, but finding all of them is a job for a computer, and it turns out that there are exactly 33 possible cutting patterns. In Chapter 8 you'll learn some counting techniques that might help you write the program to find all possible patterns.

The next question is, what's the best we can do? We have to use an integral number of 100-inch raw rolls, and we can find the total "roll-inches" ordered by multiplying the width of each standard roll by the number ordered for this width. This computation is a natural one for the matrix notation that you have learned. If W and O are 6×1 column matrices, then the total roll inches used is $W^T O$:

$$W = \begin{bmatrix} 14 \\ 17 \\ 31 \\ 33 \\ 36 \\ 45 \end{bmatrix} \quad O = \begin{bmatrix} 100 \\ 123 \\ 239 \\ 121 \\ 444 \\ 87 \end{bmatrix} \quad W^T O = 34{,}792$$

Since each raw roll is 100 inches, the best we can do is to use 348 rolls with a total width of 34,800. As a percentage of the raw material, the corresponding waste is

$$\frac{8}{34{,}800} \approx .02\%,$$

which represents very low waste. Of course, we'll only reach this target if we can lay out the cutting with perfect efficiency.

As we noted above, these integer programming problems are difficult, but many mathematical analysis and spreadsheet

*This application is based on material from the following online sources:

The Web site of the Optimization Technology Center at Northwestern University at http://www.ece.nwu.edu/OTC/. There is a link to a thorough explanation of the stock-cutting problem.

Home page of the Special Interest Group on Cutting and Packing at http://prodlog.wiwi.uni-halle.de/sicup/index.html.

The linear programming FAQ at http://www.faqs.org/faqs/linear-programming-faq/.

programs have built-in optimization routines that can handle problems of modest size. We submitted this problem to one such program, giving it the lists of orders and widths and a list of the 33 allowable cutting patterns. Figure 10 shows the seven cutting patterns chosen by the minimizer software, with a graphical representation, and the total number of times each pattern was used.

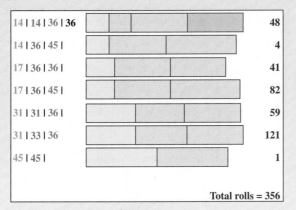

FIGURE 10

With these cutting choices we generate the following numbers of each standard width:

$$\begin{pmatrix} \text{Width:} & 14 & 17 & 31 & 33 & 36 & 45 \\ \text{Quantity produced:} & 100 & 123 & 239 & 121 & 444 & 88 \\ \text{Quantity ordered:} & 100 & 123 & 239 & 121 & 444 & 87 \end{pmatrix}$$

We figured that the minimum possible number of raw rolls was 348, so we have used only 8 more than the minimum. In the Exercises you'll figure the percentage of waste with this cutting plan.

Manufacturers of glass and sheet metal encounter a two-dimensional version of this problem: They need to cut the rectangular pieces that have been ordered from a larger rectangular piece of glass or metal, laying out the ordered sizes so as to minimize waste. Besides the extra dimension, this problem is complicated by another constraint: The typical cutting machine can make only "guillotine cuts" that go completely across the sheet being cut, so a cutting algorithm must usually begin with a few long cuts that cut the original rectangle into strips, followed by crossways cuts that begin to create the order sizes. A typical finished cutting layout might look like Figure 11.

The first cuts would be the three vertical cuts labeled 1, 2, and 3, followed by horizontal cuts in each of the four resulting strips, then vertical cuts in these new rectangles, and so on. Areas of waste are marked with X. An additional complication in designing the layout is that any given stock rectangle can be oriented in two different directions (unless it's square), so the packing problem has many alternative solutions.

In three dimensions, a comparable problem is to fill a shipping container with smaller boxes (rectangular prisms) with the minimum wasted space. These packing problems are complicated geometric versions of a basic problem called the *knapsack problem:*

Given n objects with weights $w_1, w_2, \ldots, w_n$ and cash values $v_1, v_2, \ldots, v_n$, and a knapsack that can hold a weight of at most W, choose the objects that will pack in the greatest value. In the Exercises you can try a small example, but as soon as n gets large, this problem "explodes," that is, the number of possibilities becomes too large for a trial and error solution, even with a computer to do the bookkeeping. The development of good algorithms for cutting and packing problems is an active research specialty in the field of optimization.

Exercises

1. Suppose you plan to build a raised flower bed using landscape timbers, which come in 8-ft lengths. You want the bed's outer dimensions to be 6 ft by 4 ft, and you will use three layers of timbers. The timbers are 6 in. by 6 in. in cross

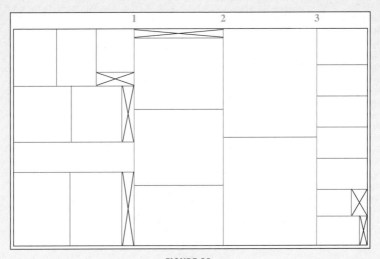

FIGURE 11

section, so if you make the bottom and top layers with 6 ft lengths on the sides and 3 ft lengths on the ends, and the middle layer with 5 ft lengths on the sides and 4 ft lengths on the ends, you could build the bed out of the following lengths:

Plan A

Length	Number needed
3 ft	4
4 ft	2
5 ft	2
6 ft	4

a. What is the smallest number of timbers you can buy to build your bed? How will you lay out the cuts? How much wood will you waste?

b. If you overlap the corners in a different way, you can build the bed with this plan:

Plan B

Length	Number needed
3 ft	2
4 ft	4
5 ft	4
6 ft	2

Does plan B allow you to build the bed with fewer 8-ft timbers?

c. What is the smallest length for the uncut timbers that would allow you to build the bed with no waste?

2. For the list of standard paper roll widths given earlier, write down four more possible cutting patterns that use at most four cuts and leave less than 14 inches of waste on the end. See if you can find ones that aren't in the list of patterns returned by the optimizer.

3. Four of the 33 possible patterns use up the raw roll with no waste, that is, the widths add up to exactly 100 inches. Find these four patterns.

4. For the computer solution of the cutting problem, figure out the percent of the 356 rolls used that is wasted.

5. In our cutting plan, we elected to use up as much as possible of each 100-inch roll with standard widths. Why might it be a better idea to allow leftover rolls that are *wider* than 14 inches?

6. The following table shows the weights of six objects and their values.

Weight	2	2.5	3	3.5	4	4.5
Value	12	11	7	13	10	11

If your knapsack holds a maximum weight of 9, what is the highest value you can pack in?

Mathematics of Finance

Buying a car usually requires both some savings for a down payment and a loan for the balance. An exercise in Section 2 calculates the regular deposits that would be needed to save up the full purchase price, and other exercises and examples in this chapter compute the payments required to amortize a loan.

- **5.1** Simple and Compound Interest

- **5.2** Future Value of an Annuity

- **5.3** Present Value of an Annuity; Amortization

Review Exercises

Extended Application: Time, Money, and Polynomials

Whether you are in a position to invest money or to borrow money, it is important for both consumers and business managers to understand *interest.* The formulas for interest are developed in this chapter.

■ 5.1 SIMPLE AND COMPOUND INTEREST

 THINK ABOUT IT
If you can borrow money at 11% interest compounded annually or at 10.8% compounded monthly, which loan would cost less?

We shall see how to make such comparisons in this section.

Simple Interest Interest on loans of a year or less is frequently calculated as **simple interest,** a type of interest that is charged (or paid) only on the amount borrowed (or invested), and not on past interest. The amount borrowed is called the **principal.** The **rate** of interest is given as a percent per year, expressed as a decimal. For example, $6\% = .06$ and $11\frac{1}{2}\% = .115$. The **time** the money is earning interest is calculated in years. Simple interest is the product of the principal, rate, and time.

SIMPLE INTEREST

$$I = Prt,$$

where p is the principal;
r is the annual interest rate;
t is the time in years.

EXAMPLE 1   Simple Interest
To buy furniture for a new apartment, Jennifer Wall borrowed $5000 at 11% simple interest for 11 months. How much interest will she pay?

Solution From the formula, $I = Prt$, with $P = 5000$, $r = .11$, and $t = 11/12$ (in years). The total interest she will pay is

$$I = 5000(.11)(11/12) = 504.17,$$

or $504.17.

A deposit of P dollars today at a rate of interest r for t years produces interest of $I = Prt$. The interest, added to the original principal P, gives

$$P + Prt = P(1 + rt).$$

This amount is called the **future value** of P dollars at an interest rate r for time t in years. When loans are involved, the future value is often called the **maturity value** of the loan. This idea is summarized as follows.

FUTURE OR MATURITY VALUE FOR SIMPLE INTEREST
The **future** or **maturity value** A of P dollars at a simple interest rate r for t years is

$$A = P(1 + rt).$$

EXAMPLE 2 Maturity Value

Find the maturity value for each of the following loans at simple interest.

(a) A loan of $2500 to be repaid in 8 months with interest of 12.1%

Solution The loan is for 8 months, or $8/12 = 2/3$ of a year. The maturity value is

$$A = P(1 + rt)$$

$$A = 2500\left[1 + .121\left(\frac{2}{3}\right)\right]$$

$$A \approx 2500(1 + .08067) \approx 2701.67,$$

or $2701.67. (The answer is rounded to the nearest cent, as is customary in financial problems.) Of this maturity value,

$$\$2701.67 - \$2500 = \$201.67$$

represents interest.

(b) A loan of $11,280 for 85 days at 11% interest

Solution It is common to assume 360 days in a year when working with simple interest. We shall usually make such an assumption in this book. The maturity value in this example is

$$A = 11,280\left[1 + .11\left(\frac{85}{360}\right)\right] \approx 11,572.97,$$

or $11,572.97.

> **CAUTION** When using the formula for future value, as well as all other formulas in this chapter, we neglect the fact that in real life, money amounts are rounded to the nearest penny. As a consequence, when the amounts are rounded, their values may differ by a few cents from the amounts given by these formulas. For instance, in Example 2(a), each monthly payment would be $2500(.121/12) \approx \$25.21$, rounded to the nearest penny. After 8 months, the total is $8(\$25.21) = \$201.68$, which is 1¢ more than we computed in the example.

In part (b) of Example 2 we assumed 360 days in a year. Interest found using a 360-day year is called *ordinary interest*, and interest found using a 365-day year is called *exact interest*.

The formula for future value has four variables, $P, r, t,$ and A. We can use the formula to find any of the quantities that these variables represent, as illustrated in the next example.

EXAMPLE 3 Simple Interest

Laurie Rosatone wants to borrow $8000 from Christine O'Brien. She is willing to pay back $8380 in 6 months. What interest rate will she pay?

Solution Use the formula for future value, with $A = 8380$, $P = 8000$, $t = 6/12 = .5$, and solve for r.

$$A = P(1 + rt)$$
$$8380 = 8000(1 + .5r)$$
$$8380 = 8000 + 4000r \quad \text{Distributive property}$$
$$380 = 4000r \quad \text{Subtract 8000.}$$
$$r = .095 \quad \text{Divide by 4000.}$$

Thus, the interest rate is 9.5%.

Compound Interest As mentioned earlier, simple interest is normally used for loans or investments of a year or less. For longer periods *compound interest* is used. With **compound interest,** interest is charged (or paid) on interest as well as on principal. For example, if $1000 is deposited at 5% interest for 1 year, at the end of the year the interest is $1000(.05)(1) = \$50$. The balance in the account is $1000 + \$50 = \$1050$. If this amount is left at 5% interest for another year, the interest is calculated on $1050 instead of the original $1000, so the amount in the account at the end of the second year is $1050 + \$1050(.05)(1) = \$1102.50$. Note that simple interest would produce a total amount of only

$$\$1000[1 + (.05)(2)] = \$1100.$$

To find a formula for compound interest, first suppose that P dollars is deposited at a rate of interest r per year. The amount on deposit at the end of the first year is found by the simple interest formula, with $t = 1$.

$$A = P(1 + r \cdot 1) = P(1 + r)$$

If the deposit earns compound interest, the interest earned during the second year is paid on the total amount on deposit at the end of the first year. Using the formula $A = P(1 + rt)$ again, with P replaced by $P(1 + r)$ and $t = 1$, gives the total amount on deposit at the end of the second year.

$$A = [P(1 + r)](1 + r \cdot 1) = P(1 + r)^2$$

In the same way, the total amount on deposit at the end of the third year is

$$P(1 + r)^3.$$

Generalizing, in t years the total amount on deposit is

$$A = P(1 + r)^t,$$

called the **compound amount.**

NOTE Compare this formula for compound interest with the formula for simple interest.

Compound interest	$A = P(1 + r)^t$
Simple interest	$A = P(1 + rt)$

The important distinction between the two formulas is that in the compound interest formula, the number of years, t, is an *exponent,* so that money grows much more rapidly when interest is compounded.

Interest can be compounded more than once per year. Common compounding periods include *semiannually* (two periods per year), *quarterly* (four periods per year), *monthly* (twelve periods per year), or *daily* (usually 365 periods per year). The *interest rate per period, i,* is found by dividing the annual interest rate, *r*, by the number of compounding periods, *m*, per year. To find the total number of compounding periods, *n*, we multiply the number of years, *t*, by the number of compounding periods per year, *m*. The following formula can be derived in the same way as the previous formula.

COMPOUND AMOUNT

$$A = P(1 + i)^n,$$

where $i = \dfrac{r}{m}$ and $n = mt$,

A is the future (maturity) value;

P is the principal;

r is the annual interest rate;

m is the number of compounding periods per year;

t is the number of years;

n is the number of compounding periods;

i is the interest rate per period.

EXAMPLE 4 Compound Interest

Suppose $1000 is deposited for 6 years in an account paying 8.31% per year compounded annually.

(a) Find the compound amount.

Solution In the formula above, $P = 1000$, $i = .0831/1$, and $n = 6(1) = 6$. The compound amount is

$$A = P(1 + i)^n$$
$$A = 1000(1.0831)^6.$$

Using a calculator, we get

$$A \approx \$1614.40,$$

the compound amount.

(b) Find the amount of interest earned.

Solution Subtract the initial deposit from the compound amount.

$$\text{Amount of interest} = \$1614.40 - \$1000 = \$614.40$$

EXAMPLE 5 Compound Interest

Find the amount of interest earned by a deposit of $2450 for 6.5 years at 5.25% compounded quarterly.

Solution Interest compounded quarterly is compounded 4 times a year. In 6.5 years, there are $6.5(4) = 26$ periods. Thus, $n = 26$. Interest of 5.25% per year is 5.25%/4 per quarter, so $i = .0525/4$. Now use the formula for compound amount.

$$A = P(1 + i)^n$$

$$A = 2450(1 + .0525/4)^{26} \approx 3438.78$$

Rounded to the nearest cent, the compound amount is $3438.78, so the interest is $3438.78 − $2450 = $988.78.

> **CAUTION** As shown in Example 5, compound interest problems involve two rates—the annual rate r and the rate per compounding period i. Be sure you understand the distinction between them. When interest is compounded annually, these rates are the same. In all other cases, $i \neq r$.

It is interesting to compare loans at the same rate when simple or compound interest is used. Figure 1 shows the graphs of the simple interest and compound interest formulas with $P = 1000$ at an annual rate of 10% from 0 to 20 years. The future value after 15 years is shown for each graph. After 15 years at compound interest, $1000 grows to $4177.25, whereas with simple interest, it amounts to $2500.00, a difference of $1677.25.

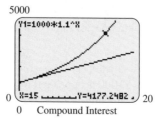
Compound Interest
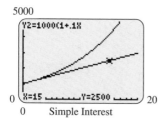
Simple Interest

FIGURE 1

Spreadsheets are ideal for performing financial calculations. Figure 2 shows a Microsoft Excel spreadsheet with the formulas for compound and simple interest used to create columns B and C, respectively, when $1000 is invested at an annual rate of 10%. Compare row 16 with the calculator results in Figure 1. For more details on the use of spreadsheets in the mathematics of finance, see *The Spreadsheet Manual* that is available with this book.

Effective Rate If $1 is deposited at 4% compounded quarterly, a calculator can be used to find that at the end of one year, the compound amount is $1.0406, an increase of 4.06% over the original $1. The actual increase of 4.06% in the money is somewhat higher than the stated increase of 4%. To differentiate between these two numbers, 4% is called the **nominal** or **stated rate** of interest, while 4.06% is called the **effective rate.*** To avoid confusion between stated rates and effective rates, we shall continue to use r for the stated rate and we will use r_e for the effective rate.

---

*When applied to consumer finance, the effective rate is called the annual percentage rate, APR, or annual percentage yield, APY.

	A	B	C
1	period	compound	simple
2	1	1100	1100
3	2	1210	1200
4	3	1331	1300
5	4	1464.1	1400
6	5	1610.51	1500
7	6	1771.561	1600
8	7	1948.7171	1700
9	8	2143.58881	1800
10	9	2357.947691	1900
11	10	2593.74246	2000
12	11	2853.116706	2100
13	12	3138.428377	2200
14	13	3452.271214	2300
15	14	3797.498336	2400
16	15	4177.248169	2500
17	16	4594.972986	2600
18	17	5054.470285	2700
19	18	5559.917313	2800
20	19	6115.909045	2900
21	20	6727.499949	3000

FIGURE 2

EXAMPLE 6 Effective Rate

Find the effective rate corresponding to a stated rate of 6% compounded semiannually.

Solution Here, $r/m = 6\%/2 = 3\%$ for $m = 2$ periods. Use a calculator to find that $(1.03)^2 \approx 1.06090$, which shows that \$1 will increase to \$1.06090, an actual increase of 6.09%. The effective rate is $r_e = 6.09\%$.

Generalizing from this example, the effective rate of interest is given by the following formula.

EFFECTIVE RATE

The effective rate corresponding to a stated rate of interest r compounded m times per year is

$$r_e = \left(1 + \frac{r}{m}\right)^m - 1.$$

EXAMPLE 7 Effective Rate

A bank pays interest of 4.9% compounded monthly. Find the effective rate.

Solution Use the formula given above with $r = .049$ and $m = 12$. The effective rate is

$$r_e = \left(1 + \frac{.049}{12}\right)^{12} - 1 = .050115575,$$

or 5.01%.

EXAMPLE 8 Effective Rate

Joe Vetere needs to borrow money. His neighborhood bank charges 11% interest compounded semiannually. A downtown bank charges 10.8% interest compounded monthly. At which bank will Joe pay the lesser amount of interest?

Solution Compare the effective rates.

$$\text{Neighborhood bank: } r_e = \left(1 + \frac{.11}{2}\right)^2 - 1 = .113025 \approx 11.3\%$$

$$\text{Downtown bank: } r_e = \left(1 + \frac{.108}{12}\right)^{12} - 1 \approx .11351 \approx 11.4\%$$

The neighborhood bank has the lower effective rate, although it has a higher stated rate.

Present Value The formula for compound interest, $A = P(1 + i)^n$, has four variables: A, P, i, and n. Given the values of any three of these variables, the value of the fourth can be found. In particular, if A (the future amount), i, and n are known, then P can be found. Here P is the amount that should be deposited today to produce A dollars in n periods.

EXAMPLE 9 Present Value

Rachel Reeve must pay a lump sum of $6000 in 5 years. What amount deposited today at 6.2% compounded annually will amount to $6000 in 5 years?

Solution Here $A = 6000$, $i = .062$, $n = 5$, and P is unknown. Substituting these values into the formula for the compound amount gives

$$6000 = P(1.062)^5$$

$$P = \frac{6000}{(1.062)^5} \approx 4441.49,$$

or $4441.49. If Rachel leaves $4441.49 for 5 years in an account paying 6.2% compounded annually, she will have $6000 when she needs it. To check your work, use the compound interest formula with $P = \$4441.49$, $i = .062$, and $n = 5$. You should get $A = \$6000.00$.

As Example 9 shows, $6000 in 5 years is approximately the same as $4441.49 today (if money can be deposited at 6.2% compounded annually). An amount that can be deposited today to yield a given sum in the future is called the *present value* of the future sum. Generalizing from Example 9, by solving $A = P(1 + i)^n$ for P, we get the following formula for present value.

PRESENT VALUE FOR COMPOUND INTEREST

The **present value** of A dollars compounded at an interest rate i per period for n periods is

$$P = \frac{A}{(1 + i)^n} \quad \text{or} \quad P = A(1 + i)^{-n}.$$

EXAMPLE 10   Present Value

Find the present value of $16,000 in 9 years if money can be deposited at 6% compounded semiannually.

Solution In 9 years there are $2 \cdot 9 = 18$ semiannual periods. A rate of 6% per year is 3% in each semiannual period. Apply the formula with $A = 16{,}000$, $i = .03$, and $n = 18$.

$$P = \frac{A}{(1 + i)^n} = \frac{16{,}000}{(1.03)^{18}} \approx 9398.31$$

A deposit of $9398.31 today, at 6% compounded semiannually, will produce a total of $16,000 in 9 years.

We can solve the compound amount formula for n also, as the following example shows.

EXAMPLE 11   Price Doubling

Suppose the general level of inflation in the economy averages 8% per year. Find the number of years it would take for the overall level of prices to double.

Solution To find the number of years it will take for $1 worth of goods or services to cost $2, find n in the equation

$$2 = 1(1 + .08)^n,$$

where $A = 2$, $P = 1$, and $i = .08$. This equation simplifies to

$$2 = (1.08)^n.$$

By trying various values of n, we find that $n = 9$ is approximately correct, because $1.08^9 = 1.99900 \approx 2$. The exact value of n can be found quickly by using logarithms, but that is beyond the scope of this chapter. Thus, the overall level of prices will double in about 9 years.

At this point, it seems helpful to summarize the notation and the most important formulas for simple and compound interest. We use the following variables.

P = principal or present value

A = future or maturity value

r = annual (stated or nominal) interest rate

t = number of years

m = number of compounding periods per year

i = interest rate per period $i = r/m$

n = total number of compounding periods $n = tm$

r_e = effective rate

Simple Interest   **Compound Interest**

$A = P(1 + rt)$ $A = P(1 + i)^n$

$$P = \frac{A}{1 + rt} \qquad P = \frac{A}{(1 + i)^n} = A(1 + i)^{-n}$$

$$r_e = \left(1 + \frac{r}{m}\right)^m - 1$$

5.1 EXERCISES

1. What is the difference between r and i? between t and n?

2. We calculated the loan in Example 2(b) assuming 360 days in a year. Find the maturity value using 365 days in a year. Which is more advantageous to the borrower?

3. What factors determine the amount of interest earned on a fixed principal?

4. In your own words, describe the *maturity value* of a loan.

5. What is meant by the *present value* of money?

Find the simple interest in Exercises 6–9.

6. $25,000 at 7% for 9 mo

7. $3850 at 9% for 8 mo

8. $1974 at 6.3% for 7 mo

9. $3724 at 8.4% for 11 mo

Find the simple interest. Assume a 360-day year.

10. $5147.18 at 10.1% for 58 days

11. $2930.42 at 11.9% for 123 days

12. Explain the difference between simple interest and compound interest.

13. In Figure 1, one graph is a straight line and the other is curved. Explain why this is, and which represents each type of interest.

Find the compound amount for each of the following deposits.

14. $1000 at 6% compounded annually for 8 yr

15. $1000 at 7% compounded annually for 10 yr

16. $470 at 10% compounded semiannually for 12 yr

17. $15,000 at 6% compounded semiannually for 11 yr

18. $6500 at 12% compounded quarterly for 6 yr

19. $9100 at 8% compounded quarterly for 4 yr

Find the amount that should be invested now to accumulate the following amounts, if the money is compounded as indicated.

20. $15,902.74 at 9.8% compounded annually for 7 yr

21. $27,159.68 at 12.3% compounded annually for 11 yr

22. $2000 at 9% compounded semiannually for 8 yr

23. $2000 at 11% compounded semiannually for 8 yr

24. $8800 at 10% compounded quarterly for 5 yr

25. $7500 at 12% compounded quarterly for 9 yr

26. How do the nominal or stated interest rate and the effective interest rate differ?

27. If interest is compounded more than once per year, which rate is higher, the stated rate or the effective rate?

Find the effective rate corresponding to each of the following nominal rates.

28. 3% compounded quarterly

29. 8% compounded quarterly

30. 8.25% compounded semiannually

31. 10.08% compounded semiannually

Applications

BUSINESS AND ECONOMICS

32. *Loan Repayment* Susan Carsten borrowed $25,900 from her father to start a flower shop. She repaid him after 11 mo, with interest of 8.4%. Find the total amount she repaid.

33. *Delinquent Taxes* An accountant for a corporation forgot to pay the firm's income tax of $725,896.15 on time. The government charged a penalty of 12.7% interest for the 34 days the money was late. Find the total amount (tax and penalty) that was paid. (Use a 365-day year.)

34. *Savings* A $100,000 certificate of deposit held for 60 days was worth $101,133.33. To the nearest tenth of a percent, what interest rate was earned?

35. *Savings* A firm of accountants has ordered 7 new IBM computers at a cost of $5104 each. The machines will not be delivered for 7 mo. What amount could the firm deposit in an account paying 6.42% to have enough to pay for the machines?

36. *Stock Growth* A stock that sold for $22 at the beginning of the year was selling for $24 at the end of the year. If the stock paid a dividend of $.50 per share, what is the simple interest rate on an investment in this stock? (*Hint:* Consider the interest to be the increase in value plus the dividend.)

37. *Bond Interest* A bond with a face value of $10,000 in 10 yr can be purchased now for $5988.02. What is the simple interest rate?

38. *Loan Interest* A small business borrows $50,000 for expansion at 12% compounded monthly. The loan is due in 4 yr. How much interest will the business pay?

39. *Wealth* An article in *The New York Times* discussed how long it will take for Bill Gates, the world's second richest person (behind the Sultan of Brunei), to become the world's first trillionaire.* His birthday is October 28, 1955, and on July 16, 1997, he was worth $42 billion. (*Note:* A trillion dollars is 1000 billion dollars.)

a. Assume that Bill Gates's fortune grows at an annual rate of 58%, the historical growth rate through 1997 of Microsoft stock, which made up most of his wealth in 1997. Find the age at which he becomes a trillionaire. (*Hint:* Use the formula for interest compounded annually, $A = P(1 + i)^n$, with $P = 42$. Graph the future value as a function of n on a graphing calculator, and find where the graph crosses the line $y = 1000$.)

b. Repeat part a using 10.9% growth, the average return on all stocks since 1926.

c. What rate of growth would be necessary for Bill Gates to become a trillionaire by the time he is eligible for Social Security on January 1, 2022, after he has turned 66?

d. An article on September 19, 1999, gave Bill Gates's wealth as roughly $90 billion.† What was the rate of growth of his wealth between the 1997 and 1999 articles?

40. *Loan Interest* A developer needs $80,000 to buy land. He is able to borrow the money at 10% per year compounded quarterly. How much will the interest amount to if he pays off the loan in 5 yr?

41. *Paying Off a Lawsuit* A company has agreed to pay $2.9 million in 5 yr to settle a lawsuit. How much must they invest now in an account paying 8% compounded monthly to have that amount when it is due?

42. *Buying a House* George Duda wants to have $20,000 available in 5 yr for a down payment on a house. He has inherited $15,000. How much of the inheritance should he invest now to accumulate $20,000, if he can get an interest rate of 8% compounded quarterly?

43. *Comparing Investments* Two partners agree to invest equal amounts in their business. One will contribute $10,000 immediately. The other plans to contribute an equivalent amount in 3 yr, when she expects to acquire a large sum of money. How much should she contribute at that time to match her partner's investment now, assuming an interest rate of 6% compounded semiannually?

44. *Comparing Investments* As the prize in a contest, you are offered $1000 now or $1210 in 5 yr. If money can be invested at 6% compounded annually, which is larger?

45. *Comparing CD Rates* A New York bank offered the following special on CD (certificate of deposit) rates. The rates are annual percentage yields, or effective rates, which are higher than the corresponding nominal rates. Assume quarterly compounding. Solve for r to approximate the corresponding nominal rates to the nearest hundredth.

Term	6 mo	1 yr	18 mo	2 yr	3 yr
APY(%)	5.00	5.30	5.45	5.68	5.75

46. *Effective Rate* An advertisement for WingspanBank.com boasted the "highest CD yield in the universe," with an APY (or effective yield) of 7.35%.‡ The actual rate was not stated. Given that interest was compounded monthly, find the actual rate.

47. *Effective Rate* According to a financial Web site, on October 16, 2000, DeepGreen Bank of Cleveland, Ohio, paid 6.96% interest, compounded daily, on a one-year CD, while Flagstar Bank of Bloomfield Hills, Michigan, paid 7.01% compounded quarterly.§ What are the effective rates for the two CDs, and which bank pays a higher effective rate?

48. *Retirement Savings* The pie graph on the next page shows the percent of baby boomers aged 46–49 who said they had investments with a total value as shown in each category.‖ Note that 30% have saved less than $10,000. Assume the money is invested at an average rate of 8% compounded quarterly. What will the top numbers in each category

*The New York Times, July 20, 1997, Sec. 4, p. 2.
†The New York Times, Sept. 19, 1999, WK Rev., p. 2.
‡The New York Times, Oct. 15, 2000, p. 39.
§www.bankrate.com
‖The New York Times, Dec. 31, 1995, Sec. 3. p. 5.

amount to in 20 yr, when this age group will be ready for retirement?

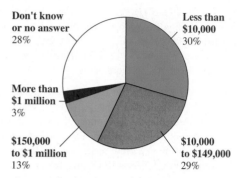

Don't know or no answer 28%

Less than $10,000 30%

More than $1 million 3%

$150,000 to $1 million 13%

$10,000 to $149,000 29%

Figures add to more than 100% because of rounding.

Doubling Time Use the ideas from Example 11 to find the time it would take for the general level of prices in the economy to double at each of the following average annual inflation rates.

49. 4%   **50.** 5%

51. *Doubling Time* The consumption of electricity has increased historically at 6% per year. If it continues to increase at this rate indefinitely, find the number of years before the electric utilities will need to double their generating capacity.

52. *Doubling Time* Suppose a conservation campaign coupled with higher rates causes the demand for electricity to increase at only 2% per year, as it has recently. Find the number of years before the utilities will need to double generating capacity.

Negative Interest Under certain conditions, Swiss banks pay negative interest: they charge you. (You didn't think all that secrecy was free?) Suppose a bank "pays" −2.4% interest compounded annually. Find the compound amount for a deposit of $150,000 after each of the following periods.

53. 4 yr   **54.** 8 yr

55. *Interest Rate* In 1995, O. G. McClain of Houston, Texas, mailed a $100 check to a descendant of Texas independence hero Sam Houston to repay a $100 debt of McClain's great-great-grandfather, who died in 1835, to Sam Houston.* A bank estimated the interest on the loan to be $420 million for the 160 yr it was due. Find the interest rate the bank was using, assuming interest is compounded annually.

56. *Investment* In the New Testament, Jesus commends a widow who contributed 2 mites to the temple treasury (Mark 12:42–44). A mite was worth roughly 1/8 of a cent. Suppose the temple invested those 2 mites at 4% interest compounded quarterly. How much would the money be worth 2000 yr later?

57. *Investments* Sun Kang borrowed $5200 from his friend Hop Fong Yee to pay for remodeling work on his house. He repaid the loan 10 mo later with simple interest at 7%. Yee then invested the proceeds in a 5-yr certificate of deposit paying 6.3% compounded quarterly. How much will he have at the end of 5 yr? (*Hint:* You need to use both simple and compound interest.)

58. *Investments* Suppose $10,000 is invested at an annual rate of 5% for 10 yr. Find the future value if interest is compounded as follows.

a. annually   **b.** quarterly

c. monthly   **d.** daily (365 days)

59. *Investments* In Exercise 58, notice that as the money is compounded more often, the compound amount becomes larger and larger. Is it possible to compound often enough so that the compound amount is $17,000 after 10 yr? Explain.

The following exercise is from an actuarial examination.[†]

60. *Savings* On January 1, 1980, Jack deposited $1000 into bank X to earn interest at a rate of j per annum compounded semiannually. On January 1, 1985, he transferred his account to bank Y to earn interest at the rate of k per annum compounded quarterly. On January 1, 1988, the balance of bank Y is $1990.76. If Jack could have earned interest at the rate of k per annum compounded quarterly from January 1, 1980, through January 1, 1988, his balance would have been $2203.76. Calculate the ratio k/j.

*The New York Times, March 30, 1995.
[†]Problem 5 from "Course 140 Examination, Mathematics of Compound Interest" of the Education and Examination Committee of The Society of Actuaries. Reprinted by permission of The Society of Actuaries.

▪ 5.2 FUTURE VALUE OF AN ANNUITY

? THINK ABOUT IT

If you deposit $1500 each year for 6 years in an account paying 8% interest compounded annually, how much will be in your account at the end of this period?

In this section and the next, we develop future value and present value formulas for such periodic payments. To develop these formulas, we must first discuss *sequences.*

Geometric Sequences   If a and r are nonzero real numbers, the infinite list of numbers $a, ar, ar^2, ar^3, ar^4, \ldots, ar^n, \ldots$ is called a **geometric sequence.** For example, if $a = 3$ and $r = -2$, we have the sequence

$$3, 3(-2), 3(-2)^2, 3(-2)^3, 3(-2)^4, \ldots,$$

or
$$3, -6, 12, -24, 48, \ldots.$$

In the sequence $a, ar, ar^2, ar^3, ar^4, \ldots,$ the number a is called the **first term** of the sequence, ar is the **second term,** ar^2 is the **third term,** and so on. Thus, for any $n \geq 1$,

ar^{n-1} **is the nth term of the sequence.**

Each term in the sequence is r times the preceding term. The number r is called the **common ratio** of the sequence.

EXAMPLE 1   Geometric Sequence
Find the seventh term of the geometric sequence $5, 20, 80, 320, \ldots.$

Solution Here, $a = 5$ and $r = 20/5 = 4$. We want the seventh term, so $n = 7$. Use ar^{n-1}, with $a = 5$, $r = 4$, and $n = 7$.

$$ar^{n-1} = (5)(4)^{7-1} = 5(4)^6 = 20{,}480$$

EXAMPLE 2   Geometric Sequence
Find the first five terms of the geometric sequence with $a = 10$ and $r = 2$.

Solution The first five terms are

$$10, 10(2), 10(2)^2, 10(2)^3, 10(2)^4,$$

or
$$10, 20, 40, 80, 160.$$

Next, we need to find the sum S_n of the first n terms of a geometric sequence, where

$$S_n = a + ar + ar^2 + ar^3 + ar^4 + \cdots + ar^{n-1}. \tag{1}$$

If $r = 1$, then

$$S_n = \underbrace{a + a + a + a + \cdots + a}_{n \text{ terms}} = na.$$

If $r \neq 1$, multiply both sides of equation (1) by r to get

$$rS_n = ar + ar^2 + ar^3 + ar^4 + \cdots + ar^n. \qquad (2)$$

Now subtract corresponding sides of equation (1) from equation (2).

$$
\begin{aligned}
rS_n &= \quad\;\; ar + ar^2 + ar^3 + ar^4 + \cdots + ar^{n-1} + ar^n \\
-S_n &= -(a + ar + ar^2 + ar^3 + ar^4 + \cdots + ar^{n-1}) \\
\hline
rS_n - S_n &= -a + ar^n \\
S_n(r - 1) &= a(r^n - 1) \\
S_n &= \frac{a(r^n - 1)}{r - 1}
\end{aligned}
$$

This result is summarized below.

SUM OF TERMS

If a geometric sequence has first term a and common ratio r, then the sum S_n of the first n terms is given by

$$S_n = \frac{a(r^n - 1)}{r - 1}, \quad r \neq 1.$$

EXAMPLE 3 Sum of a Geometric Sequence

Find the sum of the first six terms of the geometric sequence $3, 12, 48, \ldots$.

Solution Here $a = 3$ and $r = 4$. Find S_6 by the formula above.

$$
\begin{aligned}
S_6 &= \frac{3(4^6 - 1)}{4 - 1} \qquad \text{Let } n = 6, a = 3, r = 4. \\
&= \frac{3(4096 - 1)}{3} \\
&= 4095
\end{aligned}
$$

Ordinary Annuities A sequence of equal payments made at equal periods of time is called an **annuity.** If the payments are made at the end of the time period, and if the frequency of payments is the same as the frequency of compounding, the annuity is called an **ordinary annuity.** The time between payments is the **payment period,** and the time from the beginning of the first payment period to the end of the last period is called the **term** of the annuity. The **future value of the annuity,** the final sum on deposit, is defined as the sum of the compound amounts of all the payments, compounded to the end of the term.

Two common uses of annuities are to accumulate funds for some goal or to withdraw funds from an account. For example, an annuity may be used to save money for a large purchase, such as an automobile, an expensive trip, or a down payment on a home. An annuity also may be used to provide monthly payments for retirement. We explore these options in this and the next section.

For example, suppose $1500 is deposited at the end of each year for the next six years in an account paying 8% per year compounded annually. Figure 3 shows this annuity. To find the future value of the annuity, look separately at each of the $1500 payments. The first of these payments will produce a compound amount of

$$1500(1 + .08)^5 = 1500(1.08)^5.$$

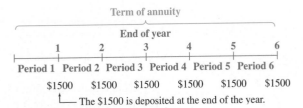

Term of annuity

FIGURE 3

Use 5 as the exponent instead of 6 since the money is deposited at the *end* of the first year and earns interest for only five years. The second payment of $1500 will produce a compound amount of $1500(1.08)^4$. As shown in Figure 4, the future value of the annuity is

$$1500(1.08)^5 + 1500(1.08)^4 + 1500(1.08)^3 + 1500(1.08)^2$$
$$+ 1500(1.08)^1 + 1500.$$

(The last payment earns no interest at all.)

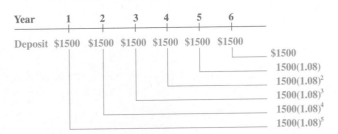

FIGURE 4

Reading this sum in reverse order, we see that it is the sum of the first six terms of a geometric sequence, with $a = 1500$, $r = 1.08$, and $n = 6$. Thus, the sum equals

$$\frac{a(r^n - 1)}{r - 1} = \frac{1500[(1.08)^6 - 1]}{1.08 - 1} \approx \$11{,}003.89.$$

To generalize this result, suppose that payments of R dollars each are deposited into an account at the end of each period for n periods, at a rate of interest i per period. The first payment of R dollars will produce a compound amount of $R(1 + i)^{n-1}$ dollars, the second payment will produce $R(1 + i)^{n-2}$ dollars, and so on; the final payment earns no interest and contributes just R dollars to the total. If S represents the future value (or sum) of the annuity, then (as shown in Figure 5 on the next page),

$$S = R(1 + i)^{n-1} + R(1 + i)^{n-2} + R(1 + i)^{n-3} + \cdots + R(1 + i) + R,$$

or, written in reverse order,

$$S = R + R(1 + i)^1 + R(1 + i)^2 + \cdots + R(1 + i)^{n-1}.$$

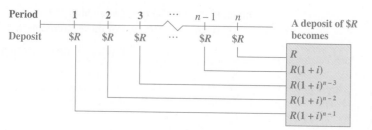

FIGURE 5

This result is the sum of the first n terms of the geometric sequence having first term R and common ratio $1 + i$. Using the formula for the sum of the first n terms of a geometric sequence,

$$S = \frac{R[(1 + i)^n - 1]}{(1 + i) - 1} = \frac{R[(1 + i)^n - 1]}{i} = R\left[\frac{(1 + i)^n - 1}{i}\right].$$

The quantity in brackets is commonly written $s_{\overline{n}|i}$ (read "*s-angle-n* at *i*"), so that

$$S = R \cdot s_{\overline{n}|i}.$$

Values of $s_{\overline{n}|i}$ can be found with a calculator.

A formula for the future value of an annuity S of n payments of R dollars each at the end of each consecutive interest period, with interest compounded at a rate i per period, follows.* Recall that this type of annuity, with payments at the *end* of each time period, is called an ordinary annuity.

FUTURE VALUE OF AN ORDINARY ANNUITY

$$S = R\left[\frac{(1 + i)^n - 1}{i}\right] \qquad \text{or} \qquad S = Rs_{\overline{n}|i}$$

where

S is the future value;

R is the payment;

i is the interest rate per period;

n is the number of periods.

A calculator will be very helpful in computations with annuities. The TI-83 graphing calculator has a special FINANCE menu that is designed to give any desired result after entering the basic information. If your calculator does not have this feature, many calculators can easily be programmed to evaluate the formulas introduced in this section and the next. We include these programs in *The Graphing Calculator Manual* available for this text.

EXAMPLE 4   Ordinary Annuity

Karen Scott is an athlete who believes that her playing career will last 7 years. To prepare for her future, she deposits $22,000 at the end of each year for 7 years in an account paying 6% compounded annually. How much will she have on deposit after 7 years?

---

*We use S for the future value here, instead of A as in the compound interest formula, to help avoid confusing the two formulas.

Solution Her payments form an ordinary annuity, with $r = 22{,}000$, $n = 7$, and $i = .06$. The future value of this annuity (by the formula on the previous page) is

$$S = 22{,}000\left[\frac{(1.06)^7 - 1}{.06}\right] \approx 184{,}664.43,$$

or $184,664.43.

Sinking Funds   A fund set up to receive periodic payments as in Example 4 is called a **sinking fund.** The periodic payments, together with the interest earned by the payments, are designed to produce a given sum at some time in the future. For example, a sinking fund might be set up to receive money that will be needed to pay off the principal on a loan at some future time. If the payments are all the same amount and are made at the end of a regular time period, they form an ordinary annuity.

EXAMPLE 5   Sinking Fund

Experts say that the baby boom generation (Americans born between 1946 and 1960) cannot count on a company pension or Social Security to provide a comfortable retirement, as their parents did. It is recommended that they start to save early and regularly. Michael Boezi, a baby boomer, has decided to deposit $200 each month for 20 years in an account that pays interest of 7.2% compounded monthly.

(a) How much will be in the account at the end of 20 years?

Solution This savings plan is an annuity with $R = 200$, $i = .072/12$, and $n = 12(20)$. The future value is

$$S = 200\left[\frac{(1 + (.072/12))^{12(20)} - 1}{.072/12}\right] \approx 106{,}752.47,$$

or $106,752.47. Figure 6 shows a calculator-generated graph of the function

$$S = 200\left[\frac{(1 + (x/12))^{12(20)} - 1}{x/12}\right]$$

where r, the annual interest rate, is designated x. The value of the function at $x = .072$, shown at the bottom of the window, agrees with our result above.

(b) Michael believes he needs to accumulate $130,000 in the 20-year period to have enough for retirement. What interest rate would provide that amount?

Solution

Method 1: Graphing Calculator

One way to answer this question is to solve the equation for S in terms of x with $S = 130{,}000$. This is a difficult equation to solve. Although trial and error could be used, it would be easier to use the graphing calculator graph in Figure 6. Adding the line $y = 130{,}000$ to the graph and then using the capability of the calculator to find the intersection point with the curve shows the annual interest rate must be at least 8.79% to the nearest hundredth. See Figure 7 on the next page.

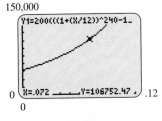

FIGURE 6

Method 2: TVM Solver

Using the TVM Solver under the FINANCE menu on the TI-83 calculator, enter 240 for N (the number of periods), 0 for PV (present value), -200 for PMT (negative because the money is being paid out), 130000 for FV (future value), and 12 for P/Y (payments per year). Put the cursor next to I% (payment) and press SOLVE. The result, shown in Figure 8, indicates that an interest rate of 8.79% is needed.

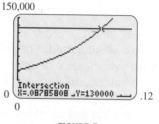

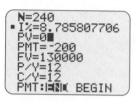

FIGURE 7   **FIGURE 8**

EXAMPLE 6 Sinking Fund

Suppose Michael, in Example 5, cannot get the higher interest rate to produce $130,000 in 20 years. To meet that goal, he must increase his monthly payment. What payment should he make each month?

Solution We start with the annuity formula

$$S = R\left[\frac{(1 + i)^n - 1}{i}\right].$$

Solve for R by multiplying both sides by $i/[(1 + i)^n - 1]$.

$$R = \frac{Si}{(1 + i)^n - 1}$$

Now substitute $S = 130,000$, $i = .072/12$, and $n = 12(20)$ to find R.

$$R = \frac{(130,000)(.072/12)}{(1 + (.072/12))^{12(20)} - 1} = 243.5540887$$

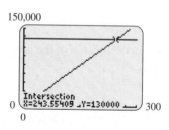

FIGURE 9

Michael will need payments of $243.56 each month for 20 years to accumulate at least $130,000. Notice that $243.55 is not quite enough, so we must round up here. Figure 9 shows the point of intersection of the graphs of

$$Y_1 = X\left[\frac{(1 + .072/12)^{12(20)} - 1}{.072/12}\right]$$

and $Y_2 = 130,000$. The result agrees with the answer we found above analytically. The table shown in Figure 9 confirms that the payment should be between $243 and $244.

We can also use a graphing calculator or spreadsheet to make a table of the amount in a sinking fund. In the formula for future value of an annuity, simply let n be a variable with values from 1 to the total number of payments. Figure 10(a)

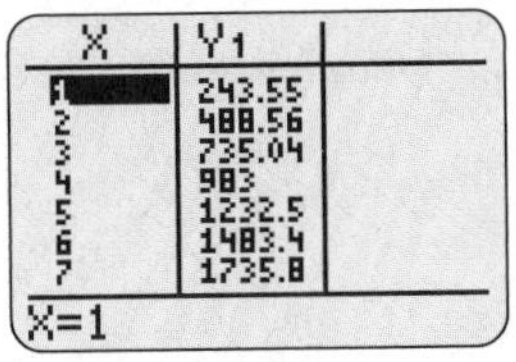

n	Amount in Fund
1	243.55
2	488.56
3	735.04
4	983.00
5	1232.45
6	1483.40
7	1735.85
8	1989.81
9	2245.30
10	2502.32
11	2760.89
12	3021.00

(a)   (b)

FIGURE 10

shows the beginning of such a table generated on a TI-83 for Example 6. Figure 10(b) shows the beginning of the same table using Microsoft Excel.

Annuities Due   The formula developed above is for *ordinary annuities*—those with payments made at the *end* of each time period. These results can be modified slightly to apply to **annuities due**—annuities in which payments are made at the *beginning* of each time period. To find the future value of an annuity due, treat each payment as if it were made at the *end* of the *preceding* period. That is, find $s_{\overline{n}|i}$ for *one additional period;* to compensate for this, subtract the amount of one payment.

Thus, the **future value of an annuity due** of n payments of R dollars each at the beginning of consecutive interest periods, with interest compounded at the rate of i per period, is

$$S = R\left[\frac{(1+i)^{n+1} - 1}{i}\right] - R \qquad or \qquad S = Rs_{\overline{n+1}|i} - R.$$

The finance feature of the TI-83 can be used to find the future value of an annuity due as well as an ordinary annuity. If this feature is not built in, you may wish to program your calculator to evaluate this formula, too.

EXAMPLE 7   Future Value of an Annuity Due

Find the future value of an annuity due if payments of $500 are made at the beginning of each quarter for 7 years, in an account paying 6% compounded quarterly.

Solution In 7 years, there are $n = 28$ quarterly periods. Add one period to get $n + 1 = 29$, and use the formula with $i = 6\%/4 = 1.5\%$.

$$S = 500\left[\frac{(1.015)^{29} - 1}{.015}\right] - 500 \approx 17{,}499.35$$

The account will have a total of $17,499.35 after 7 years.

5.2 EXERCISES

Find the fifth term of each of the following geometric sequences.

1. $a = 3$; $r = 2$

2. $a = 5$; $r = 3$

3. $a = -8$; $r = 3$

4. $a = -6$; $r = 2$

5. $a = 1$; $r = -3$

6. $a = 12$; $r = -2$

7. $a = 1024$; $r = \dfrac{1}{2}$

8. $a = 729$; $r = \dfrac{1}{3}$

Find the sum of the first four terms for each of the following geometric sequences.

9. $a = 1$; $r = 2$

10. $a = 3$; $r = 3$

11. $a = 5$; $r = \dfrac{1}{5}$

12. $a = 6$; $r = \dfrac{1}{2}$

13. $a = 128$; $r = -\dfrac{3}{2}$

14. $a = 81$; $r = -\dfrac{2}{3}$

Find each of the following values.

15. $s_{\overline{12}|.05}$

16. $s_{\overline{20}|.06}$

17. $s_{\overline{16}|.043}$

18. $s_{\overline{18}|.015}$

19. List some reasons for establishing a sinking fund.

20. Explain the difference between an ordinary annuity and an annuity due.

Find the future value of the following ordinary annuities. Interest is compounded annually.

21. $R = 100$; $i = .06$; $n = 4$

22. $R = 1000$; $i = .06$; $n = 5$

23. $R = 46{,}000$; $i = .063$; $n = 32$

24. $R = 29{,}500$; $i = .058$; $n = 15$

Find the future value of each of the following ordinary annuities, if payments are made and interest is compounded as given.

25. $R = 9200$; 10% interest compounded semiannually for 7 yr

26. $R = 3700$; 8% interest compounded semiannually for 11 yr

27. $R = 800$; 6.51% interest compounded semiannually for 12 yr

28. $R = 4600$; 8.73% interest compounded quarterly for 9 yr

29. $R = 15{,}000$; 12.1% interest compounded quarterly for 6 yr

30. $R = 42{,}000$; 10.05% interest compounded semiannually for 12 yr

Find the future value of each annuity due. Assume that interest is compounded annually.

31. $R = 600$; $i = .06$; $n = 8$

32. $R = 1400$; $i = .08$; $n = 10$

33. $R = 20{,}000$; $i = .08$; $n = 6$

34. $R = 4000$; $i = .06$; $n = 11$

Find the future value of each annuity due.

35. Payments of $1000 made at the beginning of each semiannual period for 9 yr at 8.15% compounded semiannually

36. $750 deposited at the beginning of each month for 15 yr at 5.9% compounded monthly

37. $100 deposited at the beginning of each quarter for 9 yr at 12.4% compounded quarterly

38. $1500 deposited at the beginning of each semiannual period for 11 yr at 5.6% compounded semiannually

Find the periodic payment that will amount to each of the given sums under the given conditions.

39. $S = \$10,000$; interest is 5% compounded annually; payments are made at the end of each year for 12 yr

40. $S = \$100,000$; interest is 8% compounded semiannually; payments are made at the end of each semiannual period for 9 yr

41. What is meant by a sinking fund? Give an example of a sinking fund.

Find the amount of each payment to be made into a sinking fund so that enough will be present to accumulate the following amounts. Payments are made at the end of each period.

42. $8500; money earns 8% compounded annually; 7 annual payments

43. $2000; money earns 6% compounded annually; 5 annual payments

44. $75,000; money earns 6% compounded semiannually for 4 1/2 yr

45. $25,000; money earns 5.7% compounded quarterly for 3 1/2 yr

46. $50,000; money earns 7.9% compounded quarterly for 2 1/2 yr

47. $9000; money earns 12.23% compounded monthly for 2 1/2 yr

Applications

BUSINESS AND ECONOMICS

48. *Comparing Accounts* Alex Levering deposits $12,000 at the end of each year for 9 yr in an account paying 8% interest compounded annually.

 a. Find the final amount she will have on deposit.

 b. Alex's brother-in-law works in a bank that pays 6% compounded annually. If she deposits money in this bank instead of the one above, how much will she have in her account?

 c. How much would Alex lose over 9 yr by using her brother-in-law's bank?

49. *Savings* Ron Hampton is saving for a computer. At the end of each month he puts $60 in a savings account that pays 8% interest compounded monthly. How much is in the account after 3 yr?

50. *Savings* Hassi is paid on the first day of the month and $80 is automatically deducted from his pay and deposited in a savings account. If the account pays 7.5% interest compounded monthly, how much will be in the account after 3 yr and 9 mo?

51. *Savings* A typical pack-a-day smoker spends about $55 per month on cigarettes. Suppose the smoker invests that amount each month in a savings account at 4.8% interest compounded monthly. What would the account be worth after 40 yr?

52. *Savings* A father opened a savings account for his daughter on the day she was born, depositing $1000. Each year on her birthday he deposits another $1000, making the last deposit on her twenty-first birthday. If the account pays 9.5% interest compounded annually, how much is in the account at the end of the day on the daughter's twenty-first birthday?

53. *Retirement Planning* A 45-year-old man puts $1000 in a retirement account at the end of each quarter until he reaches the age of 60 and makes no further deposits. If the account pays 8% interest compounded quarterly, how much will be in the account when the man retires at age 65?

54. *Retirement Planning* At the end of each quarter a 50-year-old woman puts $1200 in a retirement account that pays 7% interest compounded quarterly. When she reaches age 60, she withdraws the entire amount and places it in a mutual fund that pays 9% interest compounded monthly. From then on she deposits $300 in the mutual fund at the end of each month. How much is in the account when she reaches age 65?

55. *Savings* Jasspreet Kaur deposits $2435 at the beginning of each semiannual period for 8 yr in an account paying 6% compounded semiannually. She then leaves that money alone, with no further deposits, for an additional 5 yr. Find the final amount on deposit after the entire 13-yr period.

56. *Savings* Chuck Hickman deposits $10,000 at the beginning of each year for 12 yr in an account paying 5% compounded annually. He then puts the total amount on deposit in another account paying 6% compounded semiannually for another 9 yr. Find the final amount on deposit after the entire 21-yr period.

57. *Savings* Greg Tobin needs $10,000 in 8 yr.

  a. What amount should he deposit at the end of each quarter at 8% compounded quarterly so that he will have his $10,000?

  b. Find Greg's quarterly deposit if the money is deposited at 6% compounded quarterly.

58. *Buying Equipment* Harv, the owner of Harv's Meats, knows that he must buy a new deboner machine in 4 yr. The machine costs $12,000. In order to accumulate enough money to pay for the machine, Harv decides to deposit a sum of money at the end of each 6 mo in an account paying 6% compounded semiannually. How much should each payment be?

59. *Buying a Car* Susan Laferriere wants to buy an $18,000 car in 6 yr. How much money must she deposit at the end of each quarter in an account paying 5% compounded quarterly so that she will have enough to pay for her car?

Individual Retirement Accounts Suppose a 40-year-old person deposits $2000 per year in an Individual Retirement Account until age 65. Find the total in the account with the following assumptions of interest rates. (Assume semiannual compounding, with payments of $1000 made at the end of each semiannual period.)

60. 6%   **61.** 8%   **62.** 4%   **63.** 10%

In Exercises 64 and 65, use a graphing calculator to find the value of i that produces the given value of S. (See Example 5(b).)

64. *Retirement* To save for retirement, Karla Harby put $300 each month into an ordinary annuity for 20 yr. Interest was compounded monthly. At the end of the 20 yr, the annuity was worth $147,126. What annual interest rate did she receive?

65. *Rate of Return* Jennifer Wall made payments of $250 per month at the end of each month to purchase a piece of

property. At the end of 30 yr, she completely owned the property, which she sold for $330,000. What annual interest rate would she need to earn on an annuity for a comparable rate of return?

66. *Lottery* In a 1992 Virginia lottery, the jackpot was $27 million. An Australian investment firm tried to buy all possible combinations of numbers, which would have cost $7 million. In fact, the firm ran out of time and was unable to buy all combinations, but ended up with the only winning ticket anyway. The firm received the jackpot in 20 equal annual payments of $1.35 million.* Assume these payments meet the conditions of an ordinary annuity.

  a. Suppose the firm can invest money at 8% interest compounded annually. How many years would it take until the investors would be further ahead than if they had simply invested the $7 million at the same rate? (*Hint:* Experiment with different values of *n*, the number of years, or use a graphing calculator to plot the value of both investments as a function of the number of years.)

  b. How many years would it take in part a at an interest rate of 12%?

67. *Buying Real Estate* Marisa Raffaele sells some land in Nevada. She will be paid a lump sum of $60,000 in 7 yr. Until then, the buyer pays 8% simple interest quarterly.

  a. Find the amount of each quarterly interest payment on the $60,000.

  b. The buyer sets up a sinking fund so that enough money will be present to pay off the $60,000. The buyer will make semiannual payments into the sinking fund; the account pays 6% compounded semiannually. Find the amount of each payment into the fund.

68. *Buying Rare Stamps* Paul Altier bought a rare stamp for his collection. He agreed to pay a lump sum of $4000 after 5 yr. Until then, he pays 6% simple interest semiannually on the $4000.

  a. Find the amount of each semiannual interest payment.

  b. Paul sets up a sinking fund so that enough money will be present to pay off the $4000. He will make annual payments into the fund. The account pays 8% compounded annually. Find the amount of each payment.

69. *Down Payment* A conventional loan, such as for a car or a house, is similar to an annuity, but usually includes a down payment. Show that if a down payment of *D* dollars is made at the beginning of the loan period, the future value of all the payments, including the down payment, is

$$ S = D(1 + i)^n + R\left[\frac{(1 + i)^n - 1}{i}\right]. $$

*The Washington Post, March 10, 1992, p. A1.

■ 5.3 PRESENT VALUE OF AN ANNUITY; AMORTIZATION

THINK ABOUT IT

What monthly payment will pay off a $10,000 car loan in 36 monthly payments at 12% annual interest?

The answer to this question is given later in this section. We shall see that it involves finding the present value of an annuity.

Suppose that at the end of each year, for the next 10 years, $500 is deposited in a savings account paying 7% interest compounded annually. This is an example of an ordinary annuity. The **present value of an annuity** is the amount that would have to be deposited in one lump sum today (at the same compound interest rate) in order to produce exactly the same balance at the end of 10 years. We can find a formula for the present value of an annuity as follows.

Suppose deposits of R dollars are made at the end of each period for n periods at interest rate i per period. Then the amount in the account after n periods is the future value of this annuity:

$$S = R \cdot s_{\overline{n}|i} = R\left[\frac{(1 + i)^n - 1}{i}\right].$$

On the other hand, if P dollars are deposited today at the same compound interest rate i, then at the end of n periods, the amount in the account is $P(1 + i)^n$. If P is the present value of the annuity, this amount must be the same as the amount S in the formula above; that is,

$$P(1 + i)^n = R\left[\frac{(1 + i)^n - 1}{i}\right].$$

To solve this equation for P, multiply both sides by $(1 + i)^{-n}$.

$$P = R(1 + i)^{-n}\left[\frac{(1 + i)^n - 1}{i}\right]$$

Use the distributive property; also recall that $(1 + i)^{-n}(1 + i)^n = 1$.

$$P = R\left[\frac{(1 + i)^{-n}(1 + i)^n - (1 + i)^{-n}}{i}\right] = R\left[\frac{1 - (1 + i)^{-n}}{i}\right]$$

FOR REVIEW ■

Recall that for any nonzero number a, $a^0 = 1$. Also, by the product rule for exponents, $a^x \cdot a^y = a^{x+y}$. In particular, for any nonzero number a, $a^n \cdot a^{-n} = a^{n+(-n)} = a^0 = 1$.

The amount P is the *present value of the annuity*. The quantity in brackets is abbreviated as $a_{\overline{n}|i}$, so

$$a_{\overline{n}|i} = \frac{1 - (1 + i)^{-n}}{i}.$$

(The symbol $a_{\overline{n}|i}$ is read "*a*-angle-*n* at *i*." Compare this quantity with $s_{\overline{n}|i}$ in the previous section.) The formula for the present value of an annuity is summarized next.

PRESENT VALUE OF AN ANNUITY

The present value P of an annuity of n payments of R dollars each at the end of consecutive interest periods with interest compounded at a rate of interest i per period is

$$P = R\left[\frac{1 - (1 + i)^{-n}}{i}\right] \quad \text{or} \quad P = Ra_{\overline{n}|i}.$$

CAUTION Don't confuse the formula for the present value of an annuity with the one for the future value of an annuity. Notice the difference: the numerator of the fraction in the present value formula is $1 - (1 + i)^{-n}$, but in the future value formula, it is $(1 + i)^n - 1$.

The financial feature of the TI-83 calculator can be used to find the present value of an annuity by choosing that option from the menu and entering the required information. If your calculator does not have this built-in feature, it will be useful to store a program to calculate present value of an annuity in your calculator. A program is given in *The Graphing Calculator Manual* that is available with this book.

EXAMPLE 1 Present Value of an Annuity

Mr. Bryer and Ms. Gonsalez are both graduates of the Brisbane Institute of Technology. They both agree to contribute to the endowment fund of BIT. Mr. Bryer says that he will give $500 at the end of each year for 9 years. Ms. Gonsalez prefers to give a lump sum today. What lump sum can she give that will equal the present value of Mr. Bryer's annual gifts, if the endowment fund earns 7.5% compounded annually?

Solution Here, $R = 500$, $n = 9$, and $i = .075$ and we have

$$P = R \cdot a_{\overline{9}|.075} = 500\left[\frac{1 - (1.075)^{-9}}{.075}\right] \approx 3189.44.$$

Therefore, Ms. Gonsalez must donate a lump sum of $3189.44 today.

One of the most important uses of annuities is in determining the equal monthly payments needed to pay off a loan, as illustrated in the next example.

EXAMPLE 2 Car Payments

A car costs $12,000. After a down payment of $2000, the balance will be paid off in 36 equal monthly payments with interest of 12% per year on the unpaid balance. Find the amount of each payment.

Solution A single lump sum payment of $10,000 today would pay off the loan. So, $10,000 is the present value of an annuity of 36 monthly payments with interest of $12\%/12 = 1\%$ per month. Thus, $P = 10,000$, $n = 36$, $i = .01$, and we must find the monthly payment R in the formula

$$P = R\left[\frac{1 - (1 + i)^{-n}}{i}\right]$$

$$10,000 = R\left[\frac{1 - (1.01)^{-36}}{.01}\right]$$

$$R \approx 332.1430981.$$

A monthly payment of $332.14 will be needed.

Each payment in Example 2 includes interest on the unpaid balance with the remainder going to reduce the loan. For example, the first payment of $332.14 includes interest of .01($10,000) = $100 and is divided as follows.

monthly payment	interest due	to reduce the balance

$$\$332.15 - \$100 = \$232.15$$

At the end of this section, amortization schedules show that this procedure does reduce the loan to $0 after all payments are made (the final payment may be slightly different).

Amortization   A loan is **amortized** if both the principal and interest are paid by a sequence of equal periodic payments. In Example 2, a loan of $10,000 at 12% interest compounded monthly could be amortized by paying $332.14 per month for 36 months.

The periodic payment needed to amortize a loan may be found, as in Example 2, by solving the present value equation for R.

AMORTIZATION PAYMENTS

A loan of P dollars at interest rate i per period may be amortized in n equal periodic payments of R dollars made at the end of each period, where

$$R = \frac{P}{a_{\overline{n}|i}} = \frac{P}{\left[\dfrac{1 - (1 + i)^{-n}}{i}\right]} = \frac{Pi}{1 - (1 + i)^{-n}}.$$

EXAMPLE 3   Home Mortgage
The Perez family buys a house for $94,000 with a down payment of $16,000. They take out a 30-year mortgage for $78,000 at an annual interest rate of 9.6%.

(a) Find the amount of the monthly payment needed to amortize this loan.

Solution Here $P = 78,000$ and the monthly interest rate is $9.6\%/12 = .096/12 = .008$.* The number of monthly payments is $12 \cdot 30 = 360$. Therefore,

$$R = \frac{78,000}{a_{\overline{360}|.008}} = \frac{78,000}{\left[\dfrac{1 - (1.008)^{-360}}{.008}\right]} \approx 661.56.$$

Monthly payments of $661.56 are required to amortize the loan.

---

*Mortgage rates are quoted in terms of annual interest, but it is always understood that the monthly rate is 1/12 of the annual rate and that interest is compounded monthly.

(b) Find the total amount of interest paid when the loan is amortized over 30 years.

Solution The Perez family makes 360 payments of $661.56 each, for a total of $238,161.60. Since the amount of the loan was $78,000, the total interest paid is

$$\$238,161.60 - \$78,000 = \$160,161.60.$$

This large amount of interest is typical of what happens with a long mortgage. A 15-year mortgage would have higher payments, but would involve significantly less interest.

(c) Find the part of the payment that is interest and the part that is applied to reducing the debt.

Solution During the first month, the entire $78,000 is owed. Interest on this amount for 1 month is found by the formula for simple interest, with $r =$ annual interest rate and $t =$ time in years.

$$I = Prt = 78,000(.096)\frac{1}{12} = 624$$

At the end of the month, a payment of $661.56 is made; since $624 of this is interest, a total of

$$\$661.56 - \$624 = \$37.56$$

is applied to the reduction of the original debt.

It can be shown that the unpaid balance after x payments is approximately given by the function

$$y = R\left[\frac{1 - (1 + i)^{-(n-x)}}{i}\right].$$

For example, the unpaid balance in Example 3 after 1 payment is

$$y = \$661.56\left[\frac{1 - (1.008)^{-359}}{.008}\right] \approx \$77,961.87.$$

This is very close to the amount left after deducting the $37.56 applied to the loan in part (c):

$$\$78,000 - \$37.56 = \$77,962.44.$$

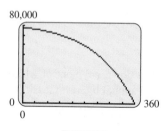

80,000

0

0   360

FIGURE 11

A calculator-generated graph of this function is shown in Figure 11.

We can find the unpaid balance after any number of payments, x, by finding the y-value that corresponds to x. For example, the remaining balance after 5 years or 60 payments is shown at the bottom of the window in Figure 12(a). You may be surprised that the remaining balance on a $78,000 loan is as large as $75,121.10. This is because most of the early payments on a loan go toward interest, as we saw in Example 3(c).

By adding the graph of $y = (1/2)78,000 = 39,000$ to the figure, we can find when half the loan has been repaid. From Figure 12(b) we see that 280 payments are required. Note that only 80 payments remain at that point, which again emphasizes the fact that the earlier payments do little to reduce the loan.

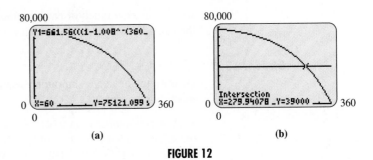

FIGURE 12

Amortization Schedules In the preceding example, 360 payments are made to amortize a $78,000 loan. The loan balance after the first payment is reduced by only $37.56, which is much less than $(1/360)(78,000) \approx \$216.67$. Therefore, even though equal *payments* are made to amortize a loan, the loan *balance* does not decrease in equal steps. This fact is very important if a loan is paid off early.

EXAMPLE 4 Early Payment

Susan Dratch borrows $1000 for 1 year at 12% annual interest compounded monthly. Verify that her monthly loan payment is $88.85. After making three payments, she decides to pay off the remaining balance all at once. How much must she pay?

Solution Since nine payments remain to be paid, they can be thought of as an annuity consisting of nine payments of $88.85 at 1% interest per period. The present value of this annuity is

$$88.85 \left[\frac{1 - (1.01)^{-9}}{.01} \right] \approx 761.09.$$

So Susan's remaining balance, computed by this method, is $761.09.

An alternative method of figuring the balance is to consider the payments already made as an annuity of three payments. At the beginning, the present value of this annuity was

$$88.85 \left[\frac{1 - (1.01)^{-3}}{.01} \right] \approx 261.31.$$

So she still owes the difference $1000 - \$261.31 = \$738.69$. Furthermore, she owes the interest on this amount for 3 months, for a total of

$$(738.69)(1.01)^3 \approx \$761.07.$$

This balance due differs from the one obtained by the first method by 2 cents because the monthly payment and the other calculations were rounded to the nearest penny.

Although most people would not quibble about a difference of 2 cents in the balance due in Example 4, the difference in other cases (larger amounts or longer terms) might be more than that. A bank or business must keep its books accurately to the nearest penny, so it must determine the balance due in such cases unambiguously and exactly. This is done by means of an **amortization schedule,** which lists how much of each payment is interest and how much goes to reduce the balance, as well as how much is owed after *each* payment.

EXAMPLE 5 Amortization Table

Determine the exact amount Susan Dratch in Example 4 owes after three monthly payments.

Solution An amortization table for the loan is shown below. It is obtained as follows. The annual interest rate is 12% compounded monthly, so the interest rate per month is $12\%/12 = 1\% = .01$. When the first payment is made, 1 month's interest—namely $.01(1000) = \$10$—is owed. Subtracting this from the $88.85 payment leaves $78.85 to be applied to repayment. Hence, the principal at the end of the first payment period is $1000 - 78.85 = \$921.15$, as shown in the "payment 1" line of the chart.

When payment 2 is made, 1 month's interest on $921.15 is owed, namely $.01(921.15) = \$9.21$. Subtracting this from the $88.85 payment leaves $79.64 to reduce the principal. Hence, the principal at the end of payment 2 is $921.15 - 79.64 = \$841.51$. The interest portion of payment 3 is based on this amount, and the remaining lines of the table are found in a similar fashion.

Payment Number	Amount of Payment	Interest for Period	Portion to Principal	Principal at End of Period
0	—	—	—	$1000.00
1	$88.85	$10.00	$78.85	$921.15
2	$88.85	$9.21	$79.64	$841.51
3	$88.85	$8.42	$80.43	$761.08
4	$88.85	$7.61	$81.24	$679.84
5	$88.85	$6.80	$82.05	$597.79
6	$88.85	$5.98	$82.87	$514.92
7	$88.85	$5.15	$83.70	$431.22
8	$88.85	$4.31	$84.54	$346.68
9	$88.85	$3.47	$85.38	$261.30
10	$88.85	$2.61	$86.24	$175.06
11	$88.85	$1.75	$87.10	$87.96
12	$88.84	$.88	$87.96	0

The schedule shows that after three payments, she still owes $761.08, an amount that differs slightly from that obtained by either method in Example 4.

The amortization schedule in Example 5 is typical. In particular, note that all payments are the same except the last one. It is often necessary to adjust the amount of the final payment to account for rounding off earlier, and to ensure that the final balance is exactly 0.

An amortization schedule also shows how the periodic payments are applied to interest and principal. The amount going to interest decreases with each payment, while the amount going to reduce the principal increases with each payment.

A graphing calculator program to produce an amortization schedule is available in *The Graphing Calculator Manual* that is available with this book. The TI-83 includes a built-in program to find the amortization payment. Spreadsheets are another useful tool for creating amortization tables. Microsoft Excel has a built-in feature for calculating monthly payments. Figure 13 shows an Excel amortization table for Example 5. For more details, see *The Spreadsheet Manual*, also available with this book.

	A	B	C	D	E	F
1	Pmt #	Payment	Interest	Principal	End Prncpl	
2	0				1000	
3	1	88.85	10.00	78.85	921.15	
4	2	88.85	9.21	79.64	841.51	
5	3	88.85	8.42	80.43	761.08	
6	4	88.85	7.61	81.24	679.84	
7	5	88.85	6.80	82.05	597.79	
8	6	88.85	5.98	82.87	514.91	
9	7	88.85	5.15	83.70	431.21	
10	8	88.85	4.31	84.54	346.67	
11	9	88.85	3.47	85.38	261.29	
12	10	88.85	2.61	86.24	175.05	
13	11	88.85	1.75	87.10	87.96	
14	12	88.85	0.88	87.97	-0.02	

FIGURE 13

5.3 EXERCISES

1. Which of the following is represented by $a_{\overline{n}|i}$?

a. $\dfrac{(1 + i)^{-n} - 1}{i}$ **b.** $\dfrac{(1 + i)^n - 1}{i}$ **c.** $\dfrac{1 - (1 + i)^{-n}}{i}$ **d.** $\dfrac{1 - (1 + i)^n}{i}$

2. Which of the choices in Exercise 1 represents $s_{\overline{n}|i}$?

Find each of the following.

3. $a_{\overline{15}|.06}$ **4.** $a_{\overline{10}|.03}$ **5.** $a_{\overline{18}|.045}$ **6.** $a_{\overline{32}|.029}$

7. Explain the difference between the present value of an annuity and the future value of an annuity. For a given annuity, which is larger? Why?

Find the present value of each ordinary annuity.

8. Payments of $890 each year for 16 yr at 8% compounded annually

9. Payments of $1400 each year for 8 yr at 8% compounded annually

10. Payments of $10,000 semiannually for 15 yr at 10% compounded semiannually

11. Payments of $50,000 quarterly for 10 yr at 8% compounded quarterly

12. Payments of $15,806 quarterly for 3 yr at 10.8% compounded quarterly

13. Payments of $18,579 every 6 mo for 8 yr at 9.4% compounded semiannually

Find the lump sum deposited today that will yield the same total amount as payments of $10,000 at the end of each year for 15 yr, at each of the given interest rates.

14. 4% compounded annually   **15.** 6% compounded annually

16. What does it mean to amortize a loan?

Find the payment necessary to amortize each of the following loans.

17. $2500, 8% compounded quarterly, 6 quarterly payments

18. $41,000, 10% compounded semiannually, 10 semiannual payments

19. $90,000, 8% compounded annually, 12 annual payments

20. $140,000, 12% compounded quarterly, 15 quarterly payments

21. $7400, 8.2% compounded semiannually, 18 semiannual payments

22. $5500, 12.5% compounded monthly, 24 monthly payments

Use the amortization table in Example 5 to answer the questions in Exercises 23–26.

23. How much of the fourth payment is interest?

24. How much of the eleventh payment is used to reduce the debt?

25. How much interest is paid in the first 4 months of the loan?

26. How much interest is paid in the last 4 months of the loan?

27. What sum deposited today at 5% compounded annually for 8 yr will provide the same amount as $1000 deposited at the end of each year for 8 yr at 6% compounded annually?

28. What lump sum deposited today at 8% compounded quarterly for 10 yr will yield the same final amount as deposits of $4000 at the end of each six-month period for 10 yr at 6% compounded semiannually?

Find the monthly house payments necessary to amortize the following loans.

29. $149,560 at 7.75% for 25 yr

30. $170,892 at 8.11% for 30 yr

31. $153,762 at 8.45% for 30 yr

32. $96,511 at 9.57% for 25 yr

Applications

BUSINESS AND ECONOMICS

33. *House Payments* Calculate the monthly payment and total amount of interest paid in Example 3 with a 15-yr loan, and then compare with the results of Example 3.

34. *Installment Buying* Stereo Shack sells a stereo system for $600 down and monthly payments of $30 for the next 3 yr. If the interest rate is 1.25% per month on the unpaid balance, find

a. the cost of the stereo system;

b. the total amount of interest paid.

35. *Car Payments* Hong Le buys a car costing $6000. He agrees to make payments at the end of each monthly period for 4 yr. He pays 12% interest, compounded monthly.

a. What is the amount of each payment?

b. Find the total amount of interest Le will pay.

36. *Land Purchase* A speculator agrees to pay $15,000 for a parcel of land; this amount, with interest, will be paid over 4 yr, with semiannual payments, at an interest rate of 10% compounded semiannually. Find the amount of each payment.

37. *Lottery Winnings* In most states, the winnings of million-dollar lottery jackpots are divided into equal payments given annually for 20 yr. (In Colorado, the results are distributed over 25 yr.)* This means that the present value of the jackpot is worth less than the stated prize, with the actual value determined by the interest rate at which the money could be invested.

a. Find the present value of a $1 million lottery jackpot distributed in equal annual payments over 20 yr, using an interest rate of 5%.

b. Find the present value of a $1 million lottery jackpot distributed in equal annual payments over 20 yr, using an interest rate of 9%.

c. Calculate the answer for part a using the 25-yr distribution time in Colorado.

---

*Gould, Lois, "Ticket to Trouble," *The New York Times Magazine,* April 23, 1995, p. 39.

d. Calculate the answer for part b using the 25-yr distribution time in Colorado.

Student Loans Student borrowers now have more options to choose from when selecting repayment plans. The standard plan repays the loan in 10 yr with equal monthly payments. The extended plan allows from 12 to 30 years to repay the loan. A student borrows $35,000 at 7.43% compounded monthly.*

38. Find the monthly payment and total interest paid under the standard plan.

39. Find the monthly payment and total interest paid under the extended plan with 20 yr to pay off the loan.

Installment Buying In Exercises 40–42, prepare an amortization schedule showing the first four payments for each loan.

40. An insurance firm pays $4000 for a new printer for its computer. It amortizes the loan for the printer in 4 annual payments at 8% compounded annually.

41. Large semitrailer trucks cost $72,000 each. Ace Trucking buys such a truck and agrees to pay for it by a loan that will be amortized with 9 semiannual payments at 10% compounded semiannually.

42. One retailer charges $1048 for a computer monitor. A firm of tax accountants buys 8 of these monitors. They make a down payment of $1200 and agree to amortize the balance with monthly payments at 12% compounded monthly for 4 yr.

43. *Investment* In 1995, Oseola McCarty donated $150,000 to the University of Southern Mississippi to establish a scholarship fund.[†] What is unusual about her is that the entire amount came from what she was able to save each month from her work as a washer woman, a job she began in 1916 at the age of 8, when she dropped out of school.

  a. How much would Ms. McCarty have to put into her savings account at the end of every 3 mo to accumulate

$150,000 over 79 yr? Assume she received an interest rate of 5.25% compounded quarterly.

  b. Answer part a using a 2% and a 7% interest rate.

44. *Loan Payments* When Nancy Hart opened her law office, she bought $14,000 worth of law books and $7200 worth of office furniture. She paid $1200 down and agreed to amortize the balance with semiannual payments for 5 yr, at 12% compounded semiannually.

  a. Find the amount of each payment.

  b. Refer to the text and Figures 11 and 12. When her loan had been reduced below $5000, Flores received a large tax refund and decided to pay off the loan. How many payments were left at this time?

45. *House Payments* Kareem Adiagbo buys a house for $285,000. He pays $60,000 down and takes out a mortgage at 9.5% on the balance. Find his monthly payment and the total amount of interest he will pay if the length of the mortgage is

  a. 15 yr;   **b.** 20 yr;   **c.** 25 yr.

  d. Refer to the text and Figures 11 and 12. When will half the 20-yr loan in part b be paid off?

46. *Inheritance* Sandi Goldstein has inherited $25,000 from her grandfather's estate. She deposits the money in an account offering 6% interest compounded annually. She wants to make equal annual withdrawals from the account so that the money (principal and interest) lasts exactly 8 yr.

  a. Find the amount of each withdrawal.

  b. Find the amount of each withdrawal if the money must last 12 yr.

47. *Charitable Trust* The trustees of a college have accepted a gift of $150,000. The donor has directed the trustees to

---

*Hansell, Saul, "Money and College," *The New York Times,* April 2, 1995, p. 28.
[†]*The New York Times,* Nov. 12, 1996, pp. A1, A22.

deposit the money in an account paying 6% per year, compounded semiannually. The trustees may make equal withdrawals at the end of each six-month period; the money must last 5 yr.

a. Find the amount of each withdrawal.

b. Find the amount of each withdrawal if the money must last 6 yr.

Amortization Prepare an amortization schedule for each of the following loans.

48. A loan of $37,947.50 with interest at 8.5% compounded annually, to be paid with equal annual payments over 10 yr.

49. A loan of $4835.80 at 9.25% interest compounded semi-annually, to be repaid in 5 yr in equal semiannual payments.

50. *Perpetuity* A *perpetuity* is an annuity in which the payments go on forever. We can derive a formula for the present value of a perpetuity by taking the formula for the present value of an annuity and looking at what happens when *n* gets larger and larger. Explain why the present value of an annuity is given by

$$P = \frac{R}{i}.$$

51. *Perpetuity* Using the result of Exercise 50, find the present value of perpetuities for each of the following.

a. Payments of $1000 a year with 4% interest compounded annually

b. Payments of $600 every 3 mo with 6% interest compounded quarterly

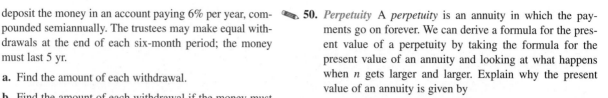

■ CHAPTER SUMMARY

This chapter introduces the mathematics of finance. Simple interest is the starting point; when interest is earned on interest previously earned, we have compound interest. In an annuity, money continues to be deposited at regular intervals, and compound interest is earned on that money. In an ordinary annuity, the compounding period is the same as the time between payments, which simplifies the calculations. An annuity due is slightly different, in that the payments are made at the beginning of each time period. A sinking fund is like an annuity; a fund is set up to receive periodic payments, so the payments plus the compound interest will produce a desired sum by a certain date. The present value of an annuity is the amount that would have to be deposited today to produce the same amount as the annuity at the end of a specified time. This idea leads to an amortization table for a loan, which shows the payments, broken down into interest and principal, for a loan to be paid back after a specified time.

A Strategy for Solving Finance Problems

We have presented a lot of new formulas in this chapter. By answering the following questions, you can decide which formula to use for a particular problem.

1. Is simple or compound interest involved?
Simple interest is normally used for investments or loans of a year or less; compound interest is normally used in all other cases.

2. If simple interest is being used, what is being sought: interest amount, future value, present value, or interest rate?

3. If compound interest is being used, does it involve a lump sum (single payment) or an annuity (sequence of payments)?
a. For a lump sum, what is being sought: present value, future value, number of periods at interest, or effective rate?
b. For an annuity,
i. Is it an ordinary annuity (payment at the end of each period) or an annuity due (payment at the beginning of each period)?
ii. What is being sought: present value, future value, or payment amount?

Once you have answered these questions, choose the appropriate formula and work the problem. As a final step, consider whether the answer you get makes sense. For instance, present value should always be less than future value. The amount of interest or the payments in an annuity should be fairly small compared to the total future value.

List of Variables

r is the annual interest rate.

i is the interest rate per period.

t is the number of years.

n is the number of periods.

m is the number of periods per year.

P is the principal or present value.

A is the future value of a lump sum.

S is the future value of an annuity.

R is the periodic payment in an annuity.

$$i = \frac{r}{m} \qquad n = tm$$

	Simple Interest	**Compound Interest**
Interest	$I = Prt$	$I = A - P$
Future Value	$A = P(1 + rt)$	$A = P(1 + i)^n$
Present Value	$P = \dfrac{A}{1 + rt}$	$P = \dfrac{A}{(1 + i)^n} = A(1 + i)^{-n}$
		Effective Rate $\quad r_e = \left(1 + \dfrac{r}{m}\right)^m - 1$

Ordinary Annuity

Future Value $\qquad S = R\left[\dfrac{(1 + i)^n - 1}{i}\right] = R \cdot s_{\overline{n}|i}$

Present Value $\qquad P = R\left[\dfrac{1 - (1 + i)^{-n}}{i}\right] = R \cdot a_{\overline{n}|i}$

Annuity Due

Future Value $\qquad S = R\left[\dfrac{(1 + i)^{n+1} - 1}{i}\right] - R$

KEY TERMS

5.1 simple interest
principal
rate
time
future value
maturity value
compound interest
compound amount

nominal (stated) rate
effective rate
present value
5.2 geometric sequence
terms
common ratio
annuity
ordinary annuity

future value of an
ordinary annuity
sinking fund
annuity due
future value of an
annuity due
5.3 present value of an
annuity

amortize a loan
amortization schedule

CHAPTER 5 REVIEW EXERCISES

Find the simple interest for the following loans.

1. $15,903 at 8% for 8 mo

2. $4902 at 9.5% for 11 mo

3. $42,368 at 5.22% for 5 mo

4. $3478 at 7.4% for 88 days
(assume a 360-day year)

5. For a given amount of money at a given interest rate for a given time period, does simple interest or compound interest produce more interest?

Find the compound amount in each of the following.

6. $2800 at 6% compounded annually for 10 yr

7. $19,456.11 at 12% compounded semiannually for 7 yr

8. $312.45 at 6% compounded semiannually for 16 yr

9. $57,809.34 at 12% compounded quarterly for 5 yr

Find the amount of interest earned by each deposit.

10. $3954 at 8% compounded annually for 12 yr

11. $12,699.36 at 10% compounded semiannually for 7 yr

12. $12,903.45 at 10.37% compounded quarterly for 29 quarters

13. $34,677.23 at 9.72% compounded monthly for 32 mo

14. What is meant by the present value of an amount A?

Find the present value of the following amounts.

15. $42,000 in 7 yr, 12% compounded monthly

16. $17,650 in 4 yr, 8% compounded quarterly

17. $1347.89 in 3.5 yr, 6.77% compounded semiannually

18. $2388.90 in 44 mo, 5.93% compounded monthly

19. Write the first five terms of the geometric sequence with $a = 2$ and $r = 3$.

20. Write the first four terms of the geometric sequence with $a = 4$ and $r = 1/2$.

21. Find the sixth term of the geometric sequence with $a = -3$ and $r = 2$.

22. Find the fifth term of the geometric sequence with $a = -2$ and $r = -2$.

23. Find the sum of the first four terms of the geometric sequence with $a = -3$ and $r = 3$.

24. Find the sum of the first five terms of the geometric sequence with $a = 8000$ and $r = -1/2$.

25. Find $s_{\overline{30}|.01}$.

26. Find $s_{\overline{20}|.05}$.

27. What is meant by the future value of an annuity?

Find the future value of each annuity.

28. $500 deposited at the end of each 6-month period for 8 yr; money earns 6% compounded semiannually

29. $1288 deposited at the end of each year for 14 yr; money earns 8% compounded annually

30. $4000 deposited at the end of each quarter for 7 yr; money earns 6% compounded quarterly

31. $233 deposited at the end of each month for 4 yr; money earns 12% compounded monthly

32. $672 deposited at the beginning of each quarter for 7 yr; money earns 8% compounded quarterly

33. $11,900 deposited at the beginning of each month for 13 mo; money earns 12% compounded monthly

34. What is the purpose of a sinking fund?

Find the amount of each payment that must be made into a sinking fund to accumulate the following amounts. (Recall, in a sinking fund, payments are made at the end of every interest period.)

35. $6500; money earns 8% compounded annually; 6 annual payments

36. $57,000; money earns 6% compounded semiannually for $8\frac{1}{2}$ yr

37. $233,188; money earns 9.7% compounded quarterly for $7\frac{3}{4}$ yr

38. $1,056,788; money earns 8.12% compounded monthly for $4\frac{1}{2}$ yr

Find the present value of each ordinary annuity.

39. Deposits of $850 annually for 4 yr at 8% compounded annually

40. Deposits of $1500 quarterly for 7 yr at 8% compounded quarterly

41. Payments of $4210 semiannually for 8 yr at 8.6% compounded semiannually

42. Payments of $877.34 monthly for 17 mo at 9.4% compounded monthly

43. Give two examples of the types of loans that are commonly amortized.

Find the amount of the payment necessary to amortize each of the following loans.

44. $80,000 loan; 8% compounded annually; 9 annual payments

45. $3200 loan; 8% compounded quarterly; 10 quarterly payments

46. $32,000 loan; 9.4% compounded quarterly; 17 quarterly payments

47. $51,607 loan; 13.6% compounded monthly; 32 monthly payments

Find the monthly house payments for the following mortgages.

48. $56,890 at 10.74% for 25 yr

49. $77,110 at 11.45% for 30 yr

A portion of an amortization table is given below for a $127,000 loan at 8.5% interest compounded monthly for 25 yr.

Payment Number	Amount of Payment	Interest for Period	Portion to Principal	Principal at End of Period
1	$1022.64	$899.58	$123.06	$126,876.94
2	$1022.64	$898.71	$123.93	$126,753.01
3	$1022.64	$897.83	$124.81	$126,628.20
4	$1022.64	$896.95	$125.69	$126,502.51
5	$1022.64	$896.06	$126.58	$126,375.93
6	$1022.64	$895.16	$127.48	$126,248.45
7	$1022.64	$894.26	$128.38	$126,120.07
8	$1022.64	$893.35	$129.29	$125,990.78
9	$1022.64	$892.43	$130.21	$125,860.57
10	$1022.64	$891.51	$131.13	$125,729.44
11	$1022.64	$890.58	$132.06	$125,597.38
12	$1022.64	$889.65	$132.99	$125,464.39

Use the table to answer the following questions.

50. How much of the fifth payment is interest?

51. How much of the twelfth payment is used to reduce the debt?

52. How much interest is paid in the first 3 months of the loan?

53. How much has the debt been reduced at the end of the first year?

Applications

BUSINESS AND ECONOMICS

54. *Personal Finance* Michael Garbin owes $5800 to his mother. He has agreed to repay the money in 10 mo at an interest rate of 10.3%. How much will he owe in 10 mo? How much interest will he pay?

55. *Business Financing* John Remington needs to borrow $9820 to buy new equipment for his business. The bank charges him 12.1% for a 7-mo loan. How much interest will he be charged? What amount must he pay in 7 mo?

56. *Business Financing* An accountant loans $28,000 at simple interest to her business. The loan is at 11.5% and earns $3255 interest. Find the time of the loan in months.

57. *Business Investment* A developer deposits $84,720 for 7 mo and earns $4055.46 in simple interest. Find the interest rate.

58. *Personal Finance* In 3 yr Joan McKee must pay a pledge of $7500 to her college's building fund. What lump sum can she deposit today, at 10% compounded semiannually, so that she will have enough to pay the pledge?

59. *Personal Finance* Tom, a graduate student, is considering investing $500 now, when he is 23, or waiting until he is 40 to invest $500. How much more money will he have at the age of 65 if he invests now, given that he can earn 5% interest compounded quarterly?

60. *Pensions* Pension experts recommend that you start drawing at least 40% of your full pension as early as possible.* Suppose you have built up a pension of $12,000-annual payments by working 10 yr for a company. When you leave to accept a better job, the company gives you the option of collecting half of the full pension when you reach age 55 or the full pension at age 65. Assume an interest rate of 8% compounded annually. By age 75, how much will each plan produce? Which plan would produce the larger amount?

61. *Business Investment* A firm of attorneys deposits $5000 of profit-sharing money at the end of each semiannual period for 7 1/2 yr. Find the final amount in the account if the deposits earn 10% compounded semiannually. Find the amount of interest earned.

62. *Business Financing* A small resort must add a swimming pool to compete with a new resort built nearby. The pool will cost $28,000. The resort borrows the money and agrees to repay it with equal payments at the end of each quarter for 6 1/2 yr at an interest rate of 12% compounded quarterly. Find the amount of each payment.

63. *Business Financing* The owner of Eastside Hallmark borrows $48,000 to expand the business. The money will be repaid in equal payments at the end of each year for 7 yr. Interest is 10%. Find the amount of each payment.

64. *Personal Finance* To buy a new computer, Mark Nguyen borrows $3250 from a friend at 9% interest compounded annually for 4 yr. Find the compound amount he must pay back at the end of the 4 yr.

65. *Effective Rate* According to a financial Web site, on October 16, 2000, Guarantee Bank of Milwaukee, Wisconsin, paid 6.90% interest, compounded quarterly, on a 1-yr CD, while Capital Crossing Bank of Boston, Massachusetts, paid 6.88% compounded monthly.[†] What are the effective rates for the two CDs, and which bank pays a higher effective rate?

66. *Home Financing* When the Lee family bought their home, they borrowed $115,700 at 10.5% compounded monthly for 25 yr. If they make all 300 payments, repaying the loan on schedule, how much interest will they pay? (Assume the last payment is the same as the previous ones.)

67. *Buying and Selling a House* The Zambrano family bought a house for $91,000. They paid $20,000 down and took out a 30-yr mortgage for the balance at 9%.

  a. Find their monthly payment.

  b. How much of the first payment is interest?

  After 180 payments, the family sells its house for $136,000. They must pay closing costs of $3700 plus 2.5% of the sale price.

  c. Estimate the current mortgage balance at the time of the sale using one of the methods from Example 4 in Section 3.

*"Pocket That Pension," *Smart Money,* Oct. 1994, p. 33.
[†]www.bankrate.com

d. Find the total closing costs.

e. Find the amount of money they receive from the sale after paying off the mortgage.

*The following exercise is from an actuarial examination.**

68. *Death Benefit* The proceeds of a $10,000 death benefit are left on deposit with an insurance company for 7 yr at an annual effective interest rate of 5%. The balance at the end of 7 yr is paid to the beneficiary in 120 equal monthly payments of *X*, with the first payment made immediately. During the payout period, interest is credited at an annual effective interest rate of 3%. Calculate *X*.

    a. 117   **b.** 118   **c.** 129   **d.** 135   **e.** 158

69. *Investment* *The New York Times* posed a scenario with two individuals, Sue and Joe, who each have $1200 a month to spend on housing and investing. Each takes out a mortgage for $140,000. Sue gets a 30-yr mortgage at a rate of 6.625%. Joe gets a 15-yr mortgage at a rate of 6.25%. Whatever money is left after the mortgage payment is invested in a mutual fund with a return of 10% annually.†

a. What annual interest rate, when compounded monthly, gives an effective annual rate of 10%?

b. What is Sue's monthly payment?

c. If Sue invests the remainder of her $1200 each month, after the payment in part b, in a mutual fund with the interest rate in part a, how much money will she have in the fund at the end of 30 yr?

d. What is Joe's monthly payment?

e. You found in part d that Joe has nothing left to invest until his mortgage is paid off. If he then invests the entire $1200 monthly in a mutual fund with the interest rate in part a, how much money will he have at the end of 30 yr (that is, after 15 yr of paying the mortgage and 15 yr of investing)?

f. Who is ahead at the end of the 30 yr, and by how much?

g. Discuss to what extent the difference found in part f is due to the different interest rates or to the different amounts of time.

---

*Problem 16 from "Course 140 Examination, Mathematics of Compound Interest" of the Education and Examination Committee of The Society of Actuaries. Reprinted by permission of The Society of Actuaries.
†*The New York Times,* Sept. 27, 1998, p. BU 10.

EXTENDED APPLICATION: Time, Money, and Polynomials*

A *time line* is often helpful for evaluating complex investments. For example, suppose you buy a $1000 CD at time t_0. After one year $2500 is added to the CD at t_1. By time t_2, after another year, your money has grown to $3851 with interest. What rate of interest, called *yield to maturity* (YTM), did your money earn? A time line for this situation is shown in Figure 14.

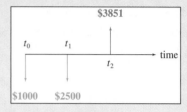

FIGURE 14

Assuming interest is compounded annually at a rate i, and using the compound interest formula, gives the following description of the YTM.

$$1000(1 + i)^2 + 2500(1 + i) = 3851$$

To determine the yield to maturity, we must solve this equation for i. Since the quantity $1 + i$ is repeated, let $x = 1 + i$ and first solve the second-degree (quadratic) polynomial equation for x.

$$1000x^2 + 2500x - 3851 = 0$$

We can use the quadratic formula with $a = 1000$, $b = 2500$, and $c = -3851$.

$$x = \frac{-2500 \pm \sqrt{2500^2 - 4(1000)(-3851)}}{2(1000)}$$

We get $x = 1.0767$ and $x = -3.5767$. Since $x = 1 + i$, the two values for i are $.0767 = 7.67\%$ and $-4.5767 = -457.67\%$. We reject the negative value because the final accumulation is greater than the sum of the deposits. In some applications, however, negative rates may be meaningful. By checking in the first equation, we see that the yield to maturity for the CD is 7.67%.

Now let us consider a more complex but realistic problem. Suppose Bill Poole has contributed for four years to a retirement fund. He contributed $6000 at the beginning of the first year. At the beginning of the next three years, he contributed $5840, $4000, and $5200, respectively. At the end of the fourth year, he had $29,912.38 in his fund. The interest rate earned by the fund varied between 21% and −3%, so Poole would like to know the YTM = i for his hard-earned retirement dollars. From a time line (see Figure 15), we set up the following equation in $1 + i$ for Poole's savings program.

$$6000(1 + i)^4 + 5840(1 + i)^3 + 4000(1 + i)^2$$
$$+ 5200(1 + i) = 29,912.38$$

Let $x = 1 + i$. We need to solve the fourth-degree polynomial equation

$$f(x) = 6000x^4 + 5840x^3 + 4000x^2 + 5200x$$
$$- 29,912.38 = 0.$$

There is no simple way to solve a fourth-degree polynomial equation, so we will use a graphing calculator.

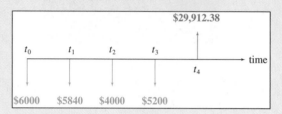

FIGURE 15

We expect that $0 < i < 1$, so that $1 < x < 2$. Let us calculate $f(1)$ and $f(2)$. If there is a change of sign, we will know that there is a solution to $f(x) = 0$ between 1 and 2. We find that

$$f(1) = -8872.38 \quad \text{and} \quad f(2) = 139,207.62.$$

Using a graphing calculator, we find that there is one positive solution to this equation, $x = 1.14$, so $i = $ YTM $= .14 = 14\%$.

Exercises

1. Brenda Bravener received $50 on her 16th birthday, and $70 on her 17th birthday, both of which she immediately invested in the bank with interest compounded annually. On her 18th birthday, she had $127.40 in her account. Draw a time line, set up a polynomial equation, and calculate the YTM.

2. At the beginning of the year. Jay Beckenstein invested $10,000 at 5% for the first year. At the beginning of the second year, he added $12,000 to the account. The total account earned 4.5% for the second year.

   a. Draw a time line for this investment.

   b. How much was in the fund at the end of the second year?

   c. Set up and solve a polynomial equation and determine the YTM. What do you notice about the YTM?

3. On January 2 each year for three years, Greg Odjakjian deposited bonuses of $1025, $2200, and $1850, respectively, in an account. He received no bonus the following year, so he made no deposit. At the end of the fourth year, there was $5864.17 in the account.

   a. Draw a time line for these investments.

---

*From *Time, Money, and Polynomials*, COMAP

b. Write a polynomial equation in x ($x = 1 + i$) and use a graphing calculator to find the YTM for these investments.

4. Pat Kelley invested yearly in a fund for his children's college education. At the beginning of the first year, he invested $1000; at the beginning of the second year, $2000; at the third through the sixth, $2500 each year, and at the beginning of the seventh, he invested $5000. At the beginning of the eighth year, there was $21,259 in the fund.

   a. Draw a time line for this investment program.

   b. Write a seventh-degree polynomial equation in $1 + i$ that gives the YTM for this investment program.

   c. Use a graphing calculator to show that the YTM is less than 5.07% and greater than 5.05%.

   d. Use a graphing calculator to calculate the solution for $1 + i$ and find the YTM.

5. People often lose money on investments. Jim Carlson invested $50 at the beginning of each of two years in a mutual fund, and at the end of two years his investment was worth $90.

   a. Draw a time line and set up a polynomial equation in $1 + i$. Solve for i.

   b. Examine each negative solution (rate of return on the investment) to see if it has a reasonable interpretation in the context of the problem. To do this, use the compound interest formula on each value of i to trace each $50 payment to maturity.

6

Logic

The rules of a game often include complex conditional statements, such as "if you roll doubles, you can roll again, but if you roll doubles twice in a row, you lose a turn." As exercises in this chapter illustrate, logical analysis of complex statements helps us clarify not only the rules of games but any precise use of language, from legal codes to medical diagnoses.

Modern digital computers are exceedingly complex machines but their design involves only two basic components: gates and memory cells. A gate is a simple electrical switch that controls the flow of data much like a floodgate controls the flow of water in a canal. A memory cell can store a single piece of data. Working together, gates and memory cells perform operations such as addition and subtraction. In 1960 only a few gates could be reliably printed onto a silicon chip. Today, over several million logic gates can be printed onto a single chip. Some experts believe that a silicon chip containing over a billion components will soon be achieved.

What is the mathematics that makes these logic gates possible? You will learn the answer to this question in this chapter.

■ 6.1 STATEMENTS AND QUANTIFIERS

THINK ABOUT IT

What message is a store trying to convey when its advertisement states, "All items are not available in all stores," as a disclaimer for a 30 percent off sale?

You will be asked to answer this question in one of the exercises at the end of this section.

This section introduces the study of *symbolic logic,* which uses letters to represent statements, and symbols for words such as *and, or, not.* One of the main applications of logic is in the study of the *truth value* (that is, the truth or falsity) of statements with many parts. The truth value of these statements depends on the components that comprise them.

Many kinds of sentences occur in ordinary language, including factual statements, opinions, commands, and questions. Symbolic logic discusses only the first type of sentence, the kind that involves facts.

Statements A **statement** is defined as a declarative sentence that is either true or false, but not both simultaneously. For example, both of the following are statements:

> Electronic mail provides a means of communication.

$$11 + 6 = 12.$$

Each one is either true or false. However, based on this definition, the following sentences are not statements:

> Access the file.
>
> Is this a great time, or what?
>
> This sentence is false.

These sentences cannot be identified as being either true or false. The first sentence is a command, and the second is a question. "This sentence is false" is a paradox; if we assume it is true, then it is false, and if we assume it is false, then it is true. Paradoxes such as this can be avoided if we do not allow statements that refer to themselves.

A **compound statement** may be formed by combining two or more statements. The statements making up a compound statement are called **component statements.** Various **logical connectives,** or simply **connectives,** can be used in forming compound statements. Words such as *and, or, not,* and *if...then* are examples of connectives. (While a statement such as "Today is not Tuesday" does not consist of two component statements, for convenience it is considered compound, since its truth value is determined by noting the truth value of a different statement, "Today is Tuesday.")

EXAMPLE 1 Compound Statements

Decide whether each statement is compound.

(a) Shakespeare wrote sonnets and the poem exhibits iambic pentameter.

Solution This statement is compound, since it is made up of the component statements "Shakespeare wrote sonnets" and "the poem exhibits iambic pentameter." The connective is *and.*

(b) You can pay me now or you can pay me later.

Solution The connective here is *or.* The statement is compound.

(c) If he said it, then it must be true.

Solution The connective here is *if...then*, discussed in more detail in a later section. The statement is compound.

(d) My pistol was made by Smith and Wesson.

Solution While the word "and" is used in this statement, it is not used as a *logical* connective, since it is part of the name of the manufacturer. The statement is not compound.

Negations The sentence "Tom Jones has a red car" is a statement; the **negation** of this statement is "Tom Jones does not have a red car." The negation of a true statement is false, and the negation of a false statement is true.

EXAMPLE 2 Negation

Form the negation of each statement.

(a) That state has a governor.

Solution To negate this statement, we introduce *not* into the sentence: "That state does not have a governor."

(b) The sun is not a star.

Solution Negation: "The sun is a star."

One way to detect incorrect negations is to check truth values. A negation must have the opposite truth value from the original statement.

EXAMPLE 3   Negation

Give a negation of each inequality.

(a) $p < 9$

> **Solution** The negation of "p is less than 9" is "p is *not* less than 9," which means "p is greater than or equal to 9," or $p \geq 9$.

(b) $7x + 11y \geq 77$

> **Solution** The negation is $7x + 11y < 77$.

FOR REVIEW ■

Recall the following inequalities from Chapter R.

Symbolism	Meaning
$a < b$	a is less than b
$a > b$	a is greater than b
$a \leq b$	a is less than or equal to b
$a \geq b$	a is greater than or equal to b

Some calculators have logic functions that allow the user to test the truth or falsity of statements involving $=$, $\neq$, $>$, $\geq$, $<$, and $\leq$. For example, functions from the TEST menu of a TI-83 calculator, illustrated in Figure 1, return a 1 when a statement is true and 0 when a statement is false. Figure 2 shows the input of $4 > 9$ and the corresponding output of 0, indicating that the statement is false. Verify that the statement $5 > 3$ has a corresponding output of 1 on a TI-83 calculator.

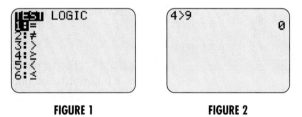

FIGURE 1   **FIGURE 2**

Symbols   To simplify work with logic, symbols are used. Statements are represented with letters, such as p, q, or r, while several symbols for connectives are shown in the following table. The table also gives the type of compound statement having the given connective.

Connective	Symbol	Type of Statement
and	$\wedge$	Conjunction
or	$\vee$	Disjunction
not	$\sim$	Negation

The symbol $\sim$ represents the connective *not*. If p represents the statement "Jimmy Carter was president in 1970" then $\sim p$ represents "Jimmy Carter was not president in 1970."

EXAMPLE 4   Symbolic Statements

Let p represent "It is 80° today," and let q represent "It is Tuesday." Write each symbolic statement in words.

(a) $p \vee q$

Solution From the table, $\vee$ symbolizes *or*; thus, $p \vee q$ represents

It is 80° today or it is Tuesday.

(b) $\sim p \wedge q$

Solution

It is not 80° today and it is Tuesday.

(c) $\sim(p \vee q)$

Solution

It is not the case that it is 80° today or it is Tuesday.

(d) $\sim(p \wedge q)$

Solution

It is not the case that it is 80° today and it is Tuesday.

The statement in part (c) of Example 4 usually is translated in English as "Neither p nor q," as in "It is neither 80° today nor is it Tuesday."

Quantifiers The words *all, each, every,* and *no(ne)* are called **universal quantifiers,** while words and phrases such as *some, there exists,* and *(for) at least one* are called **existential quantifiers.** Quantifiers are used extensively in mathematics to indicate *how many* cases of a particular situation exist. Be careful when forming the negation of a statement involving quantifiers.

The negation of a statement must be false if the given statement is true and must be true if the given statement is false, in all possible cases. Consider the statement

All girls in the group are named Mary.

Many people would write the negation of this statement as "No girls in the group are named Mary" or "All girls in the group are not named Mary." But this would not be correct. To see why, look at the three groups below.

Group I: Mary Jones, Mary Smith, Mary Jackson

Group II: Mary Johnson, Betty Parker, Margaret Westmoreland

Group III: Joycelyn Lowe, Fran Liberto, Cheryl Joslyn

These groups contain all possibilities that need to be considered. In Group I, *all* girls are named Mary; in Group II, *some* girls are named Mary (and some are not); in Group III, *no* girls are named Mary. Look at the truth values in the chart on the next page and keep in mind that "some" means "at least one (and possibly all)."

The negation of the given statement (1) must have opposite truth values in *all* cases. It can be seen that statements (2) and (3) do not satisfy this condition (for Group II), but statement (4) does. It may be concluded that the correct negation

Truth Value

	Group I	Group II	Group III
(1) All girls in the group are named Mary. (Given)	T	F	F
(2) No girls in the group are named Mary. (Possible negation)	F	F	T
(3) All girls in the group are not named Mary. (Possible negation)	F	F	T
(4) Some girls in the group are not named Mary. (Possible negation)	F	T	T

Negation

for "All girls in the group are named Mary" is "Some girls in the group are not named Mary." Other ways of stating the negation are:

Not all girls in the group are named Mary.

It is not the case that all girls in the group are named Mary.

At least one girl in the group is not named Mary.

The following table can be used to generalize the method of finding the negation of a statement involving quantifiers.

NEGATIONS OF QUANTIFIED STATEMENTS

Statement	Negation
All do.	Some do not. (Equivalently: Not all do.)
Some do.	None do. (Equivalently: All do not.)

The negation of the negation of a statement is simply the statement itself. For instance, the negations of the statements in the Negation column are simply the corresponding original statements in the Statement column. As an example, the negation of "Some do not" is "All do." Symbolically, $\sim(\sim p)$ is equivalent to p.

EXAMPLE 5 Negation

Write the negation of each statement.

(a) Some cats have fleas.

Solution Since *some* means "at least one," the statement "Some cats have fleas" is really the same as "At least one cat has fleas." The negation of this is "No cat has fleas."

(b) Some cats do not have fleas.

Solution This statement claims that at least one cat, somewhere, does not have fleas. The negation of this is "All cats have fleas."

(c) No cats have fleas.

Solution The negation is "Some cats have fleas."

For the following example we need to remind you of the definitions of some special types of numbers.

DEFINITIONS OF REAL NUMBERS

Natural or *counting numbers* are the numbers 1, 2, 3, 4,...

Whole numbers are the numbers 0, 1, 2, 3, 4,...

Integers are the numbers ..., $-3, -2, -1, 0, 1, 2, 3,...$

Rational numbers are quotients of integers, like $3/5$, $-7/9$. Any rational number can be written as a terminating decimal number, like .25, or a repeating decimal number, like .333...

Irrational numbers include all numbers that cannot be written as a quotient of integers. Some examples of irrational numbers are $\sqrt{2}$, $\sqrt[3]{4}$, and π.

Real numbers include all rational and irrational numbers.

EXAMPLE 6 Real Number Statements

Decide whether each of the following statements about numbers is *true* or *false*.

(a) There exists a whole number that is not a natural number.

Solution Because there is such a whole number (it is 0), this statement is true.

(b) Every integer is a natural number.

Solution This statement is false, because we can find at least one integer that is not a natural number. For example, -1 is an integer but is not a natural number. (There are infinitely many other choices we could have made.)

(c) Every natural number is a rational number.

Solution Since every natural number can be written as a fraction with denominator 1, this statement is true.

(d) There exists an irrational number that is not real.

Solution The real numbers include all irrational numbers. Therefore, since we cannot give an irrational number that is not real, this statement is false. (Had we been able to find at least one, the statement would have then been true.)

6.1 EXERCISES

Decide whether each of the following is a statement or is not a statement.

1. December 7, 1941, was a Sunday.

2. The ZIP code for Manistee, MI, is 49660.

3. Listen, my children, and you shall hear of the midnight ride of Paul Revere.

4. Yield to oncoming traffic.

5. $5 + 8 = 13$ and $4 - 3 = 1$

6. $5 + 8 = 12$ or $4 - 3 = 2$

7. Some numbers are negative.

8. Andrew Johnson was president of the United States in 1867.

9. Accidents are the main cause of deaths of children under the age of 8.

10. *There's Something About Mary* was the top-grossing movie of 1998.

11. Where are you going today?

12. Behave yourself and sit down.

13. Kevin "Catfish" McCarthy once took a prolonged continuous shower for 340 hours, 40 minutes.

14. One gallon of milk weighs more than 4 pounds.

Decide whether each of the following statements is compound.

15. I read the *Chicago Tribune* and I read the *New York Times*.

16. My brother got married in London.

17. Tomorrow is not Sunday.

18. Ernie St. Hilaire is younger than 29 years of age, and so is Bart Stewart.

19. Jay Beckenstein's wife loves Ben and Jerry's ice cream.

20. The sign on the back of the car read "California or bust!"

21. If Julie Ward sells her quota, then Bill Leonard will be happy.

22. If Mike is a politician, then Jerry is a crook.

Write a negation for each of the following statements.

23. Her aunt's name is Lucia.

24. The flowers are to be watered.

25. Every dog has its day.

26. No rain fell in southern California today.

27. Some books are longer than this book.

28. All students present will get another chance.

29. No computer repairman can play blackjack.

30. Some people have all the luck.

31. Everybody loves somebody sometime.

32. Everyone loves a winner.

Give a negation of each inequality.

33. $y > 12$

34. $x < -6$

35. $q \geq 5$

36. $r \leq 19$

37. Try to negate the sentence "The exact number of words in this sentence is ten" and see what happens. Explain the problem that arises.

38. Explain why the negation of "$r > 4$" is not "$r < 4$."

Let p represent the statement "She has green eyes" and let q represent the statement "He is 48 years old." Translate each symbolic compound statement into words.

39. $\sim p$

40. $\sim q$

41. $p \wedge q$

42. $p \vee q$

43. $\sim p \vee q$

44. $p \wedge \sim q$

45. $\sim p \vee \sim q$

46. $\sim p \wedge \sim q$

47. $\sim(\sim p \wedge q)$

48. $\sim(p \vee \sim q)$

Decide whether each statement is true *or* false.

49. Every whole number is an integer.

50. Every natural number is an integer.

51. There exists a rational number that is not an integer.

52. There exists an integer that is not a natural number.

53. All rational numbers are real numbers.

54. All real numbers are irrational numbers.

55. Some rational numbers are not integers.

56. Some whole numbers are not rational numbers.

57. Each whole number is a positive number.

58. Each rational number is a positive number.

59. The statement "For some real number x, $x^2 \geq 0$" is true. However, your friend does not understand why, since he claims that $x^2 \geq 0$ for *all* real numbers x (and not *some*). How would you explain his misconception to him?

60. Only one of the following statements is true. Which one is it?

   a. For some real number x, $x \geq 0$.

   b. For all real numbers x, $x^3 > 0$.

   c. For all real numbers x less than 0, x^2 is also less than 0.

   d. For some real number x, $x^2 < 0$.

61. Use the TEST menu on a graphing calculator to determine whether each statement is true or false.

   a. $\left(3\sqrt{7} - 4\sqrt{6}\right) > .5$

   b. $\dfrac{28 - 15\sqrt{2.5}}{2 - \sqrt{4.5}} > \dfrac{-\sqrt{801}}{3}$

   c. $\dfrac{\dfrac{5}{\sqrt{3}} - \sqrt{5}}{\sqrt{10} - \sqrt{5} - \sqrt{2}} < \dfrac{3}{\sqrt{2}}$

◼ Applications

BUSINESS AND ECONOMICS

62. *Advertising* Incorrect use of quantifiers often is heard in everyday language. Suppose you hear that a local electronics chain is having a 30% off sale, and the radio advertisement states, "All items are not available in all stores." Do you think that, literally translated, the ad really means what it says? What do you think really is meant? Explain your answer.

63. *Portfolios* Repeat Exercise 62 for the following: "All people don't have the time to devote to maintaining their financial portfolios properly."

Credit Cards *The following excerpts appear in a J. C. Penney credit card application.*

(a) *The Ohio laws against discrimination require that all creditors make credit equally available to all credit-* *worthy customers, and that credit reporting agencies maintain separate credit histories on each individual upon request.*

(b) *Please provide this information to us on or with this application.*

(c) *The information about the costs of the card described above is accurate as of August 2000.*

(d) *Credit card insurance will vary by state.*

64. Which of these excerpts are statements?

65. Which of these excerpts are compound statements?

66. Write the negation of excerpt d.

Income Tax The following excerpts appear in a guide for preparing income tax reports. *

(a) *You can either increase the amount of income tax withheld from your pay or make estimated tax payments to meet your 2001 tax liability.*

(b) *You can get a four-month extension by filing Form 4868 by April 16, 2001.*

(c) *Keep the dividend statements and other literature accompanying your distributions.*

(d) *Life insurance payments are not subject to income tax.*

67. Which of these excerpts are statements?

68. Which of these excerpts are compound statements?

69. Write the negation of excerpt d.

70. Define *p* and *q* to symbolically represent excerpt a.

LIFE SCIENCES

Medicine The following excerpts appear in a home medical reference book.†

(a) *Can you climb one or two flights of stairs without shortness of breath or heaviness or fatigue in your legs?*

(b) *Regularly doing exercises that concentrate on strengthening particular muscle groups and improving overall flexibility can help prevent back pain and keep you mobile.*

(c) *If you answered yes to all of the questions above, you are reasonably fit.*

(d) *Bone strength and bone density are improved by doing regular weight-bearing exercises, such as playing tennis and walking.*

(e) *A number of viral illnesses can cause swollen glands.*

71. Which of these excerpts are statements?

72. Which of these excerpts are compound statements?

73. Write the negation of excerpt e.

SOCIAL SCIENCES

Law The following excerpts appear in a guide to common laws.‡

(a) *If a mortgage lender denies you a loan or a seller refuses to sell to you and you believe that you're being discriminated against, call the Department of Housing and Urban Development to file a formal complaint.*

(b) *Don't be pressured into signing a contract.*

(c) *The court won't do it for you, and hiring an attorney is usually not cost-effective given the small amount of money involved.*

(d) *You can't marry unless you're at least 18 years old or unless you have the permission of your parents or guardian.*

(e) *Most legal problems are matters of civil law.*

74. Which of these excerpts are statements?

75. Which of these excerpts are compound statements?

76. Write the negation of excerpt e.

77. *Philosophy* Read each of the following quotes from ancient philosophers.§ Provide an argument why these quotes may or may not be called statements.

a. "A friend is a friend of someone."—Socrates

b. "Every art, and every science reduced to a teachable form, and in like manner every action and moral choice, aims, it is thought, at some good: for which reason a common and by no means a bad description of what the Chief Good is, 'that which all things aim at.'"—Aristotle

c. "Furthermore, Friendship helps the young to keep from error: the old, in respect of attention and such deficiencies in action as their weakness makes them liable to; and those who are in their prime, in respect of noble deeds, because they are thus more able to devise plans and carry them out."—Aristotle

78. *Bible* Read each of the following quotes from the biblical book, Proverbs.‖ Provide an argument why these quotes may or may not be called statements.

a. "A soft answer turneth away wrath."—Proverbs 15:1

b. "A wrathful man stirreth up strife: but he that is slow to anger appeaseth strife."—Proverbs 15:18

c. "It is better to dwell in a corner of the housetop, than with a brawling woman in a wide house."—Proverbs 21:9

*Your Income Tax 2001, J. K. Lasser Institute, John Wiley and Sons, Inc., 2000, p. xxix.
†Goldman, D. R., ed., American College of Physicians Complete Home Medical Guide, New York, DK Publishing, Inc., 1999, p. 57.
‡Ventura, John, Law for Dummies, Foster City, CA, IDG Books Worldwide, 1996.
§Frost, S. E., ed., Masterworks of Philosophy, New York, McGraw-Hill, 1946.
‖Bartlett, John, Bartlett's Familiar Quotations, 14th edition, Boston, Little, Brown and Company, 1968.

d. "A good name is rather to be chosen than great riches."—Proverbs 22:1

e. "Train up a child in the way he should go: and when he is old, he will not depart from it."—Proverbs 22:6

GENERAL INTEREST

For Exercises 79–84, let p represent the statement "Chris collects videotapes" *and let q represent the statement* "Jack plays the tuba." *Convert each of the following compound statements into symbols.*

79. Chris collects videotapes and Jack does not play the tuba.

80. Chris does not collect videotapes or Jack does not play the tuba.

81. Chris does not collect videotapes or Jack plays the tuba.

82. Jack plays the tuba and Chris does not collect videotapes.

83. Neither Chris collects videotapes nor Jack plays the tuba.

84. Either Jack plays the tuba or Chris collects videotapes, and it is not the case that both Jack plays the tuba and Chris collects videotapes.

 85. Explain the difference between the following statements:

All students did not pass the test.
Not all students passed the test.

86. Write the following statement using *every:* There is no one here who has not done that at one time or another.

Refer to the groups of art labeled A, B, and C, and identify by letter the group or groups that are satisfied by the given statements involving quantifiers.

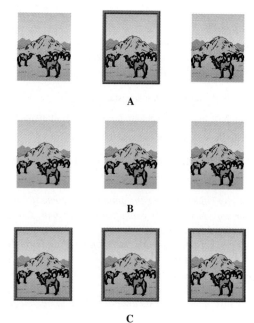

A

B

C

87. All pictures have frames.

88. No picture has a frame.

89. At least one picture does not have a frame.

90. Not every picture has a frame.

91. At least one picture has a frame.

92. No picture does not have a frame.

93. All pictures do not have frames.

94. Not every picture does not have a frame.

■ 6.2 TRUTH TABLES AND EQUIVALENT STATEMENTS

? THINK ABOUT IT When using a search engine on the Internet, you are asked to supply key words. How does the search engine connect these key words?

You will be asked to explore this question in one of the exercises at the end of this section.

In this section, the truth values of component statements are used to find the truth values of compound statements.

Conjunctions To begin, let us decide on the truth values of the **conjunction** *p and q*, symbolized $p \wedge q$. In everyday language, the connective *and* implies the idea of "both." The statement

Monday immediately follows Sunday and
March immediately follows February

is true, since each component statement is true. On the other hand, the statement

Monday immediately follows Sunday and
March immediately follows January

is false, even though part of the statement (Monday immediately follows Sunday) is true. For the conjunction $p \wedge q$ to be true, both p and q must be true. This result is summarized by a table, called a **truth table,** which shows all four of the possible combinations of truth values for the conjunction p *and* q. The truth table for *conjunction* is shown here.

TRUTH TABLE FOR THE CONJUNCTION p **and** q

p and q

p	q	$p \wedge q$
T	T	T
T	F	F
F	T	F
F	F	F

EXAMPLE 1 Truth Values

Let p represent "$5 > 3$" and let q represent "$6 < 0$." Find the truth value of $p \wedge q$.

Method 1: Truth Tables

Solution Here p is true and q is false. Looking in the second row of the conjunction truth table shows that $p \wedge q$ is false.

Method 2: Graphing Calculator

Solution This example can also be completed with the help of a TI-83 calculator. Using the LOGIC menu, illustrated in Figure 3, we input $5 > 3$ and $6 < 0$ into the calculator. The output is zero, shown in Figure 4, which means the statement is false.

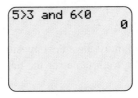

FIGURE 3

```
5>3 and 6<0
                    0
```

FIGURE 4

In some cases, the logical connective *but* is used in compound statements. For example, consider the statement

He wants to go to the mountains but she wants to go to the beach.

Here, *but* is used in place of *and* to give a different sort of emphasis to the statement. In such a case, we consider the statement as we would consider the conjunction using the word *and*. The truth table for the conjunction, given above, would apply.

Disjunctions In ordinary language, the word *or* can be ambiguous. The expression "this or that" can mean either "this or that or both," or "this or that but not both." For example, the statement

Those with a passport or driver's license will be admitted

probably has the following meaning: "Those with a passport will be admitted or those with a driver's license will be admitted or those with both a passport and a driver's license will be admitted." On the other hand, the statement

I will drive the Ford or the Nissan to the store

probably means "I will drive the Ford, or I will drive the Nissan, but I will not drive both."

The symbol $\vee$ represents the first *or* described. That is, $p \vee q$ means "*p* or *q* or both." With this meaning of *or*, $p \vee q$ is called the *inclusive disjunction,* or just the **disjunction** of *p* and *q*.

▌ **NOTE** In English, the *or* is usually interpreted as exclusive or, not the way mathematicians use it.

A disjunction is false only if both component statements are false. The truth table for *disjunction* follows.

TRUTH TABLE FOR THE DISJUNCTION *p* or *q*

p or q

p	*q*	*p* $\vee$ *q*
T	T	T
T	F	T
F	T	T
F	F	F

The inequality symbols $\leq$ and $\geq$ are examples of inclusive disjunction. For example, $x \leq 6$ is true if either $x < 6$ or $x = 6$.

EXAMPLE 2 True Statements
The following list shows several statements and the reason that each is true.

Statement	Reason That It Is True
$8 \geq 8$	$8 = 8$
$3 \geq 1$	$3 > 1$
$-5 \leq -3$	$-5 < -3$
$-4 \leq -4$	$-4 = -4$

The **negation** of a statement *p*, symbolized $\sim p$, must have the opposite truth value from the statement *p* itself. This leads to the truth table for the negation, shown here.

TRUTH TABLE FOR THE NEGATION not *p*

not p

p	$\sim p$
T	F
F	T

EXAMPLE 3 Truth Value of Compound Statements
Suppose *p* is false, *q* is true, and *r* is false. What is the truth value of the compound statement $\sim p \wedge (q \vee \sim r)$?

Solution Here parentheses are used to group q and $\sim r$ together. Work first inside the parentheses. Since r is false, $\sim r$ will be true. Since $\sim r$ is true and q is true and an *or* statement is true when either component is true, $q \vee \sim r$ must be true. An *and* statement is only true when both components are true. Since $\sim p$ is true and $q \vee \sim r$ is true, the statement $\sim p \wedge (q \vee \sim r)$ is true.

The preceding paragraph may be interpreted using a short-cut symbolic method, letting T represent a true statement and F represent a false statement:

$$\sim p \wedge (q \vee \sim r)$$
$$\sim F \wedge (T \vee \sim F)$$
$$T \wedge (T \vee T) \qquad \text{\small $\sim$F gives T.}$$
$$T \wedge T \qquad\qquad \text{\small T $\vee$ T gives T.}$$
$$T. \qquad\qquad\quad \text{\small T $\wedge$ T gives T.}$$

The T in the final row indicates that the compound statement is true.

Mathematical Statements   The next two examples show the use of truth tables to determine the truth values of mathematical statements.

EXAMPLE 4   Mathematical Statements

Let p represent the statement $3 > 2$, q represent $5 < 4$, and r represent $3 < 8$. Decide whether the following statements are *true* or *false*.

(a) $\sim p \wedge \sim q$

Solution Since p is true, $\sim p$ is false. By the *and* truth table, if one part of an "and" statement is false, the entire statement is false. This makes $\sim p \wedge \sim q$ false.

(b) $\sim (p \wedge q)$

Solution First, work within the parentheses. Since p is true and q is false, $p \wedge q$ is false by the *and* truth table. Next, apply the negation. The negation of a false statement is true, making $\sim (p \wedge q)$ a true statement.

Figure 5 shows the result of using a TI-83 calculator to solve parts a and b of this example.

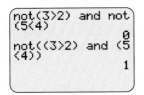

FIGURE 5

(c) $(\sim p \wedge r) \vee (\sim q \wedge \sim p)$

Solution Here p is true, q is false, and r is true. This makes $\sim p$ false and $\sim q$ true. By the *and* truth table, the statement $\sim p \wedge r$ is false, and the statement $\sim q \wedge \sim p$ is also false. Finally,

$$(\sim p \wedge r) \vee (\sim q \wedge \sim p)$$
$$\downarrow \qquad\qquad \downarrow$$
$$F \quad \vee \quad F,$$

which is false by the *or* truth table. (For an alternate solution, see Example 8(b).)

When a quantifier is used with a conjunction or a disjunction, we must be careful in determining the truth value, as shown in the following example.

EXAMPLE 5 Truth Values

Identify each statement as *true* or *false*.

(a) For some real number x, $x < 5$ and $x > 2$.

Solution Replacing x with 3 (as an example) gives $3 < 5$ and $3 > 2$. Since both $3 < 5$ and $3 > 2$ are true statements, the given statement is true by the *and* truth table. (Remember: *some* means "at least one.")

(b) For every real number b, $b > 0$ or $b < 1$.

Solution No matter which real number might be tried as a replacement for b, at least one of the statements $b > 0$ and $b < 1$ will be true. Since an *or* statement is true if one or both component statements are true, the entire statement as given is true.

(c) For all real numbers, x, $x^2 > 0$.

Solution Since the quantifier is a universal quantifier, we need only find one case in which the inequality is false to make the entire statement false. Can we find a real number whose square is not positive (that is, not greater than 0)? Yes, we can—0 itself is a real number (and the *only* real number) whose square is not positive. Therefore, this statement is false.

Truth Tables In the preceding examples, the truth value for a given statement was found by going back to the basic truth tables. In the long run, it is easier to first create a complete truth table for the given statement itself. Then final truth values can be read directly from this table. The procedure for making new truth tables is shown in the next few examples.

In this book we will use the following standard format for listing the possible truth values in compound statements involving two statements.

p	q	**Compound Statement**
T	T	
T	F	
F	T	
F	F	

EXAMPLE 6 Truth Tables

(a) Construct a truth table for $(\sim p \wedge q) \vee \sim q$.

Solution Begin by listing all possible combinations of truth values for p and q, as above. Then find the truth values of $\sim p \wedge q$. Start by listing the truth values of $\sim p$, which are the opposite of those of p.

p	q	~p
T	T	F
T	F	F
F	T	T
F	F	T

Use only the "~p" column and the "q" column, along with the *and* truth table, to find the truth values of ~p ∧ q. List them in a separate column.

p	q	~p	~p ∧ q
T	T	F	F
T	F	F	F
F	T	T	T
F	F	T	F

Next include a column for ~q.

p	q	~p	~p ∧ q	~q
T	T	F	F	F
T	F	F	F	T
F	T	T	T	F
F	F	T	F	T

Finally, make a column for the entire compound statement. To find the truth values, use *or* to combine ~p ∧ q with ~q.

p	q	~p	~p ∧ q	~q	(~p ∧ q) ∨ ~q
T	T	F	F	F	F
T	F	F	F	T	T
F	T	T	T	F	T
F	F	T	F	T	T

(b) Suppose both *p* and *q* are true. Find the truth value of (~p ∧ q) ∨ ~q.

Solution Look in the first row of the final truth table above, where both *p* and *q* have truth value T. Read across the row to find that the compound statement is false.

EXAMPLE 7 Truth Tables
Construct the truth table for p ∧ (~p ∨ ~q).

Solution Proceed as shown.

p	q	$\sim p$	$\sim q$	$\sim p \vee \sim q$	$p \wedge (\sim p \vee \sim q)$
T	T	F	F	F	F
T	F	F	T	T	T
F	T	T	F	T	F
F	F	T	T	T	F

If a compound statement involves three component statements p, q, and r, we will use the following format in setting up the truth table.

p	q	r	**Compound Statement**
T	T	T	
T	T	F	
T	F	T	
T	F	F	
F	T	T	
F	T	F	
F	F	T	
F	F	F	

EXAMPLE 8   Truth Tables

(a) Construct a truth table for $(\sim p \wedge r) \vee (\sim q \wedge \sim p)$.

Solution This statement has three component statements, p, q, and r. The truth table thus requires eight rows to list all possible combinations of truth values of p, q, and r. The final truth table, however, can be found in much the same way as the ones above.

p	q	r	$\sim p$	$\sim p \wedge r$	$\sim q$	$\sim q \wedge \sim p$	$(\sim p \wedge r) \vee (\sim q \wedge \sim p)$
T	T	T	F	F	F	F	F
T	T	F	F	F	F	F	F
T	F	T	F	F	T	F	F
T	F	F	F	F	T	F	F
F	T	T	T	T	F	F	T
F	T	F	T	F	F	F	F
F	F	T	T	T	T	T	T
F	F	F	T	F	T	T	T

(b) Suppose p is true, q is false, and r is true. Find the truth value of $(\sim p \wedge r) \vee (\sim q \wedge \sim p)$.

Solution By the third row of the truth table in part (a), the compound statement is false. (This is an alternate method for working part (c) of Example 4.)

NOTE One strategy for problem solving is noticing a pattern and using *inductive reasoning,* or reasoning that uses particular facts to find a general rule. This strategy is used in the next example.

EXAMPLE 9 Counting

If n is a counting number, and a logical statement is composed of n component statements, how many rows will appear in the truth table for the compound statement?

Solution To answer this question, let us examine some of the earlier truth tables in this section. The truth table for the negation has one statement and two rows. The truth tables for the conjunction and the disjunction have two component statements, and each has four rows. The truth table in Example 8(a) has three component statements and eight rows. Summarizing these in a table shows a pattern seen earlier.

Number of Statements	Number of Rows
1	$2 = 2^1$
2	$4 = 2^2$
3	$8 = 2^3$

Inductive reasoning leads us to the conjecture that, if a logical statement is composed of n component statements, it will have 2^n rows. This can be proved using ideas in Chapter 8.

A logical statement having n component statements will have 2^n rows in its truth table.

Alternative Method for Constructing Truth Tables After making a reasonable number of truth tables, some people prefer the shortcut method shown in Example 10, which repeats Examples 6 and 8.

EXAMPLE 10 Truth Tables

Construct the truth table for each statement.

(a) $(\sim p \wedge q) \vee \sim q$

Solution Start by inserting truth values for $\sim p$ and for q.

p	q	$(\sim p \wedge q) \vee \sim q$	
T	T	F	T
T	F	F	F
F	T	T	T
F	F	T	F

Next, use the *and* truth table to obtain the truth values of $\sim p \wedge q$.

p	q	$(\sim p$	$\wedge$	$q)$	$\vee$	$\sim q$
T	T	F	F	T		
T	F	F	F	F		
F	T	T	T	T		
F	F	T	F	F		

Now disregard the two preliminary columns of truth values for $\sim p$ and for q, and insert truth values for $\sim q$.

p	q	$(\sim p \wedge q)$	$\vee$	$\sim q$
T	T	F		F
T	F	F		T
F	T	T		F
F	F	F		T

Finally, use the *or* truth table.

p	q	$(\sim p \wedge q)$	$\vee$	$\sim q$
T	T	F	F	F
T	F	F	T	T
F	T	T	T	F
F	F	F	T	T

These steps can be summarized as follows.

p	q	$(\sim p$	$\wedge$	$q)$	$\vee$	$\sim q$
T	T	F	F	T	F	F
T	F	F	F	F	T	T
F	T	T	T	T	T	F
F	F	T	F	F	T	T
		①	②	①	④	③

The circled numbers indicate the order in which the various columns of the truth table were found.

(b) $(\sim p \wedge r) \vee (\sim q \wedge \sim p)$

Solution Work as follows.

p	q	r	$(\sim p$	$\wedge$	$r)$	$\vee$	$(\sim q$	$\wedge$	$\sim p)$
T	T	T	F	F	T	F	F	F	F
T	T	F	F	F	F	F	F	F	F
T	F	T	F	F	T	F	T	F	F
T	F	F	F	F	F	F	T	F	F
F	T	T	T	T	T	T	F	F	T
F	T	F	T	F	F	F	F	F	T
F	F	T	T	T	T	T	T	T	T
F	F	F	T	F	F	T	T	T	T
			①	②	①	⑤	③	④	③

Equivalent Statements   One application of truth tables is illustrated by show-ing that two statements are equivalent; by definition, two statements are **equiva-lent** if they have the same truth value in *every* possible situation. The columns of each truth table that were the last to be completed will be exactly the same for equivalent statements.

EXAMPLE 11   Equivalent Statements
Are the statements

$$\sim p \wedge \sim q \quad \text{and} \quad \sim(p \vee q)$$

equivalent?

Solution   To find out, make a truth table for each statement, with the follow-ing results.

p	q	$\sim p \wedge \sim q$
T	T	F
T	F	F
F	T	F
F	F	T

p	q	$\sim(p \vee q)$
T	T	F
T	F	F
F	T	F
F	F	T

Since the truth values are the same in all cases, as shown in the columns in color, the statements $\sim p \wedge \sim q$ and $\sim(p \vee q)$ are equivalent. Equivalence is written with a three-bar symbol, $\equiv$. Using this symbol, $\sim p \wedge \sim q \equiv \sim(p \vee q)$.

In the same way, the statements $\sim p \vee \sim q$ and $\sim(p \wedge q)$ are equivalent. We call these equivalences De Morgan's laws.

De MORGAN'S LAWS
For any statements p and q,

$$\sim(p \vee q) \equiv \sim p \wedge \sim q$$
$$\sim(p \wedge q) \equiv \sim p \vee \sim q.$$

66. Victoria Montoya tried to sell the book, but she was unable to do so.

67. $5 - 1 = 4$ and $9 + 12 \neq 7$

68. $3 < 10$ or $7 \neq 2$

69. Cupid or Vixen will lead Santa's sleigh next Christmas.

70. The lawyer and the client appeared in court.

Identify each of the following statements as true *or* false.

71. For every real number y, $y < 13$ or $y > 6$.

72. For every real number t, $t > 9$ or $t < 9$.

73. For some integer p, $p \geq 4$ and $p \leq 4$.

74. There exists an integer n such that $n > 0$ and $n < 0$.

75. Complete the truth table for *exclusive disjunction*. The symbol $\underline{\vee}$ represents "one or the other is true, but not both."

p	q	$p \underline{\vee} q$
T	T	
T	F	
F	T	
F	F	

Exclusive disjunction

Decide whether the following compound statements are true *or* false. *Remember from Exercise 75 that* $\underline{\vee}$ *is the exclusive disjunction; that is, assume "either p or q is true, but not both."*

76. $3 + 1 = 4 \underline{\vee} 2 + 5 = 7$

77. $3 + 1 = 4 \underline{\vee} 2 + 5 = 9$

78. $3 + 1 = 7 \underline{\vee} 2 + 5 = 7$

79. Let p represent $2\sqrt{6} - 4\sqrt{5} > -1$, q represent $\dfrac{14 - 7\sqrt{8}}{2.5 - \sqrt{5}} > -22$, and s represent $\dfrac{7 - \dfrac{5}{\sqrt{3}}}{\sqrt{8} - 2} < \dfrac{\sqrt{3}}{\sqrt{2}}$. Use the LOGIC menu on a graphing calculator to find the truth value of each of the following statements.

    a. $p \wedge q$ **b.** $\sim p \wedge q$ **c.** $\sim(p \vee q)$ **d.** $(s \wedge \sim p) \vee (\sim s \wedge q)$

Applications

BUSINESS AND ECONOMICS

80. *Credit Cards* The following statement appears in a J.C. Penney credit card application. Use one of De Morgan's laws to write the negation of this statement.

> The Ohio laws against discrimination require that all creditors make credit equally available to all credit-worthy customers, and that credit reporting agencies maintain separate credit histories on each individual upon request.

81. *Income Tax* The following statement appears in a guide for preparing income tax reports.* Use one of De Morgan's laws to write the negation of this statement.

> You can either increase the amount of income tax withheld from your pay or make estimated tax payments to meet your 2001 tax liability.

*Your Income Tax 2001, J. K. Lasser Institute, John Wiley and Sons, Inc., 2000, p. xxix.

82. *Warranty* The statement below appears in a tire manufacturer's warranty.* Describe the use of *and/or* in this situation.

> This warranty is in addition to and/or may be limited by any other applicable written warranty concerning special tires or situations you may have received.

83. *Warranty* The following statement appears in a tire manufacturer's warranty.* Negate this statement. Does the new statement make sense?

> This warranty gives you specific legal rights and you may also have other rights, which vary from state to state.

LIFE SCIENCES

84. *Medicine* The following statements appear in a home medical reference book.[†] Define *p* and *q* so that the statements can be written symbolically. Then negate each statement.

  a. Tissue samples may be taken from almost anywhere in the body, and the procedure used depends on the site.

  b. The doctor or nurse holds your tongue down with a depressor and uses a plastic stick with a sterile cotton end (swab) to collect a fluid sample from your throat.

  c. The doctor holds your tongue down with a depressor and wipes a swab over your tonsils and the back of your throat.

85. *Fisheries Management* The following statement is found in the mission of the Pennsylvania Fish and Boat Commission.[‡] Define *p* and *q* so that the statement can be written symbolically. Then negate the statement.

> The Pennsylvania Fish and Boat Commission is sensitive to the needs of the physically challenged and works to make our facilities accessible.

SOCIAL SCIENCES

86. *Law* Attorneys sometimes use the phrase "and/or." This phrase corresponds to which usage of the word *or:* inclusive or exclusive?

87. *Law* The following statement appears in a guide to common laws.[§] Define *p* and *q* so that the statement can be written symbolically. Then negate the statement.

> The court won't do it for you, and hiring an attorney is usually not cost effective.

GENERAL INTEREST

88. *Yahtzee®* The following statement appears in the instructions for the Milton Bradley game Yahtzee.®[||] Negate the statement.

> You could reroll the die again for your Large Straight or set aside the 2 Twos and roll for your Twos or for 3 of a Kind.

89. *Logic Puzzles* Raymond Smullyan is one of today's foremost writers of logic puzzles. Smullyan proposed a question, based on the classic Frank Stockton short story, in which a prisoner must make a choice between two doors: behind one is a beautiful lady, and behind the other is a hungry tiger.[#] What if each door has a sign, and the prisoner knows that only one sign is true? The sign on Door 1 reads: In this room there is a lady and in the other room there is a tiger. The sign on Door 2 reads: In one of these rooms there is a lady and in one of these rooms there is a tiger. With this information, determine what is behind each door.

90. Describe how a search engine on the Internet uses key words and logical connectives to locate information.

---

*"Passenger and Light Truck Tire Limited Warranty with Tire Maintenance and Safety Manual," Nashville, TN, Bridgestone/Firestone, Inc., 1997.

[†]Goldman, David R., ed., *American College of Physicians Complete Home Medical Guide,* New York, DK Publishing, Inc., 1999, p. 223.

[‡]"2000 Pennsylvania Summary of Fishing Regulations and Laws," Peter A. Colangelo, Executive Director, Commonwealth of Pennsylvania Fish and Boat Commission, Harrisburg, PA.

[§]Ventura, John, *Law for Dummies,* Foster City, CA, IDG Books Worldwide, 1996.

[||]Milton Bradley Company, East Longmeadow, MA, 1996.

[#]Smullyan, Raymond, *The Lady or the Tiger? And Other Logic Puzzles, Including a Mathematical Novel That Features Godel's Great Discovery,* New York, Knopf, 1982.

6.3 THE CONDITIONAL AND CIRCUITS

THINK ABOUT IT   How can logic be used in the design of electrical circuits?

This question will be answered in this section.

Conditionals

"If you build it, he will come."
—The Voice in the 1990 movie *Field of Dreams*

Ray Kinsella, an Iowa farmer in the movie *Field of Dreams,* heard a voice from the sky. Ray interpreted it as a promise that if he would build a baseball field in his cornfield, then the ghost of Shoeless Joe Jackson (a baseball star in the early days of the twentieth century) would come to play on it. The promise came in the form of a conditional statement. A **conditional** statement is a compound statement that uses the connective *if...then*. For example, here are a few conditional statements.

> *If* I read for too long, *then* I get a headache.
>
> *If* looks could kill, *then* I would be dead.
>
> *If* he doesn't get back soon, *then* you should go look for him.

In each of these conditional statements, the component coming after the word *if* gives a condition (but not necessarily the only condition) under which the statement coming after *then* will be true. For example, "If it is over 90°, then I'll go to the mountains" tells one possible condition under which I will go to the mountains—if the temperature is over 90°.

The conditional is written with an arrow, so that "if p, then q" is symbolized as

$$p \rightarrow q.$$

We read $p \rightarrow q$ as "p implies q" or "if p, then q." In the conditional $p \rightarrow q$, the statement p is the **antecedent,** while q is the **consequent.**

The conditional connective may not always be explicitly stated. That is, it may be "hidden" in an everyday expression. For example, the statement

> Big girls don't cry

can be written in *if...then* form as

> If you're a big girl, then you don't cry.

As another example, the statement

> It is difficult to study when you are distracted

can be written

> If you are distracted, then it is difficult to study.

As seen in the quote from the movie *Field of Dreams* earlier, the word "then" is sometimes not stated but understood to be there from the context of the statement. In that statement, "you build it" is the antecedent and "he will come" is the consequent.

The conditional truth table is a little harder to define than were the tables in the previous section. To see how to define the conditional truth table, let us analyze a statement made by a hypothetical politician, Senator Anne Ruppert:

If I am elected, then taxes will go down.

As before, there are four possible combinations of truth values for the two component statements. Let p represent "I am elected," and let q represent "Taxes will go down."

As we analyze the four possibilities, it is helpful to think in terms of the following: "Did Senator Ruppert lie?" If she lied, then the conditional statement is considered false; if she did not lie, then the conditional statement is considered true.

Possibility	Elected?	Taxes Go Down?	
1	Yes	Yes	p is T, q is T
2	Yes	No	p is T, q is F
3	No	Yes	p is F, q is T
4	No	No	p is F, q is F

The four possibilities are as follows:

1. In the first case assume that the senator was elected and taxes did go down (p is T, q is T). The senator told the truth, so place T in the first row of the truth table. (We do not claim that taxes went down *because* she was elected; it is possible that she had nothing to do with it at all.)

2. In the second case assume that the senator was elected and taxes did not go down (p is T, q is F). Then the senator did not tell the truth (that is, she lied). So we put F in the second row of the truth table.

3. In the third case assume that the senator was defeated, but taxes went down anyway (p is F, q is T). Senator Ruppert did not lie; she only promised a tax reduction if she were elected. She said nothing about what would happen if she were not elected. In fact, her campaign promise gives no information about what would happen if she lost. Since we cannot say that the senator lied, place T in the third row of the truth table.

4. In the last case assume that the senator was defeated but taxes did not go down (p is F, q is F). We cannot blame her, since she only promised to reduce taxes if elected. Thus, T goes in the last row of the truth table.

The completed truth table for the conditional is defined as follows.

> **TRUTH TABLE FOR THE CONDITIONAL If p, then q**
>
> *If p, then q*
>
> | p | q | $p \rightarrow q$ |
> |-----|-----|-------------------|
> | T | T | T |
> | T | F | F |
> | F | T | T |
> | F | F | T |

It must be emphasized that the use of the conditional connective in no way implies a cause-and-effect relationship. Any two statements may have an arrow placed between them to create a compound statement. For example,

If I pass mathematics, then the sun will rise the next day

is true, since the consequent is true. (See the box after Example 1.) There is, however, no cause-and-effect connection between my passing mathematics and the sun's rising. The sun will rise no matter what grade I get in a course.

EXAMPLE 1 Truth Values

Given that p, q, and r are all false, find the truth value of the statement $(p \rightarrow \sim q) \rightarrow (\sim r \rightarrow q)$.

Solution Using the short-cut method explained in Example 3 of the previous section, we can replace p, q, and r with F (since each is false) and proceed as before, using the negation and conditional truth tables as necessary.

$$(p \rightarrow \sim q) \rightarrow (\sim r \rightarrow q)$$
$$(\text{F} \rightarrow \sim\text{F}) \rightarrow (\sim\text{F} \rightarrow \text{F})$$
$$(\text{F} \rightarrow \text{T}) \rightarrow (\text{T} \rightarrow \text{F}) \qquad \text{Use the negation truth table.}$$
$$\text{T} \rightarrow \text{F} \qquad \text{Use the conditional truth table.}$$
$$\text{F}$$

The statement $(p \rightarrow \sim q) \rightarrow (\sim r \rightarrow q)$ is false when p, q, and r are all false.

The following observations come from the truth table for $p \rightarrow q$.

> **SPECIAL CHARACTERISTICS OF CONDITIONAL STATEMENTS**
>
> **1.** $p \rightarrow q$ is false only when the antecedent is *true* and the consequent is *false*.
>
> **2.** If the antecedent is *false*, then $p \rightarrow q$ is automatically *true*.
>
> **3.** If the consequent is *true*, then $p \rightarrow q$ is automatically *true*.

EXAMPLE 2 Conditional Statements

Write *true* or *false* for each statement. Here T represents a true statement, and F represents a false statement.

(a) $T \rightarrow (6 = 3)$

Solution Since the antecedent is true, while the consequent, $6 = 3$, is false, the given statement is false by the first point mentioned above.

(b) $(5 < 2) \rightarrow F$

Solution The antecedent is false, so the given statement is true by the second observation.

(c) $(3 \neq 2 + 1) \rightarrow T$

Solution The consequent is true, making the statement true by the third characteristic of conditional statements.

Truth tables for compound statements involving conditionals are found using the techniques described in the previous section. The next example shows how this is done.

EXAMPLE 3 Truth Tables

Construct a truth table for each statement.

(a) $(\sim p \rightarrow \sim q) \rightarrow (\sim p \wedge q)$

Solution First insert the truth values of $\sim p$ and of $\sim q$. Then find the truth values of $\sim p \rightarrow \sim q$.

p	q	$\sim p$	$\sim q$	$\sim p \rightarrow \sim q$
T	T	F	F	T
T	F	F	T	T
F	T	T	F	F
F	F	T	T	T

Next use $\sim p$ and q to find the truth values of $\sim p \wedge q$.

p	q	$\sim p$	$\sim q$	$\sim p \rightarrow \sim q$	$\sim p \wedge q$
T	T	F	F	T	F
T	F	F	T	T	F
F	T	T	F	F	T
F	F	T	T	T	F

Now complete the work by using the conditional truth table to find the truth values of $(\sim p \rightarrow \sim q) \rightarrow (\sim p \wedge q)$.

p	q	$\sim p$	$\sim q$	$\sim p \rightarrow \sim q$	$\sim p \wedge q$	$(\sim p \rightarrow \sim q) \rightarrow (\sim p \wedge q)$
T	T	F	F	T	F	F
T	F	F	T	T	F	F
F	T	T	F	F	T	T
F	F	T	T	T	F	F

(b) $(p \rightarrow q) \rightarrow (\sim p \vee q)$

Solution Go through steps similar to the ones above.

p	q	$p \rightarrow q$	$\sim p$	$\sim p \vee q$	$(p \rightarrow q) \rightarrow (\sim p \vee q)$
T	T	T	F	T	T
T	F	F	F	F	T
F	T	T	T	T	T
F	F	T	T	T	T

As the truth table in Example 3(b) shows, the statement $(p \rightarrow q) \rightarrow (\sim p \vee q)$ is always true, no matter what the truth values of the components. Such a statement is called a **tautology.** Other examples of tautologies (as can be checked by forming truth tables) include $p \vee \sim p$, $p \rightarrow p$, $(\sim p \vee \sim q) \rightarrow \sim (q \wedge p)$, and so on. By the way, the truth tables in Example 3 also could have been found by the alternative method shown in the previous section.

Negation of a Conditional   Suppose that someone makes the conditional statement

"If it rains, then I take my umbrella."

When will the person have lied to you? The only case in which you would have been misled is when it rains *and* the person does *not* take the umbrella. Letting p represent "it rains" and q represent "I take my umbrella," you might suspect that the symbolic statement

$$p \wedge \sim q$$

is a candidate for the negation of $p \rightarrow q$. That is,

$$\sim (p \rightarrow q) \equiv p \wedge \sim q.$$

It happens that this is indeed the case, as the next truth table indicates.

p	q	$p \rightarrow q$	$\sim(p \rightarrow q)$	$\sim q$	$p \wedge \sim q$
T	T	T	F	F	F
T	F	F	T	T	T
F	T	T	F	F	F
F	F	T	F	T	F

$$\equiv$$

NEGATION OF $p \rightarrow q$

The negation of $p \rightarrow q$ is $p \wedge \sim q$.

Since

$$\sim(p \rightarrow q) \equiv p \wedge \sim q,$$

by negating each expression we have

$$\sim[\sim(p \rightarrow q)] \equiv \sim(p \wedge \sim q).$$

The left side of the above equivalence is $p \rightarrow q$, and one of De Morgan's laws can be applied to the right side.

$$p \rightarrow q \equiv \sim p \vee \sim(\sim q)$$
$$p \rightarrow q \equiv \sim p \vee q$$

This final row indicates that a conditional may be written as a disjunction.

WRITING A CONDITIONAL AS AN *or* STATEMENT

$p \rightarrow q$ is equivalent to $\sim p \vee q$.

EXAMPLE 4   Negation

Write the negation of each statement.

(a) If you build it, he will come.

Solution If b represents "you build it" and q represents "he will come," then the given statement can be symbolized by $b \rightarrow q$. The negation of $b \rightarrow q$, as shown earlier, is $b \wedge \sim q$, so the negation of the statement is

You build it and he will not come.

(b) It must be alive if it is breathing.

Solution First, we must restate the given statement in *if...then* form:

If it is breathing, then it must be alive.

Based on our earlier discussion, the negation is

It is breathing and it is not alive.

A common error occurs when students try to write the negation of a conditional statement as another conditional statement. As seen in Example 4, the negation of a conditional statement is written as a conjunction.

EXAMPLE 5 Equivalent Statements

Write each conditional as an equivalent statement without using *if...then*.

(a) If the Cubs win the pennant, then Gwen will be happy.

Solution Since the conditional $p \rightarrow q$ is equivalent to $\sim p \lor q$, let p represent "the Cubs win the pennant" and q represent "Gwen will be happy." Restate the conditional as

> The Cubs do not win the pennant or Gwen will be happy.

(b) If it's Borden's, it's got to be good.

Solution If p represents "it's Borden's" and if q represents "it's got to be good," the conditional may be restated as

> It's not Borden's or it's got to be good.

Circuits One of the first nonmathematical applications of symbolic logic was seen in the master's thesis of Claude Shannon in 1937. Shannon showed how logic could be used as an aid in designing electrical **circuits.** His work was immediately taken up by the designers of computers. These computers, then in the developmental stage, could be simplified and built for less money using the ideas of Shannon.

To see how Shannon's ideas work, look at the electrical switch shown in Figure 6. We assume that current will flow through this switch when it is closed and not when it is open.

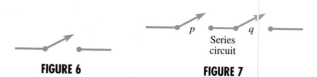

Series circuit

FIGURE 6　　　　**FIGURE 7**

Figure 7 shows two switches connected in **series;** in such a circuit, current will flow only when both switches are closed. Note how closely a series circuit corresponds to the conjunction $p \land q$. We know that $p \land q$ is true only when both p and q are true.

A circuit corresponding to the disjunction $p \lor q$ can be found by drawing a **parallel** circuit, as in Figure 8. Here, current flows if either p *or* q is closed or if both p *and* q are closed.

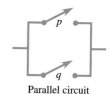

Parallel circuit

FIGURE 8

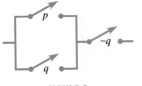

FIGURE 9

The circuit in Figure 9 corresponds to the statement $(p \lor q) \land \sim q$, which is a compound statement involving both a conjunction and a disjunction.

The way that logic is used to simplify an electrical circuit depends on the idea of equivalent statements, from Section 2. Recall that two statements are equivalent if they have exactly the same truth table final column. The symbol $\equiv$ is used to indicate that the two statements are equivalent. Some of the equivalent statements that we shall need are shown in the following box.

EQUIVALENT STATEMENTS USED TO SIMPLIFY CIRCUITS

1. $p \lor (q \land r) \equiv (p \lor q) \land (p \lor r)$ **5.** $p \lor p \equiv p$

2. $p \land (q \lor r) \equiv (p \land q) \lor (p \land r)$ **6.** $p \land p \equiv p$

3. $p \to q \equiv \sim q \to \sim p$ **7.** $\sim(p \land q) \equiv \sim p \lor \sim q$

4. $p \to q \equiv \sim p \lor q$ **8.** $\sim(p \lor q) \equiv \sim p \land \sim q$

If T represents any true statement and F represents any false statement, then

9. $p \lor T \equiv T$ **10.** $p \lor \sim p \equiv T$

11. $p \land F \equiv F$ **12.** $p \land \sim p \equiv F.$

Circuits can be used as models of compound statements, with a closed switch corresponding to T, while an open switch corresponds to F. The method for simplifying circuits is explained in the following example.

EXAMPLE 6 Circuit
Simplify the circuit in Figure 10.

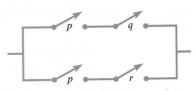

FIGURE 10

Solution At the top of Figure 10, p and q are connected in series, and at the bottom, p and r are connected in series. These are interpreted as the compound statements $p \land q$ and $p \land r$, respectively. These two conjunctions are connected in parallel, as indicated by the figure treated as a whole. Therefore, we write the disjunction of the two conjunctions:

$$(p \land q) \lor (p \land r).$$

(Think of the two switches labeled "p" as being controlled by the same handle.) By equivalent statement 2 in the box above,

$$(p \land q) \lor (p \land r) \equiv p \land (q \lor r),$$

which has the circuit of Figure 11. This new circuit is logically equivalent to the one in Figure 10, and yet contains only three switches instead of four—which might well lead to a large savings in manufacturing costs.

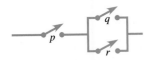

FIGURE 11

EXAMPLE 7 Circuit

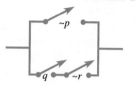

FIGURE 12

Draw a circuit for $p \rightarrow (q \wedge \sim r)$.

Solution By equivalent statement 4 on the previous page, $p \rightarrow q$ is equivalent to $\sim p \vee q$. This equivalence gives $p \rightarrow (q \wedge \sim r) \equiv \sim p \vee (q \wedge \sim r)$, which has the circuit diagram in Figure 12.

6.3 EXERCISES

In Exercises 1–8, decide whether each statement is true *or* false.

1. If the antecedent of a conditional statement is false, the conditional statement is true.

2. If the consequent of a conditional statement is true, the conditional statement is true.

3. If q is true, then $(p \wedge q) \rightarrow q$ is true.

4. If p is true, then $\sim p \rightarrow (q \vee r)$ is true.

5. The negation of "If pigs fly, I'll believe it" is "If pigs don't fly, I won't believe it."

6. The statements "If it flies, then it's a bird" and "It does not fly or it's a bird" are logically equivalent.

7. Given that $\sim p$ is true and q is false, the conditional $p \rightarrow q$ is true.

8. Given that $\sim p$ is false and q is false, the conditional $p \rightarrow q$ is true.

9. In a few sentences, explain how we determine the truth value of a conditional statement.

10. Explain why the statement "If $3 = 5$, then $4 = 6$" is true.

Tell whether each conditional is true *or* false. *Here* T *represents a true statement and* F *represents a false statement.*

11. $F \rightarrow (4 \neq 7)$

12. $T \rightarrow (6 < 3)$

13. $(6 \geq 6) \rightarrow F$

14. $F \rightarrow (3 \neq 3)$

15. $(4 = 11 - 7) \rightarrow (8 > 0)$

16. $(4^2 \neq 16) \rightarrow (4 - 4 = 8)$

Let s represent "She has a snake for a pet," *let p represent* "he trains ponies," *and let m represent* "they raise monkeys." *Express each compound statement in words.*

17. $\sim m \rightarrow p$

18. $p \rightarrow \sim m$

19. $s \rightarrow (m \wedge p)$

20. $(s \wedge p) \rightarrow m$

21. $\sim p \rightarrow (\sim m \vee s)$

22. $(\sim s \vee \sim m) \rightarrow \sim p$

Let b represent "I ride my bike," *let r represent* "it rains," *and let p represent* "the play is cancelled." *Write each compound statement in symbols.*

23. If it rains, then I ride my bike.

24. If I ride my bike, then the play is cancelled.

25. If I do not ride my bike, then it does not rain.

26. If the play is cancelled, then it does not rain.

27. I ride my bike, or if the play is cancelled then it rains.

28. The play is cancelled, and if it rains then I do not ride my bike.

29. I'll ride my bike if it doesn't rain.

30. It rains if the play is cancelled.

Find the truth value of each statement. Assume that p and r are false, and q is true.

31. $\sim r \rightarrow q$

32. $\sim p \rightarrow \sim r$

33. $q \rightarrow p$

34. $\sim r \rightarrow p$

35. $p \rightarrow q$

36. $\sim q \rightarrow r$

37. $\sim p \to (q \land r)$ **38.** $(\sim r \lor p) \to p$ **39.** $\sim q \to (p \land r)$

40. $(\sim p \land \sim q) \to (p \land \sim r)$ **41.** $(p \to \sim q) \to (\sim p \land \sim r)$ **42.** $(p \to \sim q) \land (p \to r)$

43. Explain why, if we know that p is true, we also know that

$$[r \lor (p \lor s)] \to (p \lor q)$$

is true, even if we are not given the truth values of q, r, and s.

44. Construct a true statement involving a conditional, a conjunction, a disjunction, and a negation (not necessarily in that order), that consists of component statements p, q, and r, with all of these component statements false.

Construct a truth table for each statement. Identify any tautologies.

45. $\sim q \to p$ **46.** $p \to \sim q$ **47.** $(\sim p \to q) \to p$

48. $(\sim q \to \sim p) \to \sim q$ **49.** $(p \lor q) \to (q \lor p)$ **50.** $(p \land q) \to (p \lor q)$

51. $(\sim p \to \sim q) \to (p \land q)$ **52.** $r \to (p \land \sim q)$ **53.** $[(r \lor p) \land \sim q] \to p$

54. $(\sim r \to s) \lor (p \to \sim q)$ **55.** $(\sim p \land \sim q) \to (\sim r \to \sim s)$

56. What is the minimum number of Fs that must appear in the final column of a truth table for us to be assured that the statement is not a tautology?

Write the negation of each statement. Remember that the negation of $p \to q$ is $p \land \sim q$.

57. If that is an authentic Persian rug, I'll be surprised.

58. If Ella reaches that note, she will shatter glass.

59. If the English measures are not converted to metric measures, then the spacecraft will crash on the surface of Mars.

60. If you say "I do," then you'll be happy for the rest of your life.

61. "If you want to be happy for the rest of your life, never make a pretty woman your wife." *Jimmy Soul*

62. If loving you is wrong, I don't want to be right.

Write each statement as an equivalent statement that does not use the if . . . then *connective. Remember that $p \to q$ is equivalent to $\sim p \lor q$.*

63. If you give your plants tender, loving care, they flourish.

64. If the check is in the mail, I'll be surprised.

65. If she doesn't, he will.

66. If I say yes, she says no.

67. All residents of Butte are residents of Montana.

68. All women were once girls.

Use truth tables to decide which of the pairs of statements are equivalent.

69. $p \to q$; $\sim p \lor q$ **70.** $\sim (p \to q)$; $p \land \sim q$ **71.** $p \to q$; $q \to p$

72. $q \to p$; $\sim p \to \sim q$ **73.** $p \to \sim q$; $\sim p \lor \sim q$ **74.** $p \to q$; $\sim q \to \sim p$

75. $p \land \sim q$; $\sim q \to \sim p$ **76.** $\sim p \land q$; $\sim p \to q$

Write a logical statement representing each of the following circuits. Simplify each circuit when possible.

77.

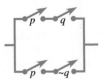

78.

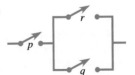

79.

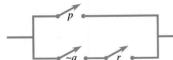

80.

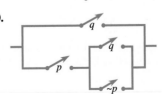

81.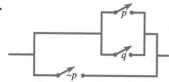

82.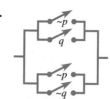

Draw circuits representing the following statements as they are given. Simplify if possible.

83. $p \wedge (q \vee \sim p)$

84. $(\sim p \wedge \sim q) \wedge \sim r$

85. $(p \vee q) \wedge (\sim p \wedge \sim q)$

86. $(\sim q \wedge \sim p) \vee (\sim p \vee q)$

87. $[(p \vee q) \wedge r] \wedge \sim p$

88. $[(\sim p \wedge \sim r) \vee \sim q] \wedge (\sim p \wedge r)$

89. $\sim q \rightarrow (\sim p \rightarrow q)$

90. $\sim p \rightarrow (\sim p \vee \sim q)$

91. Explain why the circuit

will always have exactly one open switch. What does this circuit simplify to?

92. Refer to Figures 10 and 11 in Example 6. Suppose the cost of the use of one switch for an hour is 3¢. By using the circuit in Figure 11 rather than the circuit in Figure 10, what is the savings for a year of 365 days, assuming that the circuit is in continuous use?

Applications

BUSINESS AND ECONOMICS

93. *Income Tax* The following statements appear in a guide for preparing income tax reports.* Write each of the statements in *if . . . then* form.

a. You may not file a 2000 joint return if you were divorced under a decree of divorce of separate maintenance that is final by the end of the year.

b. You may file jointly if you separated during 2000 under an interlocutory decree (provisional or temporary) or order, so long as a final divorce decree was not entered by the end of the year.

c. Where a U.S. citizen or resident is married to a nonresident alien, the couple may file a joint return if both elect to be taxed on their worldwide income.

d. To avoid current tax on pay, you may contract with your employer to defer pay to future years.

94. *Income Tax* The following statements appear in a guide for preparing income tax reports.* Rewrite each statement with an equivalent statement using "or." Then write the negation of each statement.

a. If you create an artistic work or invention for which you get a government patent or copyright, you may depreciate your costs over the life of the patent or copyright.

b. If you are married at the end of the year, you may file a joint return with your spouse.

c. If you receive your company's stock as payment for your services, you include the value of the stock as pay in the year you receive it.

LIFE SCIENCES

95. *Medicine* In making a medical diagnosis, it is common to string together a series of statements. The following statement appears in a home medical reference book with regard to difficulty swallowing.[†] Rewrite this statement replacing the *if . . . then* with an *or* statement. Then negate the statement.

If your throat is sore and food seems to stick high up in the chest and you get a burning pain in the center of the chest, then you may have gastroesophageal reflux.

Your Income Tax 2001, J. K. Lasser Institute, John Wiley and Sons, Inc., 2000.
[†]Goldman, David R., ed., *American College of Physicians Complete Home Medical Guide*, New York, DK Publishing, Inc., 1999, p. 147.

SOCIAL SCIENCES

96. *Law* The following statements appear in a guide to common laws.* Write an equivalent statement using *or*. Then negate the statement.

  a. If you are married, you can't get married again.

  b. If your job is going to cost more than $500, your contractor is legally required to put it in writing.

  c. If your application for citizenship is denied, you can appeal in federal court.

■ 6.4   MORE ON THE CONDITIONAL

 THINK ABOUT IT   Is it possible to rewrite statements in a tax guide in one or more ways?

This question will be answered in Exercise 51 of this section.

The conditional statement, introduced in the previous section, is one of the most important of all compound statements. Many mathematical properties and theorems are stated in *if...then* form. Because of their usefulness, we need to study conditional statements that are related to a statement of the form $p \rightarrow q$.

Converse, Inverse, and Contrapositive   Any conditional statement is made up of an antecedent and a consequent. If they are interchanged, negated, or both, a new conditional statement is formed. Suppose that we begin with the direct statement

If you stay, then I go,

and interchange the antecedent ("you stay") and the consequent ("I go"). We obtain the new conditional statement

If I go, then you stay.

This new conditional is called the **converse** of the given statement.

By negating both the antecedent and the consequent, we obtain the **inverse** of the given statement:

If you do not stay, then I do not go.

---

*Ventura, John, *Law for Dummies,* Foster City, CA, IDG Books Worldwide, 1996.

If the antecedent and the consequent are both interchanged *and* negated, the **contrapositive** of the given statement is formed:

If I do not go, then you do not stay.

These three related statements for the conditional $p \rightarrow q$ are summarized below. (Notice that the inverse is the contrapositive of the converse.)

RELATED CONDITIONAL STATEMENTS		
Direct Statement	$p \rightarrow q$	(If p, then q.)
Converse	$q \rightarrow p$	(If q, then p.)
Inverse	$\sim p \rightarrow \sim q$	(If not p, then not q.)
Contrapositive	$\sim q \rightarrow \sim p$	(If not q, then not p.)

EXAMPLE 1 Related Conditional Statements

Given the direct statement

If I live in Miami, then I live in Florida,

write each of the following.

(a) the converse

Solution Let p represent "I live in Miami" and q represent "I live in Florida." Then the direct statement may be written $p \rightarrow q$. The converse, $q \rightarrow p$, is

If I live in Florida, then I live in Miami.

Notice that for this statement, the converse is not necessarily true, even though the direct statement is.

(b) the inverse

Solution The inverse of $p \rightarrow q$ is $\sim p \rightarrow \sim q$. For the given statement, the inverse is

If I don't live in Miami, then I don't live in Florida,

which is again not necessarily true.

(c) the contrapositive

Solution The contrapositive, $\sim q \rightarrow \sim p$, is

If I don't live in Florida, then I don't live in Miami.

The contrapositive, like the direct statement, is true.

Example 1 shows that the converse and inverse of a true statement need not be true. They *can* be true, but they need not be. The relationship between the truth values of the direct statement, converse, inverse, and contrapositive is shown in the truth table on the next page.

As this truth table shows, the direct statement and the contrapositive always have the same truth values, making it possible to replace any statement with its contrapositive without affecting the logical meaning. Also, the converse and inverse always have the same truth values.

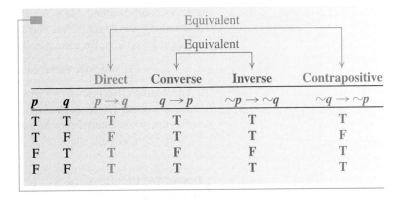

This discussion is summarized in the following sentence.

EQUIVALENCES

The direct statement and the contrapositive are equivalent, and the converse and the inverse are equivalent.

EXAMPLE 2   Related Conditional Statements

For the direct statement $\sim p \rightarrow q$, write each of the following.

(a) the converse

> **Solution** The converse of $\sim p \rightarrow q$ is $q \rightarrow \sim p$.

(b) the inverse

> **Solution** The inverse is $\sim(\sim p) \rightarrow \sim q$, which simplifies to $p \rightarrow \sim q$.

(c) the contrapositive

> **Solution** The contrapositive is $\sim q \rightarrow \sim(\sim p)$, which simplifies to $\sim q \rightarrow p$.

Alternative Forms of "if p, then q"   The conditional statement "if p, then q" can be stated in several other ways in English. For example,

> If you go to the shopping center, then you will find a place to park

can also be written

> Going to the shopping center is *sufficient* for finding a place to park.

According to this statement, going to the shopping center is enough to guarantee finding a place to park. Going to other places, such as schools or office buildings, *might* also guarantee a place to park, but at least we *know* that going to the shopping center does. Thus, $p \rightarrow q$ can be written "p is sufficient for q." Knowing that p has occurred is sufficient to guarantee that q will also occur. On the other hand,

> Turning on the set is necessary for watching television   (*)

has a different meaning. Here, we are saying that one condition that is necessary for watching television is that you turn on the set. This may not be enough; the set might be broken, for example. The statement labeled (*) could be written as

> If you watch television, then you turned on the set.

As this example suggests, $p \rightarrow q$ is the same as "q is necessary for p." In other words, if q doesn't happen, then neither will p. Notice how this idea is closely related to the idea of equivalence between the direct statement and its contrapositive.

Some common translations of $p \rightarrow q$ are summarized in the following box.

COMMON TRANSLATIONS OF $p \rightarrow q$

The conditional $p \rightarrow q$ can be translated in any of the following ways.

If p, then q.	p is sufficient for q.
If p, q.	q is necessary for p.
p implies q.	All p's are q's.
p only if q.	q if p.
q when p.	

The translation of $p \rightarrow q$ into these various word forms does not in any way depend on the truth or falsity of $p \rightarrow q$.

EXAMPLE 3 Equivalent Statements

The statement

> If you are 18, then you can vote

can be written in any of the following ways.

> You can vote if you are 18.
> You are 18 only if you can vote.
> Being able to vote is necessary for you to be 18.
> Being 18 is sufficient for being able to vote.
> All 18-year-olds can vote.
> Being 18 implies that you can vote.
> You can vote when you are 18.

EXAMPLE 4 Equivalent Statements

Write each statement in the form "if p, then q."

(a) You'll be sorry if I go.

Solution If I go, then you'll be sorry.

(b) Today is Friday only if yesterday was Thursday.

Solution If today is Friday, then yesterday was Thursday.

(c) All nurses wear white shoes.

Solution If you are a nurse, then you wear white shoes.

EXAMPLE 5 Symbolic Statements

Let p represent "A triangle is equilateral," and let q represent "A triangle has three equal sides." Write each of the following in symbols.

(a) A triangle is equilateral if it has three equal sides.

Solution

$$q \to p$$

(b) A triangle is equilateral only if it has three equal sides.

Solution

$$p \to q$$

Biconditionals The compound statement *p if and only if q* (often abbreviated *p iff q*) is called a **biconditional.** It is symbolized $p \leftrightarrow q$, and is interpreted as the conjunction of the two conditionals $p \to q$ and $q \to p$. Using symbols, this conjunction is written

$$(q \to p) \wedge (p \to q)$$

so that, by definition,

$$p \leftrightarrow q \equiv (q \to p) \wedge (p \to q).$$

Using this definition, the truth table for the biconditional $p \leftrightarrow q$ can be determined.

TRUTH TABLE FOR THE BICONDITIONAL p **if and only if** q

p if and only if *q*

p	q	$p \leftrightarrow q$
T	T	T
T	F	F
F	T	F
F	F	T

EXAMPLE 6 Biconditional Statements

Tell whether each biconditional statement is *true* or *false*.

(a) $6 + 9 = 15$ if and only if $12 + 4 = 16$

Solution Both $6 + 9 = 15$ and $12 + 4 = 16$ are true. By the truth table for the biconditional, this biconditional is true.

(b) $5 + 2 = 10$ if and only if $17 + 19 = 36$

Solution Since the first component ($5 + 2 = 10$) is false, and the second is true, the entire biconditional statement is false.

(c) $6 = 5$ if and only if $12 \neq 12$

Solution Both component statements are false, so by the last line of the truth table for the biconditional, the entire statement is true. (Understanding this might take some extra thought!)

In this section and in the previous two sections, truth tables have been derived for several important types of compound statements. The summary that follows describes how these truth tables may be remembered.

SUMMARY OF BASIC TRUTH TABLES

1. $\sim p$, the **negation** of p, has truth value opposite of p.
2. $p \wedge q$, the **conjunction,** is true only when both p and q are true.
3. $p \vee q$, the **disjunction,** is false only when both p and q are false.
4. $p \rightarrow q$, the **conditional,** is false only when p is true and q is false.
5. $p \leftrightarrow q$, the **biconditional,** is true only when p and q have the same truth value.

6.4 EXERCISES

For each given direct statement, write (**a**) *the converse,* (**b**) *the inverse, and* (**c**) *the contrapositive in* if . . . then *form. In some of the exercises, it may be helpful to restate the direct statement in* if . . . then *form.*

1. If beauty were a minute, then you would be an hour.

2. If you lead, then I will follow.

3. If the exit is ahead, then I don't see it.

4. If I had a nickel for each time that happened, I would be rich.

5. Walking in front of a moving car is dangerous to your health.

6. Milk contains calcium.

7. Birds of a feather flock together.

8. A rolling stone gathers no moss.

9. If you build it, he will come.

10. Where there's smoke, there's fire.

11. $p \rightarrow \sim q$ 12. $\sim p \rightarrow q$

13. $\sim p \rightarrow \sim q$ 14. $\sim q \rightarrow \sim p$

15. $p \rightarrow (q \vee r)$ (*Hint:* Use one of De Morgan's laws as necessary.)

16. $(r \vee \sim q) \rightarrow p$ (*Hint:* Use one of De Morgan's laws as necessary.)

17. Discuss the equivalences that exist among the direct conditional statement, the converse, the inverse, and the contrapositive.

18. State the contrapositive of "If the square of a natural number is even, then the natural number is even." The two statements must have the same truth value. Use several examples and inductive reasoning to decide whether both are true or both are false.

Write each of the following statements in the form "if p, then q."

19. If it is muddy, I'll wear my galoshes.

20. If I finish studying, I'll go to the party.

21. "17 is positive" implies that $17 + 1$ is positive.

22. "Today is Wednesday" implies that yesterday was Tuesday.

23. All integers are rational numbers.

24. All whole numbers are integers.

25. Doing crossword puzzles is sufficient for driving me crazy.

26. Being in Fort Lauderdale is sufficient for being in Florida.

27. A day's growth of beard is necessary for Greg Tobin to shave.

28. Being an environmentalist is necessary for being elected.

29. I can go from Boardwalk to Connecticut Avenue only if I pass GO.

30. The principal will hire more teachers only if the school board approves.

31. No whole numbers are not integers.

32. No integers are irrational numbers.

33. The Indians will win the pennant when their pitching improves.

34. Jesse will be a liberal when pigs fly.

35. A rectangle is a parallelogram with a right angle.

36. A parallelogram is a four-sided figure with opposite sides parallel.

37. A triangle with two sides of the same length is isosceles.

38. A square is a rectangle with two adjacent sides equal.

39. The square of a two-digit number whose units digit is 5 will end in 25.

40. An integer whose units digit is 0 or 5 is divisible by 5.

41. One of the following statements is not equivalent to all the others. Which one is it?

    a. *r* only if *s*.

    b. *r* implies *s*.

    c. If *r*, then *s*.

    d. *r* is necessary for *s*.

42. Many students have difficulty interpreting *necessary* and *sufficient.* Use the statement "Being in Canada is sufficient for being in North America" to explain why "*p* is sufficient for *q*" translates as "if *p*, then *q*."

43. Use the statement "To be an integer, it is necessary that a number be rational" to explain why "*p* is necessary for *q*" translates as "if *q*, then *p*."

44. Explain why the statement "A week has eight days if and only if December has forty days" is true.

Identify each statement as true *or* false.

45. $5 = 9 - 4$ if and only if $8 + 2 = 10$.

46. $3 + 1 \neq 6$ if and only if $8 \neq 8$.

47. $8 + 7 \neq 15$ if and only if $3 \times 5 \neq 9$.

48. $6 \times 2 = 14$ if and only if $9 + 7 \neq 16$.

49. Bill Clinton was president if and only if Jimmy Carter was not president.

50. Burger King sells Big Macs if and only if IBM manufactures computers.

Applications

BUSINESS AND ECONOMICS

 51. *Income Tax* The following statements appear in a guide for preparing income tax reports.* Write the converse, inverse, and contrapositive of each statement.

    a. If you are married at the end of the year, you may file a joint return with your spouse.

    b. If you receive your company's stock as payment for your services, you include the value of the stock as pay in the year you receive it.

52. *Credit Cards* The following statement appeared on a monthly statement of a Verizon Visa Card.[†] Write the converse, inverse, and contrapositive. Which statements are equivalent?

> If you close your account within 30 days from the date this statement was mailed, you may avoid paying the annual fee billed on this statement.

LIFE SCIENCES

53. *Medicine* In making a medical diagnosis, it is common to string together a series of statements. The following

---

*Your Income Tax 2001, J. K. Lasser Institute, John Wiley and Sons, Inc., 2000.

[†]Verizon Visa Card, Wilmington, DE.

statement appears in a home medical reference book with regard to difficulty swallowing.* Find the contrapositive of this statement.

> If your throat is sore and food seems to stick high up in the chest and you get a burning pain in the center of the chest then you may have gastroesophageal reflux.

54. *Polar Bears* The following statement is with regard to polar bear cubs.[†]

> If there are triplets, the most persistent stands to gain an extra meal and it may eat at the expense of another.

a. Use symbols to write this statement.

b. Write the contrapositive of this statement.

SOCIAL SCIENCES

55. *Law* The following statements appear in a guide to common laws.[‡] Write the converse, inverse, and contrapositive of the statements.

a. If you are married, you can't get married again.

b. If your job is going to cost more than $500, your contractor is legally required to put it (a bid) in writing.

c. If your application for citizenship is denied you can appeal in federal court.

56. *Philosophy* The following statements have been said thousands of times. Rewrite each saying as a conditional in *if . . . then* form. Then write two statements that are equivalent to the *if . . . then* statements. (*Hint:* Write the contrapositive of the statement and rewrite the conditional statement using *or*.)

a. A stitch in time saves nine.

b. A rolling stone gathers no moss.

c. Birds of a feather flock together.

57. *Philosophy* Think of some wise sayings that have been around for a long time, and state them in *if . . . then* form.

58. *Famous Quote* Aristotle once said, "If liberty and equality, as is thought by some, are chiefly to be found in democracy, they will be best attained when all persons alike share in the government to the utmost."[§] Write the contrapositive of this statement.

59. *Test of Reasoning* A test devised by psychologist Peter Wason is designed to test how people reason.[‖] As an example of this test, volunteers are given the rule, "If a card has a D on one side, then it must have a 3 on the other side." Volunteers view four cards displaying D, F, 3, and 7, respectively. They are told that each card has a letter on one side and a number on the other. Which cards do they need to turn over to determine if the rule has been violated? In Wason's experiments, fewer than one-fourth of the participants gave the correct answer.

60. *Test of Reasoning* In another example of a Wason test (see previous exercise), volunteers are given the rule, "If an employee works on the weekend, then that person gets a day off during the week." Volunteers are given four cards displaying "worked on the weekend," "did not work on the weekend," "did get a day off," and "did not get a day off." Volunteers were told that one side of the card tells whether an employee worked on the weekend, and the other side tells whether an employee got a day off. Which cards must be turned over to determine if the rule has been violated? In a set of experiments, volunteers told to take the perspective of the employees tended to give the correct answer, while volunteers told to take the perspective of employers tended to turn over the second and third card.

GENERAL INTEREST

61. *Yahtzee*® Statements similar to the ones below appear in the instructions for the Milton Bradley game, Yahtzee®[#] Rewrite each statement in the form of *if . . . then* and then write an equivalent statement using the word *or*.

a. You can score in this box only if the dice include three or more of the same number.

b. You can score in this box only if the dice show any sequence of four numbers.

c. You can score in this box only if the dice show three of one number and two of another.

*Goldman, David R., ed., *American College of Physicians Complete Home Medical Guide,* New York, DK Publishing, Inc., 1999, p. 147.
†Rosing, Norbert, "Bear Beginnings: New Life on the Ice," *National Geographic,* December, 2000, p. 33.
‡Ventura, John, *Law for Dummies,* Foster City, CA, IDG Books Worldwide, 1996.
§Bartlett, John, *Bartlett's Familiar Quotations,* 14th ed., Boston, Little, Brown and Company, 1968.
‖Bower, Bruce, "Roots of Reason," *Science News,* Vol. 145, Jan. 29, 1994, pp. 72–73.
#Milton Bradley Company, East Longmeadow, MA, 1996.

■ 6.5 ANALYZING ARGUMENTS WITH EULER DIAGRAMS

? THINK ABOUT IT   If some U.S. presidents won the popular vote and George W. Bush is the U.S. president, did he win the popular vote?

This question will be analyzed in Example 5 of this section.

There are two types of reasoning: inductive and deductive. With inductive reasoning we observe patterns to solve problems. Now, in this section and the next, we will study how deductive reasoning may be used to determine whether logical arguments are valid or invalid. A logical argument is made up of **premises** (assumptions, laws, rules, widely held ideas, or observations) and a **conclusion.** Together, the premises and the conclusion make up the argument. Also recall that *deductive* reasoning involves drawing specific conclusions from given general premises. When reasoning from the premises of an argument to obtain a conclusion, we want the argument to be valid.

VALID AND INVALID ARGUMENTS

An argument is **valid** if the fact that all the premises are true forces the conclusion to be true. An argument that is not valid is **invalid,** or a **fallacy.**

It is very important to note that *valid* and *true* are not the same—an argument can be valid even though the conclusion is false. (See Example 4.)

Several techniques can be used to check whether an argument is valid. One of these is the visual technique based on **Euler diagrams,** illustrated by the following examples. (Another is the method of truth tables, shown in the next section.) Euler (pronounced "oiler") diagrams are named after the great Swiss mathematician Leonhard Euler (1707–1783).

EXAMPLE 1 Valid Arguments
Is the following argument valid?

> All dogs are animals.
>
> Fred is a dog.
> _____
>
> Fred is an animal.

Solution Here we use the common method of placing one premise over another, with the conclusion below a line. Alternately, we could indicate that the last line is a conclusion using "therefore," as in "Therefore, Fred is an animal." To begin, draw regions to represent the first premise. One is the region for "animals." Since all dogs are animals, the region for "dogs" goes inside the region for "animals," as in Figure 13 on the next page.

The second premise, "Fred is a dog," suggests that "Fred" would go inside the region representing "dogs." Let *x* represent "Fred." Figure 14 shows that "Fred" is also inside the region for "animals." Therefore, if both premises are true, the conclusion that Fred is an animal must be true also. The argument is valid, as checked by Euler diagrams.

The method of Euler diagrams is especially convenient for arguments involving the quantifiers *all, some,* or *none.*

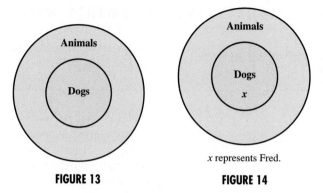

FIGURE 13

x represents Fred.

FIGURE 14

EXAMPLE 2 Valid Argument

Is the following argument valid?

> All rainy days are cloudy.
>
> Today is not cloudy.
> _____
>
> Today is not rainy.

Solution In Figure 15, the region for "rainy days" is drawn entirely inside the region for "cloudy days." Since "Today is *not* cloudy," place an *x* for "today" *outside* the region for "cloudy days." (See Figure 16.) Placing the *x* outside the region for "cloudy days" forces it also to be outside the region for "rainy days." Thus, if the first two premises are true, then it is also true that today is not rainy. The argument is valid.

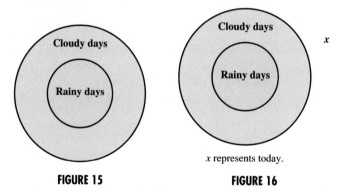

FIGURE 15

x represents today.

FIGURE 16

EXAMPLE 3 Fallacy

Is the following argument valid?

> All banana trees have green leaves.
>
> That plant has green leaves.
> _____
>
> That plant is a banana tree.

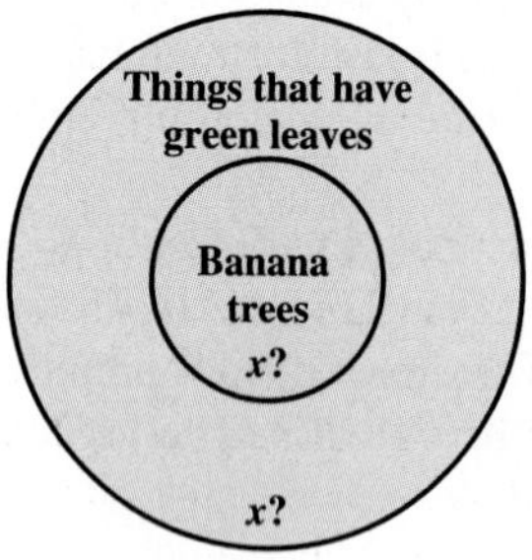

FIGURE 17

Solution The region for "banana trees" goes entirely inside the region for "things that have green leaves." (See Figure 17.) There is a choice for locating the *x* that represents "that plant." The *x* must go inside the region for "things that have green leaves," but can go either inside or outside the region for "banana trees." Even if the premises are true, we are not forced to accept that conclusion as true. This argument is invalid; it is a fallacy.

As mentioned earlier, the validity of an argument is not the same as the truth of its conclusion. The argument in Example 3 was invalid, but the conclusion "That plant is a banana tree" may or may not be true. We cannot be sure.

EXAMPLE 4   Valid Argument

Is the following argument valid?

> All expensive things are desirable.
>
> All desirable things make you feel good.
>
> All things that make you feel good make you live longer.
>
> All expensive things make you live longer.

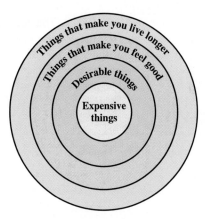

FIGURE 18

Solution A diagram for the argument is given in Figure 18. If each premise is true, then the conclusion must be true because the region for "expensive things" lies completely within the region for "things that make you live longer." Thus, the argument is valid. (This argument is an example of the fact that a *valid* argument need *not* have a true conclusion.)

Arguments with the word "some" can be tricky. One is shown in the final example of this section.

EXAMPLE 5   Invalid Argument

Is the following argument valid?

> Some U.S. presidents have won the popular vote.
>
> George W. Bush is president of the United States.
>
> George W. Bush won the popular vote.

Solution The first premise is sketched in Figure 19 on the next page. As the sketch shows, some (but not necessarily *all*) U.S. presidents have won the popular vote. Let *w* represent George W. Bush. There are two possibilities for *w*, as shown in Figure 20.

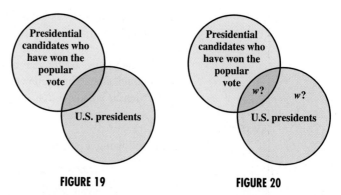

FIGURE 19   **FIGURE 20**

One possibility is that Bush won the popular vote; the other is that Bush did not win the popular vote. Since the truth of the premises does not force the conclusion to be true, the argument is invalid.

NOTE This argument is not valid regardless of whether George W. Bush won the popular vote.

6.5 EXERCISES

Decide whether each argument is valid *or* invalid.

1. All boxers wear trunks.
Steve Tomlin is a boxer.

Steve Tomlin wears trunks.

2. All amusement parks have thrill rides.
Great America is an amusement park.

Great America has thrill rides.

3. All residents of New York love Coney Island hot dogs.
Ann Stypuloski loves Coney Island hot dogs.

Ann Stypuloski is a resident of New York.

4. All politicians lie, cheat, and steal.
That man lies, cheats, and steals.

That man is a politician.

5. All contractors use cell phones.
Doug Boyle does not use a cell phone.

Doug Boyle is not a contractor.

6. All dogs love to bury bones.
Archie does not love to bury bones.

Archie is not a dog.

7. All people who apply for a loan must pay for a title search.
Cindy Herring paid for a title search.

Cindy Herring applied for a loan.

8. All residents of Minnesota know how to live in freezing temperatures.
Wendy Rockswold knows how to live in freezing temperatures.

Wendy Rockswold lives in Minnesota.

9. Some philosophers are absent-minded.
Deidre McGill is a philosopher.

Deidre McGill is absent-minded.

10. Some dinosaurs were plant-eaters.
Danny was a plant-eater.

Danny was a dinosaur.

11. Some trucks have sound systems.
Some trucks have gun racks.

Some trucks with sound systems have gun racks.

12. Some nurses wear blue uniforms.
Kim Falgout is a nurse.

Kim Falgout wears a blue uniform.

13. Refer to Example 3. If the second premise and the conclusion were interchanged, would the argument then be valid?

14. Refer to Example 4. Give a different conclusion than the one given there so that the argument is still valid.

Construct a valid argument based on the Euler diagram shown.

15.

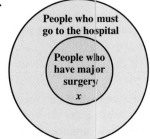

x represents Andrea Sheehan.

16.

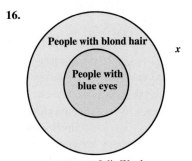

x represents Julie Ward.

As mentioned in the text, an argument can have a true conclusion yet be invalid. In these exercises, each argument has a true *conclusion. Identify each argument as* valid *or* invalid.

17. All cars have tires.
All tires are rubber.

All cars have rubber.

18. All birds fly.
All planes fly.

A bird is not a plane.

19. All chickens have beaks.
All birds have beaks.

All chickens are birds.

20. All chickens have beaks.
All hens are chickens.

All hens have beaks.

21. California is adjacent to Arizona.
Arizona is adjacent to Nevada.

California is adjacent to Nevada.

22. Quebec is northeast of Ottawa.
Quebec is northeast of Toronto.

Ottawa is northeast of Toronto.

23. A scalene triangle has a longest side.
A scalene triangle has a largest angle.

The largest angle in a scalene triangle is opposite the longest side.

24. No whole numbers are negative.

   -4 is negative.
   _____
   -4 is not a whole number.

In Exercises 25–30, the premises marked A, B, *and* C *are followed by several possible conclusions. Take each conclusion in turn, and check whether the resulting argument is* valid *or* invalid.

   A. *All people who drive contribute to air pollution.*
   B. *All people who contribute to air pollution make life a little worse.*
   C. *Some people who live in a suburb make life a little worse.*

25. Some people who live in a suburb drive.

26. Some people who live in a suburb contribute to air pollution.

27. Some people who contribute to air pollution live in a suburb.

28. Suburban residents never drive.

29. All people who drive make life a little worse.

30. Some people who make life a little worse live in a suburb.

 31. Find examples of arguments in magazine ads. Check them for validity.

 32. Find examples of arguments on television commercials. Check them for validity.

■ 6.6 ANALYZING ARGUMENTS WITH TRUTH TABLES

? THINK ABOUT IT
If a man could be in two places at one time, then I'd be with you. I am not with you so, can I conclude that a man can't be in two places at one time?

This question will be analyzed using truth tables in Example 2 of this section.

In Section 6.5 we showed how to use Euler diagrams to test the validity of arguments. While Euler diagrams often work well for simple arguments, difficulties can develop with more complex ones. These difficulties occur because Euler diagrams require a sketch showing every possible case. In complex arguments, it is hard to be sure that all cases have been considered.

In deciding whether to use Euler diagrams to test the validity of an argument, look for quantifiers such as *all, some,* or *no.* These words often indicate arguments best tested by Euler diagrams. If these words are absent, it may be better to use truth tables to test the validity of an argument.

As an example of this method, consider the following argument:

> If the floor is dirty, then I must mop it.
> The floor is dirty.
> _____
> I must mop it.

In order to test the validity of this argument, we begin by identifying the *component* statements found in the argument. They are "the floor is dirty" and "I must mop it." We shall assign the letters p and q to represent these statements:

p represents "the floor is dirty;"

q represents "I must mop it."

Now we write the two premises and the conclusion in symbols:

Premise 1: $p \rightarrow q$

Premise 2: p

Conclusion: q .

To decide if this argument is valid, we must determine whether the conjunction of both premises implies the conclusion for all possible cases of truth values for p and q. Therefore, write the conjunction of the premises as the antecedent of a conditional statement, and the conclusion as the consequent.

$$[(p \rightarrow q) \quad \wedge \quad p] \quad \rightarrow \quad q$$

premise   and   premise   implies   conclusion

Finally, construct the truth table for the conditional statement, as shown below.

p	q	$p \rightarrow q$	$(p \rightarrow q) \wedge p$	$[(p \rightarrow q) \wedge p] \rightarrow q$
T	T	T	T	T
T	F	F	F	T
F	T	T	F	T
F	F	T	F	T

Since the final column, shown in color, indicates that the conditional statement that represents the argument is true for all possible truth values of p and q, the statement is a tautology. Thus, the argument is valid.

The pattern of the argument in the floor-mopping example,

$$p \rightarrow q$$

$$p$$

$$q \quad ,$$

is a common one, and is called **modus ponens,** or the *law of detachment.*

In summary, to test the validity of an argument using a truth table, go through the steps in the box that follows.

TESTING THE VALIDITY OF AN ARGUMENT WITH A TRUTH TABLE

1. Assign a letter to represent each component statement in the argument.
2. Express each premise and the conclusion symbolically.
3. Form the symbolic statement of the entire argument by writing the *conjunction* of *all* the premises as the antecedent of a conditional statement, and the conclusion of the argument as the consequent.
4. Complete the truth table for the conditional statement formed in part 3 above. If it is a tautology, then the argument is valid; otherwise, it is invalid.

EXAMPLE 1 Invalid Argument

Determine whether the argument is *valid* or *invalid*.

> If my check arrives in time, I'll register for the fall semester.
>
> I've registered for the fall semester.
> _____
>
> My check arrived in time.

Solution Let p represent "my check arrives (arrived) in time" and let q represent "I'll register (I've registered) for the fall semester." Using these symbols, the argument can be written in the form

$$p \rightarrow q$$
$$\underline{q \qquad}$$
$$p \qquad .$$

To test for validity, construct a truth table for the statement

$$[(p \rightarrow q) \wedge q] \rightarrow p.$$

p	q	$p \rightarrow q$	$(p \rightarrow q) \wedge q$	$[(p \rightarrow q) \wedge q] \rightarrow p$
T	T	T	T	T
T	F	F	F	T
F	T	T	T	F
F	F	T	F	T

The third row of the final column of the truth table shows F, and this is enough to conclude that the argument is invalid.

If a conditional and its converse were logically equivalent, then an argument of the type found in Example 1 would be valid. Since a conditional and its converse are *not* equivalent, the argument is an example of what is sometimes called the **fallacy of the converse.**

EXAMPLE 2 Valid Argument

Determine whether the argument is *valid* or *invalid*.

> If a man could be in two places at one time, I'd be with you.
>
> I am not with you.
> _____
>
> A man can't be in two places at one time.

Solution If p represents "a man could be in two places at one time" and q represents "I'd be with you," the argument becomes

$$p \rightarrow q$$
$$\underline{\sim q \qquad}$$
$$\sim p \qquad .$$

The symbolic statement of the entire argument is

$$[(p \rightarrow q) \wedge \sim q] \rightarrow \sim p.$$

The truth table for this argument, shown below, indicates a tautology, and the argument is valid.

p	q	$p \rightarrow q$	$\sim q$	$(p \rightarrow q) \wedge \sim q$	$\sim p$	$[(p \rightarrow q) \wedge \sim q] \rightarrow \sim p$
T	T	T	F	F	F	T
T	F	F	T	F	F	T
F	T	T	F	F	T	T
F	F	T	T	T	T	T

The pattern of reasoning of this example is called **modus tollens,** or the *law of contraposition,* or *indirect reasoning.*

With reasoning similar to that used to name the fallacy of the converse, the fallacy

$$p \rightarrow q$$
$$\underline{\sim p}$$
$$\sim q$$

is often called the **fallacy of the inverse.** An example of such a fallacy is "If it rains, I get wet. It doesn't rain. Therefore, I don't get wet."

EXAMPLE 3 Valid Argument

Determine whether the argument is *valid* or *invalid.*

I'll buy a car or I'll take a vacation.

I won't buy a car.

I'll take a vacation.

Solution If p represents "I'll buy a car" and q represents "I'll take a vacation," the argument becomes

$$p \vee q$$
$$\underline{\sim p}$$
$$q \qquad .$$

We must set up a truth table for

$$[(p \vee q) \wedge \sim p] \rightarrow q.$$

p	q	$p \vee q$	$\sim p$	$(p \vee q) \wedge \sim p$	$[(p \vee q) \wedge \sim p] \rightarrow q$
T	T	T	F	F	T
T	F	T	F	F	T
F	T	T	T	T	T
F	F	F	T	F	T

The statement is a tautology and the argument is valid. Any argument of this form is valid by the law of **disjunctive syllogism.**

EXAMPLE 4   Valid Argument

Determine whether the argument is *valid* or *invalid*.

If I experience persistent crushing chest pain in the center or left side of
   my chest, then I will call an ambulance.*

If I call the ambulance then I will be taken to the emergency room.

If I experience persistent crushing chest pain in the center or left side of
   my chest, then I will be taken to the emergency room.

Solution Let p represent "I experience persistent crushing chest pain in the center
or left side of my chest," let q represent "I will call an ambulance," and let r represent "I will be taken to the emergency room." The argument takes on the general form

$$p \rightarrow q$$
$$q \rightarrow r$$
$$\overline{}$$
$$p \rightarrow r.$$

Make a truth table for the following statement:

$$[(p \rightarrow q) \wedge (q \rightarrow r)] \rightarrow (p \rightarrow r).$$

It will require eight rows.

p	q	r	$p \rightarrow q$	$q \rightarrow r$	$p \rightarrow r$	$(p \rightarrow q) \wedge (q \rightarrow r)$	$[(p \rightarrow q) \wedge (q \rightarrow r)] \rightarrow (p \rightarrow r)$
T	T	T	T	T	T	T	T
T	T	F	T	F	F	F	T
T	F	T	F	T	T	F	T
T	F	F	F	T	F	F	T
F	T	T	T	T	T	T	T
F	T	F	T	F	T	F	T
F	F	T	T	T	T	T	T
F	F	F	T	T	T	T	T

This argument is valid since the final statement is a tautology. The pattern of argument shown in this example is called **reasoning by transitivity,** or the *law of hypothetical syllogism.*

A summary of the valid and invalid forms of argument presented so far follows.

VALID ARGUMENT FORMS

Modus Ponens	Modus Tollens	Disjunctive Syllogism	Reasoning by Transitivity
$p \rightarrow q$	$p \rightarrow q$	$p \vee q$	$p \rightarrow q$
p	$\sim q$	$\sim p$	$q \rightarrow r$
q	$\sim p$	q	$p \rightarrow r$

*Goldman, David, R., ed., *American College of Physicians Complete Home Medical Guide,*
New York, DK Publishing, Inc., 1999, p. 57.

INVALID ARGUMENT FORMS (FALLACIES)

Fallacy of the Converse	**Fallacy of the Inverse**
$p \rightarrow q$	$p \rightarrow q$
$\underline{q\qquad\qquad}$	$\underline{\sim p\qquad\qquad}$
p	$\sim q$

When an argument contains three or more premises, it will be necessary to determine the truth values of the conjunction of all of them. Remember that if *at least one* premise in a conjunction of several premises is false, then the entire conjunction is false. This will be used in the next example.

EXAMPLE 5 Invalid Argument

Determine whether the argument is *valid* or *invalid.*

> If Eddie goes to town, then Mabel stays at home.
>
> If Mabel does not stay at home, then Rita will cook.
>
> Rita will not cook.
>
> Therefore, Eddie does not go to town.

Solution In an argument written in this manner, the premises are given first, and the conclusion is the statement that follows the word "Therefore." Let p represent "Eddie goes to town," let q represent "Mabel stays at home," and let r represent "Rita will cook." The symbolic form of the argument is

$$p \rightarrow q$$
$$\sim q \rightarrow r$$
$$\underline{\sim r\qquad\qquad}$$
$$\sim p\qquad .$$

To test validity, set up a truth table for the statement

$$[(p \rightarrow q) \wedge (\sim q \rightarrow r) \wedge \sim r] \rightarrow \sim p.$$

The table is shown below.

p	q	r	$p \rightarrow q$	$\sim q$	$\sim q \rightarrow r$	$\sim r$	$(p \rightarrow q) \wedge (\sim q \rightarrow r) \wedge \sim r$	$\sim p$	$[(p \rightarrow q) \wedge (\sim q \rightarrow r) \wedge \sim r] \rightarrow \sim p$
T	T	T	T	F	T	F	F	F	T
T	T	F	T	F	T	T	T	F	F
T	F	T	F	T	T	F	F	F	T
T	F	F	F	T	F	T	F	F	T
F	T	T	T	F	T	F	F	T	T
F	T	F	T	F	T	T	T	T	T
F	F	T	T	T	T	F	F	T	T
F	F	F	T	T	F	T	F	T	T

Because the final column does not contain all Ts, the statement is not a tautology. The argument is invalid.

Consider the following poem, which has been around for many years.

For want of a nail, the shoe was lost.
For want of a shoe, the horse was lost.
For want of a horse, the rider was lost.
For want of a rider, the battle was lost.
For want of a battle, the war was lost.
Therefore, for want of a nail, the war was lost.

Each line of the poem may be written as an *if...then* statement. For example, the first line may be restated as, "If a nail is lost, then the shoe is lost." Other statements may be worded similarly. The conclusion, "for want of a nail, the war was lost," follows from the premises since repeated use of the law of transitivity applies. Arguments used by Lewis Carroll* often take on a similar form. The next example comes from one of his works.

EXAMPLE 6 Valid Argument

Supply a conclusion that yields a valid argument for the following premises.

Babies are illogical.

Nobody is despised who can manage a crocodile.

Illogical persons are despised.

Solution First, write each premise in the form *if...then*.

If you are a baby, then you are illogical.

If you can manage a crocodile, then you are not despised.

If you are illogical, then you are despised.

Let p be "you are a baby," let q be "you are logical," let r be "you can manage a crocodile," and let s be "you are despised." With these letters, the statements can be written symbolically as

$$p \rightarrow \sim q$$
$$r \rightarrow \sim s$$
$$\sim q \rightarrow s.$$

Now begin with any letter that appears only once. Here p appears only once, namely, in the first statement. Notice that this statement ends with $\sim q$, which starts the third statement. The third statement ends in s. We do not have a statement starting with s, but we will if we apply the contrapositive to the statement $r \rightarrow \sim s$, yielding $s \rightarrow \sim r$. Then rearrange the three statements as follows:

$$p \rightarrow \sim q$$
$$\sim q \rightarrow s$$
$$s \rightarrow \sim r.$$

From the three statements, repeated use of reasoning by transitivity gives the conclusion

$$p \rightarrow \sim r,$$

leading to a valid argument.

In words, the conclusion is "If you are a baby, then you cannot manage a crocodile," or, as Lewis Carroll would have written it, "Babies cannot manage crocodiles."

---

*Lewis Carroll is the pseudonym for Charles Dodgson (1832–1898), mathematician and author of *Alice in Wonderland.*

6.6 EXERCISES

Each of the following arguments is either valid by one of the forms of valid arguments discussed in this section, or is a fallacy by one of the forms of invalid arguments discussed. (See the summary boxes.) Decide whether the argument is valid *or a* fallacy, *and give the form that applies.*

1. If you use binoculars, then you get a glimpse of the comet.
   If you get a glimpse of the comet, then you'll be amazed.

   If you use binoculars, then you'll be amazed.

2. If Billy Joel comes to town, then I will go to the concert.
   If I go to the concert, then I'll call in sick for work.

   If Billy Joel comes to town, then I'll call in sick for work.

3. If Kevin O'Brien sells his quota, he'll get a bonus.
   Kevin O'Brien sells his quota.

   He gets a bonus.

4. If Colleen Pollock works hard enough, she will get a promotion.
   Colleen Pollock works hard enough.

   She gets a promotion.

5. If she buys another pair of shoes, her closet will overflow.
   Her closet will overflow.

   She buys another pair of shoes.

6. If he doesn't have to get up at 5:30 A.M., he's ecstatic.
   He's ecstatic.

   He doesn't have to get up at 5:30 A.M.

7. If Patrick Roy plays, the opponent gets shut out.
   The opponent does not get shut out.

   Patrick Roy does not play.

8. If Pedro Martinez pitches, the Red Sox win.
   The Red Sox do not win.

   Pedro Martinez does not pitch.

9. "If we evolved a race of Isaac Newtons, that would not be progress."
   (quote from Aldous Huxley)
   We have not evolved a race of Isaac Newtons.

   That is progress.

10. "If I have seen farther than others, it is because I stood on the shoulders of giants."
   (quote from Sir Isaac Newton)
   I have not seen farther than others.

   I have not stood on the shoulders of giants.

11. Pat Quinlin jogs or John Remington pumps iron.
   John Remington does not pump iron.

   Pat Quinlin jogs.

12. She uses e-commerce or she pays by credit card.
   She does not pay by credit card.

   She uses e-commerce.

Use a truth table to determine whether the argument is valid *or* invalid.

13. $p \lor q$
$\underline{p}$
$\sim q$

14. $p \land \sim q$
$\underline{p}$
$\sim q$

15. $\sim p \to \sim q$
$\underline{q}$
p

16. $p \lor \sim q$
$\underline{p}$
$\sim q$

17. $p \to q$
$\underline{q \to p}$
$p \land q$

18. $\sim p \to q$
$\underline{p}$
$\sim q$

19. $p \to \sim q$
$\underline{q}$
$\sim p$

20. $p \to \sim q$
$\underline{\sim p}$
$\sim q$

21. $(\sim p \lor q) \land (\sim p \to q)$
$\underline{p}$
$\sim q$

22. $(p \to q) \land (q \to p)$
$\underline{p}$
$p \lor q$

23. $(\sim p \land r) \to (p \lor q)$
$\underline{\sim r \to p}$
$q \to r$

24. $(r \land p) \to (r \lor q)$
$\underline{q \land p}$
$r \lor p$

25. Explain in a few sentences how to determine the statement for which a truth table will be constructed so that the arguments that follow in Exercises 28–37 can be analyzed for validity.

26. Earlier we showed how to analyze arguments using Euler diagrams. Refer to Example 4 in this section, restate each premise and the conclusion using a quantifier, and then draw an Euler diagram to illustrate the relationship.

27. Cheryl Arabie made the following observation: "If I want to determine whether an argument leading to the statement

$$[(p \to q) \land \sim q] \to \sim p$$

is valid, I only need to consider the lines of the truth table which lead to T for the column headed $(p \to q) \land \sim q$." Cheryl was very perceptive. Can you explain why her observation was correct?

▪ Applications

GENERAL INTEREST

Determine whether the following arguments are valid *or* invalid.

28. *Golf* Jeff loves to play golf. If Joan likes to sew, then Jeff does not love to play golf. If Joan does not like to sew, then Brad sings in the choir. Therefore, Brad sings in the choir.

29. *Arbor Day* If that tree is infested with pine bark beetles, then it will die. If people plant a tree on Arbor Day, it will not die. Therefore, if people plant a tree on Arbor Day, then that tree is not infested with pine bark beetles.

30. *Pokémon* If the Pokémon craze continues, then Beanie Babies will remain popular. Barbie dolls continue to be favorites or Beanie Babies will remain popular. Barbie dolls do not continue to be favorites. Therefore, the Pokémon craze does not continue.

31. *Music* Christina Aguilera sings or Ricky Martin is not a teen idol. If Ricky Martin is not a teen idol, then Britney Spears does not win an American Music Award. Britney Spears wins an American Music Award. Therefore, Christina Aguilera does not sing.

32. *Love* If I've got you under my skin, then you are deep in the heart of me. If you are deep in the heart of me, then you are not really a part of me. You are deep in the heart of me or you are really a part of me. Therefore, if I've got you under my skin, then you are really a part of me.

33. *Sports* The Colts will be in the playoffs if and only if Peyton leads the league in passing. Marv loves the Colts or Peyton leads the league in passing. Marv does not love the Colts. Therefore, the Colts will not be in the playoffs.

34. *Radio* If Otis is a disc jockey, then he lives in Lexington. He lives in Lexington and he is a history buff. Therefore, if Otis is not a history buff, then he is not a disc jockey.

35. *Love* If I were your woman and you were my man, then I'd never stop loving you. I've stopped loving you. Therefore, I am not your woman or you are not my man.

36. *Equality* All men are created equal. All people who are created equal are women. Therefore, all men are women.

37. *Socrates* All men are mortal. Socrates is a man. Therefore, Socrates is mortal.

38. *Time* Suppose that you ask someone for the time and you get the following response:

> "If I tell you the time, then we'll start chatting. If we start chatting, then you'll want to meet me at a truck stop. If we meet at a truck stop, then we'll discuss my family. If we discuss my family, then you'll find out that my daughter is available for marriage. If you find out that she is available for marriage, then you'll want to marry her. If you want to marry her, then my life will be miserable since I don't want my daughter married to some fool who can't afford a $10 watch."

Use reasoning by transitivity to draw a valid conclusion.

Lewis Carroll In the arguments used by Lewis Carroll, it is helpful to restate a premise in if . . . then *form in order to more easily identify a valid conclusion. The following premises come from Lewis Carroll. Write each premise in* if . . . then *form.*

39. All my poultry are ducks.

40. None of your sons can do logic.

41. Guinea pigs are hopelessly ignorant of music.

42. No teetotalers are pawnbrokers.

43. No teachable kitten has green eyes.

44. Opium-eaters have no self-command.

45. I have not filed any of them that I can read.

46. All of them written on blue paper are filed.

Lewis Carroll The following exercises involve premises from Lewis Carroll. Write each premise in symbols, and then in the final part, give a conclusion that yields a valid argument.

47. Let *p* be "it is a duck," *q* be "it is my poultry," *r* be "one is an officer," and *s* be "one is willing to waltz."

   a. No ducks are willing to waltz.

   b. No officers ever decline to waltz.

   c. All my poultry are ducks.

   d. Give a conclusion that yields a valid argument.

48. Let *p* be "one is able to do logic," *q* be "one is fit to serve on a jury," *r* be "one is sane," and *s* be "he is your son."

   a. Everyone who is sane can do logic.

   b. No lunatics are fit to serve on a jury.

   c. None of your sons can do logic.

   d. Give a conclusion that yields a valid argument.

49. Let *p* be "one is honest," *q* be "one is a pawnbroker," *r* be "one is a promise breaker," *s* be "one is trustworthy," *t* be "one is very communicative," and *u* be "one is a wine drinker."

   a. Promise breakers are untrustworthy.

   b. Wine drinkers are very communicative.

   c. A person who keeps a promise is honest.

   d. No teetotalers are pawnbrokers. (*Hint:* Assume "teetotaler" is the opposite of "wine drinker.")

   e. One can always trust a very communicative person.

   f. Give a conclusion that yields a valid argument.

50. Let *p* be "it is a guinea pig," *q* be "it is hopelessly ignorant of music," *r* be "it keeps silent while the *Moonlight Sonata* is being played," and *s* be "it appreciates Beethoven."

   a. Nobody who really appreciates Beethoven fails to keep silent while the *Moonlight Sonata* is being played.

   b. Guinea pigs are hopelessly ignorant of music.

   c. No one who is hopelessly ignorant of music ever keeps silent while the *Moonlight Sonata* is being played.

   d. Give a conclusion that yields a valid argument.

51. Let *p* be "it begins with 'Dear Sir'," *q* be "it is crossed," *r* be "it is dated," *s* be "it is filed," *t* be "it is in black ink," *u* be "it is in the third person," *v* be "I can read it," *w* be "it is on blue paper," *x* be "it is on one sheet," and *y* be "it is written by Brown."

   a. All the dated letters are written on blue paper.

   b. None of them are in black ink, except those that are written in the third person.

   c. I have not filed any of them that I can read.

   d. None of them that are written on one sheet are undated.

   e. All of them that are not crossed are in black ink.

   f. All of them written by Brown begin with "Dear Sir."

   g. All of them written on blue paper are filed.

   h. None of them written on more than one sheet are crossed.

   i. None of them that begin with "Dear Sir" are written in the third person.

   j. Give a conclusion that yields a valid argument.

52. Let p be "he is going to a party," q be "he brushes his hair," r be "he has self-command," s be "he looks fascinating," t be "he is an opium eater," u be "he is tidy," and v be "he wears white kid gloves."

a. No one who is going to a party ever fails to brush his hair.

b. No one looks fascinating if he is untidy.

c. Opium eaters have no self-command.

d. Everyone who has brushed his hair looks fascinating.

e. No one wears white kid gloves unless he is going to a party. (*Hint:* "a unless b" $\equiv \sim b \rightarrow a$.)

f. A man is always untidy if he has no self-command.

g. Give a conclusion that yields a valid argument.

■ CHAPTER SUMMARY

In this chapter we introduced symbolic logic, which uses letters to represent statements, and symbols for words such as *and, or,* and *not.* Statements, which are declarative sentences that are either true or false, but not both simultaneously, were explored. The concept of negation was also introduced. Using logical connectives, two or more statements can be combined to form a compound statement. Truth values of various compound statements were explored using truth tables. One compound statement of particular interest is the conditional that is formed when two statements are combined using the *if. . . then* connective. Symbolic logic was then used to design circuits. Finally, arguments were analyzed with Euler diagrams and with truth tables.

■ KEY TERMS

6.1 statement	disjunction	**6.4** converse	Euler diagram
compound statement	equivalent	inverse	**6.6** modus ponens
component statement	**6.3** conditional	contrapositive	fallacy of the converse
logical connective	antecedent	biconditional	modus tollens
negation	consequent	**6.5** premise	fallacy of the inverse
universal quantifier	tautology	conclusion	disjunctive syllogism
existential quantifier	circuit	valid	reasoning by
6.2 conjunction	parallel	invalid	transitivity
truth table	series	fallacy	

■ CHAPTER 6 REVIEW EXERCISES

Write a negation for each of the following statements.

1. $6 - 3 = 3$

2. All men are created equal.

3. Some members of the class went on the field trip.

4. If that's the way you feel, then I will accept it.

5. She passed GO and collected $200.

Let p represent "You will love me" *and let q represent* "I will love you." *Write each of the following in symbols.*

6. If you won't love me, then I will love you.

7. I will love you if you will love me.

8. I won't love you if and only if you won't love me.

Using the same statements as for Exercises 6–8, write each of the following in words.

9. $\sim p \wedge q$ **10.** $\sim(p \vee \sim q)$

In each of the following, assume that p is true and that q and r are false. Find the truth value of each statement.

11. $\sim q \wedge \sim r$ **12.** $r \vee (p \wedge \sim q)$

13. $r \rightarrow (s \vee r)$ (The truth value of the statement s is unknown.)

14. $p \leftrightarrow (p \rightarrow q)$

15. Explain in your own words why, if p is a statement, the biconditional $p \leftrightarrow \sim p$ must be false.

16. State the necessary conditions for

 a. a conditional statement to be false

 b. a conjunction to be true

 c. a disjunction to be false.

Construct a truth table for each of the following.

17. $p \wedge (\sim p \vee q)$ **18.** $\sim(p \wedge q) \rightarrow (\sim p \vee \sim q)$

Decide whether each statement is true *or* false.

19. Some negative integers are whole numbers.

20. All irrational numbers are real numbers.

Write each conditional statement in if . . . then *form.*

21. All integers are rational numbers.

22. Being a rhombus is sufficient for a polygon to be a quadrilateral.

23. Being divisible by 3 is necessary for a number to be divisible by 9.

24. She digs dinosaur bones only if she is a paleontologist.

For each statement, write **(a)** *the converse,* **(b)** *the inverse, and* **(c)** *the contrapositive.*

25. If a picture paints a thousand words, the graph will help me understand it.

26. $\sim p \rightarrow (q \wedge r)$ (Use one of De Morgan's laws as necessary.)

27. Use an Euler diagram to determine whether the following argument is *valid* or *invalid.*

 All members of that video club save money.

 Pat Pearson is a member of that video club.

 Pat Pearson saves money.

28. Match each argument in parts a–d with the law that justifies its validity, or the fallacy of which it is an example.

 a. Modus ponens

 b. Modus tollens

 c. Reasoning by transitivity

 d. Disjunctive syllogism

   e. Fallacy of the converse

   f. Fallacy of the inverse

   a. If he eats liver, then he'll eat anything.

      He eats liver. _____

      He'll eat anything.

   b. If you use your seat belt, you will be safer.

      You don't use your seat belt. _____

      You won't be safer.

   c. If I hear *Come Saturday Morning,* I think of her.

      If I think of her, I get depressed. _____

      If I hear *Come Saturday Morning,* I get depressed.

   d. She sings or she dances.

      She does not sing. _____

      She dances.

Use a truth table to determine whether each argument is valid *or* invalid.

29. If I write a check, it will bounce. If the bank guarantees it, then it does not bounce. The bank guarantees it. Therefore, I don't write a check.

30. $\sim p \rightarrow \sim q$

    $\dfrac{q \rightarrow p}{p \vee q}$

Applications

BUSINESS AND ECONOMICS

Income Tax The following excerpts appear in a guide for preparing income tax reports. *

a. If you adopt a child who is not a U.S. citizen or resident at the time the adoption effort begins, a credit may not be claimed until the year the adoption becomes final.

b. A foster child who lives with you for the whole year qualifies only if the child was placed with you by an authorized placement agency or the child is your sibling or a descendant of a sibling.

c. An individual who is a nonresident alien for any part of the year is not eligible for the credit unless he or she is married and an election is made by the couple to have all of their worldwide income subject to U.S. tax.

d. Buy a savings certificate after June 30 with a maturity of one year or less.

31. Which of these excerpts are statements?

32. Which of these excerpts are compound statements?

33. Write the converse of statement a.

34. Write the contrapositive of statement b.

LIFE SCIENCES

35. *Medicine* The following statement appears in a home medical reference book with regard to pregnancy and rubella.[†] Write the contrapositive of the statement and then write an equivalent statement using the word *or.*

> If you contract rubella in early pregnancy, your baby is at serious risk of being born with abnormalities, such as congenital deafness, congenital heart disease, clouding of the lens in the eye, and the nervous system disorder cerebral palsy.

**Your Income Tax 2001,* J. K. Lasser Institute, New York, John Wiley and Sons, Inc., 2000, p. xxix.
[†]Goldman, David R., ed., *American College of Physicians Complete Home Medical Guide,* New York, DK Publishing, Inc., 1999, p. 294.

SOCIAL SCIENCES

36. *Gun Control*   The following statements appear in an article about gun control.* Negate each statement.

a. Regulations have both costs and benefits, and rules that are passed to solve a problem can sometimes make it worse.

b. Shooters overwhelmingly have problems with alcoholism and have long criminal histories, particularly arrests for violent acts.

c. They are disproportionately involved in automobile crashes and are much more likely to have had their driver's license suspended or revoked.

GENERAL INTEREST

Monopoly Junior™   *The following excerpts can be found on the box to the game, Monopoly Junior.*™†

a. Join Rich Uncle Pennybags and his nieces and nephews for a thrill-filled day at the Boardwalk Amusements—the Roller Coaster, the Magic Show, the Water Slide, the Video Arcade and more.

b. Draw a Chance card and take a ride on the Miniature Railroad, win a free Ticket Booth . . . or pay $3 to visit the Rest Rooms!

c. When someone finally runs out of money, the player with the most cash on hand wins this colorful, fast-paced junior version of the world's most popular board game.

d. Set up Ticket Booths at the Amusements and collect fees from other players who land on them.

37. Which of these excerpts are statements?

38. Which of these excerpts are compound statements?

---

*Lott, John R., "When Gun Control Costs Lives," *National Forum: The Phi Kappa Phi Journal,* Vol. 80, No. 4, Fall 2000, pp. 29–32.
†Monopoly® is Hasbro, Inc.'s trademark for its real estate trading board game equipment.

EXTENDED APPLICATION: Logic Puzzles

Some people find that logic puzzles, which appear in periodicals such as *World-Class Logic Problems* (Penny Press) and *Logic Puzzles* (Dell), provide hours of enjoyment. They are based on deductive reasoning, and players answer questions based on clues given. The following explanation on solving such problems appeared in the Autumn 1999 issue of *World-Class Logic Problems.*

How to Solve Logic Problems

Solving logic problems is entertaining and challenging. All the information you need to solve a logic problem is given in the introduction and clues, and in illustrations, when provided. If you've never solved a logic problem before, our sample should help you get started. Fill in the Sample Solving Chart in Figure 21 as you follow our explanation. We use a "•" to signify "Yes" and an "X" to signify "No."

Five couples were married last week, each on a different weekday. From the information provided, determine the woman (one is Cathy) and man (one is Paul) who make up each couple, as well as the day on which each couple was married.

1. Anne was married on Monday, but not to Wally.
2. Stan's wedding was on Wednesday. Rob was married on Friday, but not to Ida.
3. Vern (who married Fran) was married the day after Eve.

SAMPLE SOLVING CHART:

	PAUL	ROB	STAN	VERN	WALLY	MONDAY	TUESDAY	WEDNESDAY	THURSDAY	FRIDAY
ANNE										
CATHY										
EVE										
FRAN										
IDA										
MONDAY										
TUESDAY										
WEDNESDAY										
THURSDAY										
FRIDAY										

FIGURE 21

Anne was married Monday (1), so put a "•" at the intersection of Anne and Monday Put "X"s in all the other days in Anne's row and all the other names in the Monday column. (Whenever you establish a relationship, as we did here, be sure to place "X"s at the intersections of all relationships that become impossible as a result.) Anne wasn't married to Wally (1), so put an "X" at the intersection of Anne and Wally. Stan's wedding was Wednesday (2), so put a "•" at the intersection of Stan and Wednesday (don't forget the "X"s). Stan didn't marry Anne, who was married Monday, so put an "X" at the intersection of Anne and Stan. Rob was married Friday, but not to Ida (2), so put a "•" at the intersection of Rob and Friday, and "X"s at the intersections of Rob and Ida and Ida and Friday. Rob also

didn't marry Anne, who was married Monday, so put an "X" at the intersection of Anne and Rob. Now your chart should look like Figure 22.

	PAUL	ROB	STAN	VERN	WALLY	MONDAY	TUESDAY	WEDNESDAY	THURSDAY	FRIDAY
ANNE		X	X		X	•	X	X	X	X
CATHY						X				
EVE						X				
FRAN						X				
IDA		X				X				X
MONDAY		X	X							
TUESDAY		X	X							
WEDNESDAY	X	X	•	X	X					
THURSDAY		X	X							
FRIDAY	X	•	X	X	X					

FIGURE 22

Vern married Fran (3), so put a "•" at the intersection of Vern and Fran. This leaves Anne's only possible husband as Paul, so put a "•" at the intersection of Anne and Paul and Paul and Monday Vern and Fran's wedding was the day after Eve's (3), which wasn't Monday [Anne], so Vern's wasn't Tuesday. It must have been Thursday [see chart], so Eve's was Wednesday (3). Put "•"'s at the intersections of Vern and Thursday, Fran and Thursday, and Eve and Wednesday. Now your chart should look like Figure 23.

	PAUL	ROB	STAN	VERN	WALLY	MONDAY	TUESDAY	WEDNESDAY	THURSDAY	FRIDAY
ANNE	•	X	X	X	X	•	X	X	X	X
CATHY	X		X		X					
EVE	X		X		X	X	X	•	X	X
FRAN	X	X	X	•	X	X	X	X	•	X
IDA	X	X		X		X	•		X	X
MONDAY	•	X	X	X	X					
TUESDAY	X	X	X	X						
WEDNESDAY	X	X	•	X	X					
THURSDAY	X	X	X	•	X					
FRIDAY	X	•	X	X	X					

FIGURE 23

The chart shows that Cathy was married Friday, Ida was married Tuesday, and Wally was married Tuesday. Ida married Wally, and Cathy's wedding was Friday, so she married Rob. After this information is filled in, Eve could only have married Stan. You've completed the puzzle, and your chart should now look like Figure 24.

In summary: Anne and Paul, Monday; Cathy and Rob, Friday; Eve and Stan, Wednesday; Fran and Vern, Thursday; Ida and Wally, Tuesday.

In some problems, it may be necessary to make a logical guess based on facts you've established. When you do, always look for clues or other facts that disprove it. If you find that your guess is incorrect, eliminate it as a possibility.

FIGURE 24

Exercises

1. *Pumpkin-Patch Kids.** Last Saturday, each of five children, accompanied by one of his or her parents, went to Goldman's Farm to pick out a pumpkin. Once there, however, each child (including Raven) became far more fascinated with a different aspect of the farm (in one case, feeding the animals) than he or she was in the pumpkin patch! Fortunately, each parent was able to recapture his or her child's interest just long enough for them to pick out the perfect pumpkin for their Halloween jack-o'-lantern! From the information provided, match each child with his or her parent (one is Mr. Maier) and determine the aspect of the farm each found more intriguing than pumpkin picking.

   a. Neither Lauren nor Zach was the child accompanied by Ms. Reed. Tara was more interested in shopping at the country store than in picking pumpkins.

   b. Xander and his father, Mr. Morgan, didn't go on the hay ride. Ms. Fedor's child (who isn't Zach or Lauren) was fascinated by the cider-making process.

   c. Zach is neither the child who went on the hay ride nor the one who wanted to go apple picking. Mr. Hanson's child didn't go on the hay ride.

2. *Have a Ball.** With the start of the winter league just a few weeks away, the pro shop at Bowl Me Over Lanes has been packed with people picking out their new bowling balls. Just this evening, each of five bowlers purchased a ball, each a different color (one is gray) and weight (10, 12, 14, 16, or 18 pounds). After having his or her fingers sized and the holes drilled, each person's name was imprinted on his or her ball. Although their balls won't be ready for another week, these high rollers are anxious to experience life in the fast lane! From the information provided, determine the color and weight of each bowler's ball.

   a. Arlene had her name imprinted on her new orange bowling ball. Devon bought a bowling bag for his new ball, which is lighter than Silas's.

   b. Tina bowls with a 14-pound ball. The pink bowling ball is exactly 6 pounds lighter than the turquoise one.

   c. The red bowling ball, which isn't Rosetta's, weighs the most. Rosetta didn't buy the 16-pound ball.

3. *It's a Tie*[†] by Jenny Roberts. Four colleagues recently got into a discussion about some of the flamboyant patterns showing up on neckties these days. As a joke, each man arrived at work the next day sporting the most ridiculous tie he could find (no two men wore ties with the same pattern—one tie was decorated with smiling cupids). None of the men had to venture outside of his own closet, as each had received at least one such tie from a different relative! From the following clues, can you match each man with the pattern on his flamboyant tie, as well as determine the relative who presented each man with his tie?

   a. The tie with the grinning leprechauns wasn't a present from a daughter.

   b. Mr. Crow's tie features neither the dancing reindeer nor the yellow happy faces.

   c. Mr. Speigler's tie wasn't a present from his uncle.

   d. The tie with the yellow happy faces wasn't a gift from a sister.

   e. Mr. Evans and Mr. Speigler own the tie with the grinning leprechauns and the tie that was a present from a father-in-law, in some order.

---

*World-Class Logic Problems, Autumn 1999, p. 5.
[†]Dell Logic Puzzles, October 1999, p. 7.

f. Mr. Hurley received his flamboyant tie from his sister.

	Cupids	Happy faces	Leprechauns	Reindeer	Daughter	Father-in-law	Sister	Uncle
Mr. Crow								
Mr. Evans								
Mr. Hurley								
Mr. Speigler								
Daughter								
Father-in-law								
Sister								
Uncle								

4. *Imaginary Friends** by Keith King. Grantville's local library recently sponsored a writing contest for young children in the community. Each of four contestants (including Ralph) took on the task of bringing to life an imaginary friend in a short story. Each child selected a different type of animal (including a moose) to personify, and each described a different adventure involving this new friend (one story described how an imaginary friend had formed a rock band). From the following clues, can you match each young author with his or her imaginary friend and determine the adventure the two had together?

	Grizzly bear	Moose	Seal	Zebra	Circus	Rock band	Spaceship	Train
Joanne								
Lou								
Ralph								
Winnie								
Circus								
Rock band								
Spaceship								
Train								

a. The seal (who isn't the creation of either Joanne or Lou) neither rode to the moon in a spaceship nor took a trip around the world on a magic train.

b. Joanne's imaginary friend (who isn't the grizzly bear) went to the circus.

c. Winnie's imaginary friend is a zebra.

d. The grizzly bear didn't board the spaceship to the moon.

5. *Dunking Booth*† by Margaret Rounds. In order to help raise money for Dellville's annual fall festival, the coach of a local men's baseball team volunteered to sit in a dunking booth. Five of Coach Clark's players, each of whom plays a different position on the team (including first base), stepped up to the challenge of dunking him, and each player took a different number of throws (two, four, five, six, or nine) to complete his mission. From the clues below, can you determine the order in which each player (identified by full name—one surname is Winslow) dunked the coach, the number of throws it took to do so, and the position each plays?

1. The right fielder took three more throws than Josh (who wasn't the last dunker).

2. The shortstop threw immediately before Mr. Belford, who threw immediately before Geoffrey.

3. Mr. Cavallo dunked the coach with his sixth pitch.

4. The players who took an even number of throws are Scott, Mr. Hepler, and the fifth person to dunk the coach, in some order.

5. The pitcher (who wasn't the fourth dunker) took more throws than Oliver, but fewer throws than Mr. Janney.

6. The players who took an odd number of throws are Edgar and the catcher, in some order.

Order	First name	Last name	Throws	Position
_____	_____	_____	_____	_____
_____	_____	_____	_____	_____
_____	_____	_____	_____	_____
_____	_____	_____	_____	_____
_____	_____	_____	_____	_____

	Edgar	Geoffrey	Josh	Oliver	Scott	Belford	Cavallo	Hepler	Janney	Winslow	Two	Four	Five	Six	Nine	Catcher	First base	Pitcher	Right field	Shortstop
First																				
Second																				
Third																				
Fourth																				
Fifth																				
Catcher																				
First base																				
Pitcher																				
Right Field																				
Shortstop																				
Two																				
Four																				
Five																				
Six																				
Nine																				
Belford																				
Cavallo																				
Hepler																				
Janney																				
Winslow																				

**Dell Logic Puzzles,* October 1999, p. 7.

†*Dell Logic Puzzles,* October 1999.

Sets and Probability

The study of probability begins with counting. An exercise in Section 2 of this chapter counts trucks carrying different combinations of early, late, and extra late peaches from the orchard to canning facilities. You'll see trees in another context in Section 5, where we use branching tree diagrams to calculate conditional probabilities.

In this chapter and the next, we introduce the basic ideas of probability theory, a branch of mathematics that has become increasingly important in management and in the biological and social sciences. Probability theory is valuable because it provides a way to deal with uncertainty. Since the language of sets and set operations is used in the study of probability, we begin there.

7.1 SETS

THINK ABOUT IT

In how many ways can two candidates win the 50 states plus the District of Columbia in a U.S. presidential election?

Using knowledge of sets, we will answer this question in one of the exercises.

Think of a **set** as a well-defined collection of objects in which it is possible to determine if a given object is included in the collection. A set of coins might include one of each type of coin now put out by the U.S. government. Another set might be made up of all the students in your English class. In mathematics, sets are often made up of numbers. The set consisting of the numbers 3, 4, and 5 is written

$$\{3, 4, 5\},$$

with set braces, $\{\ \}$, enclosing the numbers belonging to the set. The numbers 3, 4, and 5 are called the **elements** or **members** of this set. To show that 4 is an element of the set $\{3, 4, 5\}$, we use the symbol $\in$ and write

$$4 \in \{3, 4, 5\},$$

read "4 is an element of the set containing 3, 4, and 5." Also, $5 \in \{3, 4, 5\}$. To show that 8 is *not* an element of this set, place a slash through the symbol:

$$8 \notin \{3, 4, 5\}.$$

Sets often are named with capital letters, so that if

$$B = \{5, 6, 7\},$$

then, for example, $6 \in B$ and $10 \notin B$.

A collection of people called "young adults" does not constitute a set unless the words "young adults" are defined so that membership in the set can be determined.

It is possible to have a set with no elements. Some examples are the set of counting numbers less than one, the set of female presidents of the United States, and the set of men more than 10 feet tall. A set with no elements is called the **empty set** and is written $\emptyset$.

> **CAUTION** Be careful to distinguish between the symbols 0, $\emptyset$, $\{0\}$, and $\{\emptyset\}$. The symbol 0 represents a *number*; $\emptyset$ represents a *set* with 0 elements; $\{0\}$ represents a set with one element, 0; and $\{\emptyset\}$ represents a set with one element, $\emptyset$.

Two sets are *equal* if they contain the same elements. The sets $\{5,6,7\}$, $\{7,6,5\}$, and $\{6,5,7\}$ all contain exactly the same elements and are equal. In symbols,

$$\{5,6,7\} = \{7,6,5\} = \{6,5,7\}.$$

Sets that do not contain exactly the same elements are *not equal*. For example, the sets $\{5,6,7\}$ and $\{7,8,9\}$ do not contain exactly the same elements and thus are not equal. To indicate that these sets are not equal, we write

$$\{5,6,7\} \neq \{7,8,9\}.$$

The sets $\{1,2,3\}$ and $\{1,1,2,3,3\}$, however, are equal because both sets contain the same elements.

Sometimes we are interested in a common property of the elements in a set, rather than a list of the elements. This common property can be expressed by using **set-builder notation,** for example,

$$\{x \mid x \text{ has property } P\}$$

(read "the set of all elements x such that x has property P") represents the set of all elements x having some stated property P.

EXAMPLE 1   Sets

Write the elements belonging to each of the following sets.

(a) $\{x \mid x \text{ is a natural number less than } 5\}$

Solution   The natural numbers less than 5 make up the set $\{1,2,3,4\}$.

(b) $\{x \mid x \text{ is a state that borders Florida}\}$

Solution   The states that border Florida make up the set $\{\text{Alabama, Georgia}\}$.

The **universal set** for a particular discussion is a set that includes all the objects being discussed. In elementary school arithmetic, for instance, the set of whole numbers might be the universal set, while in a college algebra class the universal set might be the set of real numbers. The universal set will be specified when necessary, or it will be clearly understandable from the context of the problem.

Subsets   Sometimes every element of one set also belongs to another set. For example, if

$$A = \{3,4,5,6\}$$

and

$$B = \{2,3,4,5,6,7,8\},$$

then every element of A is also an element of B. This is an example of the following definition.

> **SUBSET**
>
> Set A is a **subset** of set B (written $A \subseteq B$) if every element of A is also an element of B. Set A is a *proper subset* (written $A \subset B$) if $A \subseteq B$ and $A \neq B$.

To indicate that A is *not* a subset of B, we write $A \nsubseteq B$.

EXAMPLE 2 Sets

Decide whether the following statements are true or false.

(a) $\{3, 4, 5, 6\} = \{4, 6, 3, 5\}$

Solution Both sets contain exactly the same elements, so the sets are equal and the given statement is true. (The fact that the elements are listed in a different order does not matter.)

(b) $\{5, 6, 9, 12\} \subseteq \{5, 6, 7, 8, 9, 10, 11\}$

Solution The first set is not a subset of the second because it contains an element, 12, that does not belong to the second set. Therefore, the statement is false.

By the definition of subset, the empty set (which contains no elements) is a subset of every set. That is, if A is any set, and the symbol $\emptyset$ represents the empty set, then $\emptyset \subseteq A$. Also, the definition of subset can be used to show that every set is a subset of itself; that is, if A is any set, then $A \subseteq A$.

> For any set A,
>
> $$\emptyset \subseteq A \quad \text{and} \quad A \subseteq A.$$

EXAMPLE 3 Subsets

List all possible subsets for each of the following sets.

(a) $\{7, 8\}$

Solution There are 4 subsets of $\{7, 8\}$:

$$\emptyset, \quad \{7\}, \quad \{8\}, \quad \text{and} \quad \{7, 8\}.$$

(b) $\{a, b, c\}$

Solution There are 8 subsets of $\{a, b, c\}$:

$$\emptyset, \quad \{a\}, \quad \{b\}, \quad \{c\}, \quad \{a, b\}, \quad \{a, c\}, \quad \{b, c\}, \quad \text{and} \quad \{a, b, c\}.$$

In Example 3, all the subsets of $\{7, 8\}$ and all the subsets of $\{a, b, c\}$ were found by trial and error. An alternative method uses a **tree diagram,** a systematic way of listing all the subsets of a given set. Figure 1 shows tree diagrams for finding the subsets of $\{7, 8\}$ and $\{a, b, c\}$.

As Figure 1 shows, there are two possibilities for each element (either it's in the subset or it's not), so a set with 2 elements has $2 \cdot 2 = 2^2 = 4$ subsets, and a set with 3 elements has $2^3 = 8$ subsets. This idea can be extended to a set with any finite number of elements, which leads to the following conclusion.

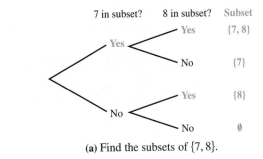

(a) Find the subsets of $\{7, 8\}$.

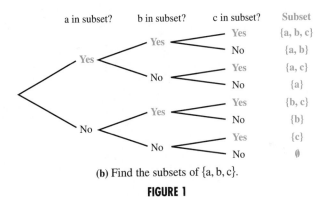

(b) Find the subsets of $\{a, b, c\}$.

FIGURE 1

A set of n distinct elements has 2^n subsets.

EXAMPLE 4   Subsets

Find the number of subsets for each of the following sets.

(a) $\{3, 4, 5, 6, 7\}$

Solution This set has 5 elements; thus, it has 2^5 or 32 subsets.

(b) $\{-1, 2, 3, 4, 5, 6, 12, 14\}$

Solution This set has 8 elements and therefore has 2^8 or 256 subsets.

(c) $\emptyset$

Solution Since the empty set has 0 elements, it has $2^0 = 1$ subset—itself.

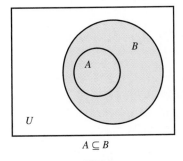

$A \subseteq B$

FIGURE 2

Figure 2 shows a set A that is a subset of set B. The rectangle represents the universal set, U. Such diagrams, called **Venn diagrams**—after the English logician John Venn (1834–1923), who invented them in 1876—are used to help illustrate relationships among sets.

Set Operations   It is possible to form new sets by combining or manipulating one or more existing sets. Given a set A and a universal set U, the set of all elements of U that do *not* belong to A is called the **complement** of set A. For example, if set A is the set of all the female students in a class, and U is the set of all students in the class, then the complement of A would be the set of all male students in the class. The complement of set A is written A', read "A-prime."

COMPLEMENT OF A SET

Let A be any set, with U representing the universal set. Then the complement of A, colored red in the figure, is

$$A' = \{x \mid x \notin A \text{ and } x \in U\}.$$

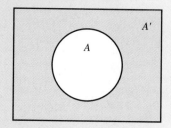

EXAMPLE 5   Set Operations

Let $U = \{1, 2, 3, 4, 5, 6, 7\}$, $A = \{1, 3, 5, 7\}$, and $B = \{3, 4, 6\}$. Find each of the following sets.

(a) A'

Solution   Set A' contains the elements of U that are not in A.

$$A' = \{2, 4, 6\}$$

(b) $B' = \{1, 2, 5, 7\}$

(c) $\emptyset' = U$ and $U' = \emptyset$

(d) $(A')' = A$

Given two sets A and B, the set of all elements belonging to *both* set A and set B is called the **intersection** of the two sets, written $A \cap B$. For example, the elements that belong to both set $A = \{1, 2, 4, 5, 7\}$ and set $B = \{2, 4, 5, 7, 9, 11\}$ are 2, 4, 5, and 7, so that

$$A \cap B = \{1, 2, 4, 5, 7\} \cap \{2, 4, 5, 7, 9, 11\} = \{2, 4, 5, 7\}.$$

INTERSECTION OF TWO SETS

The intersection of sets A and B, shown in green in the figure, is

$$A \cap B = \{x \mid x \in A \text{ and } x \in B\}.$$

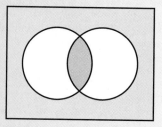

EXAMPLE 6   Set Operations

Let $A = \{3, 6, 9\}$, $B = \{2, 4, 6, 8\}$, and the universal set $U = \{0, 1, 2, \ldots, 10\}$. Find each of the following.

(a) $A \cap B$

Solution

$$A \cap B = \{3, 6, 9\} \cap \{2, 4, 6, 8\} = \{6\}$$

(b) $A \cap B'$

Solution

$$A \cap B' = \{3, 6, 9\} \cap \{0, 1, 3, 5, 7, 9, 10\} = \{3, 9\}$$

Two sets that have no elements in common are called **disjoint sets.** For example, there are no elements common to both $\{50, 51, 54\}$ and $\{52, 53, 55, 56\}$, so these two sets are disjoint, and

$$\{50, 51, 54\} \cap \{52, 53, 55, 56\} = \emptyset.$$

This result can be generalized.

DISJOINT SETS

For any sets A and B, if A and B are disjoint sets, then $A \cap B = \emptyset$.

Figure 3 shows a pair of disjoint sets.

The set of all elements belonging to set A, to set B, or to both sets is called the **union** of the two sets, written $A \cup B$. For example,

$$\{1, 3, 5\} \cup \{3, 5, 7, 9\} = \{1, 3, 5, 7, 9\}.$$

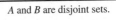

A and B are disjoint sets.

FIGURE 3

UNION OF TWO SETS

The union of sets A and B, shown in blue in the figure, is

$$A \cup B = \{x \,|\, x \in A \text{ or } x \in B\}.$$

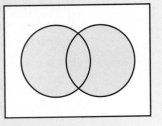

EXAMPLE 7   Union of Sets

Let $A = \{1, 3, 5, 7, 9, 11\}$, $B = \{3, 6, 9, 12\}$, $C = \{1, 2, 3, 4, 5\}$, and the universal set $U = \{0, 1, 2, \ldots, 12\}$. Find each of the following.

(a) $A \cup B$

Solution Begin by listing the elements of the first set, $\{1, 3, 5, 7, 9, 11\}$. Then include any elements from the second set *that are not already listed.* Doing this gives

$$A \cup B = \{1, 3, 5, 7, 9, 11\} \cup \{3, 6, 9, 12\} = \{1, 3, 5, 7, 9, 11, 6, 12\}$$
$$= \{1, 3, 5, 6, 7, 9, 11, 12\}.$$

(b) $(A \cup B) \cap C'$

Solution Begin with the expression in parentheses, which we calculated in part (a), and then intersect this with C'.

$$(A \cup B) \cap C' = \{1, 3, 5, 6, 7, 9, 11, 12\} \cap \{0, 6, 7, 8, 9, 10, 11, 12\}$$
$$= \{6, 7, 9, 11, 12\}$$

> **NOTE** 1. As Example 7 shows, when forming sets, do not list the same element more than once. In our final answer, we listed the elements in numerical order to make it easier to see what elements are in the set, but the set is the same, regardless of the order of the elements.
>
> 2. As shown in the definitions, an element is in the *intersection* of sets A and B if it is in A *and* B. On the other hand, an element is in the *union* of sets A and B if it is in A *or* B (or both).

EXAMPLE 8 Stocks

Table 1 gives the January 3, 2001, intraday high and low prices, the closing price, and the net change for six companies listed on the New York Stock Exchange.*

Table 1				
Stock	**high**	**low**	**close**	**change**
AT&T	20^{56}	18^{50}	20^{13}	$+1^{88}$
Chevron	85^{44}	81^{88}	82^{63}	-3^{31}
GenMills	44^{38}	42^{75}	43^{19}	-0^{81}
Hershey	65^{13}	61^{06}	61^{38}	-3^{75}
IBM	95	83^{75}	94^{63}	$+9^{81}$
PepsiCo	59^{38}	46^{38}	46^{50}	-2^{88}

Let set A include all stocks with a high price greater than $80, B all stocks with a closing price between $60 and $90, and C all stocks with a positive price change. Find the following.

(a) A'

Solution Set A' contains all the listed stocks outside set A, or those with a high price less than or equal to $80, so

$$A' = \{\text{AT\&T, GenMills, Hershey, PepsiCo}\}.$$

(b) $A \cap C$

Solution The intersection of A and C will contain those stocks with a high price greater than $80 and a positive price change.

$$A \cap C = \{\text{IBM}\}$$

The Wall Street Journal, Jan. 4, 2001, pp. C3–C5.

(c) $A \cup B$

Solution The union of A and B contains all stocks with a high price greater than $80 or a closing price between $60 and $90.

$$A \cup B = \{\text{Chevron, Hershey, IBM}\}$$

EXAMPLE 9 Employment

A department store classifies credit applicants by sex, marital status, and employment status. Let the universal set be the set of all applicants, M be the set of male applicants, S be the set of single applicants, and E be the set of employed applicants. Describe the following sets in words.

(a) $M \cap E$

Solution The set $M \cap E$ includes all applicants who are both male *and* employed; that is, employed male applicants.

(b) $M' \cup S$

Solution This set includes all applicants who are female (not male) *or* single. *All* female applicants and *all* single applicants are in this set.

(c) $M' \cap S'$

Solution These applicants are female *and* married (not single); thus, $M' \cap S'$ is the set of all married female applicants.

(d) $M \cup E'$

Solution $M \cup E'$ is the set of applicants that are male *or* unemployed. The set includes *all* male applicants and *all* unemployed applicants.

7.1 EXERCISES

In Exercises 1–9, write true *or* false *for each statement.*

1. $3 \in \{2, 5, 7, 9, 10\}$

2. $6 \in \{-2, 6, 9, 5\}$

3. $9 \notin \{2, 1, 5, 8\}$

4. $3 \notin \{7, 6, 5, 4\}$

5. $\{2, 5, 8, 9\} = \{2, 5, 9, 8\}$

6. $\{3, 7, 12, 14\} = \{3, 7, 12, 14, 0\}$

7. $\{\text{all whole numbers greater than 7 and less than 10}\} = \{8, 9\}$

8. $\{x \mid x \text{ is an odd integer}; 6 \le x \le 18\} = \{7, 9, 11, 15, 17\}$

9. $0 \in \emptyset$

10. What is set-builder notation? Give an example.

Let $A = \{2, 4, 6, 8, 10, 12\}$; $B = \{2, 4, 8, 10\}$; $C = \{4, 8, 12\}$; $D = \{2, 10\}$; $E = \{6\}$; and $U = \{2, 4, 6, 8, 10, 12, 14\}$. Insert $\subseteq$ or $\nsubseteq$ to make the statement true.

11. $A \_\_ U$

12. $E \_\_ A$

13. $A \_\_ E$

14. $B \_\_ C$

15. $\emptyset \_\_ A$

16. $\{0, 2\} \_\_ D$

17. $D \_\_ B$

18. $A \_\_ C$

19. Repeat Exercises 11–18 except insert $\subset$ or $\not\subset$ to make the statement true.

*Insert a number in each blank to make the statement true, using the sets for
Exercises 11–18.*

20. There are exactly ___ subsets of A.

21. There are exactly ___ subsets of B.

22. There are exactly ___ subsets of C.

23. There are exactly ___ subsets of D.

24. Describe the intersection and union of sets. How do they differ?

Insert ∩ or ∪ to make each statement true.

25. $\{5, 7, 9, 19\}$ ___ $\{7, 9, 11, 15\} = \{7, 9\}$

26. $\{8, 11, 15\}$ ___ $\{8, 11, 19, 20\} = \{8, 11\}$

27. $\{2, 1, 7\}$ ___ $\{1, 5, 9\} = \{1\}$

28. $\{6, 12, 14, 16\}$ ___ $\{6, 14, 19\} = \{6, 14\}$

29. $\{3, 5, 9, 10\}$ ___ $\emptyset = \emptyset$

30. $\{3, 5, 9, 10\}$ ___ $\emptyset = \{3, 5, 9, 10\}$

31. $\{1, 2, 4\}$ ___ $\{1, 2, 4\} = \{1, 2, 4\}$

32. Is it possible for two nonempty sets to have the same intersection and union? If so, give an example.

Let $U = \{2, 3, 4, 5, 7, 9\}$; $X = \{2, 3, 4, 5\}$; $Y = \{3, 5, 7, 9\}$; and $Z = \{2, 4, 5, 7, 9\}$. List the members of each of the following sets, using set braces.

33. $X \cap Y$

34. $X \cup Y$

35. X'

36. Y'

37. $X' \cap Y'$

38. $X' \cap Z$

39. $Y \cap (X \cup Z)$

40. $X' \cap (Y' \cup Z)$

41. $(X \cap Y') \cup Z'$

42. a. In Example 6, what set do you get when you calculate $(A \cap B) \cup (A \cap B')$?

b. Explain in words why $(A \cap B) \cup (A \cap B') = A$.

Let $U = \{$all students in this school$\}$; $M = \{$all students taking this course$\}$; $N = \{$all students taking accounting$\}$; $P = \{$all students taking zoology$\}$. Describe each of the following sets in words.

43. M'

44. $M \cup N$

45. $N \cap P$

46. $N' \cap P'$

47. Refer to the sets listed for Exercises 11–18. Which pairs of sets are disjoint?

48. Refer to the sets listed for Exercises 33–41. Which pairs are disjoint?

Refer to Example 8 in the text. Describe each of the sets in Exercises 49–52 in words; then list the elements of each set.

49. B'

50. $A \cap B$

51. $(A \cap B)'$

52. $(A \cup C)'$

53. Let $A = \{1, 2, 3, \{3\}, \{1, 4, 7\}\}$. Answer each of the following as true or false.

a. $1 \in A$ **b.** $\{3\} \in A$ **c.** $\{2\} \in A$ **d.** $4 \in A$ **e.** $\{\{3\}\} \subset A$

f. $\{1, 4, 7\} \in A$ **g.** $\{1, 4, 7\} \subseteq A$

54. Let $B = \{a, b, c, \{d\}, \{e, f\}\}$. Answer each of the following as true or false.

a. $a \in B$ **b.** $\{b, c, d\} \subset B$ **c.** $\{d\} \in B$ **d.** $\{d\} \subseteq B$ **e.** $\{e, f\} \in B$

f. $\{a, \{e, f\}\} \subset B$ **g.** $\{e, f\} \subset B$

Applications

BUSINESS AND ECONOMICS

Mutual Funds The following table shows the top five holdings of four major mutual funds on July 12, 2000.*

Let U be the smallest possible set that includes all the corporations listed, and V, J, F, and T be the set of top-five holdings for each mutual fund, respectively. Find each of the following sets.

55. $V \cap J$

56. $V \cap (F \cup T)$

57. $(J \cup F)'$

58. $J' \cap T'$

Vanguard 500	Janus Worldwide	Fidelity Magellan	T. Rowe Price Blue Chip Growth Fund
General Electric Inc.	China Telecom Ltd.	General Electric Inc.	Microsoft Corp.
Intel Corp.	Cisco Systems Inc.	Microsoft Corp.	Cisco Systems Inc.
Cisco Systems Inc.	Nokia Oyj	Cisco Systems Inc.	Citigroup Inc.
Microsoft Corp.	Nortel Networks	Home Depot Inc.	Intel Corp.
Exxon Mobil Corp.	NTT Mobile Corp.	Intel Corp.	Tyco International

LIFE SCIENCES

Health The table below shows some symptoms of an overactive thyroid and an underactive thyroid.[†]

Underactive Thyroid	Overactive Thyroid
Sleepiness, s	Insomnia, i
Dry hands, d	Moist hands, m
Intolerance of cold, c	Intolerance of heat, h
Goiter, g	Goiter, g

Let U be the smallest possible set that includes all the symptoms listed, N be the set of symptoms for an underactive thyroid, and O be the set of symptoms for an overactive thyroid. Find each of the following sets.

59. O'

60. N'

61. $N \cap O$

62. $N \cup O$

63. $N \cap O'$

SOCIAL SCIENCES

64. *Electoral College* U.S. presidential elections are decided by the Electoral College, in which each of the 50 states, plus the District of Columbia, gives all of its votes to a candidate.[‡] Ignoring the number of votes each state has in the Electoral College, but including all possible combinations of states that could be won by either candidate, how many outcomes are possible in the Electoral College if there are two candidates? (*Hint:* The states that can be won by a candidate form a subset of all the states.)

GENERAL INTEREST

65. *Musicians* A concert featured a cellist, a flutist, a harpist, and a vocalist. Throughout the concert, different subsets of the four musicians performed together, with at least two musicians playing each piece. How many subsets of at least two are possible?

66. *Cat Food* Suppose 9 flavors of cat food are available in a store. Euclid, the mathematical cat, could like all 9 flavors,

---

*Top 5 holdings found for each fund at www.vanguard.com, ww4.janus.com, www.fidelity.com, www.troweprice.com, respectively.

[†]*The Merck Manual of Diagnosis and Therapy,* 16th ed., Merck Research Laboratories, 1992, pp. 1075 and 1080.

[‡]The exceptions are Maine and Nebraska, which allocate their electoral college votes according to the winner in each congressional district.

or none, or any combination of selected flavors. How many possibilities are there for the set of flavors that Euclid likes? (*Hint:* Each set of flavors is a subset of the original 9 flavors.)

Cable Television *The following table lists the top five cable television networks in 1999.* Use this information for Exercises 67–72.*

Network	Subscribers (million)	Major Content
TBS	77.0	Movies, variety, sports
Discovery Channel	76.4	Scientific exploration, variety
ESPN	76.2	Sports, specials, documentaries
USA	75.8	Movies, specials, variety
C-SPAN	75.7	Specials, political broadcasts

List the elements of the following sets.

67. F, the set of networks with more than 76 million subscribers

68. G, the set of networks that feature sports events

69. M, the set of networks that feature movies

70. $F \cap G$ **71.** $M \cup G$ **72.** F'

73. *Games* In David Gale's game of Subset Takeaway, the object is for each player, at his or her turn, to pick a nonempty proper subset of a given set subject to the condition that no subset chosen earlier by either player can be a subset of the newly chosen set.[†] Consider the set $A = \{1, 2, 3\}$. Suppose Joe and Dorothy are playing the game and Dorothy goes first. If she chooses the proper subset $\{1\}$, then Joe cannot choose any subset that includes the element 1. Joe can, however, choose $\{2\}$ or $\{3\}$ or $\{2, 3\}$. Develop a strategy for Joe so that he can always win the game if Dorothy goes first.

■ 7.2 APPLICATIONS OF VENN DIAGRAMS

THINK ABOUT IT

The responses to a survey of 100 households show that 21 have a DVD player, 56 have a videocassette recorder, and 12 have both. How many have neither a DVD player nor a videocassette recorder?

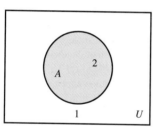

One set leads to 2 regions (numbering is arbitrary).

FIGURE 4

It is difficult to answer this question from the given information. In this section we show how a Venn diagram can be used to sort out such information, and later in this section, we are able to answer this question. Venn diagrams were used in the previous section to illustrate set union and intersection. The rectangular region of a Venn diagram represents the universal set U. Including only a single set A inside the universal set, as in Figure 4, divides U into two regions. Region 1 represents those elements of U outside set A (that is, the elements in A'), and region 2 represents those elements belonging to set A. (Our numbering of these regions is arbitrary.)

The Venn diagram in Figure 5(a) shows two sets inside U. These two sets divide the universal set into four regions. As labeled in Figure 5(a), region 1 represents the set whose elements are outside both set A and set B. Region 2 shows the set whose elements belong to A and not to B. Region 3 represents the set whose elements belong to both A and B. Which set is represented by region 4? (Again, the labeling is arbitrary.) Two other situations can arise when representing sets by Venn diagrams. If it is known that $A \cap B = \emptyset$, then the Venn diagram is drawn as in Figure 5(b). If it is known that $A \subseteq B$, then the Venn diagram is drawn as in

---

**The World Almanac and Book of Facts 2000*, p. 189.
[†]Stewart, Ian, "Mathematical Recreations: A Strategy for Subsets," *Scientific American,* Mar. 2000, pp. 96–98.

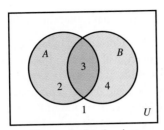

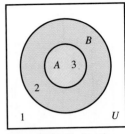

Two sets lead to 4 regions
(numbering is arbitrary).

(a)

Two sets lead to 3 regions
(numbering is arbitrary).

(b)

Two sets lead to 3 regions
(numbering is arbitrary).

(c)

FIGURE 5

Figure 5(c). For the material presented throughout this chapter we will only refer to Venn diagrams like the one in Figure 5(a), and note that some of the regions of the Venn diagram may be equal to the null set.

EXAMPLE 1 Venn Diagrams

Draw Venn diagrams similar to Figure 5(a) and shade the regions representing the following sets.

(a) $A' \cap B$

Solution Set A' contains all the elements outside set A. As labeled in Figure 5(a), A' is represented by regions 1 and 4. Set B is represented by regions 3 and 4. The intersection of sets A' and B, the set $A' \cap B$, is given by the region common to the combined regions 1 and 4 and the combined regions 3 and 4. The result is the set represented by region 4, which is blue in Figure 6. When looking for the intersection, remember to choose the area that is in one region *and* the other region.

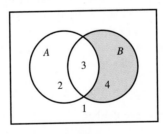

FIGURE 6

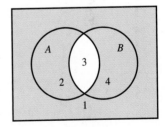

FIGURE 7

(b) $A' \cup B'$

Solution Again, set A' is represented by regions 1 and 4, and set B' by regions 1 and 2. To find $A' \cup B'$, identify the region that represents the set of all elements in A', B', or both. The result, which is blue in Figure 7, includes regions 1, 2, and 4. When looking for the union, remember to choose the area that is in one region *or* the other region (or both).

Venn diagrams also can be drawn with three sets inside U. These three sets divide the universal set into eight regions, which can be numbered (arbitrarily) as in Figure 8.

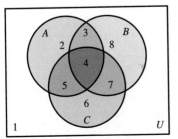

Three sets lead to 8 regions.

FIGURE 8

EXAMPLE 2 Venn Diagrams

In a Venn diagram, shade the region that represents $A' \cup (B \cap C')$.

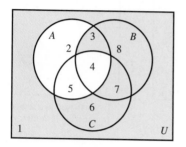

FIGURE 9

Solution First find $B \cap C'$. Set B is represented by regions 3, 4, 7, and 8, and set C' by regions 1, 2, 3, and 8. The overlap of these regions (regions 3 and 8) represents the set $B \cap C'$. Set A' is represented by regions 1, 6, 7, and 8. The union of the set represented by regions 3 and 8 and the set represented by regions 1, 6, 7, and 8 is the set represented by regions 1, 3, 6, 7, and 8, which are blue in Figure 9.

Applications We can now use a Venn diagram to answer the question posed at the beginning of this section. A researcher collecting data on 100 households finds that

21 have a DVD player;

56 have a videocassette recorder (VCR); and

12 have both.

The researcher wants to answer the following questions.

a. How many do not have a VCR?

b. How many have neither a DVD player nor a VCR?

c. How many have a DVD player but not a VCR?

Solution A Venn diagram like the one in Figure 10 will help sort out the information. In Figure 10(a), we put the number 12 in the region common to both a VCR and a DVD player, because 12 households have both. Of the 21 with a DVD player, $21 - 12 = 9$ have no VCR, so in Figure 10(b) we put 9 in the region for a DVD but no VCR. Similarly, $56 - 12 = 44$ households have a VCR but not a DVD player, so we put 44 in that region. Finally, the diagram shows that $100 - 44 - 12 - 9 = 35$ households have neither a VCR nor a DVD player. Now we can answer the questions:

a. $35 + 9 = 44$ do not have a VCR.

b. 35 have neither.

c. 9 have a DVD player but not a VCR.

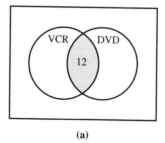

(a)

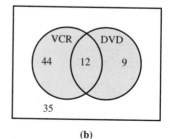

(b)

FIGURE 10

EXAMPLE 3   Magazines

A survey of 60 freshman business students at a large university produced the following results.

19 of the students read *Business Week;*

18 read *The Wall Street Journal;*

50 read *Fortune;*

13 read *Business Week* and *The Wall Street Journal;*

11 read *The Wall Street Journal* and *Fortune;*

13 read *Business Week* and *Fortune;*

 9 read all three.

Use this information to answer the following questions.

(a) How many students read none of the publications?

(b) How many read only *Fortune?*

(c) How many read *Business Week* and *The Wall Street Journal,* but not *Fortune?*

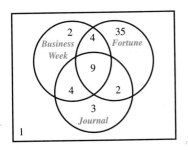

FIGURE 11

Solution Since 9 students read all three publications, begin by placing 9 in the area that belongs to all three regions, as shown in Figure 11. Of the 13 students who read *Business Week* and *Fortune,* 9 also read *The Wall Street Journal.* Therefore, only $13 - 9 = 4$ read just *Business Week* and *Fortune.* Place the number 4 in the area of Figure 11 common only to *Business Week* and *Fortune* readers. In the same way, place 4 in the region common only to *Business Week* and *The Wall Street Journal,* and 2 in the region common only to *Fortune* and *The Wall Street Journal.*

The data show that 19 students read *Business Week.* However, $4 + 9 + 4 = 17$ readers already have been placed in the region representing *Business Week.* The balance of this region will contain only $19 - 17 = 2$ students. These 2 students read *Business Week* only—not *Fortune* and not *The Wall Street Journal.* In the same way, 3 students read only *The Wall Street Journal* and 35 read only *Fortune.*

A total of $2 + 4 + 3 + 4 + 9 + 2 + 35 = 59$ students are placed in the three circles in Figure 11. Since 60 students were surveyed, $60 - 59 = 1$ student reads none of the three publications, and 1 is placed outside all three regions.

Now Figure 11 can be used to answer the questions asked above.

(a) Only 1 student reads none of the three publications.

(b) There are 35 students who read only *Fortune.*

(c) The overlap of the regions representing readers of *Business Week* and *The Wall Street Journal* shows that 4 students read *Business Week* and *The Wall Street Journal* but not *Fortune.*

CAUTION A common error in solving problems of this type is to make a circle represent one set and another circle represent its complement. In Example 3, with one circle representing those who read *Business Week,* we did not draw another for those who do not read *Business Week.* An additional circle is not only unnecessary (because those not in one set are automatically in the other) but very confusing, because the region outside or inside both circles must be empty. Similarly, if a problem involves men and women, do not draw one circle for men and another for women. Draw one circle; if you label it "women," for example, then men are automatically those outside the circle.

EXAMPLE 4 Utility Maintenance

Jeff Friedman is a section chief for an electric utility company. The employees in his section cut down trees, climb poles, and splice wire. Friedman reported the following information to the management of the utility.

"Of the 100 employees in my section,

45 can cut trees;

50 can climb poles;

57 can splice wire;

28 can cut trees and climb poles;

20 can climb poles and splice wire;

25 can cut trees and splice wire;

11 can do all three;

9 can't do any of the three (management trainees)."

The data supplied by Friedman lead to the numbers shown in Figure 12. Add the numbers from all of the regions to get the total number of employees:

$$9 + 3 + 14 + 23 + 11 + 9 + 17 + 13 = 99.$$

Friedman claimed to have 100 employees, but his data indicate only 99. The management decided that Friedman didn't qualify as a section chief, and he was reassigned as a night-shift meter reader in Guam. (*Moral:* He should have taken this course.)

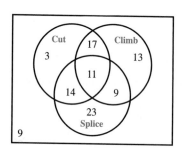

FIGURE 12

> **NOTE** In all the examples above, we started with the innermost region, the intersection of all categories. This is usually the best way to begin solving problems of this type.

We use the symbol $n(A)$ to indicate the *number* of elements in a set A. For example, if $A = \{a, b, c, d, e\}$, then $n(A) = 5$. The following statement about the number of elements in the union of two sets will be used later in our study of probability.

UNION RULE FOR SETS

$$n(A \cup B) = n(A) + n(B) - n(A \cap B)$$

To prove this statement, let $x + y$ represent $n(A)$, y represent $n(A \cap B)$, and $y + z$ represent $n(B)$, as shown in Figure 13. Then

$$n(A \cup B) = x + y + z,$$
$$n(A) + n(B) - n(A \cap B) = (x + y) + (y + z) - y = x + y + z,$$

so

$$n(A \cup B) = n(A) + n(B) - n(A \cap B).$$

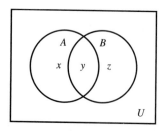

FIGURE 13

EXAMPLE 5 School Activities

A group of 10 students meet to plan a school function. All are majoring in accounting or economics or both. Five of the students are economics majors and 7 are majors in accounting. How many major in both subjects?

Solution Let A represent the set of accounting majors and B represent the set of economics majors. Use the union rule, with $n(A) = 5$, $n(B) = 7$, and $n(A \cup B) = 10$. We must find $n(A \cap B)$.

$$n(A \cup B) = n(A) + n(B) - n(A \cap B)$$
$$10 = 5 + 7 - n(A \cap B),$$

so
$$n(A \cap B) = 5 + 7 - 10 = 2.$$

When A and B are disjoint, $n(A \cap B) = 0$ and the union rule simplifies.

UNION RULE FOR DISJOINT SETS

If A and B are disjoint sets,

$$n(A \cup B) = n(A) + n(B).$$

EXAMPLE 6 Political Parties

Suppose a random sample of 200 voters was selected in the 1994 elections for the U.S. House of Representatives. The approximate numbers expected from the East, Midwest, South, or West who voted Democrat or Republican is given by the following table.*

	East (E)	Midwest (M)	South (S)	West (W)	Totals
Democrat (D)	24	22	27	26	99
Republican (R)	22	28	33	18	101
Totals	46	50	60	44	200

Using the letters given in the table, find the number of people in each of the following sets.

(a) $D \cap S$

Solution The set $D \cap S$ consists of all those who voted Democrat *and* were from the South. From the table, we see that there were 27 such people.

(b) $D \cup S$

Solution The set $D \cup S$ consists of all those who voted Democrat *or* were from the South. We include all 99 who voted Democrat, plus the 33 who were from the South and did not vote Democrat, for a total of 132. Alternatively, we could use the formula $n(D \cup S) = n(D) + n(S) - n(D \cap S) = 99 + 60 - 27 = 132$.

(c) $(E \cup W) \cap R'$

Solution Begin with the set $E \cup W$, which is everyone from the East or the West. This consists of the four categories with 24, 22, 26, and 18 people. Of this set, take those who did *not* vote Republican, for a total of $24 + 26 = 50$ people. This is the number of people from the East or the West who did not vote Republican.

---

*The New York Times, Nov. 13, 1994, p. 24.

EXAMPLE 7 Mathematics Classes

Suppose that a group of 150 students were enrolled in at least one of three mathematics classes: A, B, and C. In addition,

90 students were in class A

50 students were in class B

70 students were in class C

15 students were in classes A and C

12 students were in classes B and C

10 students were in all three classes.

Determine how many students were enrolled in both classes A and B.

Solution Since 10 students were in all three classes, begin by placing 10 in the area that belongs to all three regions, as shown in Figure 14. Of the 15 students who were in classes A and C, 10 were also in class B. Thus, only $15 - 10 = 5$ students were in the area of Figure 14 common only to classes A and C. Similarly, place a 2 in the area of Figure 14 common only to classes B and C. We usually proceed to fill in the area that is common only to A and B. In this case, however, it is currently unknown and so the variable x is placed in this area of Figure 14.

Of the 70 students who were enrolled in class C, 10 were also enrolled in classes A and B, 5 were also enrolled in class A only, and 2 were also enrolled in class B only. Thus, only $70 - 10 - 5 - 2 = 53$ were enrolled in only class C, and the number 53 is placed in this area of Figure 14. In a similar manner, we determine that $50 - 10 - 2 - x = 38 - x$ students took only class B and $90 - 10 - 5 - x = 75 - x$ students took only class A, as illustrated by Figure 14. Note that because all 150 students took at least one class, there are no elements in the region outside the three circles.

Now that the diagram is filled out, we can determine the value of x by recalling that the total number of students enrolled in at least one class was 150. Thus,

$$75 - x + 5 + 10 + x + 38 - x + 2 + 53 = 150.$$

Simplifying, we have $183 - x = 150$, implying that $x = 33$.

To complete that problem, the number of students enrolled in classes A and B is

$$33 + 10 = 43.$$

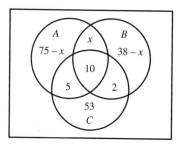

FIGURE 14

7.2 EXERCISES

Sketch a Venn diagram like the one in the figure, and use shading to show each of the following sets.

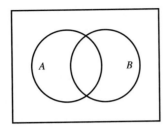

1. $B \cap A'$

2. $A \cup B'$

3. $A' \cup B$

4. $A' \cap B'$

5. $B' \cup (A' \cap B')$

6. $(A \cap B) \cup B'$

7. U'

8. $\varnothing'$

9. Three sets divide the universal set into at most _____ regions.

10. What does the notation $n(A)$ represent?

Sketch a Venn diagram like the one shown, and use shading to show each of the following sets.

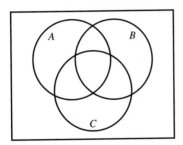

11. $(A \cap B) \cap C$

12. $(A \cap C') \cup B$

13. $A \cap (B \cup C')$

14. $A' \cap (B \cap C)$

15. $(A' \cap B') \cap C$

16. $(A \cap B') \cup C$

17. $(A \cap B') \cap C$

18. $A' \cap (B' \cup C)$

Use the union rule to answer the following questions.

19. If $n(A) = 5$, $n(B) = 8$, and $n(A \cap B) = 4$, what is $n(A \cup B)$?

20. If $n(A) = 12$, $n(B) = 27$, and $n(A \cup B) = 30$, what is $n(A \cap B)$?

21. Suppose $n(B) = 7$, $n(A \cap B) = 3$, and $n(A \cup B) = 20$. What is $n(A)$?

22. Suppose $n(A \cap B) = 5$, $n(A \cup B) = 35$, and $n(A) = 13$. What is $n(B)$?

Draw a Venn diagram and use the given information to fill in the number of elements for each region.

23. $n(U) = 38$, $n(A) = 16$, $n(A \cap B) = 12$, $n(B') = 20$

24. $n(A) = 26$, $n(B) = 10$, $n(A \cup B) = 30$, $n(A') = 17$

25. $n(A \cup B) = 17$, $n(A \cap B) = 3$, $n(A) = 8$, $n(A' \cup B') = 21$

26. $n(A') = 28$, $n(B) = 25$, $n(A' \cup B') = 45$, $n(A \cap B) = 12$

27. $n(A) = 28$, $n(B) = 34$, $n(C) = 25$, $n(A \cap B) = 14$, $n(B \cap C) = 15$, $n(A \cap C) = 11$, $n(A \cap B \cap C) = 9$, $n(U) = 59$

28. $n(A) = 54$, $n(A \cap B) = 22$, $n(A \cup B) = 85$, $n(A \cap B \cap C) = 4$, $n(A \cap C) = 15$, $n(B \cap C) = 16$, $n(C) = 44$, $n(B') = 63$

29. $n(A \cap B) = 6$, $n(A \cap B \cap C) = 4$, $n(A \cap C) = 7$, $n(B \cap C) = 4$, $n(A \cap C') = 11$, $n(B \cap C') = 8$, $n(C) = 15$, $n(A' \cap B' \cap C') = 5$

30. $n(A) = 13$, $n(A \cap B \cap C) = 4$, $n(A \cap C) = 6$, $n(A \cap B') = 6$, $n(B \cap C) = 6$, $n(B \cap C') = 11$, $n(B \cup C) = 22$, $n(A' \cap B' \cap C') = 5$

In Exercises 31–34, show that the statement is true by drawing Venn diagrams and shading the regions representing the sets on each side of the equals sign.*

31. $(A \cup B)' = A' \cap B'$

32. $(A \cap B)' = A' \cup B'$

33. $A \cap (B \cup C) = (A \cap B) \cup (A \cap C)$

34. $A \cup (B \cap C) = (A \cup B) \cap (A \cup C)$

35. Use the union rule of sets to prove that $n(A \cup B \cup C) = n(A) + n(B) + n(C) - n(A \cap B) - n(A \cap C) - n(B \cap C) + n(A \cap B \cap C)$. (*Hint:* Write $A \cup B \cup C$ as $A \cup (B \cup C)$ and use the formula from Exercise 33.)

◼ Applications

BUSINESS AND ECONOMICS

Use Venn diagrams to answer the following questions.

36. *Cooking Preferences* Jeff Friedman, of Example 4 in the text, was again reassigned, this time to the home economics department of the electric utility. He interviewed 140 people in a suburban shopping center to discover some of their cooking habits. He obtained the following results:

58 use microwave ovens;

63 use electric ranges;

58 use gas ranges;

19 use microwave ovens and electric ranges;

17 use microwave ovens and gas ranges;

4 use both gas and electric ranges;

1 uses all three;

2 use none of the three.

Should he be reassigned one more time? Why or why not?

37. *Harvesting Fruit* Toward the middle of the harvesting season, peaches for canning come in three types, early, late, and extra late, depending on the expected date of ripening. During a certain week, the following data were recorded at a fruit delivery station:

34 trucks went out carrying early peaches;

61 carried late peaches;

50 carried extra late;

25 carried early and late;

30 carried late and extra late;

8 carried early and extra late;

6 carried all three;

9 carried only figs (no peaches at all).

a. How many trucks carried only late variety peaches?

b. How many carried only extra late?

c. How many carried only one type of peach?

d. How many trucks (in all) went out during the week?

38. *Cola Consumption* Market research showed that the adult residents of a certain small town in Georgia fit the following categories of cola consumption.

Age	Drink Regular Cola (R)	Drink Diet Cola (D)	Drink No Cola (N)	Totals
21–25 (Y)	40	15	15	70
26–35 (M)	30	30	20	80
Over 35 (O)	10	50	10	70
Totals	80	95	45	220

Using the letters given in the table, find the number of people in each of the following sets.

a. $Y \cap R$

b. $M \cap D$

c. $M \cup (D \cap Y)$

d. $Y' \cap (D \cup N)$

e. $O' \cup N$

f. $M' \cap (R' \cap N')$

✎ **g.** Describe the set $M \cup (D \cap Y)$ in words.

---

*The statements in Exercises 31 and 32 are known as DeMorgan's laws. They are named for the English mathematician Augustus DeMorgan (1806–1871).

39. *Investment Habits* The following table shows the results of a survey taken by a bank in a medium-sized town in Tennessee. The survey asked questions about the investment habits of bank customers.

Age	Stocks (S)	Bonds (B)	Savings Accounts (A)	Totals
18–29 (Y)	6	2	15	23
30–49 (M)	14	5	14	33
50 or over (O)	32	20	12	64
Totals	52	27	41	120

Using the letters given in the table, find the number of people in each of the following sets.

a. $Y \cap B$ **b.** $M \cup A$ **c.** $Y \cap (S \cup B)$

d. $O' \cup (S \cup A)$ **e.** $(M' \cup O') \cap B$

f. Describe the set $Y \cap (S \cup B)$ in words.

40. *Investment Survey* Most mathematics professors love to invest their hard-earned money. A recent survey of 150 math professors revealed that

> 111 invested in stocks;
> 98 invested in bonds;
> 100 invested in certificates of deposit;
> 80 invested in stocks and bonds;
> 83 invested in bonds and certificates of deposit;
> 85 invested in stocks and certificates of deposit;
> 9 did not invest in any of the three.

How many mathematics professors invested in stocks and bonds and certificates of deposit?

LIFE SCIENCES

41. *Genetics* After a genetics experiment on 50 pea plants, the number of plants having certain characteristics was tallied, with the following results:

> 22 were tall;
> 25 had green peas;
> 39 had smooth peas;
> 9 were tall and had green peas;
> 20 had green peas and smooth peas;
> 6 had all three characteristics;
> 4 had none of the characteristics.

a. Find the number of plants that were tall and had smooth peas.

b. How many plants were tall and had peas that were neither smooth nor green?

c. How many plants were not tall but had peas that were smooth and green?

42. *Blood Antigens* Human blood can contain the A antigen, the B antigen, both the A and B antigens, or neither antigen. A third antigen, called the Rh antigen, is important in human reproduction, and again may or may not be present in an individual. Blood is called type A-positive if the individual has the A and Rh, but not the B antigen. A person having only the A and B antigens is said to have type AB-negative blood. A person having only the Rh antigen has type O-positive blood. Other blood types are defined in a similar manner. Identify the blood types of the individuals in regions (a)–(g) below. (One region has no letter assigned to it.)

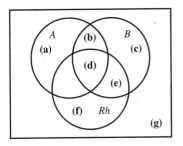

43. *Blood Antigens* (Use the diagram from Exercise 42.) In a certain hospital, the following data were recorded.

> 25 patients had the A antigen;
> 17 had the A and B antigens;
> 27 had the B antigen;
> 22 had the B and Rh antigens;
> 30 had the Rh antigen;
> 12 had none of the antigens;
> 16 had the A and Rh antigens;
> 15 had all three antigens.

How many patients

a. were represented?

b. had exactly one antigen?

c. had exactly two antigens?

d. had O-positive blood?

e. had AB-positive blood?

f. had B-negative blood?

g. had O-negative blood?

h. had A-positive blood?

44. *Mortality* The table on the next page lists the number of deaths in the U.S. during 1997 according to race and sex.*

National Vital Statistics Reports, Vol. 47, No. 19, June 30, 1999, Table I.

Use this information and the letters given to find the number of people in each set.

	White (W)	Black (B)	American Indian (I)	Asian or Pacific Islander (A)
Female (F)	1,009,509	132,410	4591	13,696
Male (M)	986,884	144,110	5985	17,060

a. F

b. $F \cap (I \cup A)$

c. $M \cup B$

d. $W' \cup I' \cup A'$

e. In words, describe the set in part b.

45. *Hockey* The following table lists the number of head and neck injuries for 319 ice hockey players wearing either a full shield or half shield in the Canadian Inter-University Athletics Union during the 1997–1998 season.* Using the letters given in the table, find the number of injuries in each set.

	Half Shield (H)	Full Shield (F)
Head and Face Injuries (A)	95	34
Concussions (B)	41	38
Neck Injuries (C)	9	7
Other Injuries (D)	202	150

a. $A \cap F$

b. $C \cap (H \cup F)$

c. $D \cup F$

d. $B' \cap C'$

SOCIAL SCIENCES

46. *Military* The number of female military personnel on active duty in September 1998 is given in the table at the bottom of the page. Use this information and the letters given to find the number of female military personnel in each of the following sets.

a. $A \cup B$

c. $E \cup (C \cup D)$

c. $O' \cap M'$

47. *Living Arrangements* In 1998, the percentage of white children under 18 years of age who lived with both parents was 74; the percentage of white children under 18 years of age who lived with their father only was 5; and the percentage of white children under 18 years of age that lived with neither parent was 3.[†] What percentage of children under age 18 lived with their mother only?

48. *Living Arrangements* In 1998, there were 3989 (in thousands) children under the age of 18 living with their grandparents. Of these children, 503 had both parents also living with them, 1827 had only their mother living with them, and 241 had only their father living with them.[‡] How many children lived with their grandparents only?

U.S. Population The U.S. population by age and race in 2000 (in millions) is given in the table at the top of the next page. Use this information in Exercises 49–54.[§]

Using the letters given in the table, find the number of people in each set.

49. $A \cap F$

50. $G \cup B$

51. $G \cup (C \cap H)$

52. $F \cap (B \cup H)$

53. $H \cup D$

54. $G' \cap (A' \cap C')$

	Army (A)	Air Force (B)	Navy (C)	Marines (D)	Total
Officers (O)	10,367	11,971	7777	854	30,969
Enlisted (E)	60,787	53,542	42,261	8928	165,518
Cadets & Midshipmen (M)	624	653	656	n.a.	1933
Total	71,778	66,166	50,694	9782	198,420

---

*Benson, Brian, Nicholas Nohtaki, M. Sarah Rose, Willem Meeuwisse, "Head and Neck Injuries Among Ice Hockey Players Wearing Full Face Shields vs. Half Face Shields," *JAMA,* Vol. 282, No. 24, Dec. 22/29, 1999, pp. 2328–2332.

[†]Brunner, Borga, ed., *Time Almanac 2000,* p. 395.

[‡]*The World Almanac and Book of Facts 2000,* p. 392.

[§]"Projections of the Total Resident Population by 5-Year Age Groups, Race, and Hispanic Origin, 1999 to 2000," from U.S. Bureau of the Census, *Current Population Reports.*

	Non-Hispanic White (A)	Hispanic (B)	African-American (C)	Asian American (D)	American Indian (E)	Totals
Under 45 (F)	120.1	25.7	24.4	7.6	1.5	179.3
45–64 (G)	47.5	4.9	6.3	2.2	.4	61.3
65 and over (H)	29.1	1.9	2.8	.8	.2	34.8
Totals	196.7	32.5	33.5	10.6	2.1	275.4

GENERAL INTEREST

55. *Chinese New Year* A survey of people attending a Lunar New Year celebration in Chinatown yielded the following results:

> 120 were women;
>
> 150 spoke Cantonese;
>
> 170 lit firecrackers;
>
> 108 of the men spoke Cantonese;
>
> 100 of the men did not light firecrackers;
>
> 18 of the non-Cantonese-speaking women lit firecrackers;
>
> 78 non-Cantonese-speaking men did not light firecrackers;
>
> 30 of the women who spoke Cantonese lit firecrackers.

a. How many attended?

b. How many of those who attended did not speak Cantonese?

c. How many women did not light firecrackers?

d. How many of those who lit firecrackers were Cantonese-speaking men?

56. *Native American Ceremonies* At a pow-wow in Arizona 75 Native Americans from all over the Southwest came to participate in the ceremonies. A coordinator of the pow-wow took a survey and found that

> 15 families brought food, costumes, and crafts;
>
> 25 families brought food and crafts;
>
> 42 families brought food;
>
> 6 families brought costumes and crafts, but not food;
>
> 4 families brought crafts, but neither food nor costumes;
>
> 10 families brought none of the three items;
>
> 18 families brought costumes but not crafts.

a. How many families brought costumes and food?

b. How many families brought costumes?

c. How many families brought crafts, but not costumes?

d. How many families did not bring crafts?

e. How many families brought food or costumes?

57. *Poultry Analysis* A chicken farmer surveyed his flock with the following results. The farmer had

> 9 fat red roosters;
>
> 2 fat red hens;
>
> 37 fat chickens;
>
> 26 fat roosters;
>
> 7 thin brown hens;
>
> 18 thin brown roosters;
>
> 6 thin red roosters;
>
> 5 thin red hens.

Answer the following questions about the flock. (Assume all chickens are thin or fat, red or brown, and hens or roosters.) How many chickens were

a. fat? **b.** red?

c. male? **d.** fat, but not male?

e. brown, but not fat? **f.** red and fat?

■ 7.3 INTRODUCTION TO PROBABILITY

THINK ABOUT IT

What is the probability that a randomly selected person in the United States is Hispanic or African American?

After introducing probability, we will answer this question in one of the exercises.

If you go to a supermarket and buy 5 pounds of peaches at 54 cents per pound, you can easily find the *exact* price of your purchase: $2.70. On the other hand, the produce manager of the market is faced with the problem of ordering peaches. The manager may have a good estimate of the number of pounds of peaches that will be sold during the day, but it is impossible to predict the *exact* amount. The number of pounds that customers will purchase during a day is *random:* the quantity cannot be predicted exactly. A great many problems that come up in applications of mathematics involve random phenomena—those for which exact prediction is impossible. The best that we can do is determine the *probability* of the possible outcomes.

Sample Spaces   In probability, an **experiment** is an activity or occurrence with an observable result. Each repetition of an experiment is called a **trial.** The possible results of each trial are called **outcomes.** The set of all possible outcomes for an experiment is the **sample space** for that experiment. A sample space for the experiment of tossing a coin is made up of the outcomes heads (h) and tails (t). If S represents this sample space, then

$$S = \{h, t\}.$$

EXAMPLE 1   Sample Space

Give the sample space for each experiment.

(a) A spinner like the one in Figure 15 is spun.

Solution The three outcomes are 1, 2, or 3, so the sample space is

$$\{1, 2, 3\}.$$

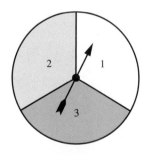

FIGURE 15

(b) For the purposes of a public opinion poll, respondents are classified as young, middle-aged, or senior, and as male or female.

Solution A sample space for this poll could be written as a set of ordered pairs:

$$\{(\text{young}, \text{male}), (\text{young}, \text{female}), (\text{middle-aged}, \text{male}),$$
$$(\text{middle-aged}, \text{female}), (\text{senior}, \text{male}), (\text{senior}, \text{female})\}.$$

(c) An experiment consists of studying the numbers of boys and girls in families with exactly 3 children. Let *b* represent *boy* and *g* represent *girl.*

Solution A three-child family can have 3 boys, written *bbb*, 3 girls, *ggg*, or various combinations, such as *bgg*. A sample space with four outcomes (not equally likely) is

$$S_1 = \{3 \text{ boys}, 2 \text{ boys and 1 girl}, 1 \text{ boy and 2 girls}, 3 \text{ girls}\}.$$

Notice that a family with 3 boys or 3 girls can occur in just one way, but a family of 2 boys and 1 girl or 1 boy and 2 girls can occur in more than one way. If the *order* of the births is considered, so that *bgg* is different from *gbg* or *ggb*, for example, another sample space is

$$S_2 = \{bbb, bbg, bgb, gbb, bgg, gbg, ggb, ggg\}.$$

The second sample space, S_2, has equally likely outcomes if we assume that boys and girls are equally likely. This assumption, while not quite true, is approximately true, so we will use it throughout this book. The outcomes in S_1 are not equally likely, since there is more than one way to get a family with 2 boys and 1 girl or a family with 2 girls and 1 boy, but only one way to get 3 boys or 3 girls.

CAUTION An experiment may have more than one sample space, as shown in Example 1(c). The most convenient sample spaces have equally likely outcomes, but it is not always possible to choose such a sample space.

Events An **event** is a subset of a sample space. If the sample space for tossing a coin is $S = \{h, t\}$, then one event is $E = \{h\}$, which represents the outcome "heads."

An ordinary die is a cube whose six different faces show the following numbers of dots: 1, 2, 3, 4, 5, and 6. If the die is fair (not "loaded" to favor certain faces over others), then any one of the faces is equally likely to come up when the die is rolled. The sample space for the experiment of rolling a single fair die is $S = \{1, 2, 3, 4, 5, 6\}$. Some possible events are listed below.

The die shows an even number: $E_1 = \{2, 4, 6\}$.

The die shows a 1: $E_2 = \{1\}$.

The die shows a number less than 5: $E_3 = \{1, 2, 3, 4\}$.

The die shows a multiple of 3: $E_4 = \{3, 6\}$.

EXAMPLE 2   Events

For the sample space S_2 in Example 1(c), write the following events.

(a) Event *H*: the family has exactly 2 girls

Solution Families with three children can have exactly 2 girls with either *bgg*, *gbg*, or *ggb*, so event *H* is

$$H = \{bgg, gbg, ggb\}.$$

(b) Event *K*: the three children are the same sex

Solution Two outcomes satisfy this condition: all boys or all girls.

$$K = \{bbb, ggg\}$$

(c) Event *J*: the family has three girls

Solution Only *ggg* satisfies this condition, so

$$J = \{ggg\}.$$

In Example 2(c), event J had only one possible outcome, ggg. Such an event, with only one possible outcome, is a **simple event.** If event E equals the sample space S, then E is called a **certain event.** If event $E = \emptyset$, then E is called an **impossible event.**

EXAMPLE 3   Events

Suppose a fair die is rolled. As shown above, the sample space is $\{1, 2, 3, 4, 5, 6\}$.

(a) The event "the die shows a 4," $\{4\}$, has only one possible outcome. It is a simple event.

(b) The event "the number showing is less than 10" equals the sample space. $S = \{1, 2, 3, 4, 5, 6\}$. This event is a certain event; if a die is rolled, the number showing (either 1, 2, 3, 4, 5, or 6) must be less than 10.

(c) The event "the die shows a 7" is the empty set $\emptyset$; this is an impossible event.

Since events are sets, we can use set operations to find unions, intersections, and complements of events. A summary of the set operations for events is given below.

SET OPERATIONS FOR EVENTS

Let E and F be events for a sample space S.
    $E \cap F$ occurs when both E and F occur;
    $E \cup F$ occurs when E or F or both occur;
    E' occurs when E does not occur.

EXAMPLE 4   Minimum Wage Workers

A study of workers earning the minimum wage grouped such workers into various categories, which can be interpreted as events when a worker is selected at random.* Consider the following events:

    E: worker is under 20;

    F: worker is white;

    G: worker is female.

Describe each of the following events in words.

(a) E'

    Solution E' is the event that the worker is 20 or over.

(b) $F \cap G'$

    Solution $F \cap G'$ is the event that the worker is white and not a female, that is, the worker is a white male.

Two events that cannot both occur at the same time, such as getting both a head and a tail on the same toss of a coin, are called **mutually exclusive events.**

---

*The New York Times, Apr. 19, 1996, p. A26.

MUTUALLY EXCLUSIVE EVENTS

Events E and F are mutually exclusive events if $E \cap F = \emptyset$.

For any event E, E and E' are mutually exclusive. By definition, mutually exclusive events are disjoint sets.

EXAMPLE 5   Mutually Exclusive Events

Let $S = \{1, 2, 3, 4, 5, 6\}$, the sample space for tossing a single die. Let $E = \{4, 5, 6\}$, and let $G = \{1, 2\}$. Then E and G are mutually exclusive events since they have no outcomes in common: $E \cap G = \emptyset$. See Figure 16.

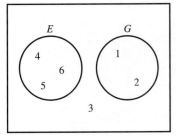

$E \cap G = \emptyset$

FIGURE 16

Probability   For sample spaces with *equally likely* outcomes, the probability of an event is defined as follows.

BASIC PROBABILITY PRINCIPLE

Let S be a sample space of equally likely outcomes, and let event E be a subset of S. Then the **probability that event E occurs** is

$$P(E) = \frac{n(E)}{n(S)}.$$

By this definition, the probability of an event is a number that indicates the relative likelihood of the event.

CAUTION   The basic probability principle only applies when the outcomes are equally likely.

EXAMPLE 6   Basic Probabilities

Suppose a single fair die is rolled. Use the sample space $S = \{1, 2, 3, 4, 5, 6\}$ and give the probability of each of the following events.

(a) E: the die shows an even number

Solution   Here, $E = \{2, 4, 6\}$, a set with three elements. Since S contains six elements,

$$P(E) = \frac{3}{6} = \frac{1}{2}.$$

(b) F: the die shows a number less than 10

Solution   Event F is a certain event, with

$$F = \{1, 2, 3, 4, 5, 6\},$$

so that

$$P(F) = \frac{6}{6} = 1.$$

(c) G: the die shows an 8

Solution   This event is impossible, so

$$P(G) = 0.$$

A standard deck of 52 cards has four suits: hearts (♥), clubs, (♣), diamonds (♦), and spades (♠), with 13 cards in each suit. The hearts and diamonds are red, and the spades and clubs are black. Each suit has an ace (A), a king (K) a queen (Q), a jack (J), and cards numbered from 2 to 10. The jack, queen, and king are called *face cards* and for many purposes can be thought of as having values 11, 12, and 13, respectively. The ace can be thought of as the low card (value 1) or the high card (value 14). See Figure 17. We will refer to this standard deck of cards often in our discussion of probability.

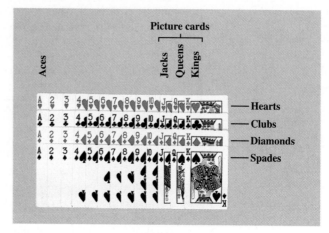

FIGURE 17

EXAMPLE 7   Playing Cards

If a single playing card is drawn at random from a standard 52-card deck, find the probability of each of the following events.

(a) Drawing an ace

Solution There are 4 aces in the deck. The event "drawing an ace" is

$$\{\text{heart ace, diamond ace, club ace, spade ace}\}.$$

Therefore,

$$P(\text{ace}) = \frac{4}{52} = \frac{1}{13}.$$

(b) Drawing a face card

Solution Since there are 12 face cards (three in each of the four suits),

$$P(\text{face card}) = \frac{12}{52} = \frac{3}{13}.$$

(c) Drawing a spade

Solution The deck contains 13 spades, so

$$P(\text{spade}) = \frac{13}{52} = \frac{1}{4}.$$

(d) Drawing a spade or a heart

Solution Besides the 13 spades, the deck contains 13 hearts, so

$$P(\text{spade or heart}) = \frac{26}{52} = \frac{1}{2}.$$

In the preceding examples, the probability of each event was a number between 0 and 1. The same thing is true in general. Any event E is a subset of the sample space S, so $0 \le n(E) \le n(S)$. Since $P(E) = n(E)/n(S)$, it follows that $0 \le P(E) \le 1$.

> For any event E, $\mathbf{0 \le P(E) \le 1.}$

EXAMPLE 8   Congressional Service

Table 2 gives the number of years of service of senators in the 106th Congress of the United States of America, which convened on January 6, 1999.*

Table 2

Years of Service	Number of Senators
0–9	51
10–19	30
20–29	13
30–39	4
40 or more	2

Find the probability that a randomly selected senator of the 106th Congress served 20–29 years when Congress convened.

Solution This probability is found by dividing the number of senators who served 20–29 years by the total number of senators. Thus,

$$P(20\text{–}29 \text{ years}) = \frac{13}{100} = .13.$$

7.3 EXERCISES

1. What is meant by a "fair" coin or die?

2. What is the sample space for an experiment?

Write sample spaces for the experiments in Exercises 3–10.

3. A month of the year is chosen for a wedding.

4. A day in April is selected for a bicycle race.

5. A student is asked how many points she earned on a recent 80-point test.

6. A person is asked the number of hours (to the nearest hour) he watched television yesterday.

7. The management of an oil company must decide whether to go ahead with a new oil shale plant or to cancel it.

The World Almanac and Book of Facts 2000, p. 81.

8. A record is kept each day for three days about whether a particular stock goes up or down.

9. A coin is tossed, and a die is rolled.

10. A box contains five balls, numbered 1, 2, 3, 4, and 5. A ball is drawn at random, the number on it recorded, and the ball replaced. The box is shaken, a second ball is drawn, and its number is recorded.

11. Define an event.

12. What is a simple event?

For the experiments in Exercises 13–18, write out the sample space, and then write the indicated events in set notation.

13. A committee of 2 people is selected from 5 executives: Alam, Bartolini, Chinn, Dickson, and Ellsberg.

a. Chinn is on the committee.

b. Dickson and Ellsberg are not both on the committee.

c. Both Alam and Chinn are on the committee.

14. A marble is drawn at random from a bowl containing 3 yellow, 4 white, and 8 blue marbles.

a. A yellow marble is drawn.   **b.** A blue marble is drawn.

c. A white marble is drawn.   **d.** A black marble is drawn.

15. Slips of paper marked with the numbers 1, 2, 3, 4, and 5 are placed in a box. After being mixed, two slips are drawn simultaneously.

a. Both slips are marked with even numbers.

b. One slip is marked with an odd number and the other is marked with an even number.

c. Both slips are marked with the same number.

16. An unprepared student takes a three-question, true/false quiz in which he guesses the answers to all three questions, so each answer is equally likely to be correct or wrong.

a. The student gets three answers wrong.

b. The student gets exactly two answers correct.

c. The student gets only the first answer correct.

17. A coin is flipped at most four times, or until two heads occur, whichever comes first.

a. The coin is tossed four times.

b. Exactly two heads are tossed.

c. No heads are tossed.

18. One jar contains four balls, labeled 1, 2, 3, and 4. A second jar contains five balls, labeled 1, 2, 3, 4, and 5. An experiment consists of taking one ball from the first jar, and then taking a ball from the second jar.

a. The number on the first ball is even.

b. The number on the second ball is even.

c. The sum of the numbers on the two balls is 5.

d. The sum of the numbers on the two balls is 1.

A single fair die is rolled. Find the probabilities of the following events.

19. Getting a 2

20. Getting an odd number

21. Getting a number less than 5

22. Getting a number greater than 2

23. Getting a 3 or a 4

24. Getting any number except 3

A card is drawn from a well-shuffled deck of 52 cards. Find the probability of drawing each of the following.

25. A 9

26. A black card

27. A black 9

28. A heart

29. The 9 of hearts

30. A face card

31. A 2 or a queen

32. A black 7 or a red 8

33. A red face card

34. A heart or a spade

A jar contains 2 white, 3 orange, 5 yellow, and 8 black marbles. If a marble is drawn at random, find the probability that it is the following.

35. White

36. Orange

37. Yellow

38. Black

39. Not black

40. Orange or yellow

41. The student sitting next to you in class concludes that the probability of the ceiling falling down on both of you before class ends is 1/2, because there are two possible outcomes—the ceiling will fall or not fall. What is wrong with this reasoning?

Applications

BUSINESS AND ECONOMICS

42. *Survey of Workers* The management of a firm wishes to check on the opinions of its assembly line workers. Before the workers are interviewed, they are divided into various categories. Define events *E*, *F*, and *G* as follows.

   E: worker is female
   F: worker has worked less than 5 years
   G: worker contributes to a voluntary retirement plan

Describe each of the following events in words.

a. E'

b. $E \cap F$

c. $E \cup G'$

d. F'

e. $F \cup G$

f. $F' \cap G'$

43. *Research Funding* According to an article in *Business Week*, in 1989 funding for university research in the United States totaled $15 billion. Support came from various sources, as shown in the table below.*

Source	Amount (in billions of dollars)
Federal government	9.0
State and local government	1.2
Institutional	2.7
Industry	1.0
Other	1.1

Find the probability that funds for a particular project came from each of the following sources.

a. Federal government

b. Industry

c. The institution

44. *Investment* As of July 2000, the Janus Mercury fund invested in equities throughout the world, as shown below.†

Region	% of Equities
Europe	18.69
Pacific Rim	6.44
United States	73.31
Latin America	1.56

Find the probability that a randomly selected equity would be from each of the following regions.

a. The Pacific Rim

b. Europe

c. The United States

45. *Investment Survey* Exercise 40 of the previous section presented a survey of 150 mathematics professors. Use the information given in that exercise to find each of the probabilities.

a. A randomly chosen professor invested in stocks and bonds.

b. A randomly chosen professor invested in stocks and bonds and certificates of deposit.

*"University Research: The Squeeze Is On," *Business Week*, May 20, 1991. Reprinted by special permission. Copyright © 1991 by McGraw-Hill, Inc.
†www.janus.com

LIFE SCIENCES

46. *Medical Survey* For a medical experiment, people are classified as to whether they smoke, have a family history of heart disease, or are overweight. Define events E, F, and G as follows.

    E: person smokes

    F: person has a family history of heart disease

    G: person is overweight

Describe each of the following events in words.

a. G' **b.** $F \cap G$ **c.** $E \cup G'$

47. *Medical Survey* Refer to Exercise 46. Describe each of the following events in words.

a. $E \cup F$ **b.** $E' \cap F$ **c.** $F' \cup G'$

48. *Causes of Death* There were 2,329,520 U.S. deaths in 1998. They are listed according to cause in the following table.* If a randomly selected person died in 1998, use this information to find the following probabilities.

Cause	Number of deaths
Heart disease	727,624
Cancer	540,702
Cerebrovascular disease	159,059
Abstructive pulmonary disease	111,823
Accidents	136,959
Pneumonia and influenza	92,048
Diabetes mellitus	63,813
All other causes	497,492

a. The probability that the cause of death was heart disease

b. The probability that the cause of death was cancer or heart disease

c. The probability that the cause of death was not an accident and was not diabetes mellitus

SOCIAL SCIENCES

49. *U.S. Population* The population of the United States by race in 2000 and the projected population by race for the year 2025 are given in the next column (in thousands).†

Race	2000	2025
White	196,700	209,900
Hispanic	32,500	56,900
African-American	33,500	44,700
Asian-American	10,600	24,000
Other	2100	2800

Find the probability of a randomly selected person being each of the following.

a. Hispanic in 2000

b. Hispanic in 2025

c. African-American in 2000

d. African-American in 2025

50. *Civil War* Estimates of the Union Army's strength and losses for the battle of Gettysburg are given in the following table, where *strength* is the number of soldiers immediately preceding the battle and *loss* indicates a soldier who was killed, wounded, captured, or missing.‡

Unit	Strength	Loss
I Corps (Reynolds)	12,222	6059
II Corps (Hancock)	11,347	4369
III Corps (Sickles)	10,675	4211
V Corps (Sykes)	10,907	2187
VI Corps (Sedgwick)	13,596	242
XI Corps (Howard)	9188	3801
XII Corps (Slocum)	9788	1082
Cavalry (Pleasonton)	11,851	610
Artillery (Tyler)	2376	242
Total	91,950	22,803

a. Find the probability that a randomly selected union soldier was from the XI Corps.

b. Find the probability that a soldier was lost in the battle.

c. Find the probability that a I Corps soldier was lost in the battle.

d. Which group had the highest probability of not being lost in the battle?

e. Which group had the highest probability of loss?

f. Explain why these probabilities vary.

---

*Brunner, Borga, ed., *Time Almanac 2000*, p. 809.

†"Projections of the Total Resident Population by 5-Year Age Groups, Race, and Hispanic Origin, 1999 to 2000," from U.S. Bureau of the Census, *Current Population Reports*.

‡Busey, John, and David Martin, *Regimental Strengths and Losses at Gettysburg*, Hightstown, N.J., Longstreet House, 1986, p. 270.

51. *Civil War* Estimates of the Confederate Army's strength and losses for the battle of Gettysburg are given in the following table, where *strength* is the number of soldiers immediately preceding the battle and *loss* indicates a soldier who was killed, wounded, captured, or missing.*

Unit	Strength	Loss
I Corps (Longstreet)	20,706	7661
II Corps (Ewell)	20,666	6603
III Corps (Hill)	22,083	8007
Cavalry (Stuart)	6621	286
Total	70,076	22,557

a. Find the probability that a randomly selected confederate soldier was from the III Corps.

b. Find the probability that a confederate soldier was lost in the battle.

c. Find the probability that a I Corps soldier was lost in the battle.

d. Which group had the highest probability of not being lost in the battle?

e. Which group had the highest probability of loss?

GENERAL INTEREST

52. *Native American Ceremonies* Exercise 56 of the previous section presented a survey of families participating in a pow-wow in Arizona. Use the information given in that exercise to find each of the following probabilities.

a. A randomly chosen family brought costumes and food.

b. A randomly chosen family brought crafts, but neither food nor costumes.

c. A randomly chosen family brought food or costumes.

53. *Chinese New Year* Exercise 55 of the previous section presented a survey of people attending a Lunar New Year celebration in Chinatown. Use the information given in that exercise to find each of the following probabilities.

a. A randomly chosen attendee speaks Cantonese.

b. A randomly chosen attendee does not speak Cantonese.

c. A randomly chosen attendee was a women that did not light a firecracker.

■ 7.4   BASIC CONCEPTS OF PROBABILITY

 THINK ABOUT IT   What is the probability that a dollar of advertising in the United States is spent on sport television or newspapers?

We determine the probability of this and other events in this section. But first we need to develop additional rules for calculating probability, beginning with the probability of a union of two events.

To determine the probability of the union of two events E and F in a sample space S, use the union rule for sets,

$$n(E \cup F) = n(E) + n(F) - n(E \cap F),$$

---

*Busey, John, and David Martin, *Regimental Strengths and Losses at Gettysburg,* Hightstown, N.J., Longstreet House, 1986, p. 270.

which was proved in Section 7.2. Dividing both sides by $n(S)$ shows that

$$\frac{n(E \cup F)}{n(S)} = \frac{n(E)}{n(S)} + \frac{n(F)}{n(S)} - \frac{n(E \cap F)}{n(S)}$$

$$P(E \cup F) = P(E) + P(F) - P(E \cap F).$$

This result is called the **union rule for probability.**

UNION RULE FOR PROBABILITY

For any events E and F from a sample space S,

$$P(E \cup F) = P(E) + P(F) - P(E \cap F).$$

(Although the union rule applies to any events E and F from any sample space, the derivation we have given is valid only for sample spaces with equally likely simple events.)

EXAMPLE 1 Probabilities with Playing Cards

If a single card is drawn from an ordinary deck of cards, find the probability that it will be a red or a face card.

Solution Let R represent the event "red card" and F the event "face card." There are 26 red cards in the deck, so $P(R) = 26/52$. There are 12 face cards in the deck, so $P(F) = 12/52$. Since there are 6 red face cards in the deck, $P(R \cap F) = 6/52$. By the union rule, the probability of the card being red or a face card is

$$P(R \cup F) = P(R) + P(F) - P(R \cap F)$$

$$= \frac{26}{52} + \frac{12}{52} - \frac{6}{52} = \frac{32}{52} = \frac{8}{13}.$$

EXAMPLE 2 Probabilities with Dice

Suppose two fair dice are rolled. Find each of the following probabilities.

(a) The first die shows a 2, or the sum of the results is 6 or 7.

Solution The sample space for the throw of two dice is shown in Figure 18, where 1-1 represents the event "the first die shows a 1 and the second die shows a 1," 1-2 represents "the first die shows a 1 and the second die shows a 2," and so on. Let A represent the event "the first die shows a 2," and B represent the event "the sum of the results is 6 or 7." These events are indicated in Figure 18. From the diagram, event A has 6 elements, B has 11 elements, and the sample space has 36 elements. Thus,

$$P(A) = \frac{6}{36}, \quad P(B) = \frac{11}{36}, \quad \text{and} \quad P(A \cap B) = \frac{2}{36}.$$

By the union rule,

$$P(A \cup B) = P(A) + P(B) - P(A \cap B)$$

$$P(A \cup B) = \frac{6}{36} + \frac{11}{36} - \frac{2}{36} = \frac{15}{36} = \frac{5}{12}.$$

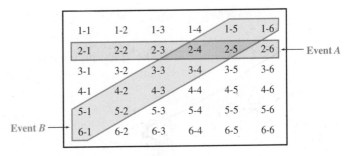

FIGURE 18

(b) The sum of the results is 11, or the second die shows a 5.

Solution $P(\text{sum is } 11) = 2/36$, $P(\text{second die shows a } 5) = 6/36$, and $P(\text{sum is } 11 \text{ and second die shows a } 5) = 1/36$, so

$$P(\text{sum is } 11 \text{ or second die shows a } 5) = \frac{2}{36} + \frac{6}{36} - \frac{1}{36} = \frac{7}{36}.$$

CAUTION You may wonder why we did not use $S = \{2, 3, 4, 5, \ldots, 12\}$ as the sample space in Example 2. Remember, we prefer to use a sample space with equally likely outcomes. The outcomes in set S above are not equally likely— a sum of 2 can occur in just one way, a sum of 3 in two ways, a sum of 4 in three ways, and so on, as shown in Figure 18.

If events E and F are mutually exclusive, then $E \cap F = \emptyset$ by definition; hence, $P(E \cap F) = 0$. Applying the union rule yields this useful fact.

UNION RULE FOR MUTUALLY EXCLUSIVE EVENTS
For mutually exclusive events E and F,

$$P(E \cup F) = P(E) + P(F).$$

By the definition of E', for any event E from a sample space S,

$$E \cup E' = S \quad \text{and} \quad E \cap E' = \emptyset.$$

Since $E \cap E' = \emptyset$, events E and E' are mutually exclusive, so that

$$P(E \cup E') = P(E) + P(E').$$

However, $E \cup E' = S$, the sample space, and $P(S) = 1$. Thus

$$P(E \cup E') = P(E) + P(E') = 1.$$

Rearranging these terms gives the following useful rule.

COMPLEMENT RULE

$$P(E) = 1 - P(E') \quad \text{and} \quad P(E') = 1 - P(E).$$

EXAMPLE 3 Complement Rule

If a fair die is rolled, what is the probability that any number but 5 will come up?

Solution If E is the event that 5 comes up, then E' is the event that any number but 5 comes up. Since $P(E) = 1/6$, we have $P(E') = 1 - 1/6 = 5/6$.

EXAMPLE 4 Complement Rule

In Example 2, find the probability that the sum of the numbers rolled is greater than 3.

Solution To calculate this probability directly, we must find the probabilities that the sum is 4, 5, 6, 7, 8, 9, 10, 11, or 12 and then add them. It is much simpler to first find the probability of the complement, the event that the sum is less than or equal to 3.

$$P(\text{sum} \leq 3) = P(\text{sum is } 2) + P(\text{sum is } 3)$$

$$= \frac{1}{36} + \frac{2}{36}$$

$$= \frac{3}{36} = \frac{1}{12}$$

Now use the fact that $P(E) = 1 - P(E')$ to get

$$P(\text{sum} > 3) = 1 - P(\text{sum} \leq 3)$$

$$= 1 - \frac{1}{12} = \frac{11}{12}.$$

Odds Sometimes probability statements are given in terms of **odds,** a comparison of $P(E)$ with $P(E')$. For example, suppose $P(E) = 4/5$. Then $P(E') = 1 - 4/5 = 1/5$. These probabilities predict that E will occur 4 out of 5 times and E' will occur 1 out of 5 times. Then we say the **odds in favor** of E are 4 to 1.

ODDS

The **odds in favor** of an event E are defined as the ratio of $P(E)$ to $P(E')$, or

$$\frac{P(E)}{P(E')}, \quad P(E') \neq 0.$$

EXAMPLE 5 Odds in Favor of Rain

Suppose the weather forecaster says that the probability of rain tomorrow is $1/3$. Find the odds in favor of rain tomorrow.

Solution Let E be the event "rain tomorrow." Then E' is the event "no rain tomorrow." Since $P(E) = 1/3$, $P(E') = 2/3$. By the definition of odds, the odds in favor of rain are

$$\frac{1/3}{2/3} = \frac{1}{2}, \quad \text{written} \quad 1 \text{ to } 2, \text{ or } 1:2.$$

On the other hand, the odds that it will *not* rain, or the *odds against* rain, are

$$\frac{2/3}{1/3} = \frac{2}{1}, \qquad \text{written} \qquad 2 \text{ to } 1, \text{ or } 2{:}1.$$

If the odds in favor of an event are, say, 3 to 5, then the probability of the event is 3/8, while the probability of the complement of the event is 5/8. (Odds of 3 to 5 indicate 3 outcomes in favor of the event out of a total of 8 possible outcomes.) This example suggests the following generalization.

If the odds favoring event E are m to n, then

$$P(E) = \frac{m}{m+n} \qquad \text{and} \qquad P(E') = \frac{n}{m+n}.$$

EXAMPLE 6   Winning Bids
The odds that a particular bid will be the low bid are 4 to 5.

(a) Find the probability that the bid will be the low bid.

Solution   Odds of 4 to 5 show 4 favorable chances out of $4 + 5 = 9$ chances altogether:

$$P(\text{bid will be low bid}) = \frac{4}{4+5} = \frac{4}{9}.$$

(b) Find the odds against that bid being the low bid.

Solution   There is a 5/9 chance that the bid will not be the low bid, so the odds against a low bid are

$$\frac{P(\text{bid will not be low})}{P(\text{bid will be low})} = \frac{5/9}{4/9} = \frac{5}{4},$$

or $5{:}4$.

EXAMPLE 7   Odds in Horse Racing
If the odds in favor of a particular horse's winning a race are 5 to 7, what is the probability that the horse will win the race?

Solution   The odds indicate chances of 5 out of 12 $(5 + 7 = 12)$ that the horse will win, so

$$P(\text{winning}) = \frac{5}{12}.$$

Race tracks generally give odds *against* a horse winning. In this case, the track would give the odds as 7 to 5.

Empirical Probability   In many real-life problems, it is not possible to establish exact probabilities for events. Instead, useful approximations are often found by drawing on past experience as a guide to the future. The next example shows one approach to such **empirical probabilities.**

EXAMPLE 8 Advertising Volume

Table 3 lists U.S. advertising volume in thousands of dollars by medium in 1998.*

Table 3

Medium	Expenditures
Magazines	13,780,249
Newspapers	16,130,928
Network television	16,271,972
Sport television	15,486,766
Syndicated television	2,691,648
Cable television	6,671,978
Network radio	824,007
Sunday magazines	1,029,447

We can find the empirical probability that a dollar of advertising is spent on each medium by first finding the total spent and then dividing the amount spent on each medium by the total. Verify that the amounts in the table sum to 72,886,995. The probability that a dollar is spent on newspapers, for example, is $P(\text{newspapers}) = 16,130,928/72,886,995 \approx .221$. Similarly, we could divide each amount by 72,886,995, with the results (rounded to three decimal places) shown in Table 4.

Table 4

Medium	Probabilities
Magazines	.189
Newspapers	.221
Network television	.223
Sport television	.212
Syndicated television	.037
Cable television	.092
Network radio	.011
Sunday magazines	.014

The numbers in Table 4 sum to .999. In theory, they should total 1.000, but this does not always occur with rounded numbers.

 The categories in the table are mutually exclusive simple events. Thus, to find the probability that an advertising dollar is spent on newspapers or sport television, we use the union rule to calculate

$$P(\text{newspapers or sport television}) = .221 + .212 = .433.$$

We could get this same result by summing the amount spent on newspapers and sport television, and dividing the total by 72,886,995.

Thus, nearly half of all advertising dollars are spent on these two media, a figure that should be of interest to both advertisers and the owners of the various media.

The World Almanac and Book of Facts 2000, p. 192.

A table of probabilities, as in Example 8, sets up a **probability distribution;** that is, for each possible outcome of an experiment, a number, called the probability of that outcome, is assigned. This assignment may be done in any reasonable way (on an empirical basis, as in Example 8, or by theoretical reasoning, as in Section 7.3), provided that it satisfies the following conditions.

PROPERTIES OF PROBABILITY

Let S be a sample space consisting of n distinct outcomes, $s_1, s_2, \ldots, s_n$. An acceptable probability assignment consists of assigning to each outcome s_i a number p_i (the probability of s_i) according to these rules.

1. The probability of each outcome is a number between 0 and 1.

$$0 \le p_1 \le 1, \quad 0 \le p_2 \le 1, \ldots, \quad 0 \le p_n \le 1$$

2. The sum of the probabilities of all possible outcomes is 1.

$$p_1 + p_2 + p_3 + \cdots + p_n = 1$$

Probability distributions are discussed further in the next chapter.

EXAMPLE 9   Clothing
Susan is a college student who receives heavy sweaters from her aunt at the first sign of cold weather. Susan has determined that the probability that a sweater is the wrong size is .47, the probability that it is a loud color is .59, and the probability that it is both the wrong size and a loud color is .31.

(a) Find the probability that the sweater is the correct size and not a loud color.

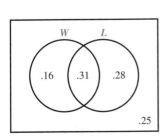

FIGURE 19

Solution Let W represent the event "wrong size," and L represent "loud color." Place the given information on a Venn diagram, starting with .31 in the intersection of the regions W and L (See Figure 19). As stated earlier, event W has probability .47. Since .31 has already been placed inside the intersection of W and L,

$$.47 - .31 = .16$$

goes inside region W, but outside the intersection of W and L. In the same way,

$$.59 - .31 = .28$$

goes inside the region for L, and outside the overlap.

Using regions W and L, the event we want is $W' \cap L'$. From the Venn diagram in Figure 19, the labeled regions have a total probability of

$$.16 + .31 + .28 = .75.$$

Since the entire region of the Venn diagram must have probability 1, the region outside W and L, or $W' \cap L'$, has probability

$$1 - .75 = .25.$$

The probability is .25 that the sweater is the correct size and not a loud color.

(b) Find the probability that the sweater is the correct size or is not loud.

Solution The corresponding region, $W' \cup L'$, has probability

$$.25 + .16 + .28 = .69.$$

7.4 EXERCISES

1. Define mutually exclusive events in your own words.

Decide whether the events in Exercises 2–7 are mutually exclusive.

2. Owning a car and owning a truck

3. Wearing glasses and wearing sandals

4. Being married and being over 30 years old

5. Being a teenager and being over 30 years old

6. Rolling a die once and getting a 4 and an odd number

7. Being a male and being a postal worker

Two dice are rolled. Find the probabilities of rolling the given sums.

8. **a.** 2   **b.** 4   **c.** 5   **d.** 6

9. **a.** 8   **b.** 9   **c.** 10   **d.** 13

10. **a.** 9 or more   **b.** Less than 7   **c.** Between 5 and 8 (exclusive)

11. **a.** Not more than 5   **b.** Not less than 8   **c.** Between 3 and 7 (exclusive)

Two dice are rolled. Find the probabilities of the following events.

12. The first die is 3 or the sum is 8.

13. The second die is 5 or the sum is 10.

14. Three unusual dice, *A*, *B*, and *C*, are constructed such that die *A* has the numbers 3, 3, 4, 4, 8, 8; die *B* has the numbers 1, 1, 5, 5, 9, 9; and die *C* has the numbers 2, 2, 6, 6, 7, 7.

    a. If dice *A* and *B* are rolled, find the probability that *B* beats *A*, that is, the number that appears on die *B* is greater than the number that appears on die *A*.

    b. If dice *B* and *C* are rolled, find the probability that *C* beats *B*.

    c. If dice *A* and *C* are rolled, find the probability that *A* beats *C*.

    d. Which die is better? Explain.

15. Laurie Rosatone, a game show contestant, could win one of two prizes: a shiny new Porsche or a shiny new penny. Laurie is given two boxes of marbles. The first box has 50 pink marbles in it and the second box has 50 blue marbles in it. The game show host will pick someone from the audience to blindfold and then draw a marble from one of the two boxes. If a pink marble is drawn, she wins the Porsche. Otherwise, Laurie wins the penny.* Can Laurie increase her chances of winning by redistributing some of the marbles from one box to the other? Explain.

One card is drawn from an ordinary deck of 52 cards. Find the probabilities of drawing the following cards.

16. **a.** A 9 or 10

    b. A red card or a 3

    c. A 9 or a black 10

    d. A heart or a black card

    e. A face card or a diamond

17. **a.** Less than a 4 (count aces as ones)

    b. A diamond or a 7

    c. A black card or an ace

    d. A heart or a jack

    e. A red card or a face card

*This problem is based on the "Puzzler of the Week: Prison Marbles" from the week of Sept. 7, 1996, on National Public Radio's *Car Talk*.

Ms. Elliott invites 10 relatives to a party: her mother, 2 aunts, 3 uncles, 2 brothers, 1 male cousin, and 1 female cousin. If the chances of any 1 guest arriving first are equally likely, find the probabilities that the first guest to arrive is as follows.

18. a. A brother or an uncle   **b.** A brother or a cousin   **c.** A brother or her mother

19. a. An uncle or a cousin   **b.** A male or a cousin   **c.** A female or a cousin

The numbers 1, 2, 3, 4, and 5 are written on slips of paper, and 2 slips are drawn at random one at a time without replacement. Find each of the probabilities in Exercises 20 and 21.

20. a. The sum of the numbers is 9.

    b. The sum of the numbers is 5 or less.

    c. The first number is 2 or the sum is 6.

21. a. Both numbers are even.

    b. One of the numbers is even or greater than 3.

    c. The sum is 5 or the second number is 2.

Use Venn diagrams to work Exercises 22 and 23.

22. Suppose $P(E) = .26$, $P(F) = .41$, and $P(E \cap F) = .14$. Find each of the following.

    a. $P(E \cup F)$ **b.** $P(E' \cap F)$ **c.** $P(E \cap F')$ **d.** $P(E' \cup F')$

23. Let $P(Z) = .42$, $P(Y) = .35$, and $P(Z \cup Y) = .61$. Find each of the following probabilities.

    a. $P(Z' \cap Y')$ **b.** $P(Z' \cup Y')$ **c.** $P(Z' \cup Y)$ **d.** $P(Z \cap Y')$

24. Define what is meant by odds.

A single fair die is rolled. Find the odds in favor of getting the results in Exercises 25–28.

25. 5

26. 3, 4, or 5

27. 1, 2, 3, or 4

28. Some number less than 2

29. A marble is drawn from a box containing 3 yellow, 4 white, and 8 blue marbles. Find the odds in favor of drawing the following.

    a. A yellow marble   **b.** A blue marble   **c.** A white marble

30. Find the odds of *not* drawing a white marble in Exercise 29.

31. Two dice are rolled. Find the odds of rolling a 7 or 11.

32. In the "Ask Marilyn" column of *Parade* magazine, a reader wrote about the following game: You and I each roll a die. If your die is higher than mine, you win. Otherwise, I win. The reader thought that the probability that each player wins is 1/2. Is this correct? If not, what is the probability that each player wins?*

33. On page 134 of Roger Staubach's autobiography, *First Down, Lifetime to Go*, Staubach makes the following statement regarding his experience in Vietnam:

> "Odds against a direct hit are very low but when your life is in danger, you don't worry too much about the odds."

Is this wording consistent with our definition of odds, for and against? How could it have been said so as to be technically correct?

Parade Magazine, Nov. 6, 1994, p. 10. Reprinted by permission of the William Morris Agency, Inc. on behalf of the author. Copyright © 1994 by Marilyn vos Savant.

34. The following table gives the odds that a particular event will occur.* Convert each odd to the probability that the event will occur.

Event	Odds for the Event
You will eat out today.	1 to 2
The next bottled water you buy will be nothing more than tap water.	1 to 4
The Earth will be struck by a huge meteor during your lifetime.	1 to 9000
You will go to Disney World this year.	1 to 9
You'll regain weight you lost by dieting.	9 to 10

Which of Exercises 35–42 are examples of empirical probability?

35. The probability of heads on 5 consecutive tosses of a coin

36. The probability that a freshman entering college will graduate with a degree

37. The probability that a person is allergic to penicillin

38. The probability of drawing an ace from a standard deck of 52 cards

39. The probability that a person will get lung cancer from smoking cigarettes

40. A weather forecast that predicts a 70% chance of rain tomorrow

41. A gambler's claim that on a roll of a fair die, $P(\text{even}) = 1/2$

42. A surgeon's prediction that a patient has a 90% chance of a full recovery

43. What is a probability distribution?

An experiment is conducted for which the sample space is $S = \{s_1, s_2, s_3, s_4, s_5\}$. Which of the probability assignments in Exercises 44–49 is possible for this experiment? If an assignment is not possible, tell why.

44.

Outcomes	s_1	s_2	s_3	s_4	s_5
Probabilities	.09	.32	.21	.25	.13

45.

Outcomes	s_1	s_2	s_3	s_4	s_5
Probabilities	.92	.03	0	.02	.03

46.

Outcomes	s_1	s_2	s_3	s_4	s_5
Probabilities	1/3	1/4	1/6	1/8	1/10

47.

Outcomes	s_1	s_2	s_3	s_4	s_5
Probabilities	1/5	1/3	1/4	1/5	1/10

48.

Outcomes	s_1	s_2	s_3	s_4	s_5
Probabilities	.64	−.08	.30	.12	.02

49.

Outcomes	s_1	s_2	s_3	s_4	s_5
Probabilities	.05	.35	.5	.2	−.3

*The Forum for Investor Advice; Krantz, Les, *What the Odds Are*, Harper Perennial, 1992; and Laudan, Larry, *Danger Ahead: The Risks You Really Face on Life's Highway*, John Wiley & Sons, 1997.

One way to solve a probability problem is to repeat the experiment many times, keeping track of the results. Then the probability can be approximated using the basic definition of the probability of an event E: $P(E) = n(E)/n(S)$, where E occurs $n(E)$ times out of $n(S)$ trials of an experiment. This is called the Monte Carlo method of finding probabilities. If physically repeating the experiment is too tedious, it may be simulated using a random number generator, available on most computers and scientific or graphing calculators. To simulate a coin toss or the roll of a die on the TI-83, change the setting to fixed decimal mode with 0 digits displayed, and enter rand *or* rand*6+.5, *respectively. For a coin toss, interpret 0 as a head and 1 as a tail. In either case, the* ENTER *key can be pressed repeatedly to perform multiple simulations.*

50. Suppose two dice are rolled. Use the Monte Carlo method with at least 50 repetitions to approximate the following probabilities. Compare with the results of Exercise 11.

a. *P*(the sum is not more than 5)   **b.** *P*(the sum is not less than 8)

51. Suppose two dice are rolled. Use the Monte Carlo method with at least 50 repetitions to approximate the following probabilities. Compare with the results of Exercise 10.

a. *P*(the sum is 9 or more)   **b.** *P*(the sum is less than 7)

52. Suppose three dice are rolled. Use the Monte Carlo method with at least 100 repetitions to approximate the following probabilities.

a. *P*(the sum is 5 or less)   **b.** *P*(neither a 1 nor a 6 is rolled)

53. Suppose a coin is tossed 5 times. Use the Monte Carlo method with at least 50 repetitions to approximate the following probabilities.

a. *P*(exactly 4 heads)   **b.** *P*(2 heads and 3 tails)

Applications

BUSINESS AND ECONOMICS

54. *Defective Merchandise* Suppose that 8% of a certain batch of calculators have a defective case, and that 11% have defective batteries. Also, 3% have both a defective case and defective batteries. A calculator is selected from the batch at random. Find the probability that the calculator has a good case and good batteries.

55. *Credit Charges* The table shows the probabilities of a person accumulating specific amounts of credit card charges over a 12-month period. Find the probabilities that a person's total charges during the period are the following.

a. $500 or more   **b.** Less than $1000

c. $500 to $2999   **d.** $3000 or more

Charges	Probability
Under $100	.31
$100–$499	.18
$500–$999	.18
$1000–$1999	.13
$2000–$2999	.08
$3000–$4999	.05
$5000–$9999	.06
$10,000 or more	.01

Customer Purchases The table below shows the probability that a customer of a department store will make a purchase in the indicated range.

Amount Spent	Probability
Below $10	.07
$10–$24.99	.18
$25–$49.99	.21
$50–$74.99	.16
$75–$99.99	.11
$100–$199.99	.09
$200–$349.99	.07
$350–$499.99	.08
$500 or more	.03

Find the probabilities that a customer makes a purchase in the following ranges.

56. a. Less than $25

   b. More than $24.99

   c. $50 to $199.99

57. a. Less than $350

   b. $75 or more

   c. $200 or more

58. *Profit* The probability that a company will make a profit this year is .74. Find the odds against the company making a profit.

LIFE SCIENCES

59. *Body Types* A study on body types gave the following results: 45% were short; 25% were short and overweight; and 24% were tall and not overweight. Find the probabilities that a person is the following.

a. Overweight

b. Short, but not overweight

c. Tall and overweight

60. *Color Blindness* Color blindness is an inherited characteristic that is more common in males than in females. If *M* represents male and *C* represents red-green color blindness, we use the relative frequencies of the incidences of males and red-green color blindness as probabilities to get

$$P(C) = .039, P(M \cap C) = .035, P(M \cup C) = .491.*$$

Find the following probabilities.

a. $P(C')$ **b.** $P(M)$ **c.** $P(M')$

d. $P(M' \cap C')$ **e.** $P(C \cap M')$ **f.** $P(C \cup M')$

61. *Genetics* Gregor Mendel, an Austrian monk, was the first to use probability in the study of genetics. In an effort to understand the mechanism of character transmittal from one generation to the next in plants, he counted the number of occurrences of various characteristics. Mendel found that the flower color in certain pea plants obeyed this scheme:

Pure red crossed with pure white produces red.

From its parents, the red offspring received genes for both red (*R*) and white (*W*), but in this case red is *dominant* and white *recessive*, so the offspring exhibits the color red. However, the offspring still carries both genes, and when two such offspring are crossed, several things can happen in the third generation. The table in the next column, which is called a *Punnet square*, shows the equally likely outcomes.

Use the fact that red is dominant over white to find each of the following. Assume that there are an equal number of red and white genes in the population.

a. *P*(a flower is red)   **b.** *P*(a flower is white)

		Second Parent	
		R	*W*
First Parent	*R*	*RR*	*RW*
	W	*WR*	*WW*

62. *Genetics* Mendel found no dominance in snapdragons, with one red gene and one white gene producing pink-flowered offspring. These second-generation pinks, however, still carry one red and one white gene, and when they are crossed, the next generation still yields the Punnet square shown above. Find each of the following probabilities.

a. *P*(red)   **b.** *P*(pink)   **c.** *P*(white)

(Mendel verified these probability ratios experimentally and did the same for many characteristics other than flower color. His work, published in 1866, was not recognized until 1890.)

63. *Genetics* In most animals and plants, it is very unusual for the number of main parts of the organism (such as arms, legs, toes, or flower petals) to vary from generation to generation. Some species, however, have *meristic variability,* in which the number of certain body parts varies from generation to generation. One researcher studied the front feet of certain guinea pigs and produced the following probabilities.[†]

$$P(\text{only four toes, all perfect}) = .77$$
$$P(\text{one imperfect toe and four good ones}) = .13$$
$$P(\text{exactly five good toes}) = .10$$

Find the probability of each of the following events.

a. No more than four good toes

b. Five toes, whether perfect or not

SOCIAL SCIENCES

64. *Earnings* The following data were gathered for 130 adult U.S. workers: 55 were women; 3 women earned more than $40,000; and 62 men earned less than $40,000. Find the probabilities that an individual is

a. a woman earning less than $40,000;

b. a man earning more than $40,000;

c. a man or is earning more than $40,000;

d. a woman or is earning less than $40,000.

---

*The probabilities of a person being male or female are from *The World Almanac and Book of Facts,* 1995. The probabilities of a male and female being color-blind are from *Parsons' Diseases of the Eye* (18th ed.) by Stephen J. H. Miller, Churchill Livingstone, 1990, p. 269. This reference gives a range of 3 to 4% for the probability of gross color blindness in men; we used the midpoint of this range.
†Wright, J. R., "An Analysis of Variability in Guinea Pigs," *Genetics,* Vol. 19, pp. 506–536. Reprinted by permission.

65. *Expenditures for Music* A survey of 100 people about their music expenditures gave the following information: 38 bought rock music; 20 were teenagers who bought rock music; and 26 were teenagers. Find the probabilities that a person is

a. a teenager who buys nonrock music;

b. someone who buys rock music or is a teenager;

c. not a teenager;

d. not a teenager, but a buyer of rock music.

66. *Refugees* In a refugee camp in southern Mexico, it was found that 90% of the refugees came to escape political oppression, 80% came to escape abject poverty, and 70% came to escape both. What is the probability that a refugee in the camp was not poor nor seeking political asylum?

67. *Community Activities* At the first meeting of a committee to plan a local Lunar New Year celebration, the persons attending are 3 Chinese men, 4 Chinese women, 3 Vietnamese women, 2 Vietnamese men, 4 Korean women, and 2 Korean men. A chairperson is selected at random. Find the probabilities that the chairperson is the following.

a. Chinese

b. Korean or a woman

c. A man or Vietnamese

d. Chinese or Vietnamese

e. Korean and a woman

68. *Elections* If the odds that a given candidate will win an election are 3 to 2, what is the probability that the candidate will lose?

69. *Military* There were 198,420 female military personnel on active duty in September 1998 in various ranks and military branches, as listed in the table below.*

	Army (A)	Air Force (B)	Navy (C)	Marines (D)
Officers (O)	10,367	11,971	7777	854
Enlisted (E)	60,787	53,542	42,261	8928
Cadets & Midshipmen (M)	624	653	656	0

a. Convert the numbers in the table to probabilities.

b. Find the probability that a randomly selected woman is enlisted in the Army.

c. Find the probability that a randomly selected woman is an officer in the Navy or Marine Corps.

d. $P(A \cup B)$

e. $P(E \cup (C \cup D))$

70. *Perceptions of Threat* Research has been carried out to measure the amount of intolerance that citizens of Russia have for left-wing Communists and right-wing Fascists, as indicated in the table at the bottom of the page. Note that the numbers are given as percents and each row sums to 100 (except for rounding).†

a. Find the probability that a randomly chosen citizen of Russia would be somewhat or extremely intolerant of right-wing Fascists.

b. Find the probability that a randomly chosen citizen of Russia would be completely tolerant of left-wing Communists.

c. Compare your answers to parts a and b and provide possible reasons for these numbers.

	None at All	Don't Know	Not Very Much	Somewhat	Extremely
Left-wing Communists	47.8	6.7	31.0	10.5	4.1
Right-wing Fascists	3.0	3.2	7.1	27.1	59.5

*Brunner, Borga, ed., *Time Almanac 2000*, p. 395.
†Gibson, J. L., "Putting Up with Fellow Russians: An Analysis of Political Tolerance in the Fledgling Russian Democracy," *Political Research Quarterly*, Vol. 51, No. 1, Mar. 1998, pp. 37–68.

71. *Perceptions of Threat* Research has been carried out to measure the amount of intolerance that U.S. citizens have for left-wing Communists and right-wing Fascists, as indicated in the table below. Note that the numbers are given as percents and each row sums to 100 (except for rounding).*

a. Find the probability that a randomly chosen U.S. citizen would have at least some intolerance of right-wing Fascists.

b. Find the probability that a randomly chosen U.S. citizen would have at least some intolerance of left-wing Communists.

c. Compare your answers to parts a and b and provide possible reasons for these numbers.

d. Compare these answers to the answers to Exercise 70.

GENERAL INTEREST

72. *Weather* If the odds that it will rain are 4 to 7, what is the probability of rain? Interpret your answer.

	None at All	Don't Know	Not Very Much	Somewhat	Extremely
Left-wing Communists	13.0	2.7	33.0	34.2	17.1
Right-wing Fascists	10.1	3.3	20.7	43.1	22.9

■ 7.5 CONDITIONAL PROBABILITY; INDEPENDENT EVENTS

THINK ABOUT IT   What is the probability that a broker who uses research picks stocks that go up?

The training manager for a large brokerage firm has noticed that some of the firm's stockbrokers use the firm's research advice, while other brokers tend to follow their own feelings of which stocks will go up. To see whether the research department performs better than brokers' feelings, the manager surveyed 100 brokers, with results as shown in the following table.

	Picked Stocks That Went Up	Didn't Pick Stocks That Went Up	Totals
Used research	30	15	45
Didn't use research	30	25	55
Totals	60	40	100

*Gibson, J. L., "Putting Up with Fellow Russians: An Analysis of Political Tolerance in the Fledgling Russian Democracy," *Political Research Quarterly*, Vol. 51, No. 1, Mar. 1998, pp. 37–68.

Letting A represent the event "picked stocks that went up," and letting B represent the event "used research," we can find the following probabilities.

$$P(A) = \frac{60}{100} = .6 \qquad P(A') = \frac{40}{100} = .4$$

$$P(B) = \frac{45}{100} = .45 \qquad P(B') = \frac{55}{100} = .55$$

To answer the question asked at the beginning of this section, suppose we want to find the probability that a broker using research will pick stocks that go up. From the table above, of the 45 brokers who use research, 30 picked stocks that went up, with

$$P(\text{broker who uses research picks stocks that go up}) = \frac{30}{45} = .667.$$

This is a different number than the probability that a broker picks stocks that go up, .6, since we have additional information (the broker uses research) that has *reduced the sample space.* It other words, we found the probability that a broker picks stocks that go up, A, given the additional information that the broker uses research, B. This is called the *conditional probability* of event A, given that event B has occurred, written $P(A|B)$. $\left(P(A|B)\right.$ may also be read as "the probability of A given B."$\left.\right)$ In the example above,

$$P(A|B) = \frac{30}{45},$$

which can be written as

$$P(A|B) = \frac{30/100}{45/100} = \frac{P(A \cap B)}{P(B)},$$

where $P(A \cap B)$ represents, as usual, the probability that both A and B will occur.

To generalize this result, assume that E and F are two events for a particular experiment. Assume that the sample space S for this experiment has n possible equally likely outcomes. Suppose event F has m elements, and $E \cap F$ has k elements ($k \leq m$). Using the fundamental principle of probability,

$$P(F) = \frac{m}{n} \qquad \text{and} \qquad P(E \cap F) = \frac{k}{n}.$$

We now want $P(E|F)$, the probability that E occurs given that F has occurred. Since we assume F has occurred, reduce the sample space to F: look only at the m elements inside F. See Figure 20. Of these m elements, there are k elements where E also occurs, since $E \cap F$ has k elements. This makes

$$P(E|F) = \frac{k}{m}.$$

Divide numerator and denominator by n to get

$$P(E|F) = \frac{\dfrac{k}{n}}{\dfrac{m}{n}} = \frac{P(E \cap F)}{P(F)}.$$

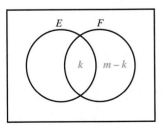

Event F has a total of m elements.

FIGURE 20

This last result motivates the definition of conditional probability.

DEFINITION OF CONDITIONAL PROBABILITY

The **conditional probability** of event E given event F, written $P(E\,|\,F)$, is

$$P(E\,|\,F) = \frac{P(E \cap F)}{P(F)}, \quad \text{where } P(F) \neq 0.$$

This definition tells us that, for equally likely outcomes, conditional probability is found by *reducing the sample space to event F*, and then finding the number of outcomes in F that are also in event E. Thus,

$$P(E\,|\,F) = \frac{n(E \cap F)}{n(F)}.$$

Although the definition of conditional probability was motivated by an example with equally likely outcomes, it is valid in all cases. For an intuitive explanation, think of the formula as giving the probability that both E and F occur compared with the entire probability of F.

EXAMPLE 1 Stocks

Use the information given in the chart at the beginning of this section to find the following probabilities.

(a) $P(B\,|\,A)$

Solution This represents the probability that the broker used research, given that the broker picked stocks that went up. Reduce the sample space to A. Then find $n(A \cap B)$ and $n(A)$.

$$P(B\,|\,A) = \frac{P(B \cap A)}{P(A)} = \frac{n(A \cap B)}{n(A)} = \frac{30}{60} = \frac{1}{2}$$

If a broker picked stocks that went up, then the probability is $1/2$ that the broker used research.

(b) $P(A'\,|\,B)$

Solution In words, this is the probability that a broker picks stocks that do not go up, even though he used research.

$$P(A'\,|\,B) = \frac{n(A' \cap B)}{n(B)} = \frac{15}{45} = \frac{1}{3}$$

(c) $P(B'\,|\,A')$

Solution Here, we want the probability that a broker who picked stocks that did not go up did not use research.

$$P(B'\,|\,A') = \frac{n(B' \cap A')}{n(A')} = \frac{25}{40} = \frac{5}{8}$$

Venn diagrams are useful for illustrating problems in conditional probability. A Venn diagram for Example 1, in which the probabilities are used to indicate the number in the set defined by each region, is shown in Figure 21. In the diagram,

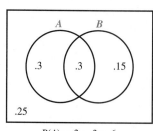

$P(A) = .3 + .3 = .6$

FIGURE 21

$P(B|A)$ is found by reducing the sample space to just set A. Then $P(B|A)$ is the ratio of the number in that part of set B which is also in A to the number in set A, or $.3/.6 = .5$.

EXAMPLE 2 Conditional Probabilities

Given $P(E) = .4$, $P(F) = .5$, and $P(E \cup F) = .7$, find $P(E|F)$.

Solution Find $P(E \cap F)$ first. By the union rule,

$$P(E \cup F) = P(E) + P(F) - P(E \cap F)$$
$$.7 = .4 + .5 - P(E \cap F)$$
$$P(E \cap F) = .2.$$

$P(E|F)$ is the ratio of the probability of that part of E which is in F to the probability of F, or

$$P(E|F) = \frac{P(E \cap F)}{P(F)} = \frac{.2}{.5} = \frac{2}{5}.$$

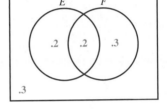

FIGURE 22

The Venn diagram in Figure 22 illustrates Example 2.

EXAMPLE 3 Tossing Coins

Two fair coins were tossed, and it is known that at least one was a head. Find the probability that both were heads.

Solution At first glance the answer to this question appears to be $1/2$. Using mathematics, however, we will see otherwise. The sample space has four equally likely outcomes, $S = \{hh, ht, th, tt\}$. Define two events:

$$E_1 = \text{at least 1 head} = \{hh, ht, th\},$$

and

$$E_2 = \text{2 heads} = \{hh\}.$$

Since there are four equally likely outcomes, $P(E_1) = 3/4$. Also, $P(E_1 \cap E_2) = 1/4$. We want the probability that both were heads, given that at least one was a head; that is, we want to find $P(E_2|E_1)$. Because of the condition that at least one coin was a head, the reduced sample space is

$$\{hh, ht, th\}.$$

Since only one outcome in this reduced sample space is 2 heads,

$$P(E_2|E_1) = \frac{1}{3}.$$

Alternatively, use the definition given above.

$$P(E_2|E_1) = \frac{P(E_2 \cap E_1)}{P(E_1)} = \frac{1/4}{3/4} = \frac{1}{3}$$

It is important not to confuse $P(A|B)$ with $P(B|A)$. For example, in a criminal trial, a prosecutor may point out to the jury that the probability of the defendant's DNA profile matching that of a sample taken at the scene of the crime, given that the defendant is innocent, is very small. What the jury must decide, however, is the probability that the defendant is innocent, given that the defendant's DNA profile matches the sample. Confusing the two is an error sometimes called "the prosecutor's fallacy," and the 1990 conviction of a rape suspect in England was overturned by a panel of judges, who ordered a retrial, because the fallacy made the original trial unfair.*

In the next section, we will see how to compute $P(A|B)$ when we know $P(B|A)$.

Product Rule

If $P(E) \neq 0$ and $P(F) \neq 0$, then the definition of conditional probability shows that

$$P(E|F) = \frac{P(E \cap F)}{P(F)} \quad \text{and} \quad P(F|E) = \frac{P(F \cap E)}{P(E)}.$$

Using the fact that $P(E \cap F) = P(F \cap E)$, and solving each of these equations for $P(E \cap F)$, we obtain the following rule.

PRODUCT RULE OF PROBABILITY

If E and F are events, then $P(E \cap F)$ may be found by either of these formulas.

$$P(E \cap F) = P(F) \cdot P(E|F) \quad \text{or} \quad P(E \cap F) = P(E) \cdot P(F|E)$$

The product rule gives a method for finding the probability that events E and F both occur, as illustrated by the next few examples.

EXAMPLE 4 Business Majors

In a class with 2/5 women and 3/5 men, 25% of the women are business majors. Find the probability that a student chosen from the class at random is a female business major.

Solution Let B and W represent the events "business major" and "woman," respectively. We want to find $P(B \cap W)$. By the product rule,

$$P(B \cap W) = P(W) \cdot P(B|W).$$

Using the given information, $P(W) = 2/5 = .4$ and $P(B|W) = .25$. Thus,

$$P(B \cap W) = .4(.25) = .10. \qquad \blacksquare$$

The next examples show how a tree diagram is used with the product rule to find the probability of a sequence of events.

EXAMPLE 5 Advertising

A company needs to hire a new director of advertising. It has decided to try to hire either person A or B, who are assistant advertising directors for its major competitor. To decide between A and B, the company does research on the campaigns managed by either A or B (no campaign is managed by both), and finds that A is in charge of twice as many advertising campaigns as B. Also, A's cam-

---

*Pringle, David, "Who's the DNA fingerprinting pointing at?" *New Scientist,* Jan. 29, 1994, pp. 51–52.

paigns have satisfactory results 3 out of 4 times, while B's campaigns have satisfactory results only 2 out of 5 times. Suppose one of the competitor's advertising campaigns (managed by A or B) is selected randomly.

We can represent this situation schematically as follows. Let A denote the event "person A does the job" and B the event "person B does the job." Let S be the event "satisfactory results" and U the event "unsatisfactory results." Then the given information can be summarized in the tree diagram in Figure 23. Since A does twice as many jobs as B, $P(A) = 2/3$ and $P(B) = 1/3$, as noted on the first-stage branches of the tree. When A does a job, the probability of satisfactory results is 3/4, and of unsatisfactory results, 1/4, as noted on the second-stage branches. Similarly, the probabilities when B does the job are noted on the remaining second-stage branches. The composite branches labeled 1–4 represent the four mutually exclusive possibilities for the running and outcome of the selected campaign.

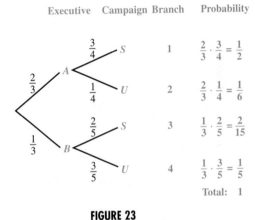

FIGURE 23

(a) Find the probability that A is in charge of the selected campaign and that it produces satisfactory results.

Solution We are asked to find $P(A \cap S)$. We know that when A does the job, the probability of success is 3/4, that is, $P(S|A) = 3/4$. Hence, by the product rule,

$$P(A \cap S) = P(A) \cdot P(S|A) = \frac{2}{3} \cdot \frac{3}{4} = \frac{1}{2}.$$

The event $A \cap S$ is represented by branch 1 of the tree, and as we have just seen, its probability is the product of the probabilities of the pieces that make up that branch.

(b) Find the probability that B runs the campaign and that it produces satisfactory results.

Solution We must find $P(B \cap S)$. The event is represented by branch 3 of the tree and, as before, its probability is the product of the probabilities of the pieces of that branch:

$$P(B \cap S) = P(B) \cdot P(S|B) = \frac{1}{3} \cdot \frac{2}{5} = \frac{2}{15}.$$

(c) What is the probability that the selected campaign is satisfactory?

Solution The event S is the union of the mutually exclusive events $A \cap S$ and $B \cap S$, which are represented by branches 1 and 3 of the diagram. By the union rule,

$$P(S) = P(A \cap S) + P(B \cap S) = \frac{1}{2} + \frac{2}{15} = \frac{19}{30}.$$

Thus, the probability of an event that appears on several branches is the sum of the probabilities of each of these branches.

(d) What is the probability that the selected campaign is unsatisfactory?

Solution $P(U)$ can be read from branches 2 and 4 of the tree.

$$P(U) = \frac{1}{6} + \frac{1}{5} = \frac{11}{30}$$

Alternatively, since U is the complement of S,

$$P(U) = 1 - P(S) = 1 - \frac{19}{30} = \frac{11}{30}.$$

EXAMPLE 6 Tree Diagram

From a box containing 3 white, 2 green, and 1 red marble, 2 marbles are drawn one at a time without replacing the first before the second is drawn. Find the probability that 1 white and 1 green marble are drawn.

Solution A tree diagram showing the various possible outcomes is given in Figure 24. In this diagram, W represents the event "drawing a white marble" and G represents "drawing a green marble." On the first draw, $P(W \text{ first}) = 3/6 = 1/2$ because 3 of the 6 marbles in the box are white. On the second draw $P(G \text{ second} \mid W \text{ first}) = 2/5$. One white marble has been removed, leaving 5, of which 2 are green.

Now we want to find the probability of drawing exactly one white marble and one green marble. This event can occur in two ways: drawing a white marble first and then a green one (branch 2 of the tree diagram), or drawing a green marble first and then a white one (branch 4). For branch 2,

$$P(W \text{ first}) \cdot P\big(G \text{ second} \mid W \text{ first}\big) = \frac{1}{2} \cdot \frac{2}{5} = \frac{1}{5}.$$

For branch 4, where the green marble is drawn first,

$$P(G \text{ first}) \cdot P\big(W \text{ second} \mid G \text{ first}\big) = \frac{1}{3} \cdot \frac{3}{5} = \frac{1}{5}.$$

Since the two events are mutually exclusive, the final probability is the sum of these two probabilities, or

$$P(1 \text{ W}, 1 \text{ G}) = P(W \text{ first}) \cdot P\big(G \text{ second} \mid W \text{ first}\big)$$

$$+ P(G \text{ first}) \cdot P\big(W \text{ second} \mid G \text{ first}\big) = \frac{2}{5}.$$

	First choice	Second choice	Branch	Probability

$$\frac{2}{5} \quad W \qquad 1 \qquad \underline{\hspace{2cm}}$$

$$\frac{2}{5} \quad G \qquad 2 \qquad \frac{1}{2} \cdot \frac{2}{5} = \frac{1}{5}$$

$$\frac{1}{5} \quad R \qquad 3 \qquad \underline{\hspace{2cm}}$$

$$\frac{3}{5} \quad W \qquad 4 \qquad \frac{1}{3} \cdot \frac{3}{5} = \frac{1}{5}$$

$$\frac{1}{5} \quad G \qquad 5 \qquad \underline{\hspace{2cm}}$$

$$\frac{1}{5} \quad R \qquad 6 \qquad \underline{\hspace{2cm}}$$

$$\frac{3}{5} \quad W \qquad 7 \qquad \underline{\hspace{2cm}}$$

$$\frac{2}{5} \quad G \qquad 8 \qquad \underline{\hspace{2cm}}$$

First choice branches: W with $\frac{1}{2}$, G with $\frac{1}{3}$, R with $\frac{1}{6}$.

FIGURE 24

The product rule is often used with *stochastic processes,* which are mathematical models that evolve over time in a probabilistic manner. For example, drawing different-colored marbles from a box (without replacing them) is such a process, in which the probabilities change with each successive draw. (Stochastic processes are studied in more detail in a later chapter.)

EXAMPLE 7 Playing Cards

Two cards are drawn from a standard deck, one after another without replacement.

FOR REVIEW ■

You may wish to refer to the picture of a deck of cards shown in Figure 17 (Section 7.3) and the description accompanying it.

(a) Find the probability that the first card is a heart and the second card is red.

Solution Start with the tree diagram in Figure 25. On the first draw, since there are 13 hearts among the 52 cards, the probability of drawing a heart is $13/52 = 1/4$. On the second draw, since a (red) heart has been drawn already, there are 25 red cards in the remaining 51 cards. Thus, the probability of drawing a red card on the second draw, given that the first is a heart, is $25/51$. By the product rule of probability,

$$P(\text{heart first and red second})$$
$$= P(\text{heart first}) \cdot P(\text{red second} \,|\, \text{heart first})$$
$$= \frac{1}{4} \cdot \frac{25}{51} = \frac{25}{204} \approx .123.$$

(b) Find the probability that the second card is red.

Solution To solve this, we need to fill out the bottom branch of the tree diagram in Figure 25. Unfortunately, if the first card is not a heart, it is not clear how to find the probability that the second card is red, because it depends upon whether the first card is red or black. To solve this problem, we divide

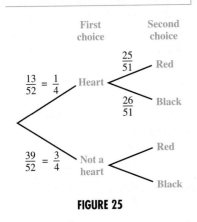

First choice / Second choice

$\frac{13}{52} = \frac{1}{4}$ Heart — $\frac{25}{51}$ Red, $\frac{26}{51}$ Black

$\frac{39}{52} = \frac{3}{4}$ Not a heart — Red, Black

FIGURE 25

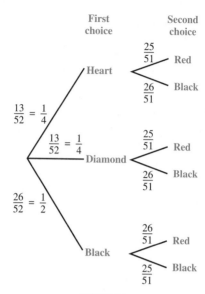

First choice Second choice

Heart $\dfrac{25}{51}$ Red $\dfrac{26}{51}$ Black

$\dfrac{13}{52} = \dfrac{1}{4}$

$\dfrac{13}{52} = \dfrac{1}{4}$ Diamond $\dfrac{25}{51}$ Red $\dfrac{26}{51}$ Black

$\dfrac{26}{52} = \dfrac{1}{2}$

Black $\dfrac{26}{51}$ Red $\dfrac{25}{51}$ Black

FIGURE 26

the bottom branch into two separate branches: diamond and black card (club or spade). The result, with the corresponding probabilities, is shown in Figure 26. Notice that if the first card is a diamond, the probabilities for the second card are the same as if the first card is a heart. On the other hand, if the first card is black, the probability that the second card is red is 26/51. The probability that the second card is red is found by multiplying the probabilities along the three branches and adding.

$$P(\text{red second}) = \frac{1}{4} \cdot \frac{25}{51} + \frac{1}{4} \cdot \frac{25}{51} + \frac{1}{2} \cdot \frac{26}{51}$$

$$= \frac{1}{2}$$

The probability is 1/2, exactly the same as the probability that any card is red. If we know nothing about the first card, there is no reason for the probability of the second card to be anything other than 1/2.

Independent Events   Suppose, in Example 7(a), that we draw the two cards *with* replacement rather than without replacement (that is, we put the first card back before drawing the second card). If the first card is a heart, then the probability of drawing a red card on the second draw is 26/52, rather than 25/51, because there are still 52 cards in the deck, 26 of them red. In this case, $P(\text{red second} \mid \text{heart first})$ is the same as $P(\text{red second})$. The value of the second card is not affected by the value of the first card. We say that the event that the second card is red is *independent* of the event that the first card is a heart since the knowledge of the first card does not influence what happens to the second card. On the other hand, when we draw without replacement, the events that the first card is a heart and that the second card is red are *dependent* events. The fact that the first card is a heart means there is one less red card in the deck, influencing the probability that the second card is red.

As another example, consider tossing a fair coin twice. If the first toss shows heads, the probability that the next toss is heads is still 1/2. Coin tosses are independent events, since the outcome of one toss does not influence the outcome of the next toss. Similarly, rolls of a fair die are independent events. On the other hand, the events "the milk is old" and "the milk is sour" are dependent events; if the milk is old, there is an increased chance that it is sour. Also, in the example at the beginning of this section, the events A (broker picked stocks that went up) and B (broker used research) are dependent events, because information about the use of research affected the probability of picking stocks that go up. That is, $P(A \mid B)$ is different from $P(A)$.

If events E and F are independent, then the knowledge that E has occurred gives no (probability) information about the occurrence or nonoccurrence of event F. That is, $P(F)$ is exactly the same as $P(F \mid E)$, or

$$P(F \mid E) = P(F).$$

This, in fact, is the formal definition of independent events.

INDEPENDENT EVENTS

E and F are **independent events** if

$$P(F|E) = P(F) \quad \text{or} \quad P(E|F) = P(E).$$

If the events are not independent, they are **dependent events.**

When E and F are independent events, then $P(F|E) = P(F)$ and the product rule becomes

$$P(E \cap F) = P(E) \cdot P(F|E) = P(E) \cdot P(F).$$

Conversely, if this equation holds, then it follows that $P(F) = P(F|E)$. Consequently, we have this useful fact:

PRODUCT RULE FOR INDEPENDENT EVENTS

E and F are independent events if and only if

$$P(E \cap F) = P(E) \cdot P(F).$$

EXAMPLE 8   Calculator

A calculator requires a keystroke assembly and a logic circuit. Assume that 99% of the keystroke assemblies are satisfactory and 97% of the logic circuits are satisfactory. Find the probability that a finished calculator will be satisfactory.

Solution If the failure of a keystroke assembly and the failure of a logic circuit are independent events, then

P(satisfactory calculator)

$\quad = P$(satisfactory keystroke assembly) $\cdot P$(satisfactory logic circuit)

$\quad = (.99)(.97) \approx .96.$

(The probability of a defective calculator is $1 - .96 = .04$.)

> **CAUTION** It is common for students to confuse the ideas of *mutually exclusive* events and *independent* events. Events E and F are mutually exclusive if $E \cap F = \emptyset$. For example, if a family has exactly one child, the only possible outcomes are $B = \{\text{boy}\}$ and $G = \{\text{girl}\}$. These two events are mutually exclusive. The events are *not* independent, however, since $P(G|B) = 0$ (if a family with only one child has a boy, the probability it has a girl is then 0). Since $P(G|B) \neq P(G)$, the events are not independent.

Of all the families with exactly two children, the events $G_1 = \{\text{first child is a girl}\}$ and $G_2 = \{\text{second child is a girl}\}$ are independent, since $P(G_2|G_1)$ equals $P(G_2)$. However, G_1 and G_2 are not mutually exclusive, since $G_1 \cap G_2 = \{\text{both children are girls}\} \neq \emptyset$.

To show that two events E and F are independent, show that $P(F|E) = P(F)$ or that $P(E|F) = P(E)$ or that $P(E \cap F) = P(E) \cdot P(F)$. Another way is to observe that knowledge of one outcome does not influence the probability of the other outcome, as we did for coin tosses.

> **NOTE** In some cases, it may not be apparent from the physical description of the problem whether two events are independent or not. For example, it is not obvious whether the event that a baseball player gets a hit tomorrow is independent of the event that he got a hit today. In such cases, it is necessary to calculate whether $P(F \mid E) = P(F)$, or, equivalently, whether $P(E \cap F) = P(E) \cdot P(F)$.

EXAMPLE 9 Snow in Manhattan

On a typical January day in Manhattan the probability of snow is .10, the probability of a traffic jam is .80, and the probability of snow or a traffic jam (or both) is .82. Are the event "it snows" and the event "a traffic jam occurs" independent?

Solution Let S represent the event "it snows" and T represent the event "a traffic jam occurs." We must determine whether

$$P(T \mid S) = P(T) \quad \text{or} \quad P(S \mid T) = P(S).$$

We know $P(S) = .10$, $P(T) = .8$, and $P(S \cup T) = .82$. We can use the union rule (or a Venn diagram) to find $P(S \cap T) = .08$, $P(T \mid S) = .8$, and $P(S \mid T) = .1$. Since

$$P(T \mid S) = P(T) = .8 \quad \text{and} \quad P(S \mid T) = P(S) = .1,$$

the events "it snows" and "a traffic jam occurs" are independent.

Although we showed $P(T \mid S) = P(T)$ and $P(S \mid T) = P(S)$ in Example 9, only one of these results is needed to establish independence. It is also important to note that independence of events does not necessarily follow intuition; it is established from the mathematical definition of independence.

7.5 EXERCISES

If a single fair die is rolled, find the probabilities of the following results.

1. A 2, given that the number rolled was odd

2. A 4, given that the number rolled was even

3. An even number, given that the number rolled was 6

If two fair dice are rolled, find the probabilities of the following results.

4. A sum of 8, given that the sum is greater than 7

5. A sum of 6, given that the roll was a "double" (two identical numbers)

6. A double, given that the sum was 9

If two cards are drawn without replacement from an ordinary deck, find the probabilities of the following results.

7. The second is a heart, given that the first is a heart.

8. The second is black, given that the first is a spade.

9. The second is a face card, given that the first is a jack.

10. The second is an ace, given that the first is not an ace.

11. A jack and a 10 are drawn.

12. An ace and a 4 are drawn.

13. Two black cards are drawn.

14. Two hearts are drawn.

15. In your own words, explain how to find the conditional probability $P(E \mid F)$.

16. Your friend asks you to explain how the product rule for independent events differs from the product rule for dependent events. How would you respond?

17. Another friend asks you to explain how to tell whether two events are dependent or independent. How would you reply? (Use your own words.)

18. A student reasons that the probability in Example 3 of both coins being heads is just the probability that the other coin is a head, that is, 1/2. Explain why this reasoning is wrong.

19. The following problem, submitted by Daniel Hahn of Blairstown, Iowa, appeared in the "Ask Marilyn" column of *Parade* magazine.*

"You discover two booths at a carnival. Each is tended by an honest man with a pair of covered coin shakers. In each shaker is a single coin, and you are allowed to bet upon the chance that both coins in that booth's shakers are heads after the man in the booth shakes them, does an inspection and can tell you that at least one of the shakers contains a head. The difference is that the man in the first booth always looks inside both of his shakers, whereas the man in the second booth looks inside only one of the shakers. Where will you stand the best chance?"

20. A student has two examinations on the same day. If the probability of passing the first test is .90 and the probability of passing the second test is .85, argue that these two events are not independent.

21. Suppose a male defendant in a court trial has a mustache, beard, tattoo, and an earring. Suppose, also, that an eyewitness has identified the perpetrator as someone with these characteristics. If the respective probabilities for the male population in this region are .35, .30, .10, and .05, is it fair to multiply these probabilities together to conclude that the probability that a person having these characteristics is .000525, or 21 in 40,000, and thus decide that the defendant must be guilty?

22. In a two-child family, if we assume that the probabilities of a male child and a female child are each .5, are the events *each child is the same sex* and *at most one male* independent? Are they independent for a three-child family?

23. Let A and B be independent events with $P(A) = \dfrac{1}{4}$ and $P(B) = \dfrac{1}{5}$. Find $P(A \cap B)$ and $P(A \cup B)$.

24. If A and B are events such that $P(A) = .5$ and $P(A \cup B) = .80$, find $P(B)$ when

   a. A and B are mutually exclusive.

   b. A and B are independent.

Applications

BUSINESS AND ECONOMICS

Banking The Midtown Bank has found that most customers at the tellers' windows either cash a check or make a deposit. The chart below indicates the transactions for one teller for one day.

	Cash Check	No Check	Totals
Make Deposit	50	20	70
No Deposit	30	10	40
Totals	80	30	110

Parade Magazine, June 12, 1994, p. 18. Reprinted by permission of the William Morris Agency, Inc. on behalf of the author. Copyright © 1994 by Marilyn vos Savant.

✎ Letting *C* represent "cashing a check" and *D* represent "making a deposit," express each of the following probabilities in words and find its value.

25. $P(C|D)$ **26.** $P(D'|C)$ **27.** $P(C'|D')$

28. $P(C'|D)$ **29.** $P[(C \cap D)']$

30. *Airline Flights* According to a booklet put out by East-west Airlines, 98% of all scheduled Eastwest flights actually take place. (The other flights are canceled due to weather, equipment problems, and so on.) Assume that the event that a given flight takes place is independent of the event that another flight takes place.

   a. Elisabeta Guervara plans to visit her company's branch offices; her journey requires 3 separate flights on Eastwest Airlines. What is the probability that all of these flights will take place?

✎ **b.** Based on the reasons we gave for a flight to be canceled, how realistic is the assumption of independence that we made?

31. *Backup Computers* Corporations where a computer is essential to day-to-day operations, such as banks, often have a second backup computer in case the main computer fails. Suppose there is a .003 chance that the main computer will fail in a given time period, and a .005 chance that the backup computer will fail while the main computer is being repaired. Assume these failures represent independent events, and find the fraction of the time that the corporation can assume it will have computer service. How realistic is our assumption of independence?

32. *ATM Transactions* Among users of automated teller machines (ATMs), 92% use ATMs to withdraw cash, and 32% use them to check their account balance.* Suppose that 96% use ATMs to either withdraw cash or check their account balance (or both). Given a woman who uses an ATM to check her account balance, what is the probability that she also uses an ATM to get cash?

Quality Control A bicycle factory runs two assembly lines, A and B. If 95% of line A's products pass inspection, while only 90% of line B's products pass inspection, and 60% of the factory's bikes come off assembly line B (the rest off A), find the probabilities that one of the factory's bikes did not pass inspection and came off the following.

33. Assembly line A   **34.** Assembly line B

35. Find the probability that one of the factory's bikes did not pass inspection.

LIFE SCIENCES

36. *Genetics* Both of a certain pea plant's parents had a gene for red and a gene for white flowers. (See Exercise 61 in Section 7.4.) If the offspring has red flowers, find the probability that it combined a gene for red and a gene for white (rather than 2 for red.)

Genetics Assuming that boy and girl babies are equally likely, fill in the remaining probabilities on the tree diagram and use the following information to find the probability that a family with three children has all girls, given the following.

37. The first is a girl.

38. The third is a girl.

39. The second is a girl.

40. At least 2 are girls.

41. At least 1 is a girl.

First child	Second child	Third child	Branch	Probability
		$\frac{1}{2}$ B	1	$\frac{1}{8}$
	$\frac{1}{2}$ B	G	2	___
$\frac{1}{2}$ B		$\frac{1}{2}$ B	3	$\frac{1}{8}$
	$\frac{1}{2}$ G	G	4	___
		B	5	___
	B	G	6	___
$\frac{1}{2}$ G		B	7	___
	G	G	8	___

42. *AIDS* The following table, based on data from the Centers for Disease Control, gives the number of new cases of the

---

*Chicago Tribune, Dec. 18, 1995, Sec. 4, p. 1.

AIDS virus for men and women in the United States in 1998 by method of transmission.*

Method of Transmission	Male	Female	Total
Homosexual contact	8388	0	8388
Heterosexual contact	1146	1806	2952
Intravenous drug use	3652	1541	5193
Other	5237	2054	7291
Total	18,423	5401	23,824

a. Find the probability that a male resident who was newly diagnosed with AIDS in 1998 contracted it via homosexual contact.

b. Find the probability that a female resident who was newly diagnosed with AIDS in 1998 contracted it via intervenous drug use.

c. Find the probability that a newly diagnosed person with AIDS is a female.

d. Find the probability that a newly diagnosed person who contracted AIDS via heterosexual contact was a female.

e. Are the events that a person with AIDS is a female and that the person contracted AIDS via heterosexual contact independent? Explain your answer.

43. *Medical Experiment* A medical experiment showed that the probability that a new medicine is effective is .75, the probability that a patient will have a certain side effect is .4, and the probability that both events occur is .3. Decide whether these events are dependent or independent.

Color Blindness *The following table shows frequencies for red-green color blindness, where M represents "person is male" and C represents "person is color-blind." Use this table to find the following probabilities. (See Exercise 60, Section 7.4.)*

	M	M'	Totals
C	.035	.004	.039
C'	.452	.509	.961
Totals	.487	.513	1.000

44. $P(M)$

45. $P(C)$

46. $P(M \cap C)$

47. $P(M \cup C)$

48. $P(M \mid C)$

49. $P(C \mid M)$

50. $P(M' \mid C)$

51. Are the events C and M, described above, dependent? What does this mean?

52. *Color Blindness* A scientist wishes to determine whether there is a relationship between color blindness (C) and deafness (D).

a. Suppose the scientist found the probabilities listed in the table. What should the findings be? (See Exercises 44–51.)

b. Explain what your answer tells us about color blindness and deafness.

	D	D'	Totals
C	.0008	.0392	.0400
C'	.0192	.9408	.9600
Totals	.0200	.9800	1.0000

53. *Obesity* A study showed that, in 1991, 31.6% of men and 35.0% of women were obese.[†] Given that 48.7% of Americans are men and 51.3% are women, find the probability that a randomly selected adult fits the following description.

a. An obese man

b. Obese

54. *Breast Cancer* To explain why the chance of a woman getting breast cancer in the next year goes up each year, while the chance of a woman getting breast cancer in her lifetime goes down, Ruma Falk made the following analogy.[‡] Suppose you are looking for a letter that you may have lost. You have 8 drawers in your desk. There is a probability of .1 that the letter is in any one of the 8 drawers, and a probability of .2 that the letter is not in any of the drawers.

a. What is the probability that the letter is in drawer 1?

b. Given that the letter is not in drawer 1, what is the probability that the letter is in drawer 2?

c. Given that the letter is not in drawer 1 or 2, what is the probability that the letter is in drawer 3?

d. Given that the letter is not in drawers 1–7, what is the probability that the letter is in drawer 8?

e. Based on your answers to parts a–d, what is happening to the probability that the letter is in the next drawer?

f. What is the probability that the letter is in some drawer?

g. Given that the letter is not in drawer 1, what is the probability that the letter is in some drawer?

Health United States, 1999, CDC, National Center for HIV, STD, and TB Prevention, Division of HIV/AIDS Prevention and *The World Almanac and Book of Facts 2000*, p. 903.
[†]*The New York Times*, July 17, 1994, p. 18.
[‡]Falk, Ruma, *Chance News*, July 23, 1995.

h. Given that the letter is not in drawer 1 or 2, what is the probability that the letter is in some drawer?

i. Given that the letter is not in drawers 1–7, what is the probability that the letter is in some drawer?

✎ **j.** Based on your answers to parts f–i, what is happening to the probability that the letter is in some drawer?

55. *Drug Screening* In searching for a new drug with commercial possibilities, drug company researchers use the ratio

$$N_S : N_A : N_P : 1.$$

That is, if the company gives preliminary screening to N_S substances, it may find that N_A of them are worthy of further study, with N_P of these surviving into full-scale development. Finally, 1 of the substances will result in a marketable drug. Typical numbers used by Smith, Kline, and French Laboratories in planning research budgets might be $2000:30:8:1$.* Use this ratio for parts a–f.

a. Suppose a compound has been chosen for preliminary screening. Find the probability that the compound will survive and become a marketable drug.

b. Find the probability that the compound will not lead to a marketable drug.

c. Suppose the number of such compounds receiving preliminary screening is a. Set up the probability that none of them produces a marketable drug. (Assume independence throughout these exercises.)

d. Use your results from part c to find the probability that at least one of the drugs will prove marketable.

e. Suppose now that N scientists are employed in the preliminary screening, and that each scientist can screen c compounds per year. Find the probability that no marketable drugs will be discovered in a year.

f. Find the probability that at least one marketable drug will be discovered.

Hockey The following table lists the number of head and neck injuries for 319 ice hockey players' exposures wearing either a full shield or half shield in the Canadian Inter-University Athletics Union during the 1997–1998 season.[†]

For a randomly selected injury, find the following probabilities.

56. $P(A)$

57. $P(C \mid F)$

58. $P(A \mid H)$

59. $P(B' \mid H')$

60. Are the events A and H independent events?

	Half Shield (H)	Full Shield (F)	Total
Head and Face Injuries (A)	95	34	129
Concussions (B)	41	38	79
Neck Injuries (C)	9	7	16
Other Injuries(D)	202	150	352
Total	347	229	576

61. *Rain Forecasts* In a letter to the journal *Nature*, Robert A. J. Matthews gives the following table of outcomes of forecast and weather over 1000 1-hour walks, based on the United Kingdom's Meteorological office's 83% accuracy in 24-hour forecasts.[‡]

	Rain	No Rain	Sum
Forecast of rain	66	156	222
Forecast of no rain	14	764	778
Sum	80	920	1000

a. Verify that the probability that the forecast called for rain, given that there was rain, is indeed 83%. Also verify that the probability that the forecast called for no rain, given that there was no rain, is also 83%.

b. Calculate the probability that there was rain, given that the forecast called for rain.

c. Calculate the probability that there was no rain, given that the forecast called for no rain.

✎ **d.** Observe that your answer to part c is higher than 83%, and that your answer to part b is much lower. Discuss which figure best describes the accuracy of the weather forecast in recommending whether or not you should carry an umbrella.

SOCIAL SCIENCES

62. *Working Women* A survey has shown that 52% of the women in a certain community work outside the home. Of these women, 64% are married, while 86% of the women who do not work outside the home are married. Find the

---

*Pyle, E. B., III, B. Douglas, G. W. Ebright, W. J. Westlake, and A. B. Bender, "Scientific Manpower Allocation to New Drug Screening Programs," *Management Science,* Vol. 19, No. 12, August 1973. Copyright © 1973 by The Institute of Management Sciences. Reprinted by permission.
[†]Benson, Brian, Nicholas Nohtaki, M. Sarah Rose, and Willem Meeuwisse, "Head and Neck Injuries Among Ice Hockey Players Wearing Full Face Shields vs. Half Face Shields," *JAMA,* Vol. 282, No. 24, Dec. 22/29, 1999, pp. 2328–2332.
[‡]Matthews, Robert A. J., *Nature,* Vol. 382, Aug. 29, 1996, p. 3.

probabilities that a woman in that community can be categorized as follows.

a. Married

b. A single woman working outside the home

GENERAL INTEREST

63. *Titanic* The following table lists the number of passengers who were on the *Titanic* and the number of passengers who survived, according to class of ticket.*

	Children		Women		Men		Totals	
	On	Survived	On	Survived	On	Survived	On	Survived
First class	6	6	144	140	175	57	325	203
Second class	24	24	165	76	168	14	357	114
Third class	79	27	93	80	462	75	634	182
Total	109	57	402	296	805	146	1316	499

Use this information to determine the following (round answers to three decimal places).

a. What is the probability that a randomly selected passenger was second class?

b. What is the overall probability of surviving?

c. What is the probability of a first-class passenger surviving?

d. What is the probability of a child who was also in the third class surviving?

e. Given that the survivor is from first class, what is the probability that she was a woman?

f. Given that a male has survived, what is the probability that he was in third class?

g. Are the events third-class survival and male survival independent events? What does this imply?

PHYSICAL SCIENCES

64. *Reliability* The probability that a key component of a space rocket will fail is .03.

a. How many such components must be used as backups to ensure that the probability of at least one of the components' working is .999999 or more?

b. Is it reasonable to assume independence here?

65. *Real Estate* A real estate agent trying to sell you an attractive beachfront house claims that it will not collapse unless it is subjected simultaneously to extremely high winds and extremely high waves. According to weather service records, there is a .001 probability of extremely high winds, and the same for extremely high waves. The real estate agent claims, therefore, that the probability of both occurring is $(.001)(.001) = .000001$. What is wrong with the agent's reasoning?

66. *Age and Loans* Suppose 20% of the population are 65 or over, 26% of those 65 or over have loans, and 53% of those under 65 have loans. Find the probabilities that a person fits into the following categories.

a. 65 or over and has a loan   **b.** Has a loan

67. *Women Joggers* In a certain area, 15% of the population are joggers and 40% of the joggers are women. If 55% of those who do not jog are women, find the probabilities that an individual from that community fits the following descriptions.

a. A woman jogger   **b.** Not a jogger   **c.** A woman

68. *Diet Soft Drinks* Two-thirds of the population are on a diet at least occasionally. Of this group, 4/5 drink diet soft drinks, while 1/2 of the rest of the population drink diet soft drinks. Find the probabilities that a person fits into the following categories.

a. Drinks diet soft drinks

b. Diets, but does not drink diet soft drinks

69. *Driver's License Test* The Motor Vehicle Department has found that the probability of a person passing the test for a driver's license on the first try is .75. The probability that an individual who fails on the first test will pass on the second try is .80, and the probability that an individual who fails the first and second tests will pass the third time is .70. Find the probabilities that an individual will do the following.

a. Fail both the first and second tests

b. Fail three times in a row

c. Require at least two tries

70. *Speeding Tickets* A smooth-talking young man has a 1/3 probability of talking a policeman out of giving him a speeding ticket. The probability that he is stopped for speeding during a given weekend is 1/2. Find the probabilities of the events in parts a and b.

---

*Takis, Sandra L., "Titanic: A Statistical Exploration," *Mathematics Teacher*, Vol. 92, No. 8, Nov. 1999, pp. 660–664.

a. He will receive no speeding tickets on a given weekend.

b. He will receive no speeding tickets on 3 consecutive weekends.

c. We have assumed that what happens on the second or third weekend is the same as what happened on the first weekend. Is this realistic? Will driving habits remain the same after getting a ticket?

71. *Luxury Cars* In one area, 4% of the population drive luxury cars. However, 17% of the CPAs drive luxury cars. Are the events "person drives a luxury car" and "person is a CPA" independent?

72. *Studying* A teacher has found that the probability that a student studies for a test is .6, the probability that a student gets a good grade on a test is .7, and the probability that both occur is .52. Are these events independent?

73. *Football* A football coach whose team is 14 points behind needs two touchdowns to win. Each touchdown is worth 6 points. After a touchdown, the coach can choose either a 1-point kick, which is almost certain to succeed, or a 2-point conversion, which is roughly half as likely to succeed. After the first touchdown, the coach must decide whether to go for 1 or 2 points. If the 2-point conversion is successful, the almost certain 1-point kick after the second touchdown will win the game. If the 2-point conversion fails, the team can try another 2-point conversion after the second touchdown to tie. Some coaches, however, prefer to go for the almost certain 1-point kick after the first touchdown, hoping that the momentum will help them get a 2-point conversion after the second touchdown and win the game. They fear that an unsuccessful 2-point conversion after the first touchdown will discourage the team, which can then at best tie.*

a. Draw a tree diagram for the 1-point kick after the first touchdown and the 2-point conversion after the second touchdown. Letting the probability of success for the 1-point kick and the 2-point conversion be k and r, respectively, show that

$$P(\text{win}) = kr,$$
$$P(\text{tie}) = r(1 - k), \quad \text{and}$$
$$P(\text{lose}) = 1 - r.$$

b. Consider the case of trying for a 2-point conversion after the first touchdown. If it succeeds, try a 1-point kick after the second touchdown. If the 2-point conversion fails, try another one after the second touchdown. Draw a tree diagram and use it to show that

$$P(\text{win}) = kr,$$
$$P(\text{tie}) = r(2 - k - r), \quad \text{and}$$
$$P(\text{lose}) = (1 - r)^2.$$

c. What can you say about the probability of winning under each strategy?

d. Given that $r < 1$, which strategy has a smaller probability of losing? What does this tell you about the value of the two strategies?

■ 7.6 BAYES' THEOREM

THINK ABOUT IT   What is the probability that a particular defective item was produced by a new machine operator?

This question and others like it are answered using Bayes' theorem, discussed in this section.

Suppose the probability that a person gets lung cancer, given that the person smokes a pack or more of cigarettes daily, is known. For a research project, it might be necessary to know the probability that a person smokes a pack or more of cigarettes daily, given that the person has lung cancer. More generally, if $P(E|F)$ is known for two events E and F, can $P(F|E)$ be found? The answer is yes, we can find $P(F|E)$ using the formula to be developed in this section. To develop this formula, we can use a tree diagram to find $P(F|E)$. Since $P(E|F)$ is known, the first outcome is either F or F'. Then for each of these outcomes, either E or E' occurs, as shown in Figure 27. The four cases have the probabilities shown on the right. By the definition of conditional probability,

$$P(F|E) = \frac{P(F \cap E)}{P(E)} = \frac{P(F) \cdot P(E|F)}{P(F) \cdot P(E|F) + P(F') \cdot P(E|F')}.$$

*Schielack, Vincent P., Jr., "The Football Coach's Dilemma: Should We Go for 1 or 2 Points First?" *The Mathematics Teacher*, Vol. 88, No. 9, Dec. 1995, pp. 731–733.

Notice that $P(E)$ is the sum of the first and third cases. This result is a special case of Bayes' theorem, which is generalized later in this section.

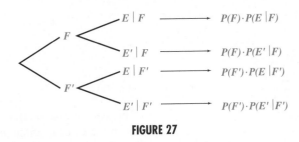

FIGURE 27

BAYES' THEOREM (SPECIAL CASE)

$$P(F|E) = \frac{P(F) \cdot P(E|F)}{P(F) \cdot P(E|F) + P(F') \cdot P(E|F')}$$

EXAMPLE 1   Worker Errors

For a fixed length of time, the probability of a worker error on a certain production line is .1, the probability that an accident will occur when there is a worker error is .3, and the probability that an accident will occur when there is no worker error is .2. Find the probability of a worker error if there is an accident.

Solution Let E represent the event of an accident, and let F represent the event of worker error. From the information above,

$$P(F) = .1, \qquad P(E|F) = .3, \qquad \text{and} \qquad P(E|F') = .2.$$

These probabilities are shown on the tree diagram in Figure 28.

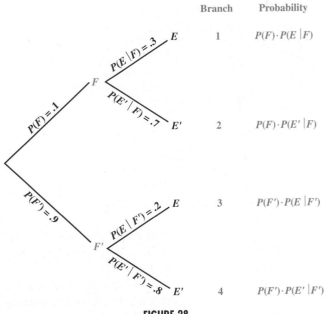

FIGURE 28

Find $P(F|E)$ by dividing the probability that both E and F occur, given by branch 1, by the probability that E occurs, given by the sum of branches 1 and 3.

$$P(F|E) = \frac{P(F) \cdot P(E|F)}{P(F) \cdot P(E|F) + P(F') \cdot P(E|F')}$$

$$= \frac{(.1)(.3)}{(.1)(.3) + (.9)(.2)} = \frac{1}{7} \approx .143$$

The special case of Bayes' theorem can be generalized to more than two events with the tree diagram in Figure 29. This diagram shows the paths that can produce an event E. We assume that the events $F_1, F_2, \ldots, F_n$ are mutually exclusive events (that is, disjoint events) whose union is the sample space, and that E is an event that has occurred. See Figure 30.

The probability $P(F_i|E)$, where $1 \le i \le n$, can be found by dividing the probability for the branch containing $P(E|F_i)$ by the sum of the probabilities of all the branches producing event E.

BAYES' THEOREM

$$P(F_i|E) = \frac{P(F_i) \cdot P(E|F_i)}{P(F_1) \cdot P(E|F_1) + P(F_2) \cdot P(E|F_2) + \cdots + P(F_n) \cdot P(E|F_n)}$$

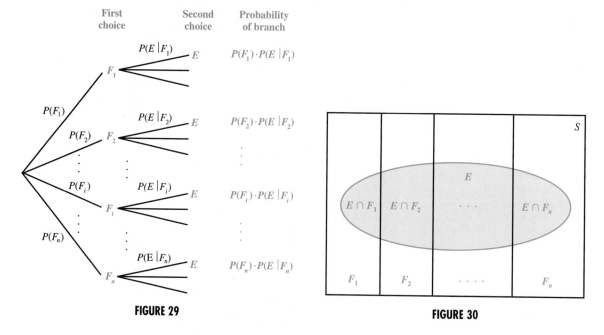

FIGURE 29

FIGURE 30

This result is known as **Bayes' theorem,** after the Reverend Thomas Bayes (1702–1761), whose paper on probability was published about three years after his death.

The statement of Bayes' theorem can be daunting. Actually, it is easier to remember the formula by thinking of the tree diagram that produced it. Go through the following steps.

USING BAYES' THEOREM

1. Start a tree diagram with branches representing $F_1, F_2, \ldots, F_n$. Label each branch with its corresponding probability.

2. From the end of each of these branches, draw a branch for event E. Label this branch with the probability of getting to it, $P(E|F_i)$.

3. You now have n different paths that result in event E. Next to each path, put its probability—the product of the probabilities that the first branch occurs, $P(F_i)$, and that the second branch occurs, $P(E|F_i)$; that is, the product $P(F_i) \cdot P(E|F_i)$, which equals $P(F_i \cap E)$.

4. $P(F_i|E)$ is found by dividing the probability of the branch for F_i by the sum of the probabilities of all the branches producing event E.

EXAMPLE 2   Machine Operators

Based on past experience, a company knows that an experienced machine operator (one or more years of experience) will produce a defective item 1% of the time. Operators with some experience (up to one year) have a 2.5% defect rate, and new operators have a 6% defect rate. At any one time, the company has 60% experienced operators, 30% with some experience, and 10% new operators. Find the probability that a particular defective item was produced by a new operator.

Solution Let E represent the event "item is defective," F_1 represent "item was made by an experienced operator," F_2 represent "item was made by an operator with some experience," and F_3 represent "item was made by a new operator." Then

$$P(F_1) = .60 \qquad P(E|F_1) = .01$$
$$P(F_2) = .30 \qquad P(E|F_2) = .025$$
$$P(F_3) = .10 \qquad P(E|F_3) = .06.$$

We need to find $P(F_3|E)$, the probability that an item was produced by a new operator, given that it is defective. First, draw a tree diagram using the given information, as in Figure 31. The steps leading to event E are shown in heavy type.

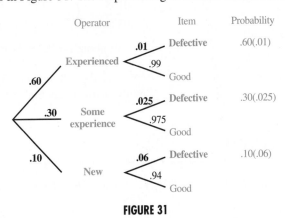

FIGURE 31

Find $P(F_3|E)$ with the bottom branch of the tree in Figure 31: divide the probability for this branch by the sum of the probabilities of all the branches leading to E, or

$$P(F_3|E) = \frac{.10(.06)}{.60(.01) + .30(.025) + .10(.06)} = \frac{.006}{.0195} = \frac{4}{13} \approx .3077.$$

In a similar way,

$$P(F_2 \mid E) = \frac{.30(.025)}{.60(.01) + .30(.025) + .10(.06)} = \frac{.0075}{.0195} = \frac{5}{13} \approx .3846.$$

Finally, $P(F_1 \mid E) = 4/13 \approx .3077$. Check that $P(F_1 \mid E) + P(F_2 \mid E) + P(F_3 \mid E) = 1$ (that is, the defective item was made by *someone*).

EXAMPLE 3 Manufacturing

A manufacturer buys items from six different suppliers. The fraction of the total number of items obtained from each supplier, along with the probability that an item purchased from that supplier is defective, are shown in the following chart.

Supplier	Fraction of Total Supplied	Probability of Defect
1	.05	.04
2	.12	.02
3	.16	.07
4	.23	.01
5	.35	.03
6	.09	.05

Find the probability that a defective item came from supplier 5.

Solution Let F_1 be the event that an item came from supplier 1, with F_2, F_3, F_4, F_5, and F_6 defined in a similar manner. Let E be the event that an item is defective. We want to find $P(F_5 \mid E)$. Use the probabilities in the table above to prepare a tree diagram, or work with the rows of the table to get

$$P(F_5 \mid E) = \frac{(.35)(.03)}{(.05)(.04) + (.12)(.02) + (.16)(.07) + (.23)(.01) + (.35)(.03) + (.09)(.05)}$$

$$= \frac{.0105}{.0329} \approx .319.$$

There is about a 32% chance that a defective item came from supplier 5.

Even though supplier 5 has only 3% defectives, his probability of being "guilty" is relatively high, about 32%, because of the large fraction supplied by 5.

7.6 EXERCISES

For two events M and N, $P(M) = .4$, $P(N \mid M) = .3$, and $P(N \mid M') = .4$. Find each of the following.

1. $P(M \mid N)$

2. $P(M' \mid N)$

For mutually exclusive events R_1, R_2, and R_3, we have $P(R_1) = .05$, $P(R_2) = .6$, and $P(R_3) = .35$. Also, $P(Q \mid R_1) = .40$, $P(Q \mid R_2) = .30$, and $P(Q \mid R_3) = .60$. Find each of the following.

3. $P(R_1 \mid Q)$

4. $P(R_2 \mid Q)$

5. $P(R_3 \mid Q)$

6. $P(R_1' \mid Q)$

Suppose you have three jars with the following contents: 2 black balls and 1 white ball in the first, 1 black ball and 2 white balls in the second, and 1 black ball and 1 white ball in the third. One jar is to be selected, and then 1 ball is to be drawn from the selected jar. If the probabilities of selecting the first, second, or third jar are 1/2, 1/3, and 1/6, respectively, find the probabilities that if a white ball is drawn, it came from the following jars.

7. The second jar

8. The third jar

Applications

BUSINESS AND ECONOMICS

9. *Employment Test* A manufacturing firm finds that 70% of its new hires turn out to be good workers and 30% become poor workers. All current workers are given a reasoning test. Of the good workers, 80% pass it; 40% of the poor workers pass it. Assume that these figures will hold true in the future. If the company makes the test part of its hiring procedure and only hires people who meet the previous requirements and also pass the test, what percent of the new hires will turn out to be good workers?

Job Qualifications Of all the people applying for a certain job, 70% are qualified and 30% are not. The personnel manager claims that she approves qualified people 85% of the time; she approves an unqualified person 20% of the time. Find each of the following probabilities.

10. A person is qualified if he or she was approved by the manager.

11. A person is unqualified if he or she was approved by the manager.

Quality Control A building contractor buys 70% of his cement from supplier A, and 30% from supplier B. A total of 90% of the bags from A arrive undamaged, while 95% of the bags from B arrive undamaged. Give the probabilities that a damaged bag is from each of the following sources.

12. Supplier A   **13.** Supplier B

Credit The probability that a customer of a local department store will be a "slow pay" is .02. The probability that a "slow pay" will make a large down payment when buying a refrigerator is .14. The probability that a person who is not a "slow pay" will make a large down payment when buying a refrigerator is .50. Suppose a customer makes a large down payment on a refrigerator. Find the probabilities that the customer is in the following categories.

14. A "slow pay"   **15.** Not a "slow pay"

Appliance Reliability Companies A, B, and C produce 15%, 40%, and 45%, respectively, of the major appliances sold in a certain area. In that area, 1% of the company A appliances, $1\frac{1}{2}$% of the company B appliances, and 2% of the company C appliances need service within the first year. Suppose a defective appliance is chosen at random; find the probabilities that it was manufactured by the following companies.

16. Company A   **17.** Company B

Television Advertising On a given weekend in the fall, a tire company can buy television advertising time for a college football game, a baseball game, or a professional football game. If the company sponsors the college football game, there is a 70% chance of a high rating, a 50% chance if they sponsor a baseball game, and a 60% chance if they sponsor a professional football game. The probabilities of the company sponsoring these various games are .5, .2, and .3, respectively. Suppose the company does get a high rating; find the probabilities that it sponsored the following.

18. A college football game

19. A professional football game

20. *Shipping Errors* The following information pertains to three shipping terminals operated by Krag Corp.*

Terminal	Percentage of Cargo Handled	Percentage of Error
Land	50	2
Air	40	4
Sea	10	14

Krag's internal auditor randomly selects one set of shipping documents, ascertaining that the set selected contains an error. Which of the following gives the probability that the error occurred in the Land Terminal?

a. .02   **b.** .10   **c.** .25   **d.** .50

---

*Uniform CPA Examination, Nov. 1989.

21. *Mortgage Defaults* A bank finds that the relationship between mortgage defaults and the size of the down payment is given by this table.

	Down Payment			
	5%	10%	20%	25%
Number of mortgages with this down payment	1260	700	560	280
Probability of default	.05	.03	.02	.01

a. If a default occurs, what is the probability that it is on a mortgage with a 5% down payment?

b. What is the probability that a mortgage that is paid to maturity has a 10% down payment?

LIFE SCIENCES

22. *Toxemia Test* A magazine article described a new test for toxemia, a disease that affects pregnant women. To perform this test, the woman lies on her left side and then rolls onto her back. The test is considered positive if there is a 20 millimeter rise in her blood pressure within one minute. The article gives the following probabilities, where T represents having toxemia at some time during the pregnancy, and N represents a negative test.

$$P(T'|N) = .90 \quad \text{and} \quad P(T|N') = .75$$

Assume that $P(N') = .11$, and find each of the following.

a. $P(N|T)$ **b.** $P(N'|T)$

23. *AIDS Testing* In 1987, it was found that one woman in 10,000 was infected with the AIDS virus. The most commonly used test always detected an infected woman, but one healthy woman out of 20,000 was incorrectly identified as being infected with the AIDS virus.*

a. Find the probability that a woman who tested positive in 1987 was not infected.

b. It has been argued that everyone should be tested for AIDS. Based on the results of part a, how useful would the results of such testing be?

24. *Hepatitis Blood Test* The probability that a person with certain symptoms has hepatitis is .8. The blood test used to confirm this diagnosis gives positive results for 90% of people with the disease and 5% of those without the disease. What is the probability that an individual who has the symptoms and who reacts positively to the test actually has hepatitis?

25. *Sensitivity and Specificity* The sensitivity of a medical test is defined as the probability that a test will be positive given that a person has a disease, written $P(T^+|D^+)$. The specificity of a test is defined as the probability that a test will be negative given that the person does not have the disease, written $P(T^-|D^-)$. For example, the sensitivity and specificity for breast cancer during a clinical breast examination by a trained expert is approximately .54 and .94, respectively.[†]

a. If 2% of U.S. women have breast cancer,[‡] find the probability that a woman who tests positive during a clinical breast examination actually has breast cancer.

b. Given that a woman tests negative during a clinical breast examination, find the probability that she does not have breast cancer.

c. Using the information above, how many false positives would you expect for every 1000 clinical breast examinations?

26. *Test for HIV* A new test for the virus that causes AIDS, developed by Octopus Diagnostics Research of Hantsport, Nova Scotia, shows the presence or absence of HIV in a drop of blood in two minutes, compared with five days for a conventional test.[§] Preliminary results indicate a false positive rate (an indication that the HIV virus is present when it is not) of less than 2%, and a false negative rate (a failure to detect the presence of the HIV virus) of up to 5%. Assume for this exercise that these rates are exactly 2% and 5%. In 1996, there were 780,000 people in North America with the HIV virus, out of a population of 295 million.[‖] Suppose a resident of North America is chosen at random and given this test. If the result is positive, what is the probability that the person actually has the HIV virus?

---

*The New York Times, Sept. 5, 1987.

†Barton, Mary B., Russell Harris, and Suzanne Fletcher, "Does This Patient Have Breast Cancer, The Screening Clinical Breast Examinations: Should It Be Done? How?" *JAMA*, Vol. 282, No. 13, Oct. 6, 1999, pp. 1270–1280.

‡The World Almanac and Book of Facts 2000, p. 902.

§Maclean's, Feb. 17, 1997, p. 70.

‖The World Almanac and Book of Facts 1997.

SOCIAL SCIENCES

27. *Binge Drinking* A 1995 study by the Harvard School of Public Health reported that 86% of male students who live in a fraternity house are binge drinkers. The figure for fraternity members who are not residents of a fraternity house is 71%, while the figure for men who do not belong to a fraternity is 45%.* Suppose that 10% of U.S. male students live in a fraternity house, 15% belong to a fraternity but do not live in a fraternity house, and 75% do not belong to a fraternity.

   a. What is the probability that a randomly selected male student is a binge drinker?

   b. If a randomly selected male student is a binge drinker, what is the probability that he lives in a fraternity house?

28. *Murder* During the murder trial of O. J. Simpson, Alan Dershowitz, an advisor to the defense team, stated on television that only about .1% of men who batter their wives actually murder them. Statistician I. J. Good observed that even if, given that a husband is a batterer, the probability he is guilty of murdering his wife is .001, what we really want to know is the probability that the husband is guilty, given that the wife was murdered.[†] Good estimates the probability of a battered wife being murdered, given that her husband is not guilty, as .001. The probability that she is murdered if her husband is guilty is 1, of course. Using these numbers and Dershowitz's .001 probability of the husband being guilty, find the probability that the husband is guilty, given that the wife was murdered.

Never-Married Adults by Age Group The following tables give the proportions of men and women in the U.S. population, and the proportions of men and women who have never married, in a recent year.[‡]

	Men	
Age	**Proportion in Population**	**Proportion Never Married**
18–24	.133	.874
25–34	.205	.397
35–44	.232	.187
45–64	.287	.075
65 or over	.142	.038

	Women	
Age	**Proportion in Population**	**Proportion Never Married**
18–24	.123	.775
25–34	.194	.298
35–44	.219	.121
45–64	.284	.062
65 or over	.181	.047

29. Find the probability that a randomly selected man who has never married is between 35 and 44 years old (inclusive).

30. Find the probability that a randomly selected woman who has been married is between 18 and 24 (inclusive).

31. Find the probability that a randomly selected woman who has never been married is between 45 and 64 (inclusive).

Seat Belt Effectiveness A federal study showed that in 1990, 49% of all those involved in a fatal car crash wore seat belts. Of those in a fatal crash who wore seat belts, 44% were injured and 27% were killed. For those not wearing seat belts, the comparable figures were 41% and 50%, respectively.[§]

32. Find the probability that a randomly selected person who was killed in a car crash was wearing a seat belt.

33. Find the probability that a randomly selected person who was unharmed in a fatal crash was not wearing a seat belt.

Voting The following table shows the proportion of people over 18 who are in various age categories, along with the probabilities that a person in a given age category will vote in a general election.

Age	**Percent of Voting Age Population**	**Probability of a Person of This Age Voting**
18–21	11.0%	.48
22–24	7.6%	.53
25–44	37.6%	.68
45–64	28.3%	.64
65 or over	15.5%	.74

---

*The New York Times, Dec. 6, 1995, p. B16.
[†]Good, I. J., "When Batterer Turns Murderer," *Nature*, Vol. 375, No. 15, June 15, 1995, p. 541.
[‡]"Marital Status and Living Arrangements of Adults 18 Years and Over," Mar. 1998, Table B, U.S. Bureau of the Census.
[§]National Highway Traffic Safety Administration, Office of Driver and Pedestrian Research: "Occupant Protection Trends in 19 Cities," Nov. 1989, and "Use of Automatic Safety Belt Systems in 19 Cities, Feb. 1991.

Suppose a voter is picked at random. Find the probabilities that the voter is in the following age categories.

34. 18–21

35. 65 or over

36. Find the probability that a person who did not vote was in the age category 45–64.

■ CHAPTER SUMMARY

We began this chapter by introducing sets, set operations, tree diagrams, and Venn diagrams. We then used these ideas to define and study probability experiments, sample spaces, mutually exclusive events, and probability distributions. These concepts were further extended to include empirical probability, conditional probability, odds, independent events, and Bayes' theorem. Many applications of probability were then explored. In the next two chapters we will employ these techniques to further our study into the fields of probability and statistics.

Probability Summary

Basic Probability Principle

Let S be a sample space of equally likely outcomes, and let event E be a subset of S. Then the probability that event E occurs is

$$P(E) = \frac{n(E)}{n(S)}.$$

Union Rule

For any events E and F from a sample space S,

$$P(E \cup F) = P(E) + P(F) - P(E \cap F).$$

For mutually exclusive events E and F,

$$P(E \cup F) = P(E) + P(F).$$

Complement Rule

$$P(E) = 1 - P(E') \quad \text{and} \quad P(E') = 1 - P(E)$$

Odds

The odds in favor of event E are $\dfrac{P(E)}{P(E')}$, $P(E') \neq 0$.

Properties of Probability

1. For any event E in sample space S, $0 \leq P(E) \leq 1$.

2. The sum of the probabilities of all possible distinct outcomes is 1.

Conditional Probability

The conditional probability of event E, given that event F has occurred, is

$$P(E \mid F) = \frac{P(E \cap F)}{P(F)}, \quad \text{where } P(F) \neq 0.$$

For equally likely outcomes, conditional probability is found by reducing the sample space to event F; then

$$P(E \mid F) = \frac{n(E \cap F)}{n(F)}.$$

Product Rule of Probability

If E and F are events, then $P(E \cap F)$ may be found by either of these formulas.

$$P(E \cap F) = P(F) \cdot P(E \mid F) \quad \text{or} \quad P(E \cap F) = P(E) \cdot P(F \mid E)$$

If E and F are independent events, then $P(E \cap F) = P(E) \cdot P(F)$.

Bayes' Theorem

$$P(F_i \mid E) = \frac{P(F_i) \cdot P(E \mid F_i)}{P(F_1) \cdot P(E \mid F_1) + P(F_2) \cdot P(E \mid F_2) + \cdots + P(F_n) \cdot P(E \mid F_n)}$$

KEY TERMS

7.1 set
element (member)
empty set
set-builder notation
universal set
subset
tree diagram
set operations
Venn diagram
complement

intersection
disjoint sets
union
7.2 union rule for sets
union rule for
disjoint sets
7.3 experiment
trial
outcome
sample space

event
simple event
certain event
impossible event
mutually exclusive
events
7.4 union rule for
probability
odds
empirical probability

probability distribution
7.5 conditional probability
product rule
independent events
dependent events
7.6 Bayes' theorem

CHAPTER 7 REVIEW EXERCISES

Write true *or* false *for each statement.*

1. $9 \in \{8, 4, -3, -9, 6\}$

2. $4 \notin \{3, 9, 7\}$

3. $2 \notin \{0, 1, 2, 3, 4\}$

4. $0 \in \{0, 1, 2, 3, 4\}$

5. $\{3, 4, 5\} \subseteq \{2, 3, 4, 5, 6\}$

6. $\{1, 2, 5, 8\} \subseteq \{1, 2, 5, 10, 11\}$

7. $\{3, 6, 9, 10\} \subseteq \{3, 9, 11, 13\}$

8. $\emptyset \subseteq \{1\}$

9. $\{2, 8\} \not\subseteq \{2, 4, 6, 8\}$

10. $0 \subseteq \emptyset$

In Exercises 11–20, let $U = \{a, b, c, d, e, f, g\}$, $K = \{c, d, f, g\}$, *and* $R = \{a, c, d, e, g\}$. *Find the following.*

11. The number of subsets of K

12. The number of subsets of R

13. K'

14. R'

15. $K \cap R$

16. $K \cup R$

17. $(K \cap R)'$

18. $(K \cup R)'$

19. $\emptyset'$

20. U'

In Exercises 21–26, let $U = \{$all employees of the K. O. Brown Company$\}$;
$\quad A = \{$employees in the accounting department$\}$;
$\quad B = \{$employees in the sales department$\}$;
$\quad C = \{$female employees$\}$;
$\quad D = \{$employees with an MBA degree$\}$.

Describe the following sets in words.

21. $A \cap C$ **22.** $B \cap D$ **23.** $A \cup D$ **24.** $A' \cap D$ **25.** $B' \cap C'$ **26.** $(B \cup C)'$

Draw a Venn diagram and shade each set in Exercises 27–30.

27. $A \cup B'$ **28.** $A' \cap B$ **29.** $(A \cap B) \cup C$ **30.** $(A \cup B)' \cap C$

Write the sample space for each of the following experiments.

31. Rolling a die

32. Drawing a card from a deck containing only the 13 spades

33. Measuring the weight of a person to the nearest half pound (the scale will not measure more than 300 lb)

34. Tossing a coin 4 times

A jar contains 5 balls labeled 3, 5, 7, 9, and 11, respectively, while a second jar contains 4 red and 2 green balls. An experiment consists of pulling 1 ball from each jar, in turn. In Exercises 35–37, write each set using set notation.

35. The sample space

36. Event E: the number on the first ball is greater than 5

37. Event F: the second ball is green

38. Are the outcomes in the sample space in Exercise 35 equally likely?

When a single card is drawn from an ordinary deck, find the probabilities that it will be the following.

39. A heart

40. A red queen

41. A face card

42. Black or a face card

43. Red, given that it is a queen

44. A jack, given that it is a face card

45. A face card, given that it is a king

46. Describe what is meant by disjoint sets

47. Describe what is meant by mutually exclusive events.

48. How are disjoint sets and mutually exclusive events related?

49. Define independent events.

50. Are independent events always mutually exclusive? Are they ever mutually exclusive?

51. An uproar has raged since September 1990 over the answer to a puzzle* published in *Parade* magazine, a supplement of the Sunday newspaper. In the "Ask Marilyn" column, Marilyn vos Savant answered the following question: "Suppose you're on a game show, and you're given the choice of three doors. Behind one door is a car; behind the others, goats. You pick a door, say number 1, and the host, who knows what's behind the other doors, opens another door, say number 3, which has a goat. He then says to you, 'Do you want to pick door number 2?' Is it to your advantage to take the switch?"

Ms. vos Savant estimates that she has since received some 10,000 letters; most of them, including many from mathematicians and statisticians, disagreed with her answer. Her answer has been debated by both professionals and amateurs, and tested in classes at all levels, from grade school to graduate school. But by performing the experiment repeatedly, it can be shown that vos Savant's answer was correct. Find the probabilities of getting the car if you switch or do not switch, and then answer the question yourself. (*Hint:* Consider the sample space.)

Find the odds in favor of a card drawn from an ordinary deck being each of the following.

52. A club

53. A black jack

54. A red face card or a queen

Find the probabilities of getting the following sums when 2 fair dice are rolled.

55. 8

56. 0

57. At least 10

58. No more than 5

59. An odd number greater than 8

60. 12, given that the sum is greater than 10

61. 7, given that at least one die shows a 4

62. At least 9, given that at least one die shows a 5

Parade Magazine, Sept. 9, 1990, p. 13. Reprinted by permission of the William Morris Agency, Inc. on behalf of the author. Copyright © 1990 by Marilyn vos Savant.

63. Suppose $P(E) = .51$, $P(F) = .37$, and $P(E \cap F) = .22$. Find each of the following.

    a. $P(E \cup F)$ **b.** $P(E \cap F')$ **c.** $P(E' \cup F)$ **d.** $P(E' \cap F')$

64. Box A contains 5 red balls and 1 black ball; box B contains 2 red balls and 3 black balls. A box is chosen, and a ball is selected from it. The probability of choosing box A is 3/8. If the selected ball is black, what is the probability that it came from box A?

65. Find the probability that the ball in Exercise 64 came from box B, given that it is red.

▨ Applications

BUSINESS AND ECONOMICS

Appliance Repairs *Of the appliance repair shops listed in the phone book, 80% are competent and 20% are not. A competent shop can repair an appliance correctly 95% of the time; an incompetent shop can repair an appliance correctly 60% of the time. Suppose an appliance was repaired correctly. Find the probabilities that it was repaired by the following.*

66. A competent shop   **67.** An incompetent shop

Suppose an appliance was repaired incorrectly. Find the probabilities that it was repaired by the following.

68. A competent shop   **69.** An incompetent shop

70. *Sales* A company sells typewriters and copiers. Let E be the event "a customer buys a typewriter," and let F be the event "a customer buys a copier." Write each of the following using $\cap$, $\cup$, or $'$ as necessary.

    a. A customer buys neither machine.

    b. A customer buys at least one of the machines.

71. *Defective Items* A sample shipment of five hair dryers is chosen at random. The probability of exactly 0, 1, 2, 3, 4, or 5 hair dryers being defective is given in the following table.

Number Defective	0	1	2	3	4	5
Probability	.31	.25	.18	.12	.08	.06

Find the probabilities that the following numbers of hair dryers are defective.

    a. No more than 3   **b.** At least 3

72. *Defective Items* A manufacturer buys items from four different suppliers. The fraction of the total number of items that is obtained from each supplier, along with the probability that an item purchased from that supplier is defective, is shown in the table in the next column.

    a. Find the probability that a defective item came from supplier 4.

    b. Find the probability that a defective item came from supplier 2.

Supplier	Fraction of Total Supplied	Probability of Defective
1	.17	.04
2	.39	.02
3	.35	.07
4	.09	.03

73. *Car Buyers* The table shows the results of a survey of buyers of a certain model of car.

Car Type	Satisfied	Not Satisfied	Totals
New	300	100	
Used	450		600
Totals		250	

    a. Complete the table.

    b. How many buyers were surveyed?

    c. How many bought a new car and were satisfied?

    d. How many were not satisfied?

    e. How many bought used cars?

    f. How many of those who were not satisfied had purchased a used car?

    g. Rewrite the event stated in part f using the expression "given that."

    h. Find the probability of the outcome in parts f and g.

    i. Find the probability that a used-car buyer is not satisfied.

    j. You should have different answers in parts h and i. Explain why.

LIFE SCIENCES

74. *Sickle Cell Anemia* The square shows the four possible (equally likely) combinations when both parents are carriers of the sickle cell anemia trait. Each carrier parent has normal cells (N) and trait cells (T).

		Second Parent	
		N_2	T_2
First Parent	N_1		$N_1 T_2$
	T_1		

a. Complete the table.

b. If the disease occurs only when two trait cells combine, find the probability that a child born to these parents will have sickle cell anemia.

c. The child will carry the trait but not have the disease if a normal cell combines with a trait cell. Find this probability.

d. Find the probability that the child is neither a carrier nor has the disease.

SOCIAL SCIENCES

75. *Elections* In the 1994 elections for the U.S. House of Representatives, 46% of men and 54% of women voted Democratic. Of those who voted, 49% were men and 51% were women.* What is the probability that a person who voted Democratic was a man?

76. *Television Viewing Habits* A telephone survey of television viewers revealed the following information.

> 20 watch situation comedies;
> 19 watch game shows;
> 27 watch movies;
> 5 watch both situation comedies and game shows;
> 8 watch both game shows and movies;
> 10 watch both situation comedies and movies;
> 3 watch all three;
> 6 watch none of these.

a. How many viewers were interviewed?

b. How many viewers watch comedies and movies but not game shows?

c. How many viewers watch only movies?

d. How many viewers watch comedies and game shows but not movies?

77. *Randomized Response Method for Getting Honest Answers to Sensitive Questions*† Basically, this is a method to guarantee that an individual who answers sensitive question will remain anonymous, thus encouraging a truthful response. This method is, in effect, an application of the formula for finding the probability of an intersection, and operates as follows. Questions A and B are posed, one of which is sensitive and the other not. The probability of receiving a "yes" to the nonsensitive question must be known. For example, one could ask

> A: Does your Social Security number end in an odd digit? (Nonsensitive)
>
> B: Have you ever intentionally cheated on your income tax? (Sensitive)

We know that $P(\text{answer yes} \mid \text{answer } A) = 1/2$. We wish to approximate $P(\text{answer yes} \mid \text{answer } B)$. The subject is asked to flip a coin and answer A if the coin comes up heads and otherwise to answer B. In this way, the interviewer does not know which question the subject is answering. Thus, a "yes" answer is not incriminating. There is no way for the interviewer to know whether the subject is saying "Yes, my Social Security number ends in an odd digit" or "Yes, I have intentionally cheated on my income taxes." The percentage of subjects in the group answering "yes" is used to approximate $P(\text{answer yes})$.

a. Use the fact that the event "answer yes" is the union of the event "answer yes and answer A" with the event "answer yes and answer B" to prove that

$$P(\text{answer yes} \mid \text{answer } B)$$
$$= \frac{P(\text{answer yes}) - P(\text{answer yes} \mid \text{answer } A) \cdot P(\text{answer } A)}{P(\text{answer } B)}.$$

b. If this technique is tried on 100 subjects and 60 answered "yes," what is the approximate probability that a person randomly selected from the group has intentionally cheated on income taxes?

PHYSICAL SCIENCES

78. *Earthquake* It has been reported that government scientists have predicted that the odds for a major earthquake occurring in the San Francisco Bay area during the next 30 years are 9 to 1.‡ What is the probability that a major earthquake will occur during the next 30 years in San Francisco?

GENERAL INTEREST

79. *Making a First Down* A first down is desirable in football—it guarantees four more plays by the team making it, assuming no score or turnover occurs in the plays. After getting a first down, a team can get another by advancing the ball at least 10 yards. During the four plays given by a first down, a team's position will be indicated by a phrase such as "third and 4," which means that the team has

---

The New York Times, Nov. 13, 1994, p. 24.
†Milton, J. S., and J. J. Corbet, *Applied Statistics with Probability.* Copyright © 1979 by Litton Educational Publishing, Inc. Reprinted by permission of Brooks/Cole Publishing Company, Monterey, California.
‡*The San Francisco Chronicle*, June 8, 1994, p. A1.

already had two of its four plays, and that 4 more yards are needed to get 10 yards necessary for another first down. An article in a management journal* offers the following results for 189 games for a recent National Football League season. "Trials" represents the number of times a team tried to make a first down, given that it was currently playing either a third or a fourth down. Here, n represents the number of yards still needed for a first down.

n	Trials	Successes	Probability of Making First Down with n Yards to Go
1	543	388	
2	327	186	
3	356	146	
4	302	97	
5	336	91	

a. Complete the table.

b. Why is the sum of the answers in the table not equal to 1?

80. *States* Of the 50 United States,

    22 are west of the Mississippi River (western states);[†]
    22 had populations of less than 3 million in the 1990 census (small states);
    26 begin with the letters A through M (early states);
    11 are western early states;
    13 are western small states;
    10 are early small states;
    5 are early small western states.

a. How many western states had populations of more than 3 million in the 1990 census and begin with the letters N through Z?

b. How many states are east of the Mississippi, had populations of more than 3 million in the 1990 census, and begin with the letters N through Z?

81. *Music* Country-western songs often emphasize three basic themes: love, prison, and trucks. A survey of the local country-western radio station produced the following data.

    12 songs were about a truckdriver who was in love while in prison;
    13 were about a prisoner in love;
    28 were about a person in love;
    18 were about a truckdriver in love;
    3 were about a truckdriver in prison who was not in love;

    2 were about a prisoner who was not in love and did not drive a truck;
    8 were about a person who was not in prison, not in love, and did not drive a truck;
    16 were about truckdrivers who were not in prison.

a. How many songs were surveyed?

Find the number of songs about

b. truckdrivers;   **c.** prisoners;

d. truckdrivers in prison; **e.** people not in prison;

f. people not in love.

 82. *Missiles* In his novel *Debt of Honor*, Tom Clancy writes the following:[‡]

"There were ten target points—missile silos, the intelligence data said, and it pleased the Colonel [Zacharias] to be eliminating the hateful things, even though the price of that was the lives of other men. There were only three of them [bombers], and his bomber, like the others, carried only eight weapons [smart bombs]. The total number of weapons carried for the mission was only twenty-four, with two designated for each silo, and Zacharias's last four for the last target. Two bombs each. Every bomb had a 95% probability of hitting within four meters of the aim point, pretty good numbers really, except that this sort of mission had precisely no margin for error. Even the paper probability was less than half a percent chance of a double miss, but that number times ten targets meant a 5% chance that [at least] one missile would survive, and that could not be tolerated."

Determine if the calculations in this quote are correct by the following steps.

a. Given that each bomb had a 95% probability of hitting the missile silo on which it was dropped, and that two bombs were dropped on each silo, what is the probability of a double miss?

b. What is the probability that a specific silo was destroyed (that is, that at least one bomb of the two bombs struck the silo)?

c. What is the probability that all ten silos were destroyed?

d. What is the probability that at least one silo survived? Does this agree with the quote?

e. What assumptions need to be made for the calculations in parts a and b to be valid? Discuss whether these assumptions seem reasonable.

*Carter, Virgil, and Robert Machols, "Optimal Strategies on Fourth Down," *Management Science*, Vol. 24, No. 16, Dec. 1978. Copyright © 1978 by The Institute of Management Sciences. Reprinted by permission.
†We count here states such as Minnesota, which has more than half of its area to the west of the Mississippi.
‡Clancy, Tom, *Debt of Honor*, G. P. Putnam's Sons, 1994, pp. 686–687. Reprinted with permission.

EXTENDED APPLICATION: Medical Diagnosis

When a patient is examined, information (typically incomplete) is obtained about his state of health. Probability theory provides a mathematical model appropriate for this situation, as well as a procedure for quantitatively interpreting such partial information to arrive at a reasonable diagnosis.*

To develop a model, we list the states of health that can be distinguished in such a way that the patient can be in one and only one state at the time of the examination. For each state of health H, we associate a number, $P(H)$, between 0 and 1 such that the sum of all these numbers is 1. This number $P(H)$ represents the probability, before examination, that a patient is in the state of health H, and $P(H)$ may be chosen subjectively from medical experience, using any information available prior to the examination. The probability may be most conveniently established from clinical records; that is, a mean probability is established for patients in general, although the number would vary from patient to patient. Of course, the more information that is brought to bear in establishing $P(H)$, the better the diagnosis.

For example, limiting the discussion to the condition of a patient's heart, suppose there are exactly three states of health, with probabilities as follows.

	State of Health, H	$P(H)$
H_1	Patient has a normal heart	.8
H_2	Patient has minor heart irregularities	.15
H_3	Patient has a severe heart condition	.05

Having selected $P(H)$, the information from the examination is processed. First, the results of the examination must be classified. The examination itself consists of observing the state of a number of characteristics of the patient. Let us assume that the examination for a heart condition consists of a stethoscope examination and a cardiogram. The outcome of such an examination, C, might be one of the following:

$C_1 =$ stethoscope shows normal heart
and cardiogram shows normal heart;

$C_2 =$ stethoscope shows normal heart
and cardiogram shows minor irregularities;

and so on.

It remains to assess for each state of health H the conditional probability $P(C|H)$ of each examination outcome C using only the knowledge that a patient is in a given state of health. (This may be based on the medical knowledge and clinical experience of the doctor.) The conditional probabilities $P(C|H)$ will not vary from patient to patient (although they

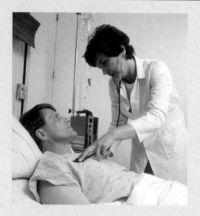

should be reviewed periodically), so that they may be built into a diagnostic system.

Suppose the result of the examination is C_1. Let us assume the following probabilities:

$$P(C_1|H_1) = .9,$$
$$P(C_1|H_2) = .4,$$
$$P(C_1|H_3) = .1.$$

Now, for a given patient, the appropriate probability associated with each state of health H, after examination, is $P(H|C)$, where C is the outcome of the examination. This can be calculated by using Bayes' theorem. For example, to find $P(H_1|C_1)$—that is, the probability that the patient has a normal heart given that the examination showed a normal stethoscope examination and a normal cardiogram—we use Bayes' theorem as follows:

$$P(H_1|C_1)$$
$$= \frac{P(C_1|H_1)P(H_1)}{P(C_1|H_1)P(H_1) + P(C_1|H_2)P(H_2) + P(C_1|H_3)P(H_3)}$$
$$= \frac{(.9)(.8)}{(.9)(.8) + (.4)(.15) + (.1)(.05)} \approx .92.$$

Hence, the probability is about .92 that the patient has a normal heart on the basis of the examination results. This means that in 8 out of 100 patients, some abnormality will be present and not be detected by the stethoscope or the cardiogram.

Exercises

1. Find $P(H_2|C_1)$.

2. Assuming the following probabilities, find $P(H_1|C_2)$.

$$P(C_2|H_1) = .2 \qquad P(C_2|H_2) = .8 \qquad P(C_2|H_3) = .3$$

3. Assuming the probabilities of Exercise 2, find $P(H_3|C_2)$.

---

*Wright, Roger, "Probabilistic Medical Diagnosis," *Some Mathematical Models in Biology*, rev. ed., Robert M. Thrall, ed., University of Michigan, 1967. Used by permission of Robert M. Thrall.

Counting Principles;
Further Probability Topics

If you have 31 ice cream flavors available, how many different three-scoop cones can you make? The answer, which is surprisingly large, involves counting permutations or combinations, the subject of the first two sections in this chapter. The counting formulas we will develop have important applications in probability theory.

In this chapter, we continue our discussion of probability theory. To use the basic definition of probability, $P(E) = n(E)/n(S)$ (where S is the sample space with equally likely outcomes), up to now we have simply listed the outcomes in S and in E. However, when S has many outcomes, listing them all becomes very tedious. In the first two sections of this chapter, we introduce methods for counting the number of outcomes in a set without actually listing them, and then we use this approach in the third section to find probabilities. In the section on binomial probability (repeated independent trials of an experiment with only two possible outcomes), we introduce a formula for finding the probability of a certain number of successes in a number of trials. The final section continues the discussion of probability distributions that we began in Chapter 7.

■ 8.1 THE MULTIPLICATION PRINCIPLE; PERMUTATIONS

THINK ABOUT IT   In how many ways can seven panelists be seated in a row of seven chairs?

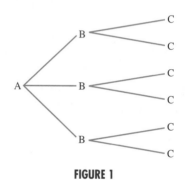

FIGURE 1

This question will be answered later in this section using *permutations*. Let us begin with a simple example. If there are 3 roads from town A to town B and 2 roads from town B to town C, in how many ways can a person travel from A to C by way of B? For each of the 3 roads from A there are 2 different routes leading from B to C, or a total of $3 \cdot 2 = 6$ different ways for the trip, as shown in Figure 1. This example illustrates a general principle of counting, called the **multiplication principle.**

MULTIPLICATION PRINCIPLE

Suppose n choices must be made, with

$$m_1 \text{ ways to make choice 1,}$$

and for each of these ways,

$$m_2 \text{ ways to make choice 2,}$$

and so on, with

$$m_n \text{ ways to make choice } n.$$

Then there are

$$m_1 \cdot m_2 \cdot \cdots \cdot m_n$$

different ways to make the entire sequence of choices.

EXAMPLE 1   Combination Lock

A certain combination lock can be set to open to any one 3-letter sequence. How many such sequences are possible?

Solution Since there are 26 letters in the alphabet, there are 26 choices for each of the 3 letters. By the multiplication principle, there are $26 \cdot 26 \cdot 26 = 17,576$ different sequences.

EXAMPLE 2   Morse Code

Morse code uses a sequence of dots and dashes to represent letters and words. How many sequences are possible with at most 3 symbols?

Solution "At most 3" means "1 or 2 or 3" here. Each symbol may be either a dot or a dash. Thus the following number of sequences are possible in each case.

Number of Symbols	Number of Sequences
1	2
2	$2 \cdot 2 = 4$
3	$2 \cdot 2 \cdot 2 = 8$

Altogether, $2 + 4 + 8 = 14$ different sequences are possible.

EXAMPLE 3   I Ching

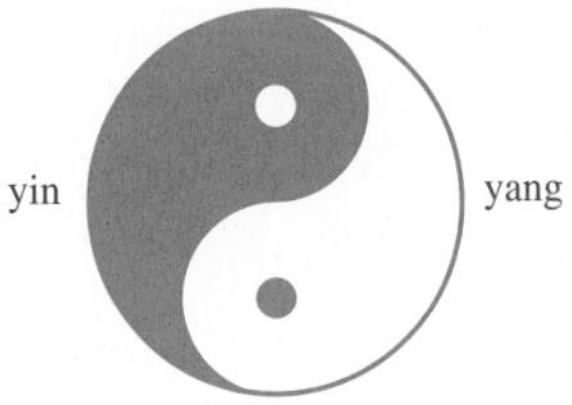

yin   yang

FIGURE 2

An ancient Chinese philosophical work known as the *I Ching (Book of Changes)* is often used as an oracle from which people can seek and obtain advice. The philosophy describes the duality of the universe in terms of two primary forces: *yin* (passive, dark, receptive) and *yang* (active, light, creative). See Figure 2. The yin energy is represented by a broken line (– –) and the yang by a solid line (—). These lines are written on top of one another in groups of three, known as *trigrams*. For example, the trigram $\equiv\equiv$ is called *Tui,* the Joyous, and has the image of a lake.

(a) How many trigrams are there altogether?

Solution Think of choosing between the 2 types of lines for each of the 3 positions in the trigram. There will be 2 choices for each position, so there are $2 \cdot 2 \cdot 2 = 8$ different trigrams.

(b) The trigrams are grouped together, one on top of the other, in pairs known as *hexagrams.* Each hexagram represents one aspect of the *I Ching* philosophy. How many hexagrams are there?

Solution For each position in the hexagram there are 8 possible trigrams, giving $8 \cdot 8 = 64$ hexagrams.

EXAMPLE 4   Books

A teacher has 5 different books that he wishes to arrange side by side. How many different arrangements are possible?

Solution Five choices will be made, one for each space that will hold a book. Any of the 5 books could be chosen for the first space. There are 4 choices for the second space, since 1 book has already been placed in the first space; there are 3 choices for the third space, and so on. By the multiplication principle, the number of different possible arrangements is $5 \cdot 4 \cdot 3 \cdot 2 \cdot 1 = 120$.

FOR REVIEW

The natural numbers, also referred to as the positive integers, are the numbers 1, 2, 3, 4, etc.

The use of the multiplication principle often leads to products such as $5 \cdot 4 \cdot 3 \cdot 2 \cdot 1$, the product of all the natural numbers from 5 down to 1. If *n* is a natural number, the symbol *n*! (read "*n factorial*") denotes the product of all the natural numbers from n down to 1. If $n = 1$, this formula is understood to give $1! = 1$.

FACTORIAL NOTATION

For any natural number n,

$$n! = n(n-1)(n-2) \cdots (3)(2)(1).$$

Also,

$$0! = 1.$$

With this symbol, the product $5 \cdot 4 \cdot 3 \cdot 2 \cdot 1$ can be written as $5!$. Also, $3! = 3 \cdot 2 \cdot 1 = 6$. The definition of $n!$ could be used to show that $n[(n-1)]! = n!$ for all natural numbers $n \geq 2$. It is helpful if this result also holds for $n = 1$. This can happen only if $0!$ equals 1, as defined above.

Some calculators have an $n!$ key. A calculator with a 10-digit display and scientific notation capability will usually give the exact value of $n!$ for $n \leq 13$, and approximate values of $n!$ for $14 \leq n \leq 69$. The value of $70!$ is approximately $1.198 \cdot 10^{100}$, which is too large for most calculators. To see how large $70!$ is, suppose a computer counted the numbers from 1 to $70!$ at a rate of 1 billion numbers per second. If the computer started when the universe began, by now it would only be done with a tiny fraction of the total.

On many graphing calculators, the factorial of a number is accessible through a menu. On the TI-83, for example, this menu is found by pressing the MATH key, and then selecting PRB (for probability).

EXAMPLE 5 Books

Suppose the teacher in Example 4 wishes to place only 3 of the 5 books on his desk. How many arrangements of 3 books are possible?

Solution The teacher again has 5 ways to fill the first space, 4 ways to fill the second space, and 3 ways to fill the third. Since he wants to use only 3 books, only 3 spaces can be filled (3 events) instead of 5, for $5 \cdot 4 \cdot 3 = 60$ arrangements.

Permutations The answer 60 in Example 5 is called the number of *permutations* of 5 things taken 3 at a time. A **permutation** of r (where $r \geq 1$) elements from a set of n elements is any specific ordering or arrangement, *without repetition,* of the r elements. Each rearrangement of the r elements is a different permutation. The number of permutations of n things taken r at a time (with $r \leq n$) is written $P(n, r)$. Based on the work in Example 5,

$$P(5, 3) = 5 \cdot 4 \cdot 3 = 60.$$

Factorial notation can be used to express this product as follows.

$$5 \cdot 4 \cdot 3 = 5 \cdot 4 \cdot 3 \cdot \frac{2 \cdot 1}{2 \cdot 1} = \frac{5 \cdot 4 \cdot 3 \cdot 2 \cdot 1}{2 \cdot 1} = \frac{5!}{2!} = \frac{5!}{(5-3)!}$$

This example illustrates the general rule of permutations, which can be stated as follows.

PERMUTATIONS

If $P(n, r)$ (where $r \leq n$) is the number of permutations of n elements taken r at a time, then

$$P(n,r) = \frac{n!}{(n-r)!}.$$

CAUTION The letter P here represents *permutations,* not *probability.* In probability notation, the quantity in parentheses describes an *event.* In permutations notation, the quantity in parentheses always comprises *two numbers.*

The proof of the permutations rule follows the discussion in Example 5. There are n ways to choose the first of the r elements, $n - 1$ ways to choose the second, and $n - r + 1$ ways to choose the rth element, so that

$$P(n,r) = n(n-1)(n-2)\cdots(n-r+1).$$

Now multiply on the right by $(n - r)!/(n - r)!$.

$$P(n,r) = n(n-1)(n-2)\cdots(n-r+1) \cdot \frac{(n-r)!}{(n-r)!}$$

$$= \frac{n(n-1)(n-2)\cdots(n-r+1)(n-r)!}{(n-r)!}$$

$$= \frac{n!}{(n-r)!}$$

To find $P(n, r)$, we can use either the permutations formula or direct application of the multiplication principle, as the following example shows.

EXAMPLE 6   Politics

Early in 1996, eight candidates sought the Republican nomination for president. In how many ways could voters rank their first, second, and third choices?

Solution

Method 1: Calculating by Hand

This is the same as finding the number of permutations of 8 elements taken 3 at a time. Since there are 3 choices to be made, the multiplication principle gives $P(8, 3) = 8 \cdot 7 \cdot 6 = 336$. Alternatively, use the permutations formula to get

$$P(8,3) = \frac{8!}{(8-3)!} = \frac{8!}{5!} = \frac{8 \cdot 7 \cdot 6 \cdot 5!}{5!} = 8 \cdot 7 \cdot 6 = 336.$$

Method 2: Graphing Calculator

Graphing calculators have the capacity to compute permutations. For example, on a TI-83, $P(8, 3)$ can be calculated by inputting 8 followed by `nPr` (found in the MATH-PRB menu), and a 3 yielding 336, as shown in Figure 3.

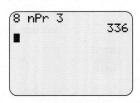

FIGURE 3

Method 3: Spreadsheet | Spreadsheets can also compute permutations. For example, in Microsoft Excel, $P(8, 3)$ can be calculated by inputting 8 and 3 in cells, say, A1 and B1, and then typing "=FACT(A1)/FACT(A1-B1)" in cell C1 or, for that matter, any other cell.

CAUTION When calculating the number of permutations with the formula, do not try to cancel unlike factorials. For example,

$$\frac{8!}{4!} \neq 2! = 2 \cdot 1 = 2.$$

$$\frac{8!}{4!} = \frac{8 \cdot 7 \cdot 6 \cdot 5 \cdot 4 \cdot 3 \cdot 2 \cdot 1}{4 \cdot 3 \cdot 2 \cdot 1} = 8 \cdot 7 \cdot 6 \cdot 5 = 1680.$$

Always write out the factors first, then cancel where appropriate.

EXAMPLE 7 Permutations
Find each of the following.

(a) The number of permutations of the letters A, B, and C

Solution By the formula for $P(n, r)$ with both n and r equal to 3,

$$P(3, 3) = \frac{3!}{(3 - 3)!} = \frac{3!}{0!} = \frac{3!}{1} = 3 \cdot 2 \cdot 1 = 6.$$

The 6 permutations (or arrangements) are

$$\text{ABC, \quad ACB, \quad BAC, \quad BCA, \quad CAB, \quad CBA.}$$

(b) The number of permutations if just 2 of the letters A, B, and C are to be used

Solution Find $P(3, 2)$.

$$P(3, 2) = \frac{3!}{(3 - 2)!} = \frac{3!}{1!} = 3! = 6$$

This result is exactly the same answer as in part (a). This is because, in the case of $P(3, 3)$, after the first 2 choices are made, the third is already determined, as shown in the table below.

First Two Letters	AB	AC	BA	BC	CA	CB
Third Letter	C	B	C	A	B	A

EXAMPLE 8 Television
A televised talk show will include 4 women and 3 men as panelists.

(a) In how many ways can the panelists be seated in a row of 7 chairs?

Solution Find $P(7, 7)$, the total number of ways to seat 7 panelists in 7 chairs.

$$P(7, 7) = \frac{7!}{(7 - 7)!} = \frac{7!}{0!} = \frac{7!}{1} = 7 \cdot 6 \cdot 5 \cdot 4 \cdot 3 \cdot 2 \cdot 1 = 5040$$

There are 5040 ways to seat the 7 panelists.

(b) In how many ways can the panelists be seated if the men and women are to be alternated?

Solution Use the multiplication principle. In order to alternate men and women, a woman must be seated in the first chair (since there are 4 women and only 3 men), any of the men next, and so on. Thus there are 4 ways to fill the first seat, 3 ways to fill the second seat, 3 ways to fill the third seat (with any of the 3 remaining women), and so on. This gives

$$4 \cdot 3 \cdot 3 \cdot 2 \cdot 2 \cdot 1 \cdot 1 = 144$$

ways to seat the panelists.

(c) In how many ways can the panelists be seated if the men must sit together, and the women must also sit together?

Solution Use the multiplication principle. We first must decide how to arrange the two groups (men and women). There are 2! ways of doing this. Next, there are 4! ways of arranging the women and 3! ways of arranging the men, for a total of

$$2! \, 4! \, 3! = 2 \cdot 24 \cdot 6 = 288$$

ways.

(d) In how many ways can one woman and one man from the panel be selected?

Solution There are 4 ways to pick the woman and 3 ways to pick the man, for a total of

$$4 \cdot 3 = 12$$

ways.

If the n objects in a permutation are not all distinguishable—that is, if there are n_1 of type 1, n_2 of type 2, and so on for r different types, then the number of **distinguishable permutations** is

$$\frac{n!}{n_1! \, n_2! \cdots n_r!}.$$

For example, suppose we want to find the number of permutations of the numbers 1, 1, 4, 4, 4. We cannot distinguish between the two 1's or among the three 4's, so using 5! would give too many distinguishable arrangements. Since the two 1's are indistinguishable and account for 2! of the permutations, we divide 5! by 2!. Similarly, we also divide by 3! to account for the three indistinguishable 4's. This gives

$$\frac{5!}{2! \, 3!} = 10$$

permutations.

EXAMPLE 9   Mississippi

In how many ways can the letters in the word *Mississippi* be arranged?

Solution This word contains 1 m, 4 i's, 4 s's, and 2 p's. To use the formula, let $n = 11, n_1 = 1, n_2 = 4, n_3 = 4$, and $n_4 = 2$ to get

$$\frac{11!}{1!\,4!\,4!\,2!} = 34{,}650$$

arrangements.

> **NOTE** If Example 9 had asked for the number of ways that the letters in a word with 11 *different* letters could be arranged, the answer would be $11! = 39{,}916{,}800$.

EXAMPLE 10   Yogurt

A student buys 3 cherry yogurts, 2 raspberry yogurts, and 2 blueberry yogurts. She puts them in her dormitory refrigerator to eat one a day for the next week. Assuming yogurts of the same flavor are indistinguishable, in how many ways can she select yogurts to eat for the next week?

Solution This problem is again one of distinguishable permutations. The 7 yogurts can be selected in 7! ways, but since the 3 cherry, 2 raspberry, and 2 blueberry yogurts are indistinguishable, the total number of distinguishable orders in which the yogurts can be selected is

$$\frac{7!}{3!\,2!\,2!} = 210.$$

8.1 EXERCISES

In Exercises 1–12, evaluate the factorial or permutation.

1. 6!

2. 7!

3. 15!

4. 16!

5. $P(13, 2)$

6. $P(12, 3)$

7. $P(38, 17)$

8. $P(33, 19)$

9. $P(n, 0)$

10. $P(n, n)$

11. $P(n, 1)$

12. $P(n, n - 1)$

13. How many different types of homes are available if a builder offers a choice of 5 basic plans, 3 roof styles, and 2 exterior finishes?

14. A menu offers a choice of 3 salads, 8 main dishes, and 5 desserts. How many different meals consisting of one salad, one main dish, and one dessert are possible?

15. A couple has narrowed down the choice of a name for their new baby to 3 first names and 5 middle names. How many different first- and middle-name arrangements are possible?

16. In a club with 15 members, how many ways can a slate of 3 officers consisting of president, vice-president, and secretary/treasurer be chosen?

17. Define *permutation* in your own words.

18. In Example 7, there are six 3-letter permutations of the letters A, B, and C. How many 3-letter subsets (unordered groups of letters) are there?

19. In Example 7, how many unordered 2-letter subsets of the letters A, B, and C are there?

20. Find the number of distinguishable permutations of the letters in each of the following words.

  a. initial   **b.** little   **c.** decreed

21. A printer has 5 A's, 4 B's, 2 C's, and 2 D's. How many different "words" are possible that use all these letters? (A "word" does not have to have any meaning here.)

22. Wing has different books to arrange on a shelf: 4 blue, 3 green, and 2 red.

    a. In how many ways can the books be arranged on a shelf?

    b. If books of the same color are to be grouped together, how many arrangements are possible?

    c. In how many distinguishable ways can the books be arranged if books of the same color are identical but need not be grouped together?

    d. In how many ways can you select 3 books, one of each color, if the order in which the books are selected does not matter?

23. A child has a set of differently shaped plastic objects. There are 3 pyramids, 4 cubes, and 7 spheres.

    a. In how many ways can she arrange the objects in a row if each is a different color?

    b. How many arrangements are possible if objects of the same shape must be grouped together?

    c. In how many distinguishable ways can the objects be arranged in a row if objects of the same shape are also the same color, but need not be grouped together?

    d. In how many ways can you select 3 objects, one of each shape, if the order in which the objects are selected does not matter?

24. Some students find it puzzling that $0! = 1$, and think that $0!$ should equal 0. If this were true, what would be the value of $P(4, 4)$ using the permutations formula?

25. If you already knew the value of $9!$, how could you find the value of $10!$ quickly?

26. When calculating $n!$, the number of ending zeros in the answer can be determined prior to calculating the actual number by finding the number of times 5 can be factored from $n!$. For example, $7!$ only has one 5 occurring in its calculation, and so there is only one ending zero in 5040. The number $10!$ has two 5's (one from the 5 and one from the 10) and so there must be two ending zeros in the answer 3,628,800. Use this idea to determine the number of zeros that occur in the following factorials, and then explain why this works.

    a. $13!$ **b.** $27!$ **c.** $75!$

27. Because of the view screen, calculators only show a fixed number of digits, often 10 digits. Thus, an approximation of a number will be shown by only including the 10 largest place values of the number. Using the ideas from the previous exercise, determine if the following numbers are correct or if they are incorrect by checking if they have the correct number of ending zeros. (*Note:* Just because a number has the correct number of zeros does not imply that it is correct.)

    a. $12! = 479,001,610$ **b.** $23! = 25,852,016,740,000,000,000,000$

    c. $15! = 1,307,643,680,000$ **d.** $14! = 87,178,291,200$

▗ Applications

BUSINESS AND ECONOMICS

28. *Automobile Manufacturing* An automobile manufacturer produces 7 models, each available in 6 different exterior colors, with 4 different upholstery fabrics and 5 interior colors. How many varieties of automobile are available?

LIFE SCIENCES

29. *Drug Sequencing* Eleven drugs have been found to be effective in the treatment of a disease. It is believed that the sequence in which the drugs are administered is important in the effectiveness of the treatment. In how many different sequences can 5 of the 11 drugs be administered?

30. *Insect Classification* A biologist is attempting to classify 52,000 species of insects by assigning 3 initials to each species. Is it possible to classify all the species in this way? If not, how many initials should be used?

31. *Genetics Experiment* In how many ways can 7 of 10 monkeys be arranged in a row for a genetics experiment?

SOCIAL SCIENCES

32. *Social Science Experiment* In an experiment on social interaction, 6 people will sit in 6 seats in a row. In how many ways can this be done?

33. *Election Ballots* In an election with 3 candidates for one office and 5 candidates for another office, how many different ballots may be printed?

GENERAL INTEREST

34. *Course Scheduling* A business school gives courses in typing, shorthand, transcription, business English, technical writing, and accounting. In how many ways can a student arrange a schedule if 3 courses are taken? Assume that the order in which courses are selected matters.

35. *Course Scheduling* If your college offers 400 courses, 20 of which are in mathematics, and your counselor arranges your schedule of 4 courses by random selection, how many schedules are possible that do not include a math course? Assume that the order in which courses are selected matters.

36. *Baseball Teams* A baseball team has 20 players. How many 9-player batting orders are possible?

37. *Union Elections* A chapter of union Local 715 has 35 members. In how many different ways can the chapter select a president, a vice-president, a treasurer, and a secretary?

38. *Programming Music* A concert to raise money for an economics prize is to consist of 5 works: 2 overtures, 2 sonatas, and a piano concerto.

   a. In how many ways can the program be arranged?

   b. In how many ways can the program be arranged if an overture must come first?

39. *Programming Music* A zydeco band from Louisiana will play 5 traditional and 3 original Cajun compositions at a concert. In how many ways can they arrange the program if

   a. they begin with a traditional piece?

   b. an original piece will be played last?

40. *Television Scheduling* The television schedule for a certain evening shows 8 choices from 8 to 9 P.M., 5 choices from 9 to 10 P.M., and 6 choices from 10 to 11 P.M. In how many different ways could a person schedule that evening of television viewing from 8 to 11 P.M.? (Assume each program that is selected is watched for an entire hour.)

41. *Radio Station Call Letters* How many different 4-letter radio station call letters can be made if

   a. the first letter must be K or W and no letter may be repeated?

   b. repeats are allowed, but the first letter is K or W?

   c. the first letter is K or W, there are no repeats, and the last letter is R?

42. *Telephone Numbers* How many 7-digit telephone numbers are possible if the first digit cannot be zero and

   a. only odd digits may be used?

   b. the telephone number must be a multiple of 10 (that is, it must end in zero)?

   c. the telephone number must be a multiple of 100?

   d. the first 3 digits are 481?

   e. no repetitions are allowed?

Telephone Area Codes *A few years ago, the United States began running out of telephone numbers. Telephone companies introduced new area codes as numbers were used up, and eventually almost all area codes were used up.*

43. a. Until recently, all area codes had a 0 or 1 as the middle digit, and the first digit could not be 0 or 1. How many area codes are there with this arrangement? How many telephone numbers does the current 7-digit sequence permit per area code? (The 3-digit sequence that follows the area code cannot start with 0 or 1. Assume there are no other restrictions.)

   b. The actual number of area codes under the previous system was 152. Explain the discrepancy between this number and your answer to part a.

44. The shortage of area codes was avoided by removing the restriction on the second digit. (This resulted in problems for some older equipment, which used the second digit to determine that a long-distance call was being made.) How many area codes are available under the new system?

45. *License Plates* For many years, the state of California used 3 letters followed by 3 digits on its automobile license plates.

   a. How many different license plates are possible with this arrangement?

   b. When the state ran out of new numbers, the order was reversed to 3 digits followed by 3 letters. How many new license plate numbers were then possible?

   c. Several years ago, the numbers described in b were also used up. The state then issued plates with 1 letter followed by 3 digits and then 3 letters. How many new license plate numbers will this provide?

46. *Social Security Numbers* A social security number has 9 digits. How many social security numbers are there? The U.S. population in 2000 was about 281 million. Is it possible for every U.S. resident to have a unique social security number? (Assume no restrictions.)

47. *Postal Zip Codes* The United States Postal Service currently uses 5-digit zip codes in most areas. How many zip codes are possible if there are no restrictions on the digits used? How many would be possible if the first number could not be 0?

48. *Postal Zip Codes* The Postal Service is encouraging the use of 9-digit zip codes in some areas, adding 4 digits after the usual 5-digit code. How many such zip codes are possible with no restrictions?

49. *Games* The game of Sets* consists of a special deck of cards. Each card has on it either one, two, or three shapes. The shapes on each card are all the same color, either green, purple, or red. The shapes on each card are the same style, either solid, shaded, or outline. There are three possible shapes: squiggle, diamond, and oval, and only one type of shape appears on a card. The deck consists of all possible combinations of shape, color, style, and number. How many cards are in a deck?

50. *Games* In the game of Scattergories,[†] the players take 12 turns. In each turn, a 20-sided die is rolled; each side has a letter. The players must then fill in 12 categories (e.g., vegetable, city, etc.) with a word beginning with the letter rolled. Considering that a game consists of 12 rolls of the 20-sided die, how many possible games are there?

 51. *Games* The game of Twenty Questions consists of asking 20 questions to determine a person, place, or thing that the other person is thinking of. The first question, "Is it an animal, vegetable, or mineral?" has three possible answers.

The other questions must be answered "Yes" or "No." How many possible objects can be distinguished in this game, assuming that all 20 questions are asked? Are 20 questions enough?

52. *Traveling Salesman* In the famous Traveling Salesman Problem, a salesman starts in any one of a set of cities, visits every city in the set once, and returns to the starting city. He would like to complete this circuit with the shortest possible distance.

   a. Suppose the salesman has 10 cities to visit. Given that it does not matter what city he starts in, how many different circuits can he take?

   b. The salesman decides to check all the different paths in part a to see which is shortest, but realizes that a circuit has the same distance whichever direction it is traveled. How many different circuits must he check?

   c. Suppose the salesman has 70 cities to visit. Would it be feasible to have a computer check all the different circuits? Explain your reasoning.

53. *Married Couples* Edouard Lucas's Problem of the Married Couples appeared in 1891 in his book, *Theorie des Nombres*. The problem is to determine the number of ways that *n* married couples can be seated around a table such that there is always one man between two women and none of the men is ever next to his own wife. This problem can be solved by calculating $2A_n \cdot n!$ where $A_3 = 1$, $A_4 = 2$, and A_n can be calculated by *Laisant's recurrence formula*,[‡]

$$(n-1)A_{n+1} = (n^2 - 1)A_n + (n+1)A_{n-1} + 4(-1)^n, \quad \text{for } n \geq 4.$$

   a. Verify that $A_5 = 13$, $A_6 = 80$, $A_7 = 579$, $A_8 = 4738$, $A_9 = 43{,}387$, $A_{10} = 439{,}792$.

   b. How many possible seating arrangements are there for this problem with 10 couples?

■ 8.2 COMBINATIONS

?	**THINK ABOUT IT**	In how many ways can a manager select 4 employees for promotion from 12 eligible employees?

As we shall see, permutations cannot be used to answer this question, but combinations will provide the answer.

In the previous section, we saw that there are 60 ways that a teacher can arrange 3 of 5 different books on his desk. That is, there are 60 permutations of 5 books taken 3 at a time. Suppose now that the teacher does not wish to arrange the books on his desk, but rather wishes to choose, without regard to order, any 3

---

*Copyright © Marsha J. Falco.
†Copyright © Milton Bradley Company.
‡Dorrie, Heinrich, *100 Great Problems of Elementary Mathematics: Their History and Solution.* New York, Dover Publishers, 1965, pp. 27–33.

of the 5 books for a book sale to raise money for his school. In how many ways can this be done?

At first glance, we might say 60 again, but this is incorrect. The number 60 counts all possible *arrangements* of 3 books chosen from 5. The following 6 arrangements, however, would all lead to the same set of 3 books being given to the book sale.

mystery-biography-textbook   biography-textbook-mystery

mystery-textbook-biography   textbook-biography-mystery

biography-mystery-textbook   textbook-mystery-biography

The list shows 6 different *arrangements* of 3 books, but only one *subset* of 3 books. A subset of items listed *without regard to order* is called a **combination.**

The number of combinations of 5 things taken 3 at a time is written $\binom{5}{3}$, and read "5 over 3" or "5 choose 3."* Since they are subsets, combinations are *not ordered.*

To evaluate $\binom{5}{3}$, start with the $5 \cdot 4 \cdot 3$ *permutations* of 5 things taken 3 at a time. Since combinations are not ordered, find the number of combinations by dividing the number of permutations by the number of ways each group of 3 can be ordered; that is, divide by 3!.

$$\binom{5}{3} = \frac{5 \cdot 4 \cdot 3}{3!} = \frac{5 \cdot 4 \cdot 3}{3 \cdot 2 \cdot 1} = 10$$

There are 10 ways that the teacher can choose 3 books for the book sale.

Generalizing this discussion gives the following formula for the number of combinations of n elements taken r at a time:

$$\binom{n}{r} = \frac{P(n,r)}{r!}.$$

Another version of this formula is found as follows.

$$\binom{n}{r} = \frac{P(n,r)}{r!}$$

$$= \frac{n!}{(n-r)!} \cdot \frac{1}{r!}$$

$$= \frac{n!}{(n-r)!\,r!}$$

The steps above lead to the following result.

COMBINATIONS

If $\binom{n}{r}$ denotes the number of combinations of n elements taken r at a time, where $r \leq n$, then

$$\binom{n}{r} = \frac{n!}{(n-r)!\,r!}.$$

---

*Other common notations for $\binom{n}{r}$ are $_nC_r$, C^n_r, and $C(n, r)$.

EXAMPLE 1   Committees

How many committees of 3 people can be formed from a group of 8 people?

Solution

Method 1: Calculating by Hand

A committee is an unordered group, so use the combinations formula for $\binom{8}{3}$.

$$\binom{8}{3} = \frac{8!}{5!\,3!} = \frac{8 \cdot 7 \cdot 6 \cdot 5 \cdot 4 \cdot 3 \cdot 2 \cdot 1}{5 \cdot 4 \cdot 3 \cdot 2 \cdot 1 \cdot 3 \cdot 2 \cdot 1} = \frac{8 \cdot 7 \cdot 6}{3 \cdot 2 \cdot 1} = 56$$

Method 2: Graphing Calculator

Graphing calculators have the capacity to compute combinations. For example, on a TI-83, $\binom{8}{3}$ can be calculated by inputting 8 followed by nCr (found in the MATH-PRB menu) and a 3 yielding 56, as shown in Figure 4.

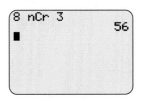

FIGURE 4

Method 3: Spreadsheet

Spreadsheets can also compute combinations. For example, in Microsoft Excel, $\binom{8}{3}$ can be calculated by inputting 8 and 3 in cells, say, A1 and B1, and then typing "=FACT(A1)/(FACT(A1-B1)*FACT(B1))" in cell C1 or, for that matter, any other cell.

Example 1 shows an alternative way to compute $\binom{n}{r}$. Take r or $n - r$, whichever is smaller. Write the factorial of this number in the denominator. In the numerator, write out a sufficient number of factors of $n!$ so there is one factor in the numerator for each factor in the denominator. For example, to calculate $\binom{8}{3}$ or $\binom{8}{5}$, write

$$\frac{8 \cdot 7 \cdot 6}{3 \cdot 2 \cdot 1} = 56.$$

The factors that are omitted (written in color in Example 1) cancel out of the numerator and denominator, so need not be included.

Notice from the previous discussion that $\binom{8}{3} = \binom{8}{5}$. (See Exercise 25 for a generalization of this idea.) One interpretation of this fact is that the number of ways to form a committee of 3 people chosen from a group of 8 is the same as the number of ways to choose the 5 people who are not on the committee.

EXAMPLE 2 Lawyers

Three lawyers are to be selected from a group of 30 to work on a special project.

(a) In how many different ways can the lawyers be selected?

Solution Here we wish to know the number of 3-element combinations that can be formed from a set of 30 elements. (We want combinations, not permutations, since order within the group of 3 doesn't matter.)

$$\binom{30}{3} = \frac{30!}{27!\,3!} = \frac{30 \cdot 29 \cdot 28 \cdot 27!}{27! \cdot 3 \cdot 2 \cdot 1}$$
$$= \frac{30 \cdot 29 \cdot 28}{3 \cdot 2 \cdot 1}$$
$$= 4060$$

There are 4060 ways to select the project group.

(b) In how many ways can the group of 3 be selected if a certain lawyer must work on the project?

Solution Since 1 lawyer already has been selected for the project, the problem is reduced to selecting 2 more from the remaining 29 lawyers.

$$\binom{29}{2} = \frac{29!}{27!\,2!} = \frac{29 \cdot 28 \cdot 27!}{27! \cdot 2 \cdot 1} = \frac{29 \cdot 28}{2 \cdot 1} = 29 \cdot 14 = 406$$

In this case, the project group can be selected in 406 ways.

(c) In how many ways can a nonempty group of at most 3 lawyers be selected from these 30 lawyers?

Solution Here, by "at most 3" we mean "1 or 2 or 3." (The number 0 is excluded because the group is nonempty.) Find the number of ways for each case.

Case	Number of Ways
1	$\binom{30}{1} = \dfrac{30!}{29!\,1!} = \dfrac{30 \cdot 29!}{29!\,(1)} = 30$
2	$\binom{30}{2} = \dfrac{30!}{28!\,2!} = \dfrac{30 \cdot 29 \cdot 28!}{28! \cdot 2 \cdot 1} = 435$
3	$\binom{30}{3} = \dfrac{30!}{27!\,3!} = \dfrac{30 \cdot 29 \cdot 28 \cdot 27!}{27! \cdot 3 \cdot 2 \cdot 1} = 4060$

The total number of ways to select at most 3 lawyers will be the sum

$$30 + 435 + 4060 = 4525.$$

EXAMPLE 3 Sales

A salesman has 10 accounts in a certain city.

(a) In how many ways can he select 3 accounts to call on?

Solution Within a selection of 3 accounts, the arrangement of the calls is not important, so there are

$$\binom{10}{3} = \frac{10!}{7!\,3!} = \frac{10 \cdot 9 \cdot 8}{3 \cdot 2 \cdot 1} = 120$$

ways he can make a selection of 3 accounts.

(b) In how many ways can he select at least 8 of the 10 accounts to use in preparing a report?

Solution "At least 8" means "8 or more," which is "8 or 9 or 10." First find the number of ways to choose in each case.

FOR REVIEW ■▬▬

Notice in Example 3 that to calculate the number of ways to select 8 or 9 or 10 accounts, we added the three numbers found. The union rule for disjoint sets from Chapter 7 says that when A and B are disjoint sets, the number of elements in A or B is the number of elements in A plus the number in B.

Case	Number of Ways
8	$\binom{10}{8} = \dfrac{10!}{2!\,8!} = \dfrac{10 \cdot 9}{2 \cdot 1} = 45$
9	$\binom{10}{9} = \dfrac{10!}{1!\,9!} = \dfrac{10}{1} = 10$
10	$\binom{10}{10} = \dfrac{10!}{0!\,10!} = 1$

He can select at least 8 of the 10 accounts in $45 + 10 + 1 = 56$ ways.

CAUTION When we are making choice 1 *and* choice 2, we *multiply* to find the total number of ways. When we are making choice 1 *or* choice 2, we *add* to find the total number of ways.

The formulas for permutations and combinations given in this section and in the previous section will be very useful in solving probability problems in the next section. Any difficulty in using these formulas usually comes from being unable to differentiate between them. Both permutations and combinations give the number of ways to choose r objects from a set of n objects. The differences between permutations and combinations are outlined in the following table.

Permutations	Combinations
Different orderings or arrangements of the r objects are different permutations. $$P(n,r) = \frac{n!}{(n-r)!}$$ Clue words: arrangement, schedule, order **Order matters!**	Each choice or subset of r objects gives one combination. Order within the group of r objects does not matter. $$\binom{n}{r} = \frac{n!}{(n-r)!\,r!}$$ Clue words: group, committee, set, sample **Order does not matter!**

In the next examples, concentrate on recognizing which formula should be applied.

EXAMPLE 4 Permutations and Combinations

For each of the following problems, tell whether permutations or combinations should be used to solve the problem.

(a) How many 4-digit code numbers are possible if no digits are repeated?

Solution Since changing the order of the 4 digits results in a different code, use permutations.

(b) A sample of 3 light bulbs is randomly selected from a batch of 15. How many different samples are possible?

Solution The order in which the 3 light bulbs are selected is not important. The sample is unchanged if the items are rearranged, so combinations should be used.

(c) In a baseball conference with 8 teams, how many games must be played so that each team plays every other team exactly once?

Solution Selection of two teams for a game is an *unordered* subset of 2 from the set of 8 teams. Use combinations again.

(d) In how many ways can 4 patients be assigned to 6 different hospital rooms so that each patient has a private room?

Solution The room assignments are an *ordered* selection of 4 rooms from the 6 rooms. Exchanging the rooms of any two patients within a selection of 4 rooms gives a different assignment, so permutations should be used.

(e) Solve the problems in parts (a)–(d) above. The answers are given in the footnote.*

EXAMPLE 5 Promotions

A manager must select 4 employees for promotion; 12 employees are eligible.

(a) In how many ways can the 4 be chosen?

Solution Since there is no reason to differentiate among the 4 who are selected, use combinations.

$$\binom{12}{4} = \frac{12!}{8!\,4!} = 495$$

(b) In how many ways can 4 employees be chosen (from 12) to be placed in 4 different jobs?

Solution In this case, once a group of 4 is selected, they can be assigned in many different ways (or arrangements) to the 4 jobs. Therefore, this problem requires permutations.

$$P(12,4) = \frac{12!}{8!} = 11{,}880$$

EXAMPLE 6 Playing Cards

In how many ways can a full house of aces and eights (3 aces and 2 eights) occur in 5-card poker?

---

*(a) 5040 **(b)** 455 **(c)** 28 **(d)** 360

FOR REVIEW ■—

Examples 6 and 7 involve a standard deck of 52 playing cards, as shown in Figure 17 in Chapter 7. Recall the discussion that accompanies the photograph.

Solution The arrangement of the 3 aces or the 2 eights does not matter, so we use combinations and the multiplication principle. There are $\binom{4}{3}$ ways to get 3 aces from the 4 aces in the deck, and $\binom{4}{2}$ ways to get 2 eights. By the multiplication principle, the number of ways to get 3 aces and 2 eights is

$$\binom{4}{3} \cdot \binom{4}{2} = 4 \cdot 6 = 24.$$

EXAMPLE 7 Playing Cards

Five cards are dealt from a standard 52-card deck.

(a) How many such hands have only face cards?

Solution The face cards are the king, queen, and jack of each suit. Since there are 4 suits, there are 12 face cards. The arrangement of the 5 cards is not important, so use combinations to get

$$\binom{12}{5} = \frac{12!}{7!\,5!} = 792.$$

(b) How many such hands have exactly 2 hearts?

Solution There are 13 hearts in the deck, so the 2 hearts will be selected from those 13 cards. The other 3 cards must come from the remaining 39 cards that are not hearts. Use combinations and the multiplication principle to get

$$\binom{13}{2}\binom{39}{3} = 78 \cdot 9139 = 712{,}842.$$

Notice that the two top numbers in the combinations add up to 52, the total number of cards, and the two bottom numbers add up to 5, the number of cards in a hand.

(c) How many such hands have cards of a single suit?

Solution The total number of ways that 5 cards of a particular suit of 13 cards can occur is $\binom{13}{5}$. Since the arrangement of the 5 cards is not important, use combinations. There are four different suits, so the multiplication principle gives

$$4 \cdot \binom{13}{5} = 4 \cdot 1287 = 5148$$

ways to deal 5 cards of the same suit.

As Example 7 shows, often both combinations and the multiplication principle must be used in the same problem.

EXAMPLE 8 Soup

To illustrate the differences between permutations and combinations in another way, suppose 2 cans of soup are to be selected from 4 cans on a shelf: noodle (N), bean (B), mushroom (M), and tomato (T). As shown in Figure 5(a), there are 12 ways to select 2 cans from the 4 cans if the order matters (if noodle first and bean second is considered different from bean, then noodle, for example). On the other hand, if order is unimportant, then there are 6 ways to choose 2 cans of soup from the 4, as illustrated in Figure 5(b).

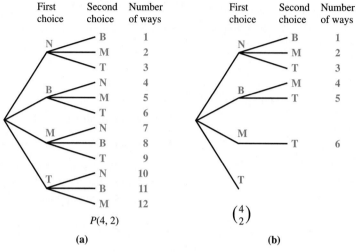

FIGURE 5

CAUTION It should be stressed that not all counting problems lend themselves to either permutations or combinations. Whenever a tree diagram or the multiplication principle can be used directly, it's best to use it.

8.2 EXERCISES

1. Define combinations in your own words.

Evaluate the following combinations.

2. $\dbinom{8}{3}$

3. $\dbinom{12}{5}$

4. $\dbinom{44}{20}$

5. $\dbinom{40}{18}$

6. $\dbinom{n}{0}$

7. $\dbinom{n}{n}$

8. $\dbinom{n}{1}$

9. $\dbinom{n}{n-1}$

Use combinations to solve Exercises 9–11.

10. In how many ways can a hand of 6 clubs be chosen from an ordinary deck?

11. Five cards are marked with the numbers 1, 2, 3, 4, and 5, then shuffled, and 2 cards are drawn.

   a. How many different 2-card combinations are possible?

   b. How many 2-card hands contain a number less than 3?

12. An economics club has 30 members.

   a. If a committee of 4 is to be selected, in how many ways can the selection be made?

   b. In how many ways can a committee of at least 1 and at most 3 be selected?

13. Use a tree diagram for the following.

   a. Find the number of ways 2 letters can be chosen from the set {L, M, N} if order is important and repetition is allowed.

   b. Reconsider part a if no repeats are allowed.

   c. Find the number of combinations of 3 elements taken 2 at a time. Does this answer differ from part a or b?

14. Repeat Exercise 13 using the set {L, M, N, P}.

15. Explain the difference between a permutation and a combination.

16. Padlocks with digit dials are often referred to as "combination locks." According to the mathematical definition of combination, is this an accurate description? Explain.

Decide whether each of the following exercises involves permutations or combinations, and then solve the problem.

17. In a club with 8 male and 11 female members, how many 5-member committees can be chosen that have

    a. all men?   **b.** all women?

    c. 3 men and 2 women?

18. In Exercise 17, how many committees can be selected that have

    a. at least 4 women?   **b.** no more than 2 men?

19. A group of 3 students is to be selected from a group of 12 students to take part in a class in cell biology.

    a. In how many ways can this be done?

    b. In how many ways can the group who will not take part be chosen?

20. In a game of musical chairs, 12 children will sit in 11 chairs arranged in a row (one will be left out). In how many ways can this happen, if we count rearrangements of the children in the chairs as different outcomes?

21. Marbles are being drawn without replacement from a bag containing 15 marbles.

    a. How many samples of 2 marbles can be drawn?

    b. How many samples of 4 marbles can be drawn?

    c. If the bag contains 3 yellow, 4 white, and 8 blue marbles, how many samples of 2 marbles can be drawn in which both marbles are blue?

22. There are 5 rotten apples in a crate of 25 apples.

    a. How many samples of 3 apples can be drawn from the crate?

    b. How many samples of 3 could be drawn in which all 3 are rotten?

    c. How many samples of 3 could be drawn in which there are two good apples and one rotten one?

23. A bag contains 5 black, 1 red, and 3 yellow jelly beans; you take 3 at random. How many samples are possible in which the jelly beans are

    a. all black?   **b.** all red?

    c. all yellow?   **d.** 2 black and 1 red?

    e. 2 black and 1 yellow?   **f.** 2 yellow and 1 black?

    g. 2 red and 1 yellow?

24. In how many ways can 5 out of 9 plants be arranged in a row on a windowsill?

25. Show that $\binom{n}{r} = \binom{n}{n-r}$.

26. The following problem was posed on National Public Radio's "Weekend Edition": In how many points can 6 circles intersect?*

    a. Find the answer for 6 circles.

    b. Find the general answer for *n* circles.

27. How many different dominoes can be formed from the numbers 0 . . . 6? (*Hint:* A domino may have the same number of dots on both halves of it or it may have a different number of dots on each half.)

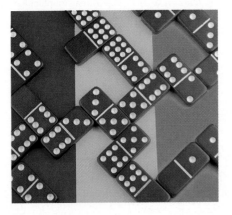

 Applications

BUSINESS AND ECONOMICS

28. *Secretarial Assignments* From a pool of 7 secretaries, 3 are selected to be assigned to 3 managers, one per manager. In how many ways can they be selected and assigned?

29. *Sales Schedules* A salesperson has the names of 6 prospects.

    a. In how many ways can she arrange her schedule if she calls on all 6?

    b. In how many ways can she arrange her schedule if she can call on only 4 of the 6?

30. *Worker Grievances* A group of 7 workers decides to send a delegation of 2 to their supervisor to discuss their grievances.

    a. How many delegations are possible?

    b. If it is decided that a particular worker must be in the delegation, how many different delegations are possible?

---

*"Weekend Edition," National Public Radio, Oct. 23, 1994.

c. If there are 2 women and 5 men in the group, how many delegations would include at least 1 woman?

31. *Hamburger Variety* Hamburger Hut sells regular hamburgers as well as a larger burger. Either type can include cheese, relish, lettuce, tomato, mustard, or catsup.

a. How many different hamburgers can be ordered with exactly three extras?

b. How many different regular hamburgers can be ordered with exactly three extras?

c. How many different regular hamburgers can be ordered with at least five extras?

32. *Assembly Line Sampling* Five items are to be randomly selected from the first 50 items on an assembly line to determine the defect rate. How many different samples of 5 items can be chosen?

LIFE SCIENCES

33. *Research Participants* From a group of 16 smokers and 20 nonsmokers, a researcher wants to randomly select 8 smokers and 8 nonsmokers for a study. In how many ways can the study group be selected?

34. *Plant Hardiness* In an experiment on plant hardiness, a researcher gathers 6 wheat plants, 3 barley plants, and 2 rye plants. She wishes to select 4 plants at random.

a. In how many ways can this be done?

b. In how many ways can this be done if exactly 2 wheat plants must be included?

SOCIAL SCIENCES

35. *Legislative Committee* A legislative committee consists of 5 Democrats and 4 Republicans. A delegation of 3 is to be selected to visit a small Pacific island republic.

a. How many different delegations are possible?

b. How many delegations would have all Democrats?

c. How many delegations would have 2 Democrats and 1 Republican?

d. How many delegations would include at least 1 Republican?

36. *Political Committee* From 10 names on a ballot, 4 will be elected to a political party committee. In how many ways can the committee of 4 be formed if each person will have a different responsibility, and different assignments of responsibility are considered different committees?

GENERAL INTEREST

37. *Bridge* How many different 13-card bridge hands can be selected from an ordinary deck?

38. *Poker* Five cards are chosen from an ordinary deck to form a hand in poker. In how many ways is it possible to get the following results?

a. 4 queens

b. No face card

c. Exactly 2 face cards

d. At least 2 face cards

e. 1 heart, 2 diamonds, and 2 clubs

39. *Baseball* If a baseball coach has 5 good hitters and 4 poor hitters on the bench and chooses 3 players at random, in how many ways can he choose at least 2 good hitters?

40. *Softball* The coach of the Morton Valley Softball Team has 6 good hitters and 8 poor hitters. He chooses 3 hitters at random.

a. In how many ways can he choose 2 good hitters and 1 poor hitter?

b. In how many ways can he choose 3 good hitters?

c. In how many ways can he choose at least 2 good hitters?

41. *Flower Selection* Five orchids from a collection of 20 are to be selected for a flower show.

a. In how many ways can this be done?

b. In how many ways can the 5 be selected if 2 special plants must be included?

42. *Ice Cream Flavors* Baskin-Robbins advertises that it has 31 flavors of ice cream.

a. How many different double-scoop cones can be made? Assume that the order of the scoops matters.

b. How many different triple-scoop cones can be made?

c. How many different double-scoop cones can be made if order doesn't matter?

43. *Lottery* A state lottery game requires that you pick 6 different numbers from 1 to 99. If you pick all 6 winning numbers, you win the jackpot.

a. How many ways are there to choose 6 numbers if order is not important?

b. How many ways are there to choose 6 numbers if order matters?

44. *Lottery* In Exercise 43, if you pick 5 of the 6 numbers correctly, you win $250,000. In how many ways can you pick exactly 5 of the 6 winning numbers without regard to order?

45. *Bingo* In the game of Bingo, each card has five columns. Column 1 has spaces for 5 numbers, chosen from 1 to 15. Column 2 similarly has 5 numbers, chosen from 16 to 30. Column 3 has a free space in the middle, plus 4 numbers chosen from 31 to 45. The 5 numbers in columns 4 and 5 are chosen from 46 to 60 and from 61 to 75, respectively.

The numbers in each card can be in any order. How many different Bingo cards are there?

46. *Pizza Varieties* A television commercial for Little Caesars pizza announced that with the purchase of two pizzas, one could receive free any combination of up to five toppings on each pizza. The commercial shows a young child waiting in line at Little Caesars who calculates that there are 1,048,576 possibilities for the toppings on the two pizzas.*

    a. Verify the child's calculation. Use the fact that Little Caesars has 11 toppings to choose from. Assume that the order of the two pizzas matters; that is, if the first pizza has combination 1 and the second pizza has combination 2, that is different from combination 2 on the first pizza and combination 1 on the second.

    b. In a letter to *The Mathematics Teacher,* Joseph F. Heiser argued that the two combinations described in part a should be counted as the same, so the child has actually overcounted. Give the number of possibilities if the order of the two pizzas doesn't matter.

47. *Cereal* The Post Corporation has introduced the cereal, *Create a Crunch*™, in which the consumers can combine ingredients to create their own unique cereal. Each box contains 8 packets of food goods. There are four types of cereal: Frosted Alpha Bits®, Cocoa Pebbles®, Fruity Pebbles®, and Honey Comb®. Also included in the box are four "Add-Ins": granola, blue rice cereal, marshmallows, and sprinkles.

    a. What is the total number of breakfasts that can be made if a breakfast is defined as any one or more cereals or add-ins?

    b. If Emily Friedrich chooses to mix one type of cereal with one add-in, how many different breakfasts can she make?

    c. If Rachel Moldovan chooses to mix two types of cereal with three add-ins, how many different breakfasts can she make?

    d. If Vincent Sonoga chooses to mix at least one type of cereal with at least one type of add-in, how many breakfasts can he make?

    e. If Ann Lombardi's favorite cereal is Fruity Pebbles®, how many different cereals can she make if each of her mixtures must include this cereal?

■ 8.3 PROBABILITY APPLICATIONS OF COUNTING PRINCIPLES

THINK ABOUT IT

If 3 engines are tested from a shipping container packed with 12 diesel engines, 2 of which are defective, what is the probability that at least 1 of the defective engines will be found (in which case the container will not be shipped)?

This problem theoretically could be solved with a tree diagram, but it would require a tree with a large number of branches. Many of the probability problems involving *dependent* events that were solved earlier by using tree diagrams can also be solved by using permutations or combinations. Permutations and combinations are especially helpful when the numbers involved are large.

To compare the method of using permutations or combinations with the method of tree diagrams used in Section 7.5, the first example repeats Example 6 from that section.

EXAMPLE 1 Marbles

From a box containing 3 white, 2 green, and 1 red marble, 2 marbles are drawn one at a time without replacement. Find the probability that 1 white and 1 green marble are drawn.

Solution Because the marbles are drawn one at a time, with one labeled as the first marble and the other as the second, we use permutations. There are two ways to draw 1 white marble and 1 green marble. The first way is to draw the white

---

*Heiser, Joseph F., "Pascal and Gauss Meet Little Caesars," *The Mathematics Teacher*, Vol. 87, Sept. 1994, p. 389.

FOR REVIEW ■■■

The use of combinations to solve probability problems depends on the basic probability principle introduced earlier and repeated here:

Let S be a sample space with equally likely outcomes, and let event E be a subset of S. Then the probability that event E occurs, written $P(E)$, is

$$P(E) = \frac{n(E)}{n(S)},$$

where $n(E)$ and $n(S)$ represent the number of elements in sets E and S.

marble followed by the green, and the second is to draw the green followed by the white. The white marble can be drawn from the 3 white marbles in $\binom{3}{1}$ ways, and the green can be drawn from the 2 green marbles in $\binom{2}{1}$ ways. By the multiplication principle and the union rule for disjoint sets, both results can occur in

$$\binom{3}{1}\binom{2}{1} + \binom{2}{1}\binom{3}{1} \text{ ways,}$$

giving the numerator of the probability fraction, $P(E) = m/n$. For the denominator, there are 6 ways to draw the first marble and 5 ways to draw the second, for a total of $6 \cdot 5$ ways. The required probability is

$$P(1 \text{ white and } 1 \text{ green}) = \frac{\binom{3}{1}\binom{2}{1} + \binom{2}{1}\binom{3}{1}}{6 \cdot 5}$$

$$= \frac{3 \cdot 2 + 2 \cdot 3}{30} = \frac{12}{30} = \frac{2}{5}.$$

This agrees with the answer found earlier.

This example can be solved more simply by observing that the probability that 1 white marble and 1 green marble are drawn should not depend upon the order in which the marbles are drawn, so we may use combinations. The numerator is simply the number of ways of drawing a white marble out of 3 white marbles and a green marble out of 2 green marbles. The denominator is just the number of ways of drawing 2 marbles out of 6. Then

$$P(1 \text{ white and } 1 \text{ green}) = \frac{\binom{3}{1}\binom{2}{1}}{\binom{6}{2}} = \frac{6}{15} = \frac{2}{5}.$$

This helps explain why combinations tend to be used more often than permutations in probability. Even if order matters in the original problem, it is sometimes possible to ignore order and use combinations. Be careful to do this only when the final result does not depend on the order of events. Order often does matter. (If you don't believe that, try getting dressed tomorrow morning and then taking your shower.)

Example 1 could also be solved using the tree diagram shown in Figure 6. Two of the branches correspond to drawing 1 white and 1 green marble. The probability for each branch is calculated by multiplying the probabilities along the branch, as we did in the previous chapter. The resulting probabilities for the two branches are then added, giving the result

$$P(1 \text{ white and } 1 \text{ green}) = \frac{3}{6} \cdot \frac{2}{5} + \frac{2}{6} \cdot \frac{3}{5} = \frac{2}{5}.$$

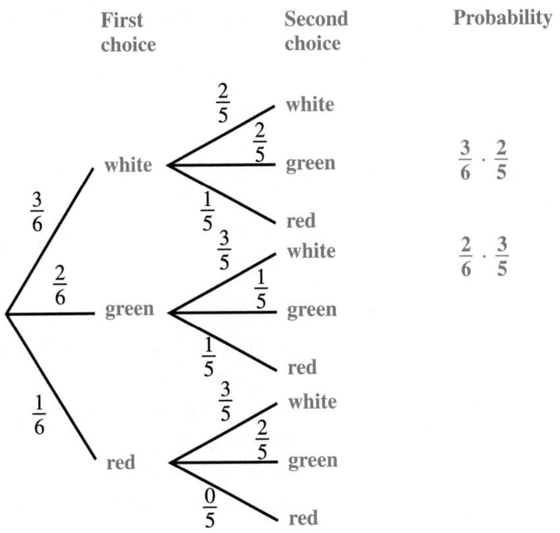

FIGURE 6

EXAMPLE 2   Nursing

From a group of 22 nurses, 4 are to be selected to present a list of grievances to management.

(a) In how many ways can this be done?

Solution Four nurses from a group of 22 can be selected in $\binom{22}{4}$ ways. (Use combinations, since the group of 4 is an unordered set.)

$$\binom{22}{4} = \frac{22!}{18!\, 4!} = \frac{(22)\,(21)\,(20)\,(19)}{(4)\,(3)\,(2)\,(1)} = 7315$$

There are 7315 ways to choose 4 people from 22.

(b) One of the nurses is Michael Branson. Find the probability that Branson will be among the 4 selected.

Solution The probability that Branson will be selected is given by m/n, where m is the number of ways the chosen group includes him, and n is the total number of ways the group of 4 can be chosen. If Branson must be one of the 4 selected, the problem reduces to finding the number of ways that the 3 additional nurses can be chosen. The 3 are chosen from 21 nurses; this can be done in

$$\binom{21}{3} = \frac{21!}{18!\, 3!} = 1330$$

ways, so $m = 1330$. Since n is the number of ways 4 nurses can be selected from 22,

$$n = \binom{22}{4} = 7315.$$

The probability that Branson will be one of the 4 chosen is

$$P(\text{Branson is chosen}) = \frac{1330}{7315} \approx .182.$$

(c) Find the probability that Branson will not be selected.

Solution The probability that he will not be chosen is $1 - .182 = .818.$ ◼

EXAMPLE 3 Diesel Engines

When shipping diesel engines abroad, it is common to pack 12 engines in one container that is then loaded on a rail car and sent to a port. Suppose that a company has received complaints from its customers that many of the engines arrive in nonworking condition. To help solve this problem, the company decides to make a spot check of containers after loading. The company will test 3 engines from a container at random; if any of the 3 are nonworking, the container will not be shipped until each engine in it is checked. Suppose a given container has 2 nonworking engines. Find the probability that the container will not be shipped.

?

Solution The container will not be shipped if the sample of 3 engines contains 1 or 2 defective engines. If $P(1 \text{ defective})$ represents the probability of exactly 1 defective engine in the sample, then

$$P(\text{not shipping}) = P(1 \text{ defective}) + P(2 \text{ defective}).$$

There are $\binom{12}{3}$ ways to choose the 3 engines for testing:

$$\binom{12}{3} = \frac{12!}{9!\,3!} = 220.$$

There are $\binom{2}{1}$ ways of choosing 1 defective engine from the 2 in the container, and for each of these ways, there are $\binom{10}{2}$ ways of choosing 2 good engines from among the 10 in the container. By the multiplication principle, there are

$$\binom{2}{1}\binom{10}{2} = \frac{2!}{1!\,1!} \cdot \frac{10!}{8!\,2!} = 90$$

ways of choosing a sample of 3 engines containing 1 defective engine with

$$P(1 \text{ defective}) = \frac{90}{220} = \frac{9}{22}.$$

There are $\binom{2}{2}$ ways of choosing 2 defective engines from the 2 defective engines in the container, and $\binom{10}{1}$ ways of choosing 1 good engine from among the 10 good engines, for

$$\binom{2}{2}\binom{10}{1} = \frac{2!}{0!\,2!} \cdot \frac{10!}{9!\,1!} = 10$$

ways of choosing a sample of 3 engines containing 2 defective engines. Finally,

$$P(2 \text{ defective}) = \frac{10}{220} = \frac{1}{22}$$

and

$$P(\text{not shipping}) = P(1 \text{ defective}) + P(2 \text{ defective})$$

$$= \frac{9}{22} + \frac{1}{22} = \frac{10}{22} \approx .455.$$

Notice that the probability is $1 - .455 = .545$ that the container will be shipped, even though it has 2 defective engines. The management must decide whether this probability is acceptable; if not, it may be necessary to test more than 3 engines from a container.

FOR REVIEW ■——

Recall that if E and E' are complements, then $P(E') = 1 - P(E)$. In Example 3, the event "0 defective in the sample" is the complement of the event "1 or 2 defective in the sample," since there are only 0 or 1 or 2 defective engines possible in the sample of 3 engines.

Instead of finding the sum $P(1 \text{ defective}) + P(2 \text{ defective})$, the result in Example 3 could be found as $1 - P(0 \text{ defective})$.

$$P(\text{not shipping}) = 1 - P(0 \text{ defective in sample})$$

$$= 1 - \frac{\binom{2}{0}\binom{10}{3}}{\binom{12}{3}}$$

$$= 1 - \frac{1(120)}{220}$$

$$= 1 - \frac{120}{220} = \frac{100}{220} \approx .455$$

EXAMPLE 4 Poker

In a common form of the card game *poker*, a hand of 5 cards is dealt to each player from a deck of 52 cards. There are a total of

$$\binom{52}{5} = \frac{52!}{47!\,5!} = 2{,}598{,}960$$

such hands possible. Find the probability of getting each of the following hands.

(a) A hand containing only hearts, called a *heart flush*

Solution There are 13 hearts in a deck, with

$$\binom{13}{5} = \frac{13!}{8!\,5!} = \frac{(13)\,(12)\,(11)\,(10)\,(9)}{(5)\,(4)\,(3)\,(2)\,(1)} = 1287$$

different hands containing only hearts. The probability of a heart flush is

$$P(\text{heart flush}) = \frac{1287}{2{,}598{,}960} \approx .000495.$$

(b) A flush of any suit (5 cards of the same suit)

Solution There are 4 suits in a deck, so

$$P(\text{flush}) = 4 \cdot P(\text{heart flush}) = 4 \cdot .000495 \approx .00198.$$

(c) A full house of aces and eights (3 aces and 2 eights)

Solution There are $\binom{4}{3}$ ways to choose 3 aces from among the 4 in the deck, and $\binom{4}{2}$ ways to choose 2 eights.

$$P(3 \text{ aces}, 2 \text{ eights}) = \frac{\binom{4}{3} \cdot \binom{4}{2}}{2,598,960} = \frac{4 \cdot 6}{2,598,960} \approx .00000923$$

(d) Any full house (3 cards of one value, 2 of another)

Solution The 13 values in a deck give 13 choices for the first value. As in part (a), there are $\binom{4}{3}$ ways to choose the 3 cards from among the 4 cards that have that value. This leaves 12 choices for the second value (order *is* important here, since a full house of 3 aces and 2 eights is not the same as a full house of 3 eights and 2 aces). From the 4 cards that have the second value, there are $\binom{4}{2}$ ways to choose 2. The probability of any full house is then

$$P(\text{full house}) = \frac{13 \cdot \binom{4}{3} \cdot 12 \cdot \binom{4}{2}}{2,598,960} \approx .00144.$$

EXAMPLE 5 Music

A music teacher has 3 violin pupils, Fred, Carl, and Helen. For a recital, the teacher selects a first violinist and a second violinist. The third pupil will play with the others, but not solo. If the teacher selects randomly, what is the probability that Helen is first violinist, Carl is second violinist, and Fred does not solo?

Solution Use *permutations* to find the number of arrangements in the sample space.

$$P(3, 3) = 3! = 6$$

(Think of this as filling the positions of first violin, second violin, and no solo.) The 6 arrangements are equally likely, since the teacher will select randomly. Thus, the required probability is $1/6$.

EXAMPLE 6 Birthdays

Suppose a group of n people is in a room. Find the probability that at least 2 of the people have the same birthday.

Solution "Same birthday" refers to the month and the day, not necessarily the same year. Also, ignore leap years, and assume that each day in the year is equally likely as a birthday. To see how to proceed, we first find the probability that *no 2 people* from among 5 people have the same birthday. There are 365 different birthdays possible for the first of the 5 people, 364 for the second (so that the people have different birthdays), 363 for the third, and so on. The number of ways the 5 people can have different birthdays is thus the number of permutations of 365 days taken 5 at a time or

$$P(365, 5) = 365 \cdot 364 \cdot 363 \cdot 362 \cdot 361.$$

The number of ways that 5 people can have the same birthday or different birthdays is

$$365 \cdot 365 \cdot 365 \cdot 365 \cdot 365 = (365)^5.$$

Finally, the *probability* that none of the 5 people have the same birthday is

$$\frac{P(365,5)}{(365)^5} = \frac{365 \cdot 364 \cdot 363 \cdot 362 \cdot 361}{365 \cdot 365 \cdot 365 \cdot 365 \cdot 365} \approx .973.$$

The probability that at least 2 of the 5 people *do* have the same birthday is $1 - .973 = .027$.

Now this result can be extended to more than 5 people. Generalizing, the probability that no 2 people among n people have the same birthday is

$$\frac{P(365,n)}{(365)^n}.$$

The probability that at least 2 of the n people *do* have the same birthday is

$$1 - \frac{P(365,n)}{(365)^n}.$$

The following table shows this probability for various values of n.

Number of People, n	Probability that Two Have the Same Birthday
5	.027
10	.117
15	.253
20	.411
22	.476
23	.507
25	.569
30	.706
35	.814
40	.891
50	.970
365	1

The probability that 2 people among 23 have the same birthday is .507, a little more than half. Many people are surprised at this result; it seems that a larger number of people should be required.

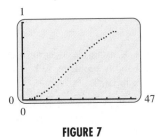

FIGURE 7

Using a graphing calculator, we can graph the probability formula in the previous example as a function of n, but care must be taken that the graphing calculator evaluates the function at integer points. Figure 7 was produced on a TI-83 by letting $Y_1 = 1 - (365 \text{ nPr } X)/365^\wedge X$ on $0 \leq x \leq 47$. (This domain ensures integer values for x.) Notice that the graph does not extend past $x = 39$. This is because $P(365,n)$ and 365^n are too large for the calculator when $n \geq 40$.

An alternative way of doing the calculations that does not run into such large numbers is based on the concept of conditional probability. The probability that the first person's birthday does not match any so far is 365/365. The probability that the second person's birthday does not match the first's is 364/365. The probability that the third person's birthday does not match the first's or the second's is 363/365. By the product rule of probability, the probability that none of the first 3 people have matching birthdays is

$$\frac{365}{365} \cdot \frac{364}{365} \cdot \frac{363}{365}.$$

Similarly, the probability that no two people in a group of 40 have the same birthday is

$$\frac{365}{365} \cdot \frac{364}{365} \cdot \frac{363}{365} \cdots \frac{326}{365}.$$

This probability can be calculated (and then subtracted from 1 to get the probability we seek) without overflowing the calculator by multiplying each fraction times the next, rather then trying to compute the entire numerator and the entire denominator. The calculations are somewhat tedious to do by hand, but can be programmed on a graphing calculator or computer.

8.3 EXERCISES

A basket contains 6 red apples and 4 yellow apples. A sample of 3 apples is drawn. Find the probabilities that the sample contains the following.

1. All red apples

2. All yellow apples

3. 2 yellow and 1 red apple

4. More red than yellow apples

Two cards are drawn at random from an ordinary deck of 52 cards.

5. How many 2-card hands are possible?

Find the probability that the 2-card hand described above contains the following.

6. 2 aces

7. At least 1 ace

8. All spades

9. 2 cards of the same suit

10. Only face cards

11. No face cards

12. No card higher than 8 (count ace as 1)

Twenty-six slips of paper are each marked with a different letter of the alphabet and placed in a basket. A slip is pulled out, its letter recorded (in the order in which the slip was drawn), and the slip is replaced. This is done 5 times. Find the probabilities that the following "words" are formed.

13. Chuck

14. A word that starts with p.

15. A word with no repetition of letters

16. A word that contains no x, y, or z

17. Discuss the relative merits of using tree diagrams versus combinations to solve probability problems. When would each approach be most appropriate?

18. Several examples in this section used the rule $P(E') = 1 - P(E)$. Explain the advantage (especially in Example 6) of using this rule.

For Exercises 19–22, refer to Example 6 in this section.

19. A total of 42 men have served as president through 2004.* Set up the probability that, if 42 men were selected at random, at least 2 have the same birthday.[†]

20. Set up the probability that at least 2 of the 100 U.S. senators have the same birthday.

21. What is the probability that at least 2 of the 435 members of the House of Representatives have the same birthday?

22. Argue that the probability that in a group of *n* people *exactly one pair* have the same birthday is

$$\binom{n}{2} \cdot \frac{P(365, n - 1)}{(365)^n}.$$

23. An elevator has 4 passengers and stops at 7 floors. It is equally likely that a person will get off at any one of the 7 floors. Find the probability that no 2 passengers leave at the same floor.

24. On National Public Radio, the "Weekend Edition" program on Sunday, September 7, 1991, posed the following probability problem: Given a certain number of balls, of which some are blue, pick 5 at random. The probability that all 5 are blue is $1/2$. Determine the original number of balls and decide how many were blue.

25. A reader wrote to the "Ask Marilyn" column[‡] in *Parade* magazine, "You have six envelopes to pick from. Two-thirds (that is, four) are empty. One-third (that is, two) contain a $100 bill. You're told to choose 2 envelopes at random. Which is more likely: (1) that you'll get at least one $100 bill, or (2) that you'll get no $100 bill at all?" Find the two probabilities.

26. After studying all night for a final exam, a bleary-eyed student randomly grabs 2 socks from a drawer containing 9 black, 6 brown, and 2 blue socks, all mixed together. What is the probability that she grabs a matched pair?

27. 3 crows, 4 blue jays, and 5 starlings sit in a random order on a section of telephone wire. Find the probability that birds of a feather flock together, that is, that all birds of the same type are sitting together.

28. If the letters l, i, t, t, l, and e are chosen at random, what is the probability that they spell the word "little"?

29. If the letters M, i, s, s, i, s, s, i, p, p, and i are chosen at random, what is the probability that they spell the word "Mississippi"?

Applications

BUSINESS AND ECONOMICS

Quality Control   A shipment of 9 typewriters contains 2 that are defective. Find the probability that a sample of each of the following sizes, drawn from the 9, will not contain a defective typewriter.

30. 1   31. 2   32. 3   33. 4

Refer to Example 3. The managers feel that the probability of .545 that a container will be shipped even though it contains 2 defective engines is too high. They decide to increase the sample size chosen. Find the probabilities that a container will be shipped even though it contains 2 defective engines, if the sample size is increased to the following.

34. 4   35. 5

---

*Although Bush is the 43rd president, the 22nd and 24th presidents were the same man: Grover Cleveland.
[†]In fact, James Polk and Warren Harding were both born on November 2.
[‡]*Parade* Magazine, Apr. 30, 1995, p. 8. Reprinted by permission of the William Morris Agency, Inc. on behalf of the author. Copyright © 1995 by Marilyn vos Savant.

SOCIAL SCIENCES

36. *Election Ballots* Five names are put on a ballot in a randomly selected order. What is the probability that they are not in alphabetical order?

37. *Native American Council* At the first meeting of a committee to plan a Northern California pow-wow, there were 3 women and 3 men from the Miwok tribe, 2 men and 3 women from the Hoopa tribe, and 4 women and 5 men from the Pomo tribe. If the ceremony subcouncil consists of 5 people, and is randomly selected, find the probabilities that the subcouncil contains

  a. 3 men and 2 women;

  b. exactly 3 Miwoks and 2 Pomos;

  c. 2 Miwoks, 2 Hoopas, and a Pomo;

  d. 2 Miwoks, 2 Hoopas, and 2 Pomos;

  e. more women than men;

  f. exactly 3 Hoopas;

  g. at least 2 Pomos.

38. *Education* A school in Bangkok requires that students take an entrance examination. After the examination, there is a drawing where 5 students are randomly selected from each group of 40 for automatic acceptance into the school, regardless of their performance on the examination. The drawing consists of placing 35 red and 5 green pieces of paper into a box. Each student picks a piece of paper from the box and then does not return the piece of paper to the box. The 5 lucky students who pick the green pieces are automatically accepted into the school.*

  a. What is the probability that the first person wins automatic acceptance?

  b. What is the probability that the last person wins automatic acceptance?

  c. If the students are chosen by the order of their seating, does this give the student who goes first a better chance of winning than the second, third,…person? (*Hint:* Imagine that the 40 pieces of paper have been mixed up and laid in a row so that the first student picks the first piece of paper, the second student picks the second piece of paper, and so on.)

GENERAL INTEREST

Poker Find the probabilities of the following hands at poker. Assume aces are either high or low.

39. Royal flush (5 highest cards of a single suit)

40. Straight flush (5 in a row in a single suit, but not a royal flush)

41. Four of a kind (4 cards of the same value)

42. Straight (5 cards in a row, not all of the same suit), with ace either high or low

43. Three of a kind (3 cards of one value, with the other cards of two different values)

44. Two pairs (2 cards of one value, 2 of another value, and 1 of a third value)

45. One pair (2 cards of one value, with the other cards of three different values)

Bridge A bridge hand is made up of 13 cards from a deck of 52. Find the probabilities that a hand chosen at random contains the following.

46. Only hearts

47. 4 aces

48. Exactly 3 aces and exactly 3 kings

49. 6 of one suit, 5 of another, and 2 of another

50. *Writers* At a conference of black writers in Detroit, special-edition books were selected to be given away in contests. There were 9 books written by Langston Hughes, 5 books by James Baldwin, and 7 books by Toni Morrison. The judge of one contest selected 6 books at random for prizes. Find the probabilities that the selection consisted of the following.

  a. 3 Hughes and 3 Morrison books

  b. exactly 4 Baldwin books

  c. 2 Hughes, 3 Baldwin, and 1 Morrison book

  d. at least 4 Hughes books

  e. exactly 4 books written by males (Morrison is female)

  f. no more than 2 books written by Baldwin

51. *Lottery* In the previous section, we found the number of ways to pick 6 different numbers from 1 to 99 in a state lottery. Assuming order is unimportant, what is the probability of picking all 6 numbers correctly to win the big prize?

52. *Lottery* In Exercise 51, what is the probability of picking exactly 5 of the 6 numbers correctly?

53. *Lottery* An article in *The New York Times* discussing the odds of winning the lottery stated, "And who cares if a game-theory professor once calculated the odds of winning as equal to a poker player's chance of drawing four royal flushes in a row, all in spades—then getting up from the card table and meeting four strangers, all with the same birthday?"[†] Calculate this probability. Does this probability seem comparable to the odds of winning the lottery? (Ignore February 29 as a birthday, and assume that all four strangers have the same birthday as each other, not necessarily the same as the poker player.)

---

*Letter to the editor, *Mathematics Teacher*, Vol. 92, No. 8, Nov. 1999.
[†]Gould, Lois, "Ticket to Trouble," *The New York Times Magazine*, Apr. 23, 1995, p. 39.

54. *Barbie* A controversy arose in 1992 over the Teen Talk Barbie doll, each of which was programmed with four sayings randomly picked from a set of 270 sayings. The controversy was over the saying, "Math class is tough," which some felt gave a negative message toward girls doing well in math. In an interview with *Science*, a spokeswoman for Mattel, the makers of Barbie, said that "There's a less than 1% chance you're going to get a doll that says math class is tough."* Is this figure correct? If not, give the correct figure.

55. *Football* During the 1988 college football season, the Big Eight Conference ended the season in a "perfect progression," as shown in the following table.†

Won	Lost	Team
7	0	Nebraska (NU)
6	1	Oklahoma (OU)
5	2	Oklahoma State (OSU)
4	3	Colorado (CU)
3	4	Iowa State (ISU)
2	5	Missouri (MU)
1	6	Kansas (KU)
0	7	Kansas State (KSU)

Someone wondered what the probability of such an outcome might be.

a. How many games do the 8 teams play?

b. Assuming no ties, how many different outcomes are there for all the games together?

c. In how many ways could the 8 teams end in a perfect progression?

d. Assuming that each team had an equally likely probability of winning each game, find the probability of a perfect progression with 8 teams.

e. Find a general expression for the probability of a perfect progression in an *n*-team league with the same assumptions.

56. *Bingo* Bingo has become popular in the United States, and it is an efficient way for many organizations to raise money. The bingo card has 5 rows and 5 columns of numbers from 1 to 75, with the center given as a free cell. Balls showing one of the 75 numbers are picked at random from a container. If the drawn number appears on a player's card, then the player covers the number. In general, the winner is the person who has a card with an entire row, column, or diagonal covered.‡

a. Find the probability that a person will win bingo after just four numbers are called.

b. An L occurs when the first column and the bottom row are both covered. Find the probability that an L will occur in the fewest number of calls.

c. An X-out occurs when both diagonals are covered. Find the probability that an X-out occurs in the fewest number of calls.

d. If bingo cards are constructed so that column one has 5 of the numbers from 1 to 15, column two has 5 of the numbers from 16 to 30, column three has 4 of the numbers from 31 to 45, column four has 5 of the numbers from 46 to 60, and column five has 5 of the numbers from 61 to 75, how many different bingo cards could be constructed? (*Hint:* Order matters!)

Science, Vol. 258, Oct. 16, 1992, p. 398.

†Madsen, Richard, "On the Probability of a Perfect Progression," *The American Statistician*, Aug. 1991, Vol. 45, No. 3, p. 214.

‡Bay, Jennifer M., Robert E. Reys, Ken Simms, and P. Mark Taylor, "Bingo Games: Turning Student Intuitions into Investigations in Probability and Number Sense," *The Mathematics Teacher*, Vol. 93, No. 3, Mar. 2000, pp. 200–206.

8.4 BINOMIAL PROBABILITY

THINK ABOUT IT What is the probability that 3 of 6 people prefer Diet Supercola over its competitors?

This question involves an experiment that is repeated 6 times. Many probability problems are concerned with experiments in which an event is repeated many times. Other examples include finding the probability of getting 7 heads in 8 tosses of a coin, of hitting a target 6 times out of 6, and of finding 1 defective item in a sample of 15 items. Probability problems of this kind are called **Bernoulli trials** problems, or **Bernoulli processes,** named after the Swiss mathematician Jakob Bernoulli (1654–1705), who is well known for his work in probability theory. In each case, some outcome is designated a success, and any other outcome is considered a failure. This labeling is arbitrary, and does not necessarily have anything to do with real success or failure. Thus, if the probability of a success in a single trial is p, the probability of failure will be $1 - p$. A Bernoulli trials problem, or **binomial experiment,** must satisfy the following conditions.

BINOMIAL EXPERIMENT

1. The same experiment is repeated several times.
2. There are only two possible outcomes, success and failure.
3. The repeated trials are independent, so that the probability of success remains the same for each trial.

EXAMPLE 1 Sleep

The chance that an American falls asleep with the TV on at least three nights a week is 1/4.* Suppose a researcher selects 5 Americans at random and is interested in the probability that all 5 are "TV sleepers."

FOR REVIEW

Recall that if A and B are independent events,

$$P(A \text{ and } B) = P(A)P(B).$$

Solution Here the experiment, selecting a person, is repeated 5 times. If selecting a TV sleeper is labeled a success, then getting a "non-TV sleeper" is labeled a failure. The 5 trials are almost independent. There is a very slight dependence; if, for example, the first person selected is a TV sleeper, then there is one less TV sleeper to choose from when we select the next person (assuming we never select the same person twice). When selecting a small sample out of a large population, however, the probability changes negligibly, so researchers consider such trials to be independent. Thus, the probability that all 5 in our sample are sleepers is

$$\frac{1}{4} \cdot \frac{1}{4} \cdot \frac{1}{4} \cdot \frac{1}{4} \cdot \frac{1}{4} = \left(\frac{1}{4}\right)^5 \approx .000977.$$

Now suppose the problem in Example 1 is changed to that of finding the probability that exactly 4 of the 5 people in the sample are TV sleepers. This out-

---

*Harper's Magazine, Mar. 1996, p. 13.

come can occur in more than one way, as shown below, where s represents a success (a TV sleeper) and f represents a failure (a non-TV sleeper).

outcome 1:	s	s	s	s	f
outcome 2:	s	s	s	f	s
outcome 3:	s	s	f	s	s
outcome 4:	s	f	s	s	s
outcome 5:	f	s	s	s	s

Keep in mind that since the probability of success is $1/4$, the probability of failure is $1 - 1/4 = 3/4$. The probability, then, of each of these 5 outcomes is

$$\left(\frac{1}{4}\right)^4\left(\frac{3}{4}\right).$$

Since the 5 outcomes represent mutually exclusive events, add the 5 identical probabilities, which is equivalent to multiplying the above probability by 5. The result is

$$P(4 \text{ of the 5 people are TV sleepers}) = 5\left(\frac{1}{4}\right)^4\left(\frac{3}{4}\right) = \frac{15}{4^5} \approx .01465.$$

In the same way, we can compute the probability of selecting 3 TV sleepers in our sample of 5. The probability of any one way of achieving 3 successes and 2 failures will be

$$\left(\frac{1}{4}\right)^3\left(\frac{3}{4}\right)^2.$$

Rather than list all the ways of achieving 3 successes out of 5 trials, we will count this number using combinations. The number of ways to select 3 elements out of a set of 5 is $\binom{5}{3} = 5!/(2!\,3!) = 10$, giving

$$P(3 \text{ of the 5 people are TV sleepers}) = 10\left(\frac{1}{4}\right)^3\left(\frac{3}{4}\right)^2 = \frac{90}{4^5} \approx .08789.$$

A similar argument works in the general case.

BINOMIAL PROBABILITY

If p is the probability of success in a single trial of a binomial experiment, the probability of x successes and $n - x$ failures in n independent repeated trials of the experiment is

$$\binom{n}{x} \cdot p^x \cdot (1 - p)^{n-x}.$$

EXAMPLE 2   Advertising

The advertising agency that handles the Diet Supercola account believes that 40% of all consumers prefer this product over its competitors. Suppose a sample of 6 people is chosen. Assume that all responses are independent of each other. Find the probability of each of the following.

(a) Exactly 3 of the 6 people prefer Diet Supercola.

Solution Think of the 6 responses as 6 independent trials. A success occurs if a person prefers Diet Supercola. Then this is a binomial experiment with

$p = P(\text{success}) = P(\text{prefer Diet Supercola}) = .4$. The sample is made up of 6 people, so $n = 6$. To find the probability that exactly 3 people prefer this drink, let $x = 3$ and use the formula in the box.

$$P(\text{exactly 3}) = \binom{6}{3}(.4)^3(1 - .4)^{6-3}$$
$$= 20(.4)^3(.6)^3$$
$$= 20(.064)(.216)$$
$$= .27648$$

(b) None of the 6 people prefer Diet Supercola.

Solution Let $x = 0$.

$$P(\text{exactly 0}) = \binom{6}{0}(.4)^0(1 - .4)^6 = 1(1)(.6)^6 \approx .0467$$

EXAMPLE 3 Coin Toss

Find the probability of getting exactly 7 heads in 8 tosses of a fair coin.

Solution The probability of success (getting a head in a single toss) is $1/2$. The probability of a failure (getting a tail) is $1 - 1/2 = 1/2$. Thus,

$$P(\text{7 heads in 8 tosses}) = \binom{8}{7}\left(\frac{1}{2}\right)^7\left(\frac{1}{2}\right)^1 = 8\left(\frac{1}{2}\right)^8 = .03125.$$

EXAMPLE 4 Defective Items

Assuming that selection of items for a sample can be treated as independent trials, and that the probability that any 1 item is defective is .01, find the following.

(a) The probability of 1 defective item in a random sample of 15 items from a production line

Solution Here, a "success" is a defective item. Since selecting each item for the sample is assumed to be an independent trial, the binomial probability formula applies. The probability of success (a defective item) is .01, while the probability of failure (an acceptable item) is .99. This makes

$$P(\text{1 defective in 15 items}) = \binom{15}{1}(.01)^1(.99)^{14}$$
$$= 15(.01)(.99)^{14}$$
$$\approx .130.$$

(b) The probability of at most 1 defective item in a random sample of 15 items from a production line

Solution "At most 1" means 0 defective items or 1 defective item. Since 0 defective items is equivalent to 15 acceptable items,

$$P(\text{0 defective}) = (.99)^{15} \approx .860.$$

Use the union rule, noting that 0 defective and 1 defective are mutually exclusive events, to get

$$P(\text{at most 1 defective}) = P(\text{0 defective}) + P(\text{1 defective})$$
$$\approx .860 + .130$$
$$= .990.$$

EXAMPLE 5   Supermarket Scanners

A survey by *Money* magazine found that supermarket scanners are overcharging customers at 30% of stores.*

(a) If you shop at 3 supermarkets that use scanners, what is the probability that you will be overcharged in at least one store?

Solution We can treat this as a binomial experiment, letting $n = 3$ and $p = .3$. At least 1 of 3 means 1 or 2 or 3. It will be simpler here to find the probability of being overcharged in none of the 3 stores, that is, $P(0 \text{ overcharges})$, and then find $1 - P(0 \text{ overcharges})$.

$$P(0 \text{ overcharges}) = \binom{3}{0}(.3)^0(.7)^3$$

$$= 1(1)(.343) = .343$$

$$P(\text{at least one}) = 1 - P(0 \text{ overcharges})$$

$$= 1 - .343 = .657$$

(b) If you shop at 3 supermarkets that use scanners, what is the probability that you will be overcharged in at most one store?

Solution "At most one" means 0 or 1, so

$$P(0 \text{ or } 1) = P(0) + P(1)$$

$$= \binom{3}{0}(.3)^0(.7)^3 + \binom{3}{1}(.3)^1(.7)^2$$

$$= 1(1)(.343) + 3(.3)(.49) = .784.$$

The triangular array of numbers shown below is called **Pascal's triangle** in honor of the French mathematician Blaise Pascal (1623–1662), who was one of the first to use it extensively. The triangle was known long be-fore Pascal's time and appears in Chinese and Islamic manuscripts from the eleventh century.

PASCAL'S TRIANGLE

```
                    1
                1       1
            1       2       1
        1       3       3       1
    1       4       6       4       1
1       5      10      10       5       1
    .       .       .       .       .
    .       .       .       .       .
    .       .       .       .       .
```

---

*O'Connell, Vanessa, "Don't Get Cheated by Supermarket Scanners," *Money,* Apr. 1993, pp. 132–138.

The array provides a quick way to find binomial probabilities. The nth row of the triangle, where $n = 0, 1, 2, 3, \ldots$, gives the coefficients $\binom{n}{r}$ for $r = 0,$ $1, 2, 3, \ldots, n$. For example, for $n = 4$, $1 = \binom{4}{0}$, $4 = \binom{4}{1}$, $6 = \binom{4}{2}$, and so on. Each number in the triangle is the sum of the two numbers directly above it. For example, in the row for $n = 4$, 1 is the sum of 1, the only number above it, 4 is the sum of 1 and 3, 6 is the sum of 3 and 3, and so on. Adding in this way gives the sixth row:

$$1 \quad 6 \quad 15 \quad 20 \quad 15 \quad 6 \quad 1.$$

Notice that Pascal's triangle tells us, for example, that $\binom{4}{1} + \binom{4}{2} = \binom{5}{2}$ (that is, $4 + 6 = 10$). Using the combinations formula, it can be shown that, in general, $\binom{n}{r} + \binom{n}{r + 1} = \binom{n + 1}{r + 1}$. This is left as an exercise.

EXAMPLE 6 Pascal's Triangle

Use Pascal's triangle to find the probability in Example 5 that if you shop at 6 supermarkets, at least 3 will overcharge you.

Solution The probability of success is .3. Since at least 3 means 3, 4, 5, or 6,

$$P(\text{at least } 3) = P(3) + P(4) + P(5) + P(6)$$

$$= \binom{6}{3}(.3)^3(.7)^3 + \binom{6}{4}(.3)^4(.7)^2$$

$$+ \binom{6}{5}(.3)^5(.7)^1 + \binom{6}{6}(.3)^6(.7)^0.$$

Use the sixth row of Pascal's triangle for the combinations to get

$$P(\text{at least } 3) = 20(.3)^3(.7)^3 + 15(.3)^4(.7)^2 + 6(.3)^5(.7)^1 + 1(.3)^6(.7)^0$$

$$= .1852 + .0595 + .0102 + .0007$$

$$= .2556.$$

EXAMPLE 7 Independent Jury

If each member of a 9-person jury acts independently of each other and makes the correct determination of guilt or innocence with probability .65, find the probability that the majority of jurors will reach a correct verdict.*

Solution

Method 1: Calculation by Hand Since the jurors in this particular situation act independently, we can treat this as a binomial experiment. Thus, the probability that the majority of the jurors will

---

*Grofman, Bernard, "A Preliminary Model of Jury Decision Making as a Function of Jury Size, Effective Jury Decision Rule, and Mean Juror Judgmental Competence," *Frontiers in Economics*, 1979, pp. 98–110.

reach the correct verdict is given by

$$P(\text{at least } 5) = \binom{9}{5}(.65)^5(.35)^4 + \binom{9}{6}(.65)^6(.35)^3 + \binom{9}{7}(.65)^7(.35)^2$$

$$+ \binom{9}{8}(.65)^8(.35)^1 + \binom{9}{9}(.65)^9.$$

$$= .2194 + .2716 + .2162 + .1004 + .0207$$

$$= .8283.$$

Method 2: Graphing Calculator Some graphing calculators provide binomial probabilities. On a TI-83, for example, the command `binompdf(9,.65,5)`, found in the menu DISTR, gives .21939, which is the probability that $x = 5$. Alternatively, the command `binomcdf(9,.65,4)` gives .17172 as the probability that 4 or fewer jurors will make the correct decision. Subtract .17172 from 1 to get .82828 as the probability that the majority of the jurors will make the correct decision. This value rounds to .8283, which is in agreement with Method 1. Often, Method 2 is more accurate than Method 1 due to the accumulation of rounding errors when doing successive calculations by hand.

Method 3: Spreadsheet Some spreadsheets also provide binomial probabilities. In Microsoft Excel, for example, the command "`=BINOMDIST(5,9,.65,0)`" gives .21939, which is the probability that $x = 5$. Alternatively, the command "`=BINOMDIST(4,9,.65,1)`" gives .17172 as the probability that 4 or fewer jurors will make the correct decision. Subtract .17172 from 1 to get .82828 as the probability that the majority of the jurors will make the correct decision. This value agrees with the value found in Methods 1 and 2.

8.4 EXERCISES

Suppose that a family has 5 children. Also, suppose that the probability of having a girl is 1/2. Find the probabilities that the family has the following children.

1. Exactly 2 girls and 3 boys

2. Exactly 3 girls and 2 boys

3. No girls

4. No boys

5. At least 4 girls

6. At least 3 boys

7. No more than 3 boys

8. No more than 4 girls

A die is rolled 12 times. Find the probabilities of rolling the following.

9. Exactly 12 ones

10. Exactly 6 ones

11. Exactly 1 one

12. Exactly 2 ones

13. No more than 3 ones

14. No more than 1 one

A coin is tossed 6 times. Find the probabilities of getting the following.

15. All heads

16. Exactly 3 heads

17. No more than 3 heads

18. At least 3 heads

19. How do you identify a probability problem that involves a binomial experiment?

20. How is Pascal's triangle used to find probabilities?

21. Using the definition of combination in Section 8.2, prove that

$$\binom{n}{r} + \binom{n}{r+1} = \binom{n+1}{r+1}.$$

(This is the formula underlying Pascal's triangle.)

In Exercises 22 and 23, argue that the use of binomial probabilities is not applicable and thus the probabilities that are computed are not correct.

22. In England, a woman was found guilty of smothering her two infant children. Much of the Crown's case against the lady was based on the testimony from a pediatrician who indicated that the chances of two crib deaths occurring in both siblings was only about 1 in 73 million. This number was calculated by assuming that the probability of a single crib death is 1 in 8500 and the probability of two crib deaths is 1 in 8500^2 (i.e., binomial).*

23. A contemporary radio station in Boston has a contest in which a caller is asked his or her date of birth. If the caller's date of birth, including the day, month, and year of birth, matches a predetermined date, the caller wins $1 million. Assuming that there were 36,525 days in the twentieth century and the contest was run 51 times on consecutive days, the probability that the grand prize will be won is

$$1 - \left(1 - \frac{1}{36,525}\right)^{51} \approx .0014.^\dagger$$

Applications

BUSINESS AND ECONOMICS

Management The survey discussed in Example 5 also found that customers overpay for 1 out of every 10 items, on average. Suppose a customer purchases 15 items. Find the following probabilities.

24. A customer overpays on 3 items.

25. A customer does not overpay for any item.

26. A customer overpays on at least one item.

27. A customer overpays on at least 2 items.

28. A customer overpays on at most 2 items.

Insurance Rating The insurance industry has found that the probability is .9 that a life insurance applicant will qualify at the regular rates. Find the probabilities that of the next 10 applicants for life insurance, the following numbers will qualify at the regular rates.

29. Exactly 10

30. Exactly 9

31. At least 9

32. Less than 8

Personnel Screening A company gives prospective workers a 6-question, multiple-choice test. Each question has 5 possible answers, so that there is a 1/5 or 20% chance of answering a question correctly just by guessing. Find the probabilities of getting the following results by chance.

33. Exactly 2 correct answers

34. No correct answers

35. At least 4 correct answers

36. No more than 3 correct answers

37. *Customer Satisfaction* Over the last decade, 10% of all clients of J. K. Loss & Company have lost their life savings. Suppose a sample of 3 of the current clients of the firm is chosen. Assuming independence, find the probability that exactly 1 of the 3 clients will lose everything.

Quality Control A factory tests a random sample of 20 transistors for defective transistors. The probability that a particular transistor will be defective has been established by past experience as .05.

38. What is the probability that there are no defective transistors in the sample?

39. What is the probability that the number of defective transistors in the sample is at most 2?

---

*Watkins, Stephen J., "Conviction by Mathematical Error?," *British Medical Journal*, Vol. 320, No. 7226, Jan. 1, 2000, pp. 2–3.

†Snell, J. Laurie, "40-Million-Dollar Thursday," *Chance News 9.04*, Mar. 7–April. 5, 2000.

40. *Quality Control* The probability that a certain machine turns out a defective item is .05. Find the probabilities that in a run of 75 items, the following results are obtained.

 a. Exactly 5 defective items

 b. No defective items

 c. At least 1 defective item

41. *Survey Results* A company is taking a survey to find out whether people like its product. Its last survey indicated that 70% of the population like the product. Based on that, in a sample of 58 people, find the probabilities of the following.

 a. All 58 like the product.

 b. From 28 to 30 (inclusive) like the product.

42. *Pecans* Pecan producers blow air through the pecans so that the lighter ones are blown out. The lighter-weight pecans are generally bad and the heavier ones tend to be better. These "blow outs" and "good nuts" are often sold to tourists along the highway. Suppose 60% of the "blow outs" are good, and 80% of the "good nuts" are good.*

 a. What is the probability that if you crack and check 20 "good nuts" you will find 8 bad ones?

 b. What is the probability that if you crack and check 20 "blow outs" you will find 8 bad ones?

 c. If we assume that 70% of the roadside stands sell "good nuts," and that out of 20 nuts we find 8 that are bad, what is the probability that the nuts are "blow outs"?

LIFE SCIENCES

Drug Effectiveness A new drug cures 70% of the people taking it. Suppose 20 people take the drug; find the probabilities of the following.

43. Exactly 18 people are cured.

44. Exactly 17 people are cured.

45. At least 17 people are cured.

46. At least 18 people are cured.

Births of Twins The probability that a birth will result in twins is .012. Assuming independence (perhaps not a valid assumption), what are the probabilities that out of 100 births in a hospital, there will be the following numbers of sets of twins?

47. Exactly 2 sets of twins

48. At most 2 sets of twins

Vitamin A Deficiency Six mice from the same litter, all suffering from a vitamin A deficiency, are fed a certain dose of carrots. If the probability of recovery under such treatment is .70, find the probabilities of the following results.

49. None of the mice recover.

50. Exactly 3 of the 6 mice recover.

51. All of the mice recover.

52. No more than 3 mice recover.

53. *Effects of Radiation* In an experiment on the effects of a radiation dose on cells, a beam of radioactive particles is aimed at a group of 10 cells. Find the probability that 8 of the cells will be hit by the beam, if the probability that any single cell will be hit is .6. (Assume independence.)

54. *Effects of Radiation* The probability of a mutation of a given gene under a dose of 1 roentgen of radiation is approximately 2.5×10^{-7}. What is the probability that in 10,000 genes, at least 1 mutation occurs?

55. *Drug Side Effects* A new drug being tested causes a serious side effect in 5 out of 100 patients. What is the probability that no side effects occur in a sample of 10 patients taking the drug?

56. *Flu Inoculations* A flu vaccine has a probability of 80% of preventing a person who is inoculated from getting the flu. A county health office inoculates 83 people. Find the probabilities of the following.

 a. Exactly 10 of the people inoculated get the flu.

 b. No more than 4 of the people inoculated get the flu.

 c. None of the people inoculated get the flu.

57. *Color Blindness* The probability that a male will be color-blind is .042. Find the probabilities that in a group of 53 men, the following will be true.

 a. Exactly 5 are color-blind.

 b. No more than 5 are color-blind.

 c. At least 1 is color-blind.

*Submitted by Professor Irvin R. Hentzel, Iowa State University.

58. *Pharmacology* In placebo-controlled trials of Pravachol®, a drug that is prescribed to lower cholesterol, 7.3% of the patients who were taking the drug experienced nausea/vomiting, whereas 7.1% of the patients who were taking the placebo experienced nausea/vomiting.*

a. If 100 patients who are taking Pravachol® are selected, what is the probability that 10 or more will experience nausea/vomiting?

b. If a second group of 100 patients receives a placebo, what is the probability that 10 or more will experience nausea/vomiting?

c. Since 7.3% is larger than 7.1%, do you believe that the Pravachol® causes more people to experience nausea/vomiting than a placebo? Explain.

59. *Genetic Fingerprinting* The use of DNA has become an integral part of many court cases. When DNA is extracted from cells and body fluids, genetic information is represented by bands of information, which look similar to a bar code at a grocery store. It is generally accepted that in unrelated people, the probability of a particular band matching is 1 in 4.†

a. If 5 bands are compared in unrelated people, what is the probability that all 5 of the bands match? (Express your answer in terms of "1 chance in ?".)

b. If 20 bands are compared in unrelated people, what is the probability that all 20 of the bands match? (Express your answer in terms of "1 chance in ?".)

c. If 20 bands are compared in unrelated people, what is the probability that 16 or more bands match? (Express your answer in terms of "1 chance in ?".)

d. If you were deciding paternity and there were 16 matches out of 20 bands compared, would you believe that the person being tested was the father? Explain.

SOCIAL SCIENCES

60. *Women Working* A recent study found that 33% of women would prefer to work part-time rather than full-time if money were not a concern.‡ Find the probability that if 10 women are selected at random, at least 3 of them would prefer to work part-time.

Testing In a 10-question, multiple-choice biology test with 5 choices for each question, an unprepared student guesses the answer to each item. Find the probabilities of the following results.

61. Exactly 6 correct answers

62. Exactly 7 correct answers

63. At least 8 correct answers

64. Fewer than 8 correct answers

65. *Community College Population* According to the state of California, 33% of all state community college students belong to ethnic minorities. Find the probabilities of the following results in a random sample of 10 California community college students.

a. Exactly 2 belong to an ethnic minority.

b. Three or fewer belong to an ethnic minority.

c. Exactly 5 do not belong to an ethnic minority.

d. Six or more do not belong to an ethnic minority.

66. *Cheating* According to a poll conducted by *U.S. News and World Report*, 84% of college students believe they need to cheat to get ahead in the world today.§

a. Do the results of this poll indicate that 84% of all college students cheat? Explain.

b. If this result is accurate and 100 college students are asked if they believe that cheating is necessary to get ahead in the world, what is the probability that 90 or more of the students will answer affirmatively to the question?

67. *Education* In the "Numbers" section of a recent *Time* magazine, it was reported that 15.2% of low-birth-weight babies graduate from high school by age 19. On the other hand, it was reported that 57.5% of the normal-birth-weight siblings graduated from high school.‖

a. If 40 low-birth-weight babies were tracked through high school, what is the probability that fewer than 15 will graduate from high school by age 19?

b. What are some of the factors that may contribute to the wide difference in high school success between these siblings? Do you believe that low birth weight is the primary cause of the difference? What other information do you need to better answer these questions?

---

*Advertisement in *Time*, July 17, 2000, for Pravachol®, developed and marketed by Bristol-Myers Squibb Company.

†"Genetic Fingerprinting Worksheet," Centre for Innovation in Mathematics Teaching, http://www.ex.ac.uk/cimt/resource/fgrprnts.htm.

‡Ferraro, Cathleen, "Feelings of the Working Women," *The Sacramento Bee*, May 11, 1995, pp. A1, A22.

§Kleiner, Carolyn, and Mary Lord, "The Cheating Game," *U.S. News and World Report*, Nov. 22, 1999, pp. 55–66.

‖"Numbers," *Time*, July 17, 2000, p. 21.

■ 8.5 PROBABILITY DISTRIBUTIONS; EXPECTED VALUE

 THINK ABOUT IT   What is the expected value of winning a prize for someone who buys one ticket in a raffle?

In this section we shall see that the *expected value* of a probability distribution is a type of average. Probability distributions were introduced briefly in the chapter on Sets and Probability. Now we take a more complete look at probability distributions. A probability distribution depends on the idea of a *random variable,* so we begin with that.

Random Variables   Suppose that the shipping manager at a company receives a package of one dozen computer monitors, of which, unknown to him, three are broken. He checks four of the monitors at random to see how many are broken in his sample of 4. The answer, which we will label x, is one of the numbers 0, 1, 2, or 3. Since the value of x is random, x is called a random variable.

> **RANDOM VARIABLE**
>
> A **random variable** is a function that assigns a real number to each outcome of an experiment.

Probability Distribution   In the example with the shipping manager, we can calculate the probability that 0, 1, 2, or 3 monitors in his sample of 4 are broken using the methods of Section 8.3. There are 3 broken monitors and 9 unbroken monitors, so the number of ways of choosing 0 broken monitors (which implies 4 unbroken monitors) is $\binom{3}{0}\binom{9}{4}$. The number of ways of choosing a sample of 4 monitors is $\binom{12}{4}$. Therefore, the probability of choosing 0 broken monitors is

$$P(0) = \frac{\binom{3}{0}\binom{9}{4}}{\binom{12}{4}} = \frac{1\left(\dfrac{9 \cdot 8 \cdot 7 \cdot 6}{4 \cdot 3 \cdot 2 \cdot 1}\right)}{\left(\dfrac{12 \cdot 11 \cdot 10 \cdot 9}{4 \cdot 3 \cdot 2 \cdot 1}\right)} = \frac{126}{495} = \frac{14}{55}.$$

Similarly, the probability of choosing 1 broken monitor is

$$P(1) = \frac{\binom{3}{1}\binom{9}{3}}{\binom{12}{4}} = \frac{3 \cdot 84}{495} = \frac{252}{495} = \frac{28}{55}.$$

The probability of choosing 2 broken monitors is

$$P(2) = \frac{\binom{3}{2}\binom{9}{2}}{\binom{12}{4}} = \frac{3 \cdot 36}{495} = \frac{108}{495} = \frac{12}{55}.$$

The probability of choosing 3 broken monitors is

$$P(3) = \frac{\binom{3}{3}\binom{9}{1}}{\binom{12}{4}} = \frac{1 \cdot 9}{495} = \frac{9}{495} = \frac{1}{55}.$$

The results can be put in a table.

Table 1

x	0	1	2	3
$P(x)$	14/55	28/55	12/55	1/55

Such a table that lists the possible values of a random variable, together with the corresponding probabilities, is called a **probability distribution.** The sum of the probabilities in a probability distribution must always equal 1. (The sum in some distributions may vary slightly from 1 because of rounding.)

Instead of writing the probability distribution as a table, we could write the same information as a set of ordered pairs:

$$\{(0, 14/55), (1, 28/55), (2, 12/55), (3, 1/55)\}.$$

There is just one probability for each value of the random variable. Thus, a probability distribution defines a function, called a **probability distribution function,** or simply a **probability function.** We shall use the terms "probability distribution" and "probability function" interchangeably.

The information in a probability distribution is often displayed graphically as a special kind of bar graph called a **histogram.** The bars of a histogram all have the same width, usually 1. The heights of the bars are determined by the probabilities. A histogram for the data in Table 1 is given in Figure 8. A histogram shows important characteristics of a distribution that may not be readily apparent in tabular form, such as the relative sizes of the probabilities and any symmetry in the distribution.

The area of the bar above $x = 0$ in Figure 8 is the product of 1 and 14/55, or $1 \cdot 14/55 = 14/55$. Since each bar has a width of 1, its area is equal to the probability that corresponds to that value of x. The probability that a particular value will occur is thus given by the area of the appropriate bar of the graph. For example, the probability that one or more monitors is broken is the sum of the areas for $x = 1$, $x = 2$, and $x = 3$. This area, shown in red in Figure 9 on the next page, corresponds to 41/55 of the total area, since

$$P(x \geq 1) = P(x = 1) + P(x = 2) + P(x = 3)$$
$$= 28/55 + 12/55 + 1/55 + 41/55.$$

EXAMPLE 1 Probability Distributions

(a) Give the probability distribution for the number of heads showing when two coins are tossed.

Solution Let x represent the random variable "number of heads." Then x can take on the values 0, 1, or 2. Now find the probability of each outcome. The results are shown in Table 2.

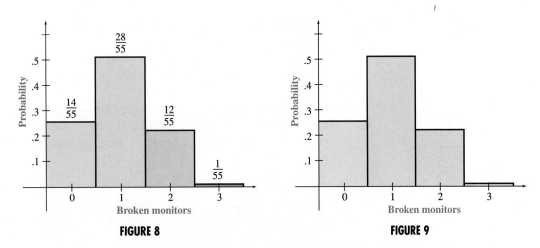

FIGURE 8   **FIGURE 9**

(b) Draw a histogram for the distribution in Table 2. Find the probability that at least one coin comes up heads.

Solution The histogram is shown in Figure 10. The portion in red represents

$$P(x \geq 1) = P(x = 1) + P(x = 2)$$

$$= \frac{3}{4}.$$

Table 2

x	0	1	2
$P(x)$	1/4	1/2	1/4

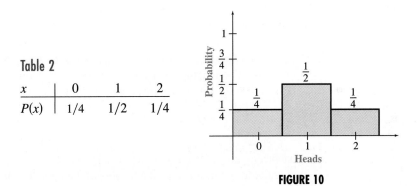

FIGURE 10

Expected Value   In working with probability distributions, it is useful to have a concept of the typical or average value that the random variable takes on. In Example 1, for instance, it seems reasonable that, on the average, one head shows when two coins are tossed. This does not tell what will happen the next time we toss two coins; we may get two heads, or we may get none. If we tossed two coins many times, however, we would expect that, in the long run, we would average about one head for each toss of two coins.

A way to solve such problems in general is to imagine flipping two coins 4 times. Based on the probability distribution in Example 1, we would expect that 1 of the 4 times we would get 0 heads, 2 of the 4 times we would get 1 head, and 1 of the 4 times we would get 2 heads. The total number of heads we would get, then, is

$$0 \cdot 1 + 1 \cdot 2 + 2 \cdot 1 = 4.$$

The expected numbers of heads per toss is found by dividing the total number of heads by the total number of tosses, or

$$\frac{0 \cdot 1 + 1 \cdot 2 + 2 \cdot 1}{4} = 0 \cdot \frac{1}{4} + 1 \cdot \frac{1}{2} + 2 \cdot \frac{1}{4} = 1.$$

Notice that the expected number of heads turns out to be the sum of the three values of the random variable x multiplied by their corresponding probabilities. We can use this idea to define the *expected value* of a random variable as follows.

EXPECTED VALUE

Suppose the random variable x can take on the n values $x_1, x_2, x_3, \ldots, x_n$. Also, suppose the probabilities that these values occur are, respectively, $p_1, p_2, p_3, \ldots, p_n$. Then the **expected value** of the random variable is

$$E(x) = x_1 p_1 + x_2 p_2 + x_3 p_3 + \cdots + x_n p_n.$$

EXAMPLE 2   Computer Monitors

In the example with the computer monitors, find the expected number of broken monitors that the shipping manager finds.

Solution Using the values in Table 1 and the definition of expected value, we find that

$$E(x) = 0 \cdot \frac{14}{55} + 1 \cdot \frac{28}{55} + 2 \cdot \frac{12}{55} + 3 \cdot \frac{1}{55} = 1.$$

On the average, the shipping manager will find 1 broken monitor in the sample of 4. On reflection, this seems natural; 3 of the 12 monitors, or 1/4 of the total, are broken. We should expect, then, that 1/4 of the sample of 4 monitors are broken.

Physically, the expected value of a probability distribution represents a balance point. If we think of the histogram in Figure 8 as a series of weights with magnitudes represented by the heights of the bars, then the system would balance if supported at the point corresponding to the expected value.

EXAMPLE 3   Symphony Orchestra

Suppose a local symphony decides to raise money by raffling a microwave oven worth $400, a dinner for two worth $80, and 2 books worth $20 each. A total of 2000 tickets are sold at $1 each. Find the expected value of winning for a person who buys one ticket in the raffle.

Solution Here the random variable represents the possible amounts of net winnings, where net winnings = amount won − cost of ticket. The net winnings of the person winning the oven are $400 (amount won) − $1 (cost of ticket) = $399. The net winnings for each losing ticket are $0 − $1 = −$1.

The net winnings of the various prizes, as well as their respective probabilities, are shown in Table 3 on the next page. The probability of winning $19 is 2/2000 because there are 2 prizes worth $20. We have not reduced the fractions in order to keep all the denominators equal. Because there are 4 winning tickets, there are 1996 losing tickets, so the probability of winning $-$1$ is $1996/2000$.

Table 3

x	$399	$79	$19	$-$1$
$P(x)$	1/2000	1/2000	2/2000	1996/2000

The expected winnings for a person buying one ticket are

$$399\left(\frac{1}{2000}\right) + 79\left(\frac{1}{2000}\right) + 19\left(\frac{2}{2000}\right) + (-1)\left(\frac{1996}{2000}\right) = -\frac{1480}{2000}$$
$$= -.74.$$

On the average, a person buying one ticket in the raffle will lose $.74, or 74¢.

It is not possible to lose 74¢ in this raffle: either you lose $1, or you win a prize worth $400, $80, or $20, minus the $1 you pay to play. But if you bought tickets in many such raffles over a long period of time, you would lose 74¢ per ticket on the average. It is important to note that the expected value of a random variable may be a number that can never occur in any one trial of the experiment.

NOTE An alternative way to compute expected value in this and other examples is to calculate the expected amount won and then subtract the cost of the ticket afterwards. The amount won is either $400 (with probability 1/2000), $80 (with probability 1/2000), $20 (with probability 2/2000), or $0 (with probability 1996/2000). The expected winnings for a person buying one ticket are then

$$400\left(\frac{1}{2000}\right) + 80\left(\frac{1}{2000}\right) + 20\left(\frac{2}{2000}\right) + 0\left(\frac{1996}{2000}\right) - 1 = -\frac{1480}{2000}$$
$$= -.74.$$

EXAMPLE 4 Friendly Wager

Each day Donna and Mary toss a coin to see who buys coffee (80 cents a cup). One tosses and the other calls the outcome. If the person who calls the outcome is correct, the other buys the coffee; otherwise the caller pays. Find Donna's expected winnings.

Solution Assume that an honest coin is used, that Mary tosses the coin, and that Donna calls the outcome. The possible results and corresponding probabilities are shown below.

	Possible Results			
Result of toss	Heads	Heads	Tails	Tails
Call	Heads	Tails	Heads	Tails
Caller wins?	Yes	No	No	Yes
Probability	1/4	1/4	1/4	1/4

Donna wins an 80¢ cup of coffee whenever the results and calls match, and she loses an 80¢ cup when there is no match. Her expected winnings are

$$(.80)\left(\frac{1}{4}\right) + (-.80)\left(\frac{1}{4}\right) + (-.80)\left(\frac{1}{4}\right) + (.80)\left(\frac{1}{4}\right) = 0.$$

On the average, over the long run, Donna neither wins nor loses.

A game with an expected value of 0 (such as the one in Example 4) is called a **fair game.** Casinos do not offer fair games. If they did, they would win (on the average) $0, and have a hard time paying the help! Casino games have expected winnings for the house that vary from 1.5 cents per dollar to 60 cents per dollar. Exercises 42–47 at the end of the section ask you to find the expected winnings for certain games of chance.

The idea of expected value can be very useful in decision making, as shown by the next example.

EXAMPLE 5 Life Insurance

At age 50, you receive a letter from Mutual of Mauritania Insurance Company. According to the letter, you must tell the company immediately which of the following two options you will choose: take $20,000 at age 60 (if you are alive, $0 otherwise) or $30,000 at age 70 (again, if you are alive, $0 otherwise). Based *only* on the idea of expected value, which should you choose?

Solution Life insurance companies have constructed elaborate tables showing the probability of a person living a given number of years into the future. From a recent such table, the probability of living from age 50 to 60 is .88, while the probability of living from age 50 to 70 is .64. The expected values of the two options are given below.

$$\text{First option: } (20,000)(.88) + (0)(.12) = 17,600$$
$$\text{Second option: } (30,000)(.64) + (0)(.36) = 19,200$$

Based strictly on expected values, choose the second option.

EXAMPLE 6 Bachelor's Degrees

According to the National Center for Education Statistics, 79% of the U.S. holders of bachelor's degrees in education in 1992–93 were women.* Suppose 5 holders of bachelor's degrees in education from 1992–93 are picked at random.

(a) Find the probability distribution for the number that are female.

Solution We first note that each of the 5 people in the sample is either female (with probability .79) or male (with probability .21). As in the previous section, we may assume that the probability for each member of the sample is independent of that of any other. Such a situation is described by binomial probability with $n = 5$ and $p = .79$, for which we use the binomial probability formula

$$\binom{n}{x} \cdot p^x \cdot (1-p)^{n-x},$$

---

*The New York Times, Jan. 7, 1996, Education Life, p. 24.

where x is the number of females in the sample. For example,

$$P(x = 0) = \binom{5}{0}(.79)^0(.21)^5 \approx .0004.$$

Similarly, we could calculate the probability that x is any value from 0 to 5, resulting in the following probability distribution (with all probabilities rounded to four places).

Table 4

x	0	1	2	3	4	5
$P(x)$	.0004	.0077	.0578	.2174	.4090	.3077

(b) Find the expected number of females in the sample of 5 people.

Solution Using the formula for expected value, we have

$$E(x) = 0(.0004) + 1(.0077) + 2(.0578) + 3(.2174)$$
$$+ 4(.4090) + 5(.3077) = 3.95.$$

On the average, 3.95 of the people in the sample of 5 will be female.

There is another way to get the answer in part (b) of the previous example. Because 79% of the U.S. holders of bachelor's degrees in education from 1992–93 are female, it is reasonable to expect 79% of our sample to be female. Thus, 79% of 5 is 5(.79) = 3.95. Notice that what we have done is to multiply n by p. It can be shown that this method always gives the expected value for binomial probability.

EXPECTED VALUE FOR BINOMIAL PROBABILITY

For binomial probability, $E(x) = np$. In other words, the expected number of successes is the number of trials times the probability of success in each trial.

EXAMPLE 7 Female Children
Suppose a family has 3 children.

(a) Find the probability distribution for the number of girls.

Solution Assuming girls and boys are equally likely, the probability distribution is binomial with $n = 3$ and $p = 1/2$. Letting x be the number of girls in the formula for binomial probability, we find, for example,

$$P(x = 0) = \binom{3}{0}\left(\frac{1}{2}\right)^0\left(\frac{1}{2}\right)^3 = \frac{1}{8}.$$

The other values are found similarly, and the results are shown in Table 5.

Table 5

x	0	1	2	3
$P(x)$	1/8	3/8	3/8	1/8

We can verify this by noticing that in the sample space S of all 3-child families, there are eight equally likely outcomes: $S = \{ggg, ggb, gbg, gbb, bgg, bgb, bbg, bbb\}$. One of the outcomes has 0 girls, three have 1 girl, three have 2 girls, and one has 3 girls.

(b) Find the expected number of girls in a 3-child family using the distribution from part (a).

Solution Using the formula for expected value, we have

$$\text{Expected number of girls} = 0\left(\frac{1}{8}\right) + 1\left(\frac{3}{8}\right) + 2\left(\frac{3}{8}\right) + 3\left(\frac{1}{8}\right)$$
$$= \frac{12}{8} = 1.5.$$

On average, a 3-child family will have 1.5 girls. This result agrees with our intuition that, on the average, half the children born will be girls.

(c) Find the expected number of girls in a 3-child family using the formula for expected value for binomial probability.

Solution Using the formula $E(x) = np$ with $n = 3$ and $p = 1/2$, we have

$$\text{Expected number of girls} = 3\left(\frac{1}{2}\right) = 1.5.$$

This agrees with our answer from part (b), as it must.

8.5 EXERCISES

For each of the experiments described below, let x determine a random variable, and use your knowledge of probability to prepare a probability distribution.

1. Four coins are tossed, and the number of heads is noted.

2. Two dice are rolled, and the total number of points is recorded.

3. Three cards are drawn from a deck. The number of aces is counted.

4. Two balls are drawn from a bag in which there are 4 white balls and 2 black balls. The number of black balls is counted.

Draw a histogram for each of the following, and shade the region that gives the indicated probability.

5. Exercise 1; $P(x \le 2)$

6. Exercise 2; $P(x \ge 11)$

7. Exercise 3; $P(\text{at least one ace})$

8. Exercise 4; $P(\text{at least one black ball})$

Find the expected value for each random variable.

9.
x	2	3	4	5
$P(x)$	.1	.4	.3	.2

10.
y	4	6	8	10
$P(y)$	.4	.4	.05	.15

11.
z	9	12	15	18	21
$P(z)$	.14	.22	.36	.18	.10

12.
x	30	32	36	38	44
$P(x)$	.31	.30	.29	.06	.04

Find the expected values for the random variables x having the probability functions graphed below.

13.

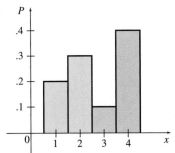

14.

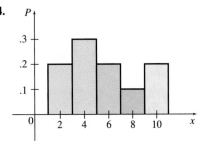

15.

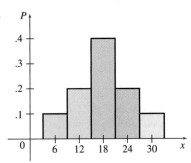

16.

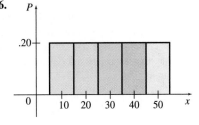

17. For the game in Example 4, find Mary's expected winnings. Is it a fair game?

18. Suppose one day Mary brings a 2-headed coin and uses it to toss for the coffee. Since Mary tosses, Donna calls.

   a. Is this still a fair game?

   b. What is Donna's expected gain if she calls heads?

   c. What is Donna's expected gain if she calls tails?

Solve each exercise. Many of these exercises require the use of combinations.

19. Suppose 3 marbles are drawn from a bag containing 3 yellow and 4 white marbles.

   a. Draw a histogram for the number of yellow marbles in the sample.

   b. What is the expected number of yellow marbles in the sample?

20. Suppose 5 apples in a barrel of 25 apples are known to be rotten.

   a. Draw a histogram for the number of rotten apples in a sample of 2 apples.

   b. What is the expected number of rotten apples in a sample of 2 apples?

21. A delegation of 3 is selected from a city council made up of 5 liberals and 4 conservatives.

   a. What is the expected number of liberals in the delegation?

   b. What is the expected number of conservatives in the delegation?

22. From a group of 2 women and 5 men, a delegation of 2 is selected. Find the expected number of women in the delegation.

23. In a club with 20 senior and 10 junior members, what is the expected number of junior members on a 3-member committee?

24. If 2 cards are drawn at one time from a deck of 52 cards, what is the expected number of diamonds?

25. Suppose someone offers to pay you $5 if you draw 2 diamonds in the game in Exercise 24. He says that you should pay 50 cents for the chance to play. Is this a fair game?

26. Your friend missed class the day probability distributions were discussed. How would you explain probability distribution to him?

27. Explain what expected value means in your own words.

Applications

BUSINESS AND ECONOMICS

28. *Complaints* A local used-car dealer gets complaints about his cars as shown in the following table.

Number of Complaints per Day	0	1	2	3	4	5	6
Probability	.01	.05	.15	.26	.33	.14	.06

Find the expected number of complaints per day.

29. *Payout on Insurance Policies* An insurance company has written 100 policies of $10,000, 500 of $5000, and 1000 of $1000 for people of age 20. If experience shows that the probability that a person will die at age 20 is .001, how much can the company expect to pay out during the year the policies were written?

30. *Rating Sales Accounts* Levi Strauss and Company* uses expected value to help its salespeople rate their accounts. For each account, a salesperson estimates potential additional volume and the probability of getting it. The product of these figures gives the expected value of the potential, which is added to the existing volume. The totals are then classified as A, B, or C, as follows: $40,000 or below, class C; from $40,000 up to and including $55,000, class B; above $55,000, class A. Complete the chart at the bottom of this page for one salesperson.

31. *Pecans* Refer to Exercise 42 in Section 8.4. Suppose that 60% of the pecan "blow outs" are good, and 80% of the "good nuts" are good.

  a. If you purchase 50 pecans, what is the expected number of good nuts you will find if you purchase "blow outs"?

  b. If you purchase 50 pecans, what is the expected number of bad nuts you will find if you have purchased "good nuts"?

LIFE SCIENCES

32. *Animal Offspring* In a certain animal species, the probability that a healthy adult female will have no offspring in a given year is .31, while the probabilities of 1, 2, 3, or 4 offspring are, respectively, .21, .19, .17, and .12. Find the expected number of offspring.

33. *Ear Infections* Otitis media, or middle ear infection, is initially treated with an antibiotic. Researchers have compared two antibiotics, amoxicillin and cefaclor, for their cost effectiveness. Amoxicillin is inexpensive, safe, and effective. Cefaclor is also safe. However, it is considerably more expensive and it is generally more effective. Use the tree diagram on the next page (where the costs are estimated as the total cost of medication, office visit, ear check, and hours of lost work) to answer the following.[†]

  a. Find the expected cost of using each antibiotic to treat a middle ear infection.

Account Number	Existing Volume	Potential Additional Volume	Probability of Getting It	Expected Value of Potential	Existing Volume + Expected Value of Potential	Class
1	$15,000	$10,000	.25	$2500	$17,500	C
2	$40,000	$0	—	—	$40,000	C
3	$20,000	$10,000	.20			
4	$50,000	$10,000	.10			
5	$5000	$50,000	.50			
6	$0	$100,000	.60			
7	$30,000	$20,000	.80			

*This example was supplied by James McDonald, Levi Strauss and Company, San Francisco.
[†]Weiss, Jeffrey, and Shoshana Melman, "Cost Effectiveness in the Choice of Antibiotics for the Initial Treatment of Otitis Media in Children: A Decision Analysis Approach," *Journal of Pediatric Infectious Disease*, Vol. 7, No. 1, 1988, pp. 23–26.

b. To minimize the total expected cost, which antibiotic should be chosen?

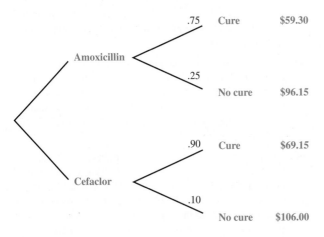

PHYSICAL SCIENCES

34. *Seeding Storms* One of the few methods that can be used in an attempt to cut the severity of a hurricane is to *seed* the storm. In this process, silver iodide crystals are dropped into the storm. Unfortunately, silver iodide crystals sometimes cause the storm to *increase* its speed. Wind speeds may also increase or decrease even with no seeding. Use the tree diagram to the right to answer the following.*

a. Find the expected amount of damage under each option, "seed" and "do not seed."

b. To minimize total expected damage, what option should be chosen?

SOCIAL SCIENCES

35. *Cheating* Recall from Exercise 66 in Section 8.4 that a poll conducted by *U.S. News and World Report* reported that 84% of college students believe they need to cheat to get ahead in the world today.[†] If 500 college students were surveyed, how many would you expect to say that they need to cheat to get ahead in the world today?

36. *Education* Recall from Exercise 67 in Section 8.4 that a *Time* magazine "Numbers" section reported that 15.2% of low-birth-weight babies graduate from high school by age 19.[‡] If 250 low-birth-weight babies are followed through high school, how many would you expect to graduate from high school?

GENERAL INTEREST

37. *Golf Tournament* At the end of play in a major golf tournament, two players, an "old pro" and a "new kid," are tied. Suppose the first prize is $80,000 and second prize is $20,000. Find the expected winnings for the old pro if

a. both players are of equal ability;

b. the new kid will freeze up, giving the old pro a 3/4 chance of winning.

38. *Cats* Kimberly Workman has four cats: Riley, Abby, Beastie, and Sylvester. Each cat has a 30% probability of climbing into the chair in which Kimberly is sitting, independent of how many cats are already in the chair with Kimberly.

*The probabilities and amounts of property damage in the tree diagram for Exercise 34 are from Howard, R. A., J. E. Matheson, and D. W. North, "The Decision to Seed Hurricanes," *Science*, Vol. 176, No. 16, June 1972, pp. 1191–1202. Copyright © 1972 by the American Association for the Advancement of Science. Reprinted by permission.
[†]Kleiner, Carolyn, and Mary Lord, "The Cheating Game," *U.S. News and World Report*, Nov. 22, 1999, pp. 55–66.
[‡]"Numbers," *Time*, July 17, 2000, p. 21.

a. Find the probability distribution for the number of cats in the chair with Kimberly.

b. Find the expected number of cats in the chair with Kimberly using the probability distribution in part a.

c. Find the expected number of cats in the chair with Kimberly using the formula for expected value of the binomial distribution.

39. *Postal Service* Mr. Statistics (a feature in *Fortune* magazine) investigated the claim of the United States Postal Service that 83% of first class mail in New York City arrives by the next day.* (The figure is 87% nationwide.) He mailed a letter to himself on 10 consecutive days; only 4 were delivered by the next day.

a. Find the probability distribution for the number of letters delivered by the next day if the overall probability of next-day delivery is 83%.

b. Using your answer to part a, find the probability that 4 or fewer out of 10 letters would be delivered by the next day.

c. Based on your answer to part b, do you think it is likely that the 83% figure is accurate? Explain.

d. Find the number of letters out of 10 that you would expect to be delivered by the next day if the 83% figure is accurate.

40. *Raffle* A raffle offers a first prize of $100 and 2 second prizes of $40 each. One ticket costs $1, and 500 tickets are sold. Find the expected winnings for a person who buys 1 ticket. Is this a fair game?

41. *Raffle* A raffle offers a first prize of $1000, 2 second prizes of $300 each, and 20 third prizes of $10 each. If 10,000 tickets are sold at 50¢ each, find the expected winnings for a person buying 1 ticket. Is this a fair game?

Find the expected winnings for the games of chance described in Exercises 42–47.

42. *Lottery* A state lottery requires you to choose 4 cards from an ordinary deck: 1 heart, 1 club, 1 diamond, and 1 spade in that order from the 13 cards in each suit. If all four choices are selected by the lottery, you win $5000. It costs $1 to play.

43. *Lottery* If exactly 3 of the 4 choices in Exercise 42 are selected, the player wins $200. (Ignore the possibility that all 4 choices are selected. It still costs $1 to play.)

44. *Roulette* In one form of roulette, you bet $1 on "even." If 1 of the 18 even numbers comes up, you get your dollar back, plus another one. If 1 of the 20 noneven (18 odd, 0, and 00) numbers comes up, you lose your dollar.

45. *Roulette* In another form of roulette, there are only 19 noneven numbers (no 00).

46. *Numbers* *Numbers* is a game in which you bet $1 on any three-digit number from 000 to 999. If your number comes up, you get $500.

47. *Keno* In one form of the game *Keno*, the house has a pot containing 80 balls, each marked with a different number from 1 to 80. You buy a ticket for $1 and mark one of the 80 numbers on it. The house then selects 20 numbers at random. If your number is among the 20, you get $3.20 (for a net winning of $2.20).

48. *Contests* A magazine distributor offers a first prize of $100,000, two second prizes of $40,000 each, and two third prizes of $10,000 each. A total of 2,000,000 entries are received in the contest. Find the expected winnings if you submit one entry to the contest. If it would cost you 50¢ in time, paper, and stamps to enter, would it be worth it?

49. *Contests* A contest at a fast-food restaurant offered the following cash prizes and probabilities of winning on one visit.

Prize	Probability
$100,000	1/176,402,500
$25,000	1/39,200,556
$5000	1/17,640,250
$1000	1/1,568,022
$100	1/282,244
$5	1/7056
$1	1/588

Suppose you spend $1 to buy a bus pass that lets you go to 25 different restaurants in the chain and pick up entry forms. Find your expected value.

50. *The Hog Game* In the hog game, each player states the number of dice that he or she would like to roll. The player then rolls that many dice. If a 1 comes up on any die, the player's score is 0. Otherwise, the player's score is the sum of the numbers rolled.[†]

a. Find the expected value of the player's score when the player rolls one die.

b. Find the expected value of the player's score when the player rolls two dice.

c. Verify that the expected nonzero score of a single die is 4, so that if a player rolls n dice that do not result in a score of 0, the expected score is $4n$.

d. Verify that if a player rolls n dice, there are 5^n possible ways to get a nonzero score, and 6^n possible ways to roll

---

*Seligman, Daniel, "Ask Mr. Statistics," *Fortune*, July 24, 1995, pp. 170–171.
[†]Bohan, James, and John Shultz, "Revisiting and Extending the Hog Game," *The Mathematics Teacher*, Vol. 89, No. 9, Dec. 1996, pp. 728–733.

the dice. Explain why the expected value, E, of the player's score when the player rolls n dice is then

$$E = \frac{5^n(4n)}{6^n}.$$

51. *Football* After a team scores a touchdown, it can either attempt to kick an extra point or attempt a two-point conversion. During the 1999–2000 NFL season, two-point conversions were successful 37% of the time and the extra-point kicks were successful 94% of the time.*

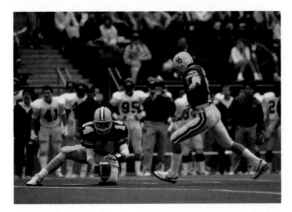

a. Calculate the expected value of each strategy.

b. Which strategy, over the long run, will maximize the number of points scored?

c. Using this information, should a team always only use one strategy? Explain.

■ CHAPTER SUMMARY

In this chapter we continued our study of probability by introducing some elementary principles of counting. Permutations were introduced to efficiently count the number of items that can occur when grouping nonrepetitive items such that order matters. Combinations were then introduced to help us determine the number of occurrences of nonrepetitive items when order does not matter. These concepts were then used to calculate various probabilities. In particular, binomial probability was defined and applied to a variety of situations. Finally, definitions of probability distribution and expected value were given. In the next chapter we will further our study of probability by introducing the field of statistics.

KEY TERMS

8.1 multiplication principle	8.2 combinations	8.5 random variable	fair game
factorial notation	8.4 Bernoulli trials	probability distribution	
permutations	binomial experiment	probability function	
distinguishable	binomial probability	histogram	
permutations	Pascal's triangle	expected value	

---

*Leonhardt, David, "In Football, 6 + 2 Often Equals 6," *The New York Times*, Sunday, Jan. 16, 2000, p. 4–2.

CHAPTER 8 REVIEW EXERCISES

1. In how many ways can 6 shuttle vans line up at the airport?

2. How many variations in first-, second-, and third-place finishes are possible in a 100-yard dash with 6 runners?

3. In how many ways can a sample of 3 oranges be taken from a bag of a dozen oranges?

4. If 2 of the oranges in Exercise 3 are rotten, in how many ways can the sample of 3 include

   a. 1 rotten orange?

   b. 2 rotten oranges?

   c. no rotten oranges?

   d. at most 2 rotten oranges?

5. In how many ways can 2 pictures, selected from a group of 5 different pictures, be arranged in a row on a wall?

6. In how many ways can the 5 pictures in Exercise 5 be arranged in a row if a certain one must be first?

7. In how many ways can the 5 pictures in Exercise 5 be arranged if 2 are landscapes and 3 are puppies and if

   a. like types must be kept together?

   b. landscapes and puppies are alternated?

8. In a Chinese restaurant the menu lists 8 items in column A and 6 items in column B.

   a. To order a dinner, the diner is told to select 3 items from column A and 2 from column B. How many dinners are possible?

   b. How many dinners are possible if the diner can select up to 3 from column A and up to 2 from column B? Assume at least one item must be included from either A or B.

9. A representative is to be selected from each of 3 departments in a small college. There are 7 people in the first department, 5 in the second department, and 4 in the third department.

   a. How many different groups of 3 representatives are possible?

   b. How many groups are possible if any number (at least 1) up to 3 representatives can form a group? (Each department is still restricted to at most one representative.)

10. Explain under what circumstances a permutation should be used in a probability problem, and under what circumstances a combination should be used.

11. Discuss under what circumstances the binomial probability formula should be used in a probability problem.

A basket contains 4 black, 2 blue, and 5 green balls. A sample of 3 balls is drawn. Find the probabilities that the sample contains the following.

12. All black balls

13. All blue balls

14. 2 black balls and 1 green ball

15. Exactly 2 black balls

16. Exactly 1 blue ball

17. 2 green balls and 1 blue ball

Suppose a family plans 6 children, and the probability that a particular child is a girl is 1/2. Find the probabilities that the 6-child family has the following children.

18. Exactly 3 girls

19. All girls

20. At least 4 girls

21. No more than 2 boys

Suppose 2 cards are drawn without replacement from an ordinary deck of 52. Find the probabilities of the following results.

22. Both cards are red.

23. Both cards are spades.

24. At least 1 card is a spade.

25. One is a face card and the other is not.

26. At least one is a face card.

27. At most one is a queen.

In Exercises 28 and 29, (a) give a probability distribution, (b) sketch its histogram, and (c) find the expected value.

28. A coin is tossed 3 times and the number of heads is recorded.

29. A pair of dice is rolled and the sum of the results for each roll is recorded.

In Exercises 30 and 31, give the probability that corresponds to the shaded region of each histogram.

30.

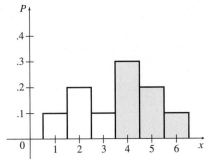

31.

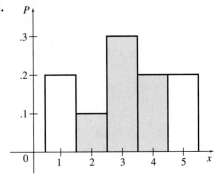

32. You pay $6 to play in a game where you will roll a die, with payoffs as follows: $8 for a 6, $7 for a 5, and $4 for any other results. What are your expected winnings? Is the game fair?

33. Find the expected number of girls in a family of 5 children.

34. Three cards are drawn from a standard deck of 52 cards.

 a. What is the expected number of aces?

 b. What is the expected number of clubs?

35. Suppose someone offers to pay you $100 if you draw 3 cards from a standard deck of 52 cards and all the cards are clubs. What should you pay for the chance to win if it is a fair game?

36. Six students will decide which of them are on a committee by flipping a coin. Each student flips the coin, and is on the committee if he or she gets a head. What is the probability that someone is on the committee, but not all 6 students?

 37. In this exercise we study the connection between sets (from Chapter 7) and combinations (from Chapter 8).

 a. Given a set with n elements, what is the number of subsets of size 0? of size 1? of size 2? of size n?

 b. Using your answer from part a, give an expression for the total number of subsets of a set with n elements.

 c. Using your answer from part b and a result from Chapter 7, explain why the following equation must be true:

$$\binom{n}{0} + \binom{n}{1} + \binom{n}{2} + \cdots + \binom{n}{n} = 2^n.$$

 d. Verify the equation in part c for $n = 4$ and $n = 5$.

 e. Explain what the equation in part c tells you about Pascal's triangle.

In the following exercise, find the digit (0 through 9) that belongs in each box. This exercise is from the 1990 University Entrance Center Examination, given in Japan to all applicants for public universities.*

 38. The numbers 1 through 9 are written individually on nine cards. Choose three cards from the nine, letting x, y, and z denote the numbers of the cards arranged in increasing order.

 a. There are □□ such x, y, and z combinations.

---

*"Japanese University Entrance Examination Problems in Mathematics," by Ling-Erl Eileen T. Wu, ed., Mathematical Association of America, 1993, p. 5.

b. The probability of having x, y and z all even is $\dfrac{\Box}{\Box\Box}$.

c. The probability of having x, y and z be consecutive numbers is $\dfrac{\Box}{\Box\Box}$.

d. The probability of having $x = 4$ is $\dfrac{\Box}{\Box\Box}$.

e. Possible values of x range from $\Box$ to $\Box$. If k is an integer such that $\Box \le k \le \Box$, the probability that $x = k$ is $\dfrac{(\Box - k)(\Box - k)}{\Box\Box\Box}$. The expected value of x is $\dfrac{\Box}{\Box}$.

Applications

BUSINESS AND ECONOMICS

Quality Control *A certain machine that is used to manufacture screws produces a defect rate of .01. A random sample of 20 screws is selected. Find the probabilities that the sample contains the following.*

39. Exactly 4 defective screws

40. Exactly 3 defective screws

41. No more than 4 defective screws

42. Set up the probability that the sample has 12 or more defective screws. (Do not evaluate.)

43. *Land Development* A developer can buy a piece of property that will produce a profit of $16,000 with probability .7, or a loss of $9000 with probability .3. What is the expected profit?

Exercises 44 and 45 are taken from actuarial examinations given by the Society of Actuaries.*

44. *Product Success* A company is considering the introduction of a new product that is believed to have probability .5 of being successful and probability .5 of being unsuccessful. Successful products pass quality control 80% of the time. Unsuccessful products pass quality control 25% of the time. If the product is successful, the net profit to the company will be $40 million; if unsuccessful, the net loss will be $15 million. Determine the expected net profit if the product passes quality control.

   a. $23 million   **b.** $24 million   **c.** $25 million

   d. $26 million   **e.** $27 million

45. *Sampling Fruit* A merchant buys boxes of fruit from a grower and sells them. Each box of fruit is either Good or Bad. A Good box contains 80% excellent fruit and will earn $200 profit on the retail market. A Bad box contains 30% excellent fruit and will produce a loss of $1000. The a priori probability of receiving a Good box of fruit is .9. Before

the merchant decides to put the box on the market, he can sample one piece of fruit to test whether it is excellent. Based on that sample, he has the option of rejecting the box without paying for it. Determine the expected value of the right to sample. (*Hint:* If the merchant samples the fruit, what are the probabilities of accepting a Good box, accepting a Bad box, and not accepting the box? What are these probabilities if he does not sample the fruit?)

   a. 0   **b.** $16   **c.** $34   **d.** $72   **e.** $80

46. *Overbooking Flights* The March 1982 issue of *Mathematics Teacher* included "Overbooking Airline Flights," an article by Joe Dan Austin. In this article, Austin developed a model for the expected income for an airline flight. With appropriate assumptions, the probability that exactly x of n people with reservations show up at the airport to buy a ticket is given by the binomial probability formula. Assume the following: 6 reservations have been accepted for 3 seats, $p = .6$ is the probability that a person with a reservation will show up, a ticket costs $100, and the airline must pay $100 to anyone with a reservation who does not get a ticket. Complete the following table.

Number Who Show Up (x)	0	1	2	3	4	5	6
Airline's Income							
$P(x)$							

a. Use the table to find $E(I)$, the expected airline income from the 3 seats.

b. Find $E(I)$ for $n = 3$, $n = 4$, and $n = 5$. Compare these answers with $E(I)$ for $n = 6$. For these values of n, how many reservations should the airline book for the 3 seats in order to maximize the expected revenue?

LIFE SCIENCES

47. *Pharmacology* In placebo-controlled trials of Prozac®, a drug that is prescribed to fight depression, 23% of the

---

*Course 130 Examination, Operations Research, Nov. 1989.

patients who were taking the drug experienced nausea, whereas 10% of the patients who were taking the placebo experienced nausea.*

a. If 50 patients who are taking Prozac® are selected, what is the probability that 10 or more will experience nausea?

b. Of the 50 patients in part a, what is the expected number of patients who will experience nausea?

c. If a second group of 50 patients receives a placebo, what is the probability that 10 or fewer will experience nausea?

d. If a patient from a study of 1000 people, who are equally divided into two groups (those taking a placebo and those taking Prozac®), is experiencing nausea, what is the probability that he/she is taking Prozac®?

e. Since .23 is more than twice as large as .10, do you think that people who take Prozac® are more likely to experience nausea than those who take a placebo? Explain.

SOCIAL SCIENCES

48. *Education* In Exercise 38 of Section 8.3, Probability Applications of Counting Principles, we saw that a school in Bangkok requires that students take an entrance examination. After the examination, 5 students are randomly drawn from each group of 40 for automatic acceptance into the school regardless of their performance on the examination. The drawing consists of placing 35 red and 5 green pieces of paper into a box. If the lottery is changed so that each student picks a piece of paper from the box and then returns the piece of paper to the box, find the probability that exactly 5 of the 40 students will choose a green piece of paper.†

GENERAL INTEREST

In Exercises 49–52, **(a)** *give a probability distribution,* **(b)** *sketch its histogram, and* **(c)** *find the expected value.*

49. *Candy* According to officials of Mars, the makers of M&M Plain Chocolate Candies, 20% of the candies in each bag are red.‡ Four candies are selected from a bag and the number of red candies is recorded.

50. *Women Athletes* In 1992, the Big 10 collegiate sports conference moved to have women compose at least 40% of its athletes within 5 years.§ Suppose they exactly achieve the 40% figure, and that 5 athletes are picked at random from Big 10 universities. The number of women is recorded.

51. *Race* In the mathematics honors society at a college, 2 of the 8 members are African American. Three members are

selected at random to be interviewed by the student newspaper, and the number of African Americans is noted.

52. *Homework* In a small class of 10 students, 3 did not do their homework. The professor selects half of the class to present solutions to homework problems on the board, and records how many of those selected did not do their homework.

53. *Lottery* A lottery has a first prize of $5000, two second prizes of $1000 each, and two $100 third prizes. A total of 10,000 tickets is sold, at $1 each. Find the expected winnings of a person buying 1 ticket.

54. *Contests* At one time, game boards for a United Airlines contest could be obtained by sending a self-addressed, stamped envelope to a certain address. The prize was a ticket for any city to which United flies. Assume that the value of the ticket was $1000 (we might as well go first-class), and that the probability that a particular game board would win was 1/4000. If the stamps to enter the contest cost 32¢ and envelopes cost 1¢ each, find the expected winnings for a person ordering 1 game board. (Notice that 2 stamps and envelopes were required to enter.)

55. *Lottery* New York has a lottery game called Quick Draw, in which the player can pick anywhere from 1 up to 10 numbers from 1 to 80. The computer then picks 20 numbers, and how much you win is based on how many of your numbers match the computer's. For simplicity, we will only consider the two cases in which you pick 4 or 5 numbers. The payoffs for each dollar that you bet are given in the table below.

| | **How many numbers match the computer's numbers** | | | | | |
	0	**1**	**2**	**3**	**4**	**5**
You pick 4	0	0	1	5	55	
You pick 5	0	0	0	2	20	300

a. According to the Quick Draw playing card, the "Overall Chances of Winning" when you pick 4 are "1:3.86," while the chances when you pick 5 are "1:10.34." Verify these figures.

b. Find the expected value when you pick 4 and when you pick 5, betting $1 each time.

c. Based on your results from parts a and b, are you better off picking 4 numbers or picking 5? Explain your reasoning.

56. *Murphy's Law* Robert Matthews wrote an article about Murphy's Law, which says that if something can go wrong,

---

*Advertisement in *The New England Journal of Medicine,* Vol. 338, No. 9, Feb. 26, 1998, for Prozac®, developed and marketed by Eli Lilly and Company.

†"Media Clips," *Mathematics Teacher,* Vol. 92, No. 8, Nov. 1999. Copyright 1999. Used with permission from the National Council of Teachers of Mathematics. All rights reserved.

‡*NCTM News Bulletin,* Feb. 1995, p. 5.

§*Chicago Tribune,* Apr. 28, 1993, p. 19.

it will.* He considers Murphy's Law of Odd Socks, which says that if an odd sock can be created it will be, in a drawer of 10 loose pairs of socks.

a. Find the probability of getting a matching pair when the following numbers of socks are selected at random from the drawer.
   i. 5 socks   **ii.** 6 socks

b. Matthews says that it is necessary to rummage through 30% of the socks to get a matching pair. Using your answers from part a, explain precisely what he means by that.

c. Matthews claims that if you lose 6 socks at random from the drawer, then it is 100 times more likely that you will be left with the worst possible outcome— 6 odd socks—than with a drawer free of odd socks. Verify this calculation by finding the probability that you will be left with 6 odd socks and the probability that you will have a drawer free of odd socks.

57. *Baseball* The number of runs scored in 16,456 half-innings of the 1986 National League Baseball season was analyzed by Hal Stern. Use the following table to answer the following questions.[†]

a. What is the probability that a given team scored 5 or more runs in any given one-half inning during the 1986 season?

b. What is the probability that a given team scored fewer than two runs in any given one-half inning of the 1986 season?

c. What is the expected number of runs that a team scored during any given half-inning of the 1986 season? Interpret this number.

Runs	Frequency	Probability
0	12,087	.7345
1	2451	.1489
2	1075	.0653
3	504	.0306
4	225	.0137
5	66	.0040
6	29	.0018
7	12	.0007
8	5	.0003
9	2	.0001

---

*Matthews, Robert, "Why Does Toast Always Land Butter-Side Down?" *Sunday Telegraph,* March 17, 1996, p. 4.
[†]J. Laurie Snell's report of Hal Stern's analysis in *Chance News 7.05,* Apr. 27–May 26, 1998.

EXTENDED APPLICATION: Optimal Inventory for a Service Truck

For many different items it is difficult or impossible to take the item to a central repair facility when service is required. Washing machines, large television sets, office copiers, and computers are only a few examples of such items. Service for items of this type is commonly performed by sending a repair person to the item, with the person driving to the location in a truck containing various parts that might be required in repairing the item. Ideally, the truck should contain all the parts that might be required. However, most parts would be needed only infrequently, so that inventory costs for the parts would be high.

An optimum policy for deciding on which parts to stock on a truck would require that the probability of not being able to repair an item without a trip back to the warehouse for needed parts be as low as possible, consistent with minimum inventory costs. An analysis similar to the one below was developed at the Xerox Corporation.*

To set up a mathematical model for deciding on the optimum truck-stocking policy, let us assume that a broken machine might require one of 5 different parts (we could assume any number of different parts—we use 5 to simplify the notation). Suppose also that the probability that a particular machine requires part 1 is p_1; that it requires part 2 is p_2; and so on. Assume also that failures of different part types are independent, and that at most one part of each type is used on a given job.

Suppose that, on the average, a repair person makes N service calls per time period. If the repair person is unable to make a repair because at least one of the parts is unavailable, there is a penalty cost, L, corresponding to wasted time for the repair person, an extra trip to the parts depot, customer unhappiness, and so on. For each of the parts carried on the truck, an average inventory cost is incurred. Let H_i be the average inventory cost for part i, where $1 \le i \le 5$.

Let M_1 represent a policy of carrying only part 1 on the repair truck, M_{24} represent a policy of carrying only parts 2 and 4,

with M_{12345} and M_0 representing policies of carrying all parts and no parts, respectively.

For policy M_{35}, carrying parts 3 and 5 only, the expected cost per time period per repair person, written $C(M_{35})$, is

$$C(M_{35}) = (H_3 + H_5) + NL[1 - (1 - p_1)(1 - p_2)(1 - p_4)].$$

(The expression in brackets represents the probability of needing at least one of the parts not carried, 1, 2, or 4 here.) As further examples,

$$C(M_{125}) = (H_1 + H_2 + H_5) + NL[1 - (1 - p_3)(1 - p_4)],$$

while

$$\begin{aligned} C(M_{12345}) &= (H_1 + H_2 + H_3 + H_4 + H_5) + NL[1 - 1] \\ &= H_1 + H_2 + H_3 + H_4 + H_5, \end{aligned}$$

and

$$C(M_0) = NL[1 - (1 - p_1)(1 - p_2)(1 - p_3)(1 - p_4)(1 - p_5)].$$

To find the best policy, evaluate $C(M_0)$, $C(M_1)$, ..., $C(M_{12345})$, and choose the smallest result. (A general solution method is in the *Management Science* paper.)

Example

Suppose that for a particular item, only 3 possible parts might need to be replaced. By studying past records of failures of the item, and finding necessary inventory costs, suppose that the following values have been found.

p_1	p_2	p_3
.09	.24	.17

H_1	H_2	H_3
$15	$40	$9

Suppose $N = 3$ and L is $54. Then, as an example,

$$\begin{aligned} C(M_1) &= H_1 + NL[1 - (1 - p_2)(1 - p_3)] \\ &= 15 + 3(54)[1 - (1 - .24)(1 - .17)] \\ &= 15 + 3(54)[1 - (.76)(.83)] \\ &\approx 15 + 59.81 = 74.81. \end{aligned}$$

Thus, if policy M_1 is followed (carrying only part 1 on the truck), the expected cost per repair person per time period is $74.81. Also,

$$\begin{aligned} C(M_{23}) &= H_2 + H_3 + NL[1 - (1 - p_1)] \\ &= 40 + 9 + 3(54)(.09) = 63.58, \end{aligned}$$

so that M_{23} is a better policy than M_1. By finding the expected values for all other possible policies (see the exercises), the optimum policy may be chosen.

*Smith, Stephen, John Chambers, and Eli Shlifer, "Optimal Inventories Based on Job Completion Rate for Repairs Requiring Multiple Items," *Management Science,* Vol, 26, No. 8, Aug. 1980. © 1980 by The Institute of Management Sciences.

Exercises

1. Refer to the example and find each of the following.

   a. $C(M_0)$ **b.** $C(M_2)$ **c.** $C(M_3)$ **d.** $C(M_{12})$

   e. $C(M_{13})$ **f.** $C(M_{123})$

2. Which policy leads to the lowest expected cost?

3. In the example, $p_1 + p_2 + p_3 = .09 + .24 + .17 = .50$. Why is it not necessary that the probabilities add up to 1?

4. Suppose an item to be repaired might need one of n different parts. How many different policies would then need to be evaluated?

Statistics

To understand the economics of large-scale farming, analysts look at historical data on the farming industry. In an exercise in Section 1 you will calculate basic descriptive statistics for U.S. wheat prices and production levels over a recent decade. Later sections in this chapter develop more sophisticated techniques for extracting useful information from this kind of data.

Statistics is a branch of mathematics that deals with the collection and summarization of data. Methods of statistical analysis make it possible for us to draw conclusions about a population based on data from a sample of the population. Statistical models have become increasingly useful in manufacturing, government, agriculture, medicine, and the social sciences, and in all types of research. In this chapter we give a brief introduction to some of the key topics from statistical theory.

9.1 FREQUENCY DISTRIBUTIONS; MEASURES OF CENTRAL TENDENCY

 THINK ABOUT IT How can the results of a survey of business executives on the number of college courses in management needed by a business major best be organized to provide useful information?

Frequency distributions can provide an answer to this question.

Often, a researcher wishes to learn something about a characteristic of a population, but because the population is very large or mobile, it is not possible to examine all of its elements. Instead, a limited sample drawn from the population is studied to determine the characteristics of the population. For these inferences to be correct, the sample chosen must be a **random sample.** Random samples are representative of the population because they are chosen so that every element of the population is equally likely to be selected. A hand dealt from a well-shuffled deck of cards is a random sample.

A random sample can be difficult to obtain in real life. For example, suppose you want to take a random sample of voters in your congressional district to see which candidate they prefer in the next election. If you do a telephone survey, you have a random sample of people who are at home to answer the telephone, underrepresenting those who work a lot of hours and are rarely home to answer the phone, or those who have an unlisted number, or those who cannot afford a telephone, or those who refuse to answer telephone surveys. Such people may have a different opinion than those you interview.

A famous example of an inaccurate poll was made by the *Literary Digest* in 1936. Their survey indicated that Alfred Landon would win the presidential election; in fact, Franklin Roosevelt won with 62% of the popular vote. The *Digest's* major error was mailing their surveys to a sample of those listed in telephone directories. During the Depression, many poor people did not have telephones, and the poor voted overwhelmingly for Roosevelt. Modern pollsters use sophisticated techniques to ensure that their sample is as random as possible.

Once a sample has been chosen and all data of interest are collected, the data must be organized so that conclusions may be more easily drawn. One method of organization is to group the data into intervals; equal intervals are usually chosen.

EXAMPLE 1 Business Executives

A survey asked a random sample of 30 business executives for their recommendations as to the number of college units in management that a business major should have. The results are shown below. Group the data into intervals and find the frequency of each interval.

3	25	22	16	0	9	14	8	34	21
15	12	9	3	8	15	20	12	28	19
17	16	23	19	12	14	29	13	24	18

Solution The highest number in the list is 34 and the lowest is 0; one convenient way to group the data is in intervals of size 5, starting with 0–4 and ending with 30–34. This gives an interval for each number in the list and results in seven equal intervals of a convenient size. Too many intervals of smaller size would not simplify the data enough, while too few intervals of larger size would conceal information that the data might provide. A rule of thumb is to use from six to fifteen intervals.

First tally the number of college units falling into each interval. Then total the tallies in each interval as in the table below. This table is an example of a **grouped frequency distribution.**

College Units	Tally	Frequency
0–4	|||	3
5–9	||||	4
10–14	⧸⧸⧸⧸ |	6
15–19	⧸⧸⧸⧸ |||	8
20–24	⧸⧸⧸⧸	5
25–29	|||	3
30–34	|	1
	Total:	30

The frequency distribution in Example 1 shows information about the data that might not have been noticed before. For example, the interval with the largest number of recommended units is 15–19, and 19 executives (more than half) recommended between 9 and 25 units. Also, the frequency in each interval increases rather evenly (up to 8) and then decreases at about the same pace. However, some information has been lost; for example, we no longer know how many executives recommended 12 units.

The information in a grouped frequency distribution can be displayed in a histogram similar to the histograms for probability distributions in the previous chapter. The intervals determine the widths of the bars; if equal intervals are used, all the bars have the same width. The heights of the bars are determined by the frequencies.

NOTE In this section, the heights of the histogram bars give the frequencies. The histograms in the previous chapter were for probability distributions, and so the heights gave the probabilities.

A **frequency polygon** is another form of graph that illustrates a grouped frequency distribution. The polygon is formed by joining consecutive midpoints of the tops of the histogram bars with straight line segments. The midpoints of the first and last bars are joined to endpoints on the horizontal axis where the next midpoint would appear.

EXAMPLE 2 Frequency Distributions

A grouped frequency distribution of college units was found in Example 1. Draw a histogram and a frequency polygon for this distribution.

Solution First draw a histogram, shown in red in Figure 1. To get a frequency polygon, connect consecutive midpoints of the tops of the bars. The frequency polygon is shown in blue.

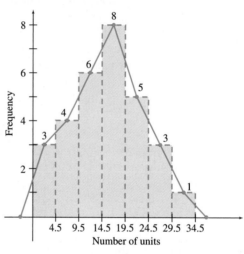

FIGURE 1

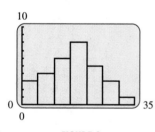

FIGURE 2

Many graphing calculators have the capability of drawing a histogram. Figure 2 shows the data of Example 1 drawn on a TI-83.

Mean The average value of a probability distribution is the expected value of the distribution. Three measures of central tendency, or "averages," are used with frequency distributions: the mean, the median, and the mode. The most important of these is the mean, which is similar to the expected value of a probability distribution. The **arithmetic mean** (the **mean**) of a set of numbers is the sum of the numbers, divided by the total number of numbers. Recall from Section 1.3 that we can write the sum of n numbers $x_1, x_2, x_3, \ldots, x_n$ in a compact way using *summation notation:*

$$x_1 + x_2 + x_3 + \cdots + x_n = \Sigma x.$$

The symbol $\bar{x}$ (read x-bar) is used to represent the mean of a sample.

MEAN

The mean of the n numbers $x_1, x_2, x_3, \ldots, x_n$ is

$$\bar{x} = \frac{\Sigma x}{n}.$$

EXAMPLE 3   Bankruptcy

The number of bankruptcy petitions (in thousands) filed in the United States in the fiscal years 1992–1997 are given in the table below.* Find the mean number of bankruptcy petitions filed annually during this period.

Year	Petitions Filed
1992	972
1993	875
1994	834
1995	927
1996	1179
1997	1404

Solution Let $x_1 = 972$, $x_2 = 875$, and so on. Here, $n = 6$, since there are 6 numbers.

$$\bar{x} = \frac{972 + 875 + 834 + 927 + 1179 + 1404}{6}$$

$$\bar{x} = \frac{6191}{6} \approx 1032$$

The mean number of bankruptcy petitions filed during the given years is about 1,032,000. Notice that this average is greater than four of the six values. This is due to the large number of petitions filed in 1997.

As another example, the mean response for the number of college units in management that a business major should have, based on the sample of 30 business executives described in Example 1, is $\bar{x} = (3 + 25 + 22 + \cdots + 18)/30 = 478/30 = 15.93$.

EXAMPLE 4   Mean for Frequency Distributions

Find the mean for the data shown in the following frequency distribution.

Value	Frequency	Value × Frequency
30	6	$30 \cdot 6 = 180$
32	9	$32 \cdot 9 = 288$
33	7	$33 \cdot 7 = 231$
37	12	$37 \cdot 12 = 444$
42	6	$42 \cdot 6 = 252$
Total:	40	Total: 1395

Solution The value 30 appears six times, 32 nine times, and so on. To find the mean, first multiply 30 by 6, 32 by 9, and so on.

*U.S. Bankruptcy Filings (Total), Department of Justice, www.usdoj.gov.

A new column, "Value $\times$ Frequency," has been added to the frequency distribution. Adding the products from this column gives a total of 1395. The total from the frequency column is 40. The mean is

$$\bar{x} = \frac{1395}{40} = 34.875.$$

The mean of grouped data is found in a similar way. For grouped data, intervals are used, rather than single values. To calculate the mean, it is assumed that all these values are located at the midpoint of the interval. The letter x is used to represent the midpoints and f represents the frequencies, as shown in the next example.

EXAMPLE 5 Business Executives

Listed below is the grouped frequency distribution for the 30 business executives described in Example 1. Find the mean from the grouped frequency distribution.

Interval	Midpoint, x	Frequency, f	Product, xf
0–4	2	3	6
5–9	7	4	28
10–14	12	6	72
15–19	17	8	136
20–24	22	5	110
25–29	27	3	81
30–34	32	1	32
		Total: 30	Total: 465

Solution A column for the midpoint of each interval has been added. The numbers in this column are found by adding the endpoints of each interval and dividing by 2. For the interval 0–4, the midpoint is $(0 + 4)/2 = 2$. The numbers in the product column on the right are found by multiplying each frequency by its corresponding midpoint. Finally, we divide the total of the product column by the total of the frequency column to get

$$\bar{x} = \frac{465}{30} = 15.5.$$

Notice that this mean is slightly different from the earlier mean of 15.93. The reason for this difference is that we have acted as if each piece of data is at the midpoint, which is not true here, and is not true in most cases. Information is always lost when the data are grouped. It is more accurate to use the original data, rather than the grouped frequency, when calculating the mean, but the original data might not be available. Furthermore, the mean based upon the grouped data is typically not too far from the mean based upon the original data, and there may be situations in which the extra accuracy is not worth the extra effort.

> **NOTE** **1.** The midpoint of the intervals in a grouped frequency distribution may be values that the data cannot take on. For example, if we grouped the data for the 30 business executives into the intervals 0–5, 6–11, 12–17, 18–23, 24–29, and 30–35, the midpoints would be 2.5, 8.5, 14.5, 20.5, 26.5, and 32.5, even though all the data are whole numbers.
>
> **2.** If we used different intervals in Example 5, the mean would come out to be a slightly different number. Verify that with the intervals 0–5, 6–11, 12–17, 18–23, 24–29, and 30–35, the mean in Example 5 is 16.1.

The formula for the mean of a grouped frequency distribution is given below.

MEAN OF A GROUPED DISTRIBUTION

The mean of a distribution, where x represents the midpoints, f the frequencies, and $n = \Sigma f$, is

$$\bar{x} = \frac{\Sigma xf}{n}.$$

The mean of a random sample is a random variable, and for this reason it is sometimes called the **sample mean.** The sample mean is a random variable because it assigns a number to the experiment of taking a random sample. If a different random sample were taken, the mean would probably have a different value, with some values more probable than others. If another set of 30 business executives were selected in Example 1, the mean number of college units in management recommended for a business major might be 13.22 or 17.69. It is unlikely that the mean would be as small as 1.21 or as large as 32.75, although these values are remotely possible.

We saw in Section 8.5 how to calculate the expected value of a random variable when we know its probability distribution. The expected value is sometimes called the **population mean,** denoted by the Greek letter μ. In other words,

$$E(x) = \mu.$$

Furthermore, it can be shown that the expected value of $\bar{x}$ is also equal to μ; that is,

$$E(\bar{x}) = \mu.$$

For instance, consider again the 30 business executives in Example 1. We found that $\bar{x} = 15.93$, but the value of μ, the average for all possible business executives, is unknown. If a good estimate of μ were needed, the best guess (based on this data) is 15.93.

Median   Asked by a reporter to give the average height of the players on his team, a Little League coach lined up his 15 players by increasing height. He picked the player in the middle and pronounced that player to be of average height. This kind of average, called the **median,** is defined as the middle entry in a set of data arranged in either increasing or decreasing order. If there is an even number of entries, the median is defined to be the mean of the two center entries.

Odd Number of Entries	Even Number of Entries
8	2
7	3
Median = 4	4
3	7 Median = $\frac{4+7}{2}$ = 5.5
1	9
	12

EXAMPLE 6 Median

Find the median for each of the following lists of numbers.

(a) 11, 12, 17, 20, 23, 28, 29

Solution The median is the middle number; in this case, 20. (Note that the numbers are already arranged in numerical order.) In this list, three numbers are smaller than 20 and three are larger.

(b) 15, 13, 7, 11, 19, 30, 39, 5, 10

Solution First arrange the numbers in numerical order, from smallest to largest.

$$5, 7, 10, 11, 13, 15, 19, 30, 39$$

The middle number, or median, can now be determined; it is 13.

(c) 47, 59, 32, 81, 74, 153

Solution Write the numbers in numerical order.

$$32, 47, 59, 74, 81, 153$$

There are six numbers here; the median is the mean of the two middle numbers.

$$\text{Median} = \frac{59 + 74}{2} = \frac{133}{2} = 66\frac{1}{2}$$

Both the mean and the median are examples of a **statistic,** which is simply a number that gives information about a sample. In some situations, the median gives a truer representation or typical element of the data than the mean. For example, suppose in an office there are 10 salespersons, 4 secretaries, the sales manager, and Ms. Daly, who owns the business. Their annual salaries are as follows: secretaries, $15,000 each; salespersons, $25,000 each; manager, $35,000; and owner, $200,000. The mean salary is

$$\bar{x} = \frac{(15,000)4 + (25,000)10 + 35,000 + 200,000}{16} = \$34,062.50.$$

However, since 14 people earn less than $34,062.50 and only 2 earn more, this does not seem very representative. The median salary is found by ranking the salaries by size: $15,000, $15,000, $15,000, $15,000, $25,000, $25,000, . . . , $200,000. Since there are 16 salaries (an even number) in the list, the mean of the eighth and ninth entries will give the value of the median. The eighth and ninth entries are both $25,000, so the median is $25,000. In this example, the median gives a truer representative element than the mean.

Mode Sue's scores on ten class quizzes include one 7, two 8's, six 9's, and one 10. She claims that her average grade on quizzes is 9, because most of her scores are 9's. This kind of "average," found by selecting the most frequent entry, is called the **mode.**

EXAMPLE 7 Mode

Find the mode for each list of numbers.

(a) 57, 38, 55, 55, 80, 87, 98

Solution The number 55 occurs more often than any other, so it is the mode. It is not necessary to place the numbers in numerical order when looking for the mode.

(b) 182, 185, 183, 185, 187, 187, 189

Solution Both 185 and 187 occur twice. This list has *two* modes.

(c) 10,708, 11,519, 10,972, 17,546, 13,905, 12,182

Solution No number occurs more than once. This list has no mode.

The mode has the advantages of being easily found and not being influenced by data that are very large or very small compared to the rest of the data. It is often used in samples where the data to be "averaged" are not numerical. A major disadvantage of the mode is that we cannot always locate exactly one mode for a set of values. There can be more than one mode, in the case of ties, or there can be no mode if all entries occur with the same frequency.

The mean is the most commonly used measure of central tendency. Its advantages are that it is easy to compute, it takes all the data into consideration, and it is reliable—that is, repeated samples are likely to give very similar means. A disadvantage of the mean is that it is influenced by extreme values, as illustrated in the salary example above.

The median can be easy to compute and is influenced very little by extremes. Like the mode, the median can be found in situations where the data are not numerical. For example, in a taste test, people are asked to rank five soft drinks from the one they like best to the one they like least. The combined rankings then produce an ordered sample, from which the median can be identified. A disadvantage of the median is the need to rank the data in order; this can be difficult when the number of items is large.

EXAMPLE 8   Seed Storage

Seeds that are dried, placed in an airtight container, and stored in a cool, dry place remain ready to be planted for a long time. The following table gives the amount of time that each type of seed can be stored and still remain viable for planting.*

Vegetable	Years
Beans	3
Cabbage	4
Carrots	1
Cauliflower	4
Corn	2
Cucumbers	5
Melons	4
Peppers	2
Pumpkin	4
Tomatoes	3

Find the mean, median, and mode of the information in the table.

Solution

Method 1: Calculating by Hand

The mean amount of time that the seeds can be stored is

$$\bar{x} = \frac{3 + 4 + 1 + 4 + 2 + 5 + 4 + 2 + 4 + 3}{10} = 3.2 \text{ years}.$$

After the numbers are arranged in order from smallest to largest, the middle number, or median, is found; it is 3.5.

The number 4 occurs more often than any other, so it is the mode.

Method 2: Graphing Calculator

Most scientific calculators have some statistical capability and can calculate the mean of a set of data; graphing calculators can often calculate the median as well. For example, Figure 3 shows the mean and the median for the data above calculated on a TI-83, where the data was stored in the list L_1. This calculator does not include a command for finding the mode.

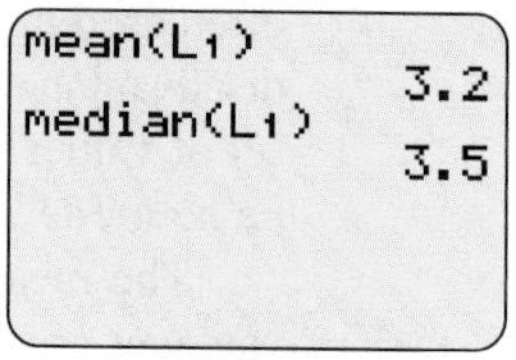

```
mean(L₁)
               3.2
median(L₁)
               3.5
```

FIGURE 3

Method 3: Spreadsheet

Using Microsoft Excel, place the data in cells A1 through A10. To find the mean of this data, type "=mean(A1..A10)" in cell A11, or any other unused cell, and then press Enter. The result of 3.2 will appear in cell A11. To find the median of this data, type "=median(A1..A10)" in cell A12, or any other unused cell, and press Enter. The result of 3.5 will appear in cell A12. To find the mode of this data, type "=mode(A1..A10)" in cell A13, or any other unused cell, and press Enter. The result of 4 will appear in cell A13.

*The Handy Science Answer Book, 2nd ed., The Carnegie Library of Pittsburgh, p. 247.

9.1 EXERCISES

For Exercises 1–4, do the following:
a. *Group the data as indicated.*
b. *Prepare a frequency distribution with a column for intervals and frequencies.*
c. *Construct a histogram.*
d. *Construct a frequency polygon.*

1. Use six intervals, starting with 0–24.

74	133	4	127	20	30
103	27	139	118	138	121
149	132	64	141	130	76
42	50	95	56	65	104
4	140	12	88	119	64

2. Use seven intervals, starting with 30–39.

79	71	78	87	69	50	63	51
60	46	65	65	56	88	94	56
74	63	87	62	84	76	82	67
59	66	57	81	93	93	54	88
55	69	78	63	63	48	89	81
98	42	91	66	60	70	64	70
61	75	82	65	68	39	77	81
67	62	73	49	51	76	94	54
83	71	94	45	73	95	72	66
71	77	48	51	54	57	69	87

3. Repeat Exercise 1 using eight intervals, starting with 0–19.

4. Repeat Exercise 2 using six intervals, starting with 39–48.

5. How does a frequency polygon differ from a histogram?

6. Discuss the advantages and disadvantages of the mean as a measure of central tendency.

Find the mean for each list of numbers. Round to the nearest tenth.

7. 8; 10; 16; 21; 25

8. 44; 41; 25; 36; 67; 51

9. 21,900; 22,850; 24,930; 29,710; 28,340; 40,000

10. 38,500; 39,720; 42,183; 21,982; 43,250

11. 9.4; 11.3; 10.5; 7.4; 9.1; 8.4; 9.7; 5.2; 1.1; 4.7

12. 30.1; 42.8; 91.6; 51.2; 88.3; 21.9; 43.7; 51.2

Find the mean for each of the following. Round to the nearest tenth.

13.

Value	Frequency
3	4
5	2
9	1
12	3

14.

Value	Frequency
9	3
12	5
15	1
18	1

Find the median for each of the following lists of numbers.

15. 12, 18, 32, 51, 58, 92, 106

16. 596, 604, 612, 683, 719

17. 100, 114, 125, 135, 150, 172

18. 1072, 1068, 1093, 1042, 1056, 1005, 1009

19. 28.4, 9.1, 3.4, 27.6, 59.8, 32.1, 47.6, 29.8

20. .6, .4, .9, 1.2, .3, 4.1, 2.2, .4, .7, .1

Use a graphing calculator or spreadsheet to calculate the mean and median for the data in each of the following exercises.

21. Exercise 1

22. Exercise 2

Find the mode or modes for each of the following lists of numbers.

23. 4, 9, 8, 6, 9, 2, 1, 3

24. 21, 32, 46, 32, 49, 32, 49

25. 74, 68, 68, 68, 75, 75, 74, 74, 70

26. 158, 162, 165, 162, 165, 157, 163

27. 6.8, 6.3, 6.3, 6.9, 6.7, 6.4, 6.1, 6.0

28. 12.75, 18.32, 19.41, 12.75, 18.30, 19.45, 18.33

29. When is the median the most appropriate measure of central tendency?

30. Under what circumstances would the mode be an appropriate measure of central tendency?

For grouped data, the modal class is the interval containing the most data values. Give the mean and modal class for each of the following collections of grouped data.

31. Use the distribution in Exercise 1.

32. Use the distribution in Exercise 2.

33. To predict the outcome of the next congressional election, you take a survey of your friends. Is this a random sample of the voters in your congressional district? Explain why or why not.

Applications

BUSINESS AND ECONOMICS

Wheat Production *U.S. wheat prices and production figures for a recent decade are given below.* *

Year	Price ($ per bushel)	Production (millions of bushels)
1989	3.72	2037
1990	2.61	2730
1991	3.00	1980
1992	3.24	2467
1993	3.26	2396
1994	3.45	2321
1995	4.55	2183
1996	4.30	2285
1997	3.38	2482
1998	2.65	2550

Find the mean and median for each of the following.

34. Price per bushel of wheat

35. Wheat production

36. *Salaries* The total pay (in thousands of dollars) for the 15 highest paid CEOs in 1997 is given in the following table.[†]

Person, Company	Total Pay
Sanford Weill, Travelers Group	230,725
Roberto Goizueta, Coca-Cola	111,832
Richard Scrushy, Health South	106,790
Ray Irani, Occidental Petroleum	101,505
Eugene Isenberg, Nabors Industries	84,547
Joseph Costelo, Cadence Design Systems	66,842
Andrew Grove, Intel	52,214
Charles McCall, HBO Co.	51,409
Philip Purcell, Morgan Stanley Dean Witter	50,807
Robert Shapiro, Monsanto	49,326
John Welch, General Electric	39,894
Harvey Golub, American Express	33,457
William Schoen, Health-Management Ass.	30,945
Charles Heimbold, Bristol-Myers Squibb	29,211
Shailesh Mehta, Providian Financial	28,365

a. Find the mean compensation for this group of people.

b. Find the median compensation for this group of people.

---

*The World Almanac and Book of Facts 2000, pp. 141–142.

[†]Wright, John W., *The American Almanac of Jobs and Salaries*, 2000–2001 ed., p. 321.

37. *Household Income* The total household income for full-time African American workers making under $100,000 in 1997 is given in the table below.*

Income Range	Midpoint Salary	Frequency (in thousands)
Under $5,000	$2500	923
$5,000–$9,999	$7500	1746
$10,000–$14,999	$12,500	1310
$15,000–$24,999	$20,000	2233
$25,000–$34,999	$30,000	1771
$35,000–$49,999	$42,500	1859
$50,000–$74,999	$62,500	1634
$75,000–$99,999	$87,500	574

Use this table to estimate the mean household income for full-time African American workers in 1997.

38. *Household Income* The total household income for full-time white American workers making under $100,000 in 1997 is given in the following table.*

Income Range	Midpoint Salary	Frequency (in thousands)
Under $5,000	$2500	2411
$5,000–$9,999	$7500	5769
$10,000–$14,999	$12,500	6716
$15,000–$24,999	$20,000	12,571
$25,000–$34,999	$30,000	11,366
$35,000–$49,999	$42,500	14,207
$50,000–$74,999	$62,500	16,188
$75,000–$99,999	$87,500	8180

a. Use this table to estimate the mean household income for full-time white American workers in 1997.

b. Compare this estimate with the estimate found in Exercise 37. Does this provide some evidence that full-time white American workers have higher household earnings than full-time African American workers?

LIFE SCIENCES

39. *Pandas* The size of the home ranges (in square kilometers) of several pandas were surveyed over a year's time, with the following results.

Home Range	Frequency
.1–.5	11
.6–1.0	12
1.1–1.5	7
1.6–2.0	6
2.1–2.5	2
2.6–3.0	1
3.1–3.5	1

Sketch a histogram and frequency polygon for the data.

40. *Blood Types* The number of recognized blood types varies by species, as indicated by the following table.[†] Find the mean, median, and mode of this data.

Animal	Number of Blood Types
Pig	16
Cow	12
Chicken	11
Horse	9
Human	8
Sheep	7
Dog	7
Rhesus Monkey	6
Mink	5
Rabbit	5
Mouse	4
Rat	4
Cat	2

SOCIAL SCIENCES

41. *Population* The histogram below shows estimates of the percent of the U.S. population in each age group in the year 2000.[‡] What percent of the population is estimated to be in each of the following age groups?

a. 10–19   **b.** 60–69

c. What age range has the largest percent of the population?

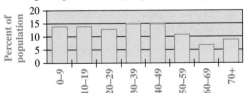

42. *Population* The histogram on the following page shows estimates of the percent of the U.S. population in each age

---

Time Almanac 2000, Time Inc., p. 829.

[†]*The Handy Science Answer Book*, Carnegie Library of Pittsburgh, Penn., p. 264.

[‡]U.S. Census Bureau, Jan. 13, 2000.

group in the year 2025.* What percent of the population is estimated to be in each of the following age groups then?

a. 20–29   **b.** 70+

c. What age group will have the smallest percent of the population?

d. Compare the histogram in Exercise 41 with the histogram below. What seems to be true of the U.S. population?

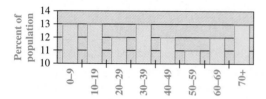

43. *Weddings* The number of weddings in each month of 1996 is given in the following table.[†]

Month	Number (in thousands)
January	110
February	155
March	118
April	172
May	241
June	242
July	235
August	239
September	225
October	231
November	171
December	184

a. Calculate the mean and median for this data.

b. Which month is closest to the mean?

GENERAL INTEREST

44. *Temperature* The following chart gives the number of days in June and July of recent years in which the temperature reached 90 degrees or higher in New York's Central Park.[‡]

a. Prepare a frequency distribution with a column for intervals and frequencies. Use six intervals, starting with 0–4.

b. Sketch a histogram and a frequency polygon, using the intervals in part a.

c. Find the mean for the original data.

d. Find the mean using the grouped data from part a.

e. Explain why your answers to parts c and d are different.

f. Find the median and the mode for the original data.

Year	Days	Year	Days	Year	Days
1970	5	1980	12	1990	6
1971	11	1981	12	1991	21
1972	11	1982	11	1992	4
1973	8	1983	20	1993	25
1974	11	1984	7	1994	16
1975	3	1985	4	1995	14
1976	8	1986	8	1996	0
1977	11	1987	14		
1978	5	1988	21		
1979	7	1989	10		

45. *Temperature* The table below gives the average monthly temperatures in degrees Fahrenheit for a certain area.

Month	Maximum	Minimum
January	39	16
February	39	18
March	44	21
April	50	26
May	60	32
June	69	37
July	79	43
August	78	42
September	70	37
October	51	31
November	47	24
December	40	20

Find the mean and median for each of the following.

a. The maximum temperature

b. The minimum temperature

46. *Olympics* The number of nations participating in the winter Olympic games, from 1968 to 1998, is given below. Find the following measures for the data.

a. Mean

b. Median

c. Mode

Year	Nations Participating
1968	37
1972	37
1976	37
1980	37
1984	49
1988	57
1992	65
1994	67
1998	72

*U.S. Census Bureau, Jan. 13, 2000.
[†]*The Amazing Almanac*, Blackbirch Press, Inc., p. 245.
[‡]*The New York Times*, July 31, 1996, p. B4.

47. *Personal Wealth Washington Post* writer John Schwartz pointed out that if Microsoft Corp. cofounder Bill Gates, who, at the time, was reportedly worth $10 billion, lived in a town with 10,000 totally penniless people, the average personal wealth in the town would make it seem as if everyone were a millionaire.*

a. Verify Schwartz's statement.

b. What would be the median personal wealth in this town?

c. What would be the mode for the personal wealth in this town?

d. In this example, which average is most representative: the mean, the median, or the mode?

48. *Baseball Salaries* According to *USA Today*,[†] the total salary in 1999 for each player on the Chicago Cubs baseball team is given in the following table.

a. Find the mean, median, and mode of the salaries.

b. Which average best describes this data?

c. Why is there such a difference between the mean and the median?

Name	Salary (in U.S. dollars)
Sammy Sosa	8,400,000
Mark Clark	5,050,000
Lance Johnson	4,900,000
Mark Grace	4,225,000
Jeff Blauser	4,000,000
Kevin Tapani	4,000,000
Rod Beck	3,600,000
Henry Rodriguez	2,000,000
Scott Servais	1,645,000
Mike Morgan	1,300,000
Jose Hernandez	725,000
Terry Mulholland	550,000
Manny Alexander	425,000
Glenallen Hill	325,000
Matt Karchner	300,000
Terry Adams	295,000
Don Wengert	240,000
Tyler Houston	225,000
Dave Stevens	225,000
Felix Heredia	225,000
Jeremi Gonzalez	210,000
Brant Brown	180,000
Sandy Martinez	174,500
Rodney Myers	174,000
Gary Gaetti	170,000
Pedro Valdes	170,000
Kerry Wood	170,000

49. *SAT I: Reasoning Test* Given the following sequence of numbers[‡]

$$1, a, a^2, a^3, \ldots, a^n,$$

where n is a positive even integer, with the *additional assumption* that a is a positive number, the median is best described as

a. greater than $a^{n/2}$ **b.** smaller than $a^{n/2}$

c. equal to $a^{n/2}$

d. The relationship cannot be determined from the information given.

*Schwartz, J., "Mean Statistics: When Is Average Best?" *The Washington Post*, Jan. 11, 1995, p. H7.
†*USA Today* Sports, www.usatoday.com/sports/baseball/mlbfs12.htm.
‡Reprinted by permission of the College Entrance Examination Board, the copyright owner. Permission to reprint SAT materials does not constitute review or endorsement by Educational Testing Service of the College Board of this publication as a whole or of any other questions or testing information it may contain. This problem appeared, minus the *additional assumption*, on an SAT in 1996. Colin Rizzio, a high school student at the time, became an instant celebrity when he noticed that the additional assumption was needed to complete the problem. *New York Times*, Feb. 7, 1997, p. A1.

9.2 MEASURES OF VARIATION

THINK ABOUT IT   How can we tell when a manufacturing process is out of control?

To answer this question, we need to understand measures of variation, which tell us how much the numbers in a sample vary from the mean.

The mean gives a measure of central tendency of a list of numbers, but tells nothing about the *spread* of the numbers in the list. For example, look at the following three samples.

I	3	5	6	3	3
II	4	4	4	4	4
III	10	1	0	0	9

Each of these three samples has a mean of 4, and yet they are quite different; the amount of dispersion or variation within the samples is different. Therefore, in addition to a measure of central tendency, another kind of measure is needed that describes how much the numbers vary.

The largest number in sample I is 6, while the smallest is 3, a difference of 3. In sample II this difference is 0; in sample III, it is 10. The difference between the largest and smallest number in a sample is called the **range,** one example of a measure of variation. The range of sample I is 3, of sample II, 0, and of sample III, 10. The range has the advantage of being very easy to compute, and gives a rough estimate of the variation among the data in the sample. It depends only on the two extremes, however, and tells nothing about how the other data are distributed between the extremes.

EXAMPLE 1   Range

Find the range for each list of numbers.

(a) 12, 27, 6, 19, 38, 9, 42, 15

Solution The highest number here is 42; the lowest is 6. The range is the difference between these numbers, or $42 - 6 = 36$.

(b) 74, 112, 59, 88, 200, 73, 92, 175

Solution

$$\text{Range} = 200 - 59 = 141$$

The most useful measure of variation is the *standard deviation*. Before defining it, however, we must find the **deviations from the mean,** the differences found by subtracting the mean from each number in a sample.

EXAMPLE 2   Deviations from the Mean

Find the deviations from the mean for the numbers

$$32, 41, 47, 53, 57.$$

Solution Adding these numbers and dividing by 5 gives a mean of 46. To find the deviations from the mean, subtract 46 from each number in the list. For example, the first deviation from the mean is $32 - 46 = -14$; the last is $57 - 46 = 11$.

Number	Deviation from Mean
32	−14
41	−5
47	1
53	7
57	11

To check your work, find the sum of these deviations. It should always equal 0. (The answer is always 0 because the positive and negative numbers cancel each other.)

To find a measure of variation, we might be tempted to use the mean of the deviations. As mentioned above, however, this number is always 0, no matter how widely the data are dispersed. One way to solve this problem is to use absolute value and find the mean of the absolute values of the deviations from the mean. Absolute value is awkward to work with algebraically, and there is an alternative approach that provides better theoretical results. In this method, the way to get a list of positive numbers is to square each deviation and then find the mean. When finding the mean of the squared deviations, most statisticians prefer to divide by $n - 1$, rather than n. We will give the reason later in this section. For the data above, this gives

$$\frac{(-14)^2 + (-5)^2 + 1^2 + 7^2 + 11^2}{5 - 1} = \frac{196 + 25 + 1 + 49 + 121}{4}$$

$$= 98.$$

This number, 98, is called the **variance** of the distribution. Since it is found by averaging a list of squares, the variance of a sample is represented by s^2.

For a sample of n numbers $x_1, x_2, x_3, \ldots, x_n$, with mean $\bar{x}$, the variance is

$$s^2 = \frac{\Sigma(x - \bar{x})^2}{n - 1}.$$

The following shortcut formula for the variance can be derived algebraically from the formula above. This is left as an exercise.

VARIANCE

The variance of a sample of n numbers $x_1, x_2, x_3, \ldots, x_n$, with mean $\bar{x}$, is

$$s^2 = \frac{\Sigma x^2 - n\bar{x}^2}{n - 1}.$$

To find the variance, we squared the deviations from the mean, so the variance is in squared units. To return to the same units as the data, we use the *square root* of the variance, called the **standard deviation.**

STANDARD DEVIATION

The standard deviation of the n numbers $x_1, x_2, x_3, \ldots, x_n$, with mean $\bar{x}$, is

$$s = \sqrt{\frac{\Sigma x^2 - n\bar{x}^2}{n - 1}}.$$

As its name indicates, the standard deviation is the most commonly used measure of variation. The standard deviation is a measure of the variation from the mean. The size of the standard deviation tells us something about how spread out the data are from the mean.

EXAMPLE 3 Standard Deviation

Find the standard deviation of the numbers

$$7, 9, 18, 22, 27, 29\ 32, 40.$$

Solution

Method 1: Calculating by Hand

The mean of the numbers is

$$\frac{7 + 9 + 18 + 22 + 27 + 29 + 32 + 40}{8} = 23.$$

Arrange the work in columns, as shown in the table.

Number	Square of the Number
7	49
9	81
18	324
22	484
27	729
29	841
32	1024
40	1600
Total:	5132

The total of the second column gives $\Sigma x^2 = 5132$. Now find the variance. The variance is

$$s^2 = \frac{\Sigma x^2 - n\bar{x}^2}{n - 1}$$

$$= \frac{5132 - 8(23)^2}{8 - 1}$$

$$\approx 128.6,$$

rounded, and the standard deviation is

$$\sqrt{128.57} \approx 11.3$$

Method 2: Graphing Calculator

The data are entered into the L_5 list on a TI-83 calculator. Figure 4 shows how the variance and standard deviation are then calculated. Figure 5 shows an alternative method, going through the STAT menu, which calculates the mean, the standard deviation using both $n - 1$ and n, and other statistics.

```
variance(L5)
         128.5714286
stdDev(L5)
         11.33893419
■
```

FIGURE 4

```
1-Var Stats
 x̄=23
 Σx=184
 Σx²=5132
 Sx=11.33893419
 σx=10.60660172
↓n=8
■
```

FIGURE 5

Method 3: Spreadsheet | The data are entered in cells A1 through A8. Then, in cell A9, type "=VAR(A1..A8)" and press Enter. The standard deviation can be calculated by either taking the square root of cell A9 or by typing "=STDEV(A1..A8)" in cell A10 and pressing Enter.

CAUTION One must be careful to divide by $n - 1$, not n, when calculating the standard deviation of a sample. Many calculators are equipped with statistical keys that compute the variance and standard deviation. Some of these calculators use $n - 1$ and others use n for these computations; some may have keys for both. Check your calculator's instruction book before using a statistical calculator for the exercises.

One way to interpret the standard deviation uses the fact that, for many populations, most of the data are within three standard deviations of the mean. (See Section 9.3.) This implies that, in Example 3, most of the population from which this sample is taken are between

$$\bar{x} - 3s = 23 - 3(11.3) = -10.9$$

and

$$\bar{x} + 3s = 23 + 3(11.3) = 56.9.$$

This has important implications for quality control. If the sample in Example 3 represents measurements of a product that the manufacturer wants to be between 5 and 45, the standard deviation is too large, even though all the numbers are within these bounds.

We saw in the previous section that the mean of a random sample is a random variable. It should not surprise you, then, to learn that the variance and standard deviation are also random variables. We will refer to the variance and standard deviation of a random sample as the **sample variance** and **sample standard deviation.** The expected value of the sample variance is called the **population variance,** denoted σ^2. The **population standard deviation** is simply σ, the square root of the population variance. (σ is the lower case version of the Greek letter sigma. You have already seen Σ, the upper case version.) In other words,

$$E(s^2) = \sigma^2$$

and

$$\sigma = \sqrt{\sigma^2}.$$

The reason most statisticians prefer $n - 1$ in the denominator of the standard deviation formula is that it makes $E(s^2) = \sigma^2$ true; that is not true if n is used in the denominator.* It may surprise you, then, that $E(s) = \sigma$ is *false,* whether n or $n - 1$ is used. This is shown in more advanced courses in statistics. If n is large, the difference between $E(s)$ and σ is slight, so in practice, the sample standard deviation s gives a good estimate of the population standard deviation σ.

For data in a grouped frequency distribution, a slightly different formula for the standard deviation is used.

FOR REVIEW ■

Recall from Section 8.5 that a random variable is a function that assigns a real number to each outcome of an experiment. When the experiment consists of drawing a random sample, the standard deviation and the variance are two real numbers assigned to each outcome. Every time the experiment is performed, the standard deviation and variance will most likely have different values, with some values more probable than others.

---

*The definition of σ^2 is $E(x - \mu)^2$. The fact that $E(s^2) = \sigma^2$ can be proven from this definition.

STANDARD DEVIATION FOR A GROUPED DISTRIBUTION

The standard deviation for a distribution with mean $\bar{x}$, where x is an interval midpoint with frequency f, and $n = \Sigma f$, is

$$s = \sqrt{\frac{\Sigma f x^2 - n\bar{x}^2}{n - 1}}.$$

The formula indicates that the product fx^2 is to be found for each interval. Then these products are summed, n times the sum of the mean is subtracted, and the difference is divided by one less than the total frequency; that is, by $n - 1$. The square root of this result is s, the standard deviation.

CAUTION In calculating the standard deviation for either a grouped or ungrouped distribution, using a rounded value for the mean may produce an inaccurate value.

EXAMPLE 4 Standard Deviation for Grouped Data

Find s for the grouped data of Example 5, Section 9.1.

Solution Begin by adding columns for x^2 (the midpoint of the interval) and fx^2. Recall from Example 5 of Section 9.1 that $\bar{x} = 15.5$.

Interval	x	x^2	f	fx^2
0–4	2	4	3	12
5–9	7	49	4	196
10–14	12	144	6	864
15–19	17	289	8	2312
20–24	22	484	5	2420
25–29	27	729	3	2187
30–34	32	1024	1	1024
		Total:	30	Total: 9015

Use the formula above with $n = 30$ to find s.

$$s = \sqrt{\frac{\Sigma f x^2 - n\bar{x}^2}{n - 1}}$$

$$= \sqrt{\frac{9015 - 30(15.5)^2}{30 - 1}}$$

$$\approx 7.89$$

Verify that the standard deviation of the original, ungrouped data in Example 1 of Section 9.1 is 7.92.

EXAMPLE 5 Quality Assurance

Statistical process control is a method of determining when a manufacturing process is out of control, producing defective items. The procedure involves taking samples of a measurement on a product over a production run and calculating the mean and standard deviation of each sample. These results are used to determine when the manufacturing process is out of control. For example, three sample measurements from a manufacturing process on each of four days are given in the table below. The mean $\bar{x}$ and standard deviation s are calculated for each sample.

Day	1			2			3			4		
Sample Number	1	2	3	1	2	3	1	2	3	1	2	3
Measurements	−3	0	4	5	−2	4	3	−1	0	4	−2	1
	0	5	3	4	0	3	−2	0	0	3	0	3
	2	2	2	3	1	4	0	1	−2	3	−1	0
$\bar{x}$	−1/3	7/3	3	4	−1/3	11/3	1/3	0	−2/3	10/3	−1	4/3
s	2.5	2.5	1	1	1.5	.6	2.5	1	1.2	.6	1	1.5

Next, the mean of the 12 sample means, $\bar{X}$, and the mean of the 12 sample standard deviations, $\bar{s}$, are found (using the formula for $\bar{x}$). Here, these measures are

$$\bar{X} = 1.3 \quad \text{and} \quad \bar{s} = 1.41.$$

The control limits for the sample means are given by

$$\bar{X} \pm k_1\bar{s},$$

where k_1 is a constant found from a manual.* For samples of size 3, $k_1 = 1.954$, so the control limits for the sample means are

$$1.3 \pm (1.954)(1.41).$$

The upper control limit is 4.06, and the lower control limit is −1.46.

Similarly, the control limits for the sample standard deviations are given by $k_2 \cdot \bar{s}$ and $k_3 \cdot \bar{s}$, where k_2 and k_3 also are values given in the same manual. Here, $k_2 = 2.568$ and $k_3 = 0$, with the upper and lower control limits for the sample standard deviations equal to 2.568(1.41) and 0(1.41), or 3.62 and 0. As long as the sample means are between −1.46 and 4.06 and the sample standard deviations are between 0 and 3.62, the process is in control.

9.2 EXERCISES

1. How are the variance and the standard deviation related?

2. Why can't we use the sum of the deviations from the mean as a measure of dispersion of a distribution?

Find the range and standard deviation for each set of numbers.

3. 42, 38, 29, 74, 82, 71, 35

4. 122, 132, 141, 158, 162, 169, 180

5. 241, 248, 251, 257, 252, 287

6. 51, 58, 62, 64, 67, 71, 74, 78, 82, 93

7. 3, 7, 4, 12, 15, 18, 19, 27, 24, 11

8. 15, 42, 53, 7, 9, 12, 28, 47, 63, 14

---

*For example, see *Statistical Process Control* by Leonard A. Doty, Industrial Press, Inc., 1991, p. 317.

Find the standard deviation for the following grouped data.

9. (From Exercise 1, Section 9.1)

Interval	Frequency
0–24	4
25–49	3
50–74	6
75–99	3
100–124	5
125–149	9

10. (From Exercise 2, Section 9.1)

Interval	Frequency
30–39	1
40–49	6
50–59	13
60–69	22
70–79	17
80–89	13
90–99	8

Chebyshev's Theorem states that for any set of numbers, the fraction that will lie within k standard deviations of the mean is at least

$$1 - \frac{1}{k^2}.$$

For example, at least $1 - 1/2^2 = 3/4$ of any set of numbers lie within 2 standard deviations of the mean. Similarly, for any probability distribution, the probability that a number will lie within k standard deviations of the mean is at least $1 - 1/k^2$. For example, if the mean is 100 and the standard deviation is 10, the probability that a number will lie within 2 standard deviations of 100, or between 80 and 120, is at least 3/4. Use Chebyshev's Theorem to find the fraction of all the numbers of a data set that must lie within the following numbers of standard deviations from the mean.

11. 3   **12.** 4   **13.** 5

In a certain distribution of numbers, the mean is 50 with a standard deviation of 6. Use Chebyshev's Theorem to tell the probability that a number lies in each of the following intervals.

14. Between 38 and 62

15. Between 32 and 68

16. Less than 38 or more than 62

17. Less than 32 or more than 68

18. Discuss what the standard deviation tells us about a distribution.

19. Explain the difference between the sample mean and standard deviation, and the population mean and standard deviation.

20. Derive the shortcut formula for the variance

$$s^2 = \frac{\Sigma x^2 - n\bar{x}^2}{n - 1}$$

from the formula

$$s^2 = \frac{\Sigma(x - \bar{x})^2}{n - 1}$$

and the following summation formulas, in which c is a constant:

$$\Sigma\, cx = c\Sigma x, \quad \Sigma c = nc, \quad \text{and} \quad \Sigma(x \pm y) = \Sigma x \pm \Sigma y.$$

(*Hint:* Multiply out $(x - \bar{x})^2$.)

Applications

BUSINESS AND ECONOMICS

21. *Battery Life* Forever Power Company analysts conducted tests on the life of its batteries and those of a competitor (Brand X). They found that their batteries had a mean life (in hours) of 26.2, with a standard deviation of 4.1. Their results for a sample of 10 Brand X batteries were as follows: 15, 18, 19, 23, 25, 25, 28, 30, 34, 38.

  a. Find the mean and standard deviation for the sample of Brand X batteries.

  b. Which batteries have a more uniform life in hours?

  c. Which batteries have the highest average life in hours?

22. *Sales Promotion* The Quaker Oats Company conducted a survey to determine whether a proposed premium, to be included in boxes of cereal, was appealing enough to generate new sales.* Four cities were used as test markets, where the cereal was distributed with the premium, and four cities as control markets, where the cereal was distributed without the premium. The eight cities were chosen on the basis of their similarity in terms of population, per capita income, and total cereal purchase volume. The results were as follows.

City		Percent Change in Average Market Share per Month
	1	+18
	2	+15
Test Cities	3	+7
	4	+10
	1	+1
	2	−8
Control Cities	3	−5
	4	0

  a. Find the mean of the change in market share for the four test cities.

  b. Find the mean of the change in market share for the four control cities.

  c. Find the standard deviation of the change in market share for the test cities.

  d. Find the standard deviation of the change in market share for the control cities.

  e. Find the difference between the means of parts a and b. This difference represents the estimate of the percent change in sales due to the premium.

  f. The two standard deviations from parts c and d were used to calculate an "error" of ± 7.95 for the estimate in part e. With this amount of error, what are the smallest and largest estimates of the increase in sales?

On the basis of the interval estimate of part f, the company decided to mass-produce the premium and distribute it nationally.

23. *Process Control* The following table gives 10 samples of three measurements, made during a production run.

				Sample Number					
1	**2**	**3**	**4**	**5**	**6**	**7**	**8**	**9**	**10**
2	3	−2	−3	−1	3	0	−1	2	0
−2	−1	0	1	2	2	1	2	3	0
1	4	1	2	4	2	2	3	2	2

Use the information in Example 5 to find the following.

  a. Find the mean $\bar{x}$ for each sample of three measurements.

  b. Find the standard deviation s for each sample of three measurements.

  c. Find the mean $\bar{X}$ of the sample means.

  d. Find the mean $\bar{s}$ of the sample standard deviations.

  e. Using $k_1 = 1.954$, find the upper and lower control limits for the sample means.

  f. Using $k_2 = 2.568$ and $k_3 = 0$, find the upper and lower control limits for the sample standard deviations.

24. *Process Control* Given the following measurements from later samples on the process in Exercise 23, decide whether the process is out of control. (*Hint:* Use the results of Exercise 23e and f.)

		Sample Number			
1	**2**	**3**	**4**	**5**	**6**
3	−4	2	5	4	0
−5	2	0	1	−1	1
2	1	1	−4	−2	−6

*This example was supplied by Jeffery S. Berman, Senior Analyst, Marketing Information, Quaker Oats Company.

25. *Washer Thickness* An assembly-line machine turns out washers with the following thicknesses (in millimeters).

1.20	1.01	1.25	2.20	2.58	2.19	1.29	1.15
2.05	1.46	1.90	2.03	2.13	1.86	1.65	2.27
1.64	2.19	2.25	2.08	1.96	1.83	1.17	2.24

Find the mean and standard deviation of these thicknesses.

26. *Unemployment* The number of unemployed workers in the United States in recent years (in millions) is given below.*

Year	Number Unemployed
1989	6.53
1990	7.05
1991	8.63
1992	9.61
1993	8.94
1994	8.00
1995	7.40
1996	7.24
1997	6.74
1998	6.21

a. Find the mean number unemployed (in millions) in this period. Which year has unemployment closest to the mean?

b. Find the standard deviation for the data.

c. In how many of these years is unemployment within 1 standard deviation of the mean?

d. In how many of these years is unemployment within 3 standard deviations of the mean?

LIFE SCIENCES

27. *Blood pH* A medical laboratory tested 21 samples of human blood for acidity on the pH scale, with the following results.

7.1	7.5	7.3	7.4	7.6	7.2	7.3
7.4	7.5	7.3	7.2	7.4	7.3	7.5
7.5	7.4	7.4	7.1	7.3	7.4	7.4

a. Find the mean and standard deviation.

b. What percent of the data is within 2 standard deviations of the mean?

28. *Blood Types* The number of recognized blood types between species is given in the following table.† In Exer-

cise 40 of the previous section, the mean was found to be 7.38.

Animal	Number of Blood Types
Pig	16
Cow	12
Chicken	11
Horse	9
Human	8
Sheep	7
Dog	7
Rhesus Monkey	6
Mink	5
Rabbit	5
Mouse	4
Rat	4
Cat	2

a. Find the variance and the standard deviation of these data.

b. How many of these animals have blood types that are within 1 standard deviation of the mean?

29. *Tumor Growth* The amount of time that it takes for various slow-growing tumors to double in size are listed in the table below.‡

Type of Cancer	Doubling time (days)
Breast cancer	84
Rectal cancer	91
Synovioma	128
Skin cancer	131
Lip cancer	143
Testicular cancer	153
Esophageal cancer	164

a. Find the mean and standard deviation of these data.

b. How many of these cancers have doubling times that are within 2 standard deviations of the mean?

c. If a person had a nonspecified tumor that was doubling every 200 days, discuss if this particular tumor is growing at a rate that would be expected.

---

*The World Almanac and Book of Facts 2000, p. 145.
†The Handy Science Answer Book, Carnegie Library of Pittsburgh, Penn., p. 264.
‡Collins, Vincent, R. Kenneth Lodffer, and Harold Tivey, "Observations on Growth Rates of Human Tumors," *American Journal of Roentgen*, Vol. 76, No. 5, Nov. 1956, pp. 988–1000.

GENERAL INTEREST

30. *Governors* In 1999, fourteen state governors earned at least $115,000 annually (not counting expense allowances) as listed below. (Salaries are given in thousands of dollars and are rounded to the nearest $1000.)*

State	Salary
California	165
Georgia	116
Illinois	140
Maryland	120
Michigan	127
Minnesota	120
Nevada	117
New York	179
Ohio	119
Texas	115
Vermont	116
Virginia	125
Washington	132
Wisconsin	116

a. Find the mean salary of these governors. Which state has the governor with the salary closest to the mean?

b. Find the standard deviation for the data.

c. What percent of the governors have salaries within 1 standard deviation of the mean?

d. What percent of the governors have salaries within 3 standard deviations of the mean?

31. *Baseball Salaries* The table in Exercise 48 in the previous section listed the total salary in 1999 for each player on the Chicago Cubs baseball team.

a. Calculate the standard deviation of these data.

b. What percent of the 1999 Chicago Cubs players have salaries that are beyond 3 standard deviations from the mean?

c. What does your answer to part b suggest?

32. *Cookies* Marie Revak and Jihan Williams performed an experiment to determine whether Oreo Double Stuf cookies contain twice as much filling as traditional Oreo cookies. The following table gives the results in grams of the amount of filling inside 49 traditional cookies and 52 Double Stuf cookies.[†]

a. Find the mean, maximum, minimum, and standard deviation of the weights for traditional Oreo cookies.

b. Find the mean, maximum, minimum, and standard deviation of the weights for Oreo Double Stuf cookies.

Traditional	Traditional	Traditional	Double Stuf	Double Stuf	Double Stuf
2.9	2.4	2.7	4.7	6.5	5.8
2.8	2.8	2.8	6.5	6.3	5.9
2.6	3.8	2.6	5.5	4.8	6.2
3.5	3.1	2.6	5.6	3.3	5.9
3.0	2.9	3.0	5.1	6.4	6.5
2.4	3.0	2.8	5.3	5.0	6.5
2.7	2.1	3.5	5.4	5.3	6.1
2.4	3.8	3.3	5.4	5.5	5.8
2.5	3.0	3.3	3.5	5.0	6.0
2.2	3.0	2.8	5.5	6.0	6.2
2.6	2.8	3.1	6.5	5.7	6.2
2.6	2.9	2.6	5.9	6.3	6.0
2.9	2.7	3.5	5.4	6.0	6.8
2.6	3.2	3.5	4.9	6.3	6.2
2.6	2.8	3.1	5.6	6.1	5.4
3.1	3.1	3.1	5.7	6.0	6.6
2.9			5.3	5.8	6.2
			6.9		

The World Almanac and Book of Facts 2000, p. 98.

[†]Revak, Marie, and Jihan Williams, "The Double Stuf Dilemma," *The Mathematics Teacher*, Vol. 92, No. 8, Nov. 1999, pp. 674–675.

c. What percent of the data of traditional Oreo cookies is within 2 standard deviations of the Double Stuf Oreo mean? (*Hint:* Use the mean and standard deviation for the Double Stuf data.)

 d. What percent of the data of traditional Oreo cookies, when multiplied by 2, is within 2 standard deviations of the Double Stuf Oreo mean? (*Hint:* Use the mean and standard deviation for the Double Stuf data.)

e. Is there evidence that Double Stuf Oreos have twice as much filling as the traditional Oreo cookie? Explain.

9.3 THE NORMAL DISTRIBUTION

? THINK ABOUT IT What is the probability that a salesperson drives between 1200 miles and 1600 miles per month?

This question can be answered by using the normal probability distribution introduced in this section.

Suppose a bank is interested in improving its services to customers. The manager decides to begin by finding the amount of time tellers spend on each transaction, rounded to the nearest minute. The times for 75 different transactions are recorded, with the results shown in the following table. The frequencies listed in the second column are divided by 75 to find the empirical probabilities.

FOR REVIEW ■

Empirical probabilities, discussed in Section 7.4, are derived from grouped data by dividing the frequency or amount for each group by the total for all the groups. This gives one example of a probability distribution, discussed further in Sections 7.4 and 8.5.

Time	Frequency	Probability
1	3	$3/75 = .04$
2	5	$5/75 \approx .07$
3	9	$9/75 = .12$
4	12	$12/75 = .16$
5	15	$15/75 = .20$
6	11	$11/75 \approx .15$
7	10	$10/75 \approx .13$
8	6	$6/75 = .08$
9	3	$3/75 = .04$
10	1	$1/75 \approx .01$

Figure 6(a) shows a histogram and frequency polygon for the data. The heights of the bars are the empirical probabilities, rather than the frequencies. The transaction times are given to the nearest minute. Theoretically at least, they could have been timed to the nearest tenth of a minute, or hundredth of a minute,

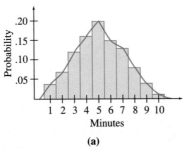

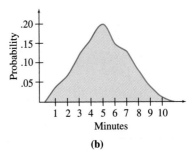

FIGURE 6

or even more precisely. In each case, a histogram and frequency polygon could be drawn. If the times are measured with smaller and smaller units, there are more bars in the histogram and the frequency polygon begins to look more and more like the curve in Figure 6(b) instead of a polygon. Actually, it is possible for the transaction times to take on any real number value greater than 0. A distribution in which the outcomes can take any real number value within some interval is a **continuous distribution.** The graph of a continuous distribution is a curve.

The distribution of heights (in inches) of college women is another example of a continuous distribution, since these heights include infinitely many possible measurements, such as 53, 58.5, 66.3, 72.666, ..., and so on. Figure 7 shows the continuous distribution of heights of college women. Here the most frequent heights occur near the center of the interval shown.

Another continuous curve, which approximates the distribution of yearly incomes in the United States, is given in Figure 8. The graph shows that the most frequent incomes are grouped near the low end of the interval. This kind of distribution, where the peak is not at the center, is called **skewed.**

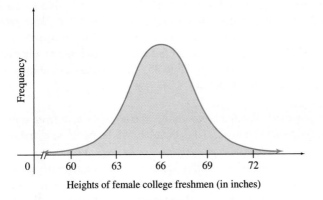

FIGURE 7

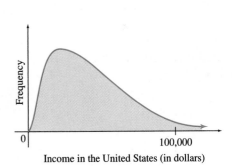

FIGURE 8

Many natural and social phenomena produce continuous probability distributions whose graphs can be approximated very well by bell-shaped curves, such as those shown in Figure 9. Such distributions are called **normal distributions** and their graphs are called **normal curves.** Examples of distributions that are approximately normal are the heights of college women and the errors made in filling 1-pound cereal boxes. We use the Greek letters μ (mu) to denote the mean, and σ (sigma) to denote the standard deviation, of a normal distribution.

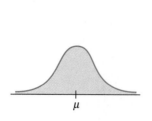

Three normal distributions

FIGURE 9

There are many normal distributions. Some of the corresponding normal curves are tall and thin and others are short and wide, as shown in Figure 9. But every normal curve has the following properties.

1. Its peak occurs directly above the mean μ.

2. The curve is symmetric about the vertical line through the mean (that is, if you fold the page along this line, the left half of the graph will fit exactly on the right half).

3. The curve never touches the x-axis—it extends indefinitely in both directions.

4. The area under the curve (and above the horizontal axis) is always 1. (This agrees with the fact that the sum of the probabilities in any distribution is 1.)

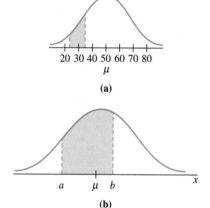

(a)

(b)

FIGURE 10

It can be shown that a normal distribution is completely determined by its mean μ and standard deviation σ.* A small standard deviation leads to a tall, narrow curve like the one in the center of Figure 9. A large standard deviation produces a flat, wide curve, like the one on the right in Figure 9.

Since the area under a normal curve is 1, parts of this area can be used to determine certain probabilities. For instance, Figure 10(a) is the probability distribution of the annual rainfall in a certain region. Calculus can be used to show that the probability that the annual rainfall will be between 25 inches and 35 inches is the area under the curve from 25 to 35. The general case, shown in Figure 10(b), can be stated as shown below.

> The area of the shaded region under the normal curve from a to b is the probability that an observed data value will be between a and b.

To use normal curves effectively, we must be able to calculate areas under portions of these curves. These calculations have already been done for the normal curve with mean $\mu = 0$ and standard deviation $\sigma = 1$ (which is called the **standard normal curve**) and are available in a table in the Appendix. The following examples demonstrate how to use the table to find such areas. Later we shall see how the standard normal curve may be used to find areas under any normal curve.

---

*As is shown in more advanced courses, its graph is the graph of the function

$$f(x) = \frac{1}{\sigma\sqrt{2\pi}} e^{-(x-\mu)^2/(2\sigma^2)},$$

where $e \approx 2.71828$ is a real number.

EXAMPLE 1   Standard Normal Curve

The horizontal axis of the standard normal curve is usually labeled z. Find the following areas under the standard normal curve.

Method 1: Using a Table   **(a)** The area to the left of $z = 1.25$

Solution Look up 1.25 in the normal curve table. (Find 1.2 in the left-hand column and .05 at the top, then locate the intersection of the corresponding row and column.) The specified area is .8944, so the shaded area shown in Figure 11 is .8944. This area represents 89.44% of the total area under the normal curve, and so the probability that $z \leq 1.25$ is

$$P(z \leq 1.25) = .8944.$$

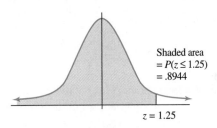

Shaded area
$= P(z \leq 1.25)$
$= .8944$

$z = 1.25$

FIGURE 11

(b) The area to the right of $z = 1.25$

Solution From part (a), the area to the left of $z = 1.25$ is .8944. The total area under the normal curve is 1, so the area to the right of $z = 1.25$ is

$$1 - .8944 = .1056.$$

See Figure 12, where the shaded area represents 10.56% of the total area under the normal curve, and the probability that $z \geq 1.25$ is .1056.

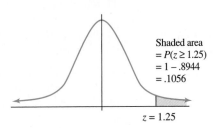

Shaded area
$= P(z \geq 1.25)$
$= 1 - .8944$
$= .1056$

$z = 1.25$

FIGURE 12

(c) Between $z = -1.02$ and $z = .92$

Solution To find this area, which is shaded in Figure 13, start with the area to the left of $z = .92$ and subtract the area to the left of $z = -1.02$. See the two shaded regions in Figure 14 on the next page. The result is

$$P(-1.02 \leq z \leq .92) = .8212 - .1539 = .6673.$$

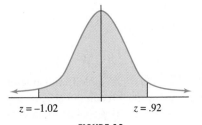

$z = -1.02$ $z = .92$

FIGURE 13

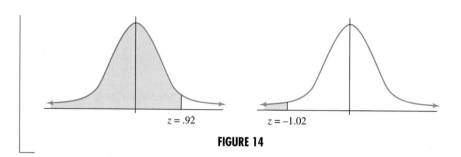

FIGURE 14

Method 2: Graphing Calculator

```
normalcdf(-1E99,
1.25,0,1)
        .894350161
normalcdf(1.25,1
E99,0,1)
        .105649839
```

FIGURE 15

Because of convenience and accuracy, graphing calculators and computers have made normal curve tables less important. Figure 15 shows how parts a and b of this example can be done on a TI-83 using the `normalcdf` command in the DISTR menu. In Figure 15, $-1E99$ stands for $-1 \cdot 10^{99}$. The area between $-1 \cdot 10^{99}$ and 1.25 is essentially the same as the area to the left of 1.25. Similarly, the area between 1.25 and $1 \cdot 10^{99}$ is essentially the same as the area to the right of 1.25. Verify the results of part c with a graphing calculator.

Method 3: Spreadsheet

Many statistical software packages are widely used today. All of these packages are set up in a way that is similar to a spreadsheet, and they all can be used to generate normal curve values. In addition, most spreadsheets can also perform a wide range of statistical calculations. For example, Microsoft Excel can be used to generate the answers to parts a, b, and c of this example. In any cell, type "=NORMDIST(1.25,0,1,1)" and press Enter. The value of .894350161 is returned. The first three input values represent the z value, mean, and standard deviation. The fourth value is always either a 0 or 1. For applications in this text, we will always place a 1 in this position to indicate that we want the area to the left of the first input value. Similarly, by typing "=1-NORMDIST(1.25,0,1,1)" and pressing Enter, we find that the area to the right of $z = 1.25$ is .105649839.

NOTE Notice in Example 1 that $P(z \leq 1.25) = P(z < 1.25)$. The area under the curve is the same, whether we include the endpoint or not. Notice also that $P(z = 1.25) = 0$, because no area is included.

CAUTION When calculating normal probability, it is wise to draw a normal curve with the mean and the z-scores every time. This will avoid confusion as to whether you should add or subtract probabilities.

EXAMPLE 2 Normal Probabilities

Find a value of z satisfying the following conditions.

Method 1: Using a Table

(a) 12.1% of the area is to the left of z.

Solution Use the table backwards. Look in the body of the table for an area of .1210, and find the corresponding value of z using the left column and the top column of the table. You should find that $z = -1.17$ corresponds to an area of .1210.

(b) 20% of the area is to the right of z.

> **Solution** If 20% of the area is to the right, 80% is to the left. Find the value of z corresponding to an area of .8000. The closest value is $z = .84$.

Method 2: Graphing Calculator Figure 16 illustrates how a TI-83 can be used to find z values for the particular probabilities given in parts a and b of this example. The command `invNorm` is found in the DISTR menu.

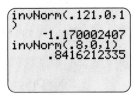

FIGURE 16

Method 3: Spreadsheet Microsoft Excel can also be used to generate the answers to parts a and b of this example. In any cell, type "=NORMINV(.121,0,1)" and press Enter. The value of -1.170002407 is returned. Similarly, by typing "=NORMINV(.8,0,1)" and pressing Enter, we find that the corresponding z value is .8416212335.

The key to finding areas under *any* normal curve is to express each number x on the horizontal axis in terms of standard deviation above or below the mean. The **z-score** for x is the number of standard deviations that x lies from the mean (positive if x is above the mean, negative if x is below the mean).

EXAMPLE 3 Z-Scores

If a normal distribution has mean 50 and standard deviation 4, find the following z-scores.

(a) The z-score for $x = 46$

> **Solution** Since 46 is 4 units below 50 and the standard deviation is 4, 46 is 1 standard deviation below the mean. So, its z-score is -1.

(b) The z-score for $x = 60$

> **Solution** The z-score is 2.5 because 60 is 10 units above the mean (since $60 - 50 = 10$), and 10 units is 2.5 standard deviations (since $10/4 = 2.5$).

In Example 3(b), we found the z-score by taking the difference between 60 and the mean and dividing this difference by the standard deviation. The same procedure works in the general case.

> If a normal distribution has mean μ and standard deviation σ, then the z-score for the number x is
> $$z = \frac{x - \mu}{\sigma}.$$

The importance of z-scores lies in the following fact.

> **AREA UNDER A NORMAL CURVE**
>
> The area under a normal curve between $x = a$ and $x = b$ is the same as the area under the standard normal curve between the z-score for a and the z-score for b.

Therefore, by converting to z-scores and using the table for the standard normal curve, we can find areas under any normal curve. Since these areas are probabilities, we can now handle a variety of applications.

EXAMPLE 4   Sales

Dixie Office Supplies finds that its sales force drives an average of 1200 miles per month per person, with a standard deviation of 150 miles. Assume that the number of miles driven by a salesperson is closely approximated by a normal distribution.

(a) Find the probability that a salesperson drives between 1200 miles and 1600 miles per month.

Solution Here $\mu = 1200$ and $\sigma = 150$, and we must find the area under the normal distribution curve between $x_1 = 1200$ and $x_2 = 1600$. We begin by finding the z-score for $x_1 = 1200$.

$$z_1 = \frac{x_1 - \mu}{\sigma} = \frac{1200 - 1200}{150} = \frac{0}{150} = 0$$

The z-score for $x_2 = 1600$ is

$$z_2 = \frac{x_2 - \mu}{\sigma} = \frac{1600 - 1200}{150} = \frac{400}{150} = 2.67.$$

From the table, the area to the left of $z_2 = 2.67$ is .9962, the area to the left of $z_1 = 0$ is .5000, and

$$.9962 - .5000 = .4962.$$

Therefore, the probability that a salesperson drives between 1200 miles and 1600 miles per month is .4962. See Figure 17.

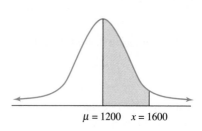

$\mu = 1200$ $x = 1600$

FIGURE 17

(b) Find the probability that a salesperson drives between 1000 miles and 1500 miles per month.

Solution As shown in Figure 18, z-scores for both $x_1 = 1000$ and $x_2 = 1500$ are needed.

For $x_1 = 1000$,

$$z_1 = \frac{1000 - 1200}{150}$$

$$= \frac{-200}{150}$$

$$z_1 \approx -1.33.$$

For $x_2 = 1500$,

$$z_2 = \frac{1500 - 1200}{150}$$

$$= \frac{300}{150}$$

$$z_2 = 2.00.$$

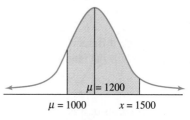

$\mu = 1000$ $x = 1500$

FIGURE 18

From the table, $z_1 = -1.33$ leads to an area of .0918, while $z_2 = 2.00$ corresponds to .9772. A total of $.9772 - .0918 = .8854$, or 88.54%, of the drivers travel between 1000 and 1500 miles per month. The probability that a driver travels between 1000 miles and 1500 miles per month is .8854.

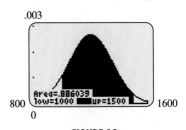

.003

800 ⟨   ⟩ 1600
0

FIGURE 19

Example 4 can also be done using a graphing calculator or computer, as described before, putting 1200 and 150 in place of 0 and 1 for the mean and standard deviation. On the TI-83, we could also use the common `Shade-Norm(1000,1500,1200,150)` for Example 4(b), with the results shown in Figure 19. The answer of .886039 is more accurate than the value of .8854 found using the normal curve table, which required rounding the z-scores to two decimal places.

NOTE The answers given to the exercises in this text are found using the normal curve table. If you use a graphing calculator or computer program, your answers will differ slightly.

As mentioned above, z-scores are the number of standard deviations from the mean, so $z = 1$ corresponds to 1 standard deviation above the mean, and so on. Looking up $z = 1.00$ and $z = -1.00$ in the table shows that

$$.8413 - .1587 = .6826,$$

or 68.3% of the area under a normal curve lies within 1 standard deviation of the mean. Also,

$$.9772 - .0228 = .9544,$$

or 95.4% of the area lies within 2 standard deviations of the mean. These results, summarized in Figure 20, can be used to get a quick estimate of results when working with normal curves.

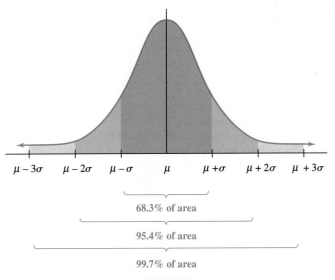

FIGURE 20

9.3 EXERCISES

1. The peak in a normal curve occurs directly above _____ .

2. The total area under a normal curve (above the horizontal axis) is _____ .

3. How are z-scores found for normal distributions where $\mu \neq 0$ or $\sigma \neq 1$?

4. How is the standard normal curve used to find probabilities for normal distributions?

Find the percent of the area under a normal curve between the mean and the given number of standard deviations from the mean.

5. 2.50   6.   .81   7. -1.71 8. -2.04

Find the percent of the total area under the standard normal curve between each pair of z-scores.

9. $z = 1.41$ and $z = 2.83$

10. $z = .64$ and $z = 2.11$

11. $z = -2.48$ and $z = -.05$

12. $z = -1.74$ and $z = -1.02$

13. $z = -3.11$ and $z = 1.44$

14. $z = -2.94$ and $z = .43$

Find a z-score satisfying the following conditions.

15. 5% of the total area is to the left of z.

16. 1% of the total area is to the left of z.

17. 15% of the total area is to the right of z.

18. 25% of the total area is to the right of z.

19. For any normal distribution, what is the value of $P(x \leq \mu)$? $P(x \geq \mu)$?

20. Compare the probability that a number will lie within 2 standard deviations of the mean of a probability distribution using Chebyshev's Theorem and using the normal distribution. (See Exercises 11–17, Section 9.2.) Explain what you observe.

21. Repeat Exercise 20 using 3 standard deviations.

Applications

In all of the following applications, assume the distributions are normal. In each case, you should consider whether this is reasonable.

BUSINESS AND ECONOMICS

Life of Light Bulbs A certain type of light bulb has an average life of 500 hr, with a standard deviation of 100 hr. The length of life of the bulb can be closely approximated by a normal curve. An amusement park buys and installs 10,000 such bulbs. Find the total number that can be expected to last for each of the following periods of time.

22. At least 500 hr

23. Less than 500 hr

24. Between 650 and 780 hr

25. Between 290 and 540 hr

26. Less than 740 hr

27. More than 300 hr

28. Find the shortest and longest lengths of life for the middle 80% of the bulbs.

Quality Control A box of oatmeal must contain 16 oz. The machine that fills the oatmeal boxes is set so that, on the average, a box contains 16.5 oz. The boxes filled by the machine have weights that can be closely approximated by a normal curve. What fraction of the boxes filled by the machine are underweight if the standard deviation is as follows?

29. .5 oz

30. .3 oz

31. .2 oz

32. .1 oz

Quality Control The chickens at Colonel Thompson's Ranch have a mean weight of 1850 g, with a standard deviation of 150 g. The weights of the chickens are closely approximated by a normal curve. Find the percent of all chickens having weights in the following ranges.

33. More than 1700 g

34. Less than 1800 g

35. Between 1750 and 1900 g

36. Between 1600 and 2000 g

37. More than 2100 g or less than 1550 g

38. Find the smallest and largest weights for the middle 95% of the chickens.

39. *Quality Control* A machine produces bolts with an average diameter of .25 inch and a standard deviation of .02 inch. What is the probability that a bolt will be produced with a diameter greater than .3 inch?

40. *Quality Control* A machine that fills quart milk cartons is set up to average 32.2 oz per carton, with a standard deviation of 1.2 oz. What is the probability that a filled carton will contain less than 32 oz of milk?

41. *Grocery Bills* At the Discount Market, the average weekly grocery bill is $52.25, with a standard deviation of $19.50. What are the largest and smallest amounts spent by the middle 50% of this market's customers?

42. *Grading Eggs* To be graded extra large, an egg must weigh at least 2.2 oz. If the average weight for an egg is 1.5 oz, with a standard deviation of .4 oz, how many eggs in a sample of five dozen would you expect to grade extra large?

LIFE SCIENCES

Vitamin Requirements In nutrition, the Recommended Daily Allowance of vitamins is a number set by the government as a guide to an individual's daily vitamin intake. Actually, vitamin needs vary drastically from person to person, but the needs are very closely approximated by a normal curve. To calculate the Recommended Daily Allowance, the government first finds the average need for vitamins among people in the population, and the standard deviation. The Recommended Daily Allowance is then defined as the mean plus 2.5 times the standard deviation.

43. What percent of the population will receive adequate amounts of vitamins under this plan?

Find the Recommended Daily Allowance for each vitamin in Exercises 44–46.

44. Mean = 1800 units; standard deviation = 140 units

45. Mean = 159 units; standard deviation = 12 units

46. Mean = 1200 units; standard deviation = 92 units

47. *Blood Clotting* The mean clotting time of blood is 7.45 sec, with a standard deviation of 3.6 sec. What is the probability that an individual's blood clotting time will be less than 7 sec or greater than 8 sec?

48. *Fish* The average size of the fish caught by anglers in Lake Amotan is 12.3 inches, with a standard deviation of 4.1 inches. Find the probability of catching a fish longer than 18 inches in Lake Amotan.

SOCIAL SCIENCES

Speed Limits New studies by Federal Highway Administration traffic engineers suggest that speed limits on many thoroughfares are set arbitrarily and often are artificially low. According to traffic engineers, the ideal limit should be the "85th percentile speed." This means the speed at or below which 85 percent of the traffic moves. Assuming speeds are normally distributed, find the 85th percentile speed for roads with the following conditions.

49. The mean speed is 50 mph with a standard deviation of 10 mph.

50. The mean speed is 30 mph with a standard deviation of 5 mph.

Education The grading system known as "grading on the curve" is based on the assumption that grades are often distributed according to the normal curve, and that a certain percent of a class should receive each grade, regardless of the performance of the class as a whole. The following is how one professor might grade on the curve.

Grade	Total Points
A	Greater than $\mu + (3/2)\sigma$
B	$\mu + (1/2)\sigma$ to $\mu + (3/2)\sigma$
C	$\mu - (1/2)\sigma$ to $\mu + (1/2)\sigma$
D	$\mu - (3/2)\sigma$ to $\mu - (1/2)\sigma$
F	Below $\mu - (3/2)\sigma$

What percent of the students receive the following grades?

51. A   **52.** B

53. C

54. Do you think this system would be more likely to be fair in a large freshman class in psychology or in a graduate seminar of five students? Why?

Education A teacher gives a test to a large group of students. The results are closely approximated by a normal curve. The mean is 74, with a standard deviation of 6. The teacher wishes to give A's to the top 8% of the students and F's to the bottom 8%. A grade of B is given to the next 15%, with D's given similarly. All other students get C's. Find the bottom cutoff (rounded to the nearest whole number) for the following grades.

55. A   **56.** B

57. C   **58.** D

59. *Standardized Tests* David Rogosa, a professor of educational statistics at Stanford University, has calculated the accuracy of tests used in California to abolish social promotion. Dr Rogosa has claimed that a fourth grader whose true reading score is exactly at reading level (50th percentile—half of all the students read worse and half read better than this student) has a 58 percent chance of either scoring above the 55th percentile or below the 45th percentile on any one test.* Assume that the results of a given test are normally distributed with mean .50 and standard deviation .09.

a. Verify that Dr Rogosa's claim is true.

b. Find the probability that this student will be above the 60th percentile or below the 40th percentile.

✎ **c.** Using the results of parts a and b, discuss problems with the use of standardized testing to prevent social promotion.

GENERAL INTEREST

60. *Christopher Columbus* Before Christopher Columbus crossed the ocean, he measured the heights of the men on his three ships and found that they were normally distributed with a mean of 69.60 inches and a standard deviation of 3.20 inches. What is the probability that a member of his crew had a height less than 66.27 inches? (The answer has another connection with Christopher Columbus!)

✎ **61.** *Lead Poisoning* Historians and biographers have collected evidence that suggests that President Andrew Jackson suffered from lead poisoning. Recently, researchers measured the amount of lead in samples of Jackson's hair in 1815. The results of this experiment showed that Jackson had a mean lead level of 130.5 ppm.†

a. If levels of lead in hair samples from that time period follow a normal distribution with mean 93 and standard deviation 16‡, find the probability that a randomly selected person from this time period would have a lead level of 130.5 ppm or higher. Does this provide evidence that Jackson suffered from lead poisoning during this time period?§

b. Today's typical lead levels follow a normal distribution with approximate mean 10 ppm and standard deviation 5 ppm.∥ By these standards, calculate the probability that a randomly selected person from today would have a lead level of 130.5 or higher. From this can we conclude that Andrew Jackson had lead poisoning? (*Note:* These standards may not be valid for this experiment.)

62. *Mercury Poisoning* Historians and biographers have also collected evidence that suggests that President Andrew Jackson suffered from mercury poisoning. Recently, researchers measured the amount of mercury in samples of Jackson's hair from 1815. The results of this experiment showed that Jackson had a mean mercury level of 6.0 ppm.†

a. If levels of mercury in hair samples from that time period follow a normal distribution with mean 6.9 and standard deviation 4.6#, find the probability that a randomly selected person from that time period would have a mercury level of 6.0 ppm or higher.

✎ **b.** Discuss whether this provides evidence that Jackson suffered from mercury poisoning during this time period.

c. Today's accepted normal mercury levels follow a normal distribution with approximate mean .6 ppm and standard deviation .3 ppm∥. By present standards, is it likely that a randomly selected person from today would have a mercury level of 6.0 ppm or higher?

✎ **d.** Discuss whether we can conclude that Andrew Jackson suffered from mercury poisoning.

63. *Barbie* The popularity and voluptuous shape of Barbie dolls have generated much discussion about the influence these dolls may have on young children, particularly with regard to normal body shape. In fact, many people have speculated as to what Barbie's measurements would be if they were scaled to a common human height. Researchers have done this and have compared Barbie's measurements

---

*Rothstein, R., "How Tests Can Drop the Ball," *The New York Times,* Sept. 13, 2000, p. B11.

†Deppisch, Lidwig, Jose Centeno, David Gemmel, and Norca Torres, "Andrew Jackson's Exposure to Mercury and Lead," *JAMA,* Vol. 282, No. 6, Aug. 11, 1999, pp. 569–571.

‡Weiss, D., B. Whitten, and D. Leddy, "Lead Content of Human Hair (1871–1971)," *Science,* Vol. 178, 1972, pp. 69–70.

§Although this provides evidence that Andrew Jackson had elevated lead levels, the authors of the paper concluded that Andrew Jackson did not die from lead poisoning.

∥Iyengar, V., and J. Woittiez, "Trace Elements in Human Clinical Specimens," *Clinical Chemistry,* Vol. 34, 1988, pp. 474–481.

#Suzuki, T., T. Hongo, M. Morita, and R. Yamamoto, "Elemental Contamination of Japanese Women's Hair from Historical Samples," *Sci. Total Environ.,* Vol. 39, 1984, pp. 81–91.

to the average 18–35 year old woman, labeled Reference, and with the average model. The table below illustrates some of the results of their research where each measurement is in centimeters.* Assume that the distributions of measurements for the models and for the reference group follow a normal distribution with the given mean and standard deviation.

Measurement	Models		Reference		Barbie
	Mean	**s.d.**	**Mean**	**s.d.**	
Head	50.0	2.4	55.3	2.0	55.0
Neck	31.0	1.0	32.7	1.4	23.9
Chest (bust)	87.4	3.0	90.3	5.5	82.3
Wrist	15.0	.6	16.1	.8	10.6
Waist	65.7	3.5	69.8	4.7	40.7

a. Find the probability of Barbie's head size or larger occurring for the reference group and for the models.

b. Find the probability of Barbie's neck size or smaller occurring for the reference group and for the models.

c. Find the probability of Barbie's bust size or larger occurring for the reference group and for the models.

d. Find the probability of Barbie's wrist size or smaller occurring for the reference group and for the models.

e. Find the probability of Barbie's waist size or smaller occurring for the reference group and for the models.

f. Compare the above values and discuss whether Barbie represents either the reference group or models. Any surprises?

64. *Ken* The same researchers from Exercise 63 wondered how the famous Ken doll measured up to average males and with Australian football players. The table below illustrates some of the results of their research, where each measurement is in centimeters.* Assume that the distributions of measurements for the football players and for the reference group follow a normal distribution with the given mean and standard deviation.

Measurement	Football		Reference		Ken
	Mean	**s.d.**	**Mean**	**s.d.**	
Head	52.1	2.3	53.7	2.9	53.0
Neck	34.6	1.8	34.2	1.9	32.1
Chest	92.3	3.5	91.2	4.8	75.0
Upper Arm	29.9	1.9	28.8	2.2	27.1
Waist	75.1	3.6	80.9	9.8	56.5

a. Find the probability of Ken's head size or larger occurring for the reference group and for the football players.

b. Find the probability of Ken's neck size or smaller occurring for the reference group and for the football players.

c. Find the probability of Ken's chest size or smaller occurring for the reference group and for the football players.

d. Find the probability of Ken's upper arm size or smaller occurring for the reference group and for the football players.

e. Find the probability of Ken's waist size or smaller occurring for the reference group and for the football players.

f. Compare the above values and discuss whether Ken's measurements are representative of either the reference group or football players. Then compare these results with the results of Exercise 63. Any surprises?

9.4 NORMAL APPROXIMATION TO THE BINOMIAL DISTRIBUTION

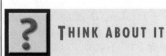

THINK ABOUT IT What is the probability that at least 40 out of 100 drivers exceed the speed limit on Interstate 10 in Texas?

This is a binomial probability problem with a large number of trials (100). In this section we will see how the normal curve can be used to approximate the binomial distribution and answer this question.

As we saw in Section 8.4 on Binomial Probability, many practical experiments have only two possible outcomes, sometimes referred to as success or failure. Such experiments are called Bernoulli trials or Bernoulli processes. Examples of Bernoulli trials include flipping a coin (with heads being a success, for instance, and tails a failure) or testing a computer chip coming off the assembly line to see whether or not it is defective. A binomial experiment consists of repeated independent Bernoulli trials, such as flipping a coin 10 times or taking a random sample of 20 computer chips from the assembly line. In Section 8.5 on Probability Distributions and Expected Value, we found the probability distribution for several binomial experiments, such as sampling five people with bachelor's degrees in education and counting how many are women. The probability distribution for a binomial experiment is known as a **binomial distribution.**

As another example, it is reported that 29% of drivers on Interstate 10 in Texas exceed the 70 mph speed limit.* Suppose a state trooper wants to verify this statistic and records the speed of 10 randomly selected drivers. The trooper finds that 4 out of 10, or 40%, exceed the speed limit. How likely is this if the 29% figure is accurate? We can answer this question with the binomial probability formula

$$\binom{n}{x} \cdot p^x \cdot (1 - p)^{n-x},$$

FOR REVIEW ■

Recall from Chapter 8 that the symbol $\binom{n}{r}$ is defined as $\dfrac{n!}{r!(n-r)!}$. For example, $\binom{10}{4} = \dfrac{10!}{4!\,6!} = \dfrac{10 \cdot 9 \cdot 8 \cdot 7}{4 \cdot 3 \cdot 2 \cdot 1} = 210.$

where n is the size of the sample (10 in this case), x is the number of speeders (4 in this example), and p is the probability that a driver is a speeder (.29). This gives

$$P(x = 4) = \binom{10}{4} \cdot .29^4 \cdot (1 - .29)^6$$
$$= 210(.007073)(.1281) \approx .1903.$$

The probability is almost 20%, so this result is not unusual.

Suppose that the state trooper takes a larger random sample of 100 drivers. What is the probability that 40 or more drivers speed if the 29% figure is accurate? Calculating $P(x = 40) + P(x = 41) + \cdots + P(x = 100)$ is a formidable task. One solution is provided by graphing calculators or computers. On the TI-83, for example, we can first calculate the probability that 39 or fewer drivers exceed the speed limit using the DISTR menu command `binomcdf(100,.29,39)`. Subtracting the answer from 1 gives a probability of .0119. But this high-tech method fails as n becomes larger; the command `binomcdf(1000,.29,300)` gives an error message. On the other hand, there

*Time, May 13, 1996, p. 34.

is a low-tech method which works regardless of the size of n. It has further interest because it connects two different distributions: the normal and the binomial. The normal distribution is continuous, since the random variable can take on any real number. The binomial distribution is *discrete,* because the random variable can only take on integer values between 0 and n. Nevertheless, the normal distribution can be used to give a good approximation to binomial probability.

In order to use the normal approximation, we first need to know the mean and standard deviation of the binomial distribution. Recall from Section 8.5 that for the binomial distribution, $E(x) = np$. In Section 9.1, we referred to $E(x)$ as μ, and that notation will be used here. It is shown in more advanced courses in statistics that the standard deviation of the binomial distribution is given by $\sigma = \sqrt{np(1 - p)}$.

MEAN AND STANDARD DEVIATION FOR THE BINOMIAL DISTRIBUTION

For the binomial distribution, the mean and standard deviation are given by

$$\mu = np \quad \text{and} \quad \sigma = \sqrt{np(1 - p)},$$

where n is the number of trials and p is the probability of success on a single trial.

EXAMPLE 1 Coin Flip

Suppose a fair coin is flipped 15 times.

(a) Find the mean and standard deviation for the number of heads.

Solution Using $n = 15$ and $p = 1/2$, the mean is

$$\mu = np = 15\left(\frac{1}{2}\right) = 7.5.$$

The standard deviation is

$$\sigma = \sqrt{np(1 - p)} = \sqrt{15\left(\frac{1}{2}\right)\left(1 - \frac{1}{2}\right)}$$

$$= \sqrt{15\left(\frac{1}{2}\right)\left(\frac{1}{2}\right)} = \sqrt{3.75} \approx 1.94.$$

We expect, on average, to get 7.5 heads out of 15 tosses. Most of the time, the number of heads will be within three standard deviations of the mean, or between $7.5 - 3(1.94) = 1.68$ and $7.5 + 3(1.94) = 13.32$.

(b) Find the probability distribution for the number of heads, and draw a histogram of the probabilities.

Solution The probability distribution is found by putting $n = 15$ and $p = 1/2$ into the formula for binomial probability. For example, the probability of 9 heads is given by

$$P(x = 9) = \binom{15}{9}\left(\frac{1}{2}\right)^9\left(1 - \frac{1}{2}\right)^6 \approx .15274.$$

Probabilities for the other values of x between 0 and 15, as well as a histogram of the probabilities, are shown in Figure 21.

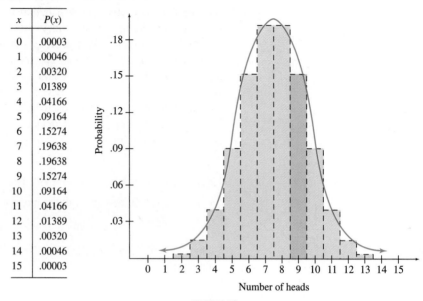

x	$P(x)$
0	.00003
1	.00046
2	.00320
3	.01389
4	.04166
5	.09164
6	.15274
7	.19638
8	.19638
9	.15274
10	.09164
11	.04166
12	.01389
13	.00320
14	.00046
15	.00003

FIGURE 21

In Figure 21, we have superimposed the normal curve with $\mu = 7.5$ and $\sigma = 1.94$ over the histogram of the distribution. Notice how well the normal distribution fits the binomial distribution. This approximation was first discovered in 1718 by Abraham De Moivre (1667–1754) for the case $p = 1/2$. The result was generalized by the French mathematician Pierre-Simon Laplace (1749–1827) in a book published in 1812. As n becomes larger and larger, a histogram for the binomial distribution looks more and more like a normal curve. Figures 22(a) and (b) show histograms of the binomial distribution with $p = .3$, using $n = 8$ and $n = 50$ in Figures 22(a) and (b), respectively.

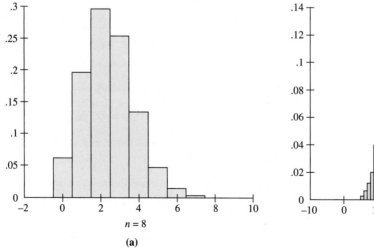

$n = 8$

(a)

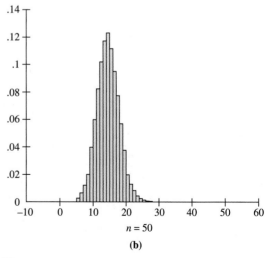

$n = 50$

(b)

FIGURE 22

The probability of getting exactly 9 heads in 15 tosses, or .15274, is the same as the area of the bar in blue in Figure 21. As the graph suggests, the area in blue is approximately equal to the area under the normal curve from $x = 8.5$ to $x = 9.5$. The normal curve is higher than the top of the bar in the left half but lower in the right half.

To find the area under the normal curve from $x = 8.5$ to $x = 9.5$, first find z-scores, as in the previous section. Use the mean and the standard deviation for the distribution, which we have already calculated, to get z-scores for $x_1 = 8.5$ and $x_2 = 9.5$.

$$
\begin{array}{ll}
\text{For } x_1 = 8.5, & \text{For } x_2 = 9.5, \\[6pt]
z_1 = \dfrac{8.5 - 7.5}{1.94} & z_2 = \dfrac{9.5 - 7.5}{1.94} \\[10pt]
= \dfrac{1.00}{1.94} & = \dfrac{2.00}{1.94} \\[10pt]
z_1 \approx .52. & z_2 \approx 1.03.
\end{array}
$$

From the table in the Appendix, $z_1 = .52$ gives an area of .6985, and $z_2 = 1.03$ gives .8485. The difference between these two numbers is the desired result.

$$P(z \le 1.03) - P(x \le .52) = .8485 - .6985 = .1500$$

This answer (.1500) is not far from the more accurate answer of .15274 found above.

> **CAUTION** The normal curve approximation to a binomial distribution is quite accurate *provided that n is large and p is not close to 0 or 1.* As a rule of thumb, the normal curve approximation can be used as long as both np and $n(1 - p)$ are at least 5.

EXAMPLE 2   Speeding

Consider the random sample discussed earlier of 100 drivers on Interstate 10 in Texas, where 29% of the drivers exceed the 70 mph speed limit.

(a) Use the normal distribution to approximate the probability that at least 40 drivers exceed the speed limit.

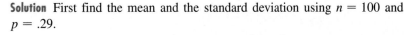

Solution First find the mean and the standard deviation using $n = 100$ and $p = .29$.

$$
\begin{array}{ll}
\mu = 100(.29) & \sigma = \sqrt{100(.29)(1 - .29)} \\[6pt]
= 29 & = \sqrt{100(.29)(.71)} \\[6pt]
& = \sqrt{20.59} \approx 4.54
\end{array}
$$

As the graph in Figure 23 on the next page shows, we need to find the area to the right of $x = 39.5$ (since we want 40 or more speeders). The z-score corresponding to $x = 39.5$ is

$$z = \frac{39.5 - 29}{4.54} \approx 2.31.$$

From the table, $z = 2.31$ leads to an area of .9896, so

$$P(z > 2.31) = 1 - .9896 = .0104.$$

This value is close to the value of .0119 found earlier with the help of a graphing calculator. Either method tells us there is roughly a 1% chance of finding 40 or more speeders out of a random sample of 100. If the trooper found this many in his sample, he might suspect that either his sample is not truly random, or that the 29% figure is too low.

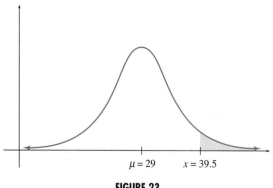

| **FIGURE 23** | **FIGURE 24** |

(b) Find the probability of finding between 30 and 35 speeders in a random sample of 100.

Solution As Figure 24 shows, we need to find the area between $x_1 = 29.5$ and $x_2 = 35.5$.

$$\text{If } x_1 = 29.5, \text{ then } z_1 = \frac{29.5 - 29}{4.54} \approx .11.$$

$$\text{If } x_2 = 35.5, \text{ then } z_2 = \frac{35.5 - 29}{4.54} \approx 1.43.$$

Use the table to find that $z_1 = .11$ gives an area of .5438, and $z_2 = 1.43$ yields .9236. The final answer is the difference of these numbers, or

$$P(.11 \le z \le 1.43) = P(z \le 1.43) - P(z \le .11)$$
$$= .9236 - .5438 = .3798.$$

The probability of finding between 30 and 35 speeders is about .3798.

9.4 EXERCISES

1. What must be known to find the mean and standard deviation of a binomial distribution?

2. What is the rule of thumb for using the normal distribution to approximate a binomial distribution?

Suppose 16 coins are tossed. Find the probability of getting each of the following results **(a)** *using the binomial probability formula, and* **(b)** *using the normal curve approximation.*

3. Exactly 8 heads

4. Exactly 7 heads

5. More than 12 tails

6. Fewer than 5 tails

For the remaining exercises in this section, use the normal curve approximation to the binomial distribution.

Suppose 1000 *coins are tossed. Find the probability of getting each of the following results.*

7. Exactly 500 heads   **8.** Exactly 510 heads   **9.** 480 heads or more   **10.** Fewer than 470 tails

A die is tossed 120 *times. Find the probability of getting each of the following results.*

11. Exactly twenty 5's

12. Exactly twenty-four 6's

13. More than eighteen 3's

14. Fewer than twenty-two 6's

15. A reader asked Mr. Statistics (a feature in *Fortune* magazine) about the game of 26 once played in the bars of Chicago.* The player chooses a number between 1 and 6, and then rolls a cup full of 10 dice 13 times. Out of the 130 numbers rolled, if the number chosen appears at least 26 times, the player wins. Calculate the probability of winning.

Applications

BUSINESS AND ECONOMICS

16. *Quality Control* Two percent of the quartz heaters produced in a certain plant are defective. Suppose the plant produced 10,000 such heaters last month. Find the probabilities that among these heaters, the following numbers were defective.

  a. Fewer than 170   **b.** More than 222

17. *Quality Control* The probability that a certain machine turns out a defective item is .05. Find the probabilities that in a run of 75 items, the following results are obtained.

  a. Exactly 5 defectives

  b. No defectives

  c. At least 1 defective

18. *Survey Results* A company is taking a survey to find out whether people like its product. Their last survey indicated that 70% of the population like the product. Based on that, of a sample of 58 people, find the probabilities of the following.

  a. All 58 like the product.

  b. From 28 to 30 (inclusive) like the product.

19. *Minimum Wage* A recent study of minimum wage earners found that 25.6% of them are teenagers.[†] Suppose a random sample of 600 minimum wage earners is selected. What is the probability that more than 160 of them are teenagers?

LIFE SCIENCES

20. *Nest Predation* For certain bird species, with appropriate assumptions, the number of nests escaping predation has a binomial distribution.[‡] Suppose the probability of success (that is, a nest escaping predation) is .3. Find the probability that at least half of 26 nests escape predation.

21. *Food Consumption* Under certain appropriate assumptions, the probability of a competing young animal eating x units of food is binomially distributed, with n equal to the maximum number of food units the animal can acquire, and p equal to the probability per time unit that an animal eats a unit of food.[§] Suppose $n = 120$ and $p = .6$.

  a. Find the probability that an animal consumes exactly 80 units of food.

  b. Suppose the animal must consume at least 70 units of food to survive. What is the probability that this happens?

22. *Coconuts* A 4-year review of trauma admissions to the Provincial Hospital, Alotau, Milne Bay Providence, reveals that 2.5% of such admissions were due to being struck by falling coconuts.[‖]

  a. Suppose 20 patients are admitted to the hospital during a certain time period. What is the probability that no more than 1 of these patients are there because they were struck by falling coconuts? Do not use the normal distribution here.

---

*Seligman, Daniel, and Patty De Llosa, "Ask Mr. Statistics," *Fortune,* May 1, 1995, p. 141.

[†]*The Chicago Tribune,* Dec. 28, 1995, pp. 1, 14.

[‡]Wilbur, H. M., *American Naturalist,* Vol. 111.

[§]deJong, G., *American Naturalist,* Vol. 110.

[‖]Barss, Peter, "Injuries Due to Falling Coconuts," *The Journal of Trauma,* Vol. 24, No. 11, 1984, pp. 990–991.

b. Suppose 2000 patients are admitted to the hospital during a longer time period. What is the approximate probability that no more than 70 of these patients are there because they were struck by falling coconuts?

23. *Kidney Dialysis* A study found that 23.6% of people on kidney dialysis in the United States die, more than in any other industrial country.* A hospital surveyed 40 of its patients who had been on dialysis that year and found that 15 died so far. What is the probability that 15 or more people out of 40 would die if this hospital's patients are typical of those in the United States?

24. *Drug Effectiveness* A new drug cures 80% of the patients to whom it is administered. It is given to 25 patients. Find the probabilities that among these patients, the following results occur.

a. Exactly 20 are cured.

b. All are cured.

c. No one is cured.

d. Twelve or fewer are cured.

25. *Flu Inoculations* A flu vaccine has a probability of 80% of preventing a person who is inoculated from getting the flu. A county health office inoculates 134 people. Find the probabilities of the following.

a. Exactly 10 of the people inoculated get the flu.

b. No more than 10 of the people inoculated get the flu.

c. None of the people inoculated get the flu.

26. *Blood Types* The blood types B− and AB− are the rarest of the eight human blood types, representing 1.5% and .6% of the population, respectively.[†]

a. If the blood types of a random sample of 1000 blood donors are recorded, what is the probability that 10 or more of the samples are AB−?

b. If the blood types of a random sample of 1000 blood donors are recorded, what is the probability that 20 to 40 inclusive of the samples are B−?

c. If a particular city had a blood drive in which 500 people gave blood and 3% of the donations were B−, would we have reason to believe that this town has a higher than normal number of donors who are B−? (*Hint:* Cal-

culate the probability of 15 or more donors being B− for a random sample of 500 and then discuss the probability obtained.)

27. *Motorcycles* In Washington state, there were 27.4 motorcycle rider fatalities per 1000 motorcycle crashes in 1989.[‡] If this rate is a true reflection of the nationwide proportion of motorcycle crashes that result in death, find the probability that for a state that has 3000 motorcycle accidents in a given year, there will be between 80 and 100 fatalities.

SOCIAL SCIENCES

28. *Straw Votes* In one state, 55% of the voters expect to vote for Edison Diest. Suppose 1400 people are asked the name of the person for whom they expect to vote. Find the probability that at least 750 people will say that they expect to vote for Diest.

29. *Weapons and Youth* A poll of 2000 teenagers found that 1 in 8 reported carrying a weapon for protection.[§] In a typical high school with 1200 students, what is the probability that more than 120 students, but fewer than 180, carry a weapon?

30. *Election 2000* As of November 16, 2000, the Florida recount gave George W. Bush 2,910,492 votes and Al Gore 2,910,192 votes.[‖] What is the likelihood of the vote being so close, even if the electorate is evenly divided? Assume that the number of votes for Bush is binomially distributed with $n = 5,820,684$ (the sum of the votes for the two candidates) and $p = .5$.

a. Using the binomial probability feature on a graphing calculator, try to calculate $P(2,910,192 \le X \le 2,910,492)$. What happens?

b. Use the normal approximation to calculate the probability in part a.

GENERAL INTEREST

31. *Homework* Only 1 out of 12 American parents requires that children do their homework before watching TV.[#] If your neighborhood is typical, what is the probability that out of 51 parents, 5 or fewer require their children to do homework before watching TV?

The New York Times, Dec, 4, 1995, p. A1.

[†]*The Handy Science Answer Book,* The Carnegie Library, Pittsburgh, Penn. 2000, p. 332.

[‡]Rowland, Jefferson, Frederick Rivara, Philip Salzberg, Robert Soderberg, Ronald Maier, and Thomas Koepsell, "Motorcycle Helmet Use and Injury Outcome and Hospitalization Costs from Crashes in Washington State," *American Journal of Public Health,* Vol. 86, No. 1, Jan. 1996, pp. 41–45.

[§]*The New York Times,* Jan. 12, 1996, p. A6.

[‖]*The New York Times,* Nov. 16, 2000, p. A28.

[#]"Harper's Index," *Harper's,* Sept. 1996, p. 15.

32. *True-False Test* A professor gives a test with 100 true-false questions. If 60 or more correct is necessary to pass, what is the probability that a student will pass by random guessing?

33. *Hole in One* In the 1989 U.S. Open, four golfers each made a hole in one on the same par-3 hole on the same day. *Sports Illustrated* writer R. Reilly stated the probability of a hole in one for a given golf pro on a given par-3 hole to be 1/3709.*

a. For a specific par-3 hole, use the binomial distribution to find the probability that 4 or more of the 156 golf pros in the tournament field shoot a hole in one.†

b. For a specific par-3 hole, use the normal approximation to the binomial distribution to find the probability that 4 or more of the 156 golf pros in the tournament field shoot a hole in one. Why must we be very cautious when using this approximation for this application?

c. If the probability of a hole in one remains constant and is 1/3709 for any par-3 hole, find the probability that in 20,000 attempts by golf pros, there will be 4 or more hole in ones. Discuss whether this assumption is reasonable.

CHAPTER SUMMARY

In this chapter we used the concepts of the previous two chapters to introduce the field of statistics. Measures of central tendency, such as mean, median, and mode, were defined. Information with regard to the spread of a group of numbers was obtained by calculating the variance and standard deviation. The normal distribution, perhaps the most important and widely used probability distribution, was defined and used to study a wide range of problems. The normal approximation of the binomial distribution was then developed, as were several important applications. The concepts that were introduced in this chapter serve only to familiarize a student to the wonderful world of statistics. Students are urged to further their study in this important area.

KEY TERMS

9.1 random sample
grouped frequency distribution
frequency polygon
(arithmetic) mean
sample mean
population mean

median
statistic
mode
9.2 range
deviations from the mean
variance

standard deviation
sample variance
sample standard deviation
population variance
population standard deviation

9.3 continuous distribution
skewed distribution
normal distribution
normal curve
standard normal curve
z-score
9.4 binomial distribution

CHAPTER 9 REVIEW EXERCISES

1. Discuss some reasons for organizing data into a grouped frequency distribution.

2. What is the rule of thumb for an appropriate interval in a grouped frequency distribution?

*Reilly, R., "King of the Hill," *Sports Illustrated,* June 1989, pp. 20–25.
†Litwiller, Bonnie, and David Duncan, "The Probability of a Hole in One," *School Science and Mathematics,* Vol. 91, No. 1, Jan. 1991, p. 30.

In Exercises 3 and 4, **(a)** *write a frequency distribution;* **(b)** *draw a histogram;* **(c)** *draw a frequency polygon.*

3. The following numbers give the sales (in dollars) for the lunch hour at a local hamburger stand for the last 20 Fridays. Use intervals 450–474, 475–499, and so on.

480	451	501	478	512	473	509	515	458	566
516	535	492	558	488	547	461	475	492	471

4. The number of units carried in one semester by students in a business mathematics class was as follows. Use intervals 9–10, 11–12, 13–14, 15–16.

10	9	16	12	13	15	13	16	15	11	13
12	12	15	12	14	10	12	14	15	15	13

Find the mean for each of the following.

5. 41, 60, 67, 68, 72, 74, 78, 83, 90, 97

6. 105, 108, 110, 115, 106, 110, 104, 113, 117

7.

Interval	Frequency
10–19	6
20–29	12
30–39	14
40–49	10
50–59	8

8.

Interval	Frequency
40–44	2
45–49	5
50–54	7
55–59	10
60–64	4
65–69	1

9. What do the mean, median, and mode of a distribution have in common? How do they differ? Describe each in a sentence or two.

Find the median and the mode (or modes) for each of the following.

10. 32, 35, 36, 44, 46, 46, 59

11. 38, 36, 42, 44, 38, 36, 48, 35

Find the modal class for the indicated distributions.

12. Exercise 7

13. Exercise 8

14. What is meant by the range of a distribution?

15. How are the variance and the standard deviation of a distribution related? What is measured by the standard deviation?

Find the range and standard deviation for each of the following distributions.

16. 14, 17, 18, 19, 32

17. 26, 43, 51, 29, 37, 56, 29, 82, 74, 93

Find the standard deviation for the following.

18. Exercise 7

19. Exercise 8

20. Describe the characteristics of a normal distribution.

21. What is meant by a skewed distribution?

Find the following areas under the standard normal curve.

22. Between $z = 0$ and $z = 1.27$

23. To the left of $z = .41$

24. Between $z = -1.88$ and $z = 2.10$

25. Between $z = 1.53$ and $z = 2.82$

26. Find a z-score such that 8% of the area under the curve is to the right of z.

27. Why is the normal distribution not a good approximation of a binomial distribution that has a value of p close to 0 or 1?

28. Suppose a card is drawn at random from an ordinary deck 1000 times with replacement.

  a. What is the probability that between 240 and 270 hearts (inclusive) are drawn?

b. Why must the normal approximation to the binomial distribution be used to solve part a?

29. Suppose four coins are flipped and the number of heads counted. This experiment is repeated 20 times. The data might look something like the following. (You may wish to try this yourself and use your own results rather than these.)

Number of Heads	Frequency
0	1
1	5
2	7
3	5
4	2

a. Calculate the sample mean $\bar{x}$ and sample standard deviation s.

b. Calculate the population mean μ and population standard deviation σ for this binomial population.

c. Compare your answer to parts a and b. What do you expect to happen?

30. Much of our work in Chapters 8 and 9 is interrelated. Note the similarities in the following parallel treatments of a frequency distribution and a probability distribution.

Frequency Distribution
Complete the table below for the following data. (Recall that x is the midpoint of the interval.)
14, 7, 1, 11, 2, 3, 11, 6, 10, 13, 11, 11, 16, 12, 9, 11, 9, 10, 7, 12, 9, 6, 4, 5, 9, 16, 12, 12, 11, 10, 14, 9, 13, 10, 15, 11, 11, 1, 12, 12, 6, 7, 8, 2, 9, 12, 10, 15, 9, 3

Probability Distribution
A binomial distribution has $n = 10$ and $p = .5$. Complete the table below.

Interval	x	Tally	f	x·f
1–3	2	ЖЖ|	6	12
4–6				
7–9				
10–12				
13–15				
16–18				

x	P(x)	x·P(x)
0	.001	
1	.010	
2	.044	
3	.117	
4		
5		
6		
7		
8		
9		
10		

a. Find the mean (or expected value) for each distribution.

b. Find the standard deviation for each distribution.

c. Use the normal approximation of the binomial probability distribution to find the interval that contains 95.44% of that distribution.

d. Why can't we use the normal distribution to answer probability questions about the frequency distribution?

Applications

BUSINESS AND ECONOMICS

31. *Stock Returns* The annual returns of two stocks for three years are given below.

Stock	1998	1999	2000
Stock I	11%	−1%	14%
Stock II	9%	5%	10%

a. Find the mean and standard deviation for each stock over the three-year period.

b. If you are looking for security (hence, less variability) with an average 8% return, which of these stocks should you choose?

32. *Quality Control* A machine that fills quart orange juice cartons is set to fill them with 32.1 oz. If the actual contents of the cartons vary normally, with a standard deviation of .1 oz, what percent of the cartons contain less than a quart (32 oz)?

33. *Quality Control* About 6% of the frankfurters produced by a certain machine are overstuffed and thus defective. Find the following probabilities for a sample of 500 frankfurters.

a. Twenty-five or fewer are overstuffed.

b. Exactly 30 are overstuffed.

c. More than 40 are overstuffed.

34. *Bankruptcy* The probability that a small business will go bankrupt in its first year is .21. For 50 such small businesses, find the following probabilities first by using the binomial probability formula, and then by using the normal approximation.

a. Exactly 8 go bankrupt.

b. No more than 2 go bankrupt.

LIFE SCIENCES

35. *Rat Diets* The weight gains of 2 groups of 10 rats fed different experimental diets were as follows.

Diet	Weight Gains									
A	1	0	3	7	1	1	5	4	1	4
B	2	1	1	2	3	2	1	0	1	0

Compute the mean and standard deviation for each group.

a. Which diet produced the greatest mean gain?

b. Which diet produced the most consistent gain?

Chemical Effectiveness White flies are devastating California crops. An area infested with white flies is to be sprayed with a chemical which is known to be 98% effective for each application. Assume a sample of 1000 flies is checked.

36. Find the approximate probability that exactly 980 of the flies are killed in one application.

37. Find the approximate probability that no more than 986 of the flies are killed in one application.

38. Find the approximate probability that at least 975 of the flies are killed in one application.

39. Find the approximate probability that between 973 and 993 (inclusive) of the flies are killed in one application.

SOCIAL SCIENCES

Commuting Times The average resident of a certain East Coast suburb spends 42 minutes per day commuting, with a standard deviation of 12 minutes. Assume a normal distribution. Find the percent of all residents of this suburb who have the following commuting times.

40. At least 50 minutes per day

41. No more than 35 minutes per day

42. Between 32 and 40 minutes per day

43. Between 38 and 60 minutes per day

44. *I.Q. Scores* On standard IQ tests, the mean is 100, with a standard deviation of 15. The results are very close to fitting a normal curve. Suppose an IQ test is given to a very large group of people. Find the percent of those people whose IQ scores are as follows.

a. More than 130   **b.** Less than 85

c. Between 85 and 115

45. *Broadway* A survey was given to 313 performers appearing in 23 Broadway companies. The percentage of performers injured during practice or a performance was 55.5%.* If a random sample of 500 Broadway performers is taken, use the normal approximation to the binomial distribution to find the approximate probability that more than 300 performers have been injured.

---

*Evans, Randolph, Richard Evans, Scott Carvajal, and Susan Perry, "A Survey of Injuries Among Broadway Performers," *American Journal of Public Health,* Vol. 86, No. 1, Jan. 1996, pp. 77–80.

46. *Broadway* In the survey described in Exercise 45, the following demographics of the Broadway performers were recorded.* Assume that all of these demographics follow a normal distribution, an assumption that always must be verified prior to using it in real situations.

   a. Find the probability that a female dancer is 35 years old or older.

   b. Find the probability that a male dancer is 35 years old or older.

   c. Compare your answers to parts a and b.

   d. Find the probability that a female performer is 1.4 meters tall or taller.

   e. Find the probability that a female performer has a career duration that is more than 1.5 standard deviations from the mean.

   f. Would a female who has more than 6 injuries during her career be considered a rare event? Explain.

	Mean	Standard Deviation
Dancer's Age (female)	28.0	5.5
Dancer's Age (male)	32.2	8.4
Height in m (female)	1.64	.08
Duration as Professional in yr (female)	11.0	8.9
Total No. of Injuries as Performer (female)	3.0	2.2

*Evans, Randolph, Richard Evans, Scott Carvajal, and Susan Perry, "A Survey of Injuries Among Broadway Performers," *American Journal of Public Health,* Vol. 86, No. 1, Jan. 1996, pp. 77–80.

EXTENDED APPLICATION: Statistics in the Law — The *Castaneda* Decision

Statistical evidence is now routinely presented in both criminal and civil cases. In this application we'll look at a famous case that established use of the binomial distribution and measurement by standard deviation as an accepted procedure.*

Defendants who are convicted in criminal cases sometimes appeal their conviction on the grounds that the jury that indicted or convicted them was drawn from a pool of jurors that does not represent the population of the district in which they live. These appeals almost always cite the Supreme Court's decision in *Castaneda v. Partida* [430 U.S. 482], a case that dealt with the selection of grand juries in the state of Texas. The decision summarizes the facts this way:

> After respondent, a Mexican-American, had been convicted of a crime in a Texas District Court and had exhausted his state remedies on his claim of discrimination in the selection of the grand jury that had indicted him, he filed a habeas corpus petition in the Federal District Court, alleging a denial of due process and equal protection under the Fourteenth Amendment, because of gross underrepresentation of Mexican-Americans on the county grand juries.

The case went to the Appeals Court, which noted that "the county population was 79% Mexican-American, but, over an 11-year period, only 39% of those summoned for grand jury service were Mexican-American," and concluded that together with other testimony about the selection process, "the proof offered by respondent was sufficient to demonstrate a prima facie case of intentional discrimination in grand jury selection...."

The state appealed to the Supreme Court, and the Supreme Court needed to decide whether the underrepresentation of Mexican-Americans on grand juries was indeed too extreme to be an effect of chance. To do so, they invoked the binomial distribution. Here is the argument:

> Given that 79.1% of the population is Mexican-American, the expected number of Mexican-Americans among the 870 persons summoned to serve as grand jurors over the 11-year period is approximately 688. The observed number is 339. Of course, in any given drawing some fluctuation from the expected number is predicted. The important point, however, is that the statistical model shows that the results of a random drawing are likely to fall in the vicinity of the expected value....
>
> The measure of the predicted fluctuations from the expected value is the standard deviation, defined for the binomial distribution as the square root of the product of the total number in the sample (here 870) times the probability of selecting a Mexican-American (.791) times

the probability of selecting a non-Mexican-American (.209).... Thus, in this case the standard deviation is approximately 12. As a general rule for such large samples, if the difference between the expected value and the observed number is greater than two or three standard deviations, then the hypothesis that the jury drawing was random would be suspect to a social scientist. The 11-year data here reflect a difference between the expected and observed number of Mexican-Americans of approximately 29 standard deviations. A detailed calculation reveals that the likelihood that such a substantial departure from the expected value would occur by chance is less than 1 in 10^{140}.

The Court decided that the statistical evidence supported the conclusion that jurors were not randomly selected, and that it was up to the state to show that its selection process did not discriminate against Mexican-Americans. The Court concluded:

> The proof offered by respondent was sufficient to demonstrate a prima facie case of discrimination in grand jury selection. Since the State failed to rebut the presumption of purposeful discrimination by competent testimony, despite two opportunities to do so, we affirm the Court of Appeals' holding of a denial of equal protection of the law in the grand jury selection process in respondent's case.

Exercises

1. Check the Court's calculation of 29 standard deviations as the difference between the expected number of Mexican-Americans and the number actually chosen.

2. Where do you think the Court's figure of 1 in 10^{140} came from?

---

*The *Castaneda* case and many other interesting applications of statistics in law are discussed in Finkelstein and Levin, *Statistics for Lawyers,* New York, Springer-Verlag, 1990. U.S. Supreme Court decisions are online at http://www.findlaw.com/casecode/supreme.html, and most states now have important state court decisions online.

3. The *Castaneda* decision also presents data from a $2\frac{1}{2}$-year period during which the State District Judge supervised the selection process. During this period, 220 persons were called to serve as grand jurors, and only 100 of these were Mexican-American.

a. Considering the 220 jurors as a random selection from a large population, what is the expected number of Mexican-Americans, using the 79.1% population figure?

b. If we model the drawing of jurors as a sequence of 220 independent Bernoulli trials, what is the standard deviation of the number of Mexican-Americans?

c. About how many standard deviations is the actual number of Mexican-Americans drawn (100) from the expected number that you calculated in a?

d. What does the normal distribution table at the back of the book tell you about this result?

4. The following information is from an appeal brought by Hy-Vee stores before the Iowa Supreme Court, appealing a ruling by the Iowa Civil Rights Commission in favor of a female employee of one of their grocery stores.

In 1985, there were 112 managerial positions in the ten Hy-Vee stores located in Cedar Rapids. Only 6 of these managers were women. During that same year there were 294 employees; 206 were men and 88 were women.

a. How far from the expected number of women in management was the actual number, assuming that gender had nothing to do with promotion? Measure the difference in standard deviations.

b. Does this look like evidence of purposeful discrimination?

CHAPTER

10

Markov Chains

In a Markov process, the next state of the system you are analyzing depends only on the current state and a set of transition probabilities. For example, the length of a waiting line in a bank a minute from now depends on the current length and the probabilities of a customer arriving or completing a transaction. This *queuing chain* model is studied in the exercises in Section 1.

In Chapter 7 we touched on *stochastic processes,* mathematical models that evolve over time in a probabilistic manner. In this chapter we study a special kind of stochastic process called a *Markov chain,* where the outcome of an experiment depends only on the outcome of the previous experiment. In other words, the next **state** of the system depends only on the present state, not on preceding states. Such experiments are common enough in applications to make their study worthwhile. Markov chains are named after the Russian mathematician A. A. Markov (1856–1922), who started the theory of stochastic processes.

■ 10.1 BASIC PROPERTIES OF MARKOV CHAINS

 THINK ABOUT IT   If we know the probability that the child of a lower-class parent becomes middle-class or upper-class, and we know similar information for the child of a middle-class or upper-class parent, what is the probability that the grandchild or great-grandchild of a lower-class parent is middle- or upper-class?

Using Markov chains, we will learn the answers to such questions.

Transition Matrix   In sociology, it is convenient to classify people by income as *lower-class, middle-class,* and *upper-class.* Sociologists have found that the strongest determinant of the income class of an individual is the income class of the individual's parents. For example, if an individual in the lower-income class is said to be in *state 1*, an individual in the middle-income class is in *state 2*, and an individual in the upper-income class is in *state 3*, then the following probabilities of change in income class from one generation to the next might apply.*

	State	**Next Generation**		
		1	**2**	**3**
Current	1	.65	.28	.07
Generation	2	.15	.67	.18
	3	.12	.36	.52

This table shows that if an individual is in state 1 (lower-income class) then there is a probability of .65 that any offspring will be in the lower-income class, a probability of .28 that offspring will be in the middle-income class, and a probability of .07 that offspring will be in the upper-income class.

The symbol p_{ij} will be used for the probability of transition from state i to state j in one generation. For example, p_{23} represents the probability that a person in state 2 will have offspring in state 3; from the table above,

$$p_{23} = .18.$$

Also from the table, $p_{31} = .12$, $p_{22} = .67$, and so on.

FOR REVIEW ■

Recall from Chapter 7 that the probability of an event is a real number between 0 and 1. The closer the probability is to 1, the more likely the event will occur. Also, in a sample space consisting of n mutually exclusive events, the sum of the probabilities of these events equals 1.

---

*For an example with actual data, see Glass, D. V., and J. R. Hall, "Social Mobility in Great Britain: A Study of Intergenerational Changes in Status," in *Social Mobility in Great Britain,* D. V. Glass, ed., Routledge & Kegan Paul, 1954. This data is analyzed using Markov chains in *Finite Markov Chains* by John G. Kemeny and J. Laurie Snell, Springer-Verlag, 1976.

The information from the table can be written in other forms. Figure 1 is a **transition diagram** that shows the three states and the probabilities of going from one state to another.

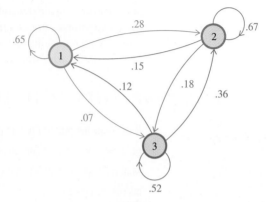

FIGURE 1

In a **transition matrix,** the states are indicated at the side and the top. If P represents the transition matrix for the table above, then

$$
\begin{array}{cc}
 & \begin{array}{ccc} 1 & 2 & 3 \end{array} \\
\begin{array}{c} 1 \\ 2 \\ 3 \end{array} &
\begin{bmatrix}
.65 & .28 & .07 \\
.15 & .67 & .18 \\
.12 & .36 & .52
\end{bmatrix} = P.
\end{array}
$$

A transition matrix has several features:

1. It is square, since all possible states must be used both as rows and as columns.

2. All entries are between 0 and 1, inclusive; this is because all entries represent probabilities.

3. The sum of the entries in any row must be 1, since the numbers in the row give the probability of changing from the state at the left to one of the states indicated across the top.

Markov Chains A transition matrix, such as matrix P above, also shows two key features of a Markov chain.

MARKOV CHAIN

A sequence of trials of an experiment is a **Markov chain** if

1. the outcome of each experiment is one of a set of discrete states;

2. the outcome of an experiment depends only on the present state, and not on any past states.

For example, in transition matrix P, a person is assumed to be in one of three discrete states (lower, middle, or upper income), with each offspring in one of these same three discrete states.

EXAMPLE 1   Dry Cleaners

A small town has only two dry cleaners, Johnson and NorthClean. Johnson's manager hopes to increase the firm's market share by conducting an extensive advertising campaign. After the campaign, a market research firm finds that there is a probability of .8 that a customer of Johnson's will bring his next batch of dirty clothes to Johnson, and a .35 chance that a NorthClean customer will switch to Johnson for his next batch. Write a transition matrix showing this information.

Solution We must assume that the probability that a customer comes to a given dry cleaner depends only on where the last batch of clothes was taken. If there is a probability of .8 that a Johnson customer will return to Johnson, then there must be a $1 - .8 = .2$ chance that the customer will switch to NorthClean. In the same way, there is a $1 - .35 = .65$ chance that a NorthClean customer will return to NorthClean. These probabilities give the following transition matrix.

$$
\begin{array}{c}
\textit{First Batch}
\end{array}
\begin{array}{c}
\\
\text{Johnson} \\
\text{NorthClean}
\end{array}
\begin{array}{c}
\textit{Second Batch} \\
\begin{array}{cc}
\text{Johnson} & \text{NorthClean}
\end{array} \\
\begin{bmatrix}
.8 & .2 \\
.35 & .65
\end{bmatrix}
\end{array}
$$

We shall come back to this transition matrix later in this section (Example 4).

Figure 2 shows a transition diagram with the probabilities of using each dry cleaner for the second batch of dirty clothes.

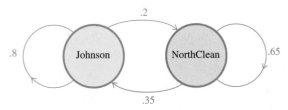

FIGURE 2

Look again at transition matrix P for income-class changes.

$$
\begin{array}{c}
1 \\
2 \\
3
\end{array}
\begin{array}{ccc}
1 & 2 & 3
\end{array}
\begin{bmatrix}
.65 & .28 & .07 \\
.15 & .67 & .18 \\
.12 & .36 & .52
\end{bmatrix} = P
$$

This matrix shows the probability of change in income class from one generation to the next. Now let us investigate the probabilities for changes in income class over *two* generations. For example, if a parent is in state 3 (the upper-income class), what is the probability that a grandchild will be in state 2?

To find out, start with a tree diagram, as shown in Figure 3; the various probabilities come from transition matrix P. The arrows point to the outcomes "grandchild in state 2"; the grandchild can get to state 2 after having had parents in either state 1, state 2, or state 3. The probability that a parent in state 3 will have a grandchild in state 2 is given by the sum of the probabilities indicated with arrows, or

$$.0336 + .2412 + .1872 = .4620.$$

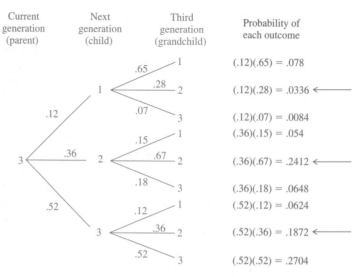

FIGURE 3

FOR REVIEW ■

Multiplication of matrices was covered in Chapter 2. To get the entry in row i, column j of a product, multiply row i of the first matrix times column j of the second matrix and add up the products. For example, to get the element in row 1, column 1 of P^2, where

$$P = \begin{bmatrix} .65 & .28 & .07 \\ .15 & .67 & .18 \\ .12 & .36 & .52 \end{bmatrix},$$

we calculate $(.65)(.65) + (.28)(.15) + (.07)(.12) = .4729 \approx .47$. To get row 3, column 2, the computation is $(.12)(.28) + (.36)(.67) + (.52)(.36) = .462 \approx .46$. You should review matrix multiplication by working out the rest of P^2 and verifying that it agrees with the result given in Example 2.

We used p_{ij} to represent the probability of changing from state i to state j in one generation. This notation can be used to write the probability that a parent in state 3 will have a grandchild in state 2:

$$p_{31} \cdot p_{12} + p_{32} \cdot p_{22} + p_{33} \cdot p_{32}.$$

This sum of products of probabilities should remind you of matrix multiplication—it is nothing more than one step in the process of multiplying matrix P by itself. In particular, it is row 3 of P times column 2 of P. If P^2 represents the matrix product $P \cdot P$, then P^2 gives the probabilities of a transition from one state to another in *two* repetitions of an experiment. Generalizing,

> P^k gives the probabilities of a transition from one state
> to another in k repetitions of an experiment.

EXAMPLE 2 Transition Matrices

For transition matrix P (income-class changes),

$$P^2 = \begin{bmatrix} .65 & .28 & .07 \\ .15 & .67 & .18 \\ .12 & .36 & .52 \end{bmatrix} \begin{bmatrix} .65 & .28 & .07 \\ .15 & .67 & .18 \\ .12 & .36 & .52 \end{bmatrix} \approx \begin{bmatrix} .47 & .39 & .13 \\ .22 & .56 & .22 \\ .19 & \mathbf{.46} & .34 \end{bmatrix}.$$

[?] (The numbers in the product have been rounded to the same number of decimal places as in matrix P.) The entry in row 3, column 2 of P^2 gives the probability that a person in state 3 will have a grandchild in state 2; that is, that an upper-class person will have a middle-class grandchild. This number, .46, is the result (rounded to two decimal places) found through using the tree diagram.

Row 1, column 3 of P^2 gives the number .13, the probability that a person in state 1 will have a grandchild in state 3; that is, that a lower-class person will have an upper-class grandchild. How would the entry .47 be interpreted? ■

EXAMPLE 3 Powers of Transition Matrices

In the same way that matrix P^2 gives the probability of income-class changes after *two* generations, the matrix $P^3 = P \cdot P^2$ gives the probabilities of change after *three* generations.

For matrix P,

$$P^3 = P \cdot P^2 = \begin{bmatrix} .65 & .28 & .07 \\ .15 & .67 & .18 \\ .12 & .36 & .52 \end{bmatrix} \begin{bmatrix} .47 & .39 & .13 \\ .22 & .56 & .22 \\ .19 & .46 & .34 \end{bmatrix} \approx \begin{bmatrix} .38 & .44 & .17 \\ .25 & .52 & .23 \\ .23 & .49 & .27 \end{bmatrix}.$$

(The rows of P^3 don't necessarily total 1 exactly because of rounding errors.) Matrix P^3 gives a probability of .25 that a person in state 2 will have a great-grandchild in state 1. The probability is .52 that a person in state 2 will have a great-grandchild in state 2.

A graphing calculator with matrix capability is useful for finding powers of a matrix. If you enter matrix A, then multiply by A, then multiply the product by A again, you get each new power in turn. You can also raise a matrix to a power just as you do with a number.

EXAMPLE 4   Dry Cleaners

Let us return to the transition matrix for the dry cleaners.

$$\begin{array}{cc} & \textit{Second Batch} \\ & \begin{array}{cc} \text{Johnson} & \text{NorthClean} \end{array} \\ \textit{First Batch} \begin{array}{c} \text{Johnson} \\ \text{NorthClean} \end{array} & \begin{bmatrix} .8 & .2 \\ .35 & .65 \end{bmatrix} \end{array}$$

As this matrix shows, there is a .8 chance that a person bringing his first batch to Johnson will also bring his second batch to Johnson, and so on. To find the probabilities for the third batch, the second stage of this Markov chain, find the square of the transition matrix. If C represents the transition matrix, then

$$C^2 = C \cdot C = \begin{bmatrix} .8 & .2 \\ .35 & .65 \end{bmatrix} \begin{bmatrix} .8 & .2 \\ .35 & .65 \end{bmatrix} = \begin{bmatrix} .71 & .29 \\ .51 & .49 \end{bmatrix}.$$

From C^2, the probability that a person bringing his first batch of clothes to Johnson will also bring his third batch to Johnson is .71; the probability that a person bringing his first batch to NorthClean will bring his third batch to NorthClean is .49.

The cube of matrix C gives the probabilities for the fourth batch, the third step in our experiment.

$$C^3 = C \cdot C^2 = \begin{bmatrix} .67 & .33 \\ .58 & .42 \end{bmatrix}$$

The probability is .58, for example, that a person bringing his first batch to NorthClean will bring his fourth batch to Johnson.

Distribution of States   Look again at the transition matrix for income-class changes:

$$P = \begin{bmatrix} .65 & .28 & .07 \\ .15 & .67 & .18 \\ .12 & .36 & .52 \end{bmatrix}.$$

Suppose the following table gives the initial distribution of people in the three income classes.

Class	State	Proportion
Lower	1	21%
Middle	2	68%
Upper	3	11%

To see how these proportions would change after one generation, use the tree diagram in Figure 4. For example, to find the proportion of people in state 2 after one generation, add the numbers indicated with arrows.

$$.0588 + .4556 + .0396 = .5540$$

In a similar way, the proportion of people in state 1 after one generation is

$$.1365 + .1020 + .0132 = .2517,$$

and the proportion of people in state 3 after one generation is

$$.0147 + .1224 + .0572 = .1943.$$

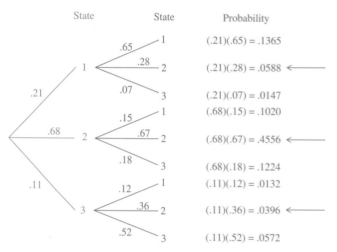

FIGURE 4

The initial distribution of states, 21%, 68%, and 11%, becomes, after one generation, 25.17% in state 1, 55.4% in state 2, and 19.43% in state 3. These distributions can be written as *probability vectors* (where the percents have been changed to decimals rounded to the nearest hundredth)

$$[.21 \quad .68 \quad .11] \quad \text{and} \quad [.25 \quad .55 \quad .19],$$

respectively. A **probability vector** is a matrix of only one row, having nonnegative entries, with the sum of the entries equal to 1.

The work with the tree diagram to find the distribution of states after one generation is exactly the work required to multiply the initial probability vector, $[.21 \quad .68 \quad .11]$, and the transition matrix P:

$$X_0 \cdot P = [.21 \quad .68 \quad .11] \begin{bmatrix} .65 & .28 & .07 \\ .15 & .67 & .18 \\ .12 & .36 & .52 \end{bmatrix} \approx [.25 \quad .55 \quad .19].$$

In a similar way, the distribution of income classes after two generations can be found by multiplying the initial probability vector and the square of P, the matrix P^2. Using P^2 from above,

$$X_0 \cdot P^2 = [.21 \quad .68 \quad .11] \begin{bmatrix} .47 & .39 & .13 \\ .22 & .56 & .22 \\ .19 & .46 & .34 \end{bmatrix} \approx [.27 \quad .51 \quad .21].$$

In the next section we will develop a long-range prediction for the proportion of the population in each income class. The work in this section is summarized below.

> Suppose a Markov chain has initial probability vector
>
> $$X_0 = [i_1 \; i_2 \; i_3 \cdots i_n]$$
>
> and transition matrix P. The probability vector after n repetitions of the experiment is
>
> $$X_0 \cdot P^n.$$

10.1 EXERCISES

Decide whether each of the matrices in Exercises 1–9 could be a probability vector.

1. $\left[\frac{2}{3} \quad \frac{1}{2}\right]$

2. $\left[\frac{1}{2} \quad 1\right]$

3. $[0 \quad 1]$

4. $[.1 \quad .1]$

5. $[.4 \quad .2 \quad 0]$

6. $\left[\frac{1}{4} \quad \frac{1}{8} \quad \frac{5}{8}\right]$

7. $[.07 \quad .04 \quad .37 \quad .52]$

8. $[.3 \quad -.1 \quad .8]$

9. $[0 \quad -.2 \quad .6 \quad .6]$

Decide whether each of the matrices in Exercises 10–15 could be a transition matrix, by definition. Sketch a transition diagram for any transition matrices.

10. $\begin{bmatrix} .5 & 0 \\ 0 & .5 \end{bmatrix}$

11. $\begin{bmatrix} \frac{2}{3} & \frac{1}{3} \\ 1 & 0 \end{bmatrix}$

12. $\begin{bmatrix} \frac{1}{4} & \frac{3}{4} \\ \frac{1}{2} & \frac{1}{2} \end{bmatrix}$

13. $\begin{bmatrix} \frac{1}{4} & \frac{3}{4} & 0 \\ 2 & 0 & 1 \\ 1 & \frac{2}{3} & 3 \end{bmatrix}$

14. $\begin{bmatrix} \frac{1}{3} & \frac{1}{3} & \frac{1}{3} \\ 0 & 1 & 0 \\ \frac{1}{2} & 0 & \frac{1}{2} \end{bmatrix}$

15. $\begin{bmatrix} \frac{1}{3} & \frac{1}{2} & 1 \\ 0 & 1 & 0 \\ \frac{1}{2} & \frac{1}{2} & 1 \end{bmatrix}$

In Exercises 16–18, write any transition diagrams as transition matrices.

16.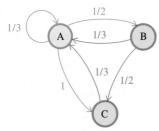

17.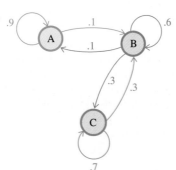

18.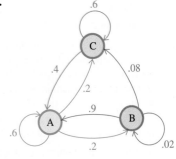

Find the first three powers of each of the transition matrices in Exercises 19–24 (for example, A, A^2, and A^3 in Exercise 19). For each transition matrix, find the probability that state 1 changes to state 2 after three repetitions of the experiment.

19. $A = \begin{bmatrix} 1 & 0 \\ .8 & .2 \end{bmatrix}$

20. $B = \begin{bmatrix} .7 & .3 \\ 0 & 1 \end{bmatrix}$

21. $C = \begin{bmatrix} .5 & .5 \\ .72 & .28 \end{bmatrix}$

22. $D = \begin{bmatrix} .3 & .2 & .5 \\ 0 & 0 & 1 \\ .6 & .1 & .3 \end{bmatrix}$

23. $E = \begin{bmatrix} .8 & .1 & .1 \\ .3 & .6 & .1 \\ 0 & 1 & 0 \end{bmatrix}$

24. $F = \begin{bmatrix} .01 & .9 & .09 \\ .72 & .1 & .18 \\ .34 & 0 & .66 \end{bmatrix}$

For each of the following transition matrices, find the first five powers of the matrix. Then find the probability that state 2 changes to state 4 after 5 repetitions of the experiment.

25. $\begin{bmatrix} .1 & .2 & .2 & .3 & .2 \\ .2 & .1 & .1 & .2 & .4 \\ .2 & .1 & .4 & .2 & .1 \\ .3 & .1 & .1 & .2 & .3 \\ .1 & .3 & .1 & .1 & .4 \end{bmatrix}$

26. $\begin{bmatrix} .3 & .2 & .3 & .1 & .1 \\ .4 & .2 & .1 & .2 & .1 \\ .1 & .3 & .2 & .2 & .2 \\ .2 & .1 & .3 & .2 & .2 \\ .1 & .1 & .4 & .2 & .2 \end{bmatrix}$

27. a. Verify that $X_0 \cdot P^n$ can be computed in two ways: (1) by first multiplying P by itself n times, then multiplying X_0 times this result; and (2) by multiplying $X_0 \cdot P$, multiplying this result by P, and continuing to multiply by P a total of n times. (*Hint:* Use the fact that matrix multiplication is associative.)

b. Which of the two methods in part a is simpler? Explain your answer.

Applications

BUSINESS AND ECONOMICS

28. *Dry Cleaning* The dry cleaning example in the text used the following transition matrix:

$$\begin{array}{cc} & \begin{array}{cc} \text{Johnson} & \text{NorthClean} \end{array} \\ \begin{array}{c} \text{Johnson} \\ \text{NorthClean} \end{array} & \begin{bmatrix} .8 & .2 \\ .35 & .65 \end{bmatrix} \end{array}.$$

Suppose now that each customer brings in one batch of clothes per week. Use various powers of the transition matrix to find the probability that a customer initially bringing a batch of clothes to Johnson also brings a batch to Johnson after the following time periods.

a. 1 wk **b.** 2 wk **c.** 3 wk **d.** 4 wk

e. What is the probability that a customer initially bringing a batch of clothes to NorthClean brings a batch to Johnson after 2 wk?

29. *Dry Cleaning* Suppose Johnson has a 40% market share initially, with NorthClean having a 60% share. Use this information to write a probability vector; use this vector, along with the transition matrix above, to find the share of the market for each firm after each of the following time periods. (As in Exercise 28, assume that customers bring in one batch of dry cleaning per week.)

a. 1 wk **b.** 2 wk **c.** 3 wk **d.** 4 wk

30. *Insurance* An insurance company classifies its drivers into three groups: G_0 (no accidents), G_1 (one accident), and G_2 (more than one accident). The probability that a G_0 driver will remain a G_0 after one year is .85, that the driver will become a G_1 is .10, and that the driver will become a G_2 is .05. A G_1 driver cannot become a G_0 (this company has a long memory). There is a .8 probability that a G_1 driver will remain a G_1. A G_2 driver must remain a G_2. Write a transition matrix using this information.

31. *Insurance* Suppose that the company in Exercise 30 accepts 50,000 new policyholders, all of whom are G_0 drivers. Find the number in each group after the following time periods.

a. 1 yr **b.** 2 yr **c.** 3 yr **d.** 4 yr

32. *Insurance* The difficulty with the mathematical model in Exercises 30 and 31 is that no "grace period" is provided; there should be a certain positive probability of moving from G_1 or G_2 back to G_0. A new system with this feature might produce the following transition matrix.

$$\begin{bmatrix} .85 & .10 & .05 \\ .15 & .75 & .10 \\ .10 & .30 & .60 \end{bmatrix}$$

Suppose that when this new policy is adopted, the company has 50,000 policyholders, all in G_0. Find the number in each group after the following time periods.

a. 1 yr   **b.** 2 yr   **c.** 3 yr

d. Write the transition matrix for a 2-yr period.

e. Use your result from part d to find the probability that a driver in G_0 is still in G_0 2 yr later.

33. *Market Share* Research done by Gulf Oil Corporation* produced the following transition matrix for the probabilities that during a given year, a person with one system of home heating will keep the same system or switch to another.

Next Year's
Heating System

		Oil	Gas	Electric
	Oil	.825	.175	0
This Year's	Gas	.060	.919	.021
Heating System	Electric	.049	0	.951

The current share of the market held by these three heating systems is given by the vector [.26 .60 .14]. Assuming that these trends continue, find the share of the market held by each heating system after the following time periods.

a. 1 yr   **b.** 2 yr   **c.** 3 yr

34. *Land Use* In one state, a Board of Realtors land use survey showed that 35% of all land was used for agricultural purposes, while 10% was urban. Ten years later, of the agricultural land, 15% had become urbanized and 80% had remained agricultural. (The remainder lay idle.) Of the idle land, 20% had become urbanized and 10% had been converted for agricultural use. Of the urban land, 90% remained urban and 10% was idle. Assume that these trends continue.

a. Write a transition matrix using this information.

b. Write a probability vector for the initial distribution of land.

Find the land use pattern after the following time periods.

c. 10 yr   **d.** 20 yr

e. Write the transition matrix for a 20-yr period.

f. Use your result from part e to find the probability that an idle plot of land is still idle 20 yr later.

35. *Queuing Chain* In the queuing chain, we assume that people are queuing up to be served by, say, a bank teller. For simplicity, let us assume that once two people are in line, no one else can enter the line. Let us further assume that once two people are in line, no one else can enter the line. Let us further assume that one person is served every minute, as long as someone is in line. Assume further that in any minute, there is a probability of $1/2$ that no one enters the line, a probability of $1/3$ that exactly one person enters the line, and a probability of $1/6$ that exactly two people enter the line, assuming there is room. If there is not enough room for two people, then the probability that one person enters the line is $1/2$. Let the state be given by the number of people in line.

a. Verify that the transition matrix is

$$
\begin{array}{c} \\ 0 \\ 1 \\ 2 \end{array}
\begin{array}{c} \begin{array}{ccc} 0 & 1 & 2 \end{array} \\ \begin{bmatrix} \frac{1}{2} & \frac{1}{3} & \frac{1}{6} \\ \frac{1}{2} & \frac{1}{3} & \frac{1}{6} \\ 0 & \frac{1}{2} & \frac{1}{2} \end{bmatrix} \end{array}.
$$

b. Find the transition matrix for a 2-minute period.

c. Use your result from part b to find the probability that a queue with no one in line has two people in line 2 minutes later.

LIFE SCIENCES

36. *Immune Response* A study of immune response in rabbits classified the rabbits into four groups, according to the strength of the response.[†] From one week to the next, the rabbits changed classification from one group to another, according to the following transition matrix.

$$
\begin{array}{c} \\ 1 \\ 2 \\ 3 \\ 4 \end{array}
\begin{array}{c} \begin{array}{cccc} 1 & 2 & 3 & 4 \end{array} \\ \begin{bmatrix} \frac{5}{7} & \frac{2}{7} & 0 & 0 \\ 0 & \frac{1}{2} & \frac{1}{3} & \frac{1}{6} \\ 0 & 0 & \frac{1}{2} & \frac{1}{2} \\ 0 & 0 & \frac{1}{4} & \frac{3}{4} \end{bmatrix} \end{array}
$$

a. What proportion of the rabbits in group 1 were still in group 1 five weeks later?

b. In the first week, there were 9 rabbits in the first group, 4 in the second, and none in the third or fourth groups. How many rabbits would you expect in each group after 4 weeks?

c. By investigating the transition matrix raised to larger and larger powers, make a reasonable guess for the long-range probability that a rabbit in group 1 or 2 will still be in group 1 or 2 after an arbitrarily long time. Explain why this answer is reasonable.

SOCIAL SCIENCES

37. *Housing Patterns* In a survey investigating changes in housing patterns in one urban area, it was found that 75%

*Ezzati, Ali, "Forecasting Market Shares of Alternative Home Heating Units," *Management Science*, Vol. 21, No. 4, Dec. 1974. Copyright © 1974 by The Institute of Management Sciences. Reprinted by permission of the author.
†McGilchrist, C. A., C. W. Aisbett, and S. Cooper, "A Markov Transition Model in the Analysis of the Immune Response," *Journal of Theoretical Biology*, Vol. 138, 1989, pp. 17–21.

of the population lived in single-family dwellings and 25% in multiple housing of some kind. Five years later, in a follow-up survey, of those who had been living in single-family dwellings, 90% still did so, but 10% had moved to multiple-family dwellings. Of those in multiple-family housing, 95% were still living in that type of housing, while 5% had moved to single-family dwellings. Assume that these trends continue.

a. Write a transition matrix for this information.

b. Write a probability vector for the initial distribution of housing.

What percent of the population can be expected in each category after the following time periods?

c. 5 yr   **d.** 10 yr

e. Write the transition matrix for a 10-yr period.

f. Use your result from part e to find the probability that someone living in a single-family dwelling is still doing so 10 yr later.

38. *Voting Trends* At the end of June in a presidential election year, 40% of the voters were registered as liberal, 45% as conservative, and 15% as independent. Over a one-month period, the liberals retained 80% of their constituency, while 15% switched to conservative and 5% to independent. The conservatives retained 70%, and lost 20% to the liberals. The independents retained 60% and lost 20% each to the conservatives and liberals. Assume that these trends continue.

a. Write a transition matrix using this information.

b. Write a probability vector for the initial distribution.

Find the percent of each type of voter at the end of each of the following months.

c. July   **d.** August   **e.** September   **f.** October

39. *Cricket* The results of cricket matches between England and Australia have been found to be modeled by a Markov chain.* The probability that England wins, loses, or draws is based on the result of the previous game, with the following transition matrix:

$$
\begin{array}{c}
\text{Wins} \\
\text{Loses} \\
\text{Draws}
\end{array}
\begin{array}{c}
\begin{array}{ccc}
\text{Wins} & \text{Loses} & \text{Draws}
\end{array} \\
\left[
\begin{array}{ccc}
.443 & .364 & .193 \\
.277 & .436 & .287 \\
.266 & .304 & .430
\end{array}
\right].
\end{array}
$$

a. Compute the transition matrix for the game after the next one, based on the result of the last game.

b. Use your answer from part a to find the probability that, if England won the last game, England will win the game after the next one.

c. Use your answer from part a to find the probability that, if Australia won the last game, England will win the game after the next one.

◼ 10.2 REGULAR MARKOV CHAINS

THINK ABOUT IT Given the transition probabilities for two dry cleaners, how can we predict the market share of each cleaner far into the future if we do not know the current proportions?

In Example 2 of this section, we will see how to answer this question.

If we start with a transition matrix P and an initial probability vector, we can use the nth power of P to find the probability vector for n repetitions of an experiment. In this section, we try to decide what happens to an initial probability vector in the long run—that is, as n gets larger and larger.

---

*Colwell, Derek, Brian Jones, and Jack Gillett, "A Markov Chain in Cricket," *The Mathematical Gazette*, June 1991.

For example, let us use the transition matrix associated with the dry cleaning example in the previous section.

$$\begin{bmatrix} .8 & .2 \\ .35 & .65 \end{bmatrix}$$

The initial probability vector, which gives the market share for each firm at the beginning of the experiment, is $[.4 \quad .6]$. The market shares shown in the following table were found by using powers of the transition matrix. (See Exercise 29 in the previous section.)

Weeks After Start	Johnson	NorthClean
0	.4	.6
1	.53	.47
2	.59	.41
3	.61	.39
4	.63	.37
5	.63	.37
12	.64	.36

The results seem to approach the numbers in the probability vector $[.64 \quad .36]$.

What happens if the initial probability vector is different from $[.4 \quad .6]$? Suppose $[.75 \quad .25]$ is used; the same powers of the transition matrix as above give the following results.

Weeks After Start	Johnson	NorthClean
0	.75	.25
1	.69	.31
2	.66	.34
3	.65	.35
4	.64	.36
5	.64	.36
6	.64	.36

The results again seem to be approaching the numbers in the probability vector $[.64 \quad .36]$, the same numbers approached with the initial probability vector $[.4 \quad .6]$. In either case, the long-range trend is for a market share of about 64% for Johnson and 36% for NorthClean. The example above suggests that this long-range trend does not depend on the initial distribution of market shares. This means that if the initial market share for Johnson was less than 64%, the advertising campaign has paid off in terms of a greater long-range market share. If the initial share was more than 64%, the campaign did not pay off.

Regular Transition Matrices One of the many applications of Markov chains is in finding long-range predictions. It is not possible to make long-range predictions with all transition matrices, but for a large set of transition matrices, long-range predictions *are* possible. Such predictions are always possible with **regular transition matrices.** A transition matrix is **regular** if some power of the matrix contains all positive entries. A Markov chain is a **regular Markov chain** if its transition matrix is regular.

EXAMPLE 1 Regular Transition Matrices

Decide whether the following transition matrices are regular.

(a) $A = \begin{bmatrix} .75 & .25 & 0 \\ 0 & .5 & .5 \\ .6 & .4 & 0 \end{bmatrix}$

Solution Square A.

$$A^2 = \begin{bmatrix} .5625 & .3125 & .125 \\ .3 & .45 & .25 \\ .45 & .35 & .2 \end{bmatrix}$$

Since all entries in A^2 are positive, matrix A is regular.

(b) $B = \begin{bmatrix} .5 & 0 & .5 \\ 0 & 1 & 0 \\ 0 & 0 & 1 \end{bmatrix}$

Solution Find various powers of B.

$$B^2 = \begin{bmatrix} .25 & 0 & .75 \\ 0 & 1 & 0 \\ 0 & 0 & 1 \end{bmatrix}; B^3 = \begin{bmatrix} .125 & 0 & .875 \\ 0 & 1 & 0 \\ 0 & 0 & 1 \end{bmatrix}; B^4 = \begin{bmatrix} .0625 & 0 & .9375 \\ 0 & 1 & 0 \\ 0 & 0 & 1 \end{bmatrix}$$

Further powers of B will still give the same zero entries, so no power of matrix B contains all positive entries. For this reason, B is not regular. ▪

NOTE If a transition matrix P has some zero entries, and P^2 does as well, you may wonder how far you must compute P^k to be certain that the matrix is not regular. The answer is that if zeros occur in the identical places in both P^k and P^{k+1} for any k, they will appear in those places for all higher powers of P, so P is not regular.

Suppose that v is any probability vector. It can be shown that for a regular Markov chain with a transition matrix P, there exists a single vector V that does not depend on v, such that $v \cdot P^n$ gets closer and closer to V as n gets larger and larger.

EQUILIBRIUM VECTOR OF A MARKOV CHAIN

If a Markov chain with transition matrix P is regular, then there is a unique vector V such that, for any probability vector v and for large values of n,

$$v \cdot P^n \approx V.$$

Vector V is called the **equilibrium vector** or the **fixed vector** of the Markov chain.

In the example with Johnson Cleaners, the equilibrium vector V is approximately [.64   .36]. Vector V can be determined by finding P^n for larger and larger values of n, and then looking for a vector that the product $v \cdot P^n$ approaches. Such an approach can be very tedious, however, and is prone to error. To find a better way, start with the fact that for a large value of n,

$$v \cdot P^n \approx V,$$

as mentioned above. From this result, $v \cdot P^n \cdot P \approx V \cdot P$, so that

$$v \cdot P^n \cdot P = v \cdot P^{n+1} \approx VP.$$

Since $v \cdot P^n \approx V$ for large values of n, it is also true that $v \cdot P^{n+1} \approx V$ for large values of n (the product $v \cdot P^n$ approaches V, so that $v \cdot P^{n+1}$ must also approach V). Thus, $v \cdot P^{n+1} \approx V$ and $v \cdot P^{n+1} \approx VP$, which suggests that

$$VP = V.$$

> If a Markov chain with transition matrix P is regular, then there exists a probability vector V such that
>
> $$VP = V.$$

This vector V gives the long-range trend of the Markov chain. Vector V is found by solving a system of linear equations, as shown in the next examples.

EXAMPLE 2   Dry Cleaners

Find the long-range trend for the Markov chain in the dry cleaning example with transition matrix

$$\begin{bmatrix} .8 & .2 \\ .35 & .65 \end{bmatrix}.$$

Solution This matrix is regular since all entries are positive. Let P represent this transition matrix, and let V be the probability vector $[v_1 \quad v_2]$. We want to find V such that

$$VP = V,$$

or

$$[v_1 \quad v_2] \begin{bmatrix} .8 & .2 \\ .35 & .65 \end{bmatrix} = [v_1 \quad v_2].$$

Use matrix multiplication on the left.

$$[.8v_1 + .35v_2 \quad .2v_1 + .65v_2] = [v_1 \quad v_2]$$

Set corresponding entries from the two matrices equal to get

$$.8v_1 + .35v_2 = v_1 \quad \text{and} \quad .2v_1 + .65v_2 = v_2.$$

Simplify each of these equations.

$$-.2v_1 + .35v_2 = 0 \quad \text{and} \quad .2v_1 - .35v_2 = 0$$

These last two equations are really the same. (The equations in the system obtained from $VP = V$ are always dependent.) To find the values of v_1 and v_2, recall that $V = [v_1 \quad v_2]$ is a probability vector, so that

$$v_1 + v_2 = 1.$$

To find v_1 and v_2, solve the system

$$-.2v_1 + .35v_2 = 0$$
$$v_1 + v_2 = 1.$$

From the second equation, $v_1 = 1 - v_2$. Substitute $1 - v_2$ for v_1 in the first equation.

$$-.2(1 - v_2) + .35v_2 = 0$$
$$-.2 + .2v_2 + .35v_2 = 0$$
$$.55v_2 = .2$$
$$v_2 = \frac{4}{11} \approx .364$$

Since $v_1 = 1 - v_2$, $v_1 = 7/11 \approx .636$, and the equilibrium vector is $V = [7/11 \quad 4/11] \approx [.636 \quad .364]$.

Some powers of the transition matrix P in Example 1 (the dry cleaning example) in the previous section (with entries rounded to two decimal places) are shown here.

$$P^2 = \begin{bmatrix} .71 & .29 \\ .51 & .49 \end{bmatrix} \qquad P^3 = \begin{bmatrix} .67 & .33 \\ .58 & .42 \end{bmatrix} \qquad P^4 = \begin{bmatrix} .65 & .35 \\ .61 & .39 \end{bmatrix}$$

$$P^5 = \begin{bmatrix} .64 & .36 \\ .62 & .38 \end{bmatrix} \qquad P^6 = \begin{bmatrix} .64 & .36 \\ .63 & .37 \end{bmatrix} \qquad P^{10} = \begin{bmatrix} .64 & .36 \\ .64 & .36 \end{bmatrix}$$

As these results suggest, higher and higher powers of the transition matrix P approach a matrix having all rows identical; these identical rows have as entries the entries of the equilibrium vector V. This agrees with the statement above: the initial state does not matter. Regardless of the initial probability vector, the system will approach a fixed vector V. This unexpected and remarkable fact is the basic property of regular Markov chains—*the limiting distribution is independent of the initial distribution.* This happens because some power of the transition matrix has all positive entries, so that all the initial probabilities are thoroughly mixed.

Let us summarize the results of this section.

PROPERTIES OF REGULAR MARKOV CHAINS

Suppose a regular Markov chain has a transition matrix P.

1. As n gets larger and larger, the product $v \cdot P^n$ approaches a unique vector V for any initial probability vector v. Vector V is called the *equilibrium vector* or *fixed vector.*
2. Vector V has the property that $VP = V$.
3. To find V, solve a system of equations obtained from the matrix equation $VP = V$, and from the fact that the sum of the entries of V is 1.
4. The powers P^n come closer and closer to a matrix whose rows are made up of the entries of the equilibrium vector V.

EXAMPLE 3 Equilibrium Vector

Find the equilibrium vector for the transition matrix

$$K = \begin{bmatrix} .2 & .6 & .2 \\ .1 & .1 & .8 \\ .3 & .3 & .4 \end{bmatrix}.$$

FOR REVIEW ■

The Gauss-Jordan process was discussed in Chapter 2. We will review it here by solving the system given in the text. Begin by ridding the matrix of decimals by multiplying by 10, and then dividing out any common factors.

$$\begin{array}{c} \\ 10R_2 \to R_2 \\ 10R_3/3 \to R_3 \\ 10R_4/2 \to R_4 \end{array} \begin{bmatrix} 1 & 1 & 1 & | & 1 \\ -8 & 1 & 3 & | & 0 \\ 2 & -3 & 1 & | & 0 \\ 1 & 4 & -3 & | & 0 \end{bmatrix}$$

Next clear column 1 by combining multiples of row 1 with multiples of the other rows.

$$\begin{array}{c} \\ 8R_1 + R_2 \to R_2 \\ -2R_1 + R_3 \to R_3 \\ -R_1 + R_4 \to R_4 \end{array} \begin{bmatrix} 1 & 1 & 1 & | & 1 \\ 0 & 9 & 11 & | & 8 \\ 0 & -5 & -1 & | & -2 \\ 0 & 3 & -4 & | & -1 \end{bmatrix}$$

Similarly, clear columns 2 and 3 using multiples of rows 2 and 3, as shown below.

$$\begin{array}{c} -R_2 + 9R_1 \to R_1 \\ \\ 5R_2 + 9R_3 \to R_3 \\ -R_2 + 3R_4 \to R_4 \end{array} \begin{bmatrix} 9 & 0 & -2 & | & 1 \\ 0 & 9 & 11 & | & 8 \\ 0 & 0 & 46 & | & 22 \\ 0 & 0 & -23 & | & -11 \end{bmatrix}$$

$$\begin{array}{c} R_3 + 23R_1 \to R_1 \\ -11R_3 + 46R_2 \to R_2 \\ R_3/2 \to R_3 \\ R_3 + 2R_4 \to R_4 \end{array} \begin{bmatrix} 207 & 0 & 0 & | & 45 \\ 0 & 414 & 0 & | & 126 \\ 0 & 0 & 23 & | & 11 \\ 0 & 0 & 0 & | & 0 \end{bmatrix}$$

Thus $v_1 = 45/207 = 5/23$; $v_2 = 126/414 = 7/23$; and $v_3 = 11/23$.

Solution Matrix K has all positive entries and thus is regular. For this reason, an equilibrium vector V must exist such that $VK = V$. Let $V = [v_1 \quad v_2 \quad v_3]$.

Then

$$[v_1 \quad v_2 \quad v_3] \begin{bmatrix} .2 & .6 & .2 \\ .1 & .1 & .8 \\ .3 & .3 & .4 \end{bmatrix} = [v_1 \quad v_2 \quad v_3].$$

Use matrix multiplication on the left.

$$[.2v_1 + .1v_2 + .3v_3 \quad .6v_1 + .1v_2 + .3v_3 \quad .2v_1 + .8v_2 + .4v_3] = [v_1 \quad v_2 \quad v_3]$$

Setting corresponding entries equal gives the following equations:

$$.2v_1 + .1v_2 + .3v_3 = v_1$$
$$.6v_1 + .1v_2 + .3v_3 = v_2$$
$$.2v_1 + .8v_2 + .4v_3 = v_3.$$

Simplifying these equations gives

$$-.8v_1 + .1v_2 + .3v_3 = 0$$
$$.6v_1 - .9v_2 + .3v_3 = 0$$
$$.2v_1 + .8v_2 - .6v_3 = 0.$$

Since V is a probability vector,

$$v_1 + v_2 + v_3 = 1.$$

This gives a system of four equations in three unknowns:

$$v_1 + v_2 + v_3 = 1$$
$$-.8v_1 + .1v_2 + .3v_3 = 0$$
$$.6v_1 - .9v_2 + .3v_3 = 0$$
$$.2v_1 + .8v_2 - .6v_3 = 0.$$

This system can be solved with the Gauss-Jordan method presented earlier. Start with the augmented matrix

$$\begin{bmatrix} 1 & 1 & 1 & | & 1 \\ -.8 & .1 & .3 & | & 0 \\ .6 & -.9 & .3 & | & 0 \\ .2 & .8 & -.6 & | & 0 \end{bmatrix}.$$

The solution of this system is $v_1 = 5/23$, $v_2 = 7/23$, $v_3 = 11/23$, and

$$V = \begin{bmatrix} \dfrac{5}{23} & \dfrac{7}{23} & \dfrac{11}{23} \end{bmatrix} \approx [.22 \quad .30 \quad .48].$$

This is a good place to use a graphing calculator that performs row operations. Refer to Section 2.2.

10.2 EXERCISES

Which of the transition matrices in Exercises 1–6 are regular?

1. $\begin{bmatrix} .2 & .8 \\ .9 & .1 \end{bmatrix}$

2. $\begin{bmatrix} .22 & .78 \\ .43 & .57 \end{bmatrix}$

3. $\begin{bmatrix} 1 & 0 \\ .6 & .4 \end{bmatrix}$

4. $\begin{bmatrix} .55 & .45 \\ 0 & 1 \end{bmatrix}$

5. $\begin{bmatrix} 0 & 1 & 0 \\ .4 & .2 & .4 \\ 1 & 0 & 0 \end{bmatrix}$

6. $\begin{bmatrix} .3 & .5 & .2 \\ 1 & 0 & 0 \\ .5 & .1 & .4 \end{bmatrix}$

Find the equilibrium vector for each transition matrix in Exercises 7–14.

7. $\begin{bmatrix} \frac{1}{4} & \frac{3}{4} \\ \frac{1}{2} & \frac{1}{2} \end{bmatrix}$

8. $\begin{bmatrix} \frac{2}{3} & \frac{1}{3} \\ \frac{1}{8} & \frac{7}{8} \end{bmatrix}$

9. $\begin{bmatrix} .3 & .7 \\ .4 & .6 \end{bmatrix}$

10. $\begin{bmatrix} .8 & .2 \\ .1 & .9 \end{bmatrix}$

11. $\begin{bmatrix} .1 & .1 & .8 \\ .4 & .4 & .2 \\ .1 & .2 & .7 \end{bmatrix}$

12. $\begin{bmatrix} .5 & .2 & .3 \\ .1 & .4 & .5 \\ .2 & .2 & .6 \end{bmatrix}$

13. $\begin{bmatrix} .25 & .35 & .4 \\ .1 & .3 & .6 \\ .55 & .4 & .05 \end{bmatrix}$

14. $\begin{bmatrix} .16 & .28 & .56 \\ .43 & .12 & .45 \\ .86 & .05 & .09 \end{bmatrix}$

Find the equilibrium vector for each transition matrix in Exercises 15–20. These matrices were first used in the exercises for Section 10.1. (Note: Not all of these transition matrices are regular, but equilibrium vectors still exist. Why doesn't this contradict the work of this section?)

15. Insurance categories (10.1 Exercise 30)

$$\begin{bmatrix} .85 & .10 & .05 \\ 0 & .80 & .20 \\ 0 & 0 & 1 \end{bmatrix}$$

16. Modified insurance categories (10.1 Exercise 32)

$$\begin{bmatrix} .85 & .10 & .05 \\ .15 & .75 & .10 \\ .10 & .30 & .60 \end{bmatrix}$$

17. Home heating systems (10.1 Exercise 33)

$$\begin{bmatrix} .825 & .175 & 0 \\ .060 & .919 & .021 \\ .049 & 0 & .951 \end{bmatrix}$$

18. Land use (10.1 Exercise 34)

$$\begin{bmatrix} .80 & .15 & .05 \\ 0 & .90 & .10 \\ .10 & .20 & .70 \end{bmatrix}$$

19. Voter registration (10.1 Exercise 38)

$$\begin{bmatrix} .80 & .15 & .05 \\ .20 & .70 & .10 \\ .20 & .20 & .60 \end{bmatrix}$$

20. Housing patterns (10.1 Exercise 37)

$$\begin{bmatrix} .90 & .10 \\ .05 & .95 \end{bmatrix}$$

21. Find the equilibrium vector for the transition matrix

$$\begin{bmatrix} p & 1-p \\ 1-q & q \end{bmatrix},$$

where $0 < p < 1$ and $0 < q < 1$. Under what conditions is this matrix regular?

22. Show that the transition matrix

$$K = \begin{bmatrix} \frac{1}{4} & 0 & \frac{3}{4} \\ 0 & 1 & 0 \\ 0 & 0 & 1 \end{bmatrix}$$

has more than one vector V such that $VK = V$. Why does this not violate the statements of this section?

23. Let

$$P = \begin{bmatrix} a_{11} & a_{12} \\ a_{21} & a_{22} \end{bmatrix}$$

be a regular matrix having *column* sums of 1. Show that the equilibrium vector for P is $[1/2 \quad 1/2]$.

24. Notice in Example 3 that the system of equations $VK = V$, with the extra equation that the sum of the elements of V must equal 1, had exactly one solution. What can you say about the number of solutions to the system $VK = V$?

Applications

BUSINESS AND ECONOMICS

25. *Quality Control* The probability that a complex assembly line works correctly depends on whether the line worked correctly the last time it was used. There is a .9 chance that the line will work correctly if it worked correctly the time before, and a .7 chance that it will work correctly if it did *not* work correctly the time before. Set up a transition matrix with this information and find the long-range probability that the line will work correctly.

26. *Quality Control* Suppose improvements are made in the assembly line of Exercise 25, so that the transition matrix becomes

	Works	Doesn't Work
Works	.95	.05
Doesn't Work	.80	.20

Find the new long-range probability that the line will work properly.

27. *Mortgage Refinancing* In 1992, many homeowners refinanced their mortgages to take advantage of lower interest rates. Most of the mortgages could be classified into three groups: 30-year fixed-rate, 15-year fixed-rate, and adjustable rate. (For this exercise, the small number of loans that are not of those three types will be classified with the adjustable rate loans.) Sometimes when a homeowner refinanced, the new loan was the same type as the old loan, and sometimes it was different. The breakdown of the percent (in decimal form) in each category is shown in the following table.*

Old loan	New loan		
	30-year fixed	**15-year fixed**	**Adjustable**
30-year fixed	.444	.479	.077
15-year fixed	.150	.802	.048
Adjustable	.463	.367	.170

If these conversion rates were to persist, find the long-range trend for the percent of loans of each type.

28. a. *Dry Cleaning* Using the initial probability vector $[.4 \quad .6]$, find the probability vector for the next 9 weeks in Example 2 (the dry cleaning example). Compute your answers to at least 6 decimal places.

b. Using your results from part a, find the difference between the equilibrium proportion of customers going to Johnson and the proportion going there for each of the first 10 weeks. Be sure to compute the equilibrium proportions to at least 6 decimal places.

c. Find the ratio between each difference calculated in part b and the difference for the previous week.

d. Using your results from part c, explain how the probability vector approaches the equilibrium vector.

e. Repeat parts a–d of this exercise using the initial probability vector $[.75 \quad .25]$.

29. *Finnish Economy* A Markov chain analysis of Finland's economy from 1988 to 1995 grouped the regions of the country into five groups according to their gross regional product per capita divided by the national average gross domestic product per capita, abbreviated as GRP.† For group 1, GRP ≤ 65. For group 2, $65 <$ GRP < 75. For group 3, $75 \leq$ GRP ≤ 85. For group 4, $85 <$ GRP < 100. For group 5, $100 \leq$ GRP. The transition matrix is as follows.

$$\begin{array}{c} \\ 1 \\ 2 \\ 3 \\ 4 \\ 5 \end{array} \begin{bmatrix} .733 & .267 & 0 & 0 & 0 \\ .278 & .555 & .167 & 0 & 0 \\ .133 & .133 & .667 & 0 & .067 \\ 0 & 0 & .412 & .294 & .294 \\ 0 & 0 & 0 & .250 & .750 \end{bmatrix}$$

a. What is the probability that a region in the second poorest group moved up by one group?

The New York Times, Nov. 18, 1992, p. D2.

†Pekkala, Sari, "Aggregate Economic Fluctuations and Regional Convergence: The Finnish Case 1988–95," *Applied Economics*, Vol. 32, No. 2, Feb. 10, 2000, pp. 211–219.

b. If these trends continue, what is the probability that a region in the middle group would be in the top group after three 7-yr periods?

c. If these trends continue, what are the long-term probabilities for each group?

LIFE SCIENCES

30. *Research with Mice* A large group of mice is kept in a cage having connected compartments A, B, and C. Mice in compartment A move to B with probability .3 and to C with probability .4. Mice in B move to A or C with probabilities of .15 and .55, respectively. Mice in C move to A or B with probabilities of .3 and .6, respectively. Find the long-range prediction for the fraction of mice in each of the compartments.

SOCIAL SCIENCES

31. *Class Mobility* The following chart shows the percent of the poor, middle class, and upper class that change into another class. The first graph shows the figures for 1967–1979, and the second for 1980–1991.* For each time period, assume the number of people who move directly from poor to affluent or from affluent to poor is essentially 0.

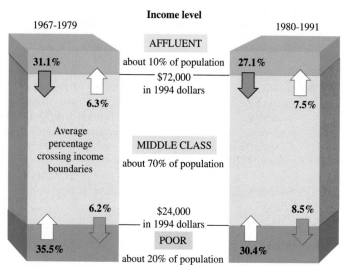

All figures are for household after-tax income, including wages, salaries and some Government assistance programs like food stamps.

a. Find the long-range percent of poor, middle class, and affluent people if the 1967–1979 trends were to continue.

b. Find the long-range percent of poor, middle class, and affluent people if the 1980–1991 trends were to continue.

32. *Criminology* A study of male criminals in Philadelphia found that the probability that one type of offense is followed by another type can be described by the following transition matrix.[†]

	Nonindex	Injury	Theft	Damage	Combination
Nonindex	.645	.099	.152	.033	.071
Injury	.611	.138	.128	.033	.090
Theft	.514	.067	.271	.030	.118
Damage	.609	.107	.178	.064	.042
Combination	.523	.093	.183	.022	.179

a. For a criminal who commits theft, what is the probability that his next crime is also a theft?

b. For a criminal who commits theft, what is the probability that his second crime after that is also a theft?

c. If these trends continue, what are the long-term probabilities for each type of crime?

33. *Education* At one liberal arts college, students are classified as humanities majors, science majors, or undecided. There is a 20% chance that a humanities major will change to a science major from one year to the next, and a 45% chance that a humanities major will change to undecided. A science major will change to humanities with probability .15, and to undecided with probability .35. An undecided will switch to humanities or science with probabilities of .5 and .3, respectively.

a. Find the long-range prediction for the fraction of students in each of these three majors.

b. Compare the result of part a with the result of Exercise 23. Make a conjecture, and describe how this conjecture, if true, would allow you to predict the answer to part a with very little computation.

34. *Rumors* The manager of the slot machines at a major casino makes a decision about whether or not to "loosen up" the slots so that the customers get a larger payback. The manager tells only one other person, a person whose word cannot be trusted. In fact, there is only a probability p, where $0 < p < 1$, that this person will tell the truth. Suppose this person tells several other people, each of

---

*The New York Times, June 4, 1995, p. E4.

[†]Stander, Julian, et al., "Markov Chain Analysis and Specialization in Criminal Careers," *The British Journal of Criminology*, Vol. 29, No. 4, Autumn 1989, pp. 317–335. The rounding was changed slightly so the rows of the transition matrix sum to 1.

whom tells several people, what the manager's decision is. Suppose there is always a probability p that the decision is passed on as heard. Find the long-range prediction for the fraction of the people who will hear the decision correctly. (*Hint:* Use a transition matrix; let the first row be $[p \quad 1 - p]$ and the second row be $[1 - p \quad p]$.)

35. *Education* A study of students taking a 20-question chemistry exam tracked their progress from one testing period to the next.* For simplicity, we have grouped students scoring from 0 to 5 in group 1, from 6 to 10 in group 2, from 11 to 15 in group 3, and from 15 to 20 in group 4. The result is the following transition matrix.

$$
\begin{array}{c} 1 \\ 2 \\ 3 \\ 4 \end{array}
\begin{bmatrix}
.065 & .585 & .34 & .01 \\
.042 & .44 & .42 & .098 \\
.018 & .276 & .452 & .254 \\
0 & .044 & .292 & .664
\end{bmatrix}
$$

a. Find the long-range prediction for the proportion of the students in each group.

b. The authors of this study were interested in the number of testing periods required before a certain proportion of the students had mastered the material. Suppose that once a student reaches group 4, the student is said to have mastered the material and is no longer tested, so the student stays in that group forever. Initially, all of the students in the study were in group 1. Find the number of testing periods you would expect for at least 70% of the students to have mastered the material. (*Hint:* Try increasing values of n in $x_0 \cdot P^n$.)

PHYSICAL SCIENCES

36. *Weather* The weather in a certain spot is classified as fair, cloudy without rain, or rainy. A fair day is followed by a fair day 60% of the time, and by a cloudy day 25% of the

time. A cloudy day is followed by a cloudy day 35% of the time, and by a rainy day 25% of the time. A rainy day is followed by a cloudy day 40% of the time, and by another rainy day 25% of the time. What proportion of days are expected to be fair, cloudy, and rainy over the long term?

GENERAL INTEREST

37. *Ehrenfest Chain* The model for the Ehrenfest chain consists of 2 boxes containing a total of n balls, where n is any integer greater than or equal to 2. In each turn, a ball is picked at random and moved from whatever box it is in to the other box. Let the state of the Markov process be the number of balls in the first box.

a. Verify that the probability of going from state i to state j is given by the following.

$$
p_{ij} = \begin{cases}
\dfrac{i}{n} & \text{if } i \geq 1 \text{ and } j = i - 1 \\[2mm]
1 - \dfrac{i}{n} & \text{if } i \leq n - 1 \text{ and } j = i + 1 \\[2mm]
1 & \text{if } i = 0 \text{ and } j = 1 \text{ or } i = n \text{ and } j = n - 1 \\[1mm]
0 & \text{otherwise.}
\end{cases}
$$

b. Verify that the transition matrix is given by

$$
\begin{array}{c} 0 \\ 1 \\ 2 \\ \cdot \\ \cdot \\ n \end{array}
\begin{bmatrix}
0 & 1 & 0 & 0 & \cdots & 0 \\
\frac{1}{n} & 0 & 1 - \frac{1}{n} & 0 & \cdots & 0 \\
0 & \frac{2}{n} & 0 & 1 - \frac{2}{n} & \cdots & 0 \\
\cdot & \cdot & \cdot & \cdot & & \cdot \\
\cdot & \cdot & \cdot & \cdot & & \cdot \\
0 & 0 & 0 & 0 & \cdots & 0
\end{bmatrix}.
$$

c. Write the transition matrix for the case $n = 2$.

d. Determine whether the transition matrix in part c is a regular transition matrix.

e. Determine an equilibrium vector for the matrix in part c. Explain what the result means.

38. *Language* One of Markov's own applications was a 1913 study of how often a vowel is followed by another vowel or a consonant by another consonant in Russian text. A similar study of a passage of English text revealed the following transition matrix.

$$
\begin{array}{c} \\ \text{Vowel} \\ \text{Consonant} \end{array}
\begin{array}{cc} \text{Vowel} & \text{Consonant} \end{array}
\begin{bmatrix}
.12 & .88 \\
.54 & .46
\end{bmatrix}
$$

Find the percent of letters in English text that are expected to be vowels.

---

*Gunzenhauser, Georg W., and Raymond G. Taylor, "Concept Mastery and First Passage Time," *National Forum of Teacher Education Journal,* Vol 1, No. 1, 1992–92, pp. 29–34.

39. *Random Walk* Many phenomena can be viewed as examples of a random walk. Consider the following simple example. A security guard can stand in front of any one of three doors 20 ft apart in front of a building, and every minute he decides whether to move to another door chosen at random. If he is at the middle door, he is equally likely to stay where he is, move to the door to the left, or move to the door to the right. If he is at the door on either end, he is equally likely to stay where he is or move to the middle door.

a. Verify that the transition matrix is given by

$$
\begin{array}{c} \\ 1 \\ 2 \\ 3 \end{array}
\begin{array}{c} \begin{array}{ccc} 1 & 2 & 3 \end{array} \\ \begin{bmatrix} \frac{1}{2} & \frac{1}{2} & 0 \\ \frac{1}{3} & \frac{1}{3} & \frac{1}{3} \\ 0 & \frac{1}{2} & \frac{1}{2} \end{bmatrix} \end{array}.
$$

b. Find the long-range trend for the fraction of time the guard spends in front of each door.

10.3 ABSORBING MARKOV CHAINS

THINK ABOUT IT If a gambler gambles until she either goes broke or wins some predetermined amount of money, what is the probability that she will eventually go broke?

Using properties of absorbing Markov chains, we will answer this question. Suppose a Markov chain has transition matrix

$$
\begin{array}{c} \\ 1 \\ 2 \\ 3 \end{array}
\begin{array}{c} \begin{array}{ccc} 1 & 2 & 3 \end{array} \\ \begin{bmatrix} .3 & .6 & .1 \\ 0 & 1 & 0 \\ .6 & .2 & .2 \end{bmatrix} \end{array} = P.
$$

The matrix shows that p_{12}, the probability of going from state 1 to state 2, is .6, and that p_{22}, the probability of staying in state 2, is 1. Thus, once state 2 is entered, it is impossible to leave. For this reason, state 2 is called an *absorbing state*. Figure 5 shows a transition diagram for this matrix. The diagram shows that it is not possible to leave state 2.

Generalizing from this example leads to the following definition.

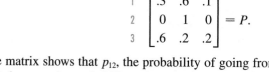

FIGURE 5

ABSORBING STATE

State i of a Markov chain is an **absorbing state** if $p_{ii} = 1$.

Using the idea of an absorbing state, we can define an absorbing Markov chain.

ABSORBING MARKOV CHAIN

A Markov chain is an **absorbing chain** if and only if the following two conditions are satisfied:

1. the chain has at least one absorbing state; and
2. it is possible to go from any nonabsorbing state to an absorbing state (perhaps in more than one step).

Note that the second condition does not mean that it is possible to go from any nonabsorbing state to *any* absorbing state, but it is possible to go to *some* absorbing state.

EXAMPLE 1 Absorbing Markov Chains

Identify all absorbing states in the Markov chains having the following matrices. Decide whether the Markov chain is absorbing.

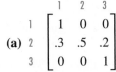

(a) $\begin{array}{c} \\ 1 \\ 2 \\ 3 \end{array} \begin{array}{ccc} 1 & 2 & 3 \\ \left[\begin{array}{ccc} 1 & 0 & 0 \\ .3 & .5 & .2 \\ 0 & 0 & 1 \end{array}\right] \end{array}$

Solution Since $p_{11} = 1$ and $p_{33} = 1$, both state 1 and state 3 are absorbing states. (Once these states are reached, they cannot be left.) The only nonabsorbing state is state 2. There is a .3 probability of going from state 2 to the absorbing state 1, and a .2 probability of going from state 2 to state 3, so that it is possible to go from the nonabsorbing state to an absorbing state. This Markov chain is absorbing. The transition diagram is shown in Figure 6.

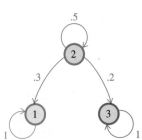

FIGURE 6

(b) $\begin{array}{c} \\ 1 \\ 2 \\ 3 \\ 4 \end{array} \begin{array}{cccc} 1 & 2 & 3 & 4 \\ \left[\begin{array}{cccc} .6 & 0 & .4 & 0 \\ 0 & 1 & 0 & 0 \\ .9 & 0 & .1 & 0 \\ 0 & 0 & 0 & 1 \end{array}\right] \end{array}$

Solution States 2 and 4 are absorbing, with states 1 and 3 nonabsorbing. From state 1, it is possible to go only to states 1 or 3; from state 3 it is possible to go only to states 1 or 3. As the transition diagram in Figure 7 shows, neither nonabsorbing state leads to an absorbing state, so that this Markov chain is nonabsorbing.

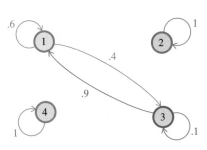

FIGURE 7

EXAMPLE 2 Gambler's Ruin

Suppose players A and B have a coin tossing game going on—a fair coin is tossed and the player predicting the toss correctly wins $1 from the other player. Suppose the players have a total of $6 between them, and that the game goes on until one player has no money (is ruined).

Let us agree that the states of this system are the amounts of money held by player A. There are seven possible states: A can have 0, 1, 2, 3, 4, 5, or 6 dollars. When either state 0 or state 6 is reached, the game is over. In any other state, the amount of money held by player A will increase by $1, or decrease by $1, with each of these events having probability $1/2$ (since we assume a fair coin). For example, in state 3 (A has $3), there is a $1/2$ chance of changing to state 2 and a $1/2$ chance of changing to state 4. Thus, $p_{32} = 1/2$ and $p_{34} = 1/2$. The probability of changing from state 3 to any other state is 0. Using this information gives the following 7×7 transition matrix.

$$\begin{array}{c} \\ 0 \\ 1 \\ 2 \\ 3 \\ 4 \\ 5 \\ 6 \end{array} \begin{array}{ccccccc} 0 & 1 & 2 & 3 & 4 & 5 & 6 \\ \left[\begin{array}{ccccccc} 1 & 0 & 0 & 0 & 0 & 0 & 0 \\ \frac{1}{2} & 0 & \frac{1}{2} & 0 & 0 & 0 & 0 \\ 0 & \frac{1}{2} & 0 & \frac{1}{2} & 0 & 0 & 0 \\ 0 & 0 & \frac{1}{2} & 0 & \frac{1}{2} & 0 & 0 \\ 0 & 0 & 0 & \frac{1}{2} & 0 & \frac{1}{2} & 0 \\ 0 & 0 & 0 & 0 & \frac{1}{2} & 0 & \frac{1}{2} \\ 0 & 0 & 0 & 0 & 0 & 0 & 1 \end{array}\right] \end{array} = P$$

Based on the rules of the game given above, states 0 and 6 are absorbing—once these states are reached, they can never be left, and the game is over. It is possible to get from one of the nonabsorbing states, 1, 2, 3, 4, or 5, to one of the absorbing states, so the Markov chain is absorbing.

For the long-term trend of the game, find various powers of the transition matrix. A computer or a graphing calculator can be used to verify the following results.

$$P^6 = \begin{bmatrix} 1.000 & .0000 & .0000 & .0000 & .0000 & .0000 & .0000 \\ .6875 & .0781 & .0000 & .1406 & .0000 & .0625 & .0313 \\ .4531 & .0000 & .2188 & .0000 & .2031 & .0000 & .1250 \\ .2188 & .1406 & .0000 & .2813 & .0000 & .1406 & .2188 \\ .1250 & .0000 & .2031 & .0000 & .2188 & .0000 & .4531 \\ .0313 & .0625 & .0000 & .1406 & .0000 & .0781 & .6875 \\ .0000 & .0000 & .0000 & .0000 & .0000 & .0000 & 1.0000 \end{bmatrix}$$

$$P^{20} = \begin{bmatrix} 1.000 & .0000 & .0000 & .0000 & .0000 & .0000 & .0000 \\ .8146 & .0094 & .0000 & .0188 & .0000 & .0094 & .1479 \\ .6385 & .0000 & .0282 & .0000 & .0282 & .0000 & .3052 \\ .4625 & .0188 & .0000 & .0375 & .0000 & .0188 & .4625 \\ .3052 & .0000 & .0282 & .0000 & .0282 & .0000 & .6385 \\ .1479 & .0094 & .0000 & .0188 & .0000 & .0094 & .8146 \\ .0000 & .0000 & .0000 & .0000 & .0000 & .0000 & 1.0000 \end{bmatrix}$$

As these results suggest, the system tends toward one of the absorbing states, so that the probability is 1 that one of the two gamblers will eventually be wiped out. ▪

EXAMPLE 3 Long-term Trend

Estimate the long-term trend for the following transition matrix.

$$P = \begin{bmatrix} .3 & .2 & .5 \\ 0 & 1 & 0 \\ 0 & 0 & 1 \end{bmatrix}$$

Solution Both states 2 and 3 are absorbing, and since it is possible to go from nonabsorbing state 1 to an absorbing state, the chain will eventually enter either state 2 or state 3. To find the long-term trend, let us find various powers of P.

$$P^2 = \begin{bmatrix} .09 & .26 & .65 \\ 0 & 1 & 0 \\ 0 & 0 & 1 \end{bmatrix} \qquad P^4 = \begin{bmatrix} .0081 & .2834 & .7085 \\ 0 & 1 & 0 \\ 0 & 0 & 1 \end{bmatrix}$$

$$P^8 = \begin{bmatrix} .0001 & .2857 & .7142 \\ 0 & 1 & 0 \\ 0 & 0 & 1 \end{bmatrix} \qquad P^{16} = \begin{bmatrix} .0000 & .2857 & .7142 \\ 0 & 1 & 0 \\ 0 & 0 & 1 \end{bmatrix}$$

Based on these powers, it appears that the transition matrix is getting closer and closer to the matrix

$$\begin{bmatrix} 0 & .29 & .71 \\ 0 & 1 & 0 \\ 0 & 0 & 1 \end{bmatrix}.$$

If the system is originally in state 1, there is no chance it will end up in state 1, but a .29 chance that it will end up in state 2 and a .71 chance it will end up in state 3. If the system was originally in state 2, it will end up in state 2; a similar statement can be made for state 3. ■

The examples suggest the following properties of absorbing Markov chains, which can be verified using more advanced methods.

1. Regardless of the original state of an absorbing Markov chain, in a finite number of steps the chain will enter an absorbing state and then stay in that state.

2. The powers of the transition matrix get closer and closer to some particular matrix.

3. The long-term trend depends on the initial state—changing the initial state can change the final result.

The third property distinguishes absorbing Markov chains from regular Markov chains, where the final result is independent of the initial state.

It would be preferable to have a method for finding the final probabilities of entering an absorbing state without finding all the powers of the transition matrix, as in Example 3. We do not really need to worry about the absorbing states (to enter an absorbing state is to stay there). Therefore, it is necessary only to work with the nonabsorbing states. To see how this is done, let us use as an example the transition matrix from the gambler's ruin problem in Example 2. Rewrite the matrix so that the rows and columns corresponding to the absorbing states come first.

$$
\begin{array}{c}
 \\
 \\
0 \\ 6 \\ 1 \\ 2 \\ 3 \\ 4 \\ 5
\end{array}
\begin{array}{cc}
\overbrace{\quad\text{Absorbing}\quad}^{} & \overbrace{\quad\quad\text{Nonabsorbing}\quad\quad}^{} \\
\begin{array}{cc} 0 & 6 \end{array} & \begin{array}{ccccc} 1 & 2 & 3 & 4 & 5 \end{array} \\
\left[\begin{array}{cc|ccccc}
1 & 0 & 0 & 0 & 0 & 0 & 0 \\
0 & 1 & 0 & 0 & 0 & 0 & 0 \\
\hline
\frac{1}{2} & 0 & 0 & \frac{1}{2} & 0 & 0 & 0 \\
0 & 0 & \frac{1}{2} & 0 & \frac{1}{2} & 0 & 0 \\
0 & 0 & 0 & \frac{1}{2} & 0 & \frac{1}{2} & 0 \\
0 & 0 & 0 & 0 & \frac{1}{2} & 0 & \frac{1}{2} \\
0 & \frac{1}{2} & 0 & 0 & 0 & \frac{1}{2} & 0
\end{array}\right] = P
\end{array}
$$

Let I_2 represent the 2×2 identity matrix in the upper left corner; let O represent the matrix of zeros in the upper right; let R represent the matrix in the lower left, and let Q represent the matrix in the lower right. Using these symbols, P can be written as

$$
P = \left[\begin{array}{c|c} I_2 & O \\ \hline R & Q \end{array}\right].
$$

The **fundamental matrix** for an absorbing Markov chain is defined as matrix F, where

$$
F = (I_n - Q)^{-1}.
$$

Here I_n is the $n \times n$ identity matrix corresponding in size to matrix Q, so that the difference $I_n - Q$ exists.

FOR REVIEW ■■■■

To find the inverse of a matrix, we first form an augmented matrix by putting the original matrix on the left and the identity matrix on the right: $[A \mid I]$. The Gauss-Jordan process is used to turn the matrix on the left into the identity. The matrix on the right is then the inverse of the original matrix: $[I \mid A^{-1}]$.

For the gambler's ruin problem, using I_5 gives

$$
F = \left(\begin{bmatrix} 1 & 0 & 0 & 0 & 0 \\ 0 & 1 & 0 & 0 & 0 \\ 0 & 0 & 1 & 0 & 0 \\ 0 & 0 & 0 & 1 & 0 \\ 0 & 0 & 0 & 0 & 1 \end{bmatrix} - \begin{bmatrix} 0 & \frac{1}{2} & 0 & 0 & 0 \\ \frac{1}{2} & 0 & \frac{1}{2} & 0 & 0 \\ 0 & \frac{1}{2} & 0 & \frac{1}{2} & 0 \\ 0 & 0 & \frac{1}{2} & 0 & \frac{1}{2} \\ 0 & 0 & 0 & \frac{1}{2} & 0 \end{bmatrix}\right)^{-1}
$$

$$
= \begin{bmatrix} 1 & -\frac{1}{2} & 0 & 0 & 0 \\ -\frac{1}{2} & 1 & -\frac{1}{2} & 0 & 0 \\ 0 & -\frac{1}{2} & 1 & -\frac{1}{2} & 0 \\ 0 & 0 & -\frac{1}{2} & 1 & -\frac{1}{2} \\ 0 & 0 & 0 & -\frac{1}{2} & 1 \end{bmatrix}^{-1}
$$

$$
\begin{array}{c} \\ 1 \\ 2 \\ = 3 \\ 4 \\ 5 \end{array}
\begin{array}{ccccc} 1 & 2 & 3 & 4 & 5 \\ \end{array}
\begin{bmatrix} \frac{5}{3} & \frac{4}{3} & 1 & \frac{2}{3} & \frac{1}{3} \\ \frac{4}{3} & \frac{8}{3} & 2 & \frac{4}{3} & \frac{2}{3} \\ 1 & 2 & 3 & 2 & 1 \\ \frac{2}{3} & \frac{4}{3} & 2 & \frac{8}{3} & \frac{4}{3} \\ \frac{1}{3} & \frac{2}{3} & 1 & \frac{4}{3} & \frac{5}{3} \end{bmatrix}.
$$

The inverse was found using techniques from Chapter 2. Recall, we also discussed finding the inverse of a matrix with a graphing calculator there.

The fundamental matrix gives the expected number of visits to each state before absorption occurs. For example, if player A currently has $2, then the second row of the fundamental matrix just computed says that she expects to have $1 an average of $1\frac{1}{3}$ times, and to have $3 twice, before quitting the game because she either runs out of money or wins $6. The total number of times that player A expects to have various amounts of money before quitting the game is the sum of entries in row 2 of F: $(4/3) + (8/3) + 2 + (4/3) + (2/3) = 8$. In other words, if player A currently has $2, she can expect to stay in the game 8 more turns before either she or player B goes broke.

To see why this is true, consider a Markov chain currently in state i. The expected number of times that the chain visits state j at this step is 1 for i and 0 for all other states. The expected number of times that the chain visits state j at the next step is given by the element in row i, column j of the transition matrix Q. The expected number of times the chain visits state j two steps from now is given by the corresponding entry in the matrix Q^2. The expected number of visits in all steps is given by $I + Q + Q^2 + Q^3 + \cdots$. To find out whether this infinite sum is the same as $(I - Q)^{-1}$, multiply the sum by $(I - Q)$:

$$
(I + Q + Q^2 + Q^3 + \cdots)(I - Q)
$$
$$
= I + Q + Q^2 + Q^3 + \cdots - Q - Q^2 - Q^3 + \cdots = I,
$$

which verifies our result.

It can be shown that

$$
P^k = \left[\begin{array}{c|c} I_m & O \\ \hline (I + Q + Q^2 + \cdots + Q^{k-1})R & Q^k \end{array}\right],
$$

where I_m is the $m \times m$ identity matrix. As $k \rightarrow \infty$, $Q^k \rightarrow O_n$, the $n \times n$ zero matrix, and

$$P^k \rightarrow \left[\begin{array}{c|c} I_m & O \\ \hline FR & O_n \end{array}\right],$$

so we see that FR gives the probabilities that if the system was originally in a nonabsorbing state, it ends up in one of the absorbing states.*

Finally, use the fundamental matrix F along with matrix R found above to get the product FR.

$$FR = \begin{bmatrix} \frac{5}{3} & \frac{4}{3} & 1 & \frac{2}{3} & \frac{1}{3} \\ \frac{4}{3} & \frac{8}{3} & 2 & \frac{4}{3} & \frac{2}{3} \\ 1 & 2 & 3 & 2 & 1 \\ \frac{2}{3} & \frac{4}{3} & 2 & \frac{8}{3} & \frac{4}{3} \\ \frac{1}{3} & \frac{2}{3} & 1 & \frac{4}{3} & \frac{5}{3} \end{bmatrix} \begin{bmatrix} \frac{1}{2} & 0 \\ 0 & 0 \\ 0 & 0 \\ 0 & 0 \\ 0 & \frac{1}{2} \end{bmatrix} = \begin{array}{c} \\ 1 \\ 2 \\ 3 \\ 4 \\ 5 \end{array} \overset{\begin{array}{cc} 0 & 6 \end{array}}{\begin{bmatrix} \frac{5}{6} & \frac{1}{6} \\ \frac{2}{3} & \frac{1}{3} \\ \frac{1}{2} & \frac{1}{2} \\ \frac{1}{3} & \frac{2}{3} \\ \frac{1}{6} & \frac{5}{6} \end{bmatrix}}$$

The product matrix FR gives the probability that if the system was originally in a nonabsorbing state, it ended up in either of the two absorbing states. For example, the probability is $2/3$ that if the system was originally in state 2, it ended up in state 0; the probability is $5/6$ that if the system was originally in state 5 it ended up in state 6, and so on. Based on the original statement of the gambler's ruin problem, if player A starts with \$2 (state 2), there is a $2/3$ chance of ending in state 0 (player A is ruined); if player A starts with \$5 (state 5) there is a $1/6$ chance of player A being ruined, and so on.

In the fundamental matrix F, the sum of the elements in row i gives the expected number of steps for the matrix to enter an absorbing state from the ith nonabsorbing state. Thus, for $i = 2$, $(4/3) + (8/3) + 2 + (4/3) + (2/3) = 8$ steps will be needed on the average for the system to go from state 2 to an absorbing state (0 or 6).

Let us summarize what we have learned about absorbing Markov chains.

PROPERTIES OF ABSORBING MARKOV CHAINS

1. Regardless of the initial state, in a finite number of steps the chain will enter an absorbing state and then stay in that state.

2. The powers of the transition matrix get closer and closer to some particular matrix.

3. The long-term trend depends on the initial state.

4. Let P be the transition matrix for an absorbing Markov chain. Rearrange the rows and columns of P so that the absorbing states come first. Matrix P will have the form

$$P = \left[\begin{array}{c|c} I_m & O \\ \hline R & Q \end{array}\right],$$

---

*We have omitted details in these steps that can be justified using advanced techniques.

where I_m is an identity matrix, with m equal to the number of absorbing states, and O is a matrix of all zeros. The fundamental matrix is defined as

$$F = (I_n - Q)^{-1},$$

where I_n has the same size as Q. The element in row i, column j of the fundamental matrix gives the number of visits to state j that are expected to occur before absorption, given that the current state is state i.

5. The product FR gives the matrix of probabilities that a particular initial nonabsorbing state will lead to a particular absorbing state.

EXAMPLE 4 Long-term Trend

Find the long-term trend for the transition matrix

$$\begin{array}{c} \\ 1 \\ 2 \\ 3 \end{array}\begin{array}{ccc} 1 & 2 & 3 \\ \begin{bmatrix} .3 & .2 & .5 \\ 0 & 1 & 0 \\ 0 & 0 & 1 \end{bmatrix} \end{array} = P$$

of Example 3.

Solution Rewrite the matrix so that absorbing states 2 and 3 come first.

$$\begin{array}{c} \\ 2 \\ 3 \\ 1 \end{array}\begin{array}{ccc} 2 & 3 & 1 \\ \left[\begin{array}{cc|c} 1 & 0 & 0 \\ 0 & 1 & 0 \\ \hline .2 & .5 & .3 \end{array}\right] \end{array}$$

Here $R = [.2 \quad .5]$ and $Q = [.3]$. Find the fundamental matrix F.

$$F = (I_1 - Q)^{-1} = [1 - .3]^{-1} = [.7]^{-1} = [1/.7] = [10/7]$$

The product FR is

$$FR = [10/7][.2 \quad .5] = [2/7 \quad 5/7] \approx [.286 \quad .714].$$

If the system starts in the nonabsorbing state 1, there is a 2/7 chance of ending up in the absorbing state 2 and a 5/7 chance of ending up in the absorbing state 3.

10.3 EXERCISES

Find all absorbing states for the transition matrices in Exercises 1–6. Which are transition matrices for absorbing Markov chains?

1. $\begin{bmatrix} .15 & .05 & .8 \\ 0 & 1 & 0 \\ .4 & .6 & 0 \end{bmatrix}$

2. $\begin{bmatrix} .1 & .5 & .4 \\ .2 & .2 & .6 \\ 0 & 0 & 1 \end{bmatrix}$

3. $\begin{bmatrix} .4 & 0 & .6 \\ 0 & 1 & 0 \\ .9 & 0 & .1 \end{bmatrix}$

4. $\begin{bmatrix} .5 & .5 & 0 \\ .8 & .2 & 0 \\ 0 & 0 & 1 \end{bmatrix}$

5. $\begin{bmatrix} .2 & .5 & .1 & .2 \\ 0 & 1 & 0 & 0 \\ .9 & .02 & .04 & .04 \\ 0 & 0 & 0 & 1 \end{bmatrix}$

6. $\begin{bmatrix} .32 & .41 & .16 & .11 \\ .42 & .30 & 0 & .28 \\ 0 & 0 & 0 & 1 \\ 1 & 0 & 0 & 0 \end{bmatrix}$

Find the fundamental matrix F for the absorbing Markov chains with the matrices in Exercises 7–14. Also, find the product matrix FR.

7. $\begin{bmatrix} 1 & 0 & 0 \\ 0 & 1 & 0 \\ .2 & .3 & .5 \end{bmatrix}$

8. $\begin{bmatrix} 1 & 0 & 0 \\ .6 & .1 & .3 \\ 0 & 0 & 1 \end{bmatrix}$

9. $\begin{bmatrix} 1 & 0 & 0 \\ 0 & 1 & 0 \\ \frac{1}{3} & \frac{1}{3} & \frac{1}{3} \end{bmatrix}$

10. $\begin{bmatrix} 1 & 0 & 0 \\ \frac{3}{8} & \frac{1}{8} & \frac{1}{2} \\ 0 & 0 & 1 \end{bmatrix}$

11. $\begin{bmatrix} 1 & 0 & 0 & 0 \\ \frac{1}{3} & 0 & \frac{2}{3} & 0 \\ 0 & 0 & 1 & 0 \\ \frac{1}{4} & \frac{1}{4} & \frac{1}{4} & \frac{1}{4} \end{bmatrix}$

12. $\begin{bmatrix} \frac{1}{4} & \frac{1}{2} & 0 & \frac{1}{4} \\ 0 & 1 & 0 & 0 \\ 0 & 0 & 1 & 0 \\ \frac{1}{2} & 0 & 0 & \frac{1}{2} \end{bmatrix}$

13. $\begin{bmatrix} 1 & 0 & 0 & 0 & 0 \\ 0 & 1 & 0 & 0 & 0 \\ .1 & .2 & .3 & .2 & .2 \\ .3 & .5 & .1 & 0 & .1 \\ 0 & 0 & 0 & 0 & 1 \end{bmatrix}$

14. $\begin{bmatrix} .4 & .2 & .3 & 0 & .1 \\ 0 & 1 & 0 & 0 & 0 \\ 0 & 0 & 1 & 0 & 0 \\ .1 & .5 & .1 & .1 & .2 \\ 0 & 0 & 0 & 0 & 1 \end{bmatrix}$

15. a. Write a transition matrix for a gambler's ruin problem when player A and player B start with a total of $4.

   b. Find matrix *F* for this transition matrix, and find the product matrix *FR*.

   c. Suppose player A starts with $1. What is the probability of ruin for A?

   d. Suppose player A starts with $3. What is the probability of ruin for A?

16. Suppose player B (Exercise 15) slips in a coin that is slightly "loaded"—such that the probability that B wins a particular toss changes from $1/2$ to $3/5$. Suppose that A and B start the game with a total of $5.

   a. If B starts with $3, find the probability that A will be ruined.

   b. If B starts with $1, find the probability that A will be ruined.

It can be shown that the probability of ruin for player A in a game such as the one described in this section is

$$x_a = \frac{b}{a+b} \text{ if } r = 1, \quad \text{and} \quad x_a = \frac{r^a - r^{a+b}}{1 - r^{a+b}} \text{ if } r \ne 1,$$

where a is the initial amount of money that player A has, b is the initial amount that player B has, $r = (1 - p)/p$, and p is the probability that player A will win on a given play.

17. Find the probability that A will be ruined if $a = 10$, $b = 30$, and $p = .49$.

18. Find the probability in Exercise 17 if *p* changes to .50.

19. Complete the following chart, assuming $a = 10$ and $b = 10$.

p	.1	.2	.3	.4	.5	.6	.7	.8	.9
x_a									

20. How can we calculate the expected total number of times a Markov chain will visit state *j* before absorption, regardless of the current state?

21. Suppose an absorbing Markov chain has only one absorbing state. What is the product *FR*?

Applications

BUSINESS AND ECONOMICS

22. *Company Training Program* A company with a new training program classified each worker into one of the following four categories: s_1, never in the program; s_2, currently in the program; s_3, discharged; s_4, completed the program. The transition matrix for this company is given below.

$$
\begin{array}{c}
\\
s_1 \\
s_2 \\
s_3 \\
s_4
\end{array}
\begin{array}{cccc}
s_1 & s_2 & s_3 & s_4 \\
\left[\begin{array}{cccc}
.4 & .2 & .05 & .35 \\
0 & .45 & .05 & .5 \\
0 & 0 & 1 & 0 \\
0 & 0 & 0 & 1
\end{array}\right]
\end{array}
$$

a. Find F and FR.

b. Find the probability that a worker originally in the program is discharged.

c. Find the probability that a worker not originally in the program goes on to complete the program.

LIFE SCIENCES

23. *Medical Prognosis* A study using Markov chains to estimate a patient's prognosis for improving under various treatment plans gives the following transition matrix as an example:*

$$
\begin{array}{c}
\\
\text{well} \\
\text{ill} \\
\text{dead}
\end{array}
\begin{array}{ccc}
\text{well} & \text{ill} & \text{dead} \\
\left[\begin{array}{ccc}
.3 & .5 & .2 \\
0 & .5 & .5 \\
0 & 0 & 1
\end{array}\right]
\end{array}
$$

a. Estimate the probability that a well person will eventually end up dead.

b. Verify your answer to part a using the matrix product FR.

c. Find the expected number of cycles that a well patient will continue to be well before dying, and the expected number of cycles that a well patient will be ill before dying.

24. *Contagion* Under certain conditions, the probability that a person will get a particular contagious disease and die from it is .05, and the probability of getting the disease and surviving is .15. The probability that a survivor will infect another person who dies from it is also .05, that a survivor will infect another person who survives it is .15, and so on. A transition matrix using the following states is given

below. A person in state 1 is one who gets the disease and dies, a person in state 2 gets the disease and survives, and a person in state 3 does not get the disease. Consider a chain of people, each of whom interacts with the previous person and may catch the disease from that individual, and then may infect the next person.

a. Verify that the transition matrix is as follows:

$$
\begin{array}{c}
\\
\\
\textit{First Person} \\
\\
\end{array}
\begin{array}{c}
\textit{Second Person} \\
\begin{array}{ccc}
1 & 2 & 3
\end{array} \\
\begin{array}{c}
1 \\
2 \\
3
\end{array}
\left[\begin{array}{ccc}
.05 & .15 & .8 \\
.05 & .15 & .8 \\
0 & 0 & 1
\end{array}\right].
\end{array}
$$

b. Find F and FR.

c. Find the probability that the disease eventually disappears.

d. Given a person who has the disease and survives, find the expected number of people in the chain who will get the disease until a person who does not get the disease is reached.

SOCIAL SCIENCES

25. *Student Retention* At a particular two-year college, a student has a probability of .25 of flunking out during a given year, a .15 probability of having to repeat the year, and a .6 probability of finishing the year. Use the states below.

State	Meaning
1	Freshman
2	Sophomore
3	Has flunked out
4	Has graduated

a. Write a transition matrix. Find F and FR.

b. Find the probability that a freshman will graduate.

c. Find the expected number of years that a freshman will be in college before graduating or flunking out.

26. *Transportation* The city of Sacramento recently completed a new light rail system to bring commuters and shoppers into the downtown area and relieve freeway congestion. City planners estimate that each year, 15% of those who drive or ride in an automobile will change to the light rail

---

*Beck, J. Robert, and Stephen G. Paukeer, "The Markov Process in Medical Prognosis," *Medical Decision Making*, Vol. 4, No. 3, 1983, pp. 419–458.

system; 80% will continue to use automobiles; and the rest will no longer go to the downtown area. Of those who use light rail, 5% will go back to using an automobile, 80% will continue to use light rail, and the rest will stay out of the downtown area. Assume those who do not go downtown will continue to stay out of the downtown area.

a. Write a transition matrix. Find F and FR.

b. Find the probability that a person who commuted by automobile ends up avoiding the downtown area.

c. Find the expected number of years until a person who commutes by automobile this year no longer enters the downtown area.

27. *Education Careers* Data has been collected on the likelihood that a teacher, or a student with a declared interest in teaching, will continue on that career path the following year.* We have simplified the classification of the original data to four groups: high school and college students, new teachers, continuing teachers, and those who have quit the profession. The transition probabilities are given in the following matrix.

	Student	New	Continuing	Quit
Student	.70	.11	0	.19
New	0	0	.86	.14
Continuing	0	0	.88	.12
Quit	0	0	0	1

a. Find the expected number of years that a student with an interest in teaching will spend as a continuing teacher.

b. Find the expected number of years that a new teacher will spend as a continuing teacher.

c. Find the expected number of additional years that a continuing teacher will spend as a continuing teacher.

d. Notice that the answer to part b is larger than the answer to part a, and the answer to part c is even larger. Explain why this is to be expected.

e. What other states might be added to this model to make it more realistic? Discuss how this would affect the transition matrix. (See the Extended Application for this chapter.)

28. *Rat Maze* A rat is placed at random in one of the compartments of the maze pictured. The probability that a rat in compartment 1 will move to compartment 2 is .3; to compartment 3 is .2; and to compartment 4 is .1. A rat in compartment 2 will move to compartments 1, 4, or 5 with probabilities of .2, .6, and .1, respectively. A rat in compartment 3 cannot leave that compartment. A rat in compartment 4 will move to 1, 2, 3, or 5 with probabilities of .1, .1, .4, and .3, respectively. A rat in compartment 5 cannot leave that compartment.

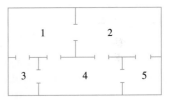

a. Set up a transition matrix using this information. Find matrices F and FR.

Find the probability that a rat ends up in compartment 5 if it was originally in the given compartment.

b. 1   **c.** 2   **d.** 3   **e.** 4

f. Find the expected number of times that a rat in compartment 1 will be in compartment 1 before ending up in compartment 3 or 5.

g. Find the expected number of times that a rat in compartment 4 will be in compartment 4 before ending up in compartment 3 or 5.

GENERAL INTEREST

29. *Gambler's Ruin* **a.** Write a transition matrix for a gambler's ruin problem, where players A and B start with a total of $7. (See Example 2.)

b. Find the probability of ruin for A if A starts with $4.

c. Find the probability of ruin for A if A starts with $5.

---

*Taylor, Raymond G., "Forecasting Teacher Shortages," *National Forum of Educational Administration and Supervision Journal*, Vol. 7, No. 2, 1990.

■ **CHAPTER SUMMARY**

Markov chains are useful for modeling situations in which there are a finite number of possibilities, known as *states*, and the probability of moving from one state to another during a specified period of time remains constant over time. The matrix containing these probabilities is called the transition matrix. To find the transition probabilities over n time periods, raise the transition matrix to the nth power. A regular Markov chain has the property that some power of the transition matrix has no zeros. In this case, an equilibrium vector can be found, which gives the long-range probability of being in each state. An absorbing Markov chain has one or more states, known as absorbing states, that are impossible to leave. The probability that a particular nonabsorbing state will lead to a particular absorbing state can be calculated by methods shown in this chapter. The applications in this chapter show the usefulness of Markov chains, whether regular or absorbing.

■ **KEY TERMS**

state	probability vector	equilibrium (or fixed)	fundamental matrix
10.1 transition diagram	10.2 regular transition	vector	
transition matrix	matrix	10.3 absorbing state	
Markov chain	regular Markov chain	absorbing chain	

■ **CHAPTER 10 REVIEW EXERCISES**

1. How can you tell by looking at a matrix whether it represents the transition matrix for a Markov chain?

2. Under what conditions is the existence of an equilibrium vector guaranteed?

Decide whether each of the matrices in Exercises 3–6 could be a transition matrix.

3. $\begin{bmatrix} .4 & .6 \\ 1 & 0 \end{bmatrix}$

4. $\begin{bmatrix} -.2 & 1.2 \\ .8 & .2 \end{bmatrix}$

5. $\begin{bmatrix} .8 & .2 & 0 \\ 0 & 1 & 0 \\ .1 & .4 & .5 \end{bmatrix}$

6. $\begin{bmatrix} .6 & .2 & .3 \\ .1 & .5 & .4 \\ .3 & .3 & .4 \end{bmatrix}$

For each of the transition matrices in Exercises 7–10, **a.** *find the first three powers; and* **b.** *find the probability that state 2 changes to state 1 after three repetitions of the experiment.*

7. $C = \begin{bmatrix} .6 & .4 \\ 1 & 0 \end{bmatrix}$

8. $D = \begin{bmatrix} .3 & .7 \\ .5 & .5 \end{bmatrix}$

9. $E = \begin{bmatrix} .2 & .5 & .3 \\ .1 & .8 & .1 \\ 0 & 1 & 0 \end{bmatrix}$

10. $F = \begin{bmatrix} .14 & .12 & .74 \\ .35 & .28 & .37 \\ .71 & .24 & .05 \end{bmatrix}$

In Exercises 11–14, use the transition matrix T, along with the given initial distribution D, to find the distribution after two repetitions of the experiment. Also, predict the long-range distribution.

11. $D = \begin{bmatrix} .3 & .7 \end{bmatrix}$; $T = \begin{bmatrix} .4 & .6 \\ .5 & .5 \end{bmatrix}$

12. $D = \begin{bmatrix} .8 & .2 \end{bmatrix}$; $T = \begin{bmatrix} .7 & .3 \\ .2 & .8 \end{bmatrix}$

13. $D = [.2 \quad .4 \quad .4]; \ T = \begin{bmatrix} .6 & .2 & .2 \\ .3 & .3 & .4 \\ .5 & .4 & .1 \end{bmatrix}$

14. $D = [.1 \quad .1 \quad .8]; \ T = \begin{bmatrix} .2 & .3 & .5 \\ .1 & .1 & .8 \\ .7 & .1 & .2 \end{bmatrix}$

15. How can you tell from the transition matrix whether a Markov chain is regular or not?

Decide whether each of the following transition matrices is regular.

16. $\begin{bmatrix} 0 & 1 \\ .2 & .8 \end{bmatrix}$

17. $\begin{bmatrix} .4 & .2 & .4 \\ 0 & 1 & 0 \\ .6 & .3 & .1 \end{bmatrix}$

18. $\begin{bmatrix} 1 & 0 & 0 \\ 0 & 1 & 0 \\ .3 & .5 & .2 \end{bmatrix}$

19. How can you tell from the transition matrix whether a Markov chain is absorbing or not?

20. How can you tell from the transition matrix where the absorbing states are in an absorbing chain?

21. Can a Markov chain be both regular and absorbing? Explain.

Find all absorbing states for the matrices in Exercises 22–24. Which are transition matrices for an absorbing Markov chain?

22. $\begin{bmatrix} 1 & 0 & 0 \\ .5 & .1 & .4 \\ 0 & 1 & 0 \end{bmatrix}$

23. $\begin{bmatrix} .2 & 0 & .8 \\ 0 & 1 & 0 \\ .7 & 0 & .3 \end{bmatrix}$

24. $\begin{bmatrix} .5 & .1 & .1 & .3 \\ 0 & 0 & 1 & 0 \\ 1 & 0 & 0 & 0 \\ .1 & .8 & .05 & .05 \end{bmatrix}$

In Exercises 25–28, find the fundamental matrix F for the absorbing Markov chains with matrices as follows. Also find the matrix FR.

25. $\begin{bmatrix} .2 & .5 & .3 \\ 0 & 1 & 0 \\ 0 & 0 & 1 \end{bmatrix}$

26. $\begin{bmatrix} 1 & 0 & 0 \\ 0 & 1 & 0 \\ .3 & .1 & .6 \end{bmatrix}$

27. $\begin{bmatrix} \frac{1}{5} & \frac{1}{5} & \frac{2}{5} & \frac{1}{5} \\ 0 & 1 & 0 & 0 \\ \frac{1}{2} & \frac{1}{4} & \frac{1}{8} & \frac{1}{8} \\ 0 & 0 & 0 & 1 \end{bmatrix}$

28. $\begin{bmatrix} .3 & .5 & .1 & .1 \\ .4 & .1 & .3 & .2 \\ 0 & 0 & 1 & 0 \\ 0 & 0 & 0 & 1 \end{bmatrix}$

Applications

BUSINESS AND ECONOMICS

Advertising Currently, 35% of all hot dogs sold in one area are made by Dogkins, and 65% are made by Long Dog. Suppose that Dogkins starts a heavy advertising campaign, with the campaign producing the following transition matrix.

		After Campaign	
		Dogkins	Long Dog
Before	Dogkins	.8	.2
Campaign	Long Dog	.4	.6

29. Find the share of the market for each company

   a. after one campaign.

   b. after three such campaigns.

30. Predict the long-range market share for Dogkins.

Credit Cards   A credit card company classifies its customers in three groups; nonusers in a given month, light users, and heavy users. The transition matrix for these states is

$$
\begin{array}{c}
 \\
\text{Nonuser} \\
\text{Light} \\
\text{Heavy}
\end{array}
\begin{array}{ccc}
\text{Nonuser} & \text{Light} & \text{Heavy}
\end{array}
\begin{bmatrix}
.8 & .15 & .05 \\
.25 & .55 & .2 \\
.04 & .21 & .75
\end{bmatrix}.
$$

Suppose the initial distribution for the three states is $[.4 \quad .4 \quad .2]$. *Find the distribution after each of the following periods.*

31. 1 mo   **32.** 2 mo   **33.** 3 mo

34. What is the long-range prediction for the distribution of users?

LIFE SCIENCES

35. *Medical Prognosis*   A study of patients at the University of North Carolina Hospitals used a Markov chain model with three categories of patients: 0 (death), 1 (unfavorable status), and 2 (favorable status).* The transition matrix for a cycle of 72 hr was as follows.

$$
\begin{array}{c}
 \\
0 \\
1 \\
2
\end{array}
\begin{array}{ccc}
0 & 1 & 2
\end{array}
\begin{bmatrix}
1 & 0 & 0 \\
.085 & .779 & .136 \\
.017 & .017 & .966
\end{bmatrix}
$$

   a. Find the fundamental matrix.

   b. For a patient with a favorable status, find the expected number of cycles that the patient will continue to have that status before dying.

   c. For a patient with an unfavorable status, find the expected number of cycles that the patient will have a favorable status before dying.

Medical Research   A medical researcher is studying the risk of heart attack in men. She first divides men into three weight categories: thin, normal, and overweight. By studying the male ancestors, sons, and grandsons of these men, the researcher comes up with the following transition matrix.

$$
\begin{array}{c}
 \\
\text{Thin} \\
\text{Normal} \\
\text{Overweight}
\end{array}
\begin{array}{ccc}
\text{Thin} & \text{Normal} & \text{Overweight}
\end{array}
\begin{bmatrix}
.3 & .5 & .2 \\
.2 & .6 & .2 \\
.1 & .5 & .4
\end{bmatrix}
$$

Find the probabilities of the following for a man of normal weight.

36. Thin son

37. Thin grandson

38. Thin great-grandson

Find the probabilities of the following for an overweight man.

39. Overweight son

40. Overweight grandson

41. Overweight great-grandson

Suppose that the distribution of men by weight is initially given by $[.2 \quad .55 \quad .25]$. *Find each of the following distributions.*

42. After 1 generation

43. After 2 generations

44. After 3 generations

45. Find the long-range prediction for the distribution of weights.

Genetics   Researchers sometimes study the problem of mating the offspring from the same two parents; two of these offspring are then mated, and so on. Let A be a dominant gene for some trait, and a the recessive gene. The original offspring can carry genes AA, Aa, or aa. There are six possible ways that these offspring can mate.

State	Mating
1	AA and AA
2	AA and Aa
3	AA and aa
4	Aa and Aa
5	Aa and aa
6	aa and aa

46. Suppose that the offspring are randomly mated with each other. Verify that the transition matrix is given by the matrix below.

$$
\begin{array}{c}
1 \\
2 \\
3 \\
4 \\
5 \\
6
\end{array}
\begin{array}{cccccc}
1 & 2 & 3 & 4 & 5 & 6
\end{array}
\begin{bmatrix}
1 & 0 & 0 & 0 & 0 & 0 \\
\frac{1}{4} & \frac{1}{2} & 0 & \frac{1}{4} & 0 & 0 \\
0 & 0 & 0 & 1 & 0 & 0 \\
\frac{1}{16} & \frac{1}{4} & \frac{1}{8} & \frac{1}{4} & \frac{1}{4} & \frac{1}{16} \\
0 & 0 & 0 & \frac{1}{4} & \frac{1}{2} & \frac{1}{4} \\
0 & 0 & 0 & 0 & 0 & 1
\end{bmatrix}
$$

---

*Chen, Pai-Lien, Estrada J. Bernard, and Pranab K. Sen, "A Markov Chain Model Used in Analyzing Disease History Applied to a Stroke Study," *Journal of Applied Statistics*, Vol. 26, No. 4, 1999, pp. 413–422.

47. Identify the absorbing states.

48. Find matrix Q.

49. Find F, and the product FR.

50. If two parents with the genes Aa are mated, find the number of pairs of offspring with these genes that can be expected before either the dominant or the recessive gene no longer appears.

51. If two parents with the genes Aa are mated, find the probability that the recessive gene will eventually disappear.

52. *Genetics* Suppose that a set of n genes includes m mutant genes. For the next generation, these genes are duplicated and a subset of n genes is selected from the $2n$ genes containing $2m$ mutant genes. Let the state be given by the number of mutant genes.

a. Verify that the transition probability from state i to state j is given by

$$p_{ij} = \frac{\binom{2i}{j}\binom{2n-2i}{n-j}}{\binom{2n}{n}},$$

where i and j are the number of mutant genes in this generation and the next, and where $\binom{n}{r}$ represents the number of combinations of n objects taken r at a time, discussed in Chapter 8. (*Hint:* Let $\binom{n}{r} = 0$ when $n < r$. If the current generation has i mutant genes, it has $n - i$ nonmutant genes. When the genes are duplicated, this results in $2i$ mutant genes and $2(n - i)$ nonmutant genes.)

b. What are the absorbing states in this chain?

c. Calculate the transition matrix for the case $n = 3$.

d. Find the fundamental matrix F and the product FR for $n = 3$.

e. If a set of 3 genes has 1 mutant gene, what is the probability that the mutant gene will eventually disappear?

f. If a set of 3 genes has 1 mutant gene, how many generations would be expected to have 1 mutant gene before either the mutant genes or the nonmutant genes disappear?

GENERAL INTEREST

53. *Monopoly* In an article on a Markov chain analysis of the game Monopoly, a simplified version of the game is presented.* The board consists of the four squares shown.

Players move clockwise by flipping a coin and moving once for heads and twice for tails. If you land on the Policeman, you go directly to jail. Therefore, landing on the Policeman is equivalent to landing on Jail, so we need only consider three states in our Markov chain. If you land on the Community Chest, you pick a card that might give or take away money, or might send you to Jail or to Go. The result is the following transition matrix.

$$
\begin{array}{c}
\text{Jail} \\
\text{CC} \\
\text{Go}
\end{array}
\begin{array}{ccc}
\text{Jail} & \text{CC} & \text{Go} \\
\end{array}
\begin{bmatrix}
\frac{17}{32} & \frac{7}{16} & \frac{1}{32} \\
\frac{1}{2} & 0 & \frac{1}{2} \\
1 & 0 & 0
\end{bmatrix}
$$

a. Explain rows 2 and 3 of the transition matrix.

b. There are 16 cards for the Community Chest. One sends you to Jail, one sends you to Go, and the others leave you on Community Chest. Use these facts to explain row 1 of the transition matrix.

c. Find the long-term probabilities for being on Jail, the Community Chest, and Go.

---

*Abbott, Stephen D., and Matt Richey, "Take a Walk on the Boardwalk," *The College Mathematics Journal*, Vol. 28, No. 3, May 1997, pp. 162–171.

EXTENDED APPLICATION: A Markov Chain Model for Teacher Retention

In an article published in the *Review of Public Personnel Administration*, Michael Reid and Raymond Taylor used a Markov chain model to describe the employment patterns for public school teachers in a New England school district.* They identified 8 possible states for teachers in the system: newly employed, continuing from the previous year, on leave without pay, on sabbatical, ill for at least 30 days during the year, resigned, retired, and deceased. Each teacher could transition to 1 of these 8 states in the following year, and the researchers recorded the transition for each teacher from year 1 of the study to year 2. The researchers also noted how each teacher in the system in year 2 transitioned to year 3. This gave them two sets of transition frequencies, which they combined in order to estimate the transition probabilities from each of the 8 possible states to each other state. The resulting transition matrix, arranged with the absorbing states in the upper left portion of the array, is given below.

Note that there are 3 irreversible transitions: resigning, retiring, and dying, so the first 3 states listed are absorbing. Once you have entered the "resigned" state, you stay there forever (at least as far as the school system is concerned). The fourth column consists of zeros, since no one already in the system can transition into the state of being a new teacher. In the exercises you will look at some of the other transition probabilities, including some of the forbidden transitions, such as from being on sabbatical to being on sabbatical again the next year.

Following the procedure outlined in Section 10.3, we can compute the fundamental matrix F, which looks like this:

$$F = \begin{bmatrix} 1 & 10.405 & .322 & .073 & .222 \\ 0 & 12.988 & .402 & .091 & .234 \\ 0 & 3.252 & 1.507 & .023 & .058 \\ 0 & 12.988 & .402 & 1.091 & .234 \\ 0 & 11.078 & .343 & .078 & 1.257 \end{bmatrix}$$

Recall that the rows and columns here represent the nonabsorbing states, in the same order in which they appear in the transition matrix. According to the properties of the fundamental matrix listed in Section 10.3, the entry in row A and column B represents the number of years during which a teacher currently in state A will be in state B before he or she exits the system into one of the absorbing states. For example, the 1 at the upper left indicates that a teacher who is new this year will spend exactly 1 year as a new teacher, which makes sense because you can only be new once. The entry 3.252 indicates that a teacher currently on leave without pay will, on the average, spend only about 3 years in the system, including the current year.

In the transition matrix, the matrix R sits under the identity matrix corresponding to the absorbing states. For the teacher transition matrix, R looks like this:

$$R = \begin{bmatrix} 0.194 & 0 & 0 \\ 0.04 & 0.016 & 0.002 \\ 0.533 & 0 & 0 \\ 0 & 0 & 0 \\ 0 & 0.139 & 0 \end{bmatrix}$$

In the exercises you will compute the product FR, which gives information about the distribution of absorbing states for each possible nonabsorbing initial state.

Exercises

1. If a teacher is currently ill, what is the probability that he or she will retire during the following year? What is the probability that he or she will be ill again in the following year?

2. The entry in row 2 and column 2 of F is 12.988. What information does this give about teachers who are currently actively teaching?

	Resigned	Retired	Deceased	New	Continuing	On Leave	On Sabbatical	Ill
Resigned	1	0	0	0	0	0	0	0
Retired	0	1	0	0	0	0	0	0
Deceased	0	0	1	0	0	0	0	0
New	.194	0	0	0	.777	0	0	.033
Continuing	.040	.016	.002	0	.896	.022	.007	.017
On leave	.533	0	0	0	.178	.289	0	0
On sabbatical	0	0	0	0	1	0	0	0
Ill	0	.139	0	0	.806	0	0	.055

*Reid, W. M., and R. G. Taylor, "An Application of Absorbing Markov Analysis to Human Resource Issues in Public Administration," *Review of Public Personnel Administration*, 1989.

3. Row 2 of F is nearly the same as row 1. Is this an accident? Can you explain the one difference?

4. What does the entry .091 in row 2 of the fundamental matrix tell you?

5. Compute the matrix FR and answer the following questions:

    a. What is the probability that a teacher now on leave without pay will eventually resign? What is the probability that a teacher now on leave will resign *the following year*?

b. Who is more likely to leave the system by retirement, a new teacher or a teacher who is currently teaching but has at least 1 year in the system?

c. What is the probability that a new teacher will die on the job?

6. The study reported here collected two sets of transition data (year 1 to year 2 and year 2 to year 3). Why was this a good idea?

Game Theory

Game theory provides a framework for making decisions with incomplete information. For example, a hospital administrator must choose nursing staffing levels even though demand is not predictable. An exercise in Section 3 models this situation as a game by assigning payoffs to the possible degrees of underutilization or overcrowding and deriving an optimal strategy.

John F. Kennedy once remarked that he had assumed that as president he would find it difficult to choose between distinct, opposite alternatives when a decision needed to be made. He said that actually, however, such decisions were easy to make; the hard decisions came when he was faced with choices that were not as clear-cut. Most decisions that we must make fall in the second category — decisions that must be made under conditions of uncertainty. *Game theory* is a branch of mathematics that provides a systematic way to attack problems of decision making when some alternatives are unclear or ambiguous.

■ 11.1 STRICTLY DETERMINED GAMES

THINK ABOUT IT

How can a football team decide which play to run when the play might have varying degrees of success, depending upon what the opposing team does?

In this section, we will learn strategies for making such decisions.

A small manufacturer of Christmas cards must decide in February what type of cards to emphasize in her fall line. She has three possible strategies: emphasize modern cards, emphasize old-fashioned cards, or emphasize a mixture of the two. Her success is dependent on the state of the economy in December. If the economy is strong, she will do well with her modern cards, while in a weak economy people long for the old days and buy old-fashioned cards. In an in-between economy, her mixture of lines would do the best.

What should the manufacturer do? She should begin by carefully defining the problem. First, she must decide on the **states of nature,** the alternatives over which she has no control. Here, there are three: a weak economy, an in-between economy, or a strong economy. Next, she should list her **strategies:** emphasize modern cards, old-fashioned cards, or a mixture of the two. The consequences of each strategy under each state of nature are called **payoffs.** They can be summarized in a **payoff matrix,** as shown below. The numbers in the matrix represent her profits in thousands of dollars.

States of Nature

		Weak Economy	In-Between	Strong Economy
	Modern	40	85	120
Strategies	Old-Fashioned	90	45	85
	Mixture	75	105	65

a. If the manufacturer is an optimist, she should aim for the biggest number on the matrix, 120 (representing \$120,000 in profit). Her strategy in this case would be to produce modern cards.

b. A pessimistic manufacturer wants to avoid the worst of all bad things that can happen. If she produces modern cards, the worst that can happen is a profit of \$40,000. For old-fashioned cards, the worst is a profit of \$45,000, while the worst that can happen from a mixture is a profit of \$65,000. Her strategy here is to use a mixture.

c. Suppose the manufacturer reads in a business magazine that leading experts think there is a 50% chance of a weak economy at Christmas, a 20% chance

FOR REVIEW ■

Recall from Section 8.5 that if a random variable x can take on the n values $x_1, x_2, x_3, \ldots, x_n$ with probabilities $p_1, p_2, p_3, \ldots, p_n$, then the expected value of the random variable is

$$E(x) = x_1 p_1 + x_2 p_2 + x_3 p_3 + \ldots + x_n p_n.$$

In the modern Christmas card example, $n = 3$, $x_1 = 40$, $x_2 = 85$, $x_3 = 120$, $p_1 = .50$, $p_2 = .20$, and $p_3 = .30$.

of an in-between economy, and a 30% chance of a strong economy. The manufacturer can now find her expected profit for each possible strategy.

Modern:	$40(.50) + 85(.20) + 120(.30) = 73$
Old-Fashioned:	$90(.50) + 45(.20) + 85(.30) = 79.5$
Mixture:	$75(.50) + 105(.20) + 65(.30) = 78$

Here the best strategy is old-fashioned cards; the expected profit is 79.5, or $79,500.

When a decision is made again and again, the meaning of the term *strategy* is more general: a strategy is then a rule for determining which choice is made each time the game is played. One possible strategy for the Christmas card manufacturer is to always choose modern cards; another is to rotate, using modern cards one year, old-fashioned the next, a mixture the third year, and so on; a third strategy is to make a random choice each year. If a decision is only made once, however, this meaning of the term *strategy* reduces to that of making a choice. In Section 11.2 on Mixed Strategies, we will see how more complex strategies can be selected.

The word *game* in the title of this chapter may have led you to think of chess, or perhaps some card game. While *game theory* does have some application to these recreational games, it was developed in the 1940s to analyze competitive situations in business, warfare, and social situations. Game theory was invented by John von Neumann (1903–1957), who, with Oskar Morgenstern, wrote the book entitled *Theory of Games and Economic Behavior* in 1944. **Game theory** is the study of how to make decisions when competing with an aggressive opponent.

A game can be set up with a payoff matrix, such as the one shown below. This game involves the two players A and B, and is called a **two-person game.** Player A can choose either row 1 or row 2, while player B can choose either column 1 or column 2. A player's choice is called a **strategy,** just as before. The payoff is at the intersection of the row and column selected. As a general agreement, a positive number represents a payoff from B to A; a negative number represents a payoff from A to B. For example, if A chooses row 2 and B chooses column 2, then B pays $4 to A.

$$\begin{array}{c} & & B \\ & & \begin{array}{cc} 1 & 2 \end{array} \\ A & \begin{array}{c} 1 \\ 2 \end{array} & \begin{bmatrix} 2 & -1 \\ -3 & 4 \end{bmatrix} \end{array}$$

EXAMPLE 1 Payoff Matrix

In the payoff matrix just shown, suppose A chooses row 1 and B chooses column 2. Who gets what?

Solution Row 1 and column 2 lead to the number -1. This number represents a payoff of $1 from A to B. ■

While the numbers in this payoff matrix represent money, they could just as easily represent goods or other property.

In the game above, no money enters the game from the outside; whenever one player wins, the other loses. Such a game is called a **zero-sum game.** The stock market is not a zero-sum game. Stocks can go up or down as a result of outside forces. Therefore, it is possible that all investors can make or lose money.

Only two-person zero-sum games are discussed in the rest of this chapter. Each player can have many different options. In particular, an $m \times n$ matrix game is one in which player A has m strategies (rows) and player B has n strategies (columns). We will always use rows for player A and columns for player B, so labels are not necessary on the matrix.

Dominated Strategies   In the rest of this section, the best possible strategy for each player is determined. Let us begin with the 3×3 game defined by the following matrix.

$$
\begin{array}{c}
1 \\
2 \\
3
\end{array}
\begin{bmatrix}
-3 & -6 & 10 \\
3 & 0 & -9 \\
5 & -4 & -8
\end{bmatrix}
$$

From B's viewpoint, strategy 2 is better than strategy 1 no matter which strategy A selects. This can be seen by comparing the two columns. If A chooses row 1, receiving $6 from A is better than receiving $3; in row 2, breaking even is better than paying $3; and in row 3, getting $4 from A is better than paying $5. Therefore, B should never select strategy 1. Strategy 2 is said to *dominate* strategy 1, and strategy 1—the **dominated strategy**—can be removed from consideration, producing the following reduced matrix.

$$
\begin{array}{c}
1 \\
2 \\
3
\end{array}
\begin{bmatrix}
-6 & 10 \\
0 & -9 \\
-4 & -8
\end{bmatrix}
$$

Either player may have dominated strategies. In fact, after a dominated strategy for one player is removed, the other player may then have a dominated strategy where there was none before.

DOMINATED STRATEGIES

A row for A **dominates** another row if every entry in the one row is *larger* than the corresponding entry in the other row. For a column for B to dominate another, each entry must be *smaller.*

In the 3×2 matrix above, neither player now has a dominated strategy. From A's viewpoint, strategy 1 is best if B chooses strategy 3, while strategy 2 is best if B chooses strategy 2. Verify that there are no dominated strategies for either player.

To find any dominated strategy, start with the first two rows. If the first row has an entry larger than the corresponding entry in the second row and another entry smaller than the corresponding entry in the second row, then neither row dominates. If this is not the case, then one row is dominated and may be removed. Continue this process for every pair of rows and every pair of columns.

EXAMPLE 2 Dominated Strategies

Find any dominated strategies in the games with the given payoff matrices.

(a)
$$\begin{array}{c} \\ 1 \\ 2 \end{array} \begin{array}{cccc} 1 & 2 & 3 & 4 \end{array} \\ \begin{bmatrix} -8 & -4 & -6 & -9 \\ -3 & 0 & -9 & 12 \end{bmatrix}$$

Solution Here every entry in column 3 is smaller than the corresponding entry in column 2. Thus, column 3 dominates column 2. (Notice that column 1 also dominates column 2.) By removing the dominated column 2, the final game is as follows.

$$\begin{array}{c} \\ 1 \\ 2 \end{array} \begin{array}{ccc} 1 & 3 & 4 \end{array} \\ \begin{bmatrix} -8 & -6 & -9 \\ -3 & -9 & 12 \end{bmatrix}$$

(b)
$$\begin{array}{c} \\ 1 \\ 2 \\ 3 \end{array} \begin{array}{cc} 1 & 2 \end{array} \\ \begin{bmatrix} 3 & -2 \\ 0 & 8 \\ 6 & 4 \end{bmatrix}$$

Solution Each entry in row 3 is greater than the corresponding entry in row 1, so that row 3 dominates row 1. Removing row 1 gives the following game.

$$\begin{array}{c} \\ 2 \\ 3 \end{array} \begin{array}{cc} 1 & 2 \end{array} \\ \begin{bmatrix} 0 & 8 \\ 6 & 4 \end{bmatrix}$$

Strictly Determined Games

Which strategies should the players choose in the following game?

$$\begin{array}{cc} & B \\ & \begin{array}{ccc} 1 & 2 & 3 \end{array} \\ A \begin{array}{c} 1 \\ 2 \\ 3 \end{array} & \begin{bmatrix} -9 & -11 & 7 \\ 2 & 3 & 5 \\ -1 & 6 & -3 \end{bmatrix} \end{array}$$

The goal of game theory is to find **optimum strategies:** those that are the most profitable to the respective players. The payoff that results from each player's choosing the optimum strategy is called the **value** of the game.

The simplest strategy for a player is to consistently choose a certain row (or column). Such a strategy is called a **pure strategy,** in contrast to strategies requiring the random choice of a row (or column); these alternative strategies are discussed in the next section.*

To choose a pure strategy in the game above, player A could choose row 1, in hopes of getting the payoff of $7. Player B would quickly discover this, however, and start playing column 2. By playing column 2, B would receive $11 from A. If A were to choose row 2 consistently, then B would again minimize outgo by choosing column 1 (a payoff of $2 by B to A is better than paying $3

*In this section we solve (find the optimum strategies for) only games that have optimum *pure* strategies.

or \$5, respectively, to A). By choosing row 3 consistently, A would cause B to choose column 3. The table shows what B will do when A chooses a given row consistently.

If A Chooses Pure Strategy:	Then B Will Choose:	With Payoff:
Row 1	Column 2	\$11 to B
Row 2	Column 1	\$2 to A
Row 3	Column 3	\$3 to B

Based on these results, A's optimum strategy is to choose row 2; in this way A will guarantee a minimum payoff of \$2 per play of the game, no matter what B does.

The optimum pure strategy in this game for A (the *row* player) is found by identifying the *smallest* number in each row of the payoff matrix; the row giving the *largest* such number gives the optimum strategy.

By going through a similar analysis for player B, we find that B should choose the column that will minimize the amount A can win. In the game above, B will pay \$2 to A if B consistently chooses column 1. By choosing column 2 consistently, B will pay \$6 to A, and by choosing column 3, player B will pay \$7 to A. The optimum strategy for B is thus to choose column 1—with each play of the game B will pay \$2 to A.

The optimum pure strategy in this game for B (the column player) is to identify the *largest* number in each column of the payoff matrix, and then choose the column producing the *smallest* such number.

In the game above, the entry 2 is both the *smallest* entry in its *row* and the *largest* entry in its *column.* Such an entry is called a **saddle point.** (The seat of a saddle is the maximum from one direction and the minimum from another direction. See Figure 1.) As Example 3(b) will show, there may be more than one such entry, but then the entries will have the same value.

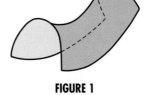

FIGURE 1

SADDLE POINT

A saddle point is the smallest entry in its row and the largest entry in its column. In a game with a saddle point, the optimum pure strategy for player A is to choose the row containing the saddle point, while the optimum pure strategy for B is to choose the column containing the saddle point.

A game with a saddle point is called a **strictly determined game.** By using these optimum strategies, A and B will ensure that the same amount always changes hands with each play of the game; this amount, given by the saddle point, is the value of the game. The value of the game above is \$2. A game having a value of 0 is a **fair game;** the game above is not fair. It is also not much fun for player B, who will lose \$2 to player A every time. Finally, it is not very interesting to play the game, because players A and B will choose strategies 2 and 1, respectively, every time. Games in which there is no optimum pure strategy are more interesting to play again and again; we will study these in the next section.

To find the saddle point easily, underline the smallest number in each row, and circle the largest number in each column. If there is a number that is both underlined and circled, it is the saddle point. In case of a tie, underline or circle all numbers that equal the smallest number in the row or the largest number in the column.

EXAMPLE 3 Saddle Points

Find the saddle points in the following games.

(a)
$$
\begin{array}{c c}
 & \begin{array}{c c} 1 & 2 \end{array} \\
\begin{array}{c} 1 \\ 2 \\ 3 \\ 4 \end{array} &
\begin{bmatrix}
2 & 2 \\
0 & 4 \\
1 & 6 \\
3 & 7
\end{bmatrix}
\end{array}
$$

Solution Underlining the smallest number in each row and circling the largest value in each column yields the following:

$$
\begin{array}{c c}
 & \begin{array}{c c} 1 & 2 \end{array} \\
\begin{array}{c} 1 \\ 2 \\ 3 \\ 4 \end{array} &
\begin{bmatrix}
\underline{2} & \underline{2} \\
\underline{0} & 4 \\
\underline{1} & 6 \\
\underline{③} & ⑦
\end{bmatrix}.
\end{array}
$$

The number that is both circled and underlined is 3. Thus, 3 is the saddle point, and the game has a value of 3. The strategies producing the saddle point can be written $(4, 1)$. (Player A's strategy is written first.)

(b)
$$
\begin{array}{c c}
 & \begin{array}{c c c c} 1 & 2 & 3 & 4 \end{array} \\
\begin{array}{c} 1 \\ 2 \end{array} &
\begin{bmatrix}
4 & 6 & 4 & 12 \\
-8 & -9 & 3 & 2
\end{bmatrix}
\end{array}
$$

Solution Underlining the smallest number in each row and circling the largest value in each column yields the following:

$$
\begin{array}{c c}
 & \begin{array}{c c c c} 1 & 2 & 3 & 4 \end{array} \\
\begin{array}{c} 1 \\ 2 \end{array} &
\begin{bmatrix}
\underline{④} & ⑥ & \underline{④} & ⑫ \\
-8 & \underline{-9} & 3 & 2
\end{bmatrix}.
\end{array}
$$

The saddle point, 4, occurs with either of two strategies, $(1, 1)$ or $(1, 3)$. The value of the game is 4. (Neither of the games in parts (a) or (b) of this example is a fair game, because neither game has a value of 0.)

(c)
$$
\begin{array}{c c}
 & \begin{array}{c c c} 1 & 2 & 3 \end{array} \\
\begin{array}{c} 1 \\ 2 \end{array} &
\begin{bmatrix}
3 & 6 & -2 \\
8 & -3 & 5
\end{bmatrix}
\end{array}
$$

Solution Underlining the smallest number in each row and circling the largest value in each column yields the following.

$$
\begin{array}{c c}
 & \begin{array}{c c c} 1 & 2 & 3 \end{array} \\
\begin{array}{c} 1 \\ 2 \end{array} &
\begin{bmatrix}
3 & ⑥ & \underline{-2} \\
⑧ & \underline{-3} & ⑤
\end{bmatrix}
\end{array}
$$

There is no number that is both the smallest number in its row and the largest number in its column, so the game has no saddle point. Since the game has no saddle point, it is not strictly determined. In the next section, methods are given for finding optimum strategies for such games.

In Example 3(a), notice that row 4 dominates all other rows. Therefore, we may as well eliminate rows 1 through 3, resulting in the matrix $[3 \quad 7]$. In this matrix, column 1 dominates column 2, so we may as well eliminate column 2, resulting in the 1×1 matrix $[3]$, whose only entry is the saddle point. Successive elimination of dominated rows and columns gives another way to find a saddle point if it exists.

11.1 EXERCISES

In the following game, decide on the payoff when the strategies of Exercises 1–6 are used.

$$
\begin{array}{c}
 & & B \\
 & & \begin{array}{ccc} 1 & 2 & 3 \end{array} \\
A \begin{array}{c} 1 \\ 2 \\ 3 \end{array} & & \begin{bmatrix} 6 & -4 & 0 \\ 3 & -2 & 6 \\ -1 & 5 & 11 \end{bmatrix}
\end{array}
$$

1. $(1, 1)$ **2.** $(1, 2)$ **3.** $(2, 2)$

4. $(2, 3)$ **5.** $(3, 1)$ **6.** $(3, 2)$

7. Does the game have any dominated strategies?

8. Does it have a saddle point?

Remove any dominated strategies in the games in Exercises 9–14. (From now on, we will save space by deleting the names of the strategies.)

9. $\begin{bmatrix} 0 & -2 & 8 \\ 3 & -1 & -9 \end{bmatrix}$ **10.** $\begin{bmatrix} 6 & 5 \\ 3 & 8 \\ -1 & -4 \end{bmatrix}$ **11.** $\begin{bmatrix} 1 & 4 \\ 4 & -1 \\ 3 & 5 \\ -4 & 0 \end{bmatrix}$

12. $\begin{bmatrix} 2 & 3 & 1 & -5 \\ -1 & 5 & 4 & 1 \\ 1 & 0 & 2 & -3 \end{bmatrix}$ **13.** $\begin{bmatrix} 8 & 12 & -7 \\ -2 & 1 & 4 \end{bmatrix}$ **14.** $\begin{bmatrix} 6 & 2 \\ -1 & 10 \\ 3 & 5 \end{bmatrix}$

For each game, when the saddle point exists, find the strategies producing it and the value of the game. Identify any games that are strictly determined.

15. $\begin{bmatrix} 3 & 5 \\ 2 & -5 \end{bmatrix}$ **16.** $\begin{bmatrix} 7 & 8 \\ -2 & 15 \end{bmatrix}$ **17.** $\begin{bmatrix} 3 & -4 & 1 \\ 5 & 3 & -2 \end{bmatrix}$ **18.** $\begin{bmatrix} -4 & 2 & -3 & -7 \\ 4 & 3 & 5 & -9 \end{bmatrix}$

19. $\begin{bmatrix} -6 & 2 \\ -1 & -10 \\ 3 & 5 \end{bmatrix}$ **20.** $\begin{bmatrix} 1 & 4 & -3 & 1 & -1 \\ 2 & 5 & 0 & 4 & 10 \\ 1 & -3 & 2 & 5 & 2 \end{bmatrix}$ **21.** $\begin{bmatrix} 2 & 3 & 1 \\ -1 & 4 & -7 \\ 5 & 2 & 0 \\ 8 & -4 & -1 \end{bmatrix}$

22. $\begin{bmatrix} 3 & 8 & -4 & -9 \\ -1 & -2 & -3 & 0 \\ -2 & 6 & -4 & 5 \end{bmatrix}$ **23.** $\begin{bmatrix} -6 & 1 & 4 & 2 \\ 9 & 3 & -8 & -7 \end{bmatrix}$ **24.** $\begin{bmatrix} 6 & -1 \\ 0 & 3 \\ 4 & 0 \end{bmatrix}$

25. Consider the matrix of Exercise 21.

   a. Remove any dominated columns, and write the resulting matrix.

   b. Remove any dominated rows from the matrix in part a, and write the resulting matrix.

   c. Remove any dominated columns from the matrix in part b, and write the resulting matrix.

   d. Remove any dominated rows from the matrix in part c, and write the resulting matrix. Verify that the only entry left is the saddle point.

26. Repeat Exercise 25, starting with the matrix of Exercise 22.

27. Suppose the payoff matrix for a game has at least three rows. Also, suppose that row 1 dominates row 2, and row 2 dominates row 3. Show that row 1 must dominate row 3.

Applications

BUSINESS AND ECONOMICS

28. *Concert Preparations* Hillsdale College has sold out all tickets for a jazz concert to be held in the stadium. If it rains, the show will have to be moved to the gym, which has a much smaller capacity. The dean must decide in advance whether to set up the seats and the stage in the gym or in the stadium, or both, just in case. The payoff matrix below shows the net profit in each case.

		States of Nature	
		Rain	No Rain
Strategies	Set Up in Stadium	$-\$1550$	$\$1500$
	Set Up in Gym	$\$1000$	$\$1000$
	Set Up in Both	$\$750$	$\$1400$

What strategy should the dean choose if she is

   a. an optimist?   **b.** a pessimist?

   c. If the weather forecaster predicts rain with a probability of .6, what strategy should the dean choose to maximize expected profit? What is the maximum expected profit?

29. *Machine Repairs* An analyst must decide what fraction of the items produced by a certain machine are defective. He has already decided that there are three possibilities for the fraction of defective items: .01, .10, and .20. He may recommend two courses of action: repair the machine or make no repairs. The payoff matrix below represents the costs to the company in each case.

		States of Nature		
		.01	.10	.20
Strategies	Repair	$-\$130$	$-\$130$	$-\$130$
	No Repair	$-\$25$	$-\$200$	$-\$500$

What strategy should the analyst recommend if he is

   a. an optimist?   **b.** a pessimist?

   c. Suppose the analyst is able to estimate probabilities for the three states of nature, as follows.

Fraction of Defective Items	Probability
.01	.70
.10	.20
.20	.10

Which strategy should he recommend? Find the expected cost to the company if this strategy is chosen.

30. *Marketing* The research department of the Allied Manufacturing Company has developed a new process that it believes will result in an improved product. Management must decide whether to go ahead and market the new product. The new product may or may not be better than the old one. If the new product is better and the company decides to market it, sales should increase by $50,000. If it is not better and they replace the old product with the new product on the market, they will lose $25,000 to competitors. If they decide not to market the new product, they will lose a total of $40,000 if it is better, and just research costs of $10,000 if it is not.

   a. Prepare a payoff matrix.

   b. If management believes there is a probability of .4 that the new product is better, find the expected profits under each strategy and determine the best action.

   c. Find any dominated strategies. Is there a saddle point?

31. *Machinery Overhaul* A businessman is planning to ship a used machine to his plant in Nigeria. He would like to use it there for the next 4 years. He must decide whether to overhaul the machine before sending it. The cost of overhaul is $2600. If the machine fails when in operation in Nigeria, it will cost him $6000 in lost production and repairs. He estimates the probability that it will fail at .3 if he does not overhaul it, and .1 if he does overhaul it. Neglect the possibility that the machine might fail more than once in the next 4 years.

a. Prepare a payoff matrix.

b. What should the businessman do to minimize his expected costs?

c. Find any dominated strategies. Is there a saddle point?

32. *Competition* Two merchants are planning to build competing stores to serve an area of three small cities. The fraction of the total population that live in each city is shown in the figure. If both merchants locate in the same city, merchant A will get 65% of the total business. If the merchants locate in different cities, each will get 80% of the business in the city it is in, and A will get 60% of the business from the city not containing B. Payoffs are measured by the number of percentage points above or below 50%. Write a payoff matrix for this game. Is this game strictly determined?

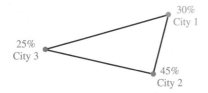

33. *Competition* In Exercise 32, if merchant A gets 55% of the total business when both merchants locate in the same city, and the population is otherwise unchanged, write the payoff matrix and find the strategies producing the saddle point and the value of the game.

SOCIAL SCIENCES

34. *Border Patrol* A person attempting an illegal crossing of the border into the United States has two options. He can cross into unoccupied territory on foot at night, or he can cross at a regular entry point if he is hidden in a vehicle. His chances of detection depend on whether extra border guards are sent to the unoccupied territory or to the regular entry point. His chances of getting across the unoccupied territory are 50% if there are no extra patrols and 40% if there are extra patrols. His chances of crossing at a regular entry point are 30% if there are no extra border guards and 20% if there are extra guards. Letting the payoffs be the probabilities of success, find the strategies producing the saddle point and the value of the game.

GENERAL INTEREST

35. *War Games* Two armies, A and B, are involved in a war game. Each army has available three different strategies,

with payoffs as shown below. These payoffs represent square kilometers of land, with positive numbers representing gains by A.

$$\begin{bmatrix} 3 & -8 & -9 \\ 0 & 6 & -12 \\ -8 & 4 & -10 \end{bmatrix}$$

Find the strategies producing the saddle point and the value of the game.

36. *Football* When a football team has the ball and is planning its next play, it can choose one of several plays or strategies. The success of the chosen play depends largely on how well the other team "reads" the chosen play. Suppose a team with the ball (team A) can choose from three plays, while the opposition (team B) has four possible strategies. The numbers shown in the following payoff matrix represent yards of gain to team A.

$$\begin{bmatrix} 9 & -3 & -4 & 16 \\ 12 & 9 & 6 & 8 \\ -5 & -2 & 3 & 18 \end{bmatrix}$$

Find the strategies producing the saddle point. Find the value of the game.

37. *Children's Game* In the children's game *rock, paper, scissors,* two players simultaneously extend one hand in the form of a fist (*rock*), a flat palm (*paper*), or two fingers extended (*scissors*). If they both make the same choice, the game is a tie. Otherwise, *rock* beats *scissors* (because rock can crush scissors), *scissors* beats *paper* (because scissors can cut paper), and *paper* beats *rock* (because paper can cover rock). Whoever wins gains one point; in the case of a tie, no one gains any points. Find the payoff matrix for this game. Is the game strictly determined?

38. *Finger Game* John and Joann play a finger matching game—each shows one or two fingers, simultaneously. If the sum of the number of fingers showing is even, Joann pays John that number of dollars; for an odd sum, John pays Joann. Find the payoff matrix for this game. Is the game strictly determined?

■ 11.2 MIXED STRATEGIES

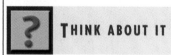

THINK ABOUT IT
How can a farmer decide whether or not to spray insecticide on his crops, given that neither strategy, when used consistently, is the best strategy?

In this section, we will learn the answer to such questions.

As mentioned earlier, not every game has a saddle point. Two-person zero-sum games still have optimum strategies, however, even if the strategy is not as simple as the ones we saw earlier. In a game with a saddle point, the optimum strategy for player A is to pick the row containing the saddle point. Such a strategy is called a *pure strategy,* since the same row is always chosen.

If there is no saddle point, then it will be necessary for both players to mix their strategies. For example, A will sometimes play row 1, sometimes row 2, and so on. If this were done in some specific pattern, the competitor would soon guess it and play accordingly.

For this reason, it is best to mix strategies according to previously determined probabilities. For example, if a player has only two strategies and has decided to play them with equal probability, the random choice could be made by tossing a fair coin, letting heads represent one strategy and tails the other. This would result in the two strategies being used about equally over the long run. On a particular play, however, it would not be possible to predetermine the strategy to be used. (Some other device, such as a spinner, is necessary if there are more than two strategies or if the probabilities are not $1/2$.)

EXAMPLE 1   Expected Value
Suppose a game has payoff matrix

$$\begin{bmatrix} -1 & 2 \\ 1 & 0 \end{bmatrix},$$

where the entries represent dollar winnings. Suppose player A chooses row 1 with probability $1/3$ and row 2 with probability $2/3$, and player B chooses each column with probability $1/2$. Find the expected value of the game.

Solution Assume that rows and columns are chosen independently, so that

$$P(\text{row } 1, \text{column } 1) = P(\text{row } 1) \cdot P(\text{column } 1) = \frac{1}{3} \cdot \frac{1}{2} = \frac{1}{6}$$

$$P(\text{row } 1, \text{column } 2) = P(\text{row } 1) \cdot P(\text{column } 2) = \frac{1}{3} \cdot \frac{1}{2} = \frac{1}{6}$$

$$P(\text{row } 2, \text{column } 1) = P(\text{row } 2) \cdot P(\text{column } 1) = \frac{2}{3} \cdot \frac{1}{2} = \frac{1}{3}$$

$$P(\text{row } 2, \text{column } 2) = P(\text{row } 2) \cdot P(\text{column } 2) = \frac{2}{3} \cdot \frac{1}{2} = \frac{1}{3}.$$

The table on the next page lists the probability of each possible outcome, along with the payoff to player A.

Outcome	Probability of Outcome	Payoff for A
Row 1, column 1	1/6	−1
Row 1, column 2	1/6	2
Row 2, column 1	1/3	1
Row 2, column 2	1/3	0

The expected value of the game is given by the sum of the products of the probabilities and the payoffs, or

$$\text{Expected value} = \frac{1}{6}(-1) + \frac{1}{6}(2) + \frac{1}{3}(1) + \frac{1}{3}(0) = \frac{1}{2}.$$

In the long run, for a great many plays of the game, the payoff to A will average $1/2$ dollar per play of the game. It is important to note that as the mixed strategies used by A and B are changed, the expected value of the game may well change. (See Example 2 below.)

To generalize the work of Example 1, let the payoff matrix for a 2×2 game be

$$M = \begin{bmatrix} a_{11} & a_{12} \\ a_{21} & a_{22} \end{bmatrix}.$$

Let player A choose row 1 with probability p_1 and row 2 with probability p_2, where $p_1 + p_2 = 1$. Write these probabilities as the row matrix

$$A = \begin{bmatrix} p_1 & p_2 \end{bmatrix}.$$

Let player B choose column 1 with probability q_1 and column 2 with probability q_2, where $q_1 + q_2 = 1$. Write this as the column matrix

$$B = \begin{bmatrix} q_1 \\ q_2 \end{bmatrix}.$$

The probability of choosing row 1 and column 1 is

$$P(\text{row}1, \text{column}1) = P(\text{row }1) \cdot P(\text{column }1) = p_1 \cdot q_1.$$

In the same way, the probabilities of each possible outcome are shown in the table below, along with the payoff for each outcome.

Outcome	Probability of Outcome	Payoff for A
Row 1, column 1	$p_1 \cdot q_1$	a_{11}
Row 1, column 2	$p_1 \cdot q_2$	a_{12}
Row 2, column 1	$p_2 \cdot q_1$	a_{21}
Row 2, column 2	$p_2 \cdot q_2$	a_{22}

The expected value for this game is

$$(p_1 \cdot q_1) \cdot a_{11} + (p_1 \cdot q_2) \cdot a_{12} + (p_2 \cdot q_1) \cdot a_{21} + (p_2 \cdot q_2) \cdot a_{22}.$$

This same result can be written as the matrix product

$$\text{Expected value} = [p_1 \quad p_2] \begin{bmatrix} a_{11} & a_{12} \\ a_{21} & a_{22} \end{bmatrix} \begin{bmatrix} q_1 \\ q_2 \end{bmatrix} = AMB.$$

The same method works for games larger than 2×2: let the payoff matrix for a game have dimension $m \times n$; call this matrix $M = [a_{ij}]$. Let the mixed strategy for player A be given by the row matrix

$$A = [p_1 \quad p_2 \quad p_3 \quad \cdots \quad p_m]$$

and the mixed strategy for player B be given by the column matrix

$$B = \begin{bmatrix} q_1 \\ q_2 \\ \vdots \\ q_n \end{bmatrix}.$$

The expected value for this game is the product

$$AMB = [p_1 \quad p_2 \quad \cdots \quad p_m] \begin{bmatrix} a_{11} & a_{12} & \cdots & a_{1n} \\ a_{21} & a_{22} & \cdots & a_{2n} \\ \vdots & \vdots & & \vdots \\ a_{m1} & a_{m2} & \cdots & a_{mn} \end{bmatrix} \begin{bmatrix} q_1 \\ q_2 \\ \vdots \\ q_n \end{bmatrix}.$$

EXAMPLE 2   Expected Value

In the game in Example 1, having payoff matrix

$$M = \begin{bmatrix} -1 & 2 \\ 1 & 0 \end{bmatrix},$$

suppose player A chooses row 1 with probability .2, and player B chooses column 1 with probability .6. Find the expected value of the game.

Solution If A chooses row 1 with probability .2, then row 2 is chosen with probability $1 - .2 = .8$, giving

$$A = [.2 \quad .8].$$

In the same way,

$$B = \begin{bmatrix} .6 \\ .4 \end{bmatrix}.$$

The expected value of this game is given by the product AMB, or

$$AMB = [.2 \quad .8] \begin{bmatrix} -1 & 2 \\ 1 & 0 \end{bmatrix} \begin{bmatrix} .6 \\ .4 \end{bmatrix}$$

$$= [.6 \quad .4] \begin{bmatrix} .6 \\ .4 \end{bmatrix}$$

$$= [.52].$$

On the average, these two strategies will produce a payoff of $.52, or 52¢, for A for each play of the game. This payoff is slightly better than the 50¢ in Example 1.

In Example 2, player B could reduce the payoff to A by changing strategy. (Check this by choosing different matrices for B.) For this reason, player A needs to develop an *optimum strategy*—a strategy that will produce the best possible payoff no matter what B does. Just as in the previous section, this is done by finding the largest of the smallest possible amounts that can be won.

To find values of p_1 and p_2 so that the probability vector $[p_1 \quad p_2]$ produces an optimum strategy, start with the payoff matrix

$$M = \begin{bmatrix} -1 & 2 \\ 1 & 0 \end{bmatrix}$$

and assume that A chooses row 1 with probability p_1. If player B chooses column 1, then player A's expectation is given by E_1, where

$$E_1 = -1 \cdot p_1 + 1 \cdot p_2 = -p_1 + p_2.$$

Since $p_1 + p_2 = 1$, we have $p_2 = 1 - p_1$, and

$$E_1 = -p_1 + 1 - p_1 = 1 - 2p_1.$$

If B chooses column 2, then A's expected value is given by E_2, where

$$E_2 = 2 \cdot p_1 + 0 \cdot p_2 = 2p_1.$$

Draw graphs of $E_1 = 1 - 2p_1$ and $E_2 = 2p_1$; see Figure 2.

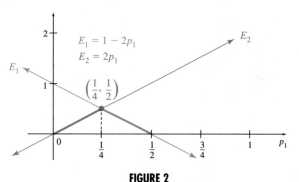

FIGURE 2

As mentioned previously, A needs to maximize the smallest amounts that can be won. On the graph, the smallest amounts that can be won are represented by the points of E_2 up to the intersection point. To the right of the intersection point, the smallest amounts that can be won are represented by the points of the line E_1. Player A can maximize the smallest amounts that can be won by choosing the point of intersection itself, the peak of the heavily shaded lines in Figure 2.

To find this point of intersection, find the simultaneous solution of the two equations. At the point of intersection, $E_1 = E_2$. Substitute $1 - 2p_1$ for E_1 and $2p_1$ for E_2.

$$E_1 = E_2$$
$$1 - 2p_1 = 2p_1$$
$$1 = 4p_1$$
$$\frac{1}{4} = p_1$$

By this result, player A should choose strategy 1 with probability 1/4, and strategy 2 with probability $1 - 1/4 = 3/4$. This will maximize A's expected winnings. To find the maximum winnings (which is also the value of the game), substitute 1/4 for p_1 in either E_1 or E_2. Choosing E_2 gives

$$E_2 = 2p_1 = 2\left(\frac{1}{4}\right) = \frac{1}{2};$$

that is, 1/2 dollar, or 50¢. Going through a similar argument for player B shows that the optimum strategy for player B is to choose each column with probability 1/2; in this case the value also turns out to be 50¢. In Example 2, A's winnings were 52¢; however, that was because B was not using his optimum strategy.

In the game above, player A can maximize expected winnings by playing row 1 with probability 1/4 and row 2 with probability 3/4. Such a strategy is called a **mixed strategy.** To actually decide which row to use on a given game, player A could use a spinner, such as the one in Figure 3, or use a random number generator on a calculator or a computer. A typical random number generator gives numbers between 0 and 1 with no apparent pattern or order. Player A could choose row 1 if the number generated were less than .25, and choose row 2 if the number is at least .25.

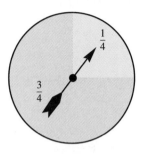

FIGURE 3

EXAMPLE 3 Boll Weevils

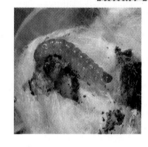

Boll weevils threaten the cotton crop near Hattiesburg. Bill Poole owns a small farm; he can protect his crop by spraying with a potent (and expensive) insecticide. He can save money by not spraying, but he risks losing his crop to the boll weevils. What should his strategy be?

Solution Bill first sets up a payoff matrix. The numbers in the matrix represent his profits.

		States of Nature	
		Boll Weevil Attack	No Attack
Strategies	Spray	$14,000	$7000
	Don't Spray	−$3000	$8000

Let p_1 represent the probability with which Bill chooses to spray, so that $1 - p_1$ is the probability with which he chooses not to spray. If nature chooses a boll weevil attack, Bill's expected value is

$$E_1 = 14,000p_1 - 3000(1 - p_1)$$
$$= 14,000p_1 - 3000 + 3000p_1$$
$$E_1 = 17,000p_1 - 3000.$$

If nature chooses not to attack, Bill has an expected value of

$$E_2 = 7000p_1 + 8000(1 - p_1)$$
$$= 7000p_1 + 8000 - 8000p_1$$
$$E_2 = 8000 - 1000p_1.$$

As suggested by the work above, to maximize his expected profit, Bill should find the value of p_1 for which $E_1 = E_2$.

$$E_1 = E_2$$
$$17{,}000p_1 - 3000 = 8000 - 1000p_1$$
$$18{,}000p_1 = 11{,}000$$
$$p_1 = 11/18$$

Thus, $p_2 = 1 - p_1 = 1 - 11/18 = 7/18$.

Bill will maximize his expected profit if he chooses to spray with probability $11/18$ and not spray with probability $7/18$. His expected profit from this mixed strategy, $[11/18 \quad 7/18]$, can be found by substituting $11/18$ for p_1 in either E_1 or E_2. If E_1 is chosen,

$$\text{Expected profit} = 17{,}000\left(\frac{11}{18}\right) - 3000 = \frac{133{,}000}{18} \approx \$7400.$$

If Bill makes this decision only once, it is meaningless to make a choice with probability $11/18$. This is because Bill either sprays or doesn't spray. Because the strategy to spray has a higher probability, he should spray, and not use the spinner discussed earlier. To make a choice with a given probability only makes sense if the decision is made many times. In game theory, the other player (nature, in this case) is thought of as trying to maximize his or her gain, which is the same as minimizing the first player's gain. It may seem strange to think of nature as choosing a strategy to minimize Bill's profit, but that is a safe and rational strategy when no other information about the state of nature is known.

To obtain a formula for the optimum strategy in a game that is not strictly determined (that is, the game has no saddle point), start with the matrix

$$M = \begin{bmatrix} a_{11} & a_{12} \\ a_{21} & a_{22} \end{bmatrix},$$

the payoff matrix of the game. Assume that A chooses row 1 with probability p_1. The expected value for A, assuming that B plays column 1, is E_1, where

$$E_1 = a_{11} \cdot p_1 + a_{21} \cdot (1 - p_1).$$

The expected value for A if B chooses column 2 is E_2, where

$$E_2 = a_{12} \cdot p_1 + a_{22} \cdot (1 - p_1).$$

As above, the optimum strategy for player A is found by letting $E_1 = E_2$.

$$a_{11} \cdot p_1 + a_{21} \cdot (1 - p_1) = a_{12} \cdot p_1 + a_{22} \cdot (1 - p_1)$$

Solve this equation for p_1.

$$a_{11} \cdot p_1 + a_{21} - a_{21} \cdot p_1 = a_{12} \cdot p_1 + a_{22} - a_{22} \cdot p_1$$
$$a_{11} \cdot p_1 - a_{21} \cdot p_1 - a_{12} \cdot p_1 + a_{22} \cdot p_1 = a_{22} - a_{21}$$
$$p_1(a_{11} - a_{21} - a_{12} + a_{22}) = a_{22} - a_{21}$$
$$p_1 = \frac{a_{22} - a_{21}}{a_{11} - a_{21} - a_{12} + a_{22}}$$

Since $p_2 = 1 - p_1$,

$$p_2 = 1 - \frac{a_{22} - a_{21}}{a_{11} - a_{21} - a_{12} + a_{22}}$$

$$= \frac{a_{11} - a_{21} - a_{12} + a_{22} - (a_{22} - a_{21})}{a_{11} - a_{21} - a_{12} + a_{22}}$$

$$= \frac{a_{11} - a_{12}}{a_{11} - a_{21} - a_{12} + a_{22}}.$$

This result is valid only if $a_{11} - a_{21} - a_{12} + a_{22} \neq 0$; this condition is satisfied if the game is not strictly determined.

There is a similar result for player B, which is included in the following summary.

OPTIMUM STRATEGIES IN A NON-STRICTLY-DETERMINED GAME

Let a non-strictly-determined game have payoff matrix

$$M = \begin{bmatrix} a_{11} & a_{12} \\ a_{21} & a_{22} \end{bmatrix}$$

and let $d = a_{11} - a_{21} - a_{12} + a_{22}$. The optimum strategy for player A is $[p_1 \quad p_2]$, where

$$p_1 = \frac{a_{22} - a_{21}}{d} \qquad \text{and} \qquad p_2 = \frac{a_{11} - a_{12}}{d}.$$

The optimum strategy for player B is $\begin{bmatrix} q_1 \\ q_2 \end{bmatrix}$, where

$$q_1 = \frac{a_{22} - a_{12}}{d} \qquad \text{and} \qquad q_2 = \frac{a_{11} - a_{21}}{d}.$$

The value of the game is

$$g = \frac{a_{11}a_{22} - a_{12}a_{21}}{d}.$$

CAUTION The optimum strategies above do not apply if the game is strictly determined. In that case, use the method of Section 11.1 on Strictly Determined Games.

NOTE (1) It is not necessary to remember all five formulas given above. Once p_1 has been calculated, it is simple to calculate $p_2 = 1 - p_1$, rather than using the above formula for p_2. Similarly, $q_2 = 1 - q_1$. Finally, g is the single element in the matrix product AMB, where $A = [p_1 \quad p_2]$ and $B = \begin{bmatrix} q_1 \\ q_2 \end{bmatrix}$.

(2) It is important to remember that the use of mixed strategies is meaningful only if a game is repeated many, many times. For one play of the game, the mixed strategy has no meaning.

EXAMPLE 4 Optimum Strategies

Suppose a game has the following payoff matrix.

$$\begin{bmatrix} 5 & -2 \\ -3 & -1 \end{bmatrix}$$

Find the optimum strategies and the value of the game.

Solution

Method 1: Formulas

Here $a_{11} = 5$, $a_{12} = -2$, $a_{21} = -3$, and $a_{22} = -1$. To find the optimum strategy for player A, first find $d = a_{11} - a_{21} - a_{12} + a_{22} = 5 - (-3) - (-2) + (-1) = 9$. Next, calculate p_1.

$$p_1 = \frac{-1 - (-3)}{9} = \frac{2}{9}$$

Player A should play row 1 with probability 2/9, and row 2 with probability $1 - 2/9 = 7/9$.

For player B,

$$q_1 = \frac{-1 - (-2)}{9} = \frac{1}{9}.$$

Player B should choose column 1 with probability 1/9, and column 2 with probability 8/9. The value of the game is

$$g = \frac{5(-1) - (-2)(-3)}{9} = -\frac{11}{9}.$$

On the average, B will receive 11/9 dollars from A per play of the game.

Method 2: Rows and Columns

There is an alternative to the formulas used in Method 1, in cases with a 2×2 matrix and no saddle point. Calculate the absolute value of the difference between elements in each row, and then swap the two numbers.

$$\begin{bmatrix} 5 & -2 \\ -3 & -1 \end{bmatrix} \begin{array}{l} \longrightarrow |5 - (-2)| = 7 \searrow 2 \\ \longrightarrow |-3 - (-1)| = 2 \nearrow 7 \end{array}$$

The sum of the two numbers calculated is $7 + 2 = 9$. The values of p_1 and p_2 are then 2/9 and 7/9, respectively.

Similarly, calculate the absolute value of the difference between elements in each column, and then swap the two numbers.

$$\begin{bmatrix} 5 & -2 \\ -3 & -1 \end{bmatrix}$$

$$\begin{array}{cc} |5 - (-3)| & |-2 - (-1)| \\ = 8 & = 1 \\ 1 & 8 \end{array}$$

The sum of the two numbers calculated is $1 + 8 = 9$, and the values of q_1 and q_2 are then 1/9 and 8/9, respectively. From these, the value of the game can be computed as

$$AMB = \begin{bmatrix} 2/9 & 7/9 \end{bmatrix} \begin{bmatrix} 5 & -2 \\ -3 & -1 \end{bmatrix} \begin{bmatrix} 1/9 \\ 8/9 \end{bmatrix} = [-11/9].$$

To see why Method 2 in Example 4 works, consider the case in which $a_{11} > a_{12}$. Then the number computed in row 1 is $|a_{11} - a_{12}| = a_{11} - a_{12}$. Since there is no saddle point, a_{22} must be greater than a_{21}, because if both elements in column 2 were the smallest number in their respective rows, there would be a saddle point in column 2. So $a_{21} < a_{22}$, and $|a_{21} - a_{22}| = a_{22} - a_{21}$. The sum of the quantities in the two rows is then $(a_{11} - a_{12}) + (a_{22} - a_{21})$, which is equivalent to $d = a_{11} - a_{21} - a_{12} + a_{22}$. This method then gives $p_1 = (a_{22} - a_{21})/d$ and $p_2 = (a_{11} - a_{12})/d$, exactly as in Method 1. The analysis is similar for q_1 and q_2, and in the case in which $a_{11} < a_{12}$.

11.2 EXERCISES

1. Suppose a game has payoff matrix $\begin{bmatrix} 3 & -4 \\ -5 & 2 \end{bmatrix}$. Suppose that player B uses the strategy $\begin{bmatrix} .3 \\ .7 \end{bmatrix}$. Find the expected value of the game if player A uses each of the following strategies.

   a. $[.5 \quad .5]$ **b.** $[.1 \quad .9]$ **c.** $[.8 \quad .2]$ **d.** $[.2 \quad .8]$

2. Suppose a game has payoff matrix $\begin{bmatrix} 0 & -4 & 1 \\ 3 & 2 & -4 \\ 1 & -1 & 0 \end{bmatrix}$. Find the expected value of the game for the following strategies for players A and B.

   a. $A = [.1 \quad .4 \quad .5]; B = \begin{bmatrix} .2 \\ .4 \\ .4 \end{bmatrix}$ **b.** $A = [.3 \quad .4 \quad .3]; B = \begin{bmatrix} .8 \\ .1 \\ .1 \end{bmatrix}$

Find the optimum strategies for player A and player B in the games in Exercises 3–14. Find the value of each game. (Be sure to look for a saddle point first.)

3. $\begin{bmatrix} 5 & 1 \\ 3 & 4 \end{bmatrix}$ 4. $\begin{bmatrix} -4 & 5 \\ 3 & -4 \end{bmatrix}$ 5. $\begin{bmatrix} -2 & 0 \\ 3 & -4 \end{bmatrix}$ 6. $\begin{bmatrix} 6 & 2 \\ -1 & 10 \end{bmatrix}$

7. $\begin{bmatrix} 4 & -3 \\ -1 & 7 \end{bmatrix}$ 8. $\begin{bmatrix} 0 & 6 \\ 4 & 0 \end{bmatrix}$ 9. $\begin{bmatrix} -2 & 1/2 \\ 0 & -3 \end{bmatrix}$ 10. $\begin{bmatrix} 6 & 3/4 \\ 2/3 & -1 \end{bmatrix}$

11. $\begin{bmatrix} 8/3 & -1/2 \\ 3/4 & -5/12 \end{bmatrix}$ 12. $\begin{bmatrix} -1/2 & 2/3 \\ 7/8 & -3/4 \end{bmatrix}$ 13. $\begin{bmatrix} -1 & 2 \\ 3 & 1 \end{bmatrix}$ 14. $\begin{bmatrix} 8 & 18 \\ -4 & 2 \end{bmatrix}$

Remove any dominated strategies, and then find the optimum strategy for each player and the value of the game.

15. $\begin{bmatrix} -4 & 9 \\ 3 & -5 \\ 8 & 7 \end{bmatrix}$ 16. $\begin{bmatrix} 3 & 4 & -1 \\ -2 & 1 & 0 \end{bmatrix}$ 17. $\begin{bmatrix} 8 & 6 & 3 \\ -1 & -2 & 4 \end{bmatrix}$

18. $\begin{bmatrix} -1 & 6 \\ 8 & 3 \\ -2 & 5 \end{bmatrix}$ 19. $\begin{bmatrix} 9 & -1 & 6 \\ 13 & 11 & 8 \\ 6 & 0 & 9 \end{bmatrix}$ 20. $\begin{bmatrix} 4 & 8 & -3 \\ 2 & -1 & 1 \\ 7 & 9 & 0 \end{bmatrix}$

21. Verify that the optimum strategy for player B in a non-strictly-determined game is as given in the text.

22. Verify that the value of a non-strictly-determined game is as given in the text when players A and B play their optimum strategies.

23. Some people claim that even if a mixed strategy is used only once, the player should use a spinner to determine what choice to make. Write a few sentences giving your reaction to this claim.

24. For a game with optimum strategy $\begin{bmatrix} \frac{1}{2} & \frac{1}{2} \end{bmatrix}$, a player could flip a coin to determine which choice to make. A nonrandom method would be to alternate between the two choices. Discuss any disadvantages of this nonrandom method. Would it make a difference whether or not the opposing player is intelligent? Explain.

25. Why doesn't the reasoning in this section apply to strictly determined games?

26. a. Prove that if player A has the optimum strategy $\begin{bmatrix} p_1 & p_2 \end{bmatrix}$ as given in the text, then $AM = \begin{bmatrix} g & g \end{bmatrix}$, where g is the value of the game.

   b. Prove that if player B has the optimum strategy $\begin{bmatrix} q_1 \\ q_2 \end{bmatrix}$ given in the text, then $MB = \begin{bmatrix} g \\ g \end{bmatrix}$, where g is the value of the game.

Applications

BUSINESS AND ECONOMICS

27. *Advertising* Suppose Allied Manufacturing Company decides to put its new product on the market with a big television and radio advertising campaign. At the same time, the company finds out that its major competitor, Bates manufacturing, also has decided to launch a big advertising campaign for a similar product. The payoff matrix shows the increased sales (in millions) for Allied, as well as the decreased sales for Bates.

$$
\begin{array}{c}
\quad\quad\quad\quad \textit{Bates} \\
\quad\quad\quad T.V. \quad Radio \\
\textit{Allied} \begin{array}{c} T.V. \\ Radio \end{array} \begin{bmatrix} 1.0 & -.7 \\ -.5 & .5 \end{bmatrix}
\end{array}
$$

Find the optimum strategy for Allied Manufacturing and the value of the game.

28. *Pricing* The payoffs in the matrix below represent the differences between Boeing Aircraft Company's profit and its competitor's profit for two prices (in millions) on commercial jet transports, with positive payoffs being in Boeing's favor. What should Boeing's price strategy be?*

$$
\begin{array}{c}
\quad\quad\quad \textit{Competitor's} \\
\quad\quad\quad \textit{Price Strategy} \\
\quad\quad\quad 4.75 \quad 4.9 \\
\textit{Boeing's Strategy} \begin{array}{c} 4.9 \\ 4.75 \end{array} \begin{bmatrix} -4 & 2 \\ 2 & 0 \end{bmatrix}
\end{array}
$$

29. *Sales The Huckster*[†] Merrill has a concession at Yankee Stadium for the sale of sunglasses and umbrellas. The business places quite a strain on him, the weather being what it is. He has observed that he can sell about 500 umbrellas when it rains, and about 100 when it is sunny; in the latter case he can also sell 1000 sunglasses. Umbrellas cost him $5 and sell for $10; sunglasses cost $2 and sell for $5. He is willing to invest $2500 in the project. Everything that is not sold is considered a total loss.

He assembles the facts regarding profit in a matrix.

$$
\begin{array}{c}
\quad\quad\quad\quad\quad \textit{Selling during:} \\
\quad\quad\quad\quad Rain \quad\quad Shine \\
\textit{Buying for:} \begin{array}{c} Rain \\ Shine \end{array} \begin{bmatrix} 2500 & -1500 \\ -1500 & 3500 \end{bmatrix}
\end{array}
$$

He immediately takes heart, for this is a mixed-strategy game, and he should be able to find a stabilizing strategy that will save him from the vagaries of the weather. Find the best mixed strategy for Merrill.

LIFE SCIENCES

30. *Choosing Medication* The number of cases of African flu has reached epidemic levels. The disease is known to have two strains with similar symptoms. Dr. De Luca has two medicines available; the first is 60% effective against the first strain and 40% effective against the second. The second medicine is completely effective against the second

---

*Brigham, Georges, "Pricing, Investment, and Games of Strategy," *Management Sciences Models and Techniques,* Vol. 1, Copyright © 1960 by Pergamon Press, Ltd. Reprinted with permission.
[†]Williams, J. D., *The Compleat Strategyst,* McGraw-Hill Book Company, 1966. Reprinted by permission from The Rand Corporation. This is an excellent nontechnical book on game theory.

strain but ineffective against the first. Use the matrix below to decide which medicine she should use and the results she can expect.

$$\begin{array}{cc} & \textit{Strain} \\ & \begin{array}{cc} 1 & 2 \end{array} \\ \text{Medicine} \begin{array}{c} 1 \\ 2 \end{array} & \begin{bmatrix} .6 & .4 \\ 0 & 1 \end{bmatrix} \end{array}$$

GENERAL INTEREST

31. *Cats* When two cats, Euclid and Jamie, play together, their game involves facing each other while several feet apart; each cat must then decide whether to pounce or to "freeze"—to stay motionless until the other cat pounces.

Euclid weighs 2 lb more than Jamie, so if they both pounce, Jamie is squashed and Euclid gains 3 points. If they both freeze, Euclid remains in control of the area and gains 2 points. If Euclid pounces while Jamie freezes, Jamie can put up a good defense, so Euclid only gains 1 point. Euclid is poor at defense, so if he freezes while Jamie pounces, he loses 2 points. Find the optimum strategy for each cat, and find the value of the game.

32. *Coin Game* In the game of matching coins, each of two players flips a coin. If both coins match (both show heads or both show tails), player A wins $1. If there is no match, player B wins $1, as in the payoff matrix below. Find the optimum strategies for the two players and the value of the game.

$$\begin{bmatrix} 1 & -1 \\ -1 & 1 \end{bmatrix}$$

33. *Finger Game* Players A and B play a game in which they show either one or two fingers at the same time. If there is a match, A wins the amount of dollars equal to the total number of fingers shown. If there is no match, B wins the amount of dollars equal to the number of fingers shown.

   a. Write the payoff matrix.

   b. Find optimum strategies for A and B and the value of the game.

34. *Finger Game* Repeat Exercise 33 if each player may show either 0 or 2 fingers with the same payoffs.

11.3 GAME THEORY AND LINEAR PROGRAMMING

THINK ABOUT IT   How can a company decide in which cities it should advertise?

Using game theory and linear programming, we will answer this and other questions.

Until now, we have not learned how to solve games in which one player has more than two choices, but for which neither player has an optimum pure strategy. In this section, we shall see how the techniques of linear programming can be used to solve games in which each player has an arbitrary number of choices.

If there is no saddle point, we must find the probabilities for mixing strategies to obtain an optimum solution and the value of the game. Let us see how linear programming can be applied to Example 1 in the previous section, for which the payoff matrix was

$$M = \begin{bmatrix} -1 & 2 \\ 1 & 0 \end{bmatrix}.$$

When player A chose row 1 with probability p_1 and row 2 with probability p_2, we found that his expectations when B chose columns 1 and 2 were as follows.

$$E_1 = -1 \cdot p_1 + 1 \cdot p_2 = -p_1 + p_2$$
$$E_2 = 2 \cdot p_1 + 0 \cdot p_2 = 2p_1$$

Player A wishes his expected gain to be as large as possible, so he should maximize the minimum of the expected gains.

Let g represent the minimum of the expected gains, so that

$$E_1 = -p_1 + p_2 \geq g$$
$$E_2 = 2p_1 \geq g.$$

We can simplify this system of linear inequalities by dividing both inequalities by g. As long as $g > 0$, which we shall show momentarily, this yields

$$E_1/g = -\left(\frac{p_1}{g}\right) + \left(\frac{p_2}{g}\right) \geq 1$$

$$E_2/g = 2\left(\frac{p_1}{g}\right) \phantom{+ \left(\frac{p_2}{g}\right)} \geq 1.$$

Then denote p_1/g by x and p_2/g by y.

$$E_1/g = -x + y \geq 1$$
$$E_2/g = 2x \geq 1$$

To determine the objective function, note that $x + y = p_1/g + p_2/g = 1/g$. Since the goal is to maximize g, which is the same as minimizing $1/g$, we can rewrite the problem as the following linear programming problem:

$$\text{Minimize} \quad w = x + y$$
$$\text{subject to:} \quad -x + y \geq 1$$
$$2x \geq 1$$
$$\text{with} \quad x \geq 0, y \geq 0,$$

where $w = 1/g$. This linear programming problem can be solved by the graphical method described in Chapter 3. The graph is shown in Figure 4. There is only one corner point, $(1/2, 3/2)$, and the value of w there is 2. Thus the value of the game is $g = 1/w = 1/2$, and the optimum strategy for player A is $p_1 = gx = (1/2)(1/2) = 1/4$, $p_2 = gy = (1/2)(3/2) = 3/4$. Notice that this agrees with the result found in the previous section.

A similar analysis can be done for player B. When B chooses columns 1 and 2 with probabilities q_1 and q_2, respectively, then her expectations when A chooses rows 1 and 2 are

$$E_1 = -1 \cdot q_1 + 2 \cdot q_2 = -q_1 + 2q_2$$
$$E_2 = 1 \cdot q_1 + 0 \cdot q_2 = q_1.$$

Player B wishes the payoff to be as small as possible, so she should minimize the maximum of the expected gains:

$$E_1 = -q_1 + 2q_2 \leq g$$
$$E_2 = q_1 \leq g.$$

FOR REVIEW ■

The graphical method of linear programming was discussed in Chapter 3. In brief, the method involves the following steps:

1. Graph the region satisfied by all the inequalities, known as the feasible region.

2. Identify the corner points of the feasible region, and find the value of the objective function at each corner point.

3. The solution is found at the corner point producing the optimum value of the objective function.

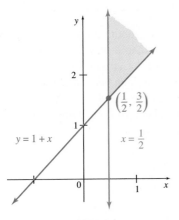

FIGURE 4

As before, we divide both inequalities by g and denote q_1/g by x and q_2/g by y, yielding:

$$E_1/g = -x + 2y \leq 1$$
$$E_2/g = x \leq 1.$$

As before, $x + y = q_1/g + q_2/g = 1/g$. Player B's goal is to minimize g, which is the same as maximizing $1/g$. We therefore have the following linear programming problem:

$$\text{Maximize} \quad z = x + y$$
$$\text{subject to:} \quad -x + 2y \leq 1$$
$$x \leq 1$$
$$\text{with} \qquad x \geq 0, y \geq 0,$$

where $z = 1/g$. This problem can also be solved by the graphical method; the reader should do this, and verify that $x = y = 1$ and $z = 2$. As before, the value of the game is $g = 1/z = 1/2$, and the optimum strategy for B is $q_1 = gx = (1/2)(1) = 1/2$, $q_2 = gy = (1/2)(1) = 1/2$. The value of the game is the same for B as for A, as must be the case, and B's strategy is the same as in the previous section.

Most significantly, notice that the linear programming problem for player A is the dual of the linear programming problem for player B. It can be shown that, under circumstances to be discussed shortly, this remarkable fact is always true. Notice further that A wants to maximize the minimum value of the game, while B wants to minimize the maximum of the game. That these two values are equal is an illustration of the **minimax principle,** which follows from the theorem of duality discussed in Chapter 4.

If the value of a zero-sum game is positive, the game can always be solved by linear programming techniques, in which case the solution of the row player is the dual of the solution of the column player. To guarantee that the value of the game is positive, find the most negative number in the payoff matrix and add its absolute value to all entries in the matrix. For example, in the previous game, -1 was the most negative entry. Adding 1 to all entries yields the payoff matrix

$$N = \begin{bmatrix} 0 & 3 \\ 2 & 1 \end{bmatrix}.$$

Solving this game as before would result in the same optimum strategies, but the value of the game would be $3/2$, that is, 1 greater than it was before. We would then subtract 1 from $3/2$ to get $1/2$, the value of the original game.

Just as linear programming problems with more than two variables could be solved by the simplex method, we can now use the simplex method to solve games involving more than two choices for each player. Before giving such an example, let us review the steps in solving a matrix game.

SOLVING ZERO-SUM GAMES

1. Set up the payoff matrix M.
2. Remove any dominated rows or columns.

3. Check to see if there is a saddle point. If so, the optimum strategies are determined by the row and column in which the saddle point is located.

4. If there is no saddle point and negative numbers are present, add a positive number to all entries in the payoff matrix so that all entries are nonnegative. Denote the elements of the remaining $m \times n$ matrix by a_{ij}.

5. Set up the linear programming problem:

$$\text{Maximize} \quad z = x_1 + \cdots + x_n$$
$$\text{subject to:} \quad a_{11}x_1 + \cdots + a_{1n}x_n \leq 1$$
$$\vdots \qquad \vdots$$
$$a_{m1}x_1 + \cdots + a_{mn}x_n \leq 1$$
$$\text{with} \quad x_1 \geq 0, \ldots, x_n \geq 0.$$

6. After finding the solution z to the linear programming problem, the value of the modified game is given by $g = 1/z$. The optimum strategy for player B is found by multiplying the solution to the linear programming problem by g: $q_i = gx_i$, $i = 1, \ldots, n$. The optimum strategy for player A is found by multiplying the solution to the dual problem by g: $p_j = gy_j$, $j = 1, \ldots, m$.

7. Subtract from g the positive number added in step 4 to find the value of the original game.

EXAMPLE 1 Optimum Strategies

Find the optimum strategies and the value of the game that was introduced in Section 11.1 on Strictly Determined Games but was never solved.

Solution The payoff matrix for this game, after deleting the column that was dominated by another column, was

$$\begin{bmatrix} -6 & 10 \\ 0 & -9 \\ -4 & -8 \end{bmatrix}.$$

Verify that this game has no saddle point. Because of the negative entries, we will add 9 to all entries to make them all nonnegative. The resulting payoff matrix is

$$\begin{bmatrix} 3 & 19 \\ 9 & 0 \\ 5 & 1 \end{bmatrix}.$$

The linear programming problem to be solved is

$$\text{Maximize} \quad z = x_1 + x_2$$
$$\text{subject to:} \quad 3x_1 + 19x_2 \leq 1$$
$$9x_1 \qquad \leq 1$$
$$5x_1 + x_2 \leq 1$$
$$\text{with} \quad x_1 \geq 0, \ x_2 \geq 0.$$

FOR REVIEW ■

The steps of the simplex algorithm can be stated briefly as follows.

1. Convert each constraint in the maximization problem into an equation by adding a slack variable.

2. Set up the initial tableau, with the negative of the payoffs in the bottom row.

3. Choose the column with the most negative indicator. Divide each element from that column into the corresponding element from the last column. The pivot is the element with the smallest quotient. Use row operations to change all elements in the pivot column other than the pivot to 0.

4. When there are no more negative numbers in the bottom row, the values of $x_1, \ldots, x_n$ can be read using the appropriate columns in combination with the last column.

5. If the z column contains a number other than 1, divide the last row by that number. The value of z, which is the same as the value of w, is given in the lower right hand corner. The solution to $y_1, \ldots, y_n$ can be read from the bottom row of the columns corresponding to the slack variables.

The initial tableau is

$$
\begin{array}{cccccc|c}
x_1 & x_2 & s_1 & s_2 & s_3 & z & \\
3 & 19 & 1 & 0 & 0 & 0 & 1 \\
9 & 0 & 0 & 1 & 0 & 0 & 1 \\
5 & 1 & 0 & 0 & 1 & 0 & 1 \\
\hline
-1 & -1 & 0 & 0 & 0 & 1 & 0
\end{array}.
$$

The first and second columns both contain -1 at the bottom; since these are equally negative, we arbitrarily choose the first column. The smallest ratio is formed by the 9 in row 2. Making this the pivot yields the following matrix.

$$
\begin{array}{l}
-R_2 + 3R_1 \to R_1 \\
\\
-5R_2 + 9R_3 \to R_3 \\
R_2 + 9R_4 \to R_4
\end{array}
\begin{array}{cccccc|c}
x_1 & x_2 & s_1 & s_2 & s_3 & z & \\
0 & 57 & 3 & -1 & 0 & 0 & 2 \\
9 & 0 & 0 & 1 & 0 & 0 & 1 \\
0 & 9 & 0 & -5 & 9 & 0 & 4 \\
0 & -9 & 0 & 1 & 0 & 9 & 1
\end{array}
$$

The next pivot is the 57 in row 1, column 2.

$$
\begin{array}{l}
\\
\\
-3R_1 + 19R_3 \to R_3 \\
3R_1 + 19R_4 \to R_4
\end{array}
\begin{array}{cccccc|c}
x_1 & x_2 & s_1 & s_2 & s_3 & z & \\
0 & 57 & 3 & -1 & 0 & 0 & 2 \\
9 & 0 & 0 & 1 & 0 & 0 & 1 \\
0 & 0 & -9 & -92 & 171 & 0 & 70 \\
0 & 0 & 9 & 16 & 0 & 171 & 25
\end{array}
$$

Since there are no more negative numbers in the bottom row, we are done pivoting. In each column with only one nonzero element, convert that element to a 1 by multiplying the corresponding row by the appropriate constant.

$$
\begin{array}{l}
\frac{1}{57}R_1 \to R_1 \\
\frac{1}{9}R_2 \to R_2 \\
\frac{1}{171}R_3 \to R_3 \\
\frac{1}{171}R_4 \to R_4
\end{array}
\begin{array}{cccccc|c}
x_1 & x_2 & s_1 & s_2 & s_3 & z & \\
0 & 1 & \frac{1}{19} & \frac{-1}{57} & 0 & 0 & \frac{2}{57} \\
1 & 0 & 0 & \frac{1}{9} & 0 & 0 & \frac{1}{9} \\
0 & 0 & \frac{-1}{19} & \frac{-92}{171} & 1 & 0 & \frac{70}{171} \\
0 & 0 & \frac{1}{19} & \frac{16}{171} & 0 & 1 & \frac{25}{171}
\end{array}
$$

From the bottom row, $z = 25/171$, so

$$
g = \frac{1}{z} = \frac{171}{25}.
$$

The values of y_1, y_2, and y_3 are read from the bottom of the columns for the three slack variables.

$$
y_1 = \frac{1}{19}, \qquad y_2 = \frac{16}{171}, \qquad y_3 = 0
$$

We find the values of p_1, p_2, and p_3 by multiplying the values of y_j by g.

$$
p_1 = \left(\frac{1}{19}\right)\left(\frac{171}{25}\right) = \frac{9}{25}, \quad p_2 = \left(\frac{16}{171}\right)\left(\frac{171}{25}\right) = \frac{16}{25}, \quad p_3 = (0)\left(\frac{171}{25}\right) = 0
$$

Next find x_1 and x_2 by using the first and second columns combined with the last column.

$$x_1 = \frac{1}{9}, \quad x_2 = \frac{2}{57}$$

Find the values of q_1 and q_2 by multiplying the values of x_1 by g.

$$q_1 = \left(\frac{1}{9}\right)\left(\frac{171}{25}\right) = \frac{19}{25}, \quad q_2 = \left(\frac{2}{57}\right)\left(\frac{171}{25}\right) = \frac{6}{25}$$

Finally, the value of the game is found by subtracting from g the 9 that was added at the beginning, yielding $(171/25) - 9 = -54/25$.

To summarize, the optimum strategy for player A is $(9/25, 16/25, 0)$, and the optimum strategy for player B is $(19/25, 6/25)$. When these strategies are used, the value of the game is $-54/25$. ■

Some applications of these techniques are given in the exercises. For many realistic games, the details of using the simplex method become very tedious, and so a graphing calculator or a computer is of great help. With the aid of a program to perform the simplex algorithm, the optimum strategies for games of virtually any size can be found.

11.3 EXERCISES

In Exercises 1–6, use the graphical method to find the optimum strategy for players A and B and the value of the game for each payoff matrix.

1. $\begin{bmatrix} 1 & 2 \\ 3 & 1 \end{bmatrix}$

2. $\begin{bmatrix} 5 & 2 \\ 0 & 3 \end{bmatrix}$

3. $\begin{bmatrix} 4 & -2 \\ -1 & 6 \end{bmatrix}$

4. $\begin{bmatrix} -1 & 5 \\ 1 & -6 \end{bmatrix}$

5. $\begin{bmatrix} 7 & -8 \\ -3 & 3 \end{bmatrix}$

6. $\begin{bmatrix} -4 & 1 \\ 5 & 0 \end{bmatrix}$

In Exercises 7–12, use the simplex method to find the optimum strategy for players A and B and the value of the game for each payoff matrix.

7. $\begin{bmatrix} 3 & -4 & 1 \\ 5 & 3 & -2 \end{bmatrix}$

8. $\begin{bmatrix} -6 & 1 & 4 & 2 \\ 9 & 3 & -8 & -7 \end{bmatrix}$

9. $\begin{bmatrix} -1 & 2 & 4 \\ 3 & -2 & 0 \end{bmatrix}$

10. $\begin{bmatrix} 1 & 0 \\ -2 & 4 \\ -1 & -1 \end{bmatrix}$

11. $\begin{bmatrix} 1 & 0 & -1 \\ -1 & 0 & 1 \\ 2 & -1 & 2 \end{bmatrix}$

12. $\begin{bmatrix} 2 & -1 & 1 \\ 0 & 2 & 3 \\ 4 & 1 & 0 \end{bmatrix}$

Applications

In Exercises 13–22, use the graphical method when the payoff matrix is a 2 × 2 matrix or can be reduced to one after removing rows or columns that are dominated. Otherwise, use the simplex method.

BUSINESS AND ECONOMICS

13. *Contractor Bidding* A contractor prepares to bid on a job. If all goes well, his bid should be $30,000, which will cover his costs plus his usual profit margin of $4500. If a threatened labor strike actually occurs, however, his bid should be $40,000 to give him the same profit. If there is a strike and he bids $30,000, he will lose $5500. If he bids $40,000 and there is no strike, his bid will be too high, and he will lose the job entirely. Find the optimum strategy for the contractor, and find the value of the game.

14. *Negotiation* In negotiating a labor contract, a labor union has considered four different approaches, based on different positions the union can take on various issues. Management can take three different approaches in the negotiations. The final contract will be regarded as either favorable to the union, favorable to the management, or neither, leading to the following payoff matrix.

Management Strategies

$$
\begin{array}{c}
\text{Labor Strategies} \\
\end{array}
\begin{array}{c}
1 \\ 2 \\ 3 \\ 4
\end{array}
\begin{bmatrix}
-1 & -1 & 1 \\
-1 & 0 & 0 \\
1 & -1 & -1 \\
1 & 1 & -1
\end{bmatrix}
$$

Find the optimum strategies for labor and management, and find the value of the game.

15. *Advertising* Marketing executives for two competing companies are trying to decide whether to advertise in Atlanta,

Boston, or Cleveland, given that each company can afford to target only one city at a time. General Items Company has a leading market share and has found that it will earn an additional profit of $10,000, $8000, or $6000 per week if they advertise in Atlanta, Boston, or Cleveland, respectively. Original Imitators, Inc., has a smaller portion of the market, but its executives can cut General Items' additional profit in half and gain that profit themselves if they run competing ads in the same city as General Items. If they run ads in a different city, their ads seem to have no impact.

a. Set up the profit matrix for this game.

b. Find the optimum strategy for each company and the value of the game.

16. *Marketing* Solve the Christmas card problem discussed in Section 11.1, for which we had the following payoff matrix.

		States of Nature	
	Weak Economy	In-Between	Strong Economy
Modern	40	85	120
Strategies Old-Fashioned	90	45	85
Mixture	75	105	65

17. *Competition* Two merchants are planning to build competing stores to serve an area of three small cities. The fraction of the total population that live in each city is shown in the figure below. If both merchants locate in the same city, merchant A will get 65% of the total business. If the merchants locate in different cities, each will get 80% of the business in the city where it is located, and 20% of the business in the city where the other merchant is located. In the city where neither is located, A will get 60% of the business and B will get 40%. Payoffs are measured by the number of percentage points above or below 50%. Find an optimum strategy for each merchant, and find the value of this game. (See Exercise 32 in Section 11.1.)

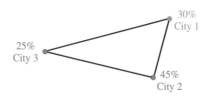

LIFE SCIENCES

18. *Nurse Staffing* Hospitals need to decide how many nurses to assign on a given day, even though they don't know the demand in advance. An article assigned the following points to various scenarios.*

Adequate staff, fully utilized	10
Moderate underutilization	8
Moderate overcrowding	6
Extreme underutilization	5
Extreme overcrowding	2
Dangerous overcrowding	1

The article gave a payoff matrix as follows, where the demand can be low, medium, or high, and anywhere from 1 to 4 nurses can be assigned.

Demand

	low	medium	high
1	5	8	10
2	8	10	6
3	10	6	2
4	6	2	1

Number of Nurses

a. Remove any dominated strategies.

b. Find the optimum strategy for the hospital.

SOCIAL SCIENCES

19. *Conflict* In an article on the sociology of conflict, the author proposes a game in which a member of a race (she calls them the green race) visits a country in which some natives are pro-green, some indifferent, and some anti-green.[†] The green visitor, unable to determine which group a native is from, has three options: act appeasing, expect civility but not fight for it, and expect civility and fight for it if he doesn't get it. The author suggests the following payoff matrix:

	pro-green	indifferent	anti-green
appease	0	0	−4
no fight	0	−3	−1
fight	−2	−1	−1

What is the visitor's optimum strategy, and what is the value of the game?

20. *Education* In an accounting class, the instructor permits the students to bring a calculator or a reference book (but not both) to an examination. The examination will emphasize either numerical problems or definitions. In trying

to decide which aid to take to an examination, a student first decides on the utilities shown in the following payoff matrix.

Exam's Emphasis

		Numbers	Definitions
Student's Choice	Calculator	50	0
	Book	10	40

What is the student's optimum strategy, and what is the value of the game?

GENERAL INTEREST

21. *Military Science* The Colonel Blotto game is a type of military strategy game.[‡] Two opposing armies are approaching two posts. Colonel Blotto has 4 regiments under his command, while his opponent, Captain Kije, has 3 regiments. Each commander must decide how many regiments to send to each post. The army that sends more regiments to a post not only captures that post but also captures the losing army's regiments. If both armies send the same number of regiments to a post, there is a stand-off, and neither army wins. The payoff is one point for capturing the post and one point for each regiment captured.

a. Set up the payoff matrix for this game. (*Hint:* Colonel Blotto has five choices, and Captain Kije has four.)

b. Find the optimum strategy for each commander and the value of the game.

c. Show that if Colonel Blotto uses the strategy found in part b, then any strategy used by Captain Kije results in the same payoff. (*Hint:* Show that $AM = (14/9)R$, where R is a row matrix consisting of all 1's, and then use the fact that $RB = [1]$.)

d. Based on the result of part c, what can you conclude about the uniqueness of the optimum strategy found by linear programming?

22. *Card Games* Player A deals Player B one of three cards—ace, king, or queen—at random, face down.[§] B looks at the card. If it is an ace, B must say "ace;" if it is a king, B can say either "king" or "ace;" if it is a queen, B can say either "queen" or "ace." If B says "ace," A can either believe him and give him $1, or ask him to show his card. If it is an ace, A must pay B $2; but if it is not, B pays A $2. If B says

---

*Duckett, Stephen, "Nurse Rostering with Game Theory," *Journal of Nursing Administration,* Jan. 1977, pp. 58–59.

[†]Bernard, Jessie, "The Theory of Games of Strategy as a Modern Sociology of Conflict," *American Journal of Sociology,* Vol. 59, No. 5, 1954, pp. 411–424.

[‡]Karlin, Samuel, *Mathematical Methods and Theory in Games, Programming, and Economics,* Addison-Wesley, 1959.

[§]Thomas, L. C., *Games, Theory, and Applications,* Halsted Press, 1986.

"king," neither side loses anything; but if he says "queen," B must pay A $1.

a. Set up the payoff matrix for this game. (*Hint:* Consider these two choices for A: (1) believe B when B says "ace;" (2) ask B to show his card when B says "ace." Consider the following four choices for B: (1) always tell the truth; (2) lie only if the card is a queen; (3) lie only if the card is a king; (4) lie if the card is a queen or a king. For each entry in the payoff matrix, find the average payoff over the three possible outcomes.)

b. Find the optimum strategy for each player and the value of the game.

23. *Children's Game* In the children's game *rock, paper, scissors*, two players simultaneously extend one hand in the form of a fist (*rock*), a flat palm (*paper*), or two fingers extended (*scissors*). If they both make the same choice, the game is a tie. Otherwise, *rock* beats *scissors* (because rock can crush scissors), *scissors* beats *paper* (because scissors can cut paper), and *paper* beats *rock* (because paper can cover rock). Whoever wins gains one point; in the case of a tie, no one gains any points. (See Exercise 37 in Section 11.1, Strictly Determined Games.)

a. Find the optimum strategy for each player and the value of the game.

b. In hindsight, how could you have guessed the answer to part a without doing any work?

CHAPTER SUMMARY

Game theory is a tool for analyzing situations in which a person, known as a player, must choose from a finite set of choices. The payoff depends not only on what the player chooses, but on what a second player chooses. In many applications, the player is not a person, but a company, an army, or some other organization. In a strictly determined game, each player has a single optimum choice. Otherwise, the best choice is a mixed strategy, in which the players randomly make a choice. For a game in which each player has only two choices, we have developed formulas giving the probability with which they should make each choice. When at least one player has more than two choices, the optimum mixed strategies can be found using the simplex method.

KEY TERMS

11.1 states of nature	two-person game	pure strategy	**11.2** mixed strategy
strategies	zero-sum game	saddle point	**11.3** minimax principle
payoff	dominated strategy	strictly determined	
payoff matrix	optimum strategy	game	
game theory	value of the game	fair game	

CHAPTER 11 REVIEW EXERCISES

1. How can you determine from the payoff matrix whether a game is strictly determined?

2. Briefly explain any advantages of removing from the payoff matrix any dominated strategies.

Use the following payoff matrix to determine the payoff if each of the strategies in Exercises 3–6 is used.

$$\begin{bmatrix} -2 & 5 & -6 & 3 \\ 0 & -1 & 7 & 5 \\ 2 & 6 & -4 & 4 \end{bmatrix}$$

3. $(1, 1)$ **4.** $(1, 4)$ **5.** $(2, 3)$ **6.** $(3, 4)$

7. Are there any dominated strategies in this game?

8. Is there a saddle point?

Remove any dominated strategies in the following games.

9. $\begin{bmatrix} -11 & 6 & 8 & 9 \\ -10 & -12 & 3 & 2 \end{bmatrix}$ **10.** $\begin{bmatrix} -1 & 9 & 0 \\ 4 & -10 & 6 \\ 8 & -6 & 7 \end{bmatrix}$ **11.** $\begin{bmatrix} -2 & 4 & 1 \\ 3 & 2 & 7 \\ -8 & 1 & 6 \\ 0 & 3 & 9 \end{bmatrix}$ **12.** $\begin{bmatrix} 3 & -1 & 4 \\ 0 & 4 & -1 \\ 1 & 2 & -3 \\ 0 & 0 & 2 \end{bmatrix}$

For the following games, find the strategies producing any saddle points. Give the value of the game. Identify any fair games.

13. $\begin{bmatrix} -2 & 3 \\ -4 & 5 \end{bmatrix}$ **14.** $\begin{bmatrix} -4 & 0 & 2 & -5 \\ 6 & 9 & 3 & 8 \end{bmatrix}$ **15.** $\begin{bmatrix} -4 & -1 \\ 6 & 0 \\ 8 & -3 \end{bmatrix}$

16. $\begin{bmatrix} 4 & -1 & 6 \\ -3 & -2 & 0 \\ -1 & -4 & 3 \end{bmatrix}$ **17.** $\begin{bmatrix} 8 & 1 & -7 & 2 \\ -1 & 4 & -3 & 3 \end{bmatrix}$ **18.** $\begin{bmatrix} 2 & -9 \\ 7 & 1 \\ 4 & 2 \end{bmatrix}$

Find the optimum strategies for each of the following games using the formulas from Section 11.2 on Mixed Strategies. Find the value of the game.

19. $\begin{bmatrix} 1 & 0 \\ -2 & 3 \end{bmatrix}$ **20.** $\begin{bmatrix} 2 & -3 \\ -3 & 5 \end{bmatrix}$ **21.** $\begin{bmatrix} -3 & 5 \\ 1 & 0 \end{bmatrix}$ **22.** $\begin{bmatrix} 8 & -3 \\ -6 & 2 \end{bmatrix}$

For each of the following games, remove any dominated strategies, then solve the game using the formulas from Section 11.2 on Mixed Strategies. Find the value of the game.

23. $\begin{bmatrix} -4 & 8 & 0 \\ -2 & 9 & -3 \end{bmatrix}$ **24.** $\begin{bmatrix} 1 & 0 & 3 & -3 \\ 4 & -2 & 4 & -1 \end{bmatrix}$ **25.** $\begin{bmatrix} 2 & -1 \\ -4 & 5 \\ -1 & -2 \end{bmatrix}$ **26.** $\begin{bmatrix} 8 & -6 \\ 4 & -8 \\ -9 & 9 \end{bmatrix}$

Find the optimum strategies for each of the following games using the graphical method of linear programming. Find the value of the game.

27. $\begin{bmatrix} -4 & 2 \\ 3 & -5 \end{bmatrix}$ **28.** $\begin{bmatrix} -2 & 2 \\ 3 & 1 \end{bmatrix}$ **29.** $\begin{bmatrix} 1 & 0 \\ -3 & 4 \end{bmatrix}$ **30.** $\begin{bmatrix} 0 & -2 \\ -1 & 3 \end{bmatrix}$

Find the optimum strategies for each of the following games using the simplex method. Find the value of the game.

31. $\begin{bmatrix} 2 & 1 & -1 \\ -3 & -2 & 0 \end{bmatrix}$ **32.** $\begin{bmatrix} 1 & -3 \\ -4 & 2 \\ -2 & 1 \end{bmatrix}$ **33.** $\begin{bmatrix} -2 & 1 & 0 \\ 2 & 0 & -2 \\ 0 & -1 & 3 \end{bmatrix}$ **34.** $\begin{bmatrix} 2 & 1 & -1 \\ 0 & 1 & 2 \\ -1 & 2 & 0 \end{bmatrix}$

35. Under what conditions is it necessary to use the simplex algorithm to solve a game?

36. Suppose you live in a country where the chance of rain is 90% every day. Should you carry an umbrella every day? Now suppose you lose 100 points if it rains and you don't carry an umbrella. Suppose further that if you carry an umbrella and it doesn't rain, you lose 1 point due to the inconvenience. Otherwise you neither gain nor lose any points. According to game theory, what should you do? Discuss any discrepancies with your first answer.

Applications

BUSINESS AND ECONOMICS

Labor Relations In labor-management relations, both labor and management can adopt either a friendly or a hostile attitude. The results are shown in the following payoff matrix. The numbers give the wage gains made by an average worker.

		Management	
		Friendly	Hostile
Labor	Friendly	$600	$800
	Hostile	$400	$950

37. Suppose the chief negotiator for labor is an optimist. What strategy should he choose?

38. What strategy should he choose if he is a pessimist?

39. The chief negotiator for labor feels that there is a 70% chance that the company will be hostile. What strategy should he adopt? What is the expected payoff?

40. Just before negotiations begin, a new management is installed in the company. There is only a 40% chance that the new management will be hostile. What strategy should be adopted by labor?

41. Find the optimum strategy for labor and management, and find the value of the game.

Investment Hector, who has inherited $10,000 from his rich grandmother, consults the firm of J. K. Knowitall & Company for advice on how to invest the sum. The company provides the following estimates of the gains he might make in 5 yr by investing the $10,000 in two different types of stocks, each dependent on the state of the economy.

		Economy	
		Inflationary	Stable
Stocks	Blue-Chip	2800	3200
	Growth	5000	−2000

42. Find the optimum strategy for Hector and the value of the game using the method found in Section 11.2 on Mixed Strategies.

43. Find the optimum strategy for Hector and the value of the game using the graphical method.

44. Find the optimum strategy for Hector and the value of the game using the simplex algorithm.

SOCIAL SCIENCES

Politics A candidate for city council can come out in favor of a new factory, be opposed to it, or waffle on the issue. The change in votes for the candidate depends on what her opponent does, with payoffs as shown.

		Opponent		
		Favors	Waffles	Opposes
	Favors	0	−1000	−4000
Candidate	Waffles	1000	0	−500
	Opposes	5000	2000	0

45. What should the candidate do if she is an optimist?

46. What should she do if she is a pessimist?

47. Suppose the candidate's campaign manager feels there is a 40% chance that the opponent will favor the plant, and a 35% chance that he will waffle. What strategy should the candidate adopt? What is the expected change in the number of votes?

48. The opponent conducts a new poll that shows strong opposition to the new factory. This changes the probability that he will favor the factory to 0, and the probability that he will waffle to .7. What strategy should our candidate adopt? What is the expected change in the number of votes now?

49. Find the optimum strategy for this candidate and her opponent, and find the value of the game.

GENERAL INTEREST

Military *The little kingdom of Ravogna has two military installations, one three times as valuable as the other. The army has the capability to successfully defend either one of the installations, but not both at the same time. Rontovia, a country which has historically been antagonistic toward Ravogna, is capable of attacking either installation, but not both. The payoff matrix below indicates the respective values of the installations to Ravogna. Installation No. 2, with a value of 1, is the lesser installation.*

Rontovia:
Attack Installation

$$\begin{array}{c} \\ \text{Ravogna:} \\ \text{Defend Installation} \end{array} \begin{array}{c} \quad\;\; 1 \quad 2 \\ \begin{array}{c} 1 \\ 2 \end{array} \left[\begin{array}{cc} 4 & 1 \\ 3 & 4 \end{array} \right] \end{array}$$

50. Find the optimum strategy for each country and the value of the game using the method described in Section 11.2 on Mixed Strategies.

51. Find the optimum strategy for each country and the value of the game using the graphical method.

52. Find the optimum strategy for each country and the value of the game using the simplex method.

53. *Theology* In his book entitled *Pensées,* the French mathematician Blaise Pascal (1623–1662) described what has since been referred to as Pascal's Wager. Pascal observed that everyone must wager on whether God exists. If you decide that God exists and in fact God does exist, you gain "an infinity of an infinitely happy life"; if you are wrong, you lose nothing of great value, having lived a good life without receiving any reward for it. If you choose not to believe in God and God actually does exist, you will have lost something of infinite value. Pascal concludes that believing in God is by far the most rational action. Write a few sentences discussing Pascal's Wager, including how the payoff matrix might be set up and what the optimum strategy would be. What is your reaction to Pascal's conclusion?

EXTENDED APPLICATION: The Prisoner's Dilemma — Non-Zero-Sum Games in Economics*

In this chapter we have looked at two-person zero-sum games, games in which the total payoff is zero so that what one player wins the other loses. When economists use game theory to model the marketplace, they encounter many *non-zero-sum games,* games in which the total payoff to all players might be positive or negative. For example, the total value of the stock market increases in periods of optimistic buying and plunges when everyone decides to sell, so the total payoff from investing is not fixed at 0. These games are more complicated to analyze, but the idea of dominated strategies introduced in Section 11.1 turns out to be useful. In this application we look at one of the simplest and most famous non-zero-sum games, the Prisoner's Dilemma.

The classic scenario goes like this: Mike and Ike are arrested and charged with committing a murder. Taken to separate cells, each has to decide whether to deny any involvement or confess. If they both deny any involvement, there is still enough evidence to convict them both of manslaughter, and they'll both get 5-year sentences. If Mike confesses and implicates Ike as the actual murderer, while Ike claims innocence, Mike will get 1 year and Ike will get 30 years. If instead Ike cuts a deal with the prosecutors, he'll get the light sentence and Mike will get the long one. If they both confess, each implicating the other, they'll both get 10-year sentences. Now *each* player has a payoff matrix in which all the payoffs are negative. Letting D stand for "deny" and C for "confess," Mike's matrix looks like this:

Mike's payoffs

Ike's strategy

$$
\text{Mike's strategy} \quad
\begin{array}{c}
C \\
D
\end{array}
\begin{bmatrix}
-10 & -1 \\
-30 & -5
\end{bmatrix}
$$

What should Mike do? Using the idea of a dominated strategy, Mike notices that each entry in his first row is larger than each entry in the second row, so D is a dominated strategy: If Ike plays C, Mike will do better with C, and if Ike plays D, Mike will also do better with C, so it looks like C is Mike's best choice. Ike's matrix is similar:

Ike's payoffs

Ike's strategy

$$
\text{Mike's strategy} \quad
\begin{array}{c}
C \\
D
\end{array}
\begin{bmatrix}
-10 & -30 \\
-1 & -5
\end{bmatrix}
$$

For Ike, the first column (strategy C) dominates the second column (D) so he also decides to play C. The result is that both prisoners get 10-year sentences; they both lose, but neither gets

the worst possible outcome, the 30-year sentence. If neither player can predict what the other will do, confessing is a sensible strategy for both players. But notice that if they could agree beforehand to deny any involvement, they would *both* do better, getting only 5-year sentences!

Game theorists call a game in which players can make binding agreements a *cooperative game.* If Mike and Ike play noncooperatively, they'll both end up with 10-year sentences, but if they play cooperatively, they can engineer a better outcome for each of them. There's just one catch: Once they have agreed to deny involvement, each player has a strong incentive to *defect* from the agreement and plead guilty, getting off with a 1-year sentence and leaving the other player with 30 years in jail. And, of course, if they *both* defect they'll end up with 10-year sentences after all.

The Prisoner's Dilemma models economic transactions in which cooperating may lead to greater gains for both players than competing, but in which the success of the cooperative strategy depends on trust. For example, suppose two art dealers have private information about a painting up for auction. They know that they'll be able to resell it for $1 million. If they bid against each other, the winning bidder will probably end up paying close to $1 million, so the profit from resale will be small. But suppose the dealers agree beforehand that only one will bid and that they will split the profits from resale. Without competition, the winning bidder may pay much less than $1 million, and both dealers will make out well. (This arrangement is called a *ring* in the auction business, and, of course, the auction houses don't like rings!) As with the Prisoner's Dilemma, trust is essential to the success of the ring. After all, the winning bidder could just deny that there was any agreement and pocket the entire profit.

In the Exercises we'll look at some variations of a two-person non-zero-sum game called the Restaurant Game. Again, though the game is simple, the issues it raises appear in many settings involving competition and negotiation, including an arms race, labor-management bargaining, and sharing of resources between species occupying the same territory. Game theory has become an important tool in economics, sociology, and evolutionary biology.

Exercises

The Restaurant Game Linda and Mel both like to eat out; Linda likes the neighborhood Chinese restaurant and Mel likes the French one. At 5 o'clock they each announce their pick for the evening's restaurant. If they pick the same restaurant, they go out, otherwise they stay home and microwave some frozen

---

*For more on the basics of Game Theory, see Morris, Peter, *Introduction to Game Theory,* Springer, 1994. Morris has a good discussion of both zero-sum and non-zero-sum games. The restaurant game presented in the exercises is a version of Morris's Battle of the Buddies.

dinners. Each player has two strategies, C for Chinese restaurant and F for French restaurant.

1. Since they like to eat out, each prefers a restaurant meal to the frozen dinner, but they enjoy their favorite food much more than the other type. Suppose that Linda and Mel have the following payoff matrices, where the numbers represent degree of enjoyment:

Linda's payoffs

Mel's strategy

$$\text{Linda's strategy} \quad \begin{matrix} & \begin{matrix} C & F \end{matrix} \\ \begin{matrix} C \\ F \end{matrix} & \begin{bmatrix} 5 & 0 \\ 0 & 2 \end{bmatrix} \end{matrix}$$

Mel's payoffs

Mel's strategy

$$\text{Linda's strategy} \quad \begin{matrix} & \begin{matrix} C & F \end{matrix} \\ \begin{matrix} C \\ F \end{matrix} & \begin{bmatrix} 2 & 0 \\ 0 & 5 \end{bmatrix} \end{matrix}$$

Does either player have a dominated strategy? How should they resolve their dilemma?

2. If Linda likes French food more than Mel likes Chinese food, their matrices might look like this:

Linda's payoffs

Mel's strategy

$$\text{Linda's strategy} \quad \begin{matrix} & \begin{matrix} C & F \end{matrix} \\ \begin{matrix} C \\ F \end{matrix} & \begin{bmatrix} 5 & 0 \\ 0 & 3 \end{bmatrix} \end{matrix}$$

Mel's payoffs

Mel's strategy

$$\text{Linda's strategy} \quad \begin{matrix} & \begin{matrix} C & F \end{matrix} \\ \begin{matrix} C \\ F \end{matrix} & \begin{bmatrix} 1 & 0 \\ 0 & 5 \end{bmatrix} \end{matrix}$$

Does either player have a dominated strategy? If they decide to cooperate, how would they pick their restaurants for maximum combined enjoyment?

3. The suspects in the Prisoner's Dilemma play their game only once, but Linda and Mel repeat the Restaurant Game each night. As we noted in Section 11.2, this gives them the option of using mixed strategies. Suppose Linda chooses Chinese with probability .8 and French with probability .2, while Mel chooses Chinese with probability .1 and French with probability .9. What is Linda's expected payoff in "enjoyment units"? What is Mel's expected payoff? Who does better?

4. Suppose Linda knows that Mel is going to stick to his strategy (Chinese with probability .1 and French with probability .9). What strategy maximizes her enjoyment? What is her expected payoff? If she plays this way, what might Mel do?

5. In repeated games, a player can use an *adaptive* strategy that changes based on past history. For example, if Mel seems to be following an "always French" strategy, Linda could decide to pick Chinese a few times in a row. What are the possible hazards and benefits of this change in strategy?

6. Suppose Linda and Mel decide to cooperate. They like eating out, so they'll eat out every night, using a spinner to determine the type of restaurant. Given that Linda likes French more than Mel likes Chinese, what do you think would be a fair way of making the choice? Consider a picture like Figure 2 in Section 11.2.

TABLES

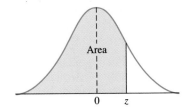

Table 1 Area Under a Normal Curve to the Left of z, Where $z = \dfrac{x - \mu}{\sigma}$

z	.00	.01	.02	.03	.04	.05	.06	.07	.08	.09
−3.4	.0003	.0003	.0003	.0003	.0003	.0003	.0003	.0003	.0003	.0002
−3.3	.0005	.0005	.0005	.0004	.0004	.0004	.0004	.0004	.0004	.0003
−3.2	.0007	.0007	.0006	.0006	.0006	.0006	.0006	.0005	.0005	.0005
−3.1	.0010	.0009	.0009	.0009	.0008	.0008	.0008	.0008	.0007	.0007
−3.0	.0013	.0013	.0013	.0012	.0012	.0011	.0011	.0011	.0010	.0010
−2.9	.0019	.0018	.0017	.0017	.0016	.0016	.0015	.0015	.0014	.0014
−2.8	.0026	.0025	.0024	.0023	.0023	.0022	.0021	.0021	.0020	.0019
−2.7	.0035	.0034	.0033	.0032	.0031	.0030	.0029	.0028	.0027	.0026
−2.6	.0047	.0045	.0044	.0043	.0041	.0040	.0039	.0038	.0037	.0036
−2.5	.0062	.0060	.0059	.0057	.0055	.0054	.0052	.0051	.0049	.0048
−2.4	.0082	.0080	.0078	.0075	.0073	.0071	.0069	.0068	.0066	.0064
−2.3	.0107	.0104	.0102	.0099	.0096	.0094	.0091	.0089	.0087	.0084
−2.2	.0139	.0136	.0132	.0129	.0125	.0122	.0119	.0116	.0113	.0110
−2.1	.0179	.0174	.0170	.0166	.0162	.0158	.0154	.0150	.0146	.0143
−2.0	.0228	.0222	.0217	.0212	.0207	.0202	.0197	.0192	.0188	.0183
−1.9	.0287	.0281	.0274	.0268	.0262	.0256	.0250	.0244	.0239	.0233
−1.8	.0359	.0352	.0344	.0336	.0329	.0322	.0314	.0307	.0301	.0294
−1.7	.0446	.0436	.0427	.0418	.0409	.0401	.0392	.0384	.0375	.0367
−1.6	.0548	.0537	.0526	.0516	.0505	.0495	.0485	.0475	.0465	.0455
−1.5	.0668	.0655	.0643	.0630	.0618	.0606	.0594	.0582	.0571	.0559
−1.4	.0808	.0793	.0778	.0764	.0749	.0735	.0722	.0708	.0694	.0681
−1.3	.0968	.0951	.0934	.0918	.0901	.0885	.0869	.0853	.0838	.0823
−1.2	.1151	.1131	.1112	.1093	.1075	.1056	.1038	.1020	.1003	.0985
−1.1	.1357	.1335	.1314	.1292	.1271	.1251	.1230	.1210	.1190	.1170
−1.0	.1587	.1562	.1539	.1515	.1492	.1469	.1446	.1423	.1401	.1379
−.9	.1841	.1814	.1788	.1762	.1736	.1711	.1685	.1660	.1635	.1611
−.8	.2119	.2090	.2061	.2033	.2005	.1977	.1949	.1922	.1894	.1867
−.7	.2420	.2389	.2358	.2327	.2296	.2266	.2236	.2206	.2177	.2148
−.6	.2743	.2709	.2676	.2643	.2611	.2578	.2546	.2514	.2483	.2451
−.5	.3085	.3050	.3015	.2981	.2946	.2912	.2877	.2843	.2810	.2776

Table 1 Area Under a Normal Curve (continued)

z	.00	.01	.02	.03	.04	.05	.06	.07	.08	.09
−.4	.3446	.3409	.3372	.3336	.3300	.3264	.3228	.3192	.3156	.3121
−.3	.3821	.3783	.3745	.3707	.3669	.3632	.3594	.3557	.3520	.3483
−.2	.4207	.4168	.4129	.4090	.4052	.4013	.3974	.3936	.3897	.3859
−.1	.4602	.4562	.4522	.4483	.4443	.4404	.4364	.4325	.4286	.4247
−.0	.5000	.4960	.4920	.4880	.4840	.4801	.4761	.4721	.4681	.4641
.0	.5000	.5040	.5080	.5120	.5160	.5199	.5239	.5279	.5319	.5359
.1	.5398	.5438	.5478	.5517	.5557	.5596	.5636	.5675	.5714	.5753
.2	.5793	.5832	.5871	.5910	.5948	.5987	.6026	.6064	.6103	.6141
.3	.6179	.6217	.6255	.6293	.6331	.6368	.6406	.6443	.6480	.6517
.4	.6554	.6591	.6628	.6664	.6700	.6736	.6772	.6808	.6844	.6879
.5	.6915	.6950	.6985	.7019	.7054	.7088	.7123	.7157	.7190	.7224
.6	.7257	.7291	.7324	.7357	.7389	.7422	.7454	.7486	.7517	.7549
.7	.7580	.7611	.7642	.7673	.7704	.7734	.7764	.7794	.7823	.7852
.8	.7881	.7910	.7939	.7967	.7995	.8023	.8051	.8078	.8106	.8133
.9	.8159	.8186	.8212	.8238	.8264	.8289	.8315	.8340	.8365	.8389
1.0	.8413	.8438	.8461	.8485	.8508	.8531	.8554	.8577	.8599	.8621
1.1	.8643	.8665	.8686	.8708	.8729	.8749	.8770	.8790	.8810	.8830
1.2	.8849	.8869	.8888	.8907	.8925	.8944	.8962	.8980	.8997	.9015
1.3	.9032	.9049	.9066	.9082	.9099	.9115	.9131	.9147	.9162	.9177
1.4	.9192	.9207	.9222	.9236	.9251	.9265	.9278	.9292	.9306	.9319
1.5	.9332	.9345	.9357	.9370	.9382	.9394	.9406	.9418	.9429	.9441
1.6	.9452	.9463	.9474	.9484	.9495	.9505	.9515	.9525	.9535	.9545
1.7	.9554	.9564	.9573	.9582	.9591	.9599	.9608	.9616	.9625	.9633
1.8	.9641	.9649	.9656	.9664	.9671	.9678	.9686	.9693	.9699	.9706
1.9	.9713	.9719	.9726	.9732	.9738	.9744	.9750	.9756	.9761	.9767
2.0	.9772	.9778	.9783	.9788	.9793	.9798	.9803	.9808	.9812	.9817
2.1	.9821	.9826	.9830	.9834	.9838	.9842	.9846	.9850	.9854	.9857
2.2	.9861	.9864	.9868	.9871	.9875	.9878	.9881	.9884	.9887	.9890
2.3	.9893	.9896	.9898	.9901	.9904	.9906	.9909	.9911	.9913	.9916
2.4	.9918	.9920	.9922	.9925	.9927	.9929	.9931	.9932	.9934	.9936
2.5	.9938	.9940	.9941	.9943	.9945	.9946	.9948	.9949	.9951	.9952
2.6	.9953	.9955	.9956	.9957	.9959	.9960	.9961	.9962	.9963	.9964
2.7	.9965	.9966	.9967	.9968	.9969	.9970	.9971	.9972	.9973	.9974
2.8	.9974	.9975	.9976	.9977	.9977	.9978	.9979	.9979	.9980	.9981
2.9	.9981	.9982	.9982	.9983	.9984	.9984	.9985	.9985	.9986	.9986
3.0	.9987	.9987	.9987	.9988	.9988	.9989	.9989	.9989	.9990	.9990
3.1	.9990	.9991	.9991	.9991	.9992	.9992	.9992	.9992	.9993	.9993
3.2	.9993	.9993	.9994	.9994	.9994	.9994	.9994	.9995	.9995	.9995
3.3	.9995	.9995	.9995	.9996	.9996	.9996	.9996	.9996	.9996	.9997
3.4	.9997	.9997	.9997	.9997	.9997	.9997	.9997	.9997	.9997	.9998

ANSWERS TO SELECTED EXERCISES

Answers to selected writing exercises are provided.

Chapter R Algebra Reference

Exercises R.1 (page xxii)

1. $-x^2 + x + 9$ **2.** $-6y^2 + 3y + 10$ **3.** $-14q^2 + 11q - 14$ **4.** $9r^2 - 4r + 19$ **5.** $-.327x^2 - 2.805x - 1.458$
6. $-2.97r^2 - 8.083r + 7.81$ **7.** $-18m^3 - 27m^2 + 9m$ **8.** $12k^2 - 20k + 3$ **9.** $25r^2 + 5rs - 12s^2$ **10.** $18k^2 - 7kq - q^2$
11. $(6/25)y^2 + (11/40)yz + (1/16)z^2$ **12.** $(15/16)r^2 - (7/12)rs - (2/9)s^2$ **13.** $144x^2 - 1$ **14.** $36m^2 - 25$
15. $27p^3 - 1$ **16.** $6p^3 - 11p^2 + 14p - 5$ **17.** $8m^3 + 1$ **18.** $12k^4 + 21k^3 - 5k^2 + 3k + 2$
19. $m^2 + mn - 2n^2 - 2km + 5kn - 3k^2$ **20.** $2r^2 - 7rs + 3s^2 + 3rt - 4st + t^2$ **21.** $x^3 + 6x^2 + 11x + 6$
22. $x^3 - 2x^2 - 5x + 6$ **23.** $9a^2 + 6ab + b^2$ **24.** $x^3 - 6x^2y + 12xy^2 - 8y^3$

Exercises R.2 (page xxv)

1. $8a(a^2 - 2a + 3)$ **2.** $3y(y^2 + 8y + 3)$ **3.** $5p^2(5p^2 - 4pq + 20q^2)$ **4.** $10m^2(6m^2 - 12mn + 5n^2)$ **5.** $(m + 7)(m + 2)$
6. $(x + 5)(x - 1)$ **7.** $(z + 4)(z + 5)$ **8.** $(b - 7)(b - 1)$ **9.** $(a - 5b)(a - b)$ **10.** $(s - 5t)(s + 7t)$
11. $(y - 7z)(y + 3z)$ **12.** $6(a - 10)(a + 2)$ **13.** $3m(m + 3)(m + 1)$ **14.** $(2x + 1)(x - 3)$ **15.** $(3a + 7)(a + 1)$
16. $(2a - 5)(a - 6)$ **17.** $(5y + 2)(3y - 1)$ **18.** $(7m + 2n)(3m + n)$ **19.** $2a^2(4a - b)(3a + 2b)$
20. $4z^3(8z + 3a)(z - a)$ **21.** $(x + 8)(x - 8)$ **22.** $(3m + 5)(3m - 5)$ **23.** $(11a + 10)(11a - 10)$ **24.** Prime
25. $(z + 7y)^2$ **26.** $(m - 3n)^2$ **27.** $(3p - 4)^2$ **28.** $(a - 6)(a^2 + 6a + 36)$ **29.** $(2r - 3s)(4r^2 + 6rs + 9s^2)$
30. $(4m + 5)(16m^2 - 20m + 25)$ **31.** $(x - y)(x + y)(x^2 + y^2)$ **32.** $(2a - 3b)(2a + 3b)(4a^2 + 9b^2)$

Exercises R.3 (page xxviii)

1. $z/2$ **2.** $5p/2$ **3.** $8/9$ **4.** $3/(t - 3)$ **5.** $2(x + 2)/x$ **6.** $4(y + 2)$ **7.** $(m - 2)/(m + 3)$ **8.** $(r + 2)/(r + 4)$
9. $(x + 4)/(x + 1)$ **10.** $(z - 3)/(z + 2)$ **11.** $(2m + 3)/(4m + 3)$ **12.** $(2y + 1)/(y + 1)$ **13.** $(3k)/5$ **14.** $(25p^2)/9$
15. $6/(5p)$ **16.** 2 **17.** $2/9$ **18.** $3/10$ **19.** $2(a + 4)/(a - 3)$ **20.** $2/(r + 2)$ **21.** $(k + 2)/(k + 3)$
22. $(m + 6)/(m + 3)$ **23.** $(m - 3)/(2m - 3)$ **24.** $(2n - 3)/(2n + 3)$ **25.** 1 **26.** $(6 + p)/(2p)$ **27.** $(8 - y)/(4y)$
28. $137/(30m)$ **29.** $(3m - 2)/[m(m - 1)]$ **30.** $(r - 12)/[r(r - 2)]$ **31.** $14/[3(a - 1)]$ **32.** $23/[20(k - 2)]$
33. $(7x + 9)/[(x - 3)(x + 1)(x + 2)]$ **34.** $y^2/[(y + 4)(y + 3)(y + 2)]$ **35.** $k(k - 13)/[(2k - 1)(k + 2)(k - 3)]$
36. $m(3m - 19)/[(3m - 2)(m + 3)(m - 4)]$ **37.** $(4a + 1)/[a(a + 2)]$ **38.** $(5x^2 + 4x - 4)/[x(x - 1)(x + 1)]$

Exercises R.4 (page xxxiv)

1. 12 **2.** $-2/7$ **3.** -12 **4.** $3/4$ **5.** $-7/8$ **6.** $-6/11$ **7.** -1 **8.** $-10/19$ **9.** $-3, -2$ **10.** $-1, 3$ **11.** 4
12. $-2, 5/2$ **13.** $-1/2, 4/3$ **14.** $2, 5$ **15.** $-4/3, 4/3$ **16.** $-4, 1/2$ **17.** $0, 4$
18. $(5 + \sqrt{13})/6 \approx 1.434, (5 - \sqrt{13})/6 \approx .232$ **19.** $(1 + \sqrt{33})/4 \approx 1.686, (1 - \sqrt{33})/4 \approx -1.186$
20. $(-1 + \sqrt{5})/2 \approx .618, (-1 - \sqrt{5})/2 \approx -1.618$ **21.** $5 + \sqrt{5} \approx 7.236, 5 - \sqrt{5} \approx 2.764$
22. $(-6 + \sqrt{26})/2 \approx -.450, (-6 - \sqrt{26})/2 \approx -5.550$ **23.** $1, 5/2$ **24.** No real number solutions
25. $(-1 + \sqrt{73})/6 \approx 1.257, (-1 - \sqrt{73})/6 \approx -1.591$ **26.** $-1, 0$ **27.** 3 **28.** 12 **29.** $-59/6$ **30.** $-11/5$
31. No real number solutions **32.** $-5/2$ **33.** $2/3$ **34.** 1 **35.** $(-13 - \sqrt{185})/4 \approx -6.650, (-13 + \sqrt{185})/4 \approx .150$
36. No solution **37.** $-15/4$

Exercises R.5 (page xxxix)

1. $(-\infty, 0)$ **2.** $[-3, \infty)$ **3.** $[1, 2)$

4. $(-5, -4]$ **5.** $(-\infty, -9)$ **6.** $[6, \infty)$

7. $-4 < x < 3$ **8.** $2 \le x < 7$ **9.** $x \le -1$ **10.** $x > 3$ **11.** $-2 \le x < 6$ **12.** $0 < x < 8$ **13.** $x \le -4$ or $x \ge 4$
14. $x < 0$ or $x \ge 3$ **15.** $(-\infty, -1]$ **16.** $(-\infty, 1)$

17. $(-1, \infty)$ **18.** $(-\infty, 1]$ **19.** $(1/5, \infty)$

20. $(1/3, \infty)$ **21.** $(-5, 6)$ **22.** $[7/3, 4]$

23. $[-11/2, 7/2]$ **24.** $[-1, 2]$

25. $[-17/7, \infty)$ **26.** $(-\infty, 50/9)$

27. $(-2, 4)$ **28.** $(-\infty, -6] \cup [1, \infty)$

29. $(1, 2)$ **30.** $(-\infty, -4) \cup (1/2, \infty)$

31. $[1, 6]$ **32.** $[-3/2, 5]$

33. $(-\infty, -1/2) \cup (1/3, \infty)$ **34.** $[-1/2, 2/5]$

35. $[-3, 1/2]$ **36.** $(-\infty, -2) \cup (5/3, \infty)$

37. $[-5, 5]$ **38.** $(-\infty, 0) \cup (16, \infty)$ **39.** $(-5, 3]$

40. $(-\infty, -1) \cup (1, \infty)$ **41.** $(-\infty, -2)$ **42.** $(-2, 3/2)$ **43.** $[-8, 5)$ **44.** $(-\infty, -3/2) \cup [-13/9, \infty)$ **45.** $(-2, \infty)$
46. $(-\infty, -1)$ **47.** $(-\infty, -1) \cup (-1/2, 1) \cup (2, \infty)$ **48.** $(-4, -2) \cup (0, 2)$ **49.** $(1, 3/2]$ **50.** $(-\infty, -2) \cup (-2, 2) \cup [4, \infty)$

Exercises R.6 (page xliv)

1. $1/64$ **2.** $1/81$ **3.** 1 **4.** 1 **5.** $-1/9$ **6.** $1/9$ **7.** $49/4$ **8.** $27/64$ **9.** $1/3^6$ **10.** 8^5 **11.** $1/10^8$ **12.** 5
13. x^2 **14.** y^3 **15.** $2^3 k^3$ **16.** $1/(3z^7)$ **17.** $x^2/(2y)$ **18.** $m^3/5^4$ **19.** $a^3 b^6$ **20.** $d^6/(2^2 c^4)$ **21.** x^4/y^4 **22.** b/a^3
23. $(a + b)/(ab)$ **24.** $(1 - ab^2)/b^2$ **25.** $2(m - n)/[mn(m + n^2)]$ **26.** $(3n^2 + 4m)/(mn^2)$ **27.** $xy/(y - x)$
28. $x^4 y^4/(x^2 + y^2)^2$ **29.** 9 **30.** 3 **31.** 4 **32.** -25 **33.** $2/3$ **34.** $4/3$ **35.** $1/32$ **36.** $1/5$ **37.** $4/3$
38. $1000/1331$ **39.** 2^2 **40.** $27^{1/3}$ **41.** 4^2 **42.** 1 **43.** r **44.** $12^3/y^8$ **45.** $1/(2^2 \cdot 3k^{5/2})$ or $1/(12k^{5/2})$ **46.** $1/(2p^2)$
47. $a^{2/3} b^2$ **48.** $y/(x^{4/3} z^{1/2})$ **49.** $h^{1/3} t^{1/5}/k^{2/5}$ **50.** $m^3 p/n$ **51.** $4x(x^2 + 2)(-x^3 + 6x - 1)$ **52.** $6x(x^3 + 7)(-2x^3 - 5x + 7)$
53. $3x^3(x^2 - 1)^{-1/2}$ **54.** $3(6x + 2)^{-1/2}(27x + 5)$ **55.** $(2x + 5)(x^2 - 4)^{-1/2}(4x^2 + 5x - 8)$
56. $(4x^2 + 1)(2x - 1)^{-1/2}(36x^2 - 16x + 1)$

Exercises R.7 (page xlviii)

1. 5 **2.** 6 **3.** -5 **4.** $5\sqrt{2}$ **5.** $20\sqrt{5}$ **6.** $4y^2\sqrt{2y}$ **7.** $7\sqrt{2}$ **8.** $9\sqrt{3}$ **9.** $2\sqrt{5}$ **10.** $-2\sqrt{7}$ **11.** $5\sqrt[3]{2}$
12. $7\sqrt[3]{3}$ **13.** $3\sqrt[3]{4}$ **14.** $xyz^2\sqrt{2x}$ **15.** $7rs^2 t^5\sqrt{2r}$ **16.** $2x^2yz\sqrt[3]{2x^2yz^2}$ **17.** $x^2yz^2\sqrt[4]{y^3z^3}$ **18.** $ab\sqrt{ab}(b - 2a^2 + b^3)$
19. $p^2\sqrt{pq}(pq - q^4 + p^2)$ **20.** $5\sqrt{7}/7$ **21.** $-2\sqrt{3}/3$ **22.** $-\sqrt{3}/2$ **23.** $\sqrt{2}$ **24.** $-3(1 + \sqrt{5})/4$
25. $-5(2 + \sqrt{6})/2$ **26.** $-2(\sqrt{3} + \sqrt{2})$ **27.** $(\sqrt{10} - \sqrt{3})/7$ **28.** $(\sqrt{r} + \sqrt{3})/(r - 3)$ **29.** $5(\sqrt{m} + \sqrt{5})/(m - 5)$
30. $\sqrt{y} + \sqrt{5}$ **31.** $\sqrt{z} + \sqrt{11}$ **32.** $-2x - 2\sqrt{x(x + 1)} - 1$ **33.** $[p^2 + p + 2\sqrt{p(p^2 - 1)} - 1]/(-p^2 + p + 1)$
34. $-1/[2(1 - \sqrt{2})]$ **35.** $-2/[3(1 + \sqrt{3})]$ **36.** $-1/[2x - 2\sqrt{x(x + 1)} + 1]$
37. $(-p^2 + p + 1)/[p^2 + p - 2\sqrt{p(p^2 - 1)} - 1]$ **38.** $|4 - x|$ **39.** $|2y + 1|$ **40.** Cannot be simplified
41. Cannot be simplified

Chapter 1 Linear Functions

Exercises 1.1 (page 15)

1. $3/5$ **3.** Not defined   **5.** 2   **7.** $5/9$ **9.** Not defined   **11.** 0   **13.** 2   **15.** $y = -2x + 5$ **17.** $y = 1$
19. $y = -(1/3)x + 10/3$ **21.** $y = -(3/5)x + 59/30$ **23.** $x = -8$ **25.** $y = (2/3)x - 2$ **27.** $x = -6$
29. $y = -(1/3)x + 11/3$ **31.** $y = x - 7$ **33.** $y = (2/3)x + 2$ **35.** No   **39.** a   **41.** -4 **45.**

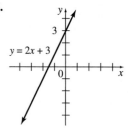

47.

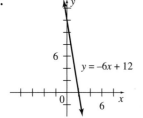

49.

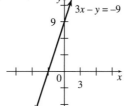

51.

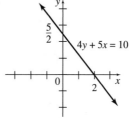

53.

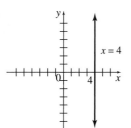

55.

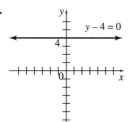

57.

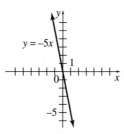

59.

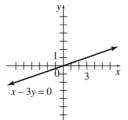

61. a.

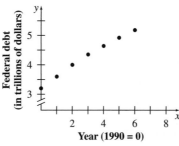

The debt is increasing and the data appear to be nearly linear.

b.

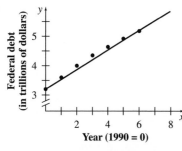

$y = .3292x + 3.207$. The slope .3292 indicates that the federal debt was increasing at a rate of .3292 trillion dollars per year.

c. 1997: 5.511 trillion dollars. 1998: 5.841 trillion dollars. For both years, the results are greater than the actual debt.

d.

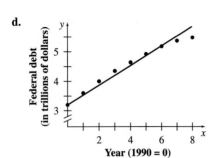

The linear trend does not continue past 1996, as the debt increased at a slower rate. A linear approximation is not necessarily valid over domains that extend past the original data.

63. a. $y \approx .053x - .043$ **b.** About 10.2 years   **65.** Approximately 4.3 m/second   **67.** 23   **69. a.** $y = .19x + 1.1$
b. About 7.4 million   **71. a.** $y = -9.3x + 504$ **b.** About \$439   **c.** The average payment is decreasing by about \$9.30 per month.   **73. a.** There appears to be a linear relationship.   **b.** $y = 76.9x$ **c.** About 780 megaparsecs (about 1.5×10^{22} miles)
d. About 12.4 billion years   **75. a.**

Yes.   **b.** $y = 633.6x - 47,831.6$; the slope of 633.6 indicates that tuition and fees have increased approximately \$634 per year.

c. The year 2020 is too far in the future to make accurate predictions based on the given data. Many factors could affect college costs by then.

Exercises 1.2 (page 27)

1. True   **3.** True   **7.** If $C(x)$ is the cost of renting a saw for x hours, then $C(x) = 12 + x$.   **9.** If $P(x)$ is the cost (in cents) of parking for x half-hours, then $P(x) = 35x + 50$.   **11.** $C(x) = 30x + 100$ **13.** $C(x) = 90x + 2500$ **17. a.** \$16   **b.** \$11
c. \$6   **d.** 800   **e.** 400   **f.** 0   **g.**   **h.** 0   **i.** About 1333   **j.** About 2667   **k.** See part g.
l. 800; \$6

19. a.

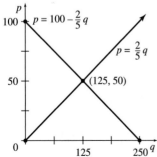

b. 125; \$50   **21. a.** $C(x) = 3.50x + 90$ **b.** 17   **c.** 108

23. a. $C(x) = .097x + 1.32$ **b.** \$98.32   **c.** \$98.417   **d.** \$.097, or 9.7¢   **e.** \$.097, or 9.7¢   **25. a.** 5,100,000
b. $(1, 100,000), (6, 5,100,000)$ **c.** $f(x) = 1,000,000x - 900,000$ **d.** In 2011   **27. a.** 2 units   **b.** \$980   **c.** 52 units

29. Break-even quantity is 45 units; don't produce; $P(x) = 20x - 900$ **31.** Break-even quantity is -50 units; impossible to make a profit here since $C(x) > R(x)$ for all positive x; $P(x) = -10x - 500$ (always a loss!)   **33. a.** 14.4°C   **b.** -28.9°C   **c.** 122°F   **35.** -40°

Exercises 1.3 (page 37)

3. c.   **5. a.** $Y = 7.09x - 530$ **b.** About 158,000   **c.** In 1999   **d.** $r = .943$, which indicates that the points on the line give good approximations of the data points.   **7. a.**

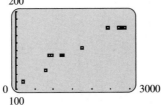

Yes   **b.** .972; yes   **c.** $y = 113 + .0243x$; 2.43¢ per mile

9. a. $Y = .129x + 2.15$; $r = .547$ **b.** $Y = .106x + 4.12$; $r = .436$ **11. a.** $Y = .212x - .309$ **b.** 15.2   **c.** 86.4°   **d.** .835
13. a.

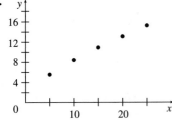

Yes   **b.** .9978; yes   **c.** $y = .4809x + 3.622$ **d.** About $25,260

15. a. $Y = -.0067x + 14.75$ **b.** 12   **c.** 11   **d.** $-.13$ **e.** There is no linear relationship.
17. a.   **b.** $Y = .366x + .803$; the line seems to fit the data.

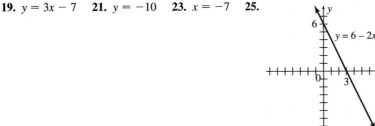

c. $r = .995$ indicates a good fit, which confirms the conclusion in part b.   **19. a.** $-.995$; yes   **b.** $Y = -.0769x + 5.91$
c. 2.07 points   **21. a.** 3.16 miles per hour   **b.**   Yes   **c.** $y = 3.06x + 4.54$ **d.** .9964; yes
e. 3.06 miles per hour

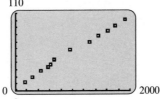

Chapter 1 Review Exercises (page 44)

3. 1/3   **5.** $-2/11$ **7.** $-2/3$ **9.** 0   **11.** -3 **13.** $y = (2/3)x - 13/3$ **15.** $y = -(5/4)x + 17/4$ **17.** $x = -1$
19. $y = 3x - 7$ **21.** $y = -10$ **23.** $x = -7$ **25.**   **27.**

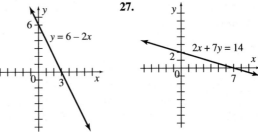

29.

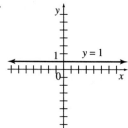

31.

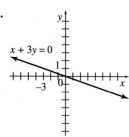

33. a. 7/6; 9/2   **b.** 2; 2   **c.** 5/2; 1/2

d.

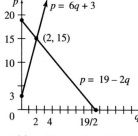

e. $15   **f.** 2   **35.** $D(q) = -.5q + 72.50$ **37.** $C(x) = 30x + 60$

39. $C(x) = 30x + 85$ **41. a.** 5 cartons   **b.** $2000   **43.** $y = 4.43x - 381.84$ **45.** $y = 9000x + 102,000$
47. a. .8286; yes, but not as close as in some other examples   **b.**

Yes

c. $y = .0131x + 33.8$ **d.** About 75 years, which is slightly lower than the actual figure   **49.** $y = -.8x + 187.7$
51. a. .494; not very closely   **b.**   150

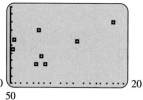

c. $y = 1.70x + 87.8$ **d.** About $170

Chapter 2 Systems of Linear Equations and Matrices

Exercises 2.1 (page 57)

1. $(3, 6)$ **3.** $(-1, 4)$ **5.** $(-2, 0)$ **7.** $(1, 3)$ **9.** $(4, -2)$ **11.** $(2, -2)$ **13.** No solution   **15.** $((y + 9)/4, y)$
17. No solution   **19.** $(12, 6)$ **21.** $(7, -2)$ **23.** $(1, 2, -1)$ **25.** $(2, 0, 3)$ **29.** $((-2z - 7)/5, (11z + 21)/5, z)$
31. $((-4z + 28)/5, (z - 7)/5, z)$ **35.** $27   **37.** 400 main floor, 200 balcony   **39.** Not possible; inconsistent system
41. $3000 at 6.5%, $6000 at 6%, $1000 at 5%   **43.** Either 10 buffets, 5 chairs, and no tables, or 11 buffets, 1 chair, and 1 table
45. a. .056057, 1.06657   **b.** 228 ft

Exercises 2.2 (page 70)

1. $\begin{bmatrix} 2 & 3 & | & 11 \\ 1 & 2 & | & 8 \end{bmatrix}$ **3.** $\begin{bmatrix} 2 & 1 & 1 & | & 3 \\ 3 & -4 & 2 & | & -7 \\ 1 & 1 & 1 & | & 2 \end{bmatrix}$ **5.** $x = 2, y = 3$ **7.** $x = 2, y = 3, z = -2$ **9.** Row operations

11. $\begin{bmatrix} 2 & 3 & 8 & | & 20 \\ 0 & -5 & -4 & | & -4 \\ 0 & 3 & 5 & | & 10 \end{bmatrix}$ **13.** $\begin{bmatrix} 1 & 0 & -18 & | & -47 \\ 0 & 1 & 5 & | & 14 \\ 0 & 3 & 8 & | & 16 \end{bmatrix}$ **15.** $\begin{bmatrix} 1 & 0 & 0 & | & 6 \\ 0 & 5 & 0 & | & 9 \\ 0 & 0 & 4 & | & 8 \end{bmatrix}$ **17.** $(2, 3)$ **19.** $(5/2, -1)$

21. No solution   **23.** $((3y + 1)/6, y)$ **25.** $(3, 2, -4)$ **27.** No solution   **29.** $(-1, 23, 16)$
31. $((-9z + 5)/23, (10z - 3)/23, z)$ **33.** $((-2z + 80)/35, (3z + 13)/7, z)$ **35.** $((9 - 3y - z)/2, y, z)$ **37.** $(0, 2, -2, 1)$ The
answers are given in the order x, y, z, w.   **39.** $(-w - 3, -4w - 19, -3w - 2, w)$ **41.** $(28.9436, 36.6326, 9.6390, 37.1036)$
43. Garcia, 20 hours; Wong, 15 hours   **45.** 2000 chairs, 1600 cabinets, 2500 buffets   **47. a.** 5 trucks, 2 vans, 3 station wagons
b. Three possibilities: 7 trucks, no vans, 2 station wagons; or 7 trucks, 1 van, 1 station wagon; or 7 trucks, 2 vans, no station
wagons   **49. a.** \$10,000 at 13%, \$7000 at 14%, \$8000 at 12%   **b.** It must be between \$1000 and \$13,000. \$16,000 is borrowed
at 13% and \$4000 at 14%.   **c.** No   **51.** The manufacturer should purchase 40 units for Roseville from the first supplier, 35 units
for Akron from the first supplier, 0 units for Roseville from the second supplier, and 40 units for Akron from the second supplier.
53. a. About 51.8 weeks; about \$959,091   **55. a.** $r = 175,000, b = 375,000$ **57. a.** 400/9 grams of A; 400/3 grams of B;
2000/9 grams of C   **b.** For any (positive) amount of C, A must be C grams less than 800/3 grams and B must be 400/3 grams.
c. No   **59.** 81 kg of the first chemical, 382.286 kg of the second, and 286.714 kg of the third   **61.** 244 of A, 39 of B, and 101
of C (rounded)   **63. b.** 10,366,482 white bulls, 7,460,514 black bulls, 7,358,060 spotted bulls, 7,206,360 white cows, 4,893,246
black cows, 3,515,820 spotted cows, 5,439,213 brown cows   **65. a.** 24 balls, 57 dolls, and 19 cars   **b.** None   **c.** 48
d. 5 balls, 95 dolls, and 0 cars   **e.** 52 balls, 1 doll, and 47 cars   **67. a.** Push every button one time.   **b.** Push the button in
the first row, second column, and push the button in the second row, first column.

Exercises 2.3 (page 81)
1. False; not all corresponding elements are equal.   **3.** True   **5.** True   **7.** 2×2; square; $\begin{bmatrix} 4 & -8 \\ -2 & -3 \end{bmatrix}$

9. 3×4; $\begin{bmatrix} 6 & -8 & 0 & 0 \\ -4 & -1 & -9 & -2 \\ -3 & 5 & -7 & -1 \end{bmatrix}$ **11.** 2×1; column; $\begin{bmatrix} -2 \\ -4 \end{bmatrix}$ **13.** The $n \times m$ zero matrix   **15.** $x = 2, y = 4, z = 8$

17. $x = -15, y = 5, k = 3$ **19.** $z = 18, r = 3, s = 3, p = 3, a = 3/4$ **21.** $\begin{bmatrix} 9 & 12 & 0 & 2 \\ 1 & -1 & 2 & -4 \end{bmatrix}$ **23.** Not possible

25. $\begin{bmatrix} 1 & 5 & 6 & -9 \\ 5 & 7 & 2 & 1 \\ -7 & 2 & 2 & -7 \end{bmatrix}$ **27.** $\begin{bmatrix} 3 & 4 \\ 4 & 8 \end{bmatrix}$ **29.** $\begin{bmatrix} 3 & 12 \\ -6 & 3 \end{bmatrix}$ **31.** $\begin{bmatrix} -12x + 8y & -x + y \\ x & 8x - y \end{bmatrix}$ **33.** $\begin{bmatrix} -x & -y \\ -z & -w \end{bmatrix}$

39. a. Chicago: $\begin{bmatrix} 4.05 & 7.01 \\ 3.27 & 3.51 \end{bmatrix}$, Seattle: $\begin{bmatrix} 4.40 & 6.90 \\ 3.54 & 3.76 \end{bmatrix}$ **b.** $\begin{bmatrix} 4.42 & 7.43 \\ 3.38 & 3.62 \end{bmatrix}$ **41. a.** $\begin{bmatrix} 2 & 1 & 2 & 1 \\ 3 & 2 & 2 & 1 \\ 4 & 3 & 2 & 1 \end{bmatrix}$ **b.** $\begin{bmatrix} 5 & 0 & 7 \\ 0 & 10 & 1 \\ 0 & 15 & 2 \\ 10 & 12 & 8 \end{bmatrix}$

c. $\begin{bmatrix} 8 \\ 4 \\ 5 \end{bmatrix}$ **43. a.** 8   **b.** 3   **c.** $\begin{bmatrix} 85 & 15 \\ 27 & 73 \end{bmatrix}$ **d.** Yes   **45. a.** $\begin{bmatrix} 60.0 & 68.3 \\ 63.8 & 72.5 \\ 64.5 & 73.6 \\ 67.2 & 74.7 \end{bmatrix}$ **b.** $\begin{bmatrix} 68.0 & 75.6 \\ 70.7 & 78.1 \\ 72.7 & 79.4 \\ 74.3 & 79.9 \end{bmatrix}$

c. $\begin{bmatrix} -8.0 & -7.3 \\ -6.9 & -5.6 \\ -8.2 & -5.8 \\ -7.1 & -5.2 \end{bmatrix}$ **47. a.** $\begin{bmatrix} 18.2 & 2.8 \\ 30.1 & 4.2 \\ 50.8 & 8.4 \\ 65.8 & 11.9 \\ 73.4 & 13.6 \\ 75.2 & 13.9 \end{bmatrix}$ **b.** $\begin{bmatrix} 21.8 & 3.3 \\ 32.5 & 4.6 \\ 51.5 & 8.3 \\ 66.5 & 10.8 \\ 74.1 & 12.9 \\ 76.7 & 15.4 \end{bmatrix}$ **c.** $\begin{bmatrix} -3.6 & -.5 \\ -2.4 & -.4 \\ -.7 & .1 \\ -.7 & 1.1 \\ -.7 & .7 \\ -1.5 & -1.5 \end{bmatrix}$

Exercises 2.4 (page 92)
1. $\begin{bmatrix} -4 & 8 \\ 0 & 6 \end{bmatrix}$ **3.** $\begin{bmatrix} 24 & -8 \\ -16 & 0 \end{bmatrix}$ **5.** $\begin{bmatrix} -22 & -6 \\ 20 & -12 \end{bmatrix}$ **7.** 2×2; 2×2 **9.** 3×2; BA does not exist.

11. *AB* does not exist; 3×2 **13.** Columns; rows   **15.** $\begin{bmatrix} 13 \\ 25 \end{bmatrix}$ **17.** $\begin{bmatrix} 5 \\ -21 \end{bmatrix}$ **19.** $\begin{bmatrix} 16 & -5 & 2 \\ 9 & -2 & 6 \end{bmatrix}$ **21.** $\begin{bmatrix} -2 & 10 \\ 0 & 8 \end{bmatrix}$

23. $\begin{bmatrix} 13 & 5 \\ 25 & 15 \end{bmatrix}$ **25.** $\begin{bmatrix} 13 \\ 29 \end{bmatrix}$ **27.** $\begin{bmatrix} 110 \\ 40 \\ -50 \end{bmatrix}$ **29.** $\begin{bmatrix} 22 & -8 \\ 11 & -4 \end{bmatrix}$ **31. a.** $\begin{bmatrix} 16 & 22 \\ 7 & 19 \end{bmatrix}$ **b.** $\begin{bmatrix} 5 & -5 \\ 0 & 30 \end{bmatrix}$ **c.** No   **d.** No

39. a. $\begin{bmatrix} 6 & 106 & 158 & 222 & 28 \\ 120 & 139 & 64 & 75 & 115 \\ -146 & -2 & 184 & 144 & -129 \\ 106 & 94 & 24 & 116 & 110 \end{bmatrix}$ **b.** Cannot be found   **c.** No

41. a. $\begin{bmatrix} -1 & 5 & 9 & 13 & -1 \\ 7 & 17 & 2 & -10 & 6 \\ 18 & 9 & -12 & 12 & 22 \\ 9 & 4 & 18 & 10 & -3 \\ 1 & 6 & 10 & 28 & 5 \end{bmatrix}$ **b.** $\begin{bmatrix} -2 & -9 & 90 & 77 \\ -42 & -63 & 127 & 62 \\ 413 & 76 & 180 & -56 \\ -29 & -44 & 198 & 85 \\ 137 & 20 & 162 & 103 \end{bmatrix}$ **c.** $\begin{bmatrix} -56 & -1 & 1 & 45 \\ -156 & -119 & 76 & 122 \\ 315 & 86 & 118 & -91 \\ -17 & -17 & 116 & 51 \\ 118 & 19 & 125 & 77 \end{bmatrix}$

d. $\begin{bmatrix} 54 & -8 & 89 & 32 \\ 114 & 56 & 51 & -60 \\ 98 & -10 & 62 & 35 \\ -12 & -27 & 82 & 34 \\ 19 & 1 & 37 & 26 \end{bmatrix}$ **e.** $\begin{bmatrix} -2 & -9 & 90 & 77 \\ -42 & -63 & 127 & 62 \\ 413 & 76 & 180 & -56 \\ -29 & -44 & 198 & 85 \\ 137 & 20 & 162 & 103 \end{bmatrix}$ **f.** Yes

43. a.
	A	B
Dept. 1	57	70
Dept. 2	41	54
Dept. 3	27	40
Dept. 4	39	40

b. Supplier A: \$164; Supplier B: \$204; Supplier A   **45. a.** $\begin{bmatrix} 4.24 & 6.95 \\ 3.42 & 3.64 \end{bmatrix}$ **b.** $\begin{bmatrix} 4.41 & 7.17 \\ 3.46 & 3.69 \end{bmatrix}$

47. a. $\begin{bmatrix} 80 & 40 & 120 \\ 60 & 30 & 150 \end{bmatrix}$ **b.** $\begin{bmatrix} 1/2 & 1/5 \\ 1/4 & 1/5 \\ 1/4 & 3/5 \end{bmatrix}$ **c.** $PF = \begin{bmatrix} 80 & 96 \\ 75 & 108 \end{bmatrix}$ The rows give the average price per pair of footwear sold by

each store, and the columns give the state.   **49. a.** $\begin{bmatrix} .038 & .014 \\ .022 & .008 \\ .023 & .007 \\ .014 & .009 \\ .010 & .011 \end{bmatrix} ; \begin{bmatrix} 283 & 1627 & 218 & 199 & 425 \\ 360 & 2038 & 286 & 227 & 460 \\ 468 & 2498 & 362 & 252 & 484 \\ 621 & 2987 & 443 & 278 & 498 \\ 798 & 3451 & 523 & 306 & 508 \end{bmatrix}$

b.
	Births	Deaths
1960	58.598	24.97
1970	72.872	30.449
1980	89.434	36.662
1990	108.373	43.671
2000	127.639	50.783

51. $\begin{bmatrix} 2.00 & 3.06 & 4.12 & 6.20 \\ 1.45 & 2.13 & 3.72 & 4.6 \\ .67 & .54 & .45 & .47 \end{bmatrix}$ **53. a.** 3819; 3824; 3763; 3698; 3638   **b.** Extinction

c. 4017; 4154; 4280; 4399; 4524; does not become extinct

Exercises 2.5 (page 105)

1. Yes   **3.** No   **5.** No   **7.** Yes   **9.** No; the row of all zeros makes it impossible to get all the 1s in the diagonal of the

identity matrix, no matter what matrix is used as an inverse.   **11.** $\begin{bmatrix} 0 & 1/2 \\ -1 & 1/2 \end{bmatrix}$ **13.** $\begin{bmatrix} 2 & 1 \\ 5 & 3 \end{bmatrix}$ **15.** No inverse

17. $\begin{bmatrix} 1 & 0 & 0 \\ 0 & -1 & 0 \\ -1 & 0 & 1 \end{bmatrix}$ **19.** $\begin{bmatrix} 15 & 4 & -5 \\ -12 & -3 & 4 \\ -4 & -1 & 1 \end{bmatrix}$ **21.** No inverse **23.** $\begin{bmatrix} 7/4 & 5/2 & 3 \\ -1/4 & -1/2 & 0 \\ -1/4 & -1/2 & -1 \end{bmatrix}$

25. $\begin{bmatrix} 1/2 & 1/2 & -1/4 & 1/2 \\ -1 & 4 & -1/2 & -2 \\ -1/2 & 5/2 & -1/4 & -3/2 \\ 1/2 & -1/2 & 1/4 & 1/2 \end{bmatrix}$ **27.** $(-1, 4)$ **29.** $(2, 1)$ **31.** $(2, 3)$ **33.** No inverse, $(-8y - 12, y)$ **35.** $(-8, 6, 1)$

37. $(15, -5, -1)$ **39.** No inverse, no solution for system **41.** $(-7, -34, -19, 7)$

51. Entries are rounded to four places. $\begin{bmatrix} -.0447 & -.0230 & .0292 & .0895 & -.0402 \\ .0921 & .0150 & .0321 & .0209 & -.0276 \\ -.0678 & .0315 & -.0404 & .0326 & .0373 \\ .0171 & -.0248 & .0069 & -.0003 & .0246 \\ -.0208 & .0740 & .0096 & -.1018 & .0646 \end{bmatrix}$

53. Entries are rounded to four places $\begin{bmatrix} .0394 & -.0880 & .0033 & .0530 & -.1499 \\ -.1492 & .0289 & .0187 & .1033 & .1668 \\ -.1330 & -.0543 & .0356 & .1768 & .1055 \\ .1407 & .0175 & -.0453 & -.1344 & .0655 \\ .0102 & -.0653 & .0993 & .0085 & -.0388 \end{bmatrix}$ **55.** Yes **57.** $\begin{bmatrix} 1.51482 \\ .053479 \\ -.637242 \\ .462629 \end{bmatrix}$

59. a. $\begin{bmatrix} 72 \\ 48 \\ 60 \end{bmatrix}$ **b.** $\begin{bmatrix} 2 & 4 & 2 \\ 2 & 1 & 2 \\ 2 & 1 & 3 \end{bmatrix} \begin{bmatrix} x_1 \\ x_2 \\ x_3 \end{bmatrix} = \begin{bmatrix} 72 \\ 48 \\ 60 \end{bmatrix}$ **c.** 8 type I, 8 type II, and 12 type III **61. a.** \$12,000 at 6%, \$7000 at 7%, and

\$6000 at 10% **b.** \$10,000 at 6%, \$15,000 at 7%, and \$5000 at 10% **c.** \$20,000 at 6%, \$10,000 at 7%, and \$10,000 at 10%
63. a. 50 Super Vim, 75 Multitab, and 100 Mighty Mix **b.** 75 Super Vim, 50 Multitab, and 60 Mighty Mix **c.** 80 Super Vim,

100 Multitab, and 50 Mighty Mix **65. a.** $\begin{bmatrix} -1/15 & -1/15 & 13/3 \\ 4/15 & 1/15 & -40/3 \\ -1/5 & 0 & 10 \end{bmatrix}$ **b.** $\begin{bmatrix} 0 & 1 & 40 \\ -1 & 0 & 60 \\ 0 & 0 & 1 \end{bmatrix}$ **c.** The shape is a sideways T
whose vertical and horizontal intersection is at the mark $(60, 10)$.

Exercises 2.6 (page 113)

1. $\begin{bmatrix} 10.67 \\ 8.33 \end{bmatrix}$ **3.** $\begin{bmatrix} 6.43 \\ 26.12 \end{bmatrix}$ **5.** $\begin{bmatrix} 6.67 \\ 20 \\ 10 \end{bmatrix}$ **7.** $33:47:23$ **9.** $\begin{bmatrix} 7697 \\ 4205 \\ 6345 \\ 4106 \end{bmatrix}$ (rounded) **11.** 1079 metric tons of wheat and

1428 metric tons of oil **13.** 1285 units of agriculture, 1455 units of manufacturing, and 1202 units of transportation
15. 3077 units of agriculture, 2564 units of manufacturing, and 3179 units of transportation **17. a.** 7/4 bushels of yams and
$15/8 \approx 2$ pigs **b.** 167.5 bushels of yams and $153.75 \approx 154$ pigs **19.** 848 units of agriculture, 516 units of manufacturing, and
2970 units of households **21.** 195 million pounds of agriculture, 26 million pounds of manufacturing, and 13.6 million pounds of
energy **23.** In millions of dollars, the amounts are 532 for natural resources, 481 for manufacturing, 805 for trade and services,

and 1185 for personal consumption. **25. a.** $\begin{bmatrix} 1.67 & .56 & .56 \\ .19 & 1.17 & .06 \\ 3.15 & 3.27 & 4.38 \end{bmatrix}$ **b.** These multipliers imply that if the demand for one

community's output increases by \$1 then the output in the other community will increase by the amount in the row and column
of that matrix. For example, if the demand for Hermitage's output increases by \$1 then output from Sharon will increase by \$.56,
Farrell by \$.06, and Hermitage by \$4.38. **27.** 9 units of steel to every 10 units of coal **29.** 6 units of mining to every 8 units of
manufacturing and 5 units of communication

Chapter 2 Review Exercises (page 117)

3. $(-4, 6)$ **5.** $(-1, 2, 3)$ **7.** $(-9, 3)$ **9.** $(7, -9, -1)$ **11.** $(6 - 7z/3, 1 + z/3, z)$ **13.** $3 \times 2; a = 2, x = -1, y = 4,$
$p = 5, z = 7$ **15.** 3×3 (square); $a = -12, b = 1, k = 9/2, c = 3/4, d = 3, l = -3/4, m = -1, p = 3, q = 9$

17. $\begin{bmatrix} 8 & -6 \\ -10 & -16 \end{bmatrix}$ **19.** Not possible **21.** $\begin{bmatrix} 26 & 86 \\ -7 & -29 \\ 21 & 87 \end{bmatrix}$ **23.** $\begin{bmatrix} 6 & 18 & -24 \\ 1 & 3 & -4 \\ 0 & 0 & 0 \end{bmatrix}$ **25.** $\begin{bmatrix} 15 \\ 16 \\ 1 \end{bmatrix}$ **27.** $\begin{bmatrix} -7/19 & 4/19 \\ 3/19 & 1/19 \end{bmatrix}$

29. No inverse **31.** $\begin{bmatrix} -1/4 & 1/6 \\ 0 & 1/3 \end{bmatrix}$ **33.** No inverse **35.** $\begin{bmatrix} 1/4 & 1/2 & 1/2 \\ 1/4 & -1/2 & 1/2 \\ 1/8 & -1/4 & -1/4 \end{bmatrix}$ **37.** No inverse **39.** Matrix A

has no inverse. Solution: $(-2y + 5, y)$ **41.** $X = \begin{bmatrix} 6 \\ 15 \\ 16 \end{bmatrix}$ **43.** $(34, -9)$ **45.** $(1, 2, 3)$ **47.** $\begin{bmatrix} 725.7 \\ 305.9 \\ 166.7 \end{bmatrix}$ **49.** 8000 standard,

6000 extra large **51.** 5 blankets, 3 rugs, 8 skirts **53.** $\begin{bmatrix} .14 & 11 & 333,675 & 20.13 & 1.88 \\ .64 & 39 & 390,591 & 47.81 & 4.06 \\ .08 & 41 & 436,351 & 15.19 & 1.88 \\ .54 & 17 & 27,077 & 23.13 & -1.50 \end{bmatrix}$

55. a. $\begin{array}{c} \\ c \\ g \end{array} \begin{array}{cc} c & g \\ \end{array} \begin{bmatrix} 0 & 1/2 \\ 2/3 & 0 \end{bmatrix}$ **b.** 1200 units of cheese; 1600 units of goats **57.** $\begin{bmatrix} 8 & 8 & 8 \\ 10 & 5 & 9 \\ 7 & 10 & 7 \\ 8 & 9 & 7 \end{bmatrix}$ **59. a.** No **b. (i)** .23, .37, .42;

A is healthy; B and D are tumorous, C is bone. **(ii)** .33, .27, .32; A and C are tumorous, B could be healthy or tumorous, D is
bone. **c.** .2, .4, .45, .3; A is healthy, B and C are bone, D is tumorous. **d.** One example is to choose beams 1, 2, 3, and 6.
61. $W_1 = W_2 = 100\sqrt{3}/3 \approx 58$ lb **63. a.** $C = .0125t^2 + .8t + 316$ **b.** 2089 **65.** There are y girls and $2500 - 1.5y$ boys,
where y is any even integer between 0 and 1666.

Chapter 3 Linear Programming: The Graphical Method

Exercises 3.1 (page 131)

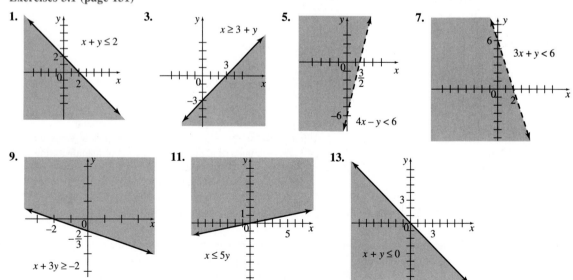

15.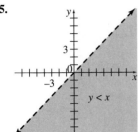

$y < x$

17.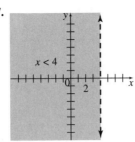

$x < 4$

19.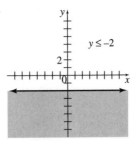

$y \leq -2$

21.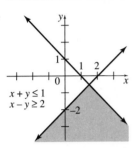

$x + y \leq 1$
$x - y \geq 2$

23.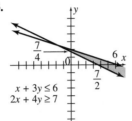

$x + 3y \leq 6$
$2x + 4y \geq 7$

25.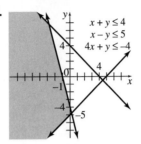

$x + y \leq 4$
$x - y \leq 5$
$4x + y \leq -4$

27.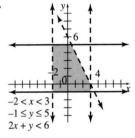

$-2 < x < 3$
$-1 \leq y \leq 5$
$2x + y < 6$

29. $2y + x \geq -5$
$y \leq 3 + x$
$x \geq 0$
$y \geq 0$

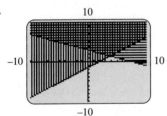

31.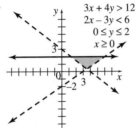

$3x + 4y > 12$
$2x - 3y < 6$
$0 \leq y \leq 2$
$x \geq 0$

33.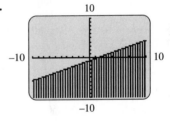

35.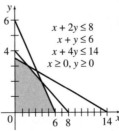

37. B: $\leq, \leq, \leq$; C: $\geq, \geq, \leq$; D: $\leq, \geq, \leq$; E: $\leq, \leq, \geq$; F: $\leq, \geq, \geq$; G: $\geq, \geq, \geq$

39. a.

	Shawls	Afghans	Total
Number Made	x	y	
Spinning Time	1	2	$\leq$ 8
Dyeing Time	1	1	$\leq$ 6
Weaving Time	1	4	$\leq$ 14

b.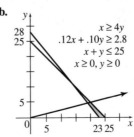

$x + 2y \leq 8$
$x + y \leq 6$
$x + 4y \leq 14$
$x \geq 0, y \geq 0$

c. Yes; no

41. a. $x \geq 4y$; $.12x + .10y \geq 2.8$; $x + y \leq 25$; $x \geq 0$; $y \geq 0$

b.

$x \geq 4y$
$.12x + .10y \geq 2.8$
$x + y \leq 25$
$x \geq 0, y \geq 0$

43. a. $x \le (1/2)y$; $x + y \le 1000$; $x \ge 0$; $y \ge 0$ **b.**

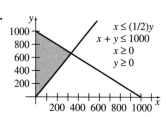

45. a. $x + y \ge 7$; $2x + y \ge 10$; $x + y \le 9$; $x \ge 0$; $y \ge 0$ **b.**

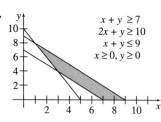

Exercises 3.2 (page 138)

1. Maximum of 65 at $(5, 10)$; minimum of 8 at $(1, 1)$ **3.** Maximum of 9 at $(0, 12)$; minimum of 0 at $(0, 0)$ **5. a.** No maximum; minimum of 16 at $(0, 8)$ **b.** No maximum; minimum of 18 at $(3, 4)$ **c.** No maximum; minimum of 21 at $(13/2, 2)$ **d.** No maximum; minimum of 12 at $(12, 0)$ **7.** Minimum of $42/5$ when $x = 6/5$, $y = 6/5$ **9.** Maximum of 30 when $x = 27/4$, $y = 33/4$, as well as when $x = 10$, $y = 5$, and all points in between **11.** Maximum of $235/4$ when $x = 105/8$, $y = 25/8$ **13. a.** Maximum of 204 when $x = 18$ and $y = 2$ **b.** Maximum of $588/5$ when $x = 12/5$ and $y = 39/5$ **c.** Maximum of 102 when $x = 0$ and $y = 17/2$ **15. b.**

Exercises 3.3 (page 144)

1. Let x be the number of product A made and y be the number of product B. Then $2x + 3y \le 45$. **3.** Let x be the number of green pills and y be the number of red pills. Then $4x + y \ge 25$. **5.** Let x be the number of pounds of \$6 coffee and y be the number of pounds of \$5 coffee. Then $x + y \ge 50$. **7.** 56 to plant I and 27 to plant II for a minimum cost of \$2065 **9. a.** 6 units of policy A and 16 units of policy B for a minimum premium cost of \$940 **b.** 30 units of policy A and 0 units of policy B for a minimum premium cost of \$750 **11.** 800 Type 1 and 1600 Type 2 for maximum revenue of \$272 **13. a.** 150 kg half-and-half mix and 75 kg other mix for maximum revenue of \$1260 **b.** 200 kg half-and-half mix and 0 kg other mix for maximum revenue of \$1600 **15.** 40 gal from dairy I and 60 gal from dairy II for maximum butterfat of 3.4% **17.** \$22 million in bonds and \$18 million in mutual funds, or \$24 million in bonds and \$14 million in mutual funds (or any solution on the line in between these two points), for a maximum annual interest of \$3.72 million **19. a.** **21.** Three #1 pills and two #2 pills for a minimum cost of \$1.05 per day **23.** 3 3/4 servings of A and 1 7/8 servings of B for a minimum cost of \$1.69 **25.** 0 plants and 18 animals, for a minimum of 270 hours

Chapter 3 Review Exercises (page 147)

3.

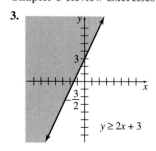

$y \ge 2x + 3$

5.

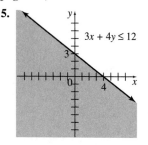

$3x + 4y \le 12$

7.

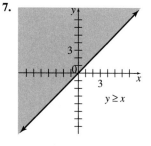

$y \ge x$

9.

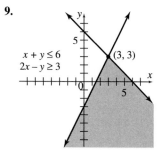

$x + y \le 6$
$2x - y \ge 3$
$(3, 3)$

11.
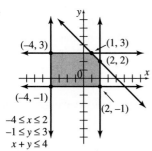

$-4 \le x \le 2$
$-1 \le y \le 3$
$x + y \le 4$

13.
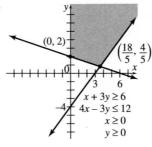
$x + 3y \ge 6$
$4x - 3y \le 12$
$x \ge 0$
$y \ge 0$

15. Minimum of 8 at $(2, 1)$; maximum of 40 at $(6, 7)$

17. Maximum of 24 at $(0, 6)$ **19.** Minimum of 40 at any point on the segment connecting $(0, 20)$ and $(10/3, 40/3)$

23. a.
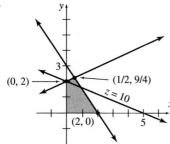
b.
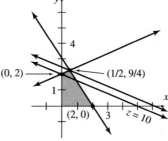

25. Let x = number of batches of cakes and y = number of batches of cookies. Then $x \ge 0$, $y \ge 0$, and $2x + (3/2)y \le 15$, $3x + (2/3)y \le 13$.

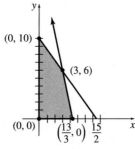

27. 3 batches of cakes and 6 of cookies for maximum profit of \$210 **29.** 7 packages of gardening mixture and 2 packages of potting mixture, for maximum income of \$31 **31.** Produce no runs of type I and 7 runs of type II for a minimum cost of \$42,000.
33. 0 acres for millet and 2 acres for wheat, for a maximum harvest of 1600 lbs

Chapter 4 Linear Programming: The Simplex Method

Exercises 4.1 (page 158)
1. $x_1 + 2x_2 + s_1 = 6$ **3.** $2x_1 + 4x_2 + 3x_3 + s_1 = 100$
5. a. 3 **b.** s_1, s_2, s_3 **c.** $4x_1 + 2x_2 + s_1 = 20$; $5x_1 + x_2 + s_2 = 50$; $2x_1 + 3x_2 + s_3 = 25$ **7. a.** 2 **b.** s_1, s_2
c. $7x_1 + 6x_2 + 8x_3 + s_1 = 118$; $4x_1 + 5x_2 + 10x_3 + s_2 = 220$ **9.** $x_1 = 0, x_2 = 0, x_3 = 20, s_1 = 0, s_2 = 15, z = 10$
11. $x_1 = 0, x_2 = 0, x_3 = 8, s_1 = 0, s_2 = 6, s_3 = 7, z = 12$ **13.** $x_1 = 0, x_2 = 20, x_3 = 0, s_1 = 16, s_2 = 0, z = 60$ **15.** $x_1 = 0$,
$x_2 = 0, x_3 = 12, s_1 = 0, s_2 = 9, s_3 = 8, z = 36$ **17.** $x_1 = 0, x_2 = 0, x_3 = 50, s_1 = 10, s_2 = 0, s_3 = 50, z = 100$

19.
$$\begin{array}{ccccc|c}
x_1 & x_2 & s_1 & s_2 & z & \\
2 & 3 & 1 & 0 & 0 & 6 \\
4 & 1 & 0 & 1 & 0 & 6 \\
\hline
-5 & -1 & 0 & 0 & 1 & 0
\end{array}$$

21.
$$\begin{array}{cccccc|c}
x_1 & x_2 & s_1 & s_2 & s_3 & z & \\
1 & 1 & 1 & 0 & 0 & 0 & 10 \\
5 & 2 & 0 & 1 & 0 & 0 & 20 \\
1 & 2 & 0 & 0 & 1 & 0 & 36 \\
\hline
-1 & -3 & 0 & 0 & 0 & 1 & 0
\end{array}$$

23.
$$\begin{array}{ccccc|c}
x_1 & x_2 & s_1 & s_2 & z & \\
3 & 1 & 1 & 0 & 0 & 12 \\
1 & 1 & 0 & 1 & 0 & 15 \\
\hline
-2 & -1 & 0 & 0 & 1 & 0
\end{array}$$

25. If x_1 is the number of simple figures, x_2 is the number of figures with additions, and x_3 is the number of computer-drawn sketches, find $x_1 \geq 0$, $x_2 \geq 0$, $x_3 \geq 0$, $s_1 \geq 0$, $s_2 \geq 0$, $s_3 \geq 0$, $s_4 \geq 0$ so that $20x_1 + 35x_2 + 60x_3 + s_1 = 2200$, $x_1 + x_2 + x_3 + s_2 = 400$, $-x_1 - x_2 + x_3 + s_3 = 0$, $-x_1 + 2x_2 + s_4 = 0$, and $z = 95x_1 + 200x_2 + 325x_3$ is maximized.

$$\begin{array}{ccccccccc} x_1 & x_2 & x_3 & s_1 & s_2 & s_3 & s_4 & z & \\ \left[\begin{array}{cccccccc|c} 20 & 35 & 60 & 1 & 0 & 0 & 0 & 0 & 2200 \\ 1 & 1 & 1 & 0 & 1 & 0 & 0 & 0 & 400 \\ -1 & -1 & 1 & 0 & 0 & 1 & 0 & 0 & 0 \\ -1 & 2 & 0 & 0 & 0 & 0 & 1 & 0 & 0 \\ -95 & -200 & -325 & 0 & 0 & 0 & 0 & 1 & 0 \end{array}\right] \end{array}$$

27. If x_1 is the number of redwood tables made, x_2 is the number of stained Douglas fir tables made, and x_3 is the number of stained white spruce tables made, find $x_1 \geq 0$, $x_2 \geq 0$, $x_3 \geq 0$, $s_1 \geq 0$, $s_2 \geq 0$, $s_3 \geq 0$ so that $8x_1 + 7x_2 + 8x_3 + s_1 = 720$, $2x_2 + 2x_3 + s_2 = 480$, $159x_1 + 138.85x_2 + 129.35x_3 + s_3 = 15{,}000$, and $z = x_1 + x_2 + x_3$ is maximized.

$$\begin{array}{ccccccc} x_1 & x_2 & x_3 & s_1 & s_2 & s_3 & z \\ \left[\begin{array}{cccccc|c} 8 & 7 & 8 & 1 & 0 & 0 & 720 \\ 0 & 2 & 2 & 0 & 1 & 0 & 480 \\ 159 & 138.85 & 129.35 & 0 & 0 & 1 & 15{,}000 \\ -1 & -1 & -1 & 0 & 0 & 0 & 0 \end{array}\right] \end{array}$$

29. If x_1 is the number of newspaper ads run, x_2 is the number of radio ads run, and x_3 is the number of TV ads run, find $x_1 \geq 0$, $x_2 \geq 0$, $x_3 \geq 0$, $s_1 \geq 0$, $s_2 \geq 0$, $s_3 \geq 0$, $s_4 \geq 0$ so that $400x_1 + 200x_2 + 1200x_3 + s_1 = 8000$, $x_1 + s_2 = 20$, $x_2 + s_3 = 30$, $x_3 + x_4 = 6$, and $z = 2000x_1 + 1200x_2 + 10{,}000x_3$ is maximized.

$$\begin{array}{cccccccc} x_1 & x_2 & x_3 & s_1 & s_2 & s_3 & s_4 & z \\ \left[\begin{array}{ccccccc|c} 400 & 200 & 1200 & 1 & 0 & 0 & 0 & 0 & 8000 \\ 1 & 0 & 0 & 0 & 1 & 0 & 0 & 0 & 20 \\ 0 & 1 & 0 & 0 & 0 & 1 & 0 & 0 & 30 \\ 0 & 0 & 1 & 0 & 0 & 0 & 1 & 0 & 6 \\ -2000 & -1200 & -10{,}000 & 0 & 0 & 0 & 0 & 1 & 0 \end{array}\right] \end{array}$$

Exercises 4.2 (page 168)

1. Maximum is 20 when $x_1 = 0$, $x_2 = 4$, $x_3 = 0$, $s_1 = 0$, and $s_2 = 2$.   **3.** Maximum is 8 when $x_1 = 4$, $x_2 = 0$, $s_1 = 8$, $s_2 = 2$, and $s_3 = 0$.   **5.** Maximum is 264 when $x_1 = 16$, $x_2 = 4$, $x_3 = 0$, $s_1 = 0$, $s_2 = 16$, and $s_3 = 0$.   **7.** Maximum is 22 when $x_1 = 5.5$, $x_2 = 0$, $s_1 = 0$, and $s_2 = .5$.   **9.** Maximum is 120 when $x_1 = 0$, $x_2 = 10$, $s_1 = 0$, $s_2 = 40$, and $s_3 = 4$.   **11.** Maximum is 944 when $x_1 = 118$, $x_2 = 0$, $x_3 = 0$, $s_1 = 0$, and $s_2 = 102$.   **13.** Maximum is 250 when $x_1 = 0$, $x_2 = 0$, $x_3 = 0$, $x_4 = 50$, $s_1 = 0$, and $s_2 = 50$.   **17.** 6 churches and 2 labor unions for a maximum of $1000 per month   **19.** 2 jazz CDs, 3 blues CDs, and 6 reggae CDs for a maximum weekly profit of $10.60   **21. a.** No racing or touring bicycles and 2700 mountain bicycles   **b.** Maximum profit is $59,400   **23. a.** 4 radio ads, 6 TV ads, no newspaper ads, for a maximum exposure of 64,800   **25. a.** 3   **b.** 4   **c.** 3   **27.** $200, $66.67, $300, $100   **29.** 163.6 kg of food P, none of Q, 1090.9 kg of R, 145.5 kg of S; maximum is 87,454.5   **31.** 12 minutes to the senator, 9 minutes to the congresswoman, and 6 minutes to the governor, for a maximum of 1,320,000 viewers

Exercises 4.3 (page 180)

1. $\begin{bmatrix} 1 & 3 & 1 \\ 2 & 2 & 10 \\ 3 & 1 & 0 \end{bmatrix}$ **3.** $\begin{bmatrix} -1 & 13 & -2 \\ 4 & 25 & -1 \\ 6 & 0 & 11 \\ 12 & 4 & 3 \end{bmatrix}$ **5.** Minimize $w = 5y_1 + 4y_2 + 15y_3$ subject to $y_1 + y_2 + 2y_3 \geq 4$, $y_1 + y_2 + y_3 \geq 3$, $y_1 + 3y_3 \geq 2$, $y_1 \geq 0$, $y_2 \geq 0$, and $y_3 \geq 0$.   **7.** Maximize $z = 50x_1 + 100x_2$ subject to $x_1 + 3x_2 \leq 1$, $x_1 + x_2 \leq 2$, $x_1 + 2x_2 \leq 1$, $x_1 + x_2 \leq 5$, $x_1 \geq 0$, and $x_2 \geq 0$.   **9.** Minimum is 14 when $y_1 = 0$ and $y_2 = 7$.   **11.** Minimum is 40 when $y_1 = 10$ and $y_2 = 0$.   **13.** Minimum is 100 when $y_1 = 0$, $y_2 = 100$, and $y_3 = 0$.   **15.** a   **17. a.** Maximize $x_1 + 1.5x_2 = z$ subject to: $x_1 + 2x_2 \leq 200$, $4x_1 + 3x_2 \leq 600$, $0 \leq x_2 \leq 90$, with $x_1 \geq 0$.   **b.** Make 120 monkeys and 40 bears, for a maximum profit of $180.   **c.** Minimize $w = 200y_1 + 600y_2 + 90y_3$ subject to $y_1 + 4y_2 \geq 1$, $2y_1 + 3y_2 + y_3 \geq 1.5$, $y_1 \geq 0$, $y_2 \geq 0$, and $y_3 \geq 0$.   **d.** $y_1 = .6$, $y_2 = .1$, $y_3 = 0$, $w = 180$ **e.** $186   **f.** $179   **19. a.** 1400 small and 700 large for a minimum cost of $294   **b.** $322   **21.** 0 gm of soybean meal, 8 gm of meat byproducts, and 3.6 gm of grain, for a minimum cost of $1.08   **23.** Make 16 large bowls, no small bowls, and 6 pots, for a minimum time of 104 hours.   **25.** 15/4 servings of A and 15/8 of B, for a minimum cost of $1.69

Exercises 4.4 (page 189)

1. $2x_1 + 3x_2 + s_1 = 8$; $x_1 + 4x_2 - s_2 = 7$ **3.** $x_1 + x_2 + x_3 + s_1 = 100$; $x_1 + x_2 + x_3 - s_2 = 75$; $x_1 + x_2 - s_3 = 27$
5. Change the objective function to maximize $z = -4y_1 - 3y_2 - 2y_3$. The constraints are not changed.   **7.** Change the objective
function to maximize $z = -y_1 - 2y_2 - y_3 - 5y_4$. The constraints are not changed.   **9.** Maximum is 480 when $x_1 = 40$, $x_2 = 0$.
11. Maximum is 750 when $x_1 = 0$, $x_2 = 150$, $x_3 = 0$.   **13.** Maximum is 300 when $x_1 = 0$, $x_2 = 100$.   **15.** Minimum is 20
when $y_1 = 5$, $y_2 = 0$.   **17.** Maximum is 400/3 when $x_1 = 100/3$, $x_2 = 50/3$.   **19.** Minimum is 512 when $y_1 = 6$, $y_2 = 8$.
23. Ship 200 barrels of oil from supplier S_1 to distributor D_1; ship 2800 barrels of oil from supplier S_2 to distributor D_1, ship
2800 barrels of oil from supplier S_1 to distributor D_2; ship 2200 barrels of oil from supplier S_2 to distributor D_2. Minimum cost is
$180,400.   **25.** Make $3,000,000 in commercial loans and $22,000,000 in home loans for a maximum return of $2,940,000.
27. Use 1000 lb of bluegrass, 2400 lb of rye, and 1600 lb of Bermuda for a minimum cost of $560.   **29.** Ship 2 computers from
W_1 to D_1, 20 computers from W_1 to D_2, 30 computers from W_2 to D_1, and 0 computers from W_2 to D_2, for a minimum cost of $628.
31. 5/3 oz of I, 20/3 oz of II, 5/3 oz of III, for a minimum cost of $1.55 per gallon; 10 oz of the additive should be used per
gallon of gasoline.

Chapter 4 Review Exercises (page 192)

1. When the problem has more than two variables   **3. a.** $2x_1 + 5x_2 + s_1 = 50$; $x_1 + 3x_2 + s_2 = 25$; $4x_1 + x_2 + s_3 = 18$;
$x_1 + x_2 + s_4 = 12$ **b.**

x_1	x_2	s_1	s_2	s_3	s_4	z	
2	5	1	0	0	0	0	50
1	3	0	1	0	0	0	25
4	1	0	0	1	0	0	18
1	1	0	0	0	1	0	12
-5	-3	0	0	0	0	1	0

5. a. $x_1 + x_2 + x_3 + s_1 = 90$; $2x_1 + 5x_2 + x_3 + s_2 = 120$; $x_1 + 3x_2 - s_3 = 80$ **b.**

x_1	x_2	x_3	s_1	s_2	s_3	z	
1	1	1	1	0	0	0	90
2	5	1	0	1	0	0	120
1	3	0	0	0	-1	0	80
-5	-8	-6	0	0	0	1	0

7. Maximum is 82.4 when $x_1 = 13.6$, $x_2 = 0$, $x_3 = 4.8$, $s_1 = 0$, and $s_2 = 0$.   **9.** Maximum is 76.67 when $x_1 = 6.67$, $x_2 = 0$,
$x_3 = 21.67$, $s_1 = 0$, $s_2 = 0$, and $s_3 = 35$.   **11. Dual Method** Solve the dual problem: Maximize $17x_1 + 42x_2$ subject to
$x_1 + 5x_2 \le 10$, $x_1 + 8x_2 \le 15$. **Method of Section 4.4** Change the objective function to maximize $z = -10y_1 - 15y_2$. The
constraints are not changed.   **13. Dual Method** Solve the dual problem: Maximize $48x_1 + 12x_2 + 10x_3 + 30x_4$ subject to
$x_1 + x_2 + 3x_4 \le 7$, $x_1 + x_2 \le 2$, $2x_1 + x_3 + x_4 \le 3$. **Method of Section 4.4** Change the objective function to maximize
$z = -7y_1 - 2y_2 - 3y_3$. The constraints are not changed.   **15.** Minimum of 62 when $y_1 = 8$, $y_2 = 12$, $s_1 = 0$, $s_2 = 1$, $s_3 = 0$, and
$s_4 = 2$.   **17.** Problems with constraints involving "$\le$" can be solved using slack variables, while those involving "$\ge$" or "$=$" can
be solved using surplus and artificial variables, respectively.   **19. a.** Maximize $z = 6x_1 + 7x_2 + 5x_3$, subject to
$4x_1 + 2x_2 + 3x_3 \le 9$, $5x_1 + 4x_2 + x_3 \le 10$, with $x_1 \ge 0$, $x_2 \ge 0$, $x_3 \ge 0$.   **b.** The first constraint would be $4x_1 + 2x_2 + 3x_3 \ge 9$.
c. $x_1 = 0$, $x_2 = 2.1$, $x_3 = 1.6$, and $z = 22.7$ **d.** Minimize $w = 9y_1 + 10y_2$, subject to $4y_1 + 5y_2 \ge 6$, $2y_1 + 4y_2 \ge 7$,
$3y_1 + y_2 \ge 5$, with $y_1 \ge 0$, $y_2 \ge 0$.   **e.** $y_1 = 1.3$, $y_2 = 1.1$ and $w = 22.7$ **21. a.** Let $x_1 =$ number of item A, $x_2 =$ number of
item B, and $x_3 =$ number of item C she should buy.   **b.** $z = 4x_1 + 3x_2 + 3x_3$ **c.** $5x_1 + 3x_2 + 6x_3 \le 1200$;
$x_1 + 2x_2 + 2x_3 \le 800$; $2x_1 + x_2 + 5x_3 \le 500$ **23. a.** Let $x_1 =$ number of gallons of fruity wine and $x_2 =$ number of gallons
of crystal wine to be made.   **b.** $z = 12x_1 + 15x_2$ **c.** $2x_1 + x_2 \le 110$; $2x_1 + 3x_2 \le 125$; $2x_1 + x_2 \le 90$ **25.** None of A,
400 of B, and none of C, for maximum profit of $1200   **27.** 36.25 gal of fruity and 17.5 gal of crystal, for maximum profit of
$697.50   **29. a. and b.** Produce 660 cases of corn, 0 cases of beans, and 340 cases of carrots, for a minimum cost of $15,100.

Chapter 5 Mathematics of Finance

Exercises 5.1 (page 208)

1. r is the interest rate per year, while i is the interest rate per compounding period. t is number of years, while n is the number
of periods.   **3.** The interest rate and number of compounding periods   **7.** $231.00   **9.** $286.75   **11.** $119.15
15. $1967.15   **17.** $28,741.55   **19.** $12,492.35   **21.** $7581.36   **23.** $849.16   **25.** $2587.74   **27.** The effective rate
29. 8.243%   **31.** 10.334%   **33.** $734,483.60   **35.** $34,438.29   **37.** 6.7%   **39. a.** 48   **b.** 72   **c.** 13.9%   **d.** 41.9%
41. About $1.946 million   **43.** $11,940.52   **45.** 4.91, 5.20, 5.34, 5.56, 5.63   **47.** 7.21% and 7.20%; DeepGreen Bank
49. About 18 years   **51.** About 12 years   **53.** $136,110.16   **55.** 10.00%   **57.** $7522.50

Exercises 5.2 (page 218)

1. 48 **3.** −648 **5.** 81 **7.** 64 **9.** 15 **11.** 156/25 **13.** −208 **15.** 15.91713 **17.** 22.35633 **21.** $437.46 **23.** $4,427,846.13 **25.** $180,307.41 **27.** $28,438.21 **29.** $518,017.56 **31.** $6294.79 **33.** $158,456.07 **35.** $26,874.97 **37.** $6655.99 **39.** $628.25 **43.** $354.79 **45.** $1626.16 **47.** $257.99 **49.** $2432.13 **51.** $79,679.68 **53.** $169,474.59 **55.** $67,940.98 **57. a.** $226.11 **b.** $245.77 **59.** $647.76 **61.** $152,667.08 **63.** $209,348 **65.** 7.397% **67. a.** $1200 **b.** $3511.58

Exercises 5.3 (page 227)

1. c 3. 9.71225 **5.** 12.15999 **9.** $8045.30 **11.** $1,367,774 **13.** $205,724.40 **15.** $97,122.49 **17.** $446.31 **19.** $11,942.55 **21.** $589.31 **23.** $7.61 **25.** $35.24 **27.** $6699 **29.** $1129.67 **31.** $1176.85 **33.** $819.21; $69,457.80; The payments are $157.65 more than for the 30-year loan, but the total interest paid is $90,703.80 less. **35. a.** $158 **b.** $1584 **37. a.** $623,110.52 **b.** $456,427.28 **c.** $563,757.78 **d.** $392,903.18 **39.** $280.46; $32,310.40

41.

Payment Number	Amount of Payment	Interest for Period	Portion to Principal	Principal at End of Period
0	—	—	—	$72,000.00
1	$10,129.69	$3600.00	$6529.69	$65,470.31
2	$10,129.69	$3273.52	$6856.17	$58,614.14
3	$10,129.69	$2930.71	$7198.98	$51,415.16
4	$10,129.69	$2570.76	$7558.93	$43,856.23

43. a. $32.49 **b.** $195.52; $10.97 **45. a.** $2349.51; $197,911.80 **b.** $2097.30; $278,352 **c.** $1965.82; $364,746 **d.** After 170 payments **47. a.** $17,584.58 **b.** $15,069.31

49.

Payment Number	Amount of Payment	Interest for Period	Portion to Principal	Principal at End of Period
0	—	—	—	$4835.80
1	$614.90	$223.66	$391.24	$4444.56
2	$614.90	$205.56	$409.34	$4035.22
3	$614.90	$186.63	$428.27	$3606.95
4	$614.90	$166.82	$448.08	$3158.87
5	$614.90	$146.10	$468.80	$2690.07
6	$614.90	$124.42	$490.48	$2199.59
7	$614.90	$101.73	$513.17	$1686.42
8	$614.90	$78.00	$536.90	$1149.52
9	$614.90	$53.17	$561.73	$587.79
10	$614.98	$27.19	$587.79	$0.00

51. a. $25,000 **b.** $40,000

Chapter 5 Review Exercises (page 232)

1. $848.16 **3.** $921.50 **5.** Compound interest **7.** $43,988.40 **9.** $104,410.10 **11.** $12,444.50 **13.** $10,213.85 **15.** $18,207.65 **17.** $1067.71 **19.** 2, 6, 18, 54, 162 **21.** −96 **23.** −120 **25.** 34.78489 **29.** $31,188.82 **31.** $14,264.87 **33.** $165,974.31 **35.** $886.05 **37.** $5132.48 **39.** $2815.31 **41.** $47,988.11 **43.** A home loan and an auto loan **45.** $356.24 **47.** $1931.82 **49.** $760.67 **51.** $132.99 **53.** $1535.61 **55.** $693.13; $10,513.13 **57.** 8.21% **59.** $2298.58 **61.** $107,892.82; $32,892.32 **63.** $9859.46 **65.** 7.08% and 7.10%; Capital Crossing Bank **67. a.** $571.28 **b.** $532.50 **c.** Method 1: $56,324.44; Method 2: $56,325.43 **d.** $7100 **e.** Method 1: $72,575.56; Method 2: $72,574.57 **69. a.** 9.569% **b.** $896.44 **c.** $626,200.88 **d.** $1200.39 **e.** $478,134.14 **f.** Sue is ahead by $148,066.74.

Chapter 6 Logic

Exercises 6.1 (page 244)

1. Statement **3.** Not a statement **5.** Statement **7.** Statement **9.** Statement **11.** Not a statement **13.** Statement **15.** Compound **17.** Compound **19.** Not compound **21.** Compound **23.** Her aunt's name is not Lucia. **25.** At least one dog does not have its day. **27.** No book is longer than this book. **29.** At least one computer repairman can play blackjack. **31.** Someone does not love somebody sometime. **33.** $y \leq 12$ **35.** $q < 5$ **39.** She does not have green eyes. **41.** She has green eyes and he is 48 years old. **43.** She does not have green eyes or he is 48 years old. **45.** She does not have green eyes or he is not 48 years old. **47.** It is not the case that she does not have green eyes and he is 48 years old. **49.** True **51.** True

53. True **55.** True **57.** False **61. a.** False **b.** False **c.** True **65.** a **67.** a, b, d **69.** Life insurance payments are subject to income tax. **71.** b, c, d, e **73.** Viral illnesses cannot cause swollen glands. **75.** c, d **79.** $p \wedge \sim q$ **81.** $\sim p \vee q$ **83.** $\sim(p \vee q)$ or, equivalently, $\sim p \wedge \sim q$ **87.** Group C **89.** Groups A and B **91.** Groups A and C **93.** Group B

Exercises 6.2 (page 258)

1. False **3.** True **5.** Both components are false. **7.** True **9.** True **11.** False **13.** True **15.** True **17.** True **19.** Disjunction **21.** True **23.** False **25.** True **27.** True **29.** False **31.** True **33.** True **35.** True **37.** 4 **39.** 16 **41.** 128 **43.** 6

45.

p	q	$\sim p$	$\sim p \wedge q$
T	T	F	F
T	F	F	F
F	T	T	T
F	F	T	F

47.

p	q	$p \wedge q$	$\sim(p \wedge q)$
T	T	T	F
T	F	F	T
F	T	F	T
F	F	F	T

49.

p	q	$\sim p$	$\sim q$	$q \vee \sim p$	$(q \vee \sim p) \vee \sim q$
T	T	F	F	T	T
T	F	F	T	F	T
F	T	T	F	T	T
F	F	T	T	T	T

In Exercises 51–59 we are using the alternative method to save space, filling in columns in the order indicated by the numbers. Observe that columns with the same number are combined (by the logical definition of the connective) to get the next numbered column. Note that this is different than the way the numbered columns are used in the textbook. Remember that the last column (highest numbered column) completed yields the truth value for the complete compound statement. Be sure to align truth values under the appropriate logical connective or simple statement.

51.

p	q	$\sim q$	$\wedge$	$(\sim p \vee q)$
T	T	F	F	F T T
T	F	T	F	F F F
F	T	F	F	T T T
F	F	T	T	T T F
		2	3	1 2 1

53.

p	q	$(p \vee \sim q)$	$\wedge$	$(p \wedge q)$
T	T	T T F	T	T T T
T	F	T T T	F	T F F
F	T	F F F	F	F F T
F	F	F T T	F	F F F
		1 2 1	3	1 2 1

55.

p	q	r	$(\sim p \wedge q)$	$\wedge$	r
T	T	T	F F T	F	T
T	T	F	F F T	F	F
T	F	T	F F F	F	T
T	F	F	F F F	F	F
F	T	T	T T T	T	T
F	T	F	T T T	F	F
F	F	T	T F F	F	T
F	F	F	T F F	F	F
			1 2 1	3	2

57.

p	q	r	$(\sim p \wedge \sim q)$	$\vee$	$(\sim r \vee \sim p)$
T	T	T	F F F	F	F F F
T	T	F	F F F	T	T T F
T	F	T	F F T	F	F F F
T	F	F	F F T	T	T T F
F	T	T	T F F	T	F T T
F	T	F	T F F	T	T T T
F	F	T	T T T	T	F T T
F	F	F	T T T	T	T T T
			1 2 1	3	1 2 1

59.

p	q	r	s	$\sim(\sim p \wedge \sim q)$	$\vee$	$(\sim r \vee \sim s)$
T	T	T	T	T F F F	T	F F F
T	T	T	F	T F F F	T	F T T
T	T	F	T	T F F F	T	T T F
T	T	F	F	T F F F	T	T T T
T	F	T	T	T F F T	T	F F F
T	F	T	F	T F F T	T	F T T
T	F	F	T	T F F T	T	T T F
T	F	F	F	T F F T	T	T T T
F	T	T	T	T T F F	T	F F F
F	T	T	F	T T F F	T	F T T
F	T	F	T	T T F F	T	T T F
F	T	F	F	T T F F	T	T T T
F	F	T	T	F T T T	F	F F F
F	F	T	F	F T T T	T	F T T
F	F	F	T	F T T T	T	T T F
F	F	F	F	F T T T	T	T T T
				3 1 2 1	4	2 3 2

61. You can't pay me now and you can't pay me later.   **63.** It is not summer or there is snow.   **65.** I did not say yes or she did not say no.   **67.** $5 - 1 \neq 4$ or $9 + 12 = 7$ **69.** Neither Cupid nor Vixen will lead Santa's sleigh next Christmas.   **71.** True
73. True   **75.**

p	q	$p \bigvee q$
T	T	F
T	F	T
F	T	T
F	F	F

77. True   **79. a.** False   **b.** True   **c.** False   **d.** True   **81.** You cannot increase the
amount of income tax withheld from your pay and cannot make estimated tax payments to meet your 2001 tax liability.
83. Neither does this warranty give you specific legal rights nor may you have other rights, which vary from state to state.   **85.** p:
The Pennsylvania Fish and Boat Commission is sensitive to the needs of the physically challenged. q: The Pennsylvania Fish and
Boat Commission works to make our facilities accessible. *Negation:* Either the Pennsylvania Fish and Boat Commission is not
sensitive to the needs of the physically challenged or it does not work to make our facilities accessible.   **87.** p: The court won't do
it for you. q: Hiring an attorney is usually not cost effective. *Negation:* Either the court will do it for you or hiring an attorney is
often cost effective.   **89.** The lady is behind Door 2 and the tiger is behind Door 1.

Exercises 6.3 (page 270)

1. True   **3.** True   **5.** False   **7.** True   **11.** True   **13.** False   **15.** True   **17.** If they do not raise monkeys, then he trains
ponies.   **19.** If she has a snake for a pet, then they raise monkeys and he trains ponies.   **21.** If he does not train ponies, then
they do not raise monkeys or she has a snake for a pet.   **23.** $r \to b$ **25.** $\sim b \to \sim r$ **27.** $b \bigvee (p \to r)$ **29.** $\sim r \to b$
31. True   **33.** False   **35.** True   **37.** False   **39.** True   **41.** True   **45.**

p	q	$\sim q$	$\to$	p
T	T	F	T	T
T	F	T	T	T
F	T	F	T	F
F	F	T	F	F
		1	2	1

47.

p	q	$(\sim p \to q)$	$\to$	p
T	T	F T T	T	T
T	F	F T F	T	T
F	T	T T T	F	F
F	F	T F F	T	F
		1 2 1	3	2

49.

p	q	$(p \vee q)$	$\to$	$(q \vee p)$
T	T	T T T	T	T T T
T	F	T T F	T	F T T
F	T	F T T	T	T T F
F	F	F F F	T	F F F
		1 2 1	3	1 2 1

Since this statement is always true (column 3), it is a tautology.

51.

p	q	$(\sim p \to \sim q)$	$\to$	$(p \wedge q)$
T	T	F T F	T	T T T
T	F	F T T	F	T F F
F	T	T F F	T	F F T
F	F	T T T	F	F F F
		1 2 1	3	1 2 1

53.

p	q	r	$[(r \vee p)$	$\wedge$	$\sim q]$	$\to$	p
T	T	T	T T T	F	F	T	T
T	T	F	F T T	F	F	T	T
T	F	T	T T T	T	T	T	T
T	F	F	F T T	T	T	T	T
F	T	T	T T F	F	F	T	F
F	T	F	F F F	F	F	T	F
F	F	T	T T F	T	T	F	F
F	F	F	F F F	F	T	T	F
			1 2 1	3	2	4	3

55.

p	q	r	s	$(\sim p \wedge \sim q)$			$\rightarrow$	$(\sim r \rightarrow \sim s)$		
T	T	T	T	F	F	F	T	F	T	F
T	T	T	F	F	F	F	T	F	T	T
T	T	F	T	F	F	F	T	T	F	F
T	T	F	F	F	F	F	T	T	T	T
T	F	T	T	F	F	T	T	F	T	F
T	F	T	F	F	F	T	T	F	T	T
T	F	F	T	F	F	T	T	T	F	F
T	F	F	F	F	F	T	T	T	T	T
F	T	T	T	T	F	F	T	F	T	F
F	T	T	F	T	F	F	T	F	T	T
F	T	F	T	T	F	F	T	T	F	F
F	T	F	F	T	F	F	T	T	T	T
F	F	T	T	T	T	T	T	F	T	F
F	F	T	F	T	T	T	T	F	T	T
F	F	F	T	T	T	T	F	T	F	F
F	F	F	F	T	T	T	T	T	T	T
				1	2	1	3	1	2	1

57. That is an authentic Persian rug and I won't be surprised.

59. The English measures are not converted to metric measures and the spacecraft does not crash on the surface of Mars.
61. You want to be happy for the rest of your life and you make a pretty woman your wife.   **63.** You do not give your plants tender, loving care or they flourish.   **65.** She does or he will.   **67.** The person is not a resident of Butte or is a resident of Montana.   **69.** Equivalent   **71.** Not equivalent   **73.** Equivalent   **75.** Not equivalent   **77.** $(p \wedge q) \vee (p \wedge \sim q) \equiv p$
79. $p \vee (\sim q \wedge r)$ **81.** $\sim p \vee (p \vee q) \equiv$ T (tautology)   **83.**

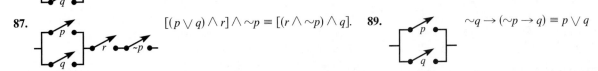

$p \wedge (q \vee \sim p) \equiv p \wedge q$

85.

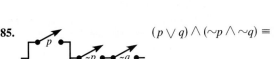

$(p \vee q) \wedge (\sim p \wedge \sim q) \equiv$ F

87.

$[(p \vee q) \wedge r] \wedge \sim p \equiv [(r \wedge \sim p) \wedge q].$ **89.**

$\sim q \rightarrow (\sim p \rightarrow q) \equiv p \vee q$

93. a. If you were divorced under a decree of divorce of separate maintenance that is final by the end of the year, you may not file a 2000 joint return.   **b.** If you separated during 2000 under an interlocutory decree (provisional or temporary) or order, so long as a final divorce decree was not entered by the end of the year, you may file jointly.   **c.** If both a U.S. citizen or resident and his/her nonresident alien spouse elect to be taxed on their worldwide income, the couple may file a joint return.   **d.** If you wish to avoid current tax on pay, then you may contract with your employer to defer pay to future years.   **95.** Your throat is not sore or food does not seem to stick high up in the chest or you do not get a burning pain in the center of the chest or you may have gastroesophageal reflux. *Negation:* Your throat is sore and food seems to stick high up in the chest and you get a burning pain in the center of the chest and you do not have gastroesophageal reflux.

Exercises 6.4 (page 278)

1. a. *Converse:* If you were an hour, then beauty would be a minute.   **b.** *Inverse:* If beauty were not a minute, then you would not be an hour.   **c.** *Contrapositive:* If you were not an hour, then beauty would not be a minute.   **3. a.** *Converse:* If I don't see it, then the exit is ahead.   **b.** *Inverse:* If the exit is not ahead, then I see it.   **c.** *Contrapositive:* If I see it, then the exit is not ahead.
5. a. *Converse:* If it is dangerous to your health, then you walk in front of a moving car.   **b.** *Inverse:* If you do not walk in front of a moving car, then it is not dangerous to your health.   **c.** *Contrapositive:* If it is not dangerous to your health, then you do not walk in front of a moving car.   **7. a.** *Converse:* If they flock together, then they are birds of a feather.   **b.** *Inverse:* If they are

not birds of a feather, then they do not flock together. **c.** *Contrapositive:* If they do not flock together, then they are not birds of a feather. **9. a.** *Converse:* If he comes, then you built it. **b.** *Inverse:* If you don't build it, then he won't come. **c.** *Contrapositive:* If he doesn't come, then you didn't build it. **11. a.** *Converse:* $\sim q \rightarrow p$. **b.** *Inverse:* $\sim p \rightarrow q$. **c.** *Contrapositive:* $q \rightarrow \sim p$. **13. a.** *Converse:* $\sim q \rightarrow \sim p$. **b.** *Inverse:* $p \rightarrow q$. **c.** *Contrapositive:* $q \rightarrow p$. **15. a.** *Converse:* $(q \vee r) \rightarrow p$. **b.** *Inverse:* $\sim p \rightarrow \sim(q \vee r)$ or $\sim p \rightarrow (\sim q \wedge \sim r)$. **c.** *Contrapositive:* $(\sim q \wedge \sim r) \rightarrow \sim p$. **19.** If it is muddy, then I'll wear my galoshes. **21.** If 17 is positive, then $17 + 1$ is positive. **23.** If a number is an integer, then it is a rational number. **25.** If I do crossword puzzles, then I am driven crazy. **27.** If Greg Tobin is to shave, then he must have a day's growth of beard. **29.** If I go from Boardwalk to Connecticut, then I pass GO. **31.** If a number is a whole number, then it is an integer. **33.** If their pitching improves, then the Indians will win the pennant. **35.** If the figure is a rectangle, then it is a parallelogram with a right angle. **37.** If a triangle has two sides of the same length, then it is isosceles. **39.** If a two-digit number whose units digit is 5 is squared, then it will end in 25. **41.** d **45.** True **47.** True **49.** False **51. a.** *Converse:* If you may file a joint return with your spouse, then you are married at the end of the year. *Inverse:* If you are not married at the end of the year, you may not file a joint return with your spouse. *Contrapositive:* If you may not file a joint return with your spouse, then you were not married at the end of the year. **b.** *Converse:* If you include the value of the stock as pay in the year you receive it, then you receive your company's stock as payment for your services. *Inverse:* If you do not receive your company's stock as payment for your services, then you do not include the value of the stock as pay in the year you receive it. *Contrapositive:* If you do not include the value of the stock as pay in the year you receive it, then you did not receive your company's stock as payment for your services. **53.** If you do not have gastroesophageal reflux then your throat is not sore or your food does not stick high up in the chest or you don't get a burning pain in the center of the chest. **55. a.** *Converse:* If you can't get married again, then you are married. *Inverse:* If you aren't married, then you can get married again. *Contrapositive:* If you can get married again, then you are not married. **b.** *Converse:* If your contractor is legally required to put it in writing, then your job is going to cost more than $500. *Inverse:* If your job is not going to cost more than $500, then your contractor is not legally required to put it in writing. *Contrapositive:* If your contractor is not legally required to put it in writing, then the job is not going to cost more than $500. **c.** *Converse:* If you can appeal in federal court, then your application for citizenship is denied. *Inverse:* If your application for citizenship is not denied, then you cannot appeal in federal court. *Contrapositive:* If you cannot appeal in federal court, then your application for citizenship has not been denied. **59.** D, 7 **61. a.** If you can score in this box then the dice include three or more of the same number. You cannot score in this box or the dice include three or more of the same number. **b.** If you can score in this box, then the dice show any sequence of four numbers. You cannot score in this box or the dice show any sequences of four numbers. **c.** If you can score in this box, then the dice show three of one number and two of another. You cannot score in this box or the dice show three of one number and two of another.

Exercises 6.5 (page 284)

1. Valid **3.** Invalid **5.** Valid **7.** Invalid **9.** Invalid **11.** Invalid **13.** Yes **15.** People who have major surgery must go to the hospital. Andrea Sheehan is having major surgery. Therefore, Andrea Sheehan must go to the hospital. **17.** Valid **19.** Invalid **21.** Invalid **23.** Invalid **25.** Invalid **27.** Invalid **29.** Valid

Exercises 6.6 (page 293)

1. Valid **3.** Valid **5.** Fallacy **7.** Valid **9.** Fallacy **11.** Valid **13.** Invalid **15.** Valid **17.** Invalid **19.** Valid **21.** Invalid **23.** Invalid **29.** Valid **31.** Invalid **33.** Invalid **35.** Valid **37.** Valid **39.** If it is my poultry, then it is a duck. **41.** If it is a guinea pig, then it is hopelessly ignorant of music. **43.** If it is a teachable kitten, then it does not have green eyes. **45.** If I can read it, then I have not filed it. **47. a.** $p \rightarrow \sim s$ **b.** $r \rightarrow s \equiv \sim s \rightarrow \sim r$ **c.** $q \rightarrow p$ **d.** $q \rightarrow \sim r$, *Conclusion:* If it is my poultry, then it is not an officer. **49. a.** $r \rightarrow \sim s \equiv s \rightarrow \sim r$ **b.** $u \rightarrow t$ **c.** $\sim r \rightarrow p$ **d.** $\sim u \rightarrow \sim q \equiv q \rightarrow u$ **e.** $t \rightarrow s$ **f.** $q \rightarrow p$, *Conclusion:* If one is a pawnbroker, then one is honest. **51. a.** $r \rightarrow w$ **b.** $\sim u \rightarrow \sim t$ **c.** $v \rightarrow \sim s \equiv s \rightarrow \sim v$ **d.** $x \rightarrow r$ **e.** $\sim q \rightarrow t \equiv \sim t \rightarrow q$ **f.** $y \rightarrow p$ (begin with y since it appears only once) **g.** $w \rightarrow s$ **h.** $\sim x \rightarrow \sim q$ **i.** $p \rightarrow \sim u$ **j.** $y \rightarrow \sim v$, *Conclusion:* If it is written by Brown, then I can't read it.

Chapter 6 Review Exercises (page 296)

1. $6 - 3 \neq 3$ **3.** No members of the class went on the field trip. **5.** She did not pass GO or did not collect $200. **7.** $p \rightarrow q$ **9.** You won't love me and I will love me. **11.** True **13.** True **17.**

p	q	p	$\wedge$	$(\sim p \vee q)$
T	T	T	T	F T T
T	F	T	F	F F F
F	T	F	F	T T T
F	F	F	F	T T F
		2	3	1 2 1

19. False

21. If the number is an integer, then it is a rational number.   **23.** If a number is divisible by 9, then it is divisible by 3.
25. *Converse:* If the graph will help me understand it, then a picture paints a thousand words. *Inverse:* If a picture doesn't paint a thousand words, the graph won't help me understand it. *Contrapositive:* If the graph doesn't help me understand it, then a picture doesn't paint a thousand words.   **27.**

Valid argument   **29.** Let p represent "I write a check," q

represent "It will bounce," and r represent "The bank guarantees it." Valid

p	q	r	$\{[(p \to q) \land (r \to \sim q)] \land r\} \to (\sim p)$
T	T	T	T F T F F F T T F
T	T	F	T T F T F F F T F
T	F	T	F F T T T F T T F
T	F	F	F F F T T F F T F
F	T	T	T F T F F F T T T
F	T	F	T T F T F F F T T
F	F	T	T T T T T T T T T
F	F	F	T T F T T F F T T
			2 3 1 2 1 4 3 5 4

31. a, b, c   **33.** *Converse:* If a credit may not be claimed until the year the adoption becomes final, then you adopt a child who is not a U.S. citizen or resident at the time the adoption effort begins.   **35.** *Contrapositive:* If your baby is not at serious risk of being born with abnormalities, such as congenital deafness, congenital heart disease, clouding of the lens in the eye, and the nervous system disorder cerebral palsy, then you did not contract rubella in early pregnancy. Or *statement:* You did not contract rubella in early pregnancy or your baby is at serious risk of being born with abnormalities, such as congenital deafness, congenital heart disease, clouding of the lens in the eye, and the nervous system disorder cerebral palsy.   **37.** C

Chapter 7 Sets and Probability

Exercises 7.1 (page 311)

1. False   **3.** True   **5.** True   **7.** True   **9.** False   **11.** $\subseteq$ **13.** $\not\subseteq$ **15.** $\subseteq$ **17.** $\subseteq$ **19.** $\subset$; $\subset$; $\not\subset$; $\not\subset$; $\subset$; $\not\subset$; $\subset$; $\not\subset$
21. 16   **23.** 4   **25.** $\cap$ **27.** $\cap$ **29.** $\cap$ **31.** $\cup$ or $\cap$ **33.** $\{3,5\}$ **35.** $\{7,9\}$ **37.** $\emptyset$ **39.** Y or $\{3,5,7,9\}$
41. $\{2,3,4\}$ **43.** All students in this school not taking this course   **45.** All students in this school taking accounting and zoology   **47.** C and D, B and E, C and E, D and E **49.** B' is the set of all stocks on the list with a closing price below $60 or above $90; $B' = \{$AT&T, GenMills, IBM, PepsiCo$\}$.   **51.** $(A \cap B)'$ is the set of all stocks on the list that do not have a high price greater than $80 and a last price between $60 and $90; $(A \cap B)' = \{$AT&T, GenMills, Hershey, IBM, PepsiCo$\}$.   **53. a.** True
b. True   **c.** False   **d.** False   **e.** True   **f.** True   **g.** False   **55.** $\{$Cisco Systems Inc.$\}$
57. $\{$Exxon Mobil Corp., Citigroup Inc., Tyco International$\}$ **59.** $\{s,d,c\}$ **61.** $\{g\}$ **63.** $\{s,d,c\}$ **65.** 11
67. $\{$TBS, The Discovery Channel, ESPN$\}$ **69.** $\{$TBS, USA$\}$ **71.** $\{$TBS, ESPN, USA$\}$ **73.** Joe should always first choose the complement of what Dorothy chose.

Exercises 7.2 (page 321)

1.

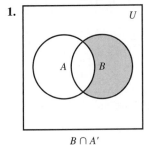

$B \cap A'$

3.

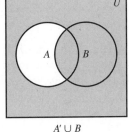

$A' \cup B$

5.

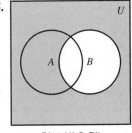

$B' \cup (A' \cap B')$

7.

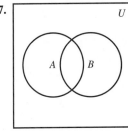

$U' = \emptyset$

9. 8   **11.**

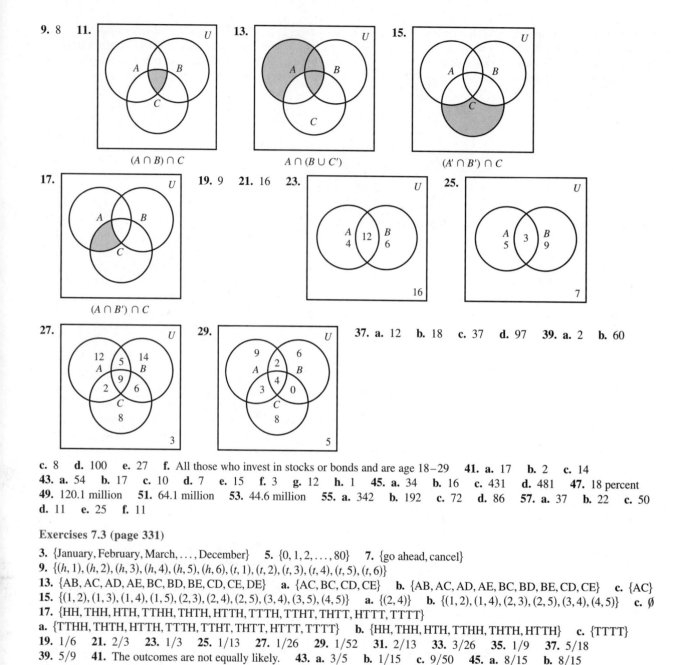

$(A \cap B) \cap C$

13.

$A \cap (B \cup C')$

15.

$(A' \cap B') \cap C$

17.

$(A \cap B') \cap C$

19. 9   **21.** 16   **23.**

25.

27.

29.

37. a. 12   **b.** 18   **c.** 37   **d.** 97   **39. a.** 2   **b.** 60

c. 8   **d.** 100   **e.** 27   **f.** All those who invest in stocks or bonds and are age 18–29   **41. a.** 17   **b.** 2   **c.** 14
43. a. 54   **b.** 17   **c.** 10   **d.** 7   **e.** 15   **f.** 3   **g.** 12   **h.** 1   **45. a.** 34   **b.** 16   **c.** 431   **d.** 481   **47.** 18 percent
49. 120.1 million   **51.** 64.1 million   **53.** 44.6 million   **55. a.** 342   **b.** 192   **c.** 72   **d.** 86   **57. a.** 37   **b.** 22   **c.** 50
d. 11   **e.** 25   **f.** 11

Exercises 7.3 (page 331)

3. {January, February, March, . . . , December}   **5.** {0, 1, 2, . . . , 80}   **7.** {go ahead, cancel}
9. {(h, 1), (h, 2), (h, 3), (h, 4), (h, 5), (h, 6), (t, 1), (t, 2), (t, 3), (t, 4), (t, 5), (t, 6)}
13. {AB, AC, AD, AE, BC, BD, BE, CD, CE, DE}   **a.** {AC, BC, CD, CE}   **b.** {AB, AC, AD, AE, BC, BD, BE, CD, CE}   **c.** {AC}
15. {(1, 2), (1, 3), (1, 4), (1, 5), (2, 3), (2, 4), (2, 5), (3, 4), (3, 5), (4, 5)}   **a.** {(2, 4)}   **b.** {(1, 2), (1, 4), (2, 3), (2, 5), (3, 4), (4, 5)}   **c.** ∅
17. {HH, THH, HTH, TTHH, THTH, HTTH, TTTH, TTHT, THTT, HTTT, TTTT}
a. {TTHH, THTH, HTTH, TTTH, TTHT, THTT, HTTT, TTTT}   **b.** {HH, THH, HTH, TTHH, THTH, HTTH}   **c.** {TTTT}
19. 1/6   **21.** 2/3   **23.** 1/3   **25.** 1/13   **27.** 1/26   **29.** 1/52   **31.** 2/13   **33.** 3/26   **35.** 1/9   **37.** 5/18
39. 5/9   **41.** The outcomes are not equally likely.   **43. a.** 3/5   **b.** 1/15   **c.** 9/50   **45. a.** 8/15   **b.** 8/15
47. a. Person smokes or has a family history of heart disease.   **b.** Person does not smoke and has a family history of heart
disease.   **c.** Person does not have a family history of heart disease or is not overweight.   **49. a.** .118   **b.** .168   **c.** .122
d. .132   **51. a.** About .32   **b.** About .32   **c.** About .37   **d.** Cavalry   **e.** I Corps   **53. a.** 25/57   **b.** 32/57
c. 4/19

Exercises 7.4 (page 342)

3. No   **5.** Yes   **7.** No   **9. a.** 5/36   **b.** 1/9   **c.** 1/12   **d.** 0   **11. a.** 5/18   **b.** 5/12   **c.** 1/3   **13.** 2/9   **15.** Yes,
Laurie should put 1 pink marble in one box and the other 49 in the box with the 50 blue marbles.   **17. a.** 3/13   **b.** 4/13
c. 7/13   **d.** 4/13   **e.** 8/13   **19. a.** 1/2   **b.** 7/10   **c.** 1/2   **21. a.** 1/10   **b.** 9/10   **c.** 7/20   **23. a.** .39   **b.** .84
c. .74   **d.** .26   **25.** 1 to 5   **27.** 2 to 1   **29. a.** 1 to 4   **b.** 8 to 7   **c.** 4 to 11   **31.** 2 to 7   **35.** Not empirical

37. Empirical   **39.** Empirical   **41.** Not empirical   **45.** Possible   **47.** Not possible; the sum of the probabilities is greater than 1.   **49.** Not possible; a probability cannot be negative.   (Note: for 50–53, theoretical answers are given; actual answers will vary.)   **51. a.** .2778   **b.** .4167   **53. a.** .15625   **b.** .3125   **55. a.** .51   **b.** .67   **c.** .39   **d.** .12   **57. a.** .89 **b.** .38   **c.** .18   **59. a.** .56   **b.** .20   **c.** .31   **61. a.** 3/4   **b.** 1/4   **63. a.** .90   **b.** .23   **65. a.** .06   **b.** .44   **c.** .74 **d.** .18   **67. a.** 7/18   **b.** 13/18   **c.** 5/9   **d.** 2/3   **e.** 2/9   **69. a.**   **b.** .306

	A	B	C	D
O	.052	.060	.039	.004
E	.306	.270	.213	.045
M	.003	.003	.003	0

c. About .04   **d.** About .7   **e.** About .88   **71. a.** .867   **b.** .843

Exercises 7.5 (page 358)

1. 0   **3.** 1   **5.** 1/6   **7.** 4/17   **9.** 11/51   **11.** .012   **13.** .245   **19.** The second booth, for which the probability that both are heads is 1/2. The probability is 1/3 for the man in the first booth.   **21.** No, these events are not independent.   **23.** 1/20; 2/5   **25.** The probability of a customer cashing a check, given that the customer made a deposit, is 5/7.   **27.** The probability of a customer not cashing a check, given that the customer did not make a deposit, is 1/4.   **29.** The probability of a customer not both cashing a check and making a deposit is 6/11.   **31.** .999985; fairly realistic   **33.** .02   **35.** .08   **37.** 1/4   **39.** 1/4 **41.** 1/7   **43.** They are independent.   **45.** .039   **47.** .491   **49.** .072   **51.** Yes   **53. a.** .154   **b.** .333   **55. a.** .0005 **b.** .9995   **c.** $(1999/2000)^a$ **d.** $1 - (1999/2000)^a$ **e.** $(1999/2000)^{Nc}$ **f.** $1 - (1999/2000)^{Nc}$ **57.** .031   **59.** .834 **61. b.** .30   **c.** .98   **63. a.** .271   **b.** .379   **c.** .625   **d.** .342   **e.** .690   **f.** .514   **g.** Not independent   **65.** The events are probably dependent, and the agent has used the product rule for independent events.   **67. a.** .060   **b.** .85   **c.** .5275 **69. a.** .05   **b.** .015   **c.** .25   **71.** No   **73. c.** They are the same.   **d.** The 2-point first strategy has a smaller probability of losing.

Exercises 7.6 (page 368)

1. 1/3   **3.** 2/41   **5.** 21/41   **7.** 8/17   **9.** 82.4%   **11.** .092   **13.** .176   **15.** .994   **17.** .364   **19.** 2/7   **21. a.** .643 **b.** .251   **23. a.** .3333   **25. a.** about .155   **b.** about .990   **c.** about 59   **27. a.** .53   **b.** .1623   **29.** .162   **31.** .086 **33.** .244   **35.** .178

Chapter 7 Review Exercises (page 373)

1. False   **3.** False   **5.** True   **7.** False   **9.** False   **11.** 16   **13.** {a, b, e}   **15.** {c, d, g}   **17.** {a, b, e, f}   **19.** U **21.** All female employees in the accounting department   **23.** All employees who are in the accounting department or who have MBA degrees   **25.** All male employees who are not in the sales department   **27.**

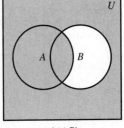

$A \cup B'$

29.

$(A \cap B) \cup C$

31. {1, 2, 3, 4, 5, 6}   **33.** {0, .5, 1, 1.5, 2, . . . , 299.5, 300}

35. {(3, R), (3, G), (5, R), (5, G), (7, R), (7, G), (9, R), (9, G), (11, R), (11, G)}   **37.** {(3, G), (5, G), (7, G), (9, G), (11, G)}   **39.** 1/4

41. 3/13   **43.** 1/2   **45.** 1   **51.** Vos Savant's answer: The contestant should switch doors.   **53.** 1 to 25   **55.** 5/36 ≈ .139
57. 1/6 ≈ .167   **59.** 1/6 ≈ .167   **61.** 2/11 ≈ .182   **63. a.** .66   **b.** .29   **c.** .71   **d.** .34   **65.** 4/9 ≈ .444
67. 3/22 ≈ .136   **69.** 2/3   **71. a.** .86   **b.** .26   **73. a.** Row 1: 400; row 2: 150; row 3: 750, 1000   **b.** 1000   **c.** 300
d. 250   **e.** 600   **f.** 150   **g.** Those who purchased a used car given that the buyer is not satisfied   **h.** 3/5   **i.** 1/4
75. .45   **77. b.** 7/10   **79. a.** .715; .569; .410; .321; .271   **81. a.** 51   **b.** 31   **c.** 18   **d.** 15   **e.** 33   **f.** 23

Chapter 8 Counting Principles; Further Probability Topics

Exercises 8.1 (page 386)

1. 720   **3.** 1.308×10^{12} **5.** 156   **7.** 1.024×10^{25} **9.** 1   **11.** n **13.** 30   **15.** 15   **19.** 3   **21.** 540,540
23. a. 8.7178291×10^{10} **b.** 4,354,560   **c.** 120,120   **d.** 84   **25.** Multiply by 10   **27. a.** incorrect   **b.** incorrect
c. incorrect   **d.** correct   **29.** 55,440   **31.** 604,800   **33.** 1440   **35.** 2.05237×10^{10} **37.** 1,256,640   **39. a.** 25,200
b. 15,120   **41. a.** 27,600   **b.** 35,152   **c.** 1104   **43. a.** 160; 8,000,000   **b.** Some numbers, such as 911, 800, and 900,
are reserved for special purposes.   **45. a.** 17,576,000   **b.** 17,576,000   **c.** 456,976,000   **47.** 100,000; 90,000   **49.** 81
51. 1,572,864; no   **53. b.** 3,191,834,419,200

Exercises 8.2 (page 396)

3. 792   **5.** 1.134×10^{11} **7.** 1   **9.** n **11. a.** 10   **b.** 7   **13. a.** 9   **b.** 6   **c.** 3; yes, from both   **17.** Combinations;
a. 56   **b.** 462   **c.** 3080   **19.** Combinations;   **a.** 220   **b.** 220   **21.** Combinations;   **a.** 105   **b.** 1365   **c.** 28
23. Combinations;   **a.** 10   **b.** 0   **c.** 1   **d.** 10   **e.** 30   **f.** 15   **g.** 0   **27.** 28   **29. a.** 720   **b.** 360   **31. a.** 40
b. 20   **c.** 7   **33.** 1,621,233,900   **35. a.** 84   **b.** 10   **c.** 40   **d.** 74   **37.** 635,013,559,600   **39.** 50   **41. a.** 15,504
b. 816   **43. a.** 1,120,529,256   **b.** 806,781,064,320   **45.** 5.524×10^{26} **47. a.** 255   **b.** 16   **c.** 24   **d.** 225   **e.** 128

Exercises 8.3 (page 406)

1. 1/6   **3.** 3/10   **5.** 1326   **7.** 33/221 ≈ .149   **9.** 52/221 ≈ .235   **11.** 130/221 ≈ .588   **13.** 8.42×10^{-8}
15. 18,975/28,561 ≈ .664   **19.** $1 - P(365, 42)/(365)^{42}$ **21.** 1   **23.** .3499   **25.** 3/5 and 2/5   **27.** 2.165×10^{-4}
29. .0000289   **31.** 7/12   **33.** 5/18   **35.** 7/22 ≈ .318   **37. a.** .348   **b.** .046   **c.** .087   **d.** 0   **e.** 1/2   **f.** .068
g. .779   **39.** 1.54×10^{-6} **41.** 2.40×10^{-4} **43.** .02113   **45.** .4226   **47.** .002641   **49.** .006511   **51.** 8.9×10^{-10}
53. 4.51×10^{-34}; no   **55. a.** 28   **b.** 268,435,456   **c.** 40,320   **d.** 1.502×10^{-4} **e.** $n!/2^{n(n-1)/2}$

Exercises 8.4 (page 415)

1. 5/16 ≈ .313   **3.** 1/32 ≈ .031   **5.** 3/16 ≈ .188   **7.** 13/16 ≈ .813   **9.** 4.6×10^{-10} **11.** .269   **13.** .875
15. 1/64 ≈ .016   **17.** 21/32 ≈ .656   **23.** The potential callers are not likely to have birthdates that are distributed evenly
throughout the twentieth century.   **25.** .2059   **27.** .4510   **29.** .349   **31.** .736   **33.** .246   **35.** .017   **37.** .243   **39.** .925
41. a. 1.037×10^{-4} **b.** .002438   **43.** .028   **45.** .107   **47.** .218   **49.** .000729   **51.** .118   **53.** .121   **55.** .599
57. a. .047822   **b.** .976710   **c.** .897110   **59. a.** 1 chance in 1024   **b.** About 1 chance in 1.1×10^{12} **c.** About 1 chance
in 2.6×10^6 **61.** .0055   **63.** .000078   **65. a.** .199   **b.** .568   **c.** .133   **d.** .794   **67. a.** .9995

Exercises 8.5 (page 426)

1.

Number of Heads	0	1	2	3	4
Probability	1/16	1/4	3/8	1/4	1/16

3.

Number of Aces	0	1	2	3
Probability	.7826	.2042	.0130	.0002

5.

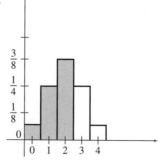

7.

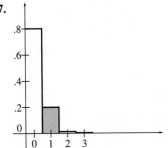

9. 3.6   **11.** 14.64   **13.** 2.7   **15.** 18   **17.** 0; yes

19. a.

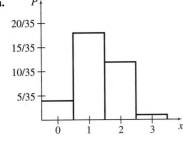

b. $9/7 \approx 1.286$ **21. a.** $5/3 \approx 1.667$ **b.** $4/3 \approx 1.333$ **23.** 1

25. No, the expected value is about $-21¢$ **29.** \$4500   **31. a.** 30   **b.** 10   **33. a.** Amoxicillin \$68.51; Cefaclor \$72.84
b. Amoxicillin   **35.** 420   **37. a.** \$50,000   **b.** \$65,000
39. a.

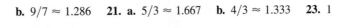

Number	0	1	2	3	4	5	6	7	8	9	10
Probability	.0000	.0000	.0000	.0003	.0024	.0141	.0573	.1600	.2929	.3178	.1552

b. .0027   **d.** 8.3   **41.** $-32¢$; no   **43.** $-66¢$ **45.** $-2.7¢$ **47.** $-20¢$ **49.** $-87.8¢$ **51. a.** .74; .94   **b.** Extra-point kick

Chapter 8 Review Exercises (page 432)

1. 720   **3.** 220   **5.** 20   **7. a.** 24   **b.** 12   **9. a.** 140   **b.** 239   **13.** 0   **15.** $14/55 \approx .255$ **17.** $4/33 \approx .121$
19. $1/64 \approx .016$ **21.** $11/32 \approx .344$ **23.** $1/17 \approx .059$ **25.** .3620   **27.** .9955
29. a.

Number	2	3	4	5	6	7	8	9	10	11	12
Probability	1/36	1/18	1/12	1/9	5/36	1/6	5/36	1/9	1/12	1/18	1/36

b.

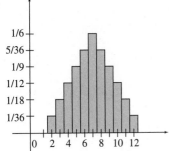

c. 7   **31.** .6   **33.** 2.5   **35.** \$1.29   **37. a.** $\binom{n}{0}$, or 1; $\binom{n}{1}$, or n; $\binom{n}{2}$; $\binom{n}{n}$, or 1

b. $\binom{n}{0} + \binom{n}{1} + \binom{n}{2} + \cdots + \binom{n}{n}$ **e.** The sum of the elements in row n of Pascal's triangle is 2^n.   **39.** .00004   **41.** 1.0000
43. \$8500   **45.** (c)   **47. a.** About .74   **b.** About 12   **c.** About .99   **d.** About .70

49. a.

Number	0	1	2	3	4
Probability	.4096	.4096	.1536	.0256	.0016

b.

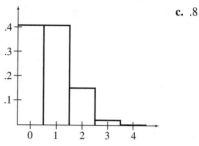

c. .8

51. a.

Number	0	1	2
Probability	10/28	15/28	3/28

b.

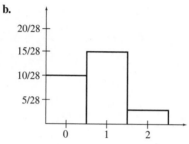

c. 3/4   **53.** −28¢

55. b. −$.4026, −$.3968   **57. a.** .0069   **b.** .8834   **c.** .4651

Chapter 9 Statistics

Exercises 9.1 (page 449)

1. a.–b.

Interval	Frequency
0–24	4
25–49	3
50–74	6
75–99	3
100–124	5
125–149	9

c.–d.

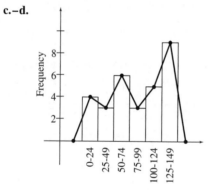

3. a.–b.

Interval	Frequency
0–19	3
20–39	3
40–59	3
60–79	5
80–99	2
100–119	4
120–139	7
140–159	3

c.–d.

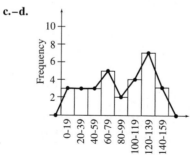

7. 16   **9.** 27,955   **11.** 7.7   **13.** 6.7   **15.** 51   **17.** 130   **19.** 29.1   **21.** 85.5, 91.5   **23.** 9   **25.** 68 and 74   **27.** 6.3

31. 86.2; 125–149   **35.** 2343 million bushels, 2359 million bushels   **37.** $29,952.28   **39.**

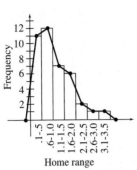

41. a. 14%   **b.** 7%   **c.** 30–49 years   **43. a.** 193,583 weddings; 204,500 weddings   **b.** December   **45. a.** 55.5°F, 50.5°F
b. 28.9°F, 28.5°F   **47. a.** The mean wealth is $999,900.01, or roughly $1 million.   **b.** $0   **c.** $0   **d.** Median or mode
49. c

Exercises 9.2 (page 459)

1. The standard deviation is the square root of the variance.   **3.** 53; 21.8   **5.** 46; 16.1   **7.** 24; 8.1   **9.** 45.2   **11.** 8/9
13. 24/25   **15.** At least .889   **17.** No more than .111   **21. a.** Mean = 25.5 hours; standard deviation = 7.2 hours
b. Forever Power   **c.** Forever Power   **23. a.** 1/3; 2; −1/3; 0; 5/3; 7/3; 1; 4/3; 7/3; 2/3   **b.** 2.1; 2.6; 1.5; 2.6; 2.5; .6; 1.0;
2.1; .6; 1.2   **c.** 1.13   **d.** 1.68   **e.** 4.41; −2.15   **f.** 4.31; 0   **25.** Mean = 1.8158 mm; standard deviation = .4451 mm
27. a. Mean = 7.3571; standard deviation = .1326   **b.** 100%   **29. a.** 127.71 days; 30.16 days   **b.** 7   **31. a.** $2,162,747.73
b. About 4%

Exercises 9.3 (page 472)

1. the mean   **3.** z-scores are found with the formula $z = (x - \mu)/\sigma$ **5.** 49.38%   **7.** 45.64%   **9.** 7.7%   **11.** 47.35%
13. 92.42%   **15.** −1.64 or −1.65   **17.** 1.04   **19.** .5, .5   **21.** .889; .997   **23.** 5000   **25.** 6375   **27.** 9772   **29.** .1587
31. .0062   **33.** 84.13%   **35.** 37.79%   **37.** 7.03%   **39.** .0062   **41.** $65.32 and $39.19   **43.** 99.38%   **45.** 189 units
47. .8887   **49.** 60.4 mph   **51.** 6.68%   **53.** 38.3%   **55.** 82   **57.** 70   **59. b.** 27%   **61. a.** About .01; yes
b. Essentially 0; yes   **63. a.** .5596; .0188   **b.** essentially 0; essentially 0   **c.** .9265; .9554   **d.** essentially 0; essentially 0
e. essentially 0; essentially 0

Exercises 9.4 (page 480)

1. The number of trials and the probability of success on each trial   **3. a.** .1964   **b.** .1974   **5. a.** .0106   **b.** .0122
7. .0240   **9.** .9032   **11.** .0956   **13.** .6443   **15.** .1841   **17. a.** .1684   **b.** .0305   **c.** .9573   **19.** .2578   **21. a.** .0237
b. .6808   **23.** .0301   **25. a.** .0001   **b.** .0002   **c.** .0000   **27.** .5977   **29.** .9898   **31.** .7357   **33. a.** 1.2139×10^{-7}
b. essentially 0   **c.** .7910

Chapter 9 Review Exercises (page 483)

3. a.

Sales	Frequency
450–474	5
475–499	6
500–524	5
525–549	2
550–574	2

b.–c.

5. 73   **7.** 34.9   **11.** 38; 36 and 38   **13.** 55–59

17. 67; 23.9   **19.** 6.2   **21.** A skewed distribution has the largest frequency at one end.   **23.** .6591   **25.** .0606
27. Because the histogram is skewed, not close to the shape of a normal distribution   **29. a.** 2.1, 1.07   **b.** 2, 1   **c.** Answers
to parts a and b should be close to each other.   **31. a.** Stock I: 8%, 7.9%; Stock II: 8%, 2.6%   **b.** Stock II   **33. a.** .1977
b. .0718   **c.** .0239   **35.** Diet A: $\bar{x} = 2.7$, $s = 2.26$; Diet B: $\bar{x} = 1.3$, $s = .95$ **a.** Diet A   **b.** Diet B   **37.** .9292
39. .9534   **41.** 28.10%   **43.** 56.25%   **45.** .0192

Chapter 10 Markov Chains

Exercises 10.1 (page 497)

1. No   **3.** Yes   **5.** No   **7.** Yes   **9.** No   **11.** Yes   **13.** No   **15.** No

17. Yes; $\begin{bmatrix} .9 & .1 & 0 \\ .1 & .6 & .3 \\ 0 & .3 & .7 \end{bmatrix}$ **19.** $A = \begin{bmatrix} 1 & 0 \\ .8 & .2 \end{bmatrix}$; $A^2 = \begin{bmatrix} 1 & 0 \\ .96 & .04 \end{bmatrix}$; $A^3 = \begin{bmatrix} 1 & 0 \\ .992 & .008 \end{bmatrix}$; 0   **21.** $C = \begin{bmatrix} .5 & .5 \\ .72 & .28 \end{bmatrix}$;

$C^2 = \begin{bmatrix} .61 & .39 \\ .5616 & .4384 \end{bmatrix}$; $C^3 = \begin{bmatrix} .5858 & .4142 \\ .596448 & .403552 \end{bmatrix}$; .4142   **23.** $E = \begin{bmatrix} .8 & .1 & .1 \\ .3 & .6 & .1 \\ 0 & 1 & 0 \end{bmatrix}$; $E^2 = \begin{bmatrix} .67 & .24 & .09 \\ .42 & .49 & .09 \\ .3 & .6 & .1 \end{bmatrix}$;

$E^3 = \begin{bmatrix} .608 & .301 & .091 \\ .483 & .426 & .091 \\ .42 & .49 & .09 \end{bmatrix}$; .301   **25.** The first power is the given transition matrix; $\begin{bmatrix} .2 & .15 & .17 & .19 & .29 \\ .16 & .2 & .15 & .18 & .31 \\ .19 & .14 & .24 & .21 & .22 \\ .16 & .19 & .16 & .2 & .29 \\ .16 & .19 & .14 & .17 & .34 \end{bmatrix}$;

$\begin{bmatrix} .17 & .178 & .171 & .191 & .29 \\ .171 & .178 & .161 & .185 & .305 \\ .18 & .163 & .191 & .197 & .269 \\ .175 & .174 & .164 & .187 & .3 \\ .167 & .184 & .158 & .182 & .309 \end{bmatrix}$; $\begin{bmatrix} .1731 & .175 & .1683 & .188 & .2956 \\ .1709 & .1781 & .1654 & .1866 & .299 \\ .1748 & .1718 & .1753 & .1911 & .287 \\ .1712 & .1775 & .1667 & .1875 & .2971 \\ .1706 & .1785 & .1641 & .1858 & .301 \end{bmatrix}$; $\begin{bmatrix} .17193 & .17643 & .1678 & .18775 & .29609 \\ .17167 & .17689 & .16671 & .18719 & .29754 \\ .17293 & .17488 & .17007 & .18878 & .29334 \\ .17192 & .17654 & .16713 & .18741 & .297 \\ .17142 & .17726 & .16629 & .18696 & .29807 \end{bmatrix}$; .18719

29. a. 53% for Johnson and 47% for NorthClean   **b.** 58.85% for Johnson and 41.15% for NorthClean   **c.** 61.48% for Johnson and 38.52% for NorthClean   **d.** 62.667% for Johnson and 37.333% for NorthClean   **31. a.** 42,500; 5000; 2500   **b.** 36,125; 8250; 5625   **c.** 30,706; 10,213; 9081   **d.** 26,100; 11,241; 12,659   **33. a.** [.257   .597   .146]   **b.** [.255   .594   .151]

c. [.254   .590   .156]   **35. b.** $\begin{bmatrix} 5/12 & 13/36 & 2/9 \\ 5/12 & 13/36 & 2/9 \\ 1/4 & 5/12 & 1/3 \end{bmatrix}$ **c.** 2/9   **37. a.** $\begin{array}{c} \\ \text{Single} \\ \text{Multiple} \end{array} \begin{array}{cc} S & M \\ \begin{bmatrix} .90 & .10 \\ .05 & .95 \end{bmatrix} \end{array}$ **b.** [.75   .25]   **c.** 68.8%

single-family and 31.3% multiple-family   **d.** 63.4% single-family and 36.6% multiple-family   **e.** $\begin{bmatrix} .815 & .185 \\ .0925 & .9075 \end{bmatrix}$ **f.** .815

39. a. $\begin{bmatrix} .348 & .379 & .273 \\ .320 & .378 & .302 \\ .316 & .360 & .323 \end{bmatrix}$ **b.** .348   **c.** .320

Exercises 10.2 (page 506)

1. Regular   **3.** Not regular   **5.** Regular   **7.** [2/5   3/5]   **9.** [4/11   7/11]   **11.** [14/83   19/83   50/83]
13. [170/563   197/563   196/563]   **15.** [0   0   1]   **17.** [81/331   175/331   75/331]   **19.** [1/2   7/20   3/20]
21. [(1 − q)/(2 − p − q)   (1 − p)/(2 − q − p)]

25. $\begin{array}{c} \\ \text{Works} \\ \text{Doesn't work} \end{array} \begin{array}{cc} \text{Works} & \begin{array}{c}\text{Doesn't}\\\text{work}\end{array} \\ \begin{bmatrix} .9 & .1 \\ .7 & .3 \end{bmatrix} \end{array}$; long-range probability of line working correctly is 7/8.   **27.** 24.0% 30-year fixed rate, 69.7%
15-year fixed-rate, 6.3% adjustable   **29. a.** .167   **b.** .107   **c.** .402 for group 1, .297 for group 2, .186 for group 3, .030 for group 4, and .086 for group 5.   **31. a.** 12.7% poor, 72.6% middle class, 14.7% affluent   **b.** 18.0% poor, 64.3% middle class, 17.8% affluent   **33. a.** [1/3   1/3   1/3]   **35. a.** 1.80% in group 1, 23.68% in group 2, 38.47% in group 3, and 36.04% in

group 4   **b.** 8   **37. c.** $\begin{bmatrix} 0 & 1 & 0 \\ 1/2 & 0 & 1/2 \\ 0 & 1 & 0 \end{bmatrix}$ **d.** Not a regular matrix   **e.** $[1/4 \quad 1/2 \quad 1/4]$ **39. b.** The guard spends 3/7 of the time in front of the middle door and 2/7 of the time in front of each of the other doors.

Exercises 10.3 (page 516)

1. State 2 is absorbing; matrix is that of an absorbing Markov chain.   **3.** State 2 is absorbing; matrix is not that of an absorbing Markov chain.   **5.** States 2 and 4 are absorbing; matrix is that of an absorbing Markov chain.

7. $F = [2]$; $FR = [.4 \quad .6]$ **9.** $F = [3/2]$; $FR = [1/2 \quad 1/2]$ **11.** $F = \begin{bmatrix} 1 & 0 \\ 1/3 & 4/3 \end{bmatrix}$; $FR = \begin{bmatrix} 1/3 & 2/3 \\ 4/9 & 5/9 \end{bmatrix}$

13. $F = \begin{bmatrix} 25/17 & 5/17 \\ 5/34 & 35/34 \end{bmatrix}$; $FR = \begin{bmatrix} 4/17 & 15/34 & 11/34 \\ 11/34 & 37/68 & 9/68 \end{bmatrix}$ **15. a.** $\begin{bmatrix} 1 & 0 & 0 & 0 & 0 \\ 1/2 & 0 & 1/2 & 0 & 0 \\ 0 & 1/2 & 0 & 1/2 & 0 \\ 0 & 0 & 1/2 & 0 & 1/2 \\ 0 & 0 & 0 & 0 & 1 \end{bmatrix}$

b. $F = \begin{bmatrix} 3/2 & 1 & 1/2 \\ 1 & 2 & 1 \\ 1/2 & 1 & 3/2 \end{bmatrix}$; $FR = \begin{bmatrix} 3/4 & 1/4 \\ 1/2 & 1/2 \\ 1/4 & 3/4 \end{bmatrix}$ **c.** 3/4   **d.** 1/4   **17.** .8756

19.

p	.1	.2	.3	.4	.5	.6	.7	.8	.9
x_a	.9999999997	.99999905	.99979	.98295	.5	.017046	.000209	.00000095	.0000000003

21. a column matrix of all 1's   **23. a.** 1   **c.** 10/7, 10/7   **25. a.** $P = \begin{bmatrix} .15 & .6 & .25 & 0 \\ 0 & .15 & .25 & .6 \\ 0 & 0 & 1 & 0 \\ 0 & 0 & 0 & 1 \end{bmatrix}$; $F = \begin{bmatrix} 20/17 & 240/289 \\ 0 & 20/17 \end{bmatrix}$;

$FR = \begin{bmatrix} 145/289 & 144/289 \\ 5/17 & 12/17 \end{bmatrix}$ **b.** 144/289   **c.** $580/289 \approx 2.007$ years   **27. a.** 2.63   **b.** 7.17   **c.** 8.33

29. a.
$$\begin{array}{c|cccccccc} & 0 & 1 & 2 & 3 & 4 & 5 & 6 & 7 \\ \hline 0 & 1 & 0 & 0 & 0 & 0 & 0 & 0 & 0 \\ 1 & 1/2 & 0 & 1/2 & 0 & 0 & 0 & 0 & 0 \\ 2 & 0 & 1/2 & 0 & 1/2 & 0 & 0 & 0 & 0 \\ 3 & 0 & 0 & 1/2 & 0 & 1/2 & 0 & 0 & 0 \\ 4 & 0 & 0 & 0 & 1/2 & 0 & 1/2 & 0 & 0 \\ 5 & 0 & 0 & 0 & 0 & 1/2 & 0 & 1/2 & 0 \\ 6 & 0 & 0 & 0 & 0 & 0 & 1/2 & 0 & 1/2 \\ 7 & 0 & 0 & 0 & 0 & 0 & 0 & 0 & 1 \end{array}$$
   b. 3/7   **c.** 2/7

Chapter 10 Review Exercises (page 520)

3. Yes   **5.** Yes   **7. a.** $C = \begin{bmatrix} .6 & .4 \\ 1 & 0 \end{bmatrix}$; $C^2 = \begin{bmatrix} .76 & .24 \\ .6 & .4 \end{bmatrix}$; $C^3 = \begin{bmatrix} .696 & .304 \\ .76 & .24 \end{bmatrix}$ **b.** .76   **9. a.** $E = \begin{bmatrix} .2 & .5 & .3 \\ .1 & .8 & .1 \\ 0 & 1 & 0 \end{bmatrix}$;

$E^2 = \begin{bmatrix} .09 & .8 & .11 \\ .1 & .79 & .11 \\ .1 & .8 & .1 \end{bmatrix}$; $E^3 = \begin{bmatrix} .098 & .795 & .107 \\ .099 & .792 & .109 \\ .1 & .79 & .11 \end{bmatrix}$ **b.** .099   **11.** $[.453 \quad .547]$; $[5/11 \quad 6/11]$ or $[.455 \quad .545]$

13. $[.48 \quad .28 \quad .24]$; $[47/95 \quad 26/95 \quad 22/95]$ or $[.495 \quad .274 \quad .232]$ **17.** Not regular   **23.** Matrix is not that of an absorbing Markov chain; state 2 is absorbing.   **25.** $F = [5/4]$; $FR = [5/8 \quad 3/8]$ **27.** $F = \begin{bmatrix} 7/4 & 4/5 \\ 1 & 8/5 \end{bmatrix}$; $FR = \begin{bmatrix} .55 & .45 \\ .6 & .4 \end{bmatrix}$

29. a. [.54 .46]   **b.** [.6464 .3536]   **31.** [.428 .322 .25]   **33.** [.431 .284 .285]   **35. a.** $\begin{bmatrix} 6.536 & 26.144 \\ 3.268 & 42.484 \end{bmatrix}$

b. 42.484   **c.** 26.144   **37.** .2   **39.** .4   **41.** .256   **43.** [.1945 .5555 .25]   **45.** [7/36 5/9 1/4]   **47.** States 1 and 6

49. $F = \begin{bmatrix} 8/3 & 1/6 & 4/3 & 2/3 \\ 4/3 & 4/3 & 8/3 & 4/3 \\ 4/3 & 1/3 & 8/3 & 4/3 \\ 2/3 & 1/6 & 4/3 & 8/3 \end{bmatrix}$; $FR = \begin{bmatrix} 3/4 & 1/4 \\ 1/2 & 1/2 \\ 1/2 & 1/2 \\ 1/4 & 3/4 \end{bmatrix}$ **51.** 1/2   **53. c.** 16/27, 7/27, 4/27

Chapter 11 Game Theory

Exercises 11.1 (page 533)

1. $6 from B to A   **3.** $2 from A to B   **5.** $1 from A to B   **7.** Yes; column 2 dominates column 3.   **9.** $\begin{bmatrix} -2 & 8 \\ -1 & -9 \end{bmatrix}$

11. $\begin{bmatrix} 4 & -1 \\ 3 & 5 \end{bmatrix}$ **13.** $\begin{bmatrix} 8 & -7 \\ -2 & 4 \end{bmatrix}$ **15.** (1, 1); 3; strictly determined   **17.** No saddle point; not strictly determined   **19.** (3, 1); 3; strictly determined   **21.** (1, 3); 1; strictly determined   **23.** No saddle point; not strictly determined

25. a. $\begin{bmatrix} 3 & 1 \\ 4 & -7 \\ 2 & 0 \\ -4 & -1 \end{bmatrix}$ **b.** $\begin{bmatrix} 3 & 1 \\ 4 & -7 \end{bmatrix}$ **c.** $\begin{bmatrix} 1 \\ -7 \end{bmatrix}$ **d.** [1]   **29. a.** Don't repair   **b.** Repair   **c.** Don't repair; −$107.50

31. a.

	Fails	Doesn't Fail
Overhaul	−$8600	−$2600
Don't Overhaul	−$6000	$0

b. Do not overhaul before shipping.   **c.** Column 1 dominates column 2 and row 2 dominates row 1. The saddle point is −$6000.

33.
A
	B 1	2	3
1	5	−2	6
2	7	5	9
3	3	−3	5

; saddle point is 5 at (2, 2); 5

35. (1, 3); −9   **37.**

	Rock	Paper	Scissors
Rock	0	−1	1
Paper	1	0	−1
Scissors	−1	1	0

; no

Exercises 11.2 (page 544)

1. a. −1   **b.** −.28   **c.** −1.54   **d.** −.46   **3.** Player A: 1: 1/5, 2: 4/5; player B: 1: 3/5; 2: 2/5; value 17/5   **5.** Player A: 1: 7/9, 2: 2/9; player B: 1: 4/9, 2: 5/9; value −8/9   **7.** Player A: 1: 8/15, 2: 7/15; player B: 1: 2/3, 2: 1/3; value 5/3   **9.** Player A: 1: 6/11, 2: 5/11; player B: 1: 7/11, 2: 4/11; value −12/11   **11.** Strictly determined; saddle point at (2, 2); value −5/12   **13.** Player A: 1: 2/5, 2: 3/5; player B: 1: 1/5, 2: 4/5; value 7/5   **15.** Player A: 1: 1/14, 2: 0, 3: 13/14; player B: 1: 1/7, 2: 6/7; value 50/7   **17.** Player A: 1: 2/3, 2: 1/3; player B: 1: 0, 2: 1/9, 3: 8/9; value 10/3   **19.** Player A: 1: 0, 2: 3/4, 3: 1/4; player B: 1: 0, 2: 1/12, 3: 11/12; value 33/4   **27.** Allied should use TV with probability 10/27 and use radio with probability 17/27. The value of the game is 1/18, which represents increased sales of $55,556.   **29.** He should invest in rainy day goods about 5/9 of the time and in sunny day goods about 4/9 of the time, for a steady profit of $722.22.   **31.** Euclid pounces with probability 2/3 and freezes with probability 1/3. Jamie pounces with probability 1/6 and freezes with probability 5/6. The value of the game is 4/3.   **33. a.** $\begin{bmatrix} 2 & -3 \\ -3 & 4 \end{bmatrix}$ **b.** Player A: 1: 7/12, 2: 5/12; player B: 1: 7/12, 2: 5/12; value −1/12

Exercises 11.3 (page 551)

1. Player A: 1: 2/3, 2: 1/3; player B: 1: 1/3, 2: 2/3; value 5/3 **3.** Player A: 1: 7/13, 2: 6/13; player B: 1: 8/13, 2: 5/13; value 22/13 **5.** Player A: 1: 2/7, 2: 5/7; player B: 1: 11/21, 2: 10/21; value −1/7 **7.** Player A: 1: 1/2, 2: 1/2; player B: 1: 0, 2: 3/10, 3: 7/10; value −1/2 **9.** Player A: 1: 5/8, 2: 3/8; player B: 1: 1/2, 2: 1/2, 3: 0; value 1/2 **11.** Player A: 1: 1/2, 2: 1/2, 3: 0; player B: 1: 1/6, 2: 2/3, 3: 1/6; value 0 **13.** The contractor should bid $30,000 with probability 9/29 and bid $40,000 with

probability 20/29. The value of the game is $1396.55. **15. a.** $\begin{bmatrix} 5000 & 10,000 & 10,000 \\ 8000 & 4000 & 8000 \\ 6000 & 6000 & 3000 \end{bmatrix}$ **b.** General Items should advertise in

Atlanta with probability 4/9, in Boston with probability 5/9, and never in Cleveland. Original Imitators should advertise in Atlanta with probability 2/3, in Boston with probability 1/3, and never in Cleveland. The value of the game is $6666.67. **17.** Merchant A should locate in city 1, 2, and 3 with probability 27/101, 129/202, and 19/202, respectively. Merchant B should locate in city 1, 2, and 3 with probability 39/101, 9/101, and 53/101, respectively. The value of the game is 885/101 ≈ 8.76 percentage points. **19.** Act appeasing with probability 1/9, expect civility but don't fight for it with probability 2/9, and expect civility and fight for it if it isn't given with probability 2/3. The value of the game is −4/3.

21. a. Blotto

$$\begin{array}{c} & \textit{Kije} \\ & \begin{array}{cccc} (3,0) & (0,3) & (2,1) & (1,2) \end{array} \\ \begin{array}{c} (4,0) \\ (0,4) \\ (3,1) \\ (1,3) \\ (2,2) \end{array} & \begin{bmatrix} 4 & 0 & 2 & 1 \\ 0 & 4 & 1 & 2 \\ 1 & -1 & 3 & 0 \\ -1 & 1 & 0 & 3 \\ -2 & -2 & 2 & 2 \end{bmatrix} \end{array}$$

b. Blotto uses strategies (4, 0) and (0, 4) with probability 4/9 each,

strategy (2, 2) with probability 1/9, and never sends 3 regiments to one post and 1 to the other. Kije used strategy (3, 0) with probability 1/30, strategy (0, 3) with probability 7/90, strategy (2, 1) with probability 8/15, and strategy (1, 2) with probability 16/45. The value of the game is 14/9. **23. a.** Each player uses each strategy 1/3 of the time, and the value of the game is 0. **b.** The game is symmetric in that neither player has an advantage, and each choice is as strong as every other choice.

Chapter 11 Review Exercises (page 554)

3. $2 from A to B **5.** $7 from B to A **7.** Row 3 dominates row 1; column 1 dominates column 4. **9.** $\begin{bmatrix} -11 & 6 \\ -10 & -12 \end{bmatrix}$

11. $\begin{bmatrix} -2 & 4 \\ 3 & 2 \\ 0 & 3 \end{bmatrix}$ **13.** (1, 1); value −2 **15.** (2, 2); value 0; fair game **17.** (2, 3); value −3 **19.** Player A: 1: 5/6, 2: 1/6;

player B: 1: 1/2, 2: 1/2; value 1/2 **21.** Player A: 1: 1/9, 2: 8/9; player B: 1: 5/9, 2: 4/9; value 5/9 **23.** Player A: 1: 1/5, 2: 4/5; player B: 1: 3/5, 2: 0, 3: 2/5; value −12/5 **25.** Player A: 1: 3/4, 2: 1/4, 3: 0; player B: 1: 1/2, 2: 1/2; value 1/2 **27.** Player A: 1: 4/7, 2: 3/7; Player B: 1: 1/2, 2: 1/2; value −1 **29.** Player A: 1: 7/8, 2: 1/8; player B: 1: 1/2, 2: 1/2; value 1/2 **31.** Player A: 1: 1/2, 2: 1/2; player B: 1: 1/6, 2: 0, 3: 5/6; value −1/2 **33.** Player A: 1: 10/29, 2: 11/29, 3: 8/29; player B: 1: 7/29, 2: 16/29, 3: 6/29; value 2/29 **37.** Hostile **39.** Hostile; $785 **41.** Labor and management should both always be friendly. The value of the game is $600. **43.** Hector should invest in blue-chip stocks with probability 35/37 and growth stocks with probability 2/37. The value of the game is $2918.92. **45.** Oppose **47.** Oppose; gain of 2700 votes **49.** Each candidate should oppose the factory. The value of the game is 0. **51.** Ravogna should defend installation No. 1 with probability 1/4 and installation No. 2 with probability 3/4. Rontovia should attack installation No. 1 with probability 3/4 and installation No. 2 with probability 1/4. The value of the game is 13/4.

PHOTO ACKNOWLEDGMENTS

page xix © FPG   **page 1** © Poulides/Thatcher/Stone   **page 18** © PhotoDisc   **page 22** © PhotoDisc   **page 25** © PhotoDisc   **page 39** © David Young-Wolff/PhotoEdit   **page 42** © Kevin Fleming/CORBIS   **page 44** © PhotoDisc   **page 48** © Kelly-Mooney Photography/CORBIS   **page 73** © Robert Landau/CORBIS   **page 84** © Bob Daemmrich/ The Image Works   **page 85** © Karla Harby   **page 96** © Kevin Schafer/CORBIS   **page 121** © Paul A. Souders/CORBIS   **page 124** © PhotoDisc   **page 132** © Joseph Sohm; ChromoSohm Inc./CORBIS   **page 141** © Corbis Corporation/Tom Stewart, 2001   **page 145** © Tony Arruza/CORBIS   **page 149** © PhotoDisc   **page 149** © Corbis Corporation/John Medere, 2001   **page 151** © John Garrett/CORBIS   **page 159** © AP/Wide World Photos   **page 169** © AP/Wide World Photos   **page 171** © Bob Rowan; Progressive Image/CORBIS   **page 182** © Corbis Corporation/Tom Stewart, 2001   **page 186** © AP/Wide World Photos   **page 199** © IT Int'l/eStock Photography/PictureQuest   **page 220** © PhotoDisc   **page 229** © AP/Wide World Photos   **page 229** © AP/Wide World Photos   **page 235** © PhotoDisc   **page 238** Copyright © David Young-Wolff/PhotoEdit   **page 246** Copyright © Amy Etra/PhotoEdit   **page 273** © PhotoDisc   **page 280** © Corbis Corporation/Ron Sanford, 2001   **page 295** © PhotoDisc   **page 303** © Ed Young/CORBIS   **page 313** © Corbis Corporation/William Whitehurst, 2001   **page 325** © Keren Su/CORBIS   **page 330** © Mary Boucher   **page 335** © Corbis Corporation/Lester Lefkowitz, 2001   **page 360** © Corbis Corporation/Ariel Skelley, 2001   **page 378** Copyright © Billy E. Barnes/PhotoEdit   **page 379** © CORBIS   **page 397** © CORBIS   **page 409** © David H. Wells/CORBIS   **page 417** © Nik Wheeler/CORBIS   **page 431** © Kevin R. Morris/CORBIS   **page 437** © Bob Daemmrich/The Image Works   **page 439** © PhotoDisc   **page 453** © AP/Wide World Photos   **page 464** © Rita Maas/The Image Bank   **page 475** © AP/Wide World Photos   **page 487** © AP/Wide World Photos   **page 488** © David Frazier/Stone   **page 490** © Corbis Corporation/Chuck Savage, 2001   **page 500** © AP/Wide World Photos   **page 509** © Corbis Corporation/Ted Horowitz, 2001   **page 519** © W. Cody/CORBIS   **page 521** © AP/Wide World Photos   **page 526** © Charles Thatcher/Stone   **page 535** © Corbis Corporation/Jim Cummins, 2001   **page 540** © George Lepp/CORBIS   **page 552** © Leif Skoogfors/CORBIS   **page 556** Copyright © Robert Brenner/PhotoEdit

INDEX

6.2, 6.3 Truth Tables

The following truth table defines the logical operators in this chapter.

p	q	$\sim p$	$p \wedge q$	$p \vee q$	$p \rightarrow q$
T	T	F	T	T	T
T	F	F	F	T	F
F	T	T	F	T	T
F	F	T	F	F	T

7.3 Basic Probability Principle

Let S be a sample space of equally likely outcomes, and let event E be a subset of S. Then the probability that event E occurs is

$$P(E) = \frac{n(E)}{n(S)}.$$

7.4 Union Rule

For any two events E and F from a sample space S,

$$P(E \cup F) = P(E) + P(F) - P(E \cap F).$$

7.4 Odds

If $P(E') \neq 0$, the odds in favor of an event E are defined as the ratio of $P(E)$ to $P(E')$, or

$$\frac{P(E)}{P(E')}.$$

7.4 Properties of Probability

Let S be a sample space consisting of n distinct outcomes, $s_1, s_2, \ldots, s_n$. An acceptable probability assignment consists of assigning to each outcome s_i a number p_i (the probability of s_i) according to these rules.

1. The probability of each outcome is a number between 0 and 1.

$$0 \le p_1 \le 1, \quad 0 \le p_2 \le 1, \ldots, \quad 0 \le p_n \le 1$$

2. The sum of the probabilities of all possible outcomes is 1.

$$p_1 + p_2 + p_3 + \cdots + p_n = 1$$

7.5 Product Rule

If E and F are events, then $P(E \cap F)$ may be found by either of these formulas.

$$P(E \cap F) = P(F) \cdot P(E|F) \quad \text{or} \quad P(E \cap F) = P(E) \cdot P(F|E)$$

7.6 Bayes' Theorem

$$P(F_i|E) = \frac{P(F_i) \cdot P(E|F_i)}{P(F_1) \cdot P(E|F_1) + P(F_2) \cdot P(E|F_2) + \cdots + P(F_n) \cdot P(E|F_n)}$$

8.1 Multiplication Principle

Suppose n choices must be made, with

$$m_1 \text{ ways to make choice 1,}$$

and for each of these ways,

$$m_2 \text{ ways to make choice 2,}$$

and so on, with

$$m_n \text{ ways to make choice } n.$$